TECHNOLOGICAL INVENTIONS

Mario D. Rucci

First Edition (2019)

RUC BOOKS – Montreal, Canada

RUCCIAN ACTIVITIES BOOKS

(Books by Mario D. Rucci)

MAGNETIC EVOLUTION

TECHNOLOGICAL INVENTIONS

VISUAL ARTS

VERBAL ARTS

TRICHOTOMOUS ASSOCIATION

UNIVERSAL VISUALIZATION

CLASSIFICATION ...
Explanation 1: The first 2 books are about Development; the middle 2 about Representation; and, the last 2 about Management.
Explanation 2: The first 2 books are about Imagination; the middle 2 about Perception; and, the last 2 about Conception.
Explanation 3: The first 2 books are of Fundamental significance; the middle 2 of Particular significance; and, the last 2 of General significance.

DEDICATION

To
The Inventive Minds
of my Ruccian Family

RUCCIAN TECHNOLOGICAL INVENTIONS

by Mario D. Rucci

Published by:
RUC™ BOOKS
Post Office Box 20002
Montreal, QC H1H 5W5 Canada
Orders from RucBooks.com

ISBN, electronic ed. 978-1-989504-17-8
ISBN, print ed. 978-1-989504-16-1

First Printing 2019

TABLE OF CONTENTS

ABOUT THE AUTHOR

Mario D. Rucci is the 5th child of Nicola Rucci and Cristina Maria Del Russo. He was born in 1948 in Sepino, Campobasso, Italy. He moved to Canada with family in 1966, when he was 18 years old. He was a teacher for 30 years: taught fine arts in High School and all subjects in Elementary School. He retired from teaching in 2006, at the age of 58, to work full time on his writing. He worked on his writing for another 12 years, until 2018 (age 70). After that came the harder parts: correcting and editing.

Over the course of his life **Mario D. Rucci** has been a painter, writer, teacher, computer expert, house builder, inventor, theoretical physicist, and philosopher – and he has excelled in whatever he happened to tackle.

INTRODUCTION

1. WORLD ACTIVITIES

WORLDS

There are 3 worlds to be considered or thought about: **the Natural/Earthly/Human/Familiar WORLD**, the Para-natural/Extra-terrestrial/Other-dimensional/Alien WORLD, and the Supernatural/Heavenly/Devine/Spiritual WORLD. Our goal here is to analyze, clarify, and classify the Natural World.
* world: existence, being, reality

[–] WORLD TYPES

	World:			
Natural	earthly, human	tangible	familiar	known
Para-natural	extra-terrestrial, other-dimensional	semi-tangible	alien	semi-known
Super-natural	heavenly, divine	intangible	spiritual	unknown

WORLDS AWARENESS as to SUBSTANCE, ACTIVITY, AWARENESS
In relation to **Substance**, the natural world is tangible, the para-natural semi-tangible, and the super-natural intangible. In relation to **Activity**, the natural world is familiar, the para-natural alien, and the super-natural spiritual. And, in relation to **Awareness**, the natural world is known, the para-natural semi-known, and the supernatural unknown.

WORLDS AWARENESS as to SUBSTANCE Position, Extent, Form
Known World has to do with Substance **Position**: the inanimate, living, adapted worlds are known. The Semi-Known World has to do with Substance **Extent**: the celestial (very large and very far) and the minute/minuscule worlds are semi-known. The Unknown World has to do with Substance **Form**: the superfluid (very fluid) world is unknown.

NATURAL WORLD IN GENERAL

IN GENERAL, the Natural World is Substance, Activity, and Awareness. We have the following order: **First**, we have Substance; **second**, we have Activity of substance; and, **third**, we have Awareness of substance and activity. In other words, the order is Physical, Social, Mental.

[–] NATURAL WORLD GENERALIZATION

	World:
Physical	Substance
Social	Activity
Mental	Awareness

NATURAL WORLD IN PARTICULAR

IN PARTICULAR, the world is Physical, Social, Mental } Substance, Activity, Awareness.

[+] NATURAL WORLD INDIVIDUATION

	Substance:	**Activity:**	**Awareness:**
Physical	inanimate	magnetic/natural	perceived

Social	living	organic	decided/chosen
Mental	adapted	synthetic/artificial	conceived

NATURAL WORLD ACTIVITY IN ESSENCE

IN ESSENCE, the world activity is Magnetic, Organic, and Synthetic.

Magnetic Activity is about the rotation, revolution, and vibration by material bodies. Organic Activity is about the transfer (transportation*, transaction, and transmission) by living organisms. Synthetic Activity is about the development, management, and representation by human beings.

* transportation or transformation/alteration like destructive metabolism/fission and constructive metabolism/fusion

[+] NATURAL WORLD ACTIVITY FOUNDATION

	Magnetic Activity:	**Organic Activity:**	**Synthetic Activity:**
Physical	solid	feeding	developmental/progressive
Social	fluid	defending	managerial
Mental	superfluid	strategizing	representational

SYNTHETIC ACTIVITY IN PARTICULAR

We have 3 Synthetic Activities: Development, Management, and Representation. Development is thru evolution, invention, and understanding. Management is thru adaptation, legislation, and classification. And, Representation is thru idea, art, and symbolization.

[+] SYNTHETIC ACTIVITY INDIVIDUATION

	Development:	**Management:**	**Representation:**
Physical	evolution	adaptation	visualization/idea
Social	invention	legislation	art
Mental	understanding	association	symbolization

SYNTHETIC ACTIVITY IN ESSENCE (Ruccian Activities are underlined.)

P Development: Evolution: magnetic (theoretical physics) , biological, theological
S Development: Invention/Device: technological, cultural, recreational
M Development: Understanding: analytical, synthetical/artificial/scoped, theoretical
P Management: Adaptation: material, behavioral, attitudinal*
S Management: Legislation: economic, political, ethical
M Management: Association: unitary*/hierarchical, binary/comparative, tertiary/trichotomous*
P Representation: Representative: substitute/stand-in, middleman/go-between, spokesperson/envoy
S Representation: Art/Illustration: visual, performing, verbal
M Representation: Visualization*/Idea*: universal/general, particular, fundamental

* P = Physical, S = Social, M = Mental
* material adaptation/modification thru roads-building, housing, clothing
*** Association: to see categories as a union/relation/hierarchy, a division/distinction/branching, and a group/set/trichotomy**
* unitary is listable/hierarchical, binary is comparative/branching, and tertiary is **trichotomous/tabular**
* visualization: idea/mental image, concept/conception, thought
* idea: signed, phonetic/alphabetic, ideographic } universal/general, particular, fundamental

[+] DEVELOPMENT

	P Development: Evolution:	S Development: Invention:	M Development: Understanding:
Physical	magnetic	technological	analytical
Social	biological	cultural	synthetical/scoped
Mental	theological	recreational	theoretical

[+] MANAGEMENT

	P Management: Adaptation:	S Management: Legislation:	M Management: Association:
Physical	material	economic	unitary/hierarchical
Social	behavioral	political	binary/comparative
Mental	attitudinal	ethical	tertiary/trichotomous

[+] REPRESENTATION

	P Representation: Representative/Agent:	S Representation: Art:	M Representation: Visualization/Idea:
Physical	substitute, stand-in	visual	universal/general
Social	middleman, go-between	performing	particular
Mental	spokesperson, envoy	verbal	fundamental

2. RUCCIAN ACTIVITIES

This series of books is about Activities I tackled or dealt-with:

PHYSICAL ACTIVITIES
Magnetic Evolution
Technological Inventions
SOCIAL ACTIVITIES
Visual Arts
Verbal Arts
MENTAL ACTIVITIES
Trichotomous Associations
Universal Visualization

3. THIS BOOK ACTIVITY

This book is about the ideas I came up with during the years I spent on TECHNOLOGICAL INVENTIONS. If you are looking for ideas you would like to develop, this is a book for you. There might be here some idea that you would like to bring to fruition. And, if not, these ideas might inspire you towards a similar idea. But, do not forget to acknowledge where your similar idea originated. If, on the other hand, you decide to use one of my ideas as is, know that these ideas are copyrighted and you have to get permission to use what is protected by copyright laws. Said copyright is easily given, because, like you, I would love to have one of my ideas come to fruition.

The purpose of TECHNOLOGICAL INVENTIONS is the control P Substance/Matter, S Activity/ Energy, and M Awareness/Information. What follows now is the classification of Technological Inventions.

TECHNOLOGICAL INVENTIONS COLLECTION (In General)

P Substance Machine: **Developer**
S Activity Machine: **Manager***
M Awareness Machine: **Representative***
* manager: administrator, director, controller, regulator
* representative/**rep**: agent, **executor**, medium, instrument, vehicle

[–] COLLECTIVE TECHNOLOGICAL INVENTIONS

	Technological Invention/Machine/Device:
P Substance	MATTER / Developmental MACHINE, Developer
S Activity	ENERGY / Managerial MACHINE, Manager
M Awareness	INFORMATION / Representational MACHINE, Representative

TECHNOLOGICAL INVENTIONS INDIVIDUATION (In Particular)

[+] INDIVIDUAL TECHNOLOGICAL INVENTIONS

	Developer:	**Manager:**	**Rep/Executor:**
Matter	preparer, smelter	conformer, complier	builder, connector
Energy	motor (motive force)	regulator	vehicle, appliance, bulb
Information*	recorder	editor, program	broadcaster, record player

Matter is **compact** matter (wood, stone, metal), **moldable** matter (mud, clay, concrete, mortar, plaster), and **fibrous** matter (wool, cotton, polyester).
Energy is from **solid** matter (wood, coal, uranium), **fluid** matter (steam, gasoline/petroleum, gas), and **superfluid** matter (gravity/waterfall, heat/wind, sunlight).
Information is from **image**, **text** symbols, and **speech** sounds.

TECHNOLOGICAL INVENTIONS FOUNDATION (In Essence)

Ruccian Inventions are <u>underlined</u>.

[+] MATTER/SUBSTANCE MACHINES

	Matter Developer:	**Matter Manager:**	**Matter Rep/Executor:**
	Preparer:	**Conformer:**	**Builder:**
Compact	straightener*, lengthener	<u>wire press</u>, <u>sculptor</u>	<u>framer</u>*
Moldable	crusher, pulverizer	mixer, mold	<u>tile setter</u>, brick layer
Fibrous	spindle, loom	scissors	<u>sewing machine</u>

[+] ENERGY/ACTIVITY MACHINES

	Energy Developer: Motor:	**Energy Manager: Regulator:**	**Energy Rep/Executor: Vehicle:**
Solid	muscles, nuclear reactor	heat regulator	lever, steam vehicle*
Fluid	combustible: gas engine	gas regulator	gas vehicle, IFO**,** gun
Superfluid	elec. atm. magn. motor	voltage regulator	electric vehicle, appliance

[+] INFORMATION/AWARENESS MACHINES

	Information Developer:	**Information Manager:**	**Information Rep/**
Executor:			
	Recorder:	**Editor:**	**Broadcaster:**
Image	camera	image editor, sizer	TV, computer
Text	printing press, keyboard	text editor or word processor	newspaper, book
Speech	tape recorder*	sound editor	telephone, radio

* straightener: straightening raw material (wood, stone, metal) so it is ready for use
* framer: frame building with rods, glass panes, walls
* gun: due to explosive device (bomb, shell, missile)
* recorder: **camera** (recording image), **keyboard** (recording text), **sound recorder** (recording sound)
NOTE: All machines can be **manually-guided, motor-powered, computer-controlled.**

RUCCIAN TECHNOLOGICAL INVENTIONS

MATTER MACHINES
1. Shaper: **Wire Press, Sculptor**
2. Shelter: **Framer**, **Tile Setter, Level**
3. Clothing: **Sewing Machine**
ENERGY MACHINES
4. Levers, Inclined Plane, Pendulum
5. Fuel/Gas Engine
6. Vehicle, Transmission; IFO
7. Gun
8. Atmospheric, Magnetic, Atm-Mag, Electric MOTOR
INFORMATION MACHINES
9. Motion Camera, Portrait Camera
10. CAD, Measuring Telescope, Measuring Pen, Figure Abstraction
11. Computer
12. Keyboard
13. Other Ideas

1A. WIRE PRESS

My idea to mass produce the **wire horse** was to use a **wire press** operated by hand.

SHAPING 3-DIMENTIONALLY ALREADY SHAPED WIRE ENCLOSING A 2-DIMENTIONAL AREA

e.g. HORSE'S HEAD

TOP VIEW

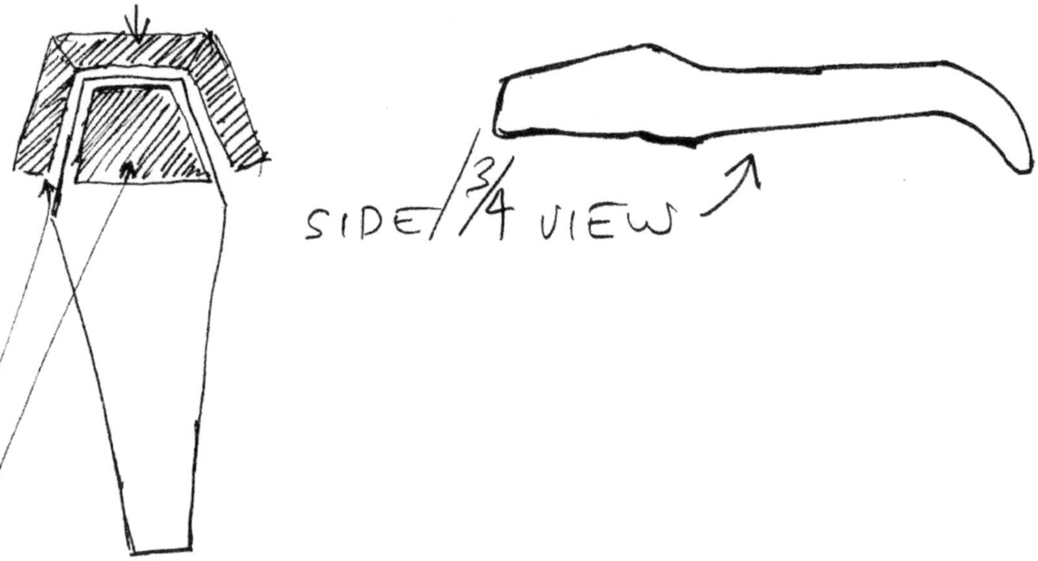

MAKE MOLD WITH SOFT WOOD

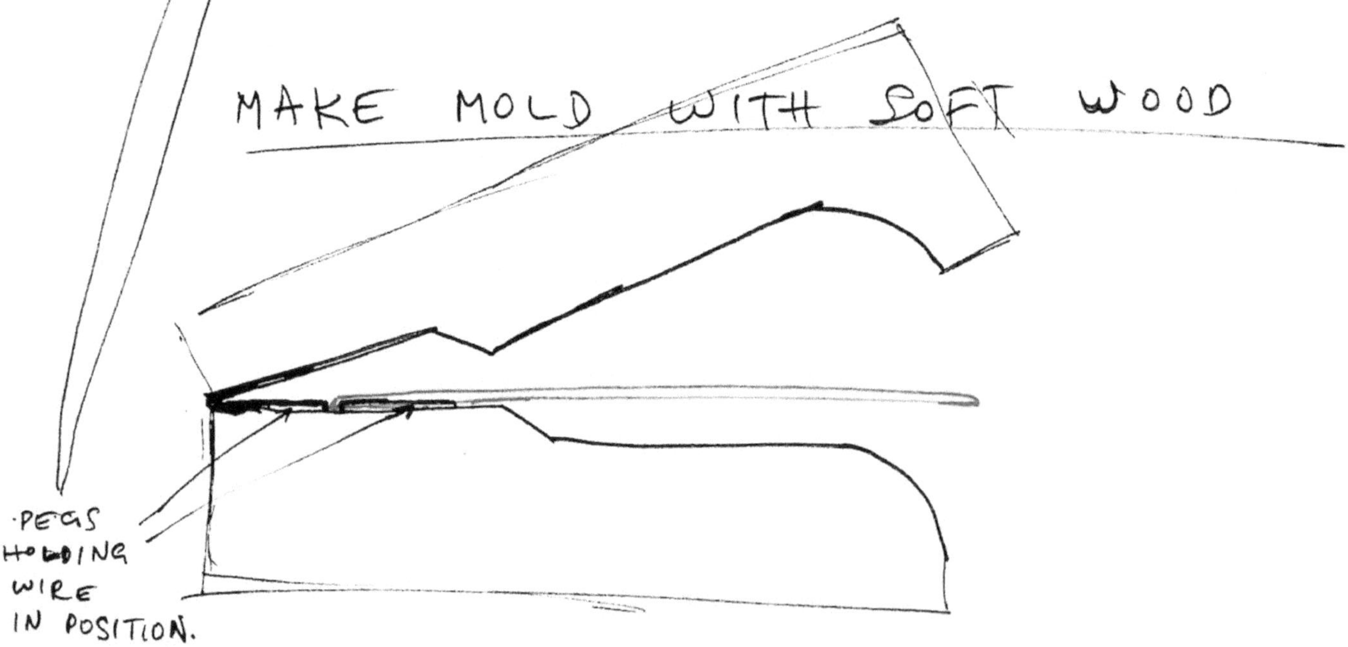

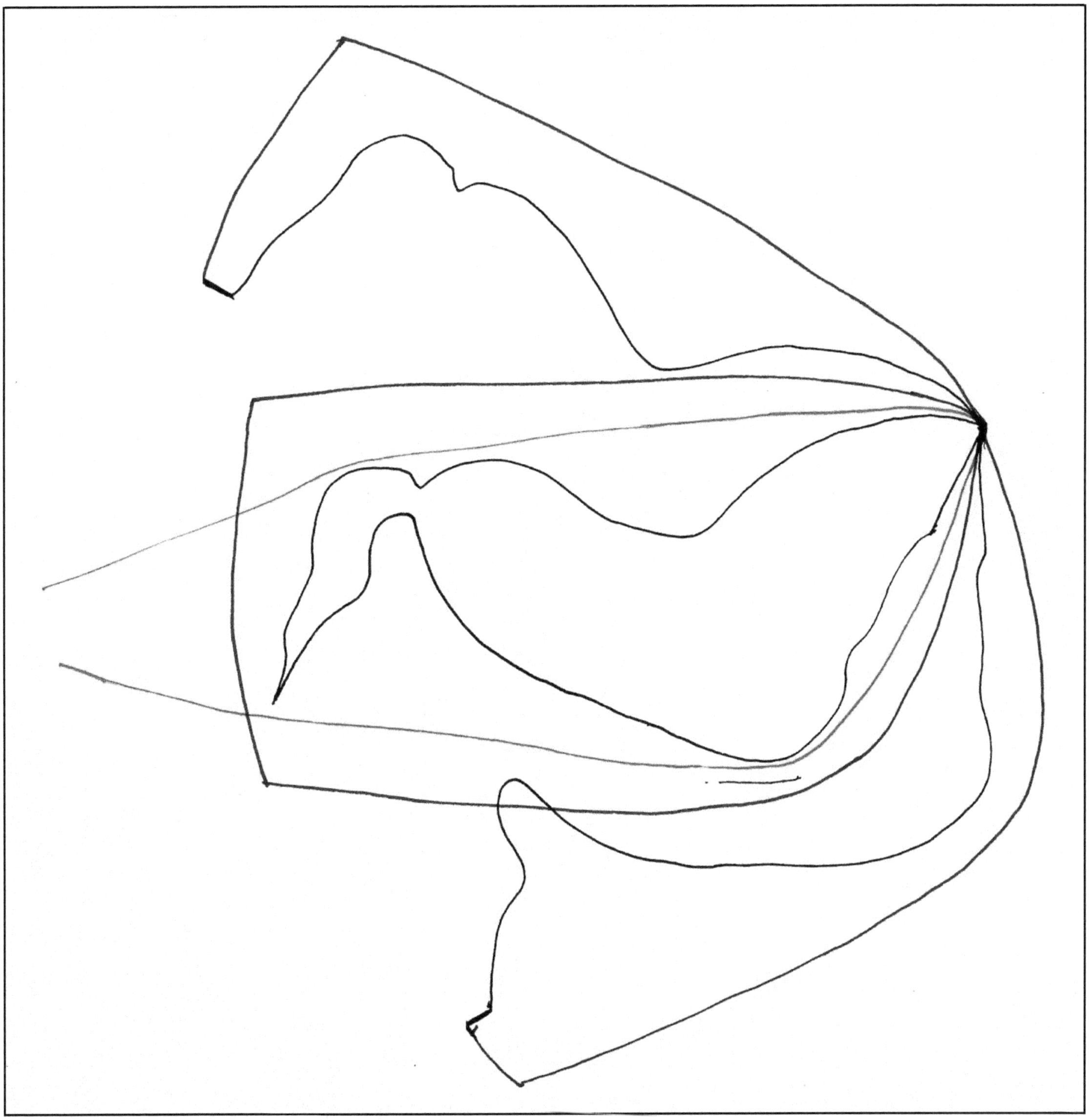

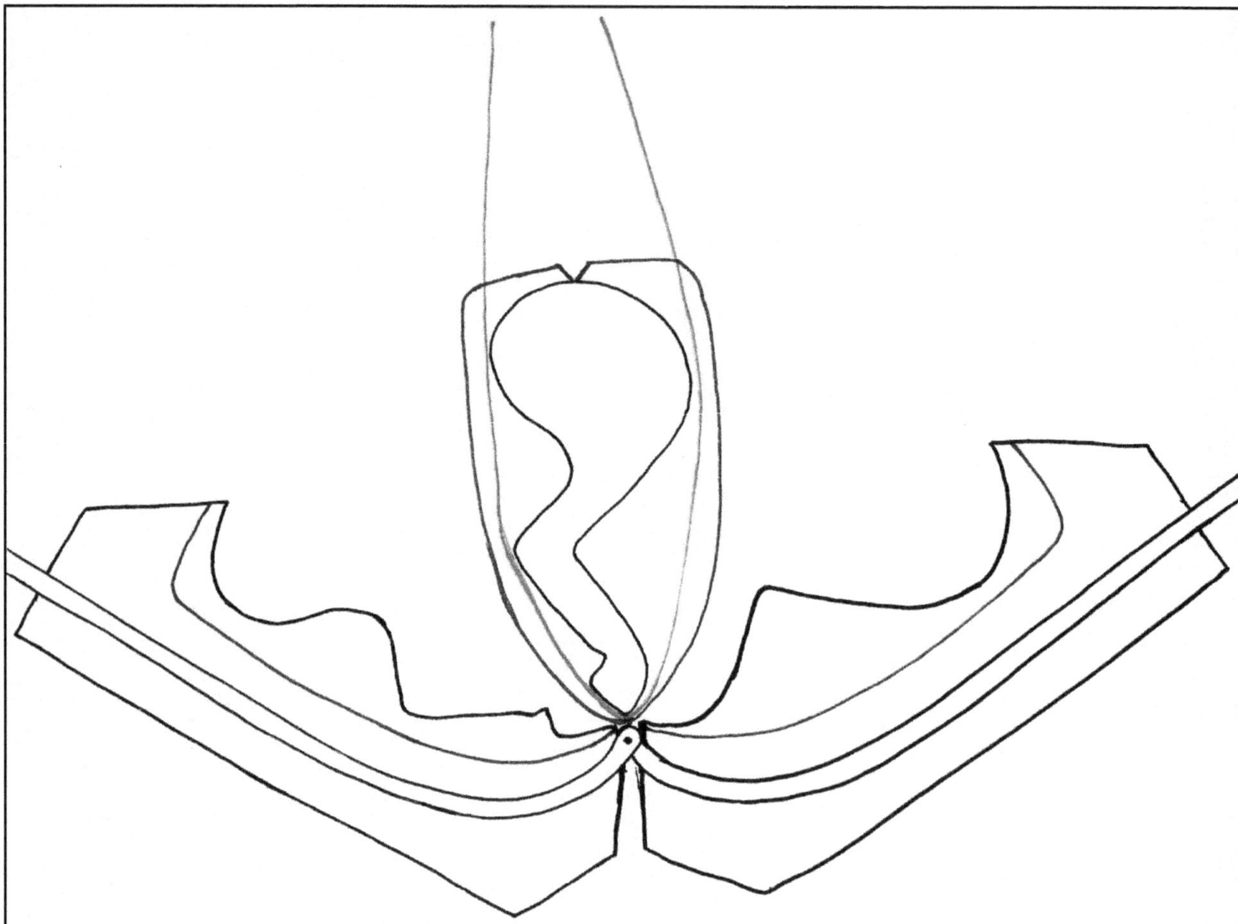

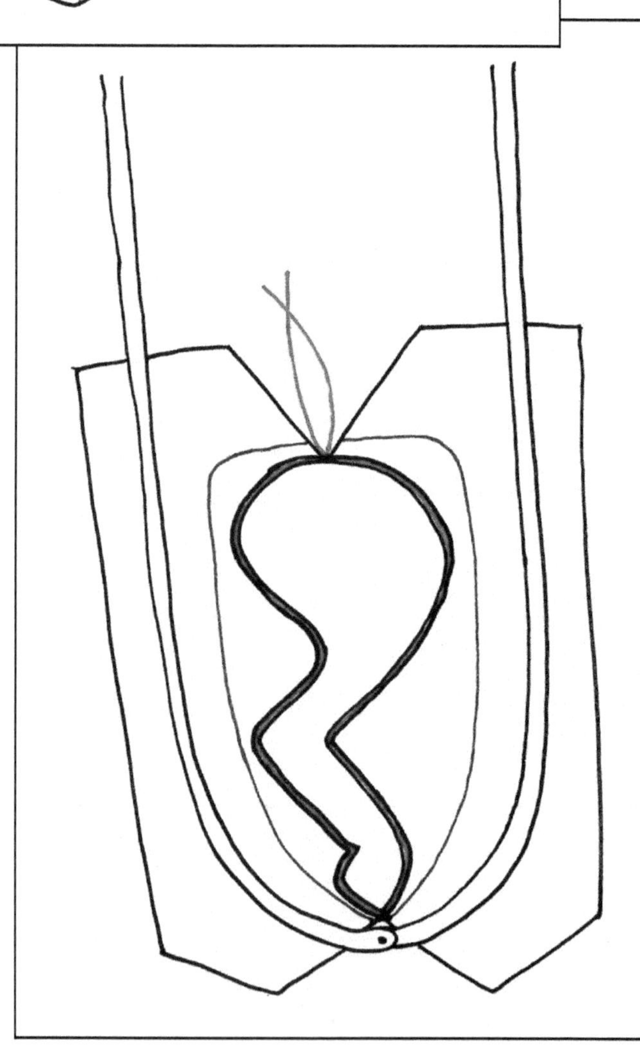

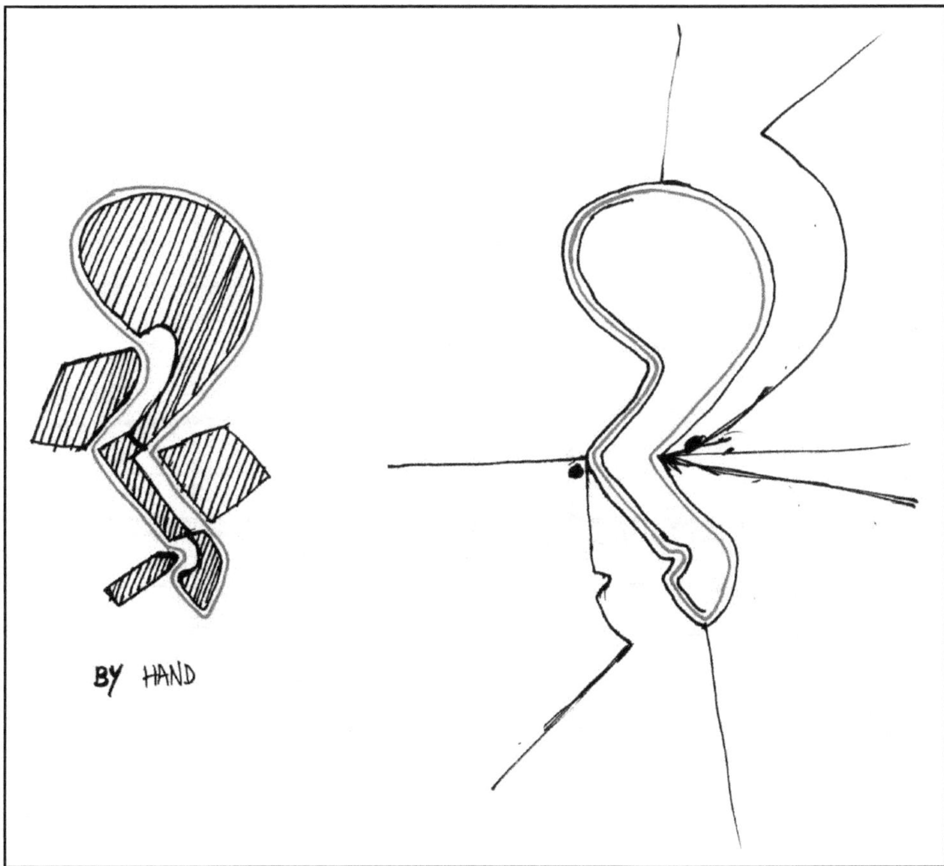
BY HAND

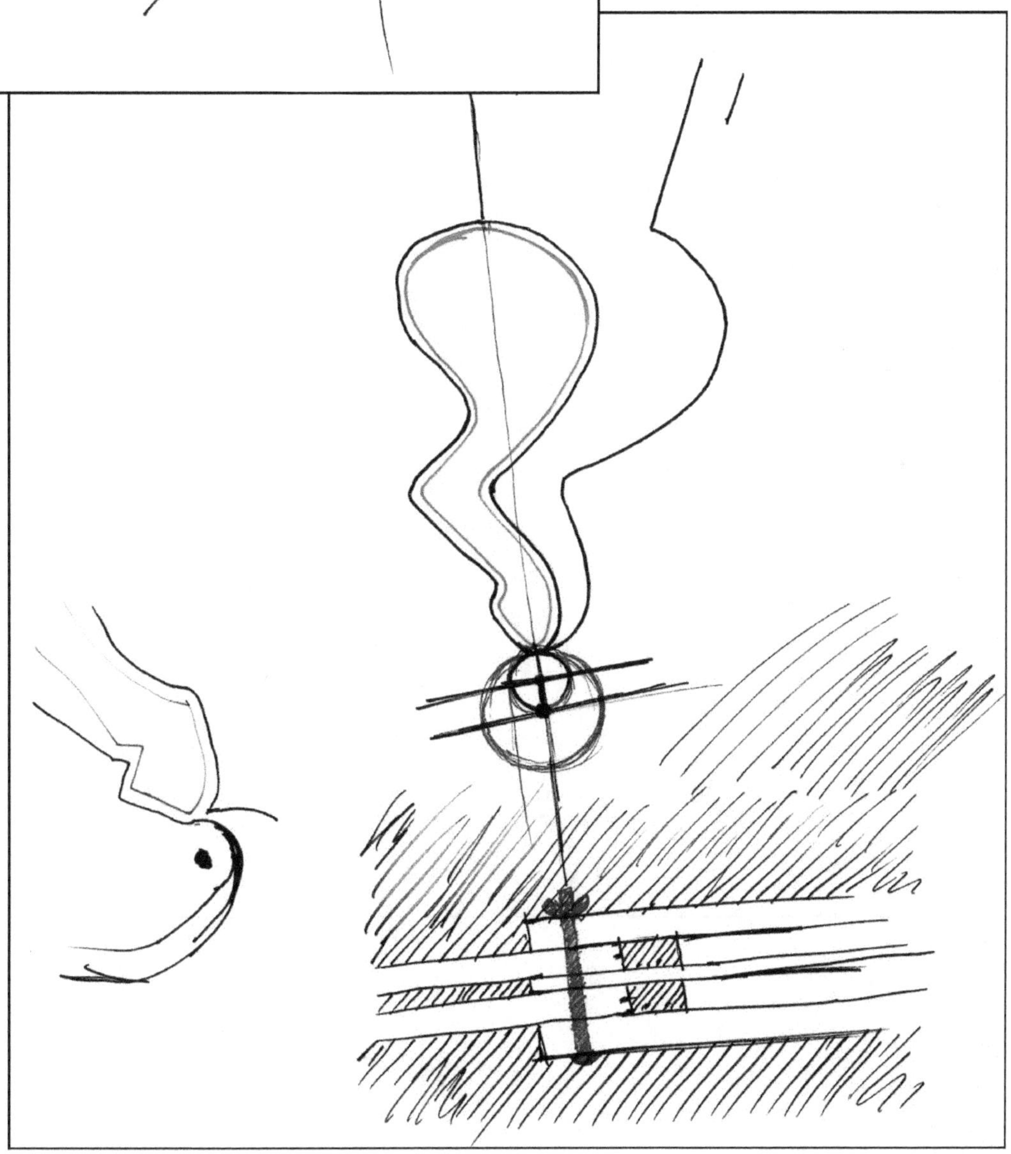

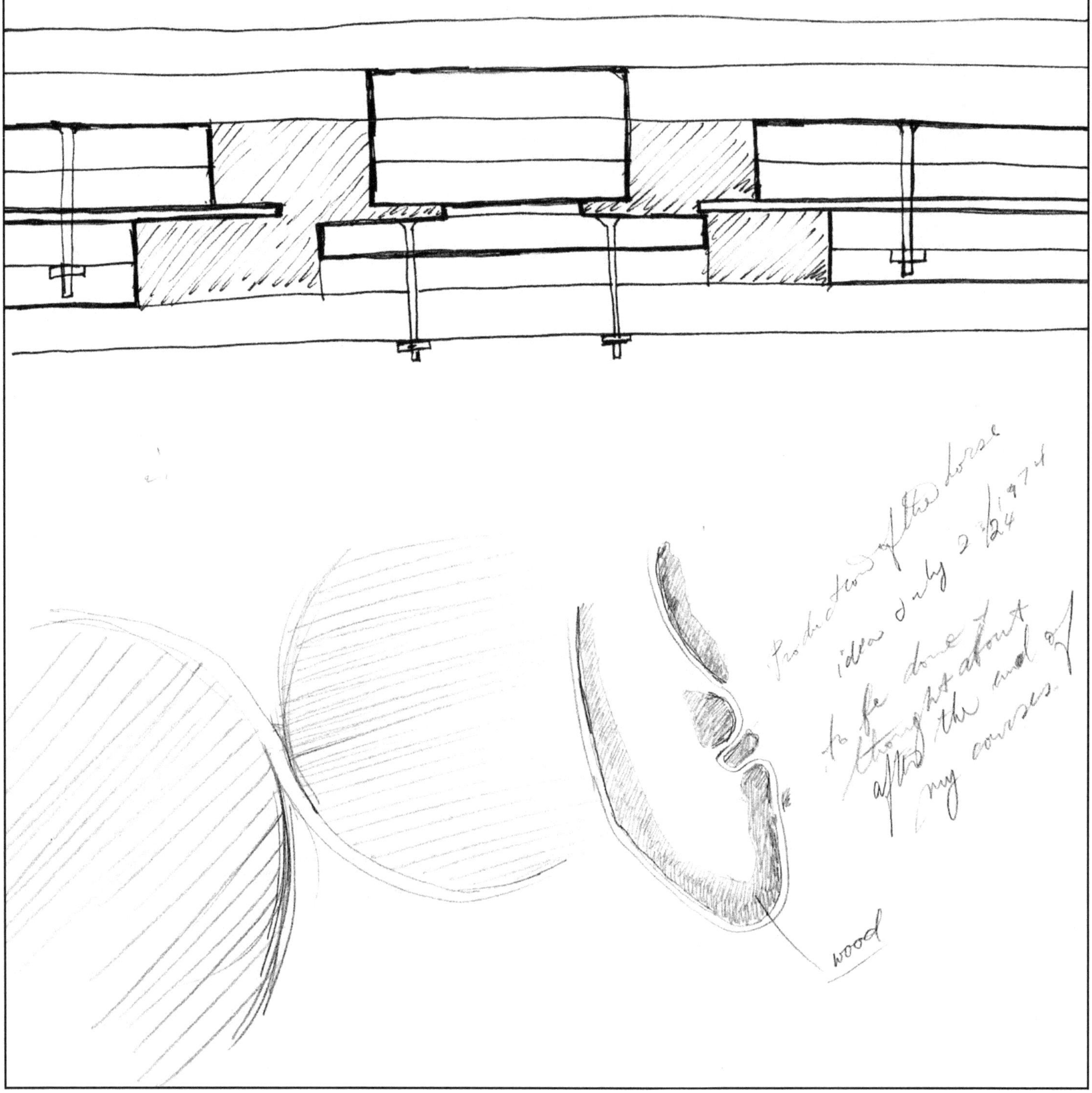
Production of the horse
idea July 2 3/24 1974
to be done
thought about
after the end of
my courses.
wood

1B. SCULPTOR

A Sculptor consists of the following parts:

1. **Cage/Enclosure/Guider**: Cubic Frame
2. **3-D Movement** of Pantograph Arm and Router
3. **Router** or **Laser Gun**
4. **Linkage**: **Pantograph** or **Computer Program**
5. **Vise** securing/holding workpiece
6. **Chuck** securing drill/router bit
7. **Transmission**: Speed Scaler & Gear Selector

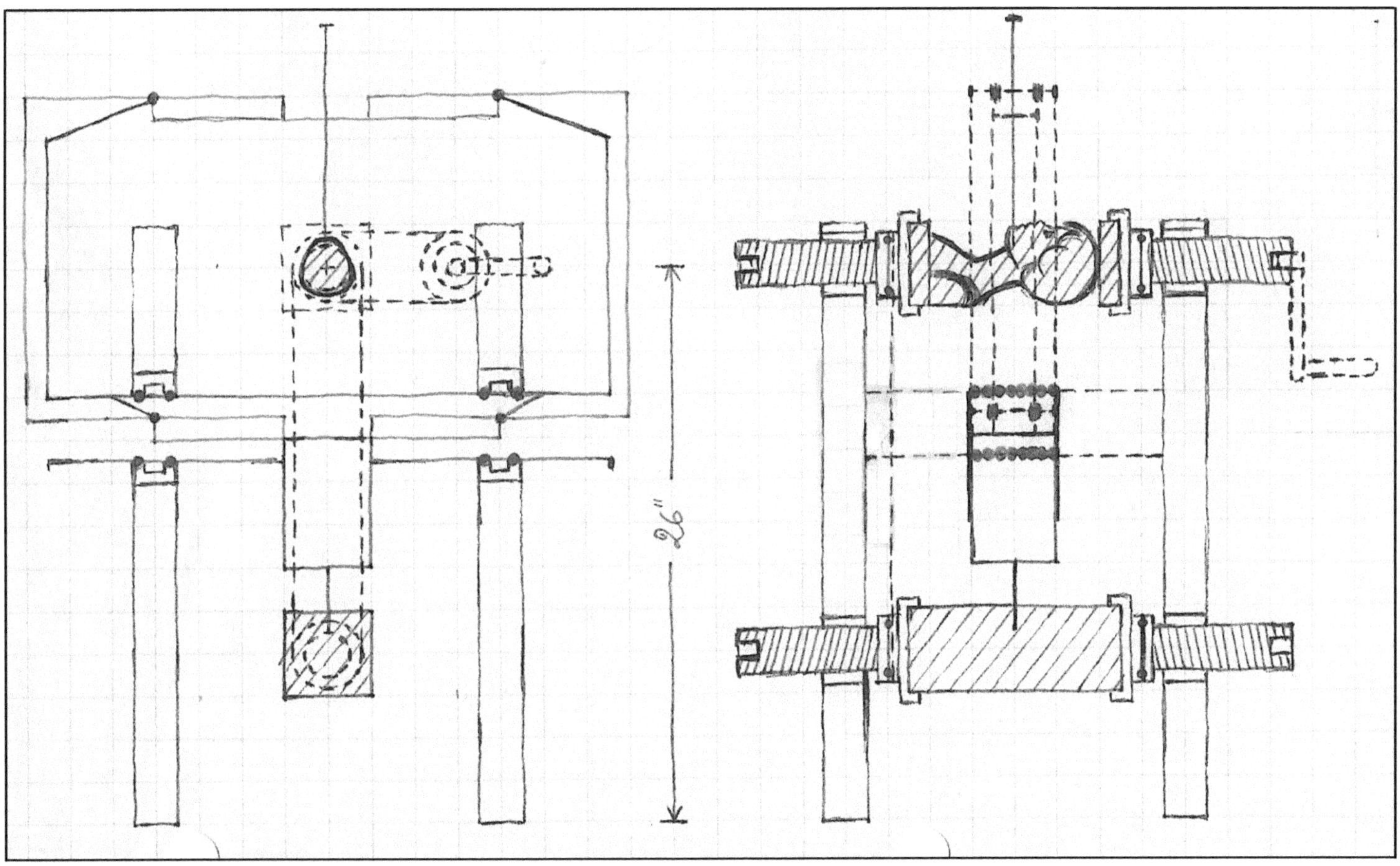

MANUAL SCULPTOR

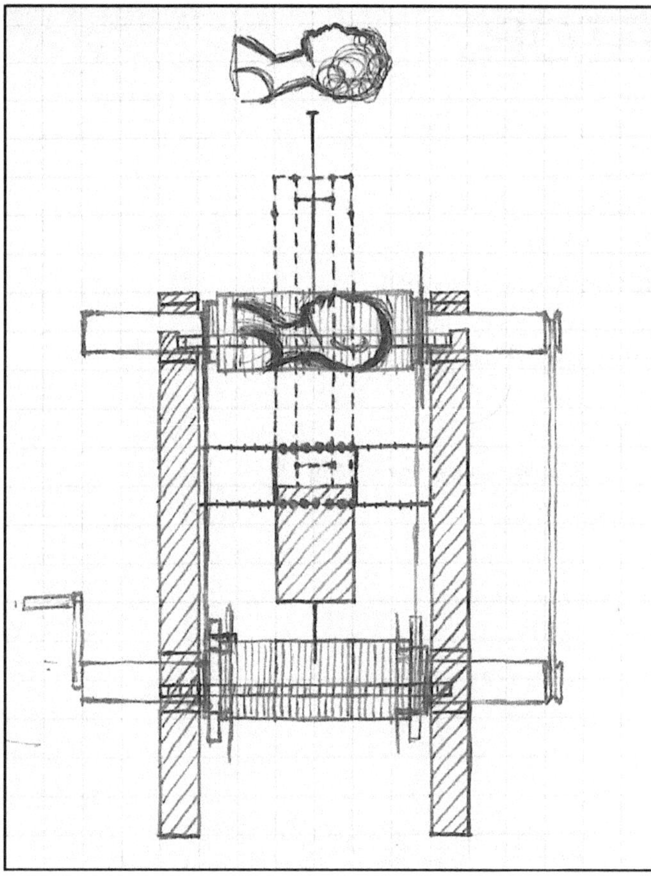

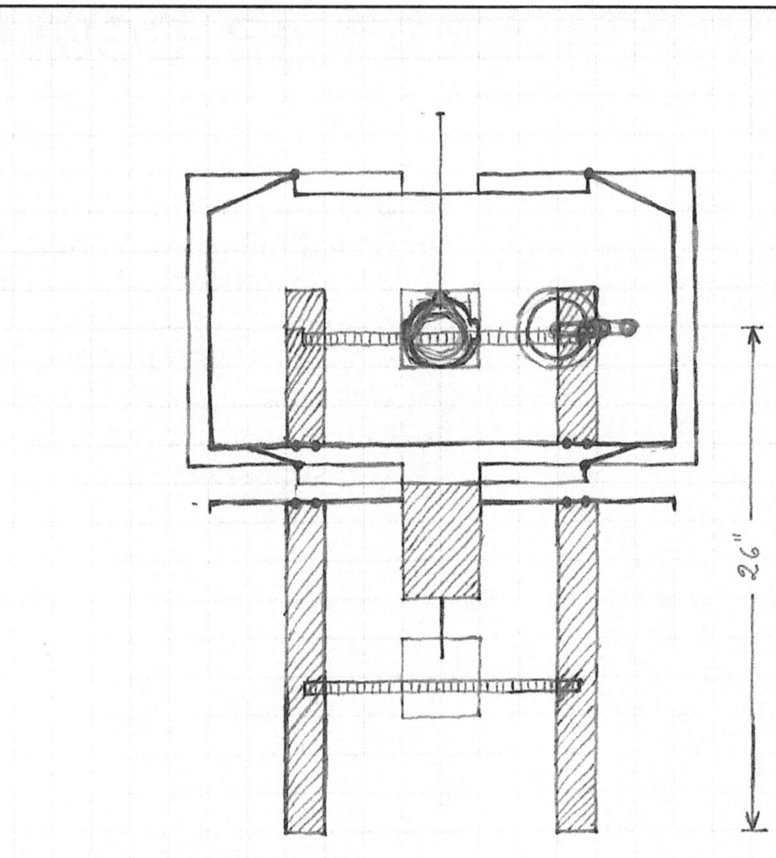
26"

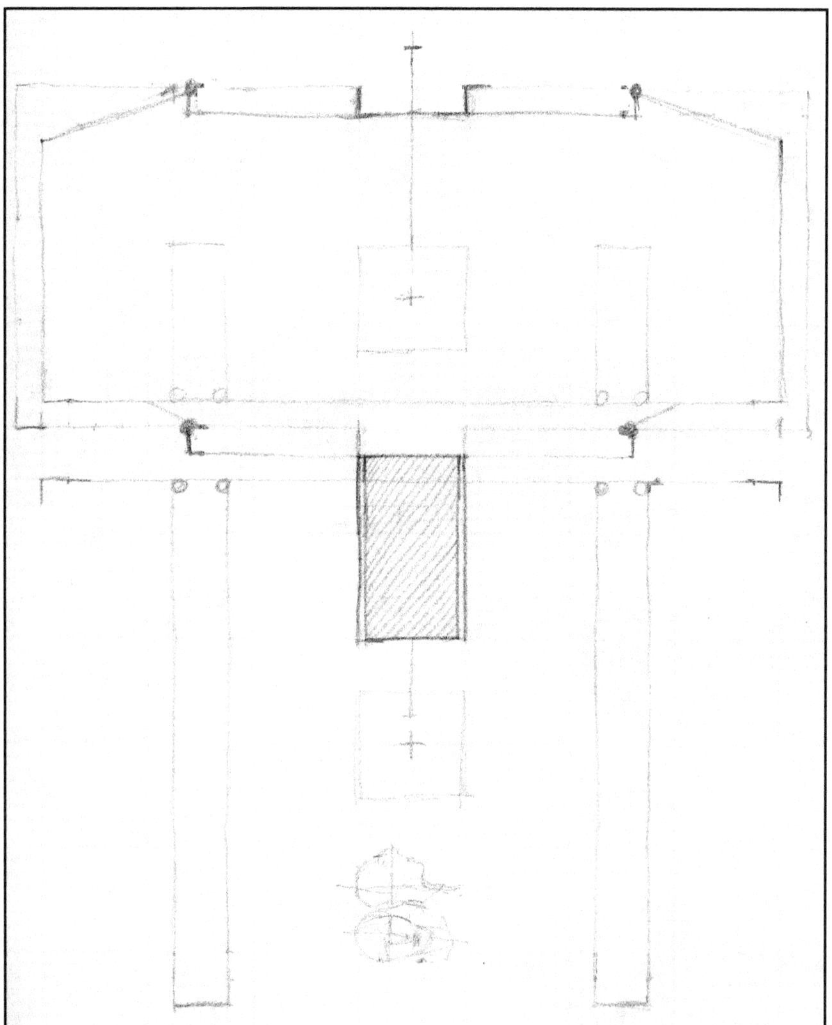

Cage

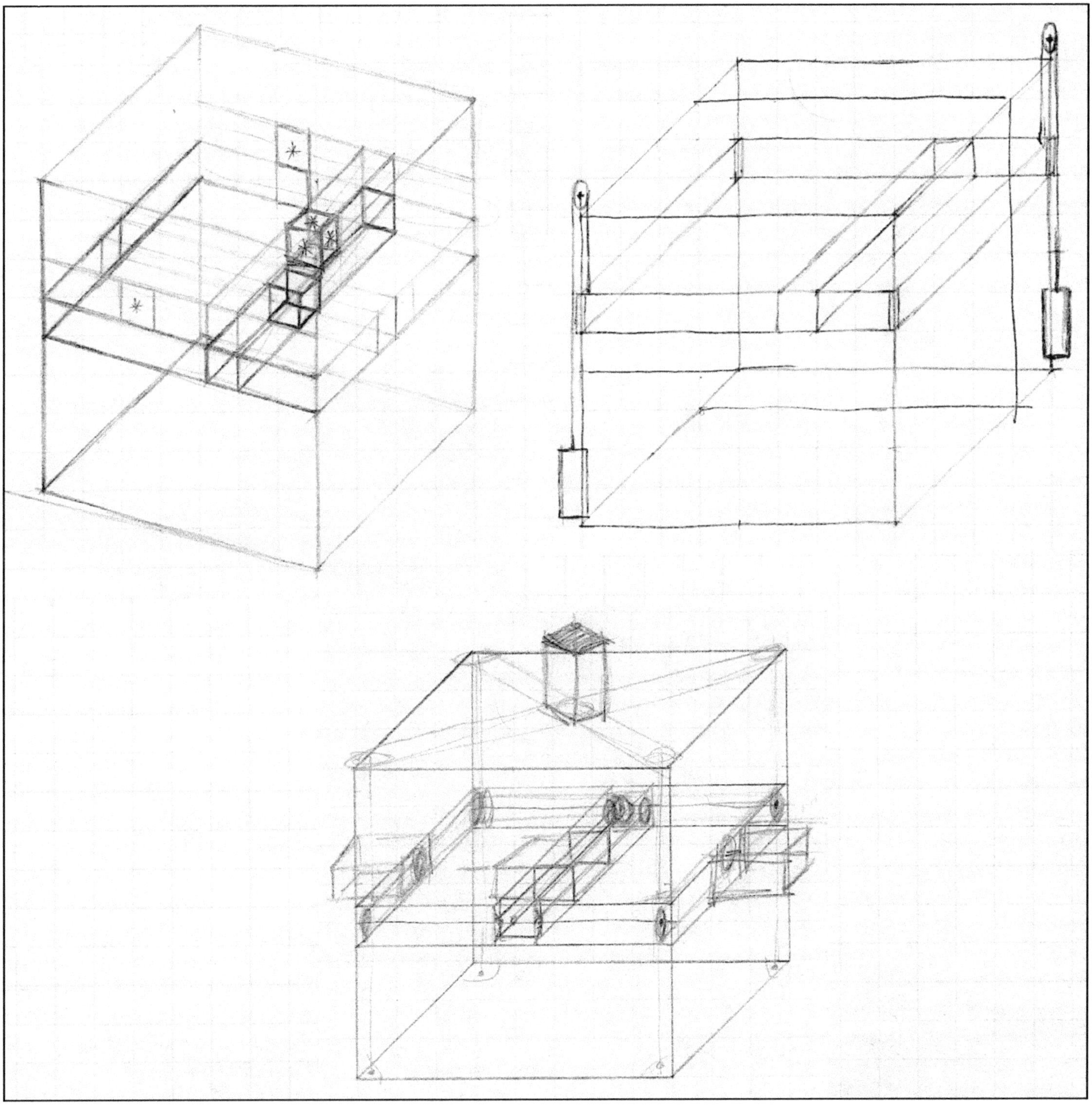

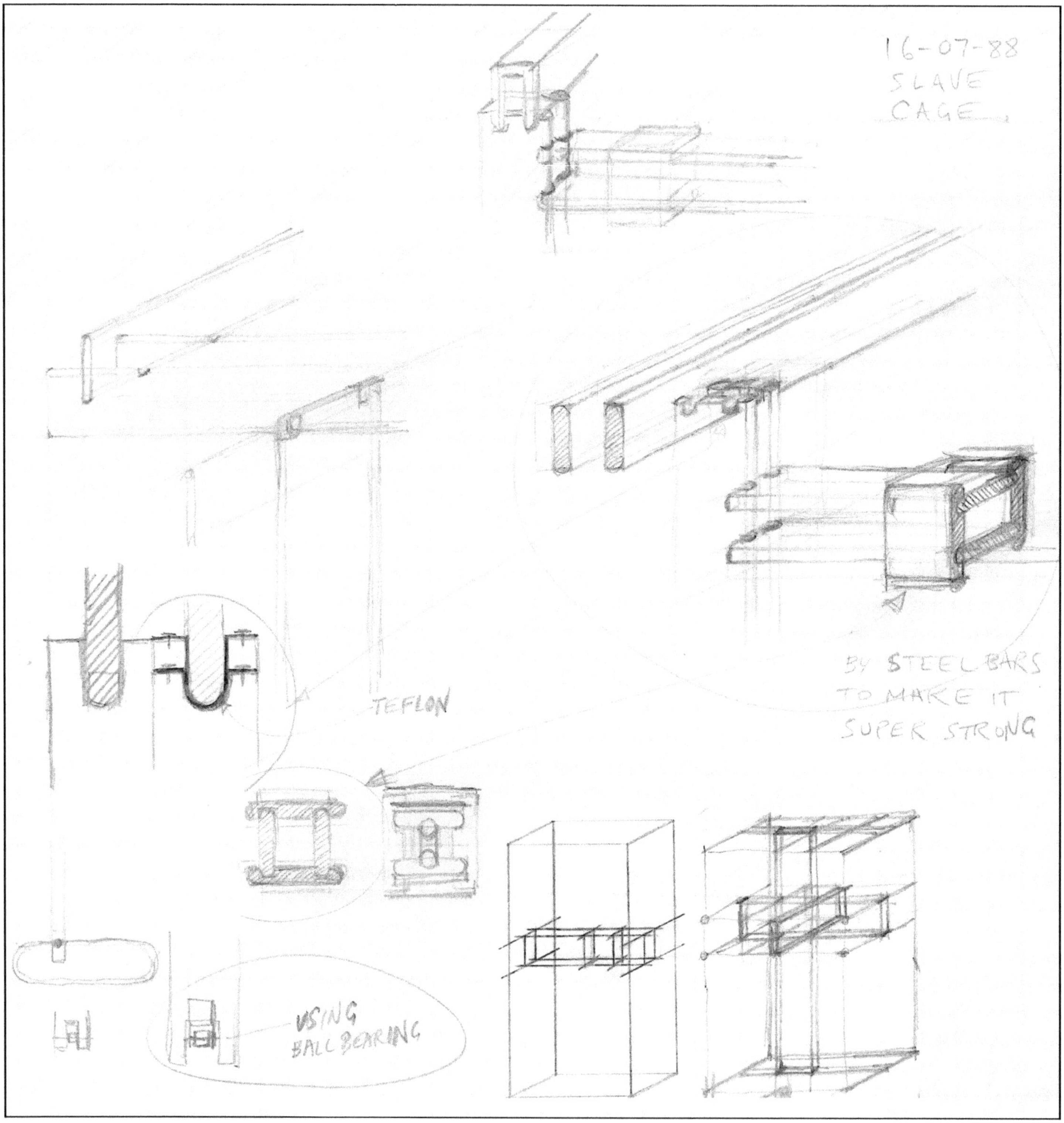
16-07-88
SLAVE
CAGE
TEFLON
BY STEEL BARS
TO MAKE IT
SUPER STRONG
USING
BALL BEARING

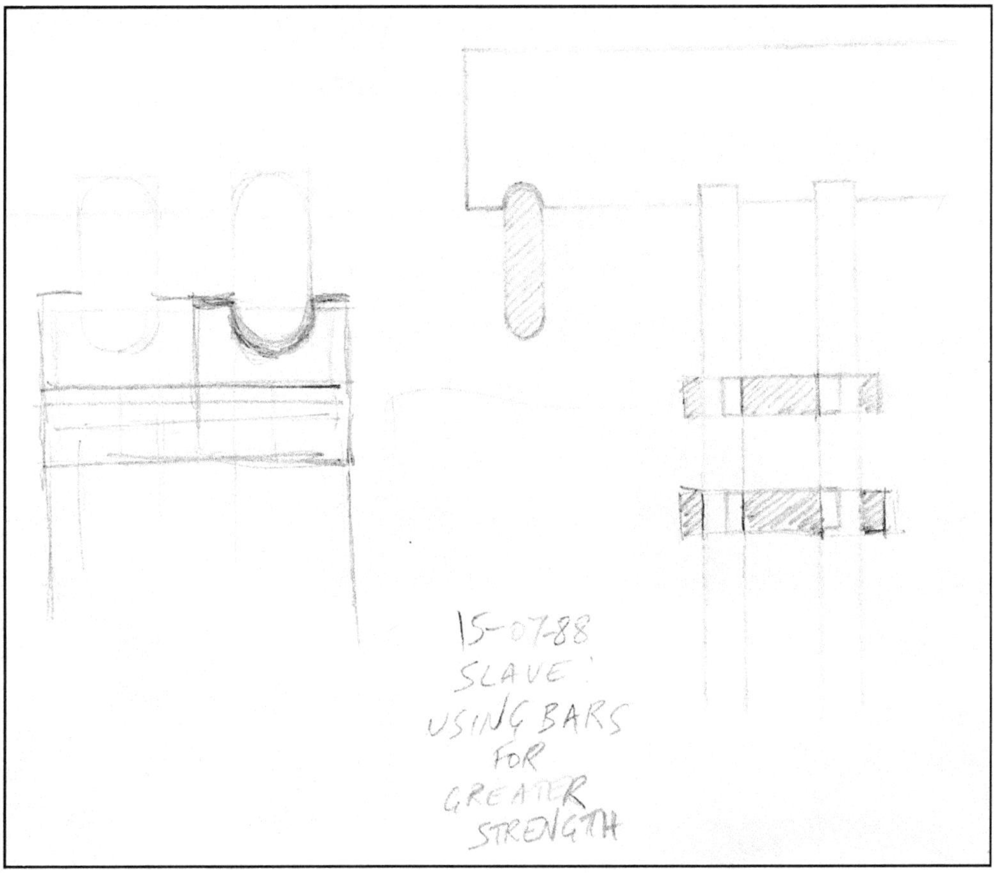
15-07-88
SLAVE
USING BARS
FOR
GREATER
STRENGTH

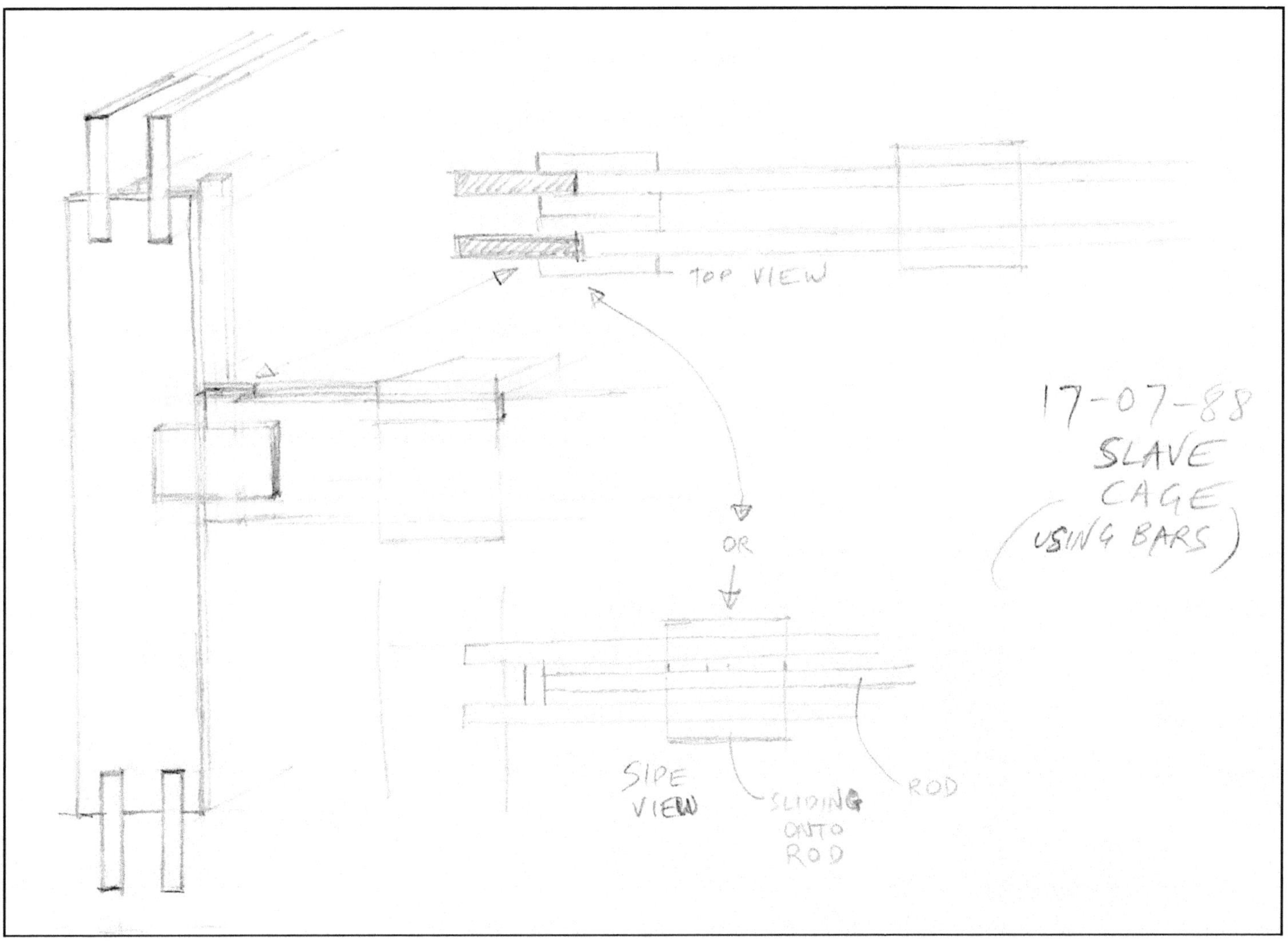
TOP VIEW
OR
17-07-88
SLAVE
CAGE
(USING BARS)
SIDE VIEW
SLIDING
ONTO
ROD
ROD

3-D MOVEMENT

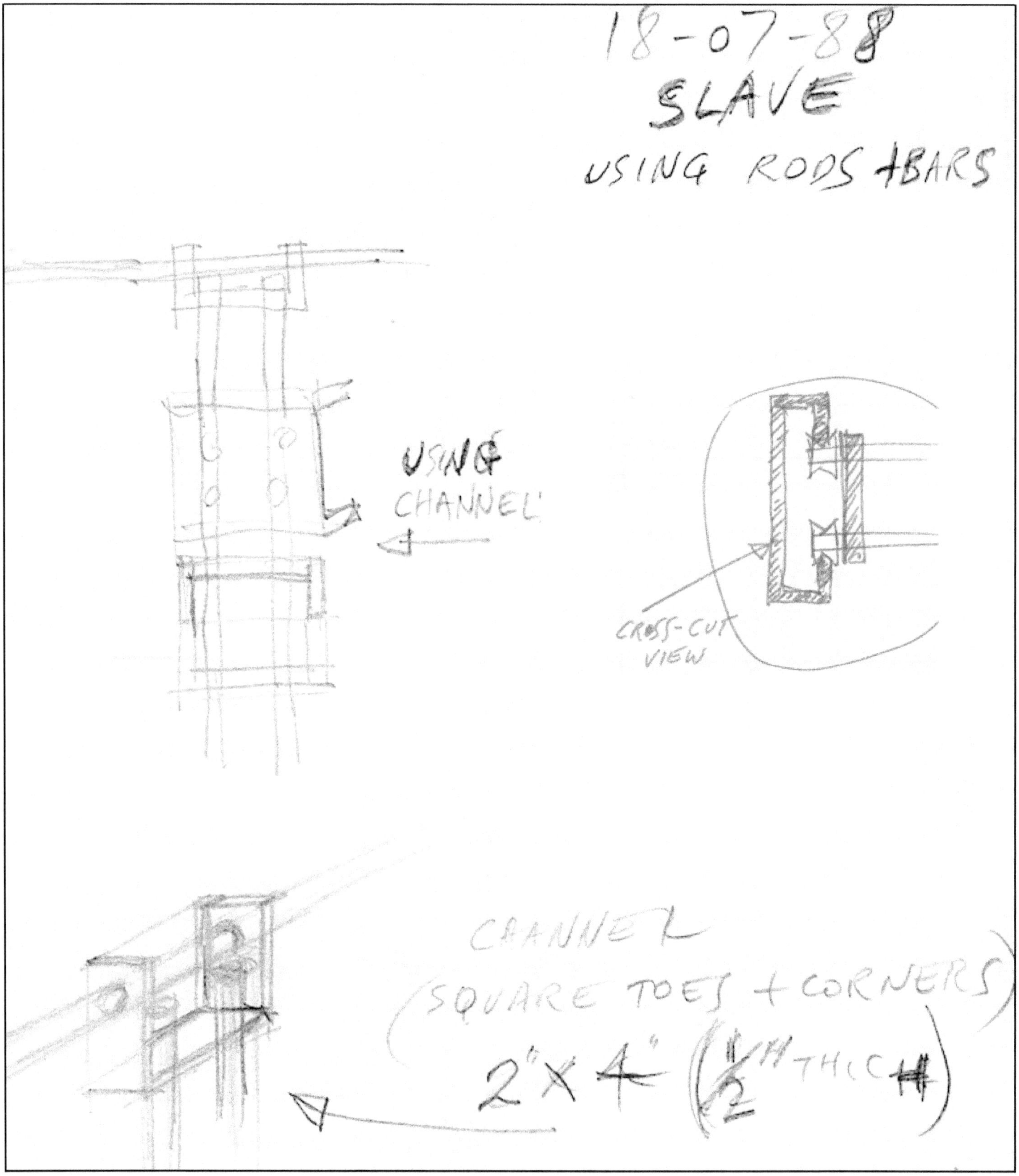
18-07-88
SLAVE
USING RODS + BARS
USING CHANNEL
CROSS-CUT VIEW
CHANNEL
(SQUARE TOES + CORNERS)
2" X 4" (1/2" THICK)

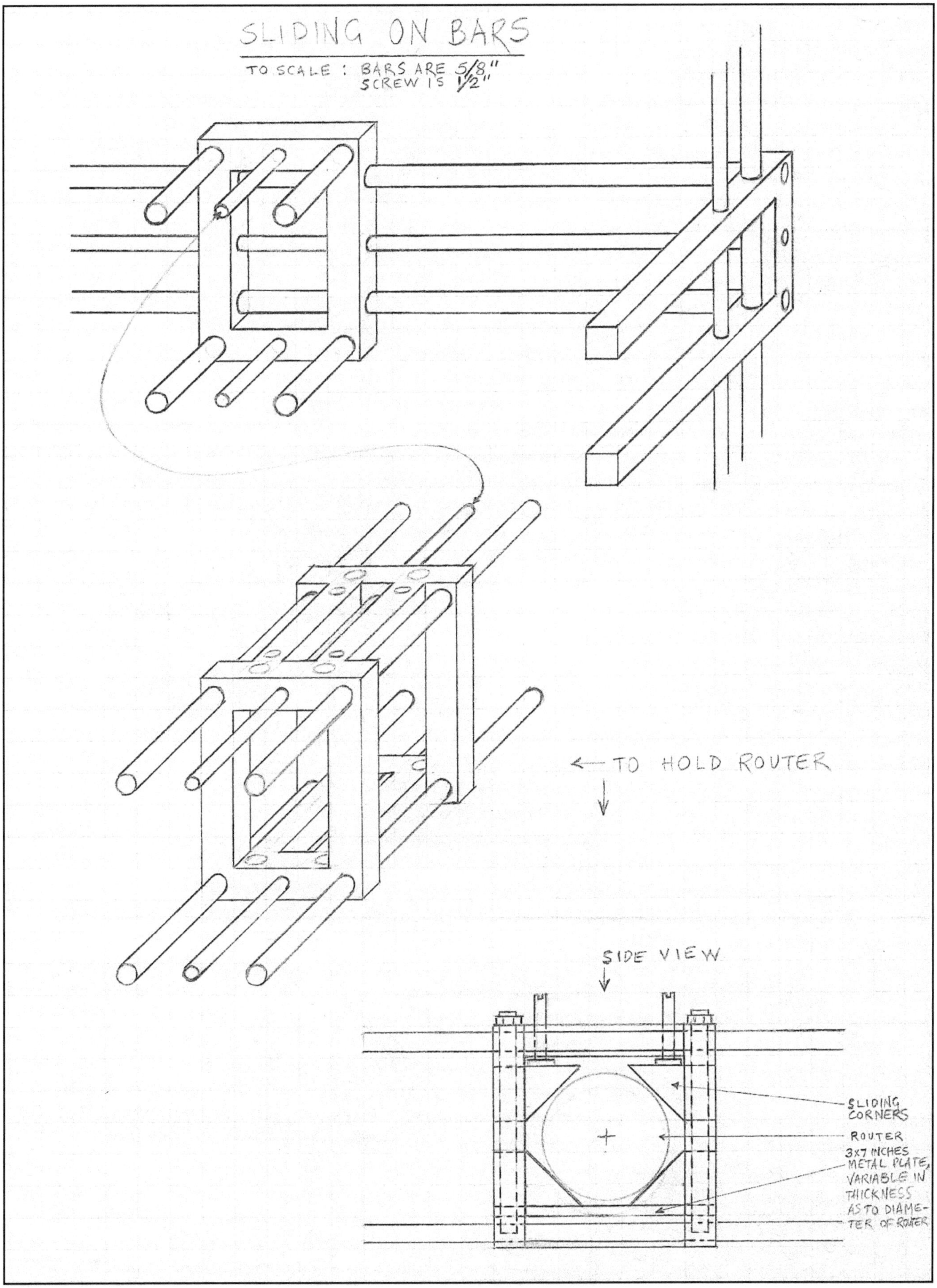
SLIDING ON BARS
TO SCALE : BARS ARE 5/8"
SCREW IS 1/2"
TO HOLD ROUTER
SIDE VIEW
SLIDING CORNERS
ROUTER
3x7 INCHES METAL PLATE, VARIABLE IN THICKNESS AS TO DIAMETER OF ROUTER

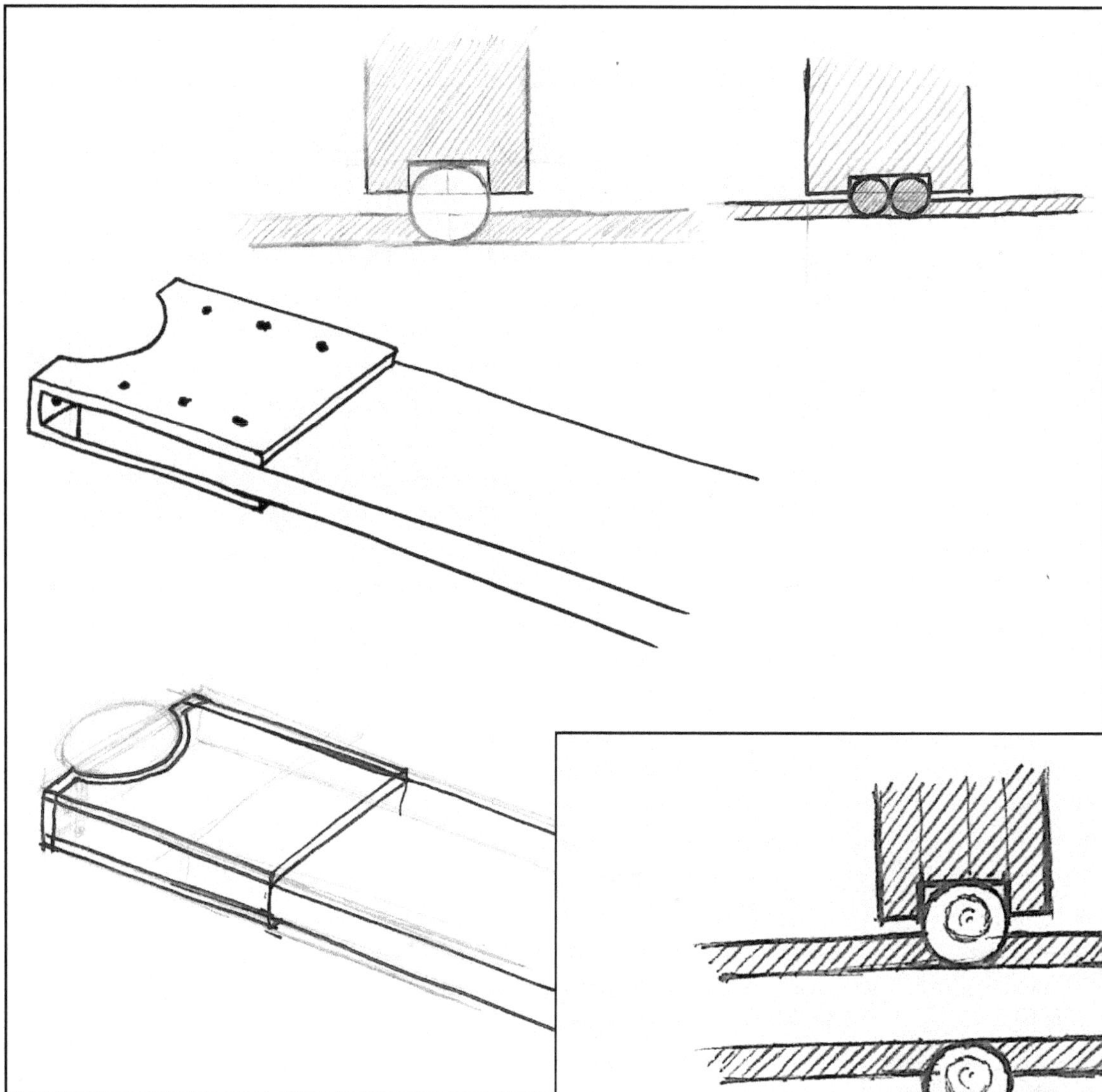

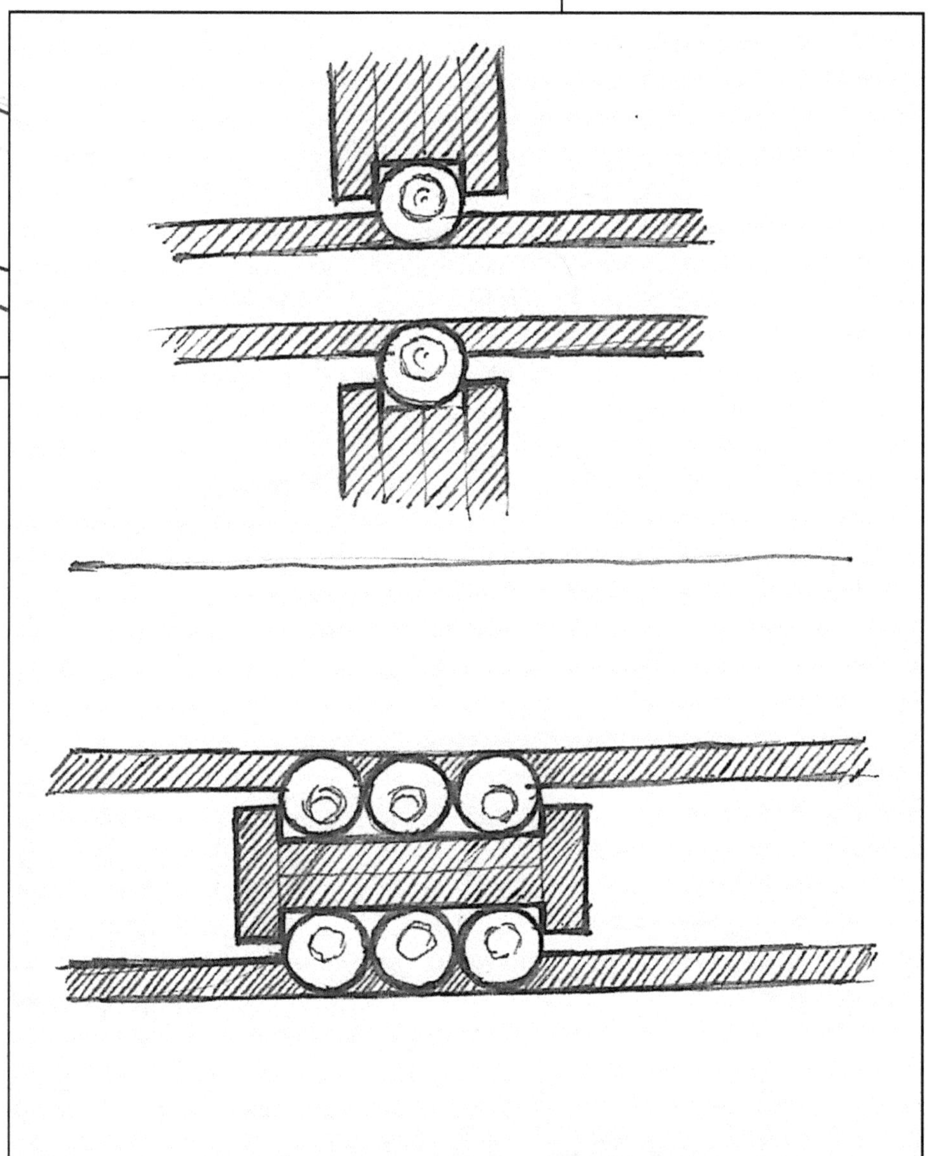

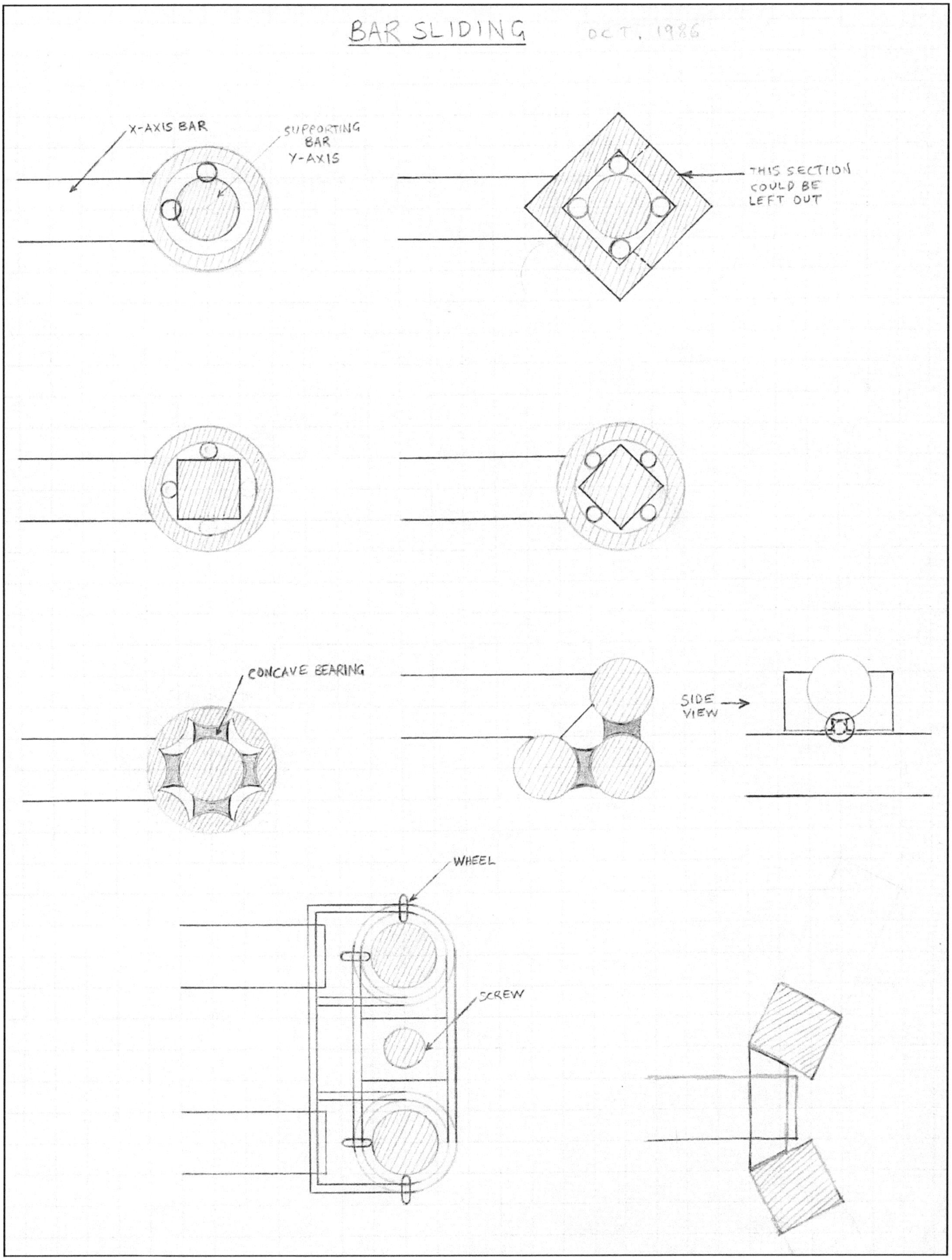
BAR SLIDING
OCT. 1986
X-AXIS BAR
SUPPORTING BAR Y-AXIS
THIS SECTION COULD BE LEFT OUT
CONCAVE BEARING
SIDE VIEW
WHEEL
SCREW

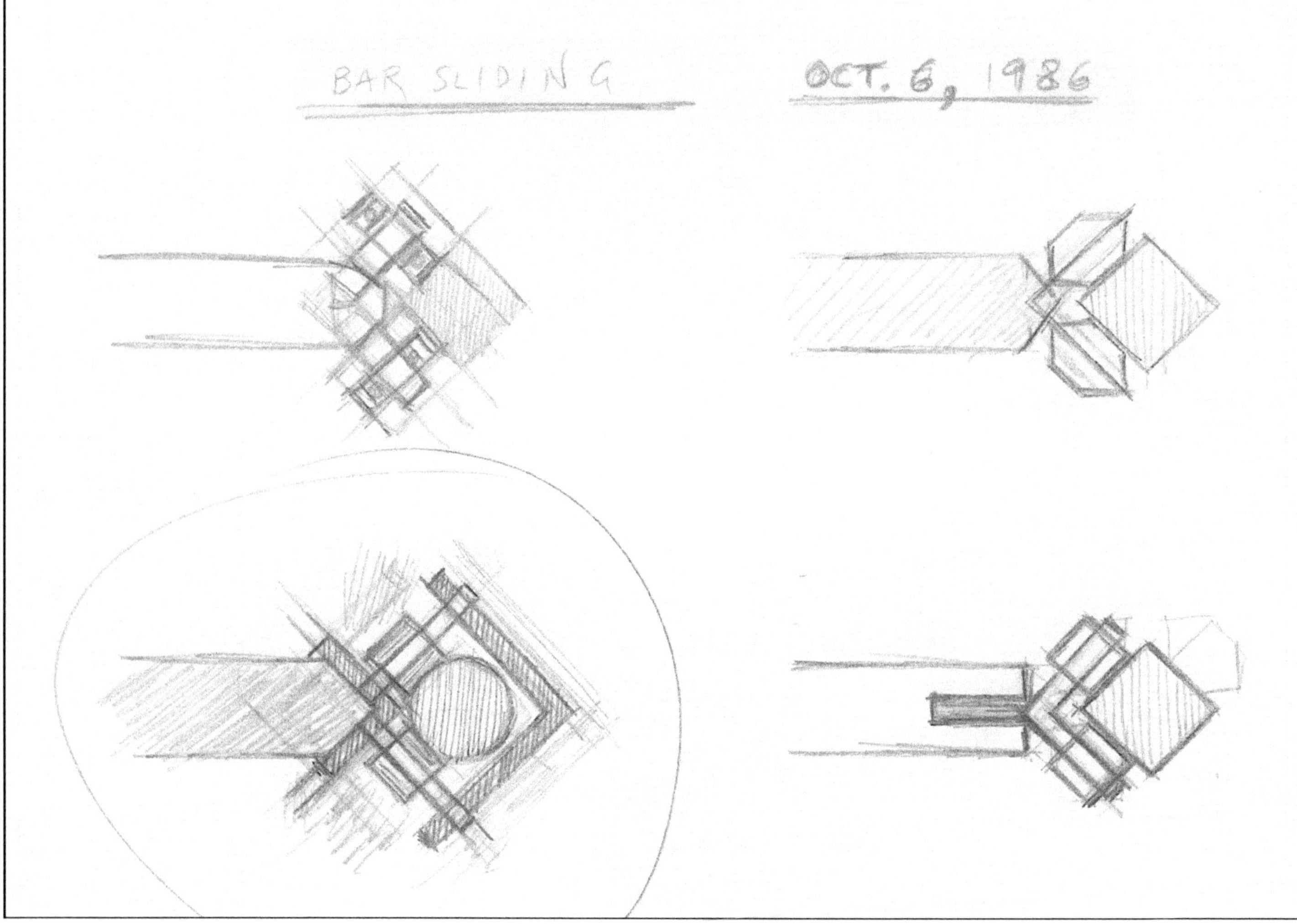
BAR SLIDING
OCT. 6, 1986

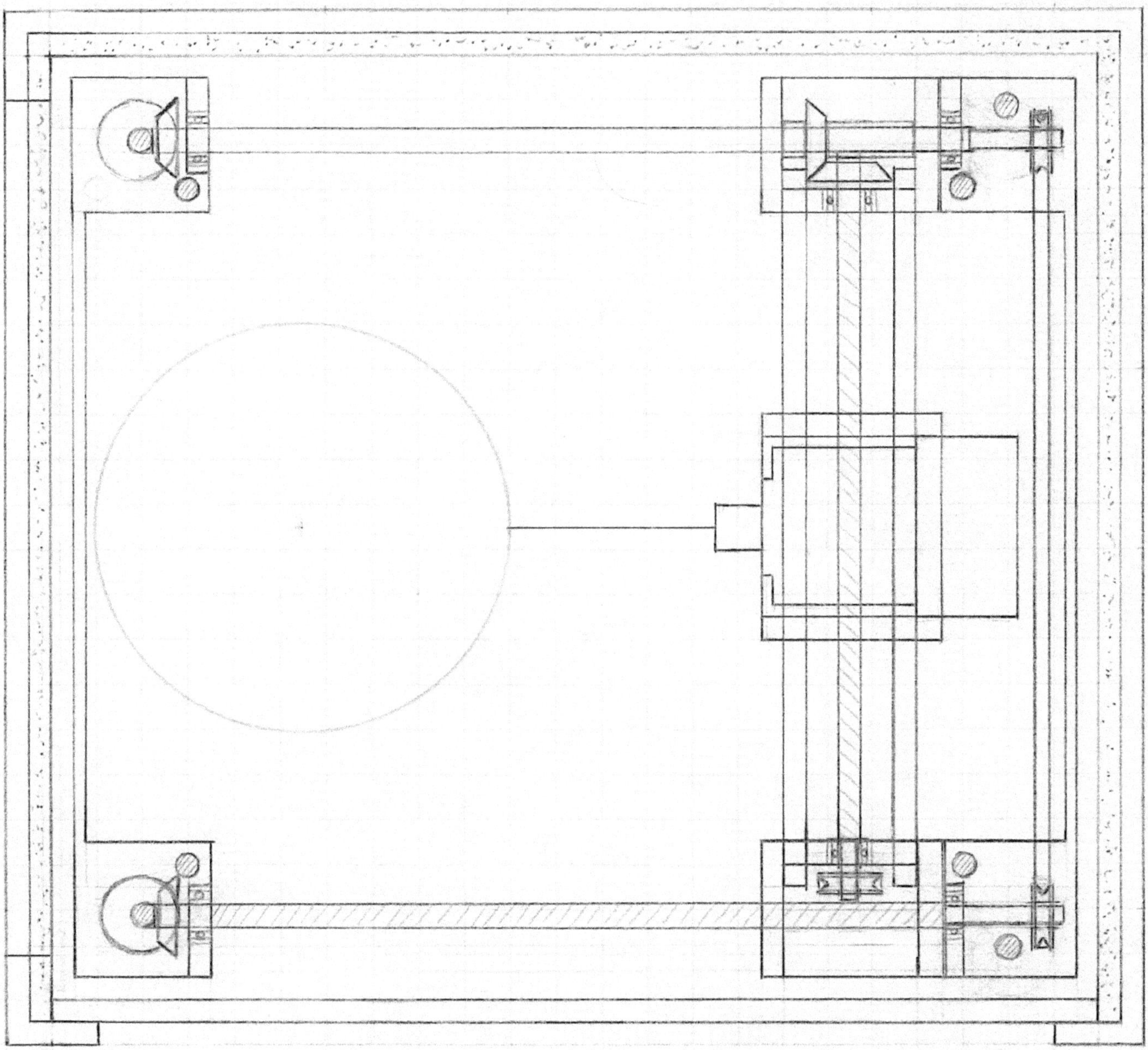

23 X 25 X 21

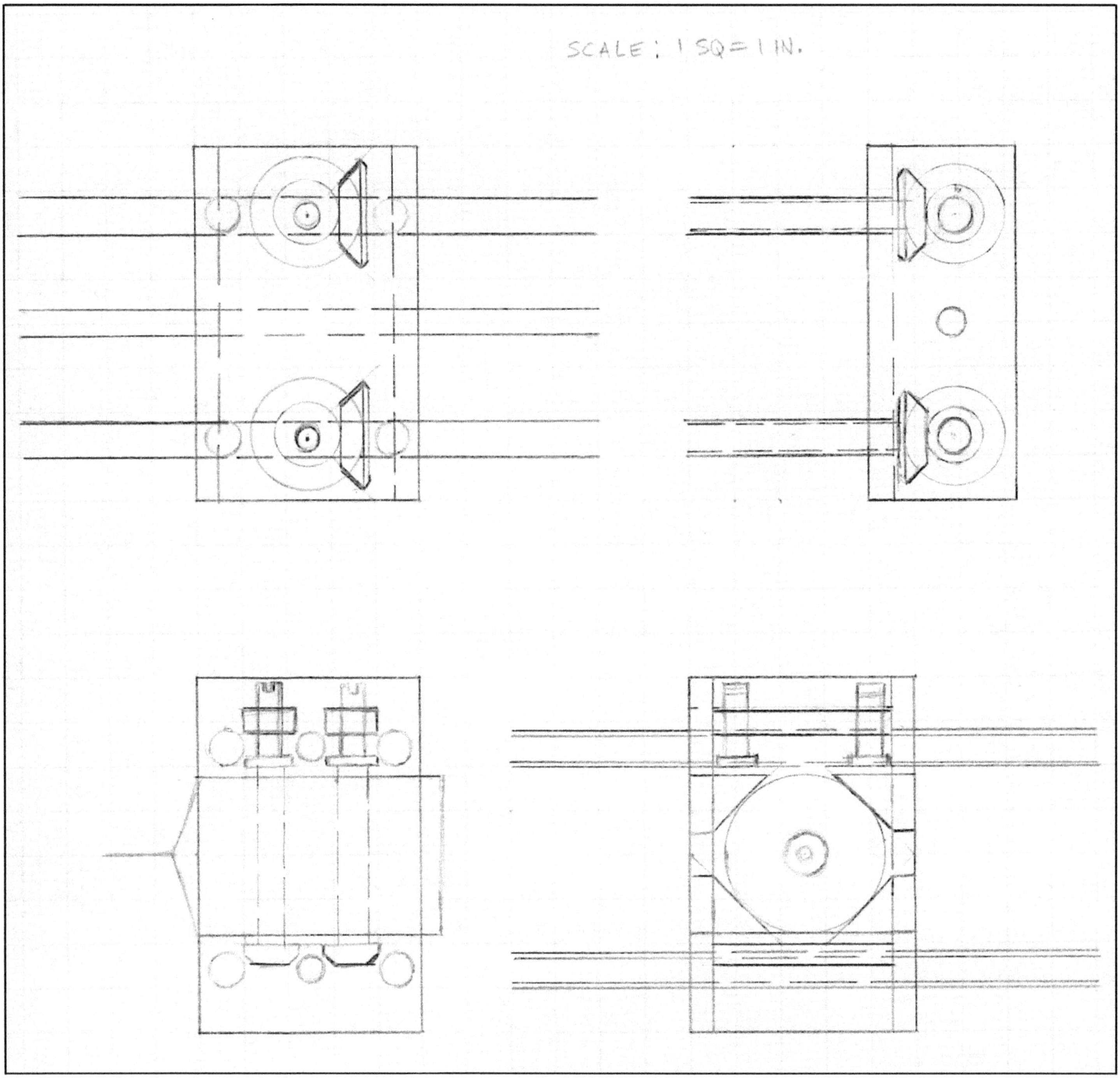
SCALE: 1 SQ = 1 IN.

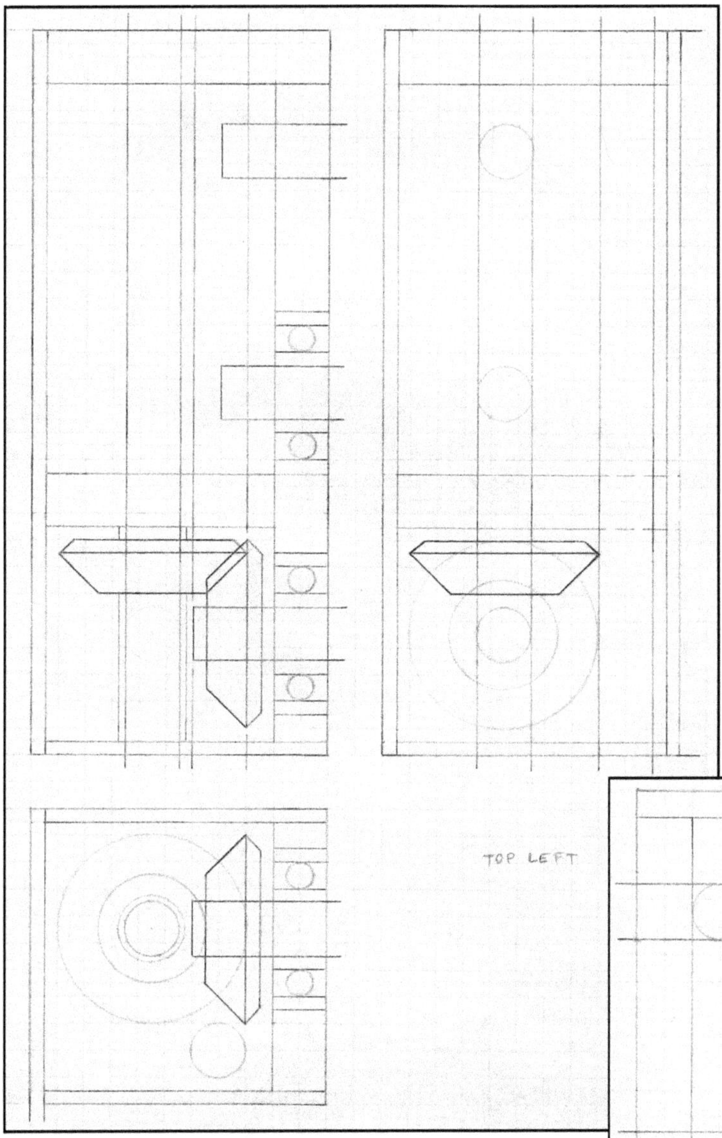
TOP LEFT

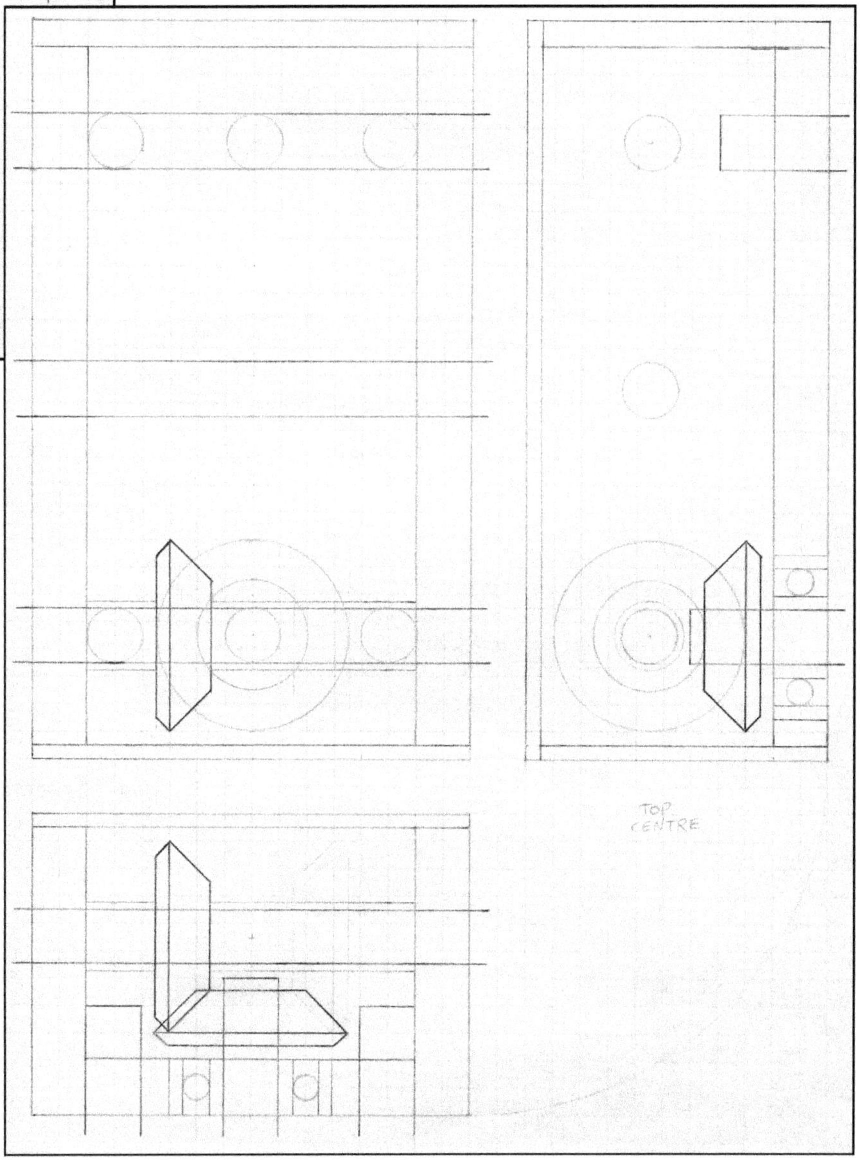
TOP CENTRE

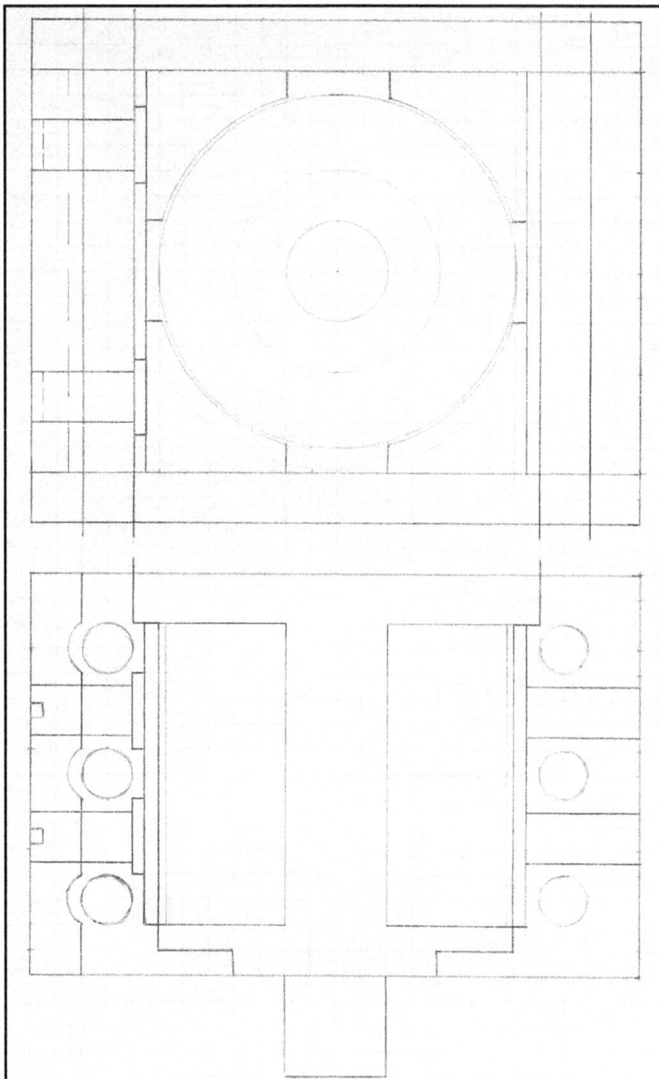

ROUTER

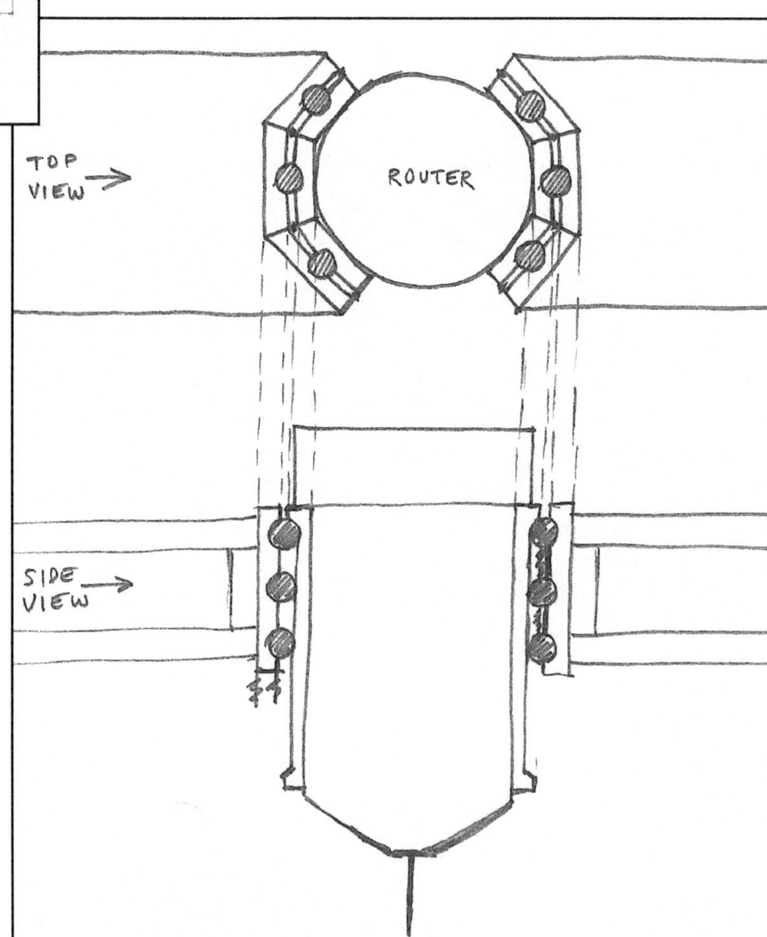

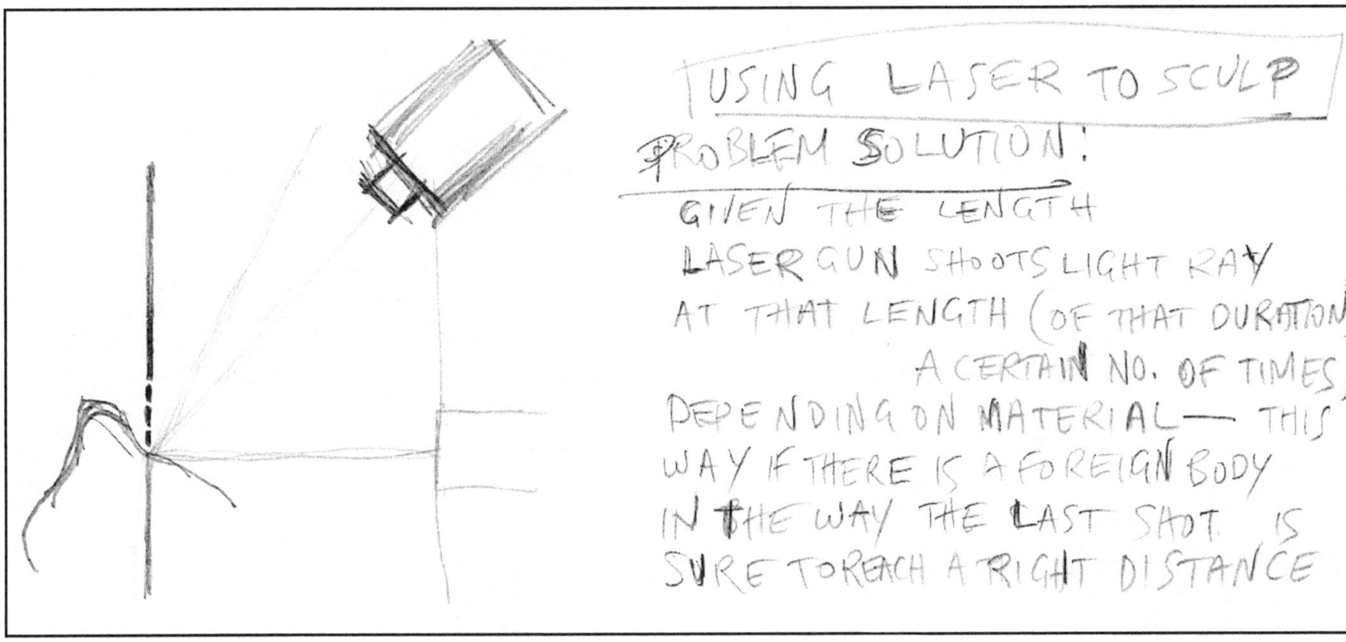
USING LASER TO SCULP
PROBLEM SOLUTION:
GIVEN THE LENGTH
LASERGUN SHOOTS LIGHT RAY
AT THAT LENGTH (OF THAT DURATION)
A CERTAIN NO. OF TIMES,
DEPENDING ON MATERIAL — THIS
WAY IF THERE IS A FOREIGN BODY
IN THE WAY THE LAST SHOT IS
SURE TO REACH A RIGHT DISTANCE

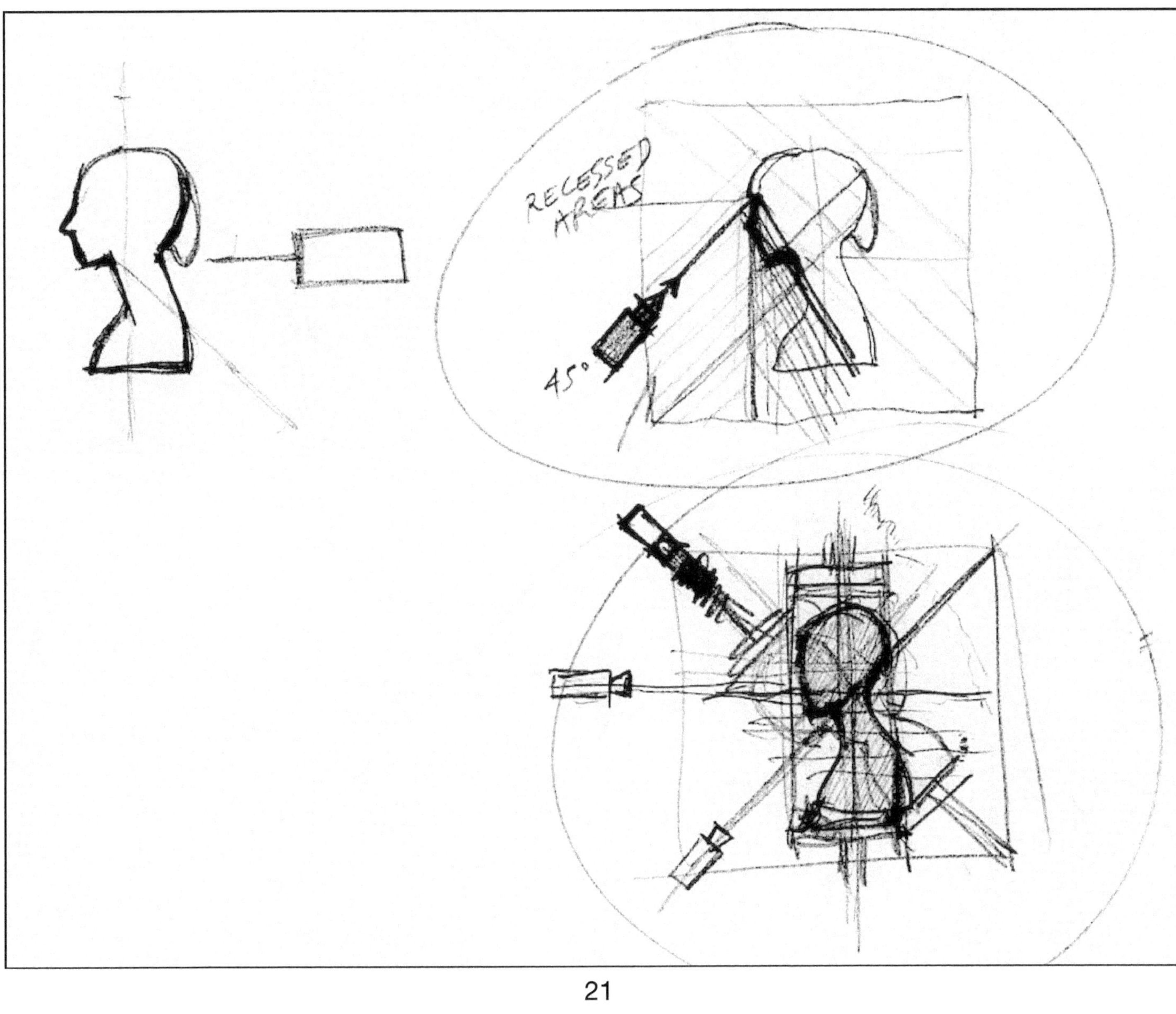
RECESSED AREAS
45°

LINKAGE - PANTOGRAPH

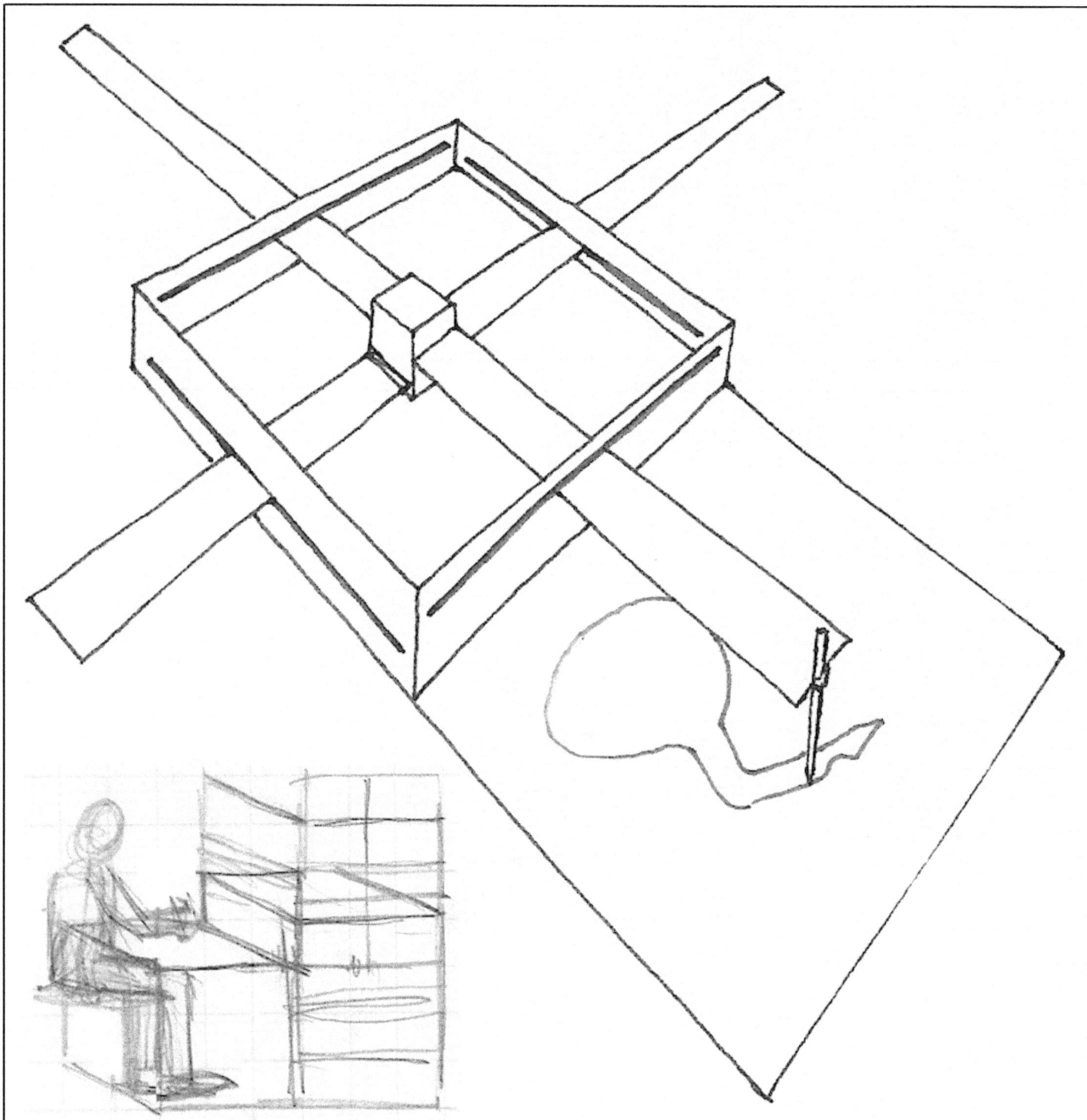

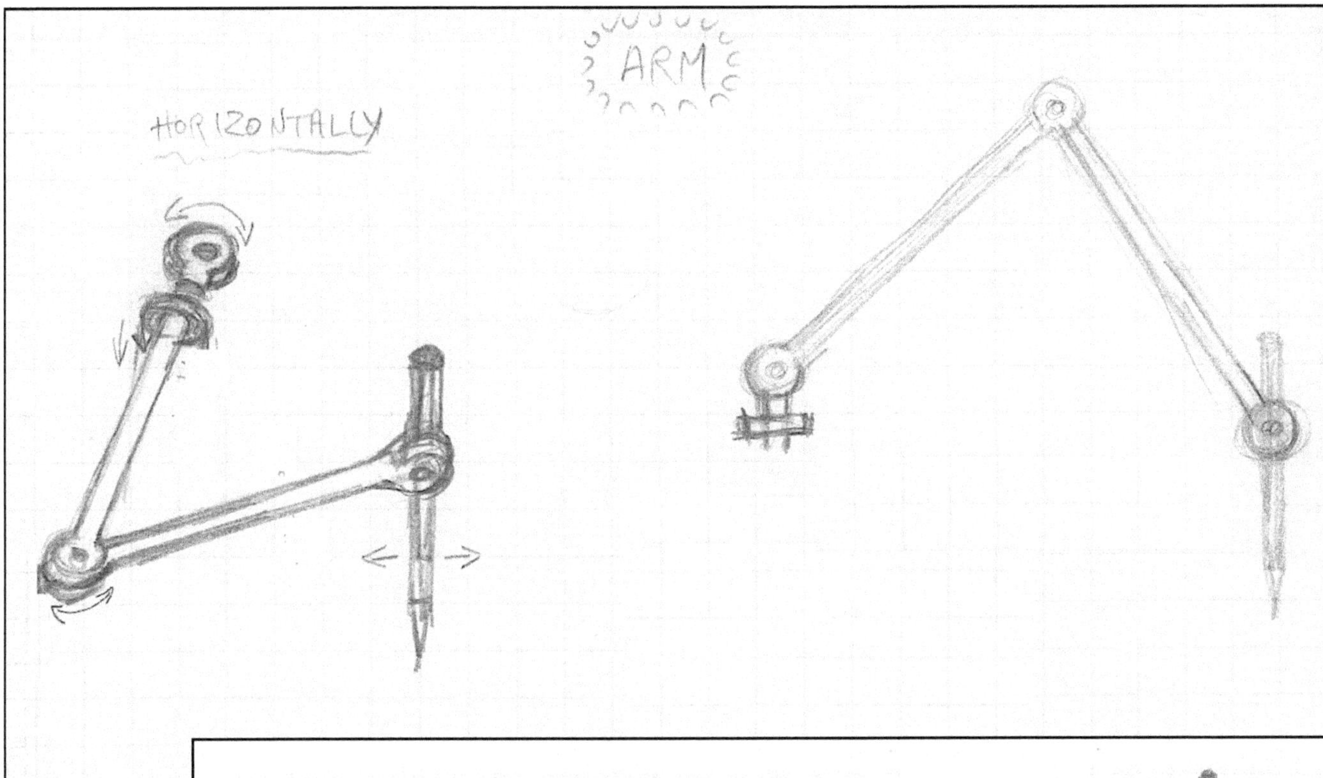
ARM
HORIZONTALLY

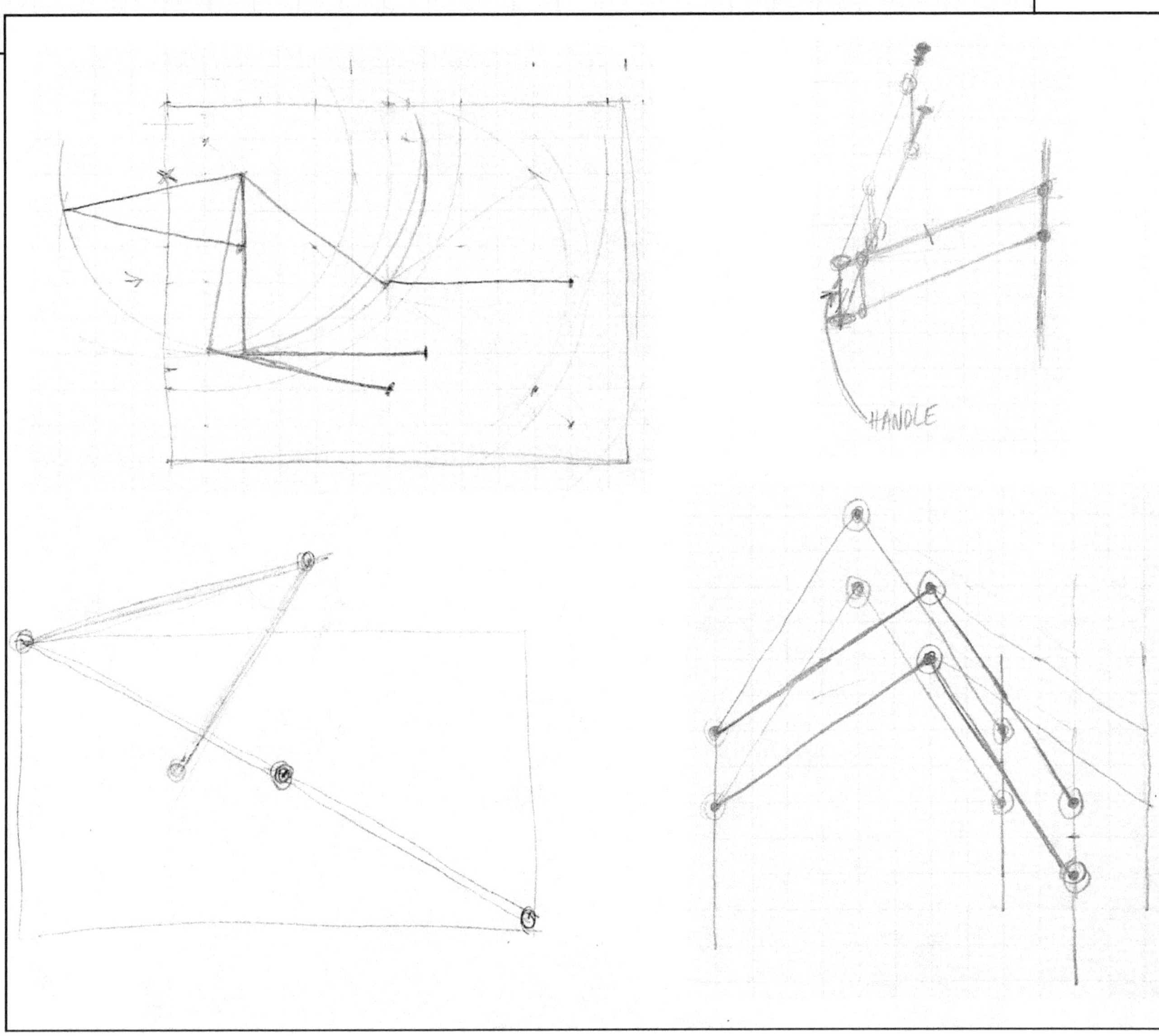
HANDLE

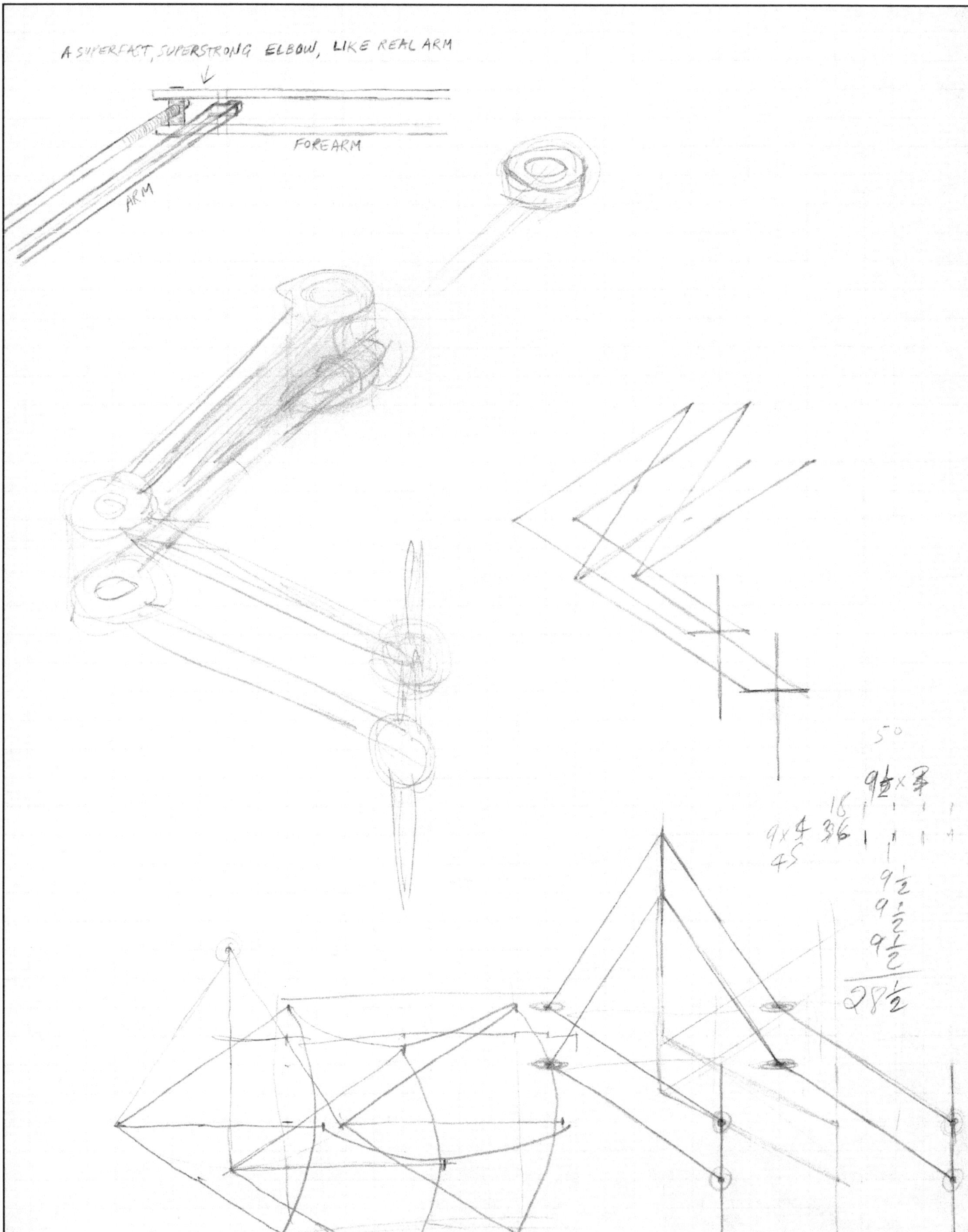
A SUPERFAST, SUPERSTRONG ELBOW, LIKE REAL ARM
FOREARM
ARM
50
9½ × 3
18
9×4 36
45
9½
9½
9½
28½

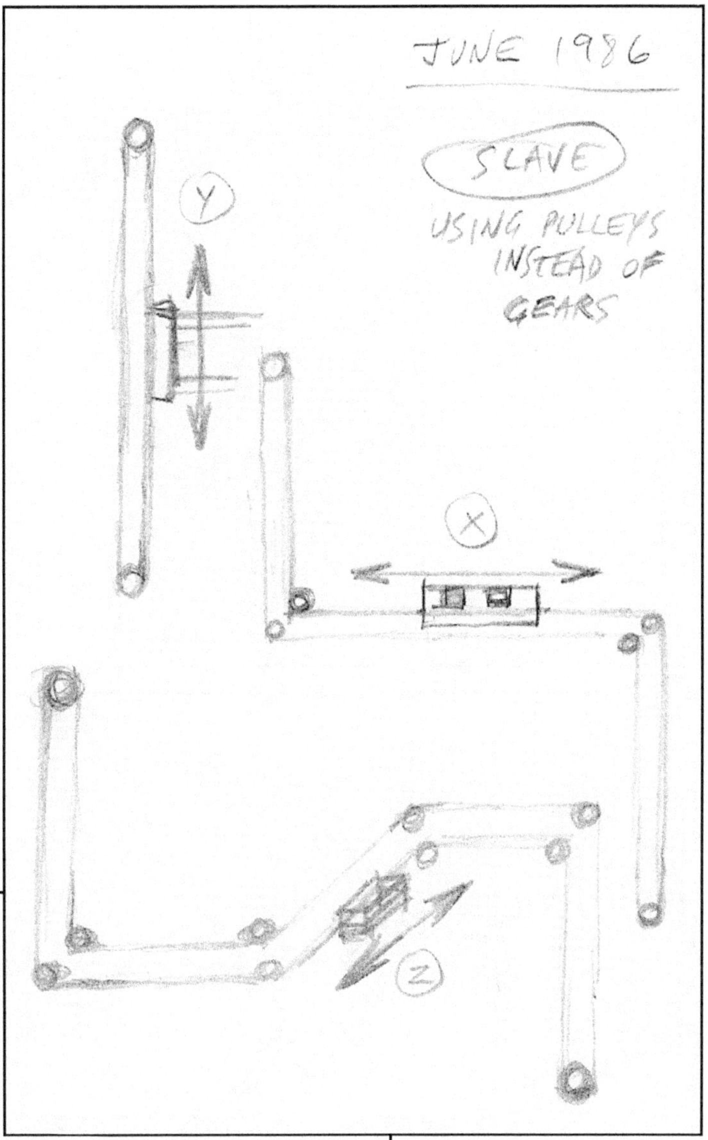
JUNE 1986
SLAVE
USING PULLEYS INSTEAD OF GEARS
Y
X
Z

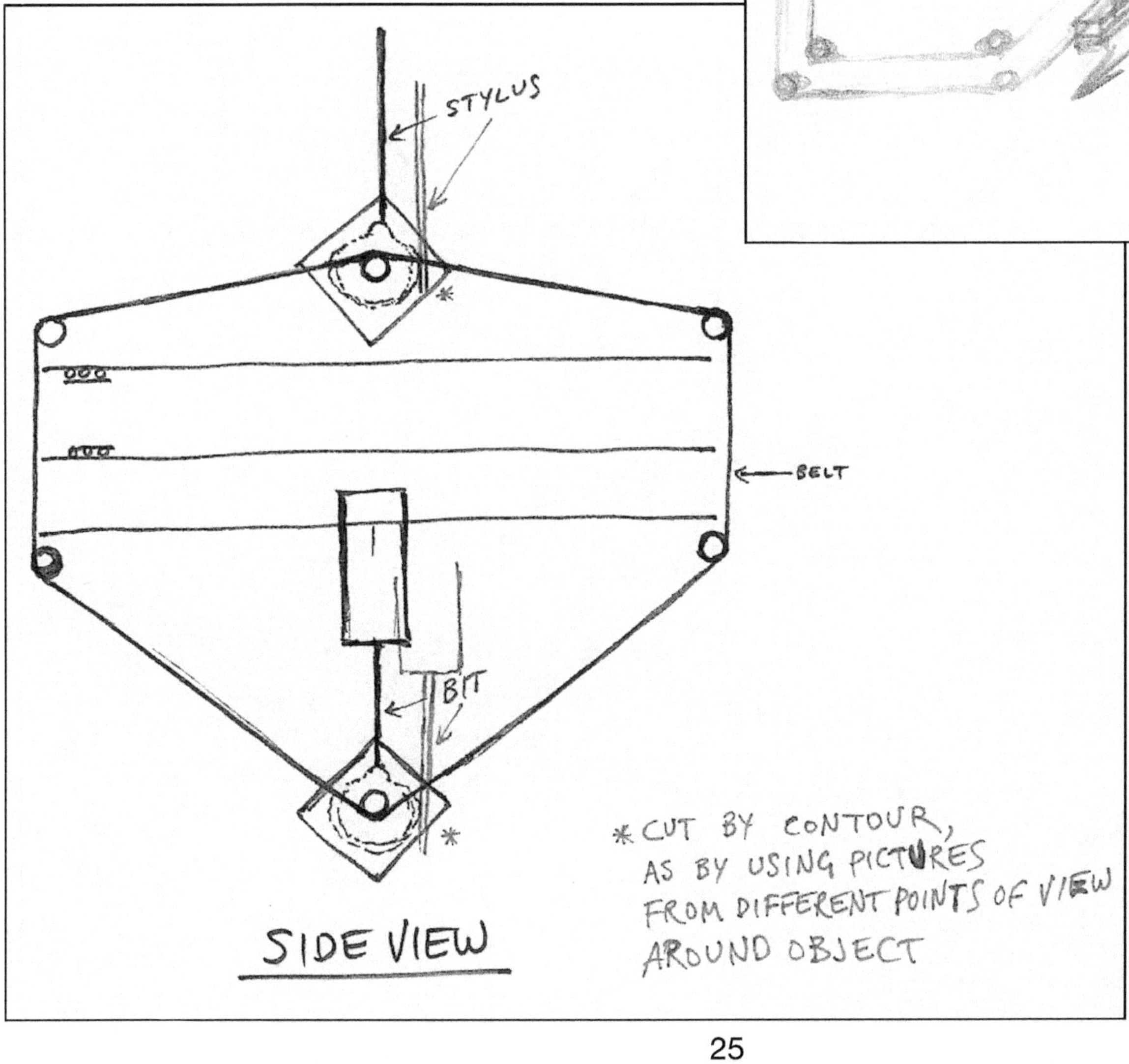
STYLUS
BELT
BIT
* CUT BY CONTOUR, AS BY USING PICTURES FROM DIFFERENT POINTS OF VIEW AROUND OBJECT
SIDE VIEW

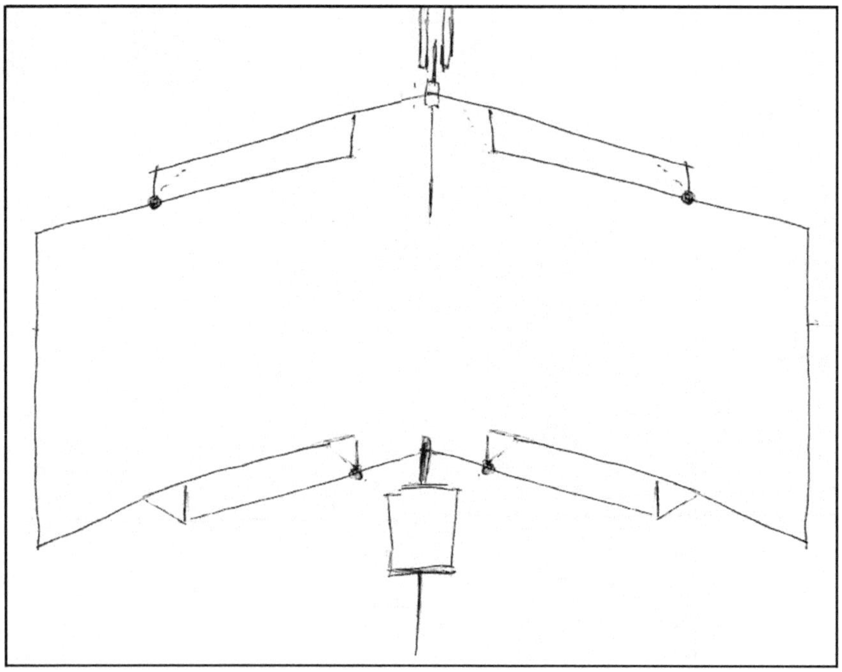

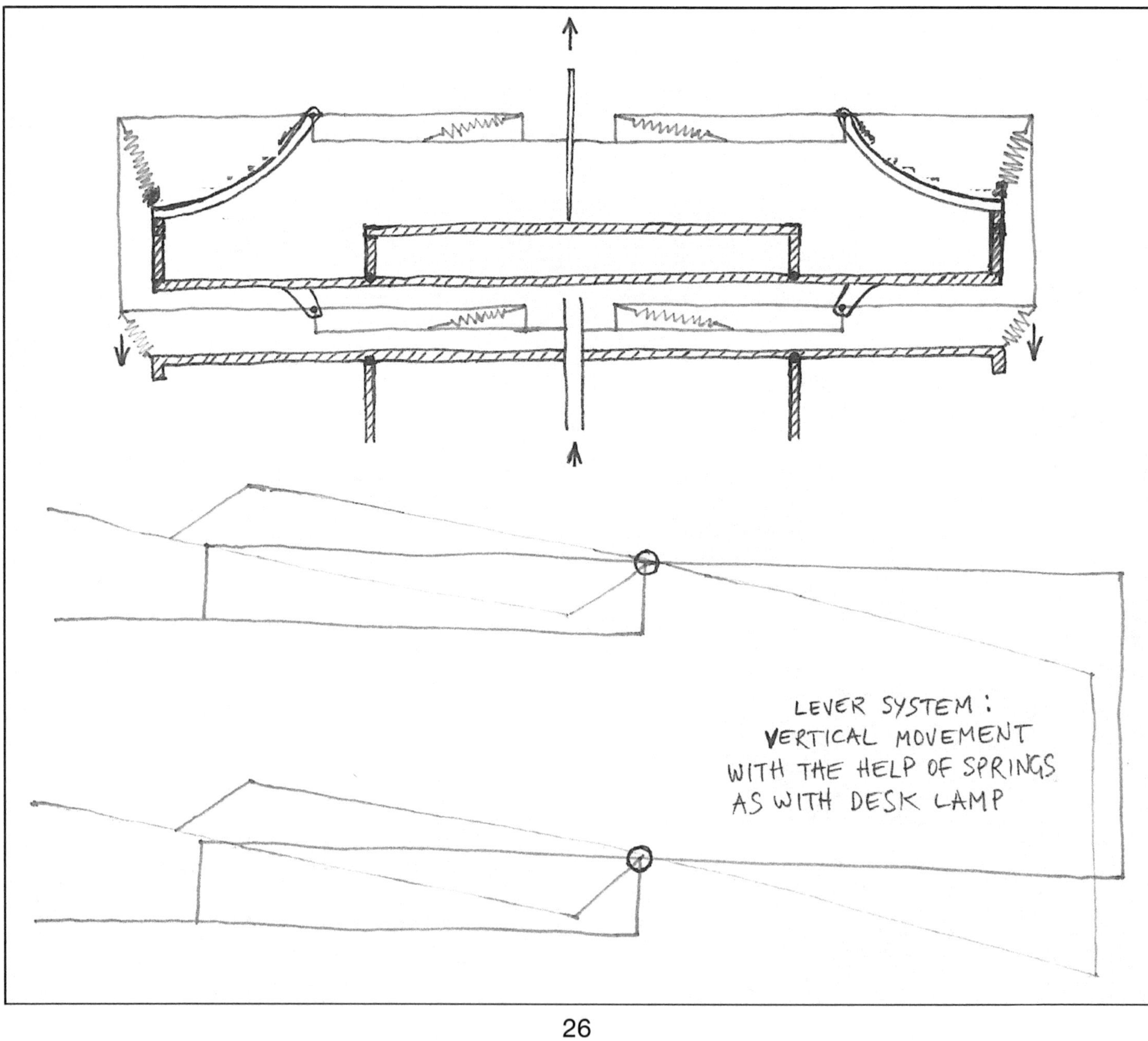
LEVER SYSTEM:
VERTICAL MOVEMENT
WITH THE HELP OF SPRINGS
AS WITH DESK LAMP

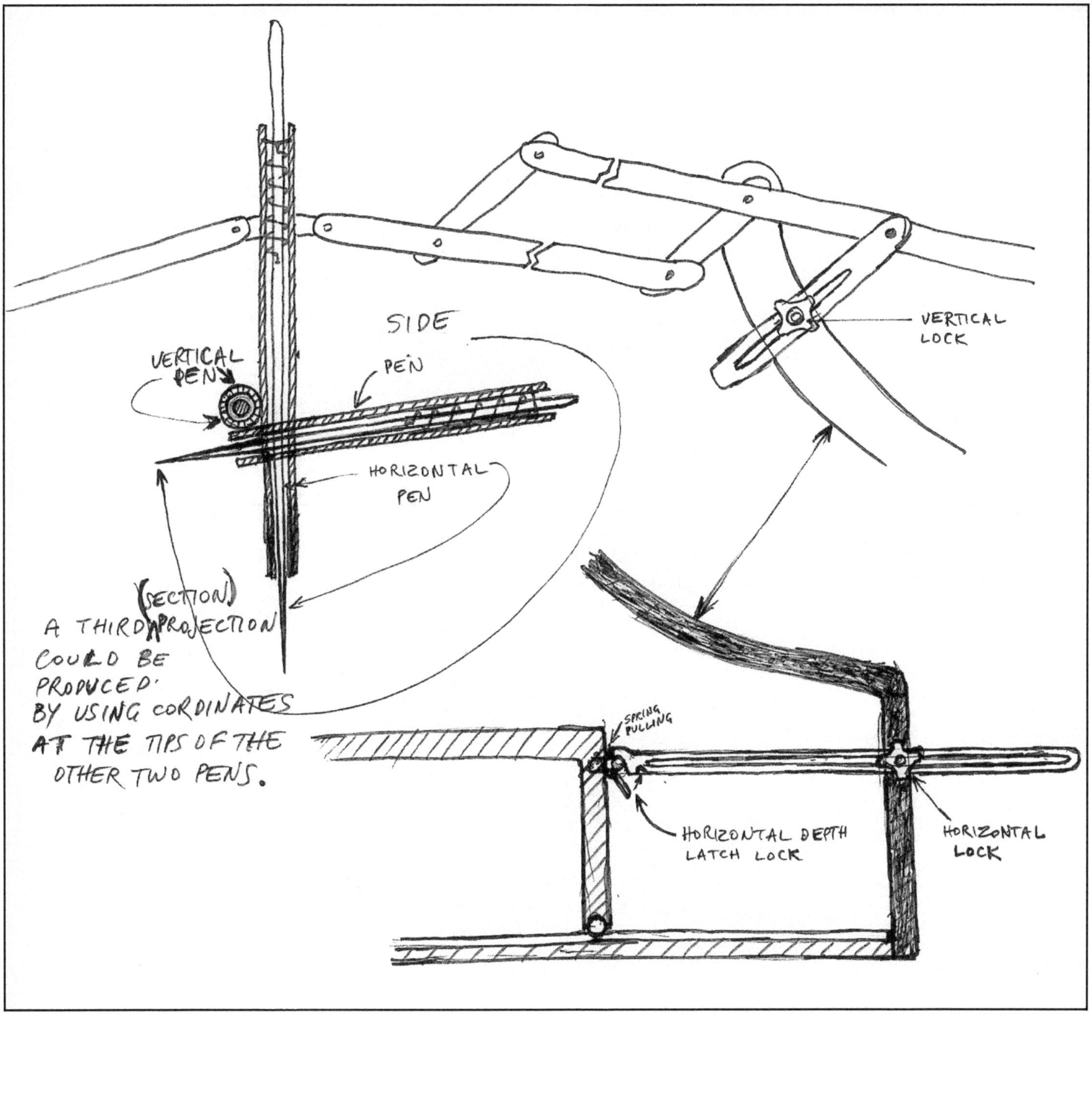
SIDE
VERTICAL PEN
PEN
HORIZONTAL PEN
VERTICAL LOCK
(SECTION)
A THIRD PROJECTION COULD BE PRODUCED BY USING CORDINATES AT THE TIPS OF THE OTHER TWO PENS.
SPRING PULLING
HORIZONTAL DEPTH LATCH LOCK
HORIZONTAL LOCK

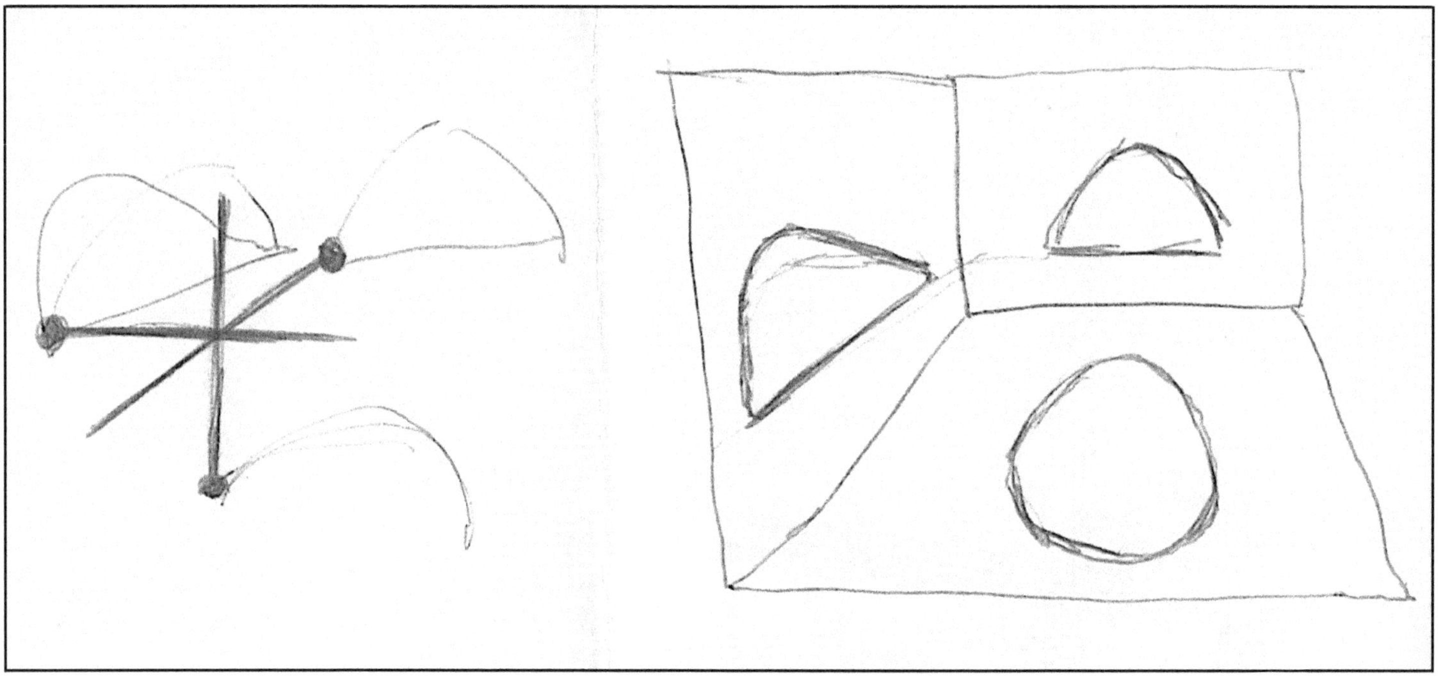

A 3-pen pantograph can produce a 3-D sculpture from 2-D orthogonal pictures.

LINKAGE - COMPUTER PROGRAM

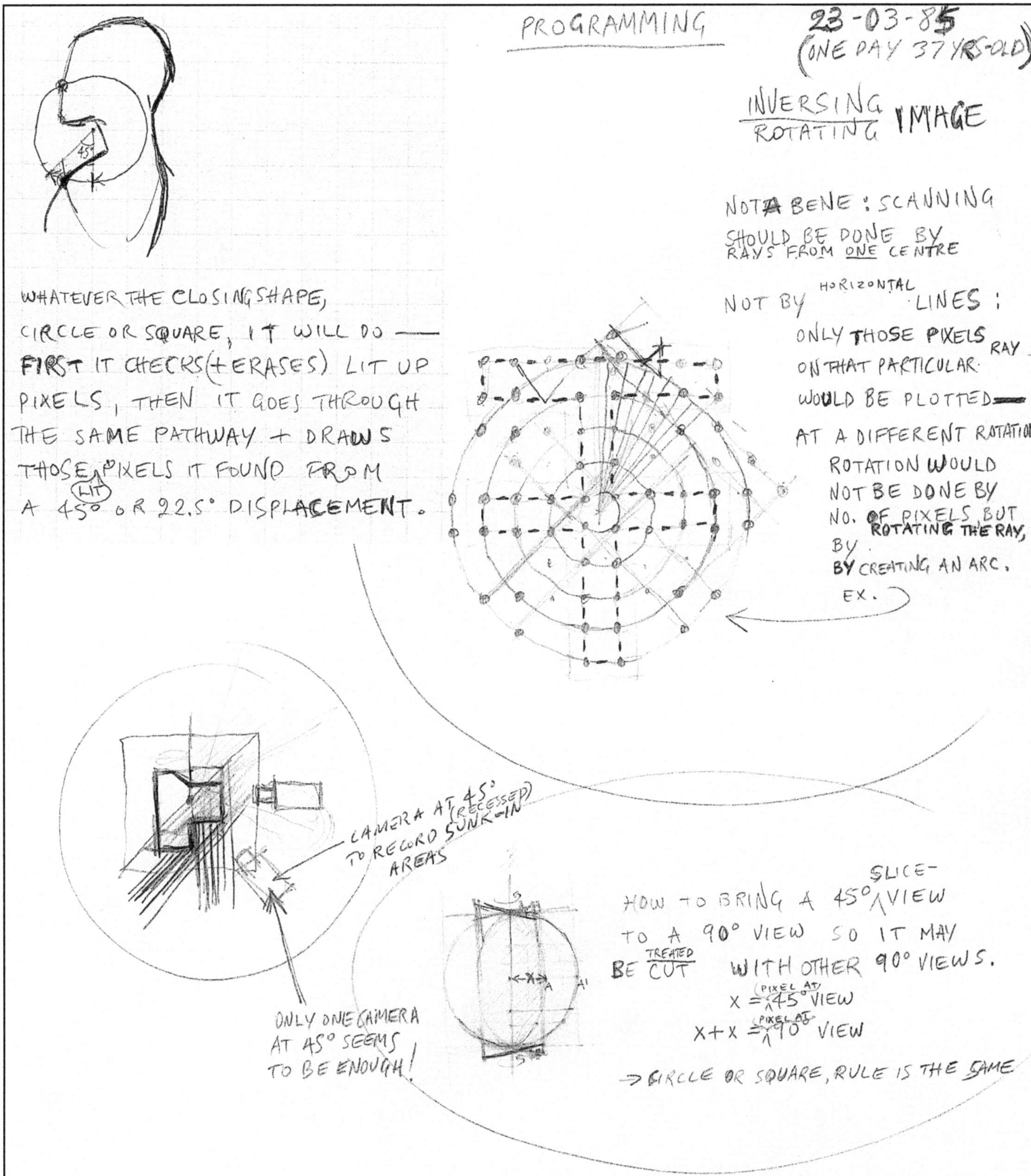

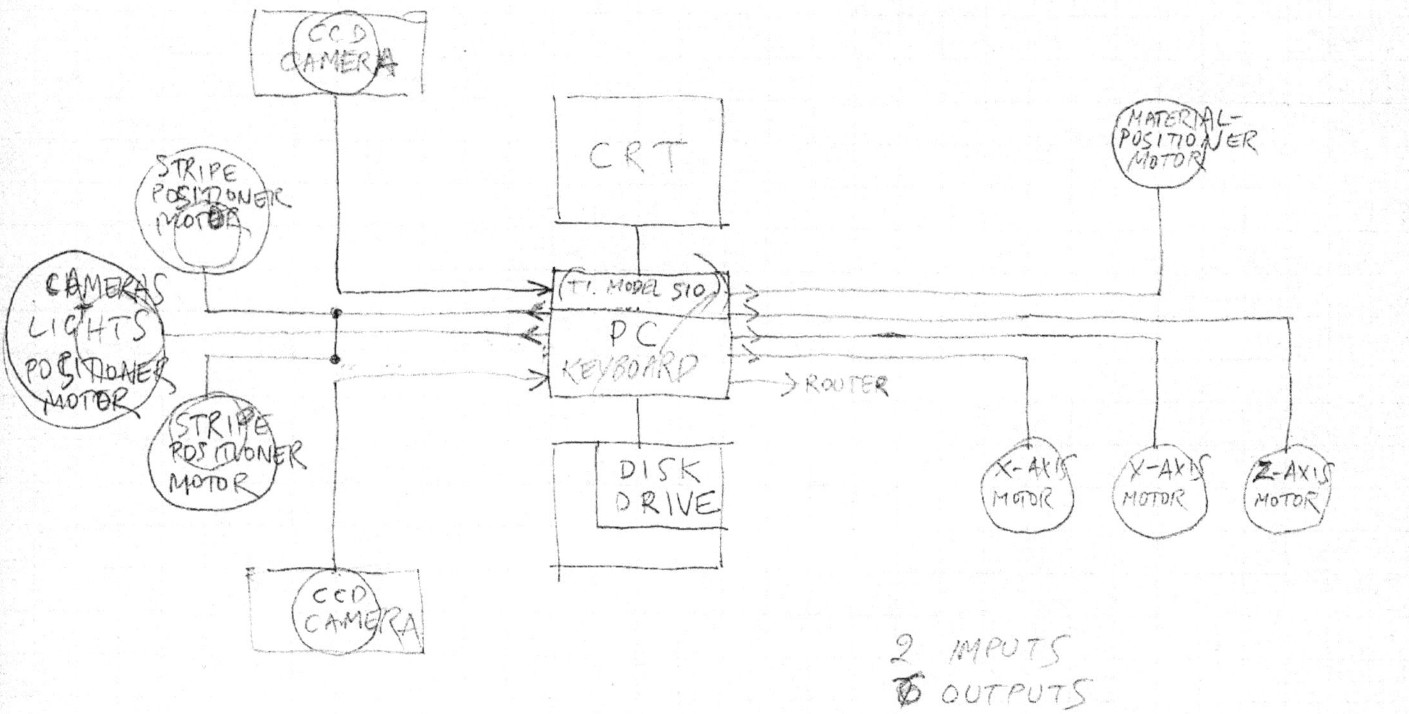

PROGRAMS

- POSITIONING LIGHTS + CAMERAS
- COMING TO LINE + RETRACTING
- FOLLOWING CONTOUR OF LINE
- GENERATING MIDLINE
- SCALING (DETERMINING SIZE)

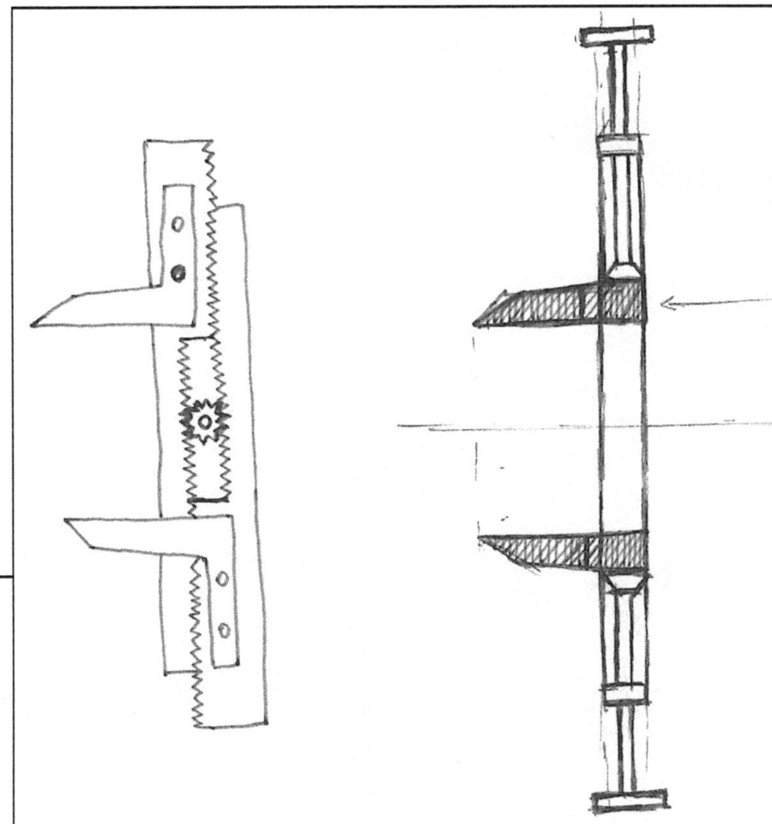

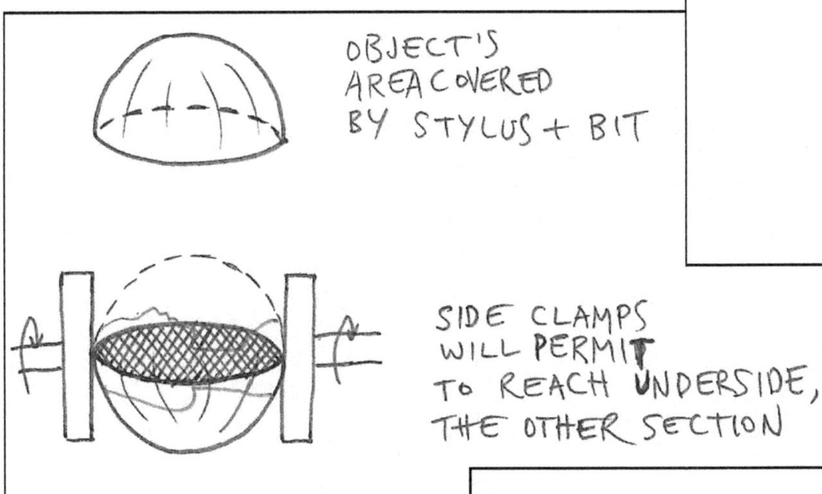
OBJECT'S
AREA COVERED
BY STYLUS + BIT
SIDE CLAMPS
WILL PERMIT
TO REACH UNDERSIDE,
THE OTHER SECTION

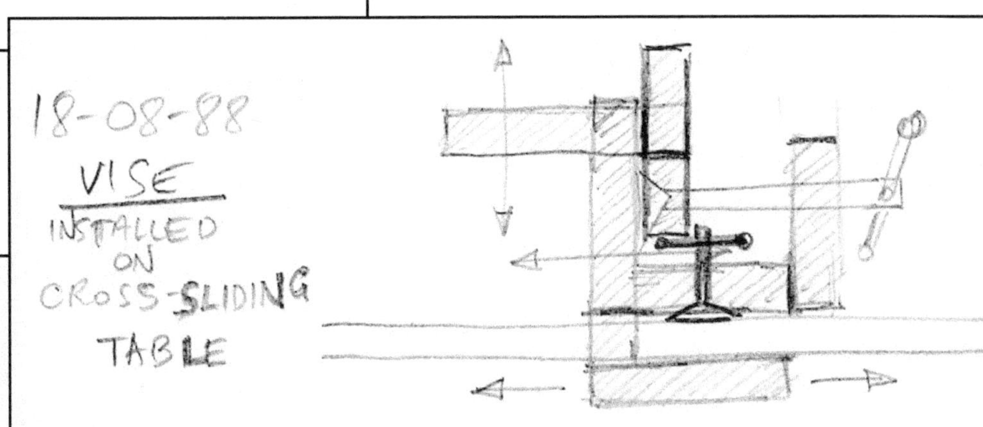
18-08-88
VISE
INSTALLED
ON
CROSS-SLIDING
TABLE

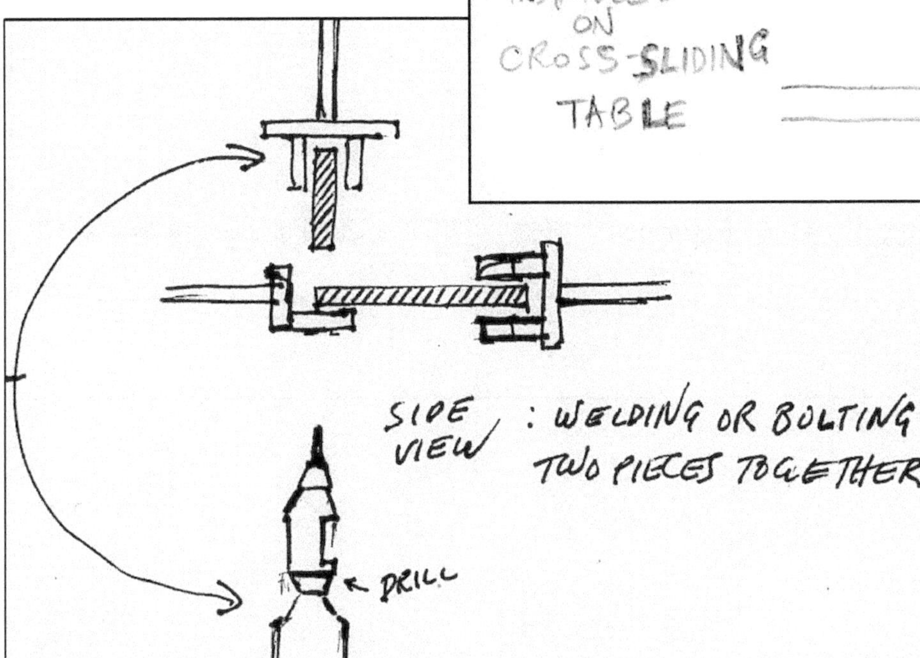
SIDE
VIEW
: WELDING OR BOLTING
TWO PIECES TOGETHER
DRILL

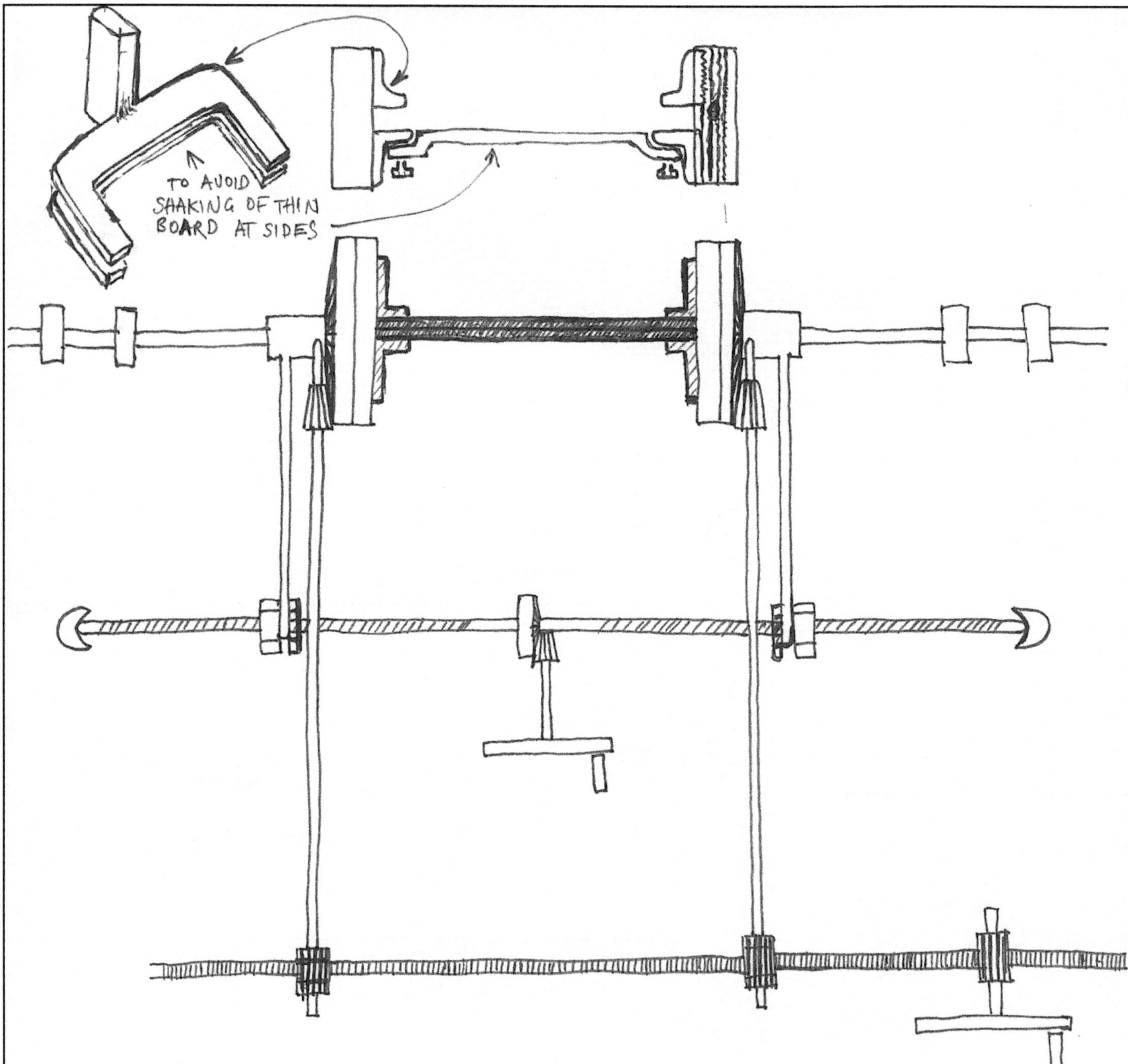
TO AVOID
SHAKING OF THIN
BOARD AT SIDES

18-01-89

MILLING MACHINE

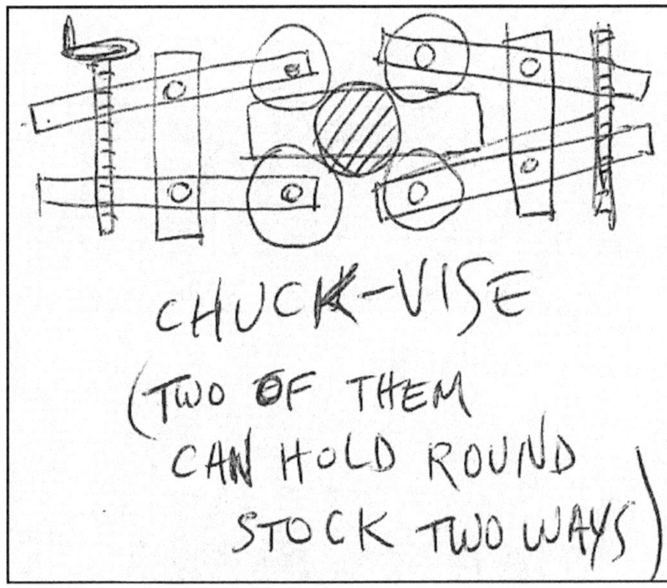

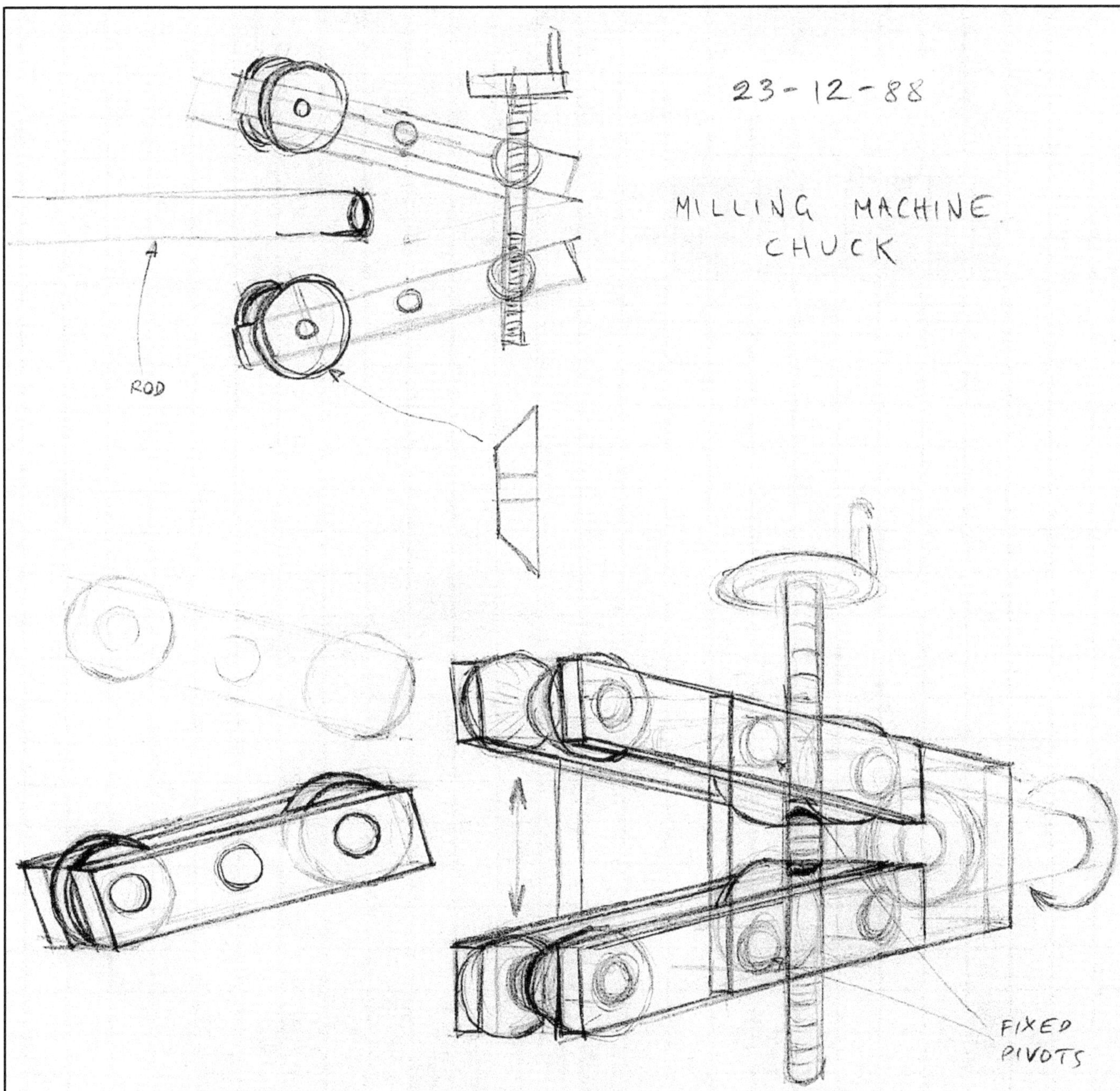

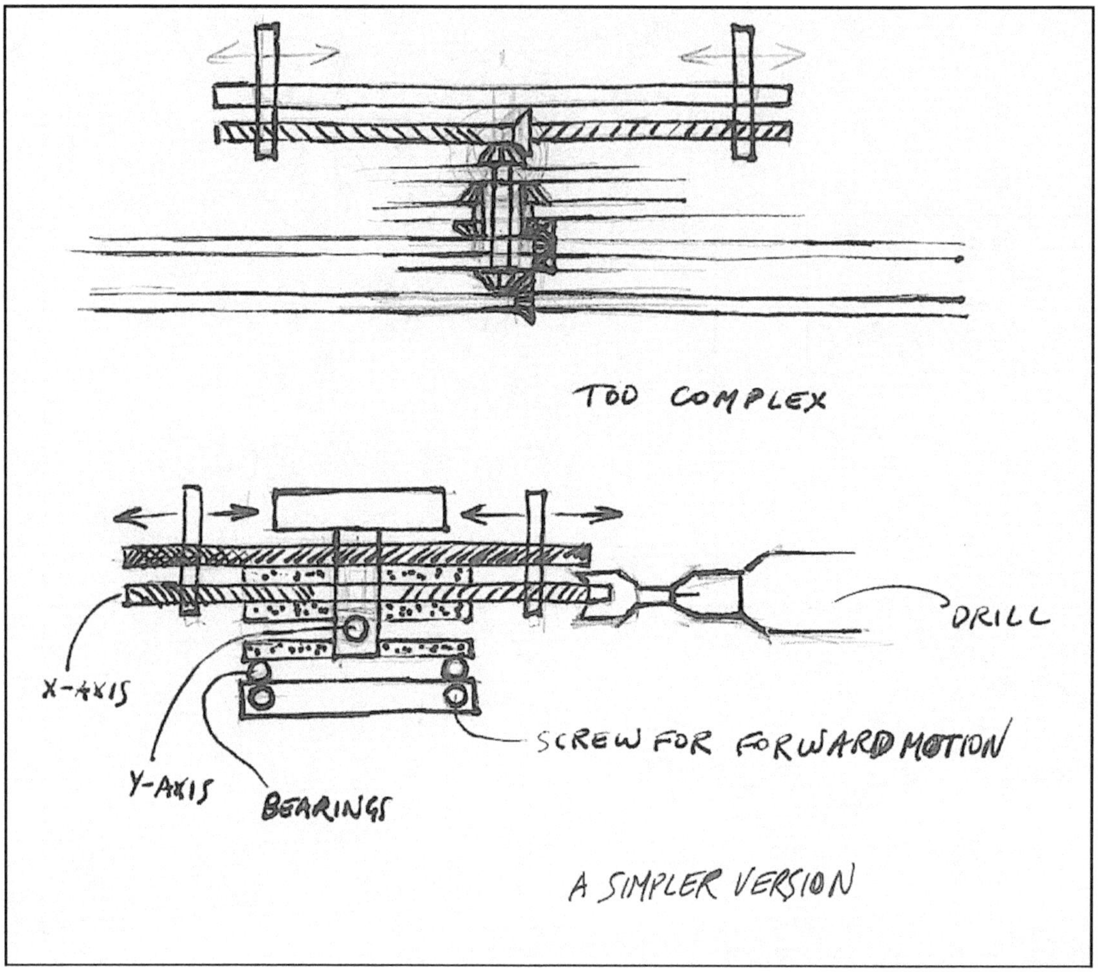
TOO COMPLEX
DRILL
X-AXIS
Y-AXIS
BEARINGS
SCREW FOR FORWARD MOTION
A SIMPLER VERSION

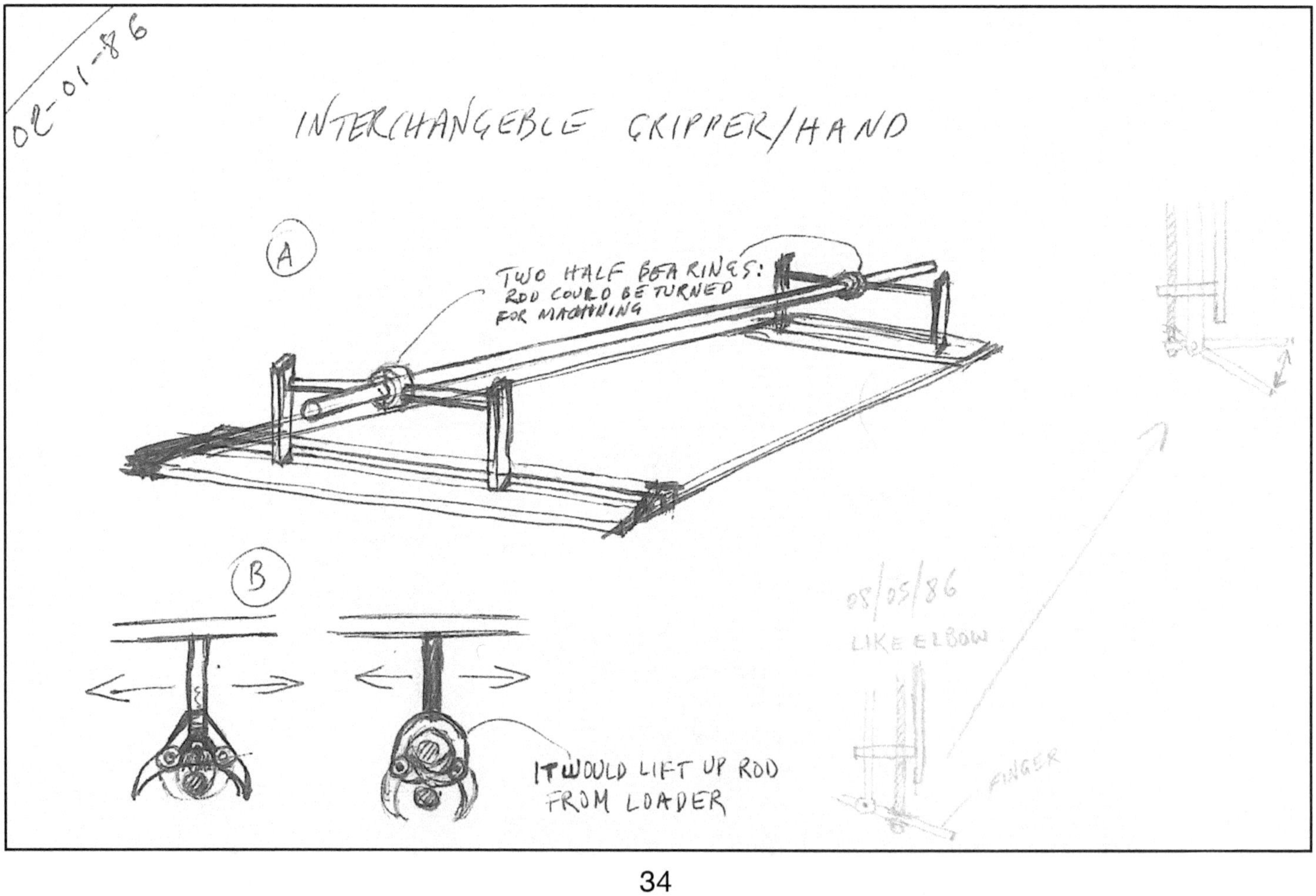
02-01-86
INTERCHANGEBLE GRIPPER/HAND
A
TWO HALF BEARINGS:
ROD COULD BE TURNED
FOR MACHINING
B
IT WOULD LIFT UP ROD
FROM LOADER
08/05/86
LIKE ELBOW
FINGER

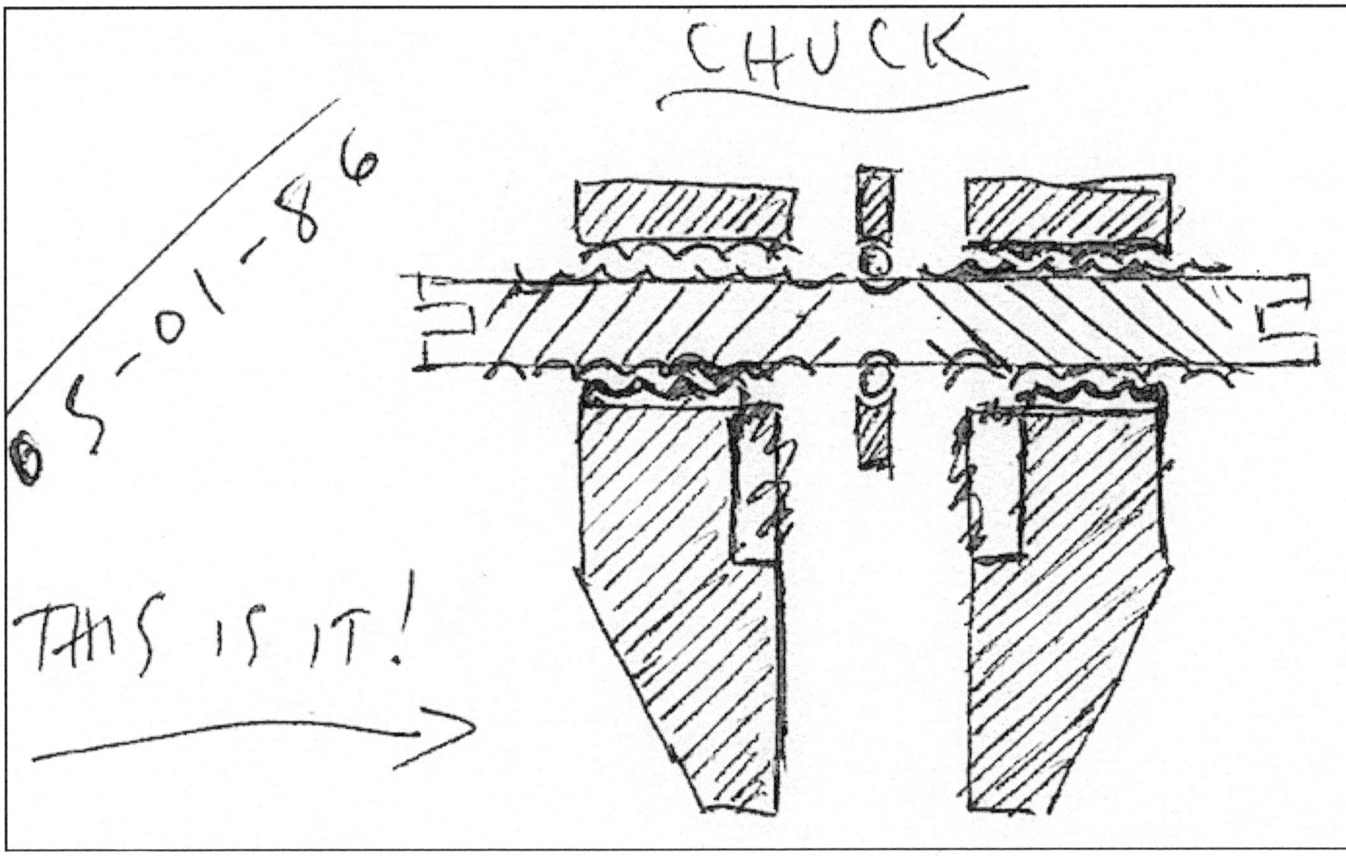
CHUCK
05-01-86
THIS IS IT!

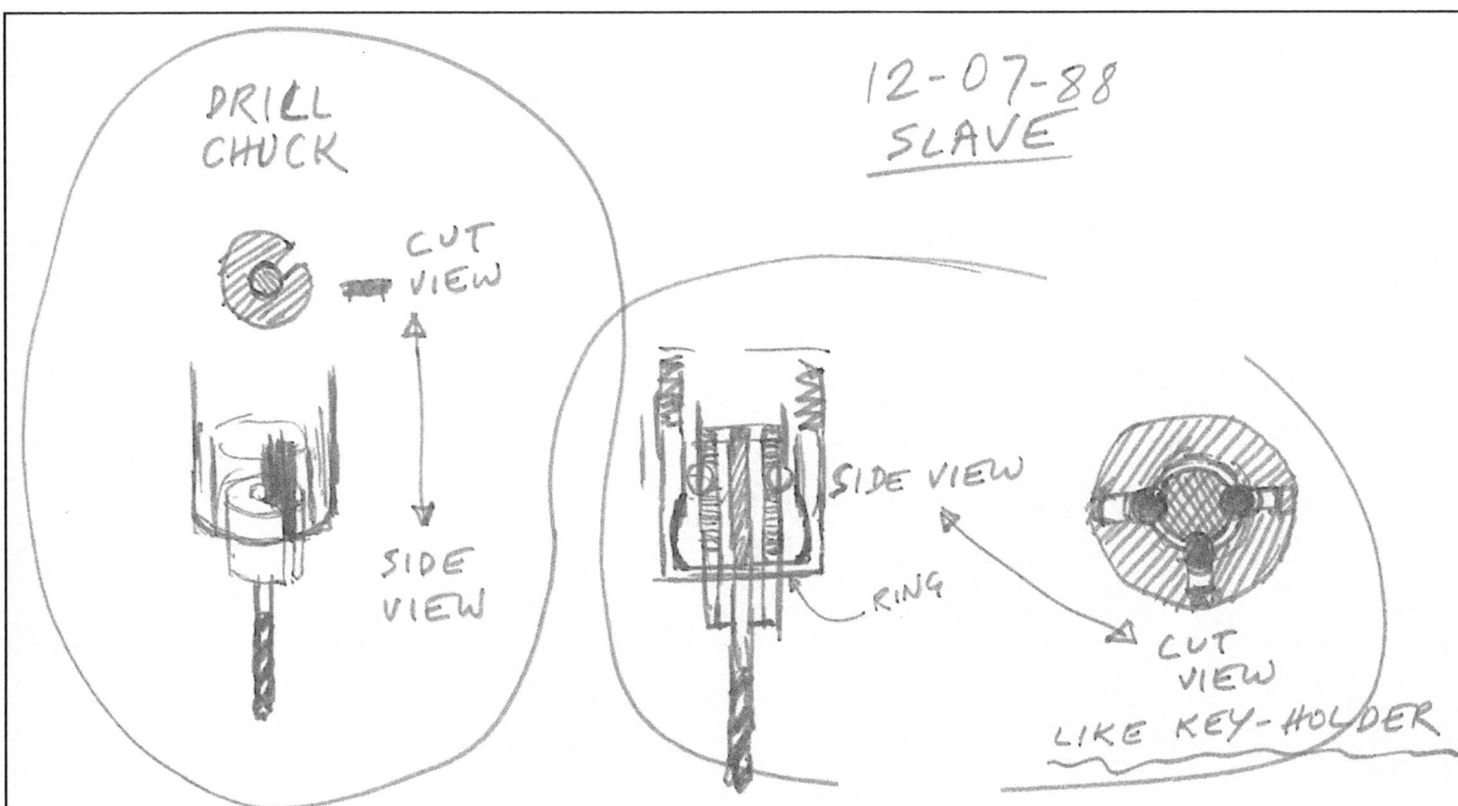
DRILL CHUCK
CUT VIEW
SIDE VIEW
12-07-88
SLAVE
SIDE VIEW
RING
CUT VIEW
LIKE KEY-HOLDER

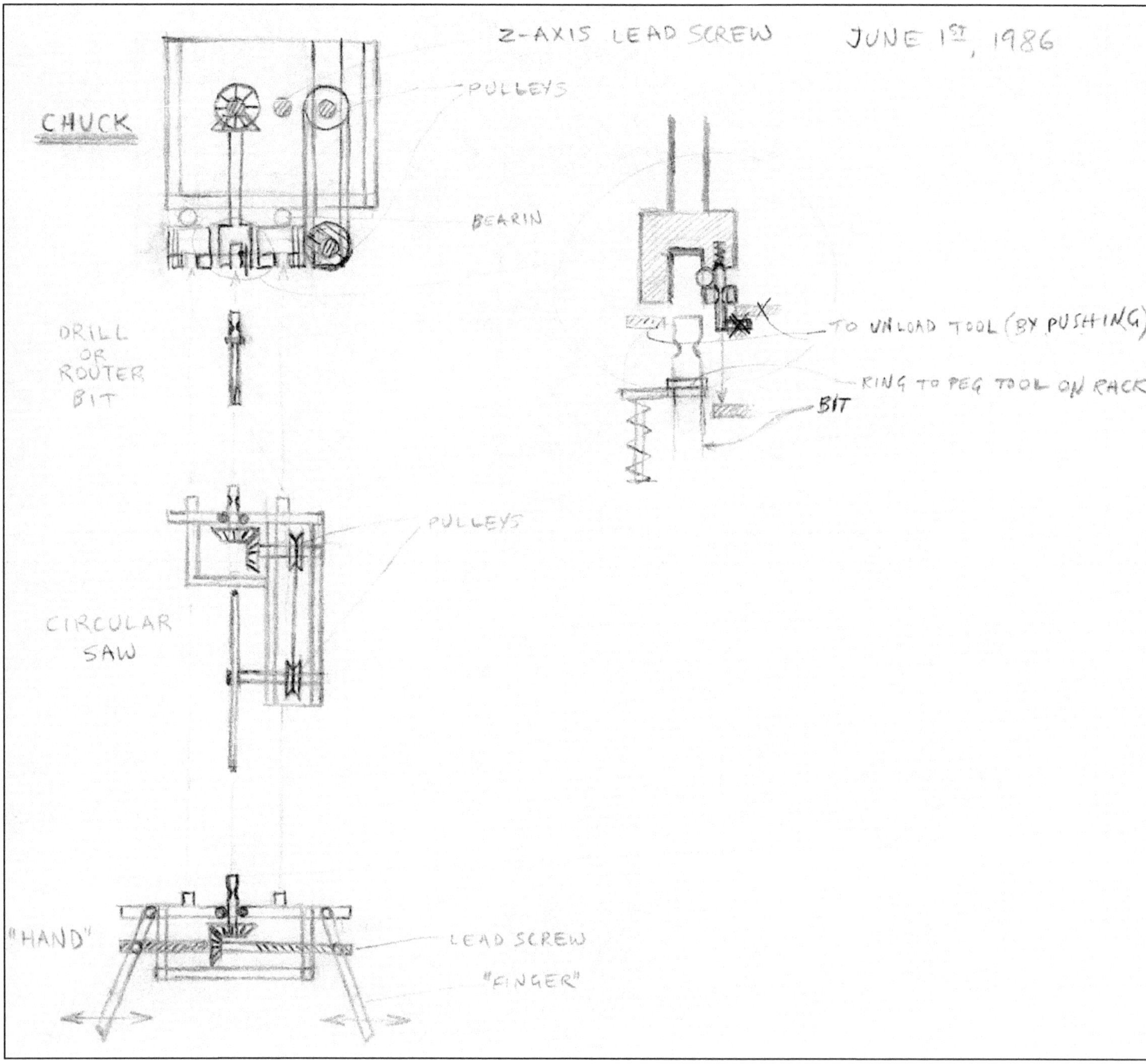
Z-AXIS LEAD SCREW
JUNE 1ST, 1986
CHUCK
PULLEYS
BEARIN
DRILL OR ROUTER BIT
TO UNLOAD TOOL (BY PUSHING)
RING TO PEG TOOL ON RACK
BIT
PULLEYS
CIRCULAR SAW
"HAND"
LEAD SCREW
"FINGER"

23-12-88

→ MILLING MACHINE CHUCK: TO AVOID CENTRE DRILLING OF ROUND STOCK FOR TAIL STOCK: (3) VERSIONS:

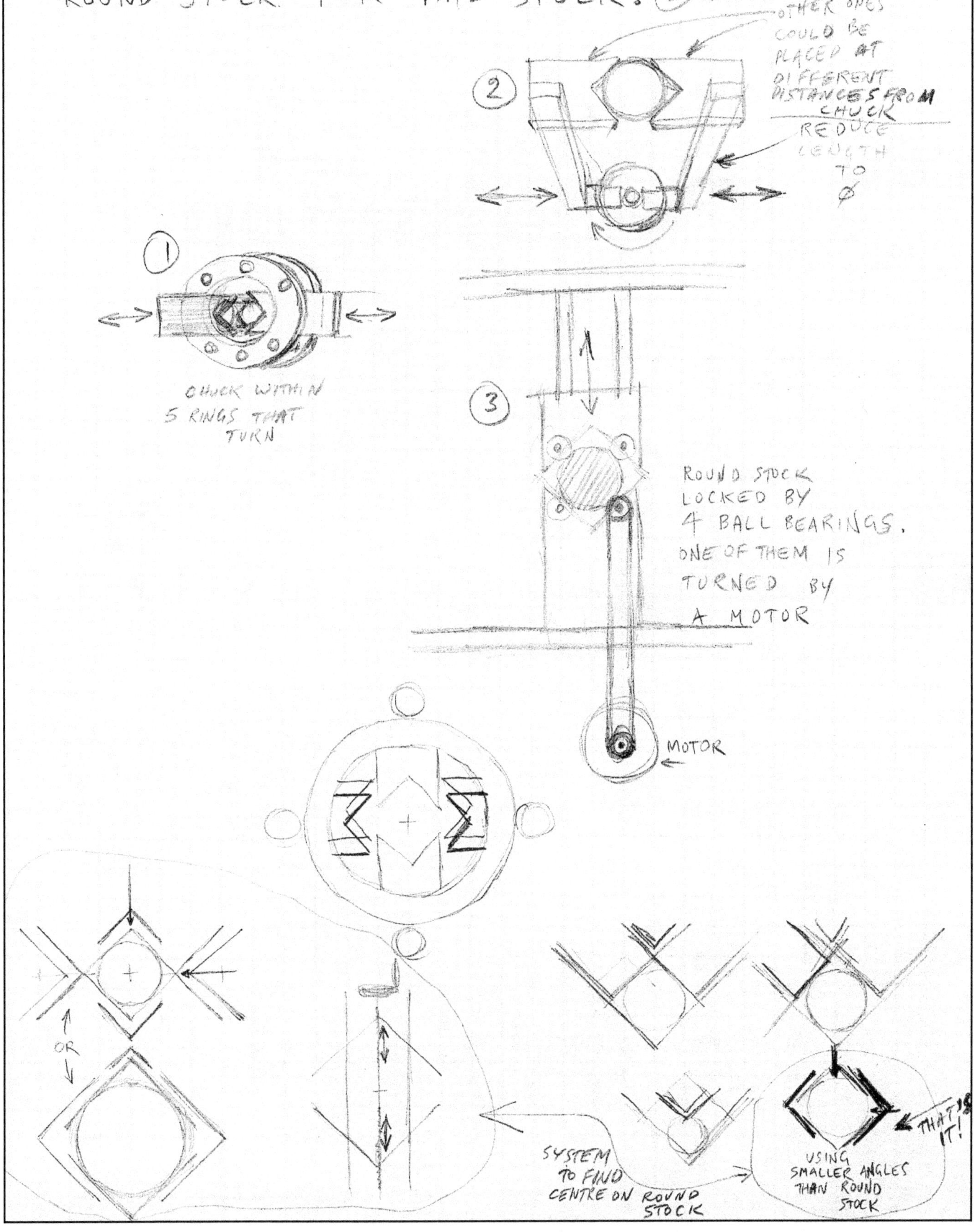

WORKPIECE ROTATION

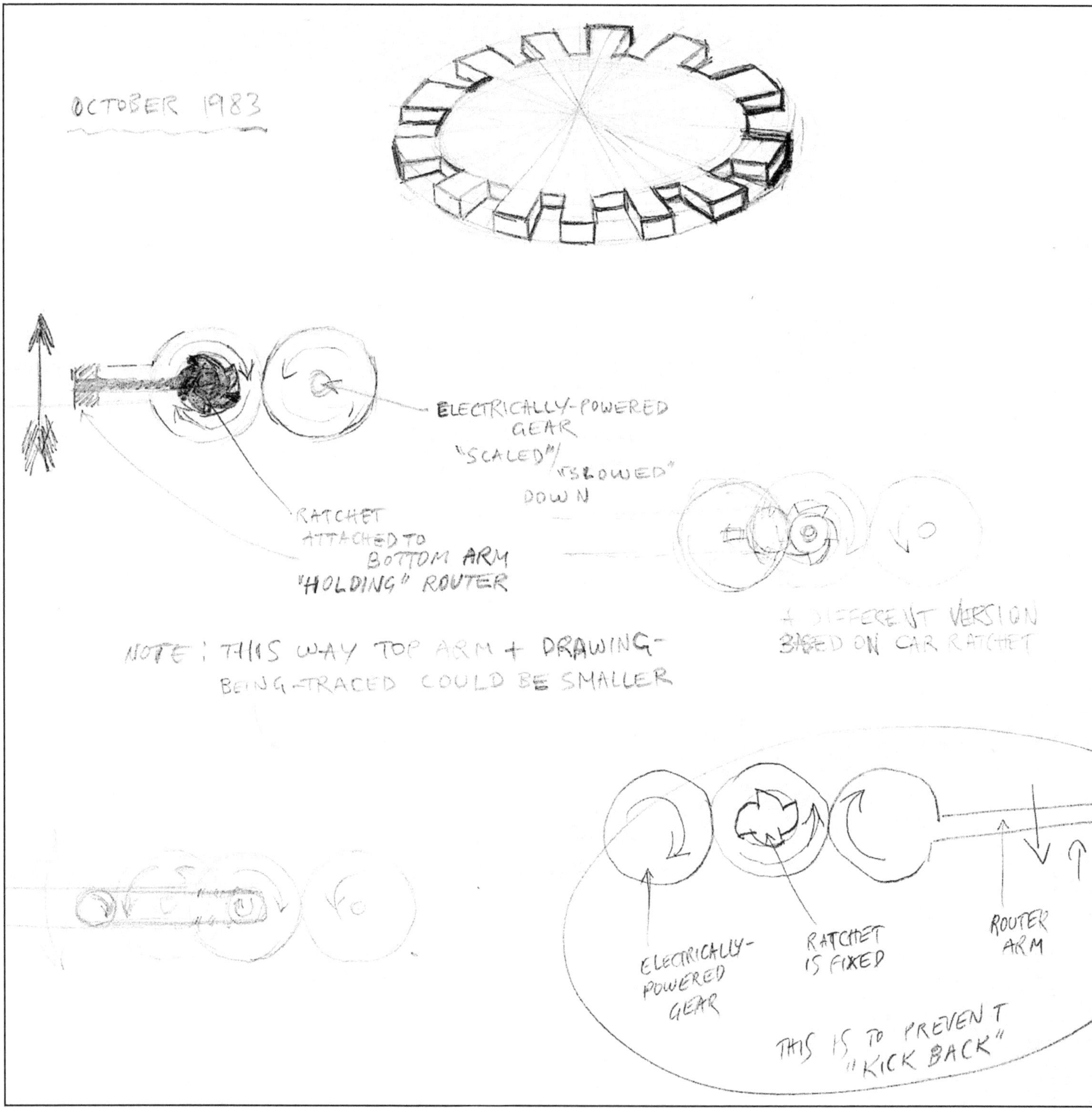

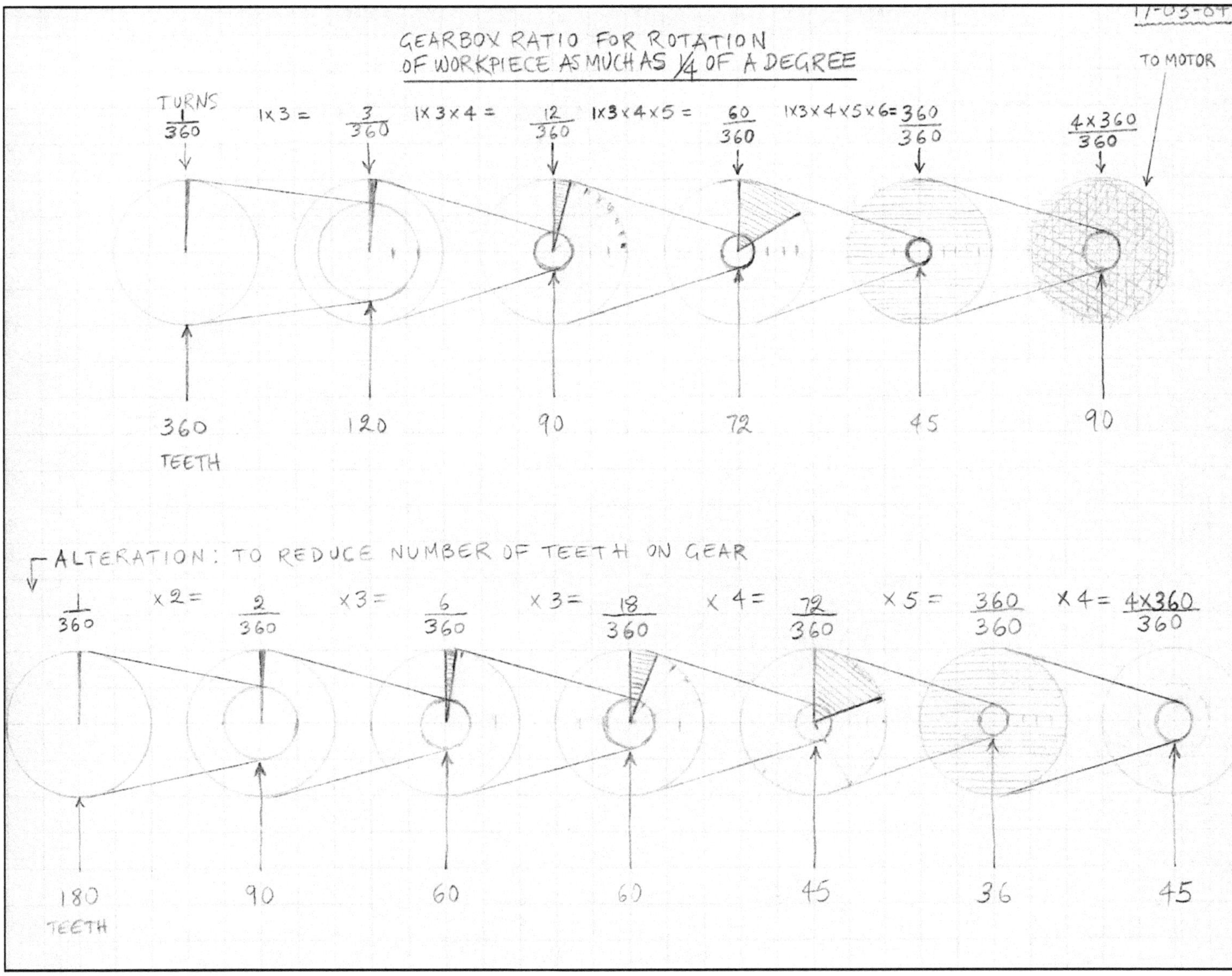
17-03-04
GEARBOX RATIO FOR ROTATION
OF WORKPIECE AS MUCH AS 1/4 OF A DEGREE
TO MOTOR
TURNS
1/360
1x3 = 3/360
1x3x4 = 12/360
1x3x4x5 = 60/360
1x3x4x5x6= 360/360
4x360/360
360
TEETH
120
90
72
45
90
ALTERATION: TO REDUCE NUMBER OF TEETH ON GEAR
1/360
x2= 2/360
x3= 6/360
x3= 18/360
x4= 72/360
x5= 360/360
x4= 4x360/360
180
TEETH
90
60
60
45
36
45

ROTATING WORKPIECE 360°

(1) USING WORM GEAR

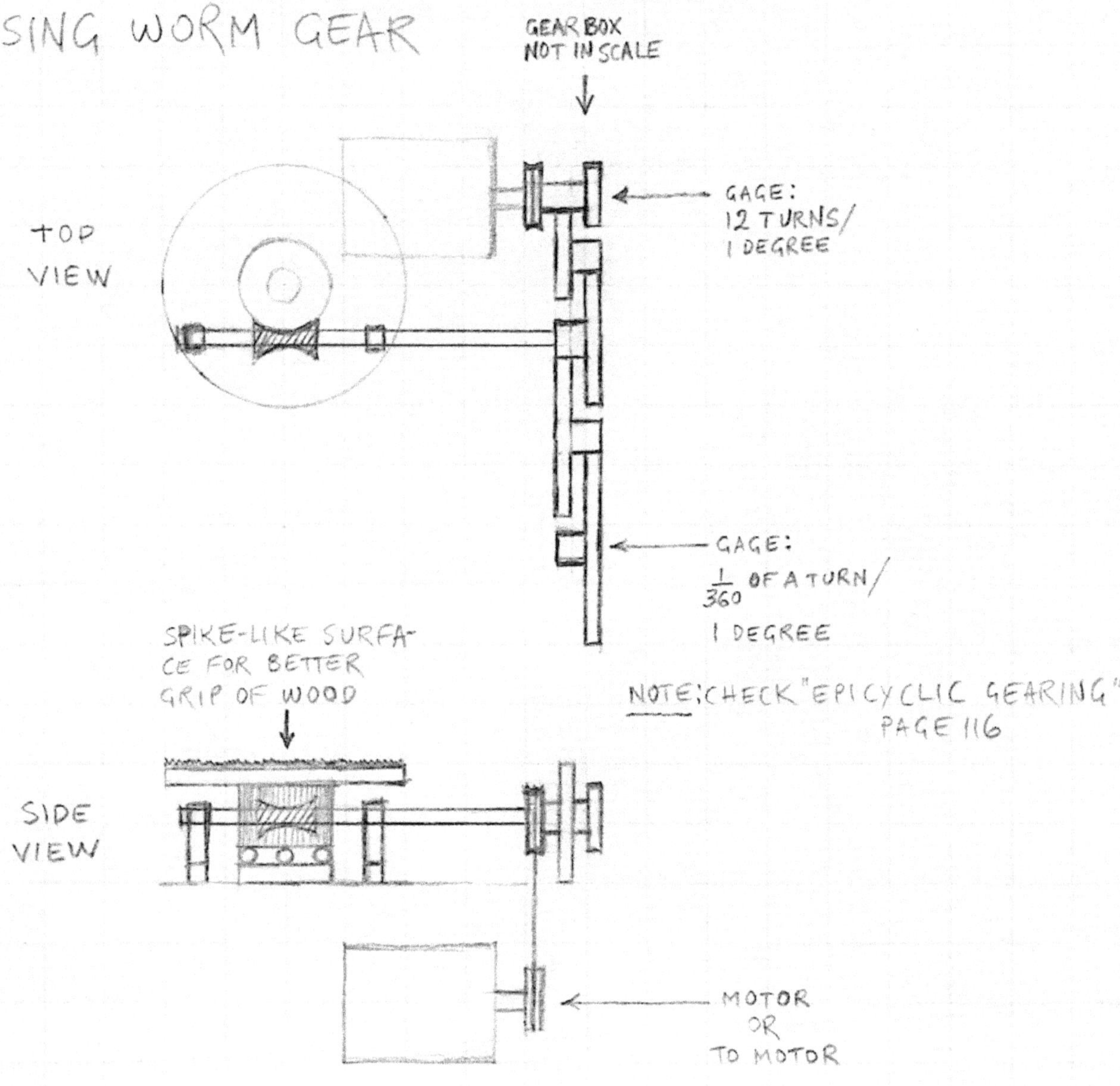

(2) USING BELT + SCREW

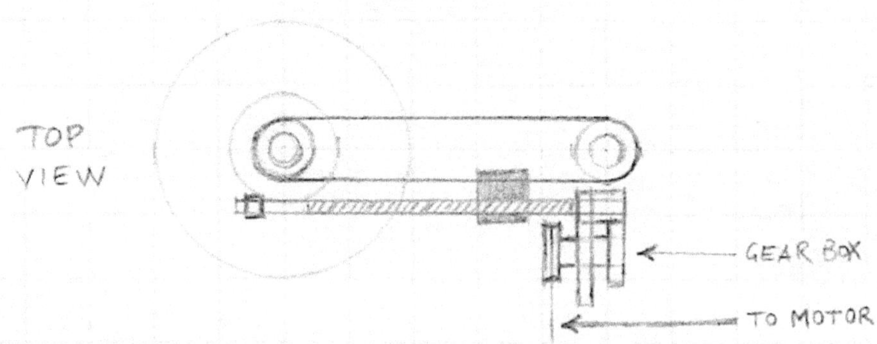

2A. FRAMER

FRAMING or FRAME BUILDING with ...
1. **Rods** as with **plastic car tents & domes**
2. **Panes** as with **glass doors & windows**
3. **Walls** as with **wood house walls**

ROD FRAMING

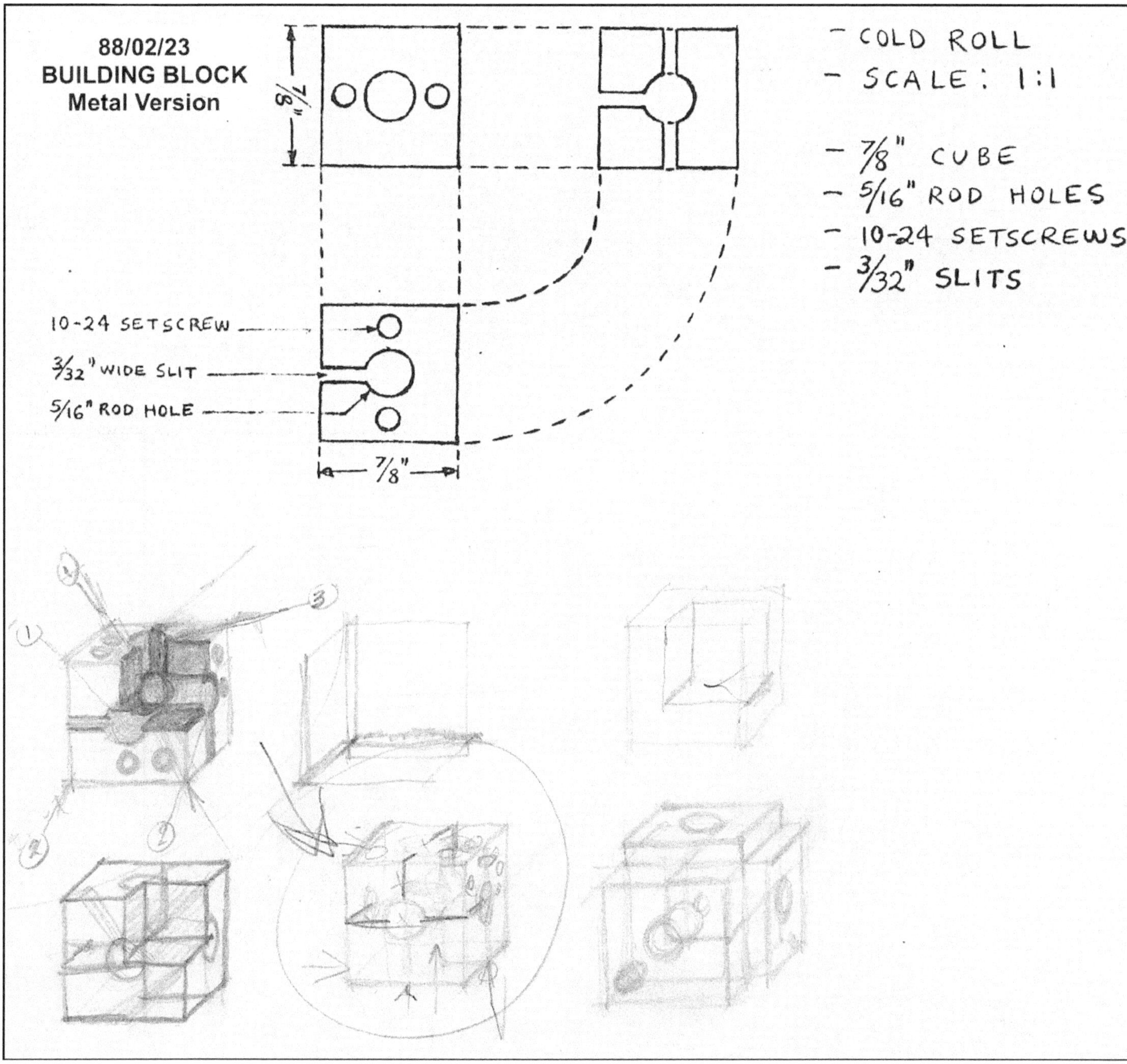

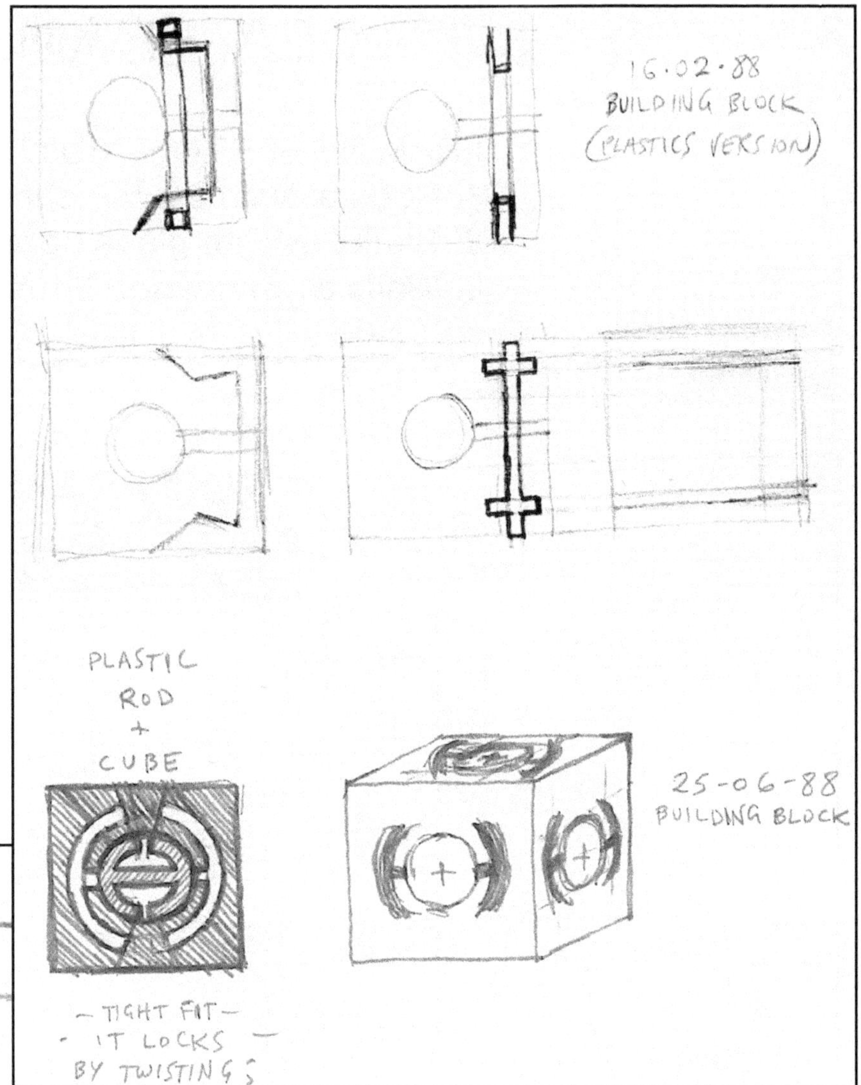
16·02·88
BUILDING BLOCK
(PLASTICS VERSION)
PLASTIC
ROD
+
CUBE
25-06-88
BUILDING BLOCK
— TIGHT FIT —
- IT LOCKS -
BY TWISTING;

ROD
BLOCK
ROD
18·02·88
BUILDING BLOCK
(PLASTICS VERSION)
BLOCK
18·02·88
SCHOOL. TWIST TO INSERT.
PUSH OR PULL AND IT'S LOCKED
ALTERNATING DIRECTION
OF ROWS

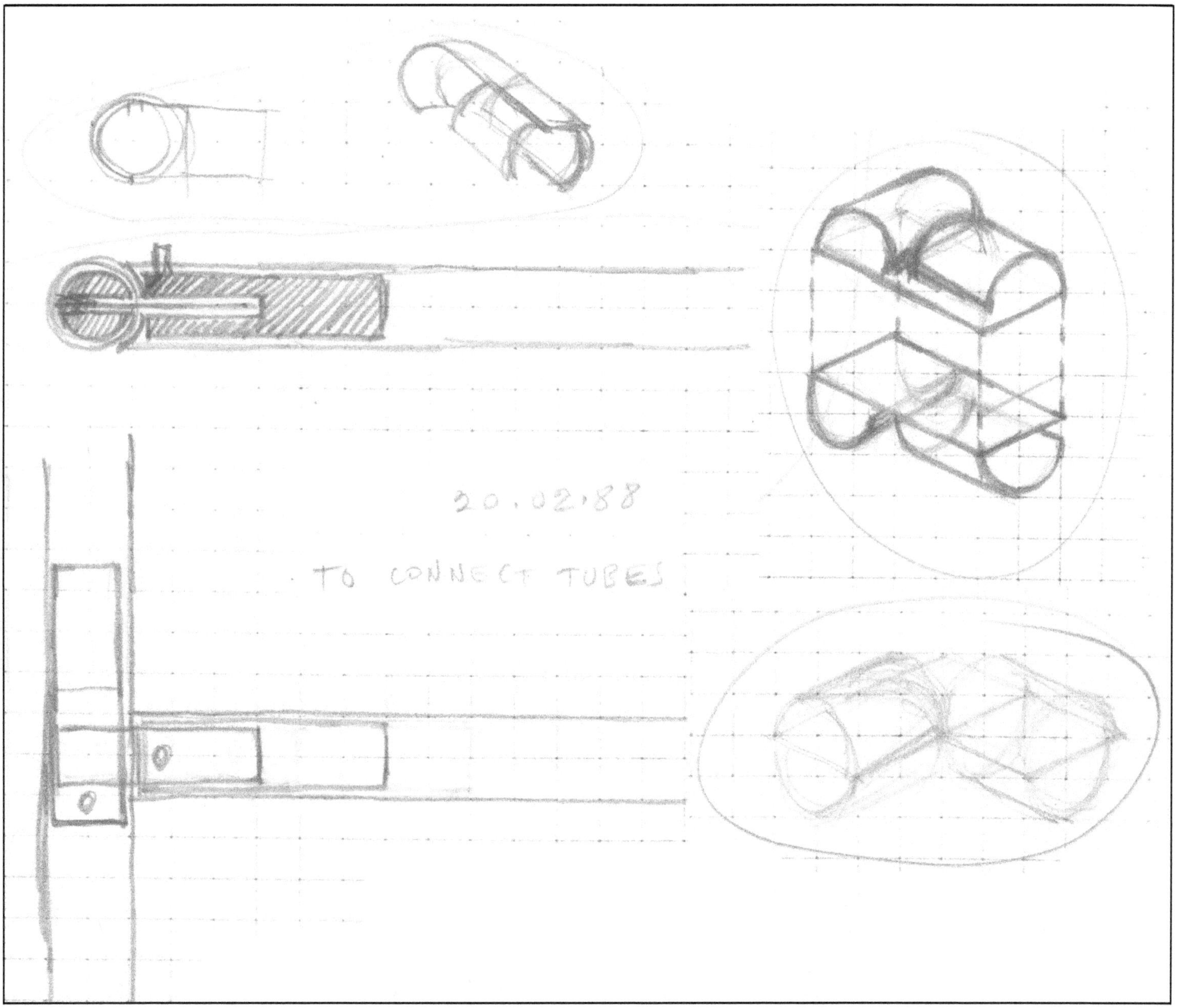
20.02.88
TO CONNECT TUBES

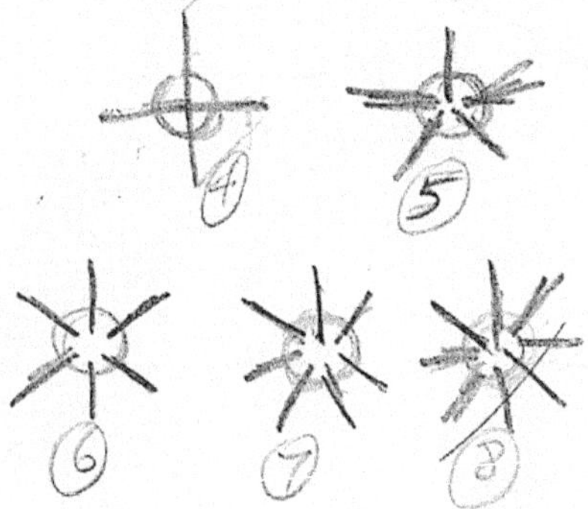

BLOCKS FOR A CERTAIN NUMBER OF RODS

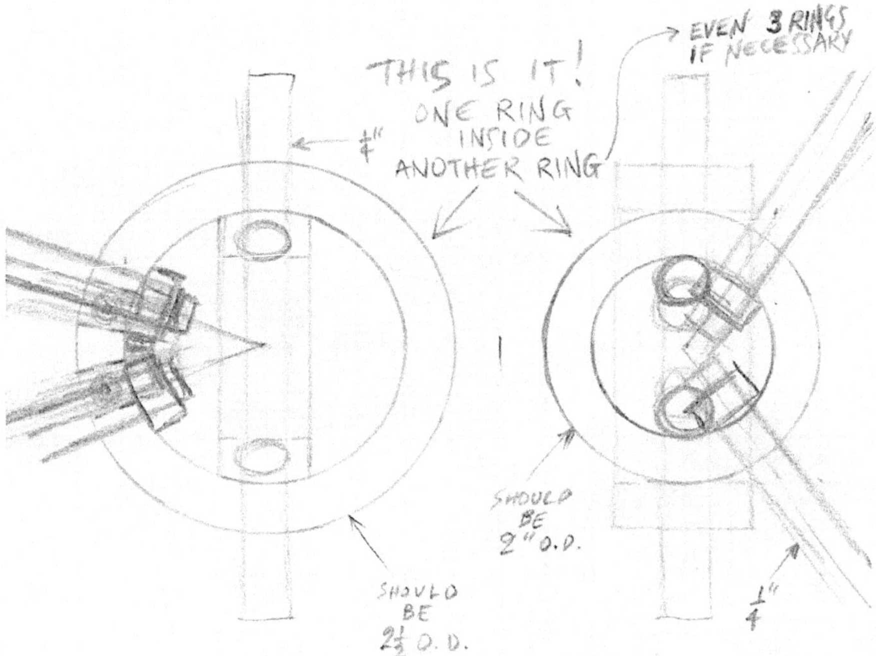

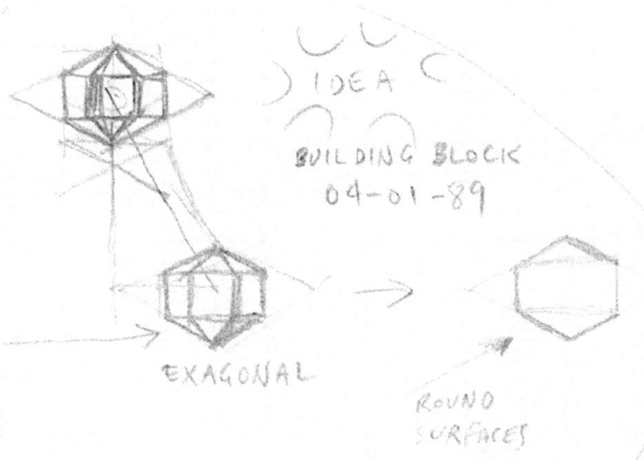

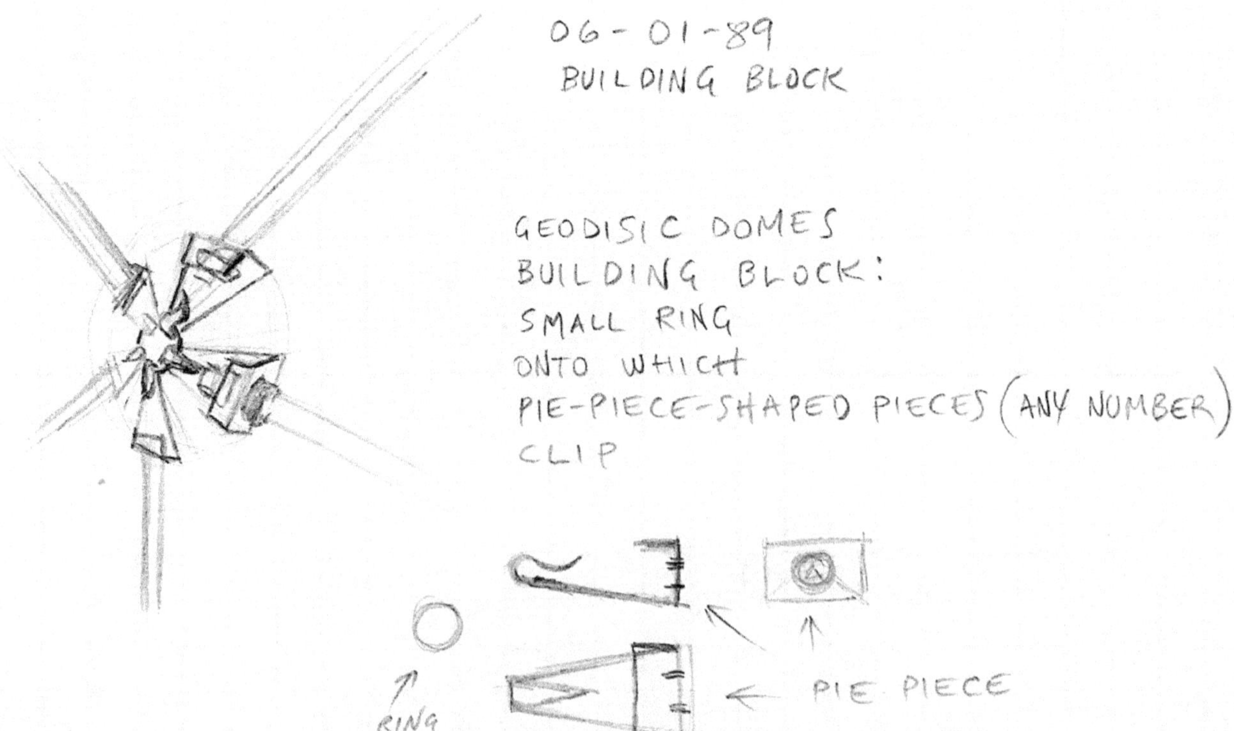

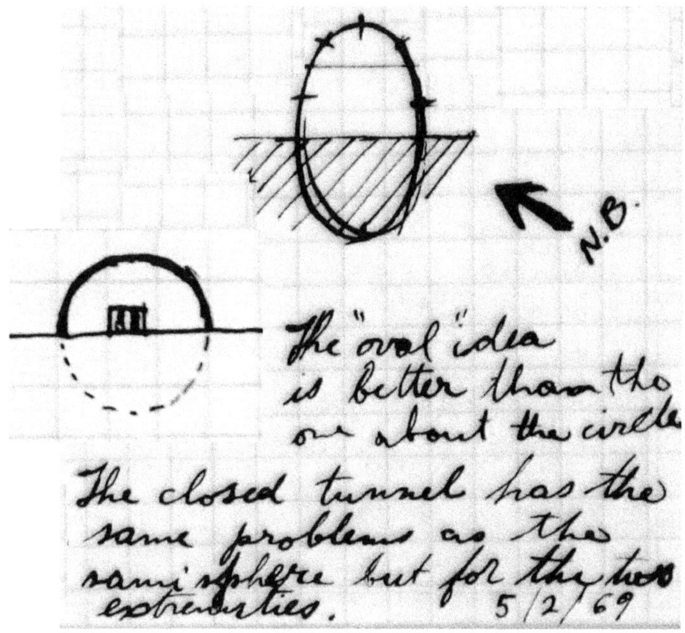

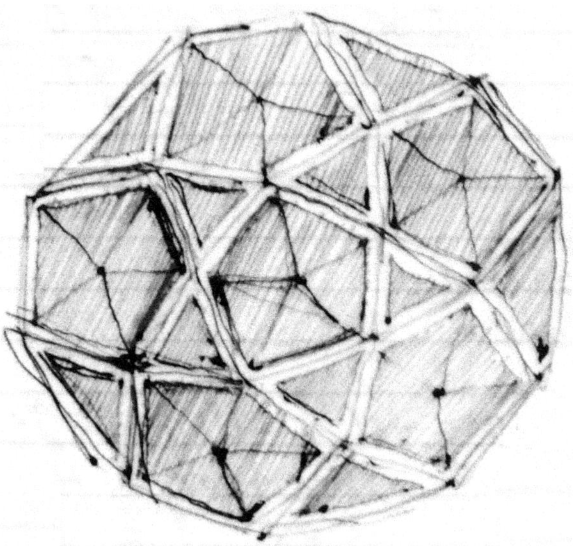

My fascination with the dome.

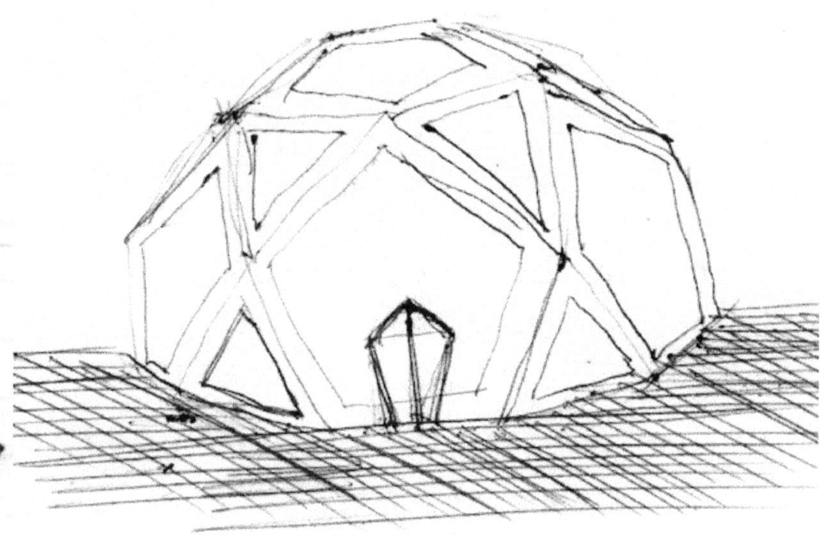

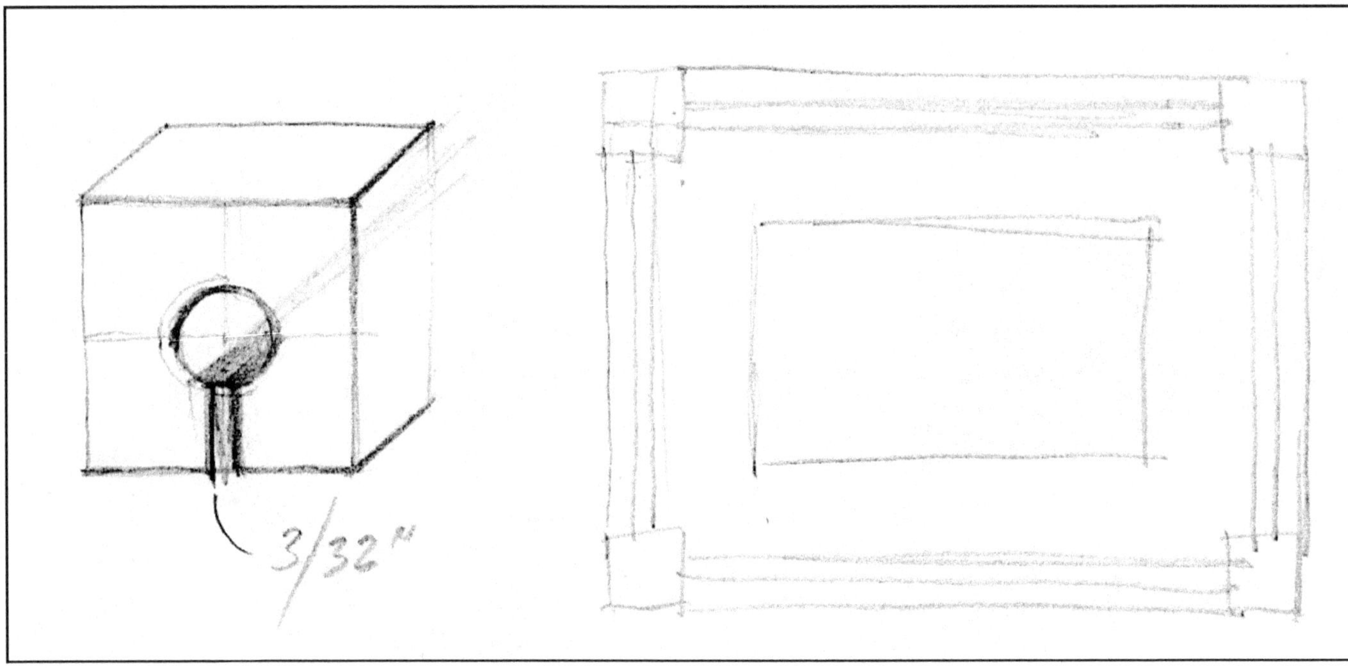

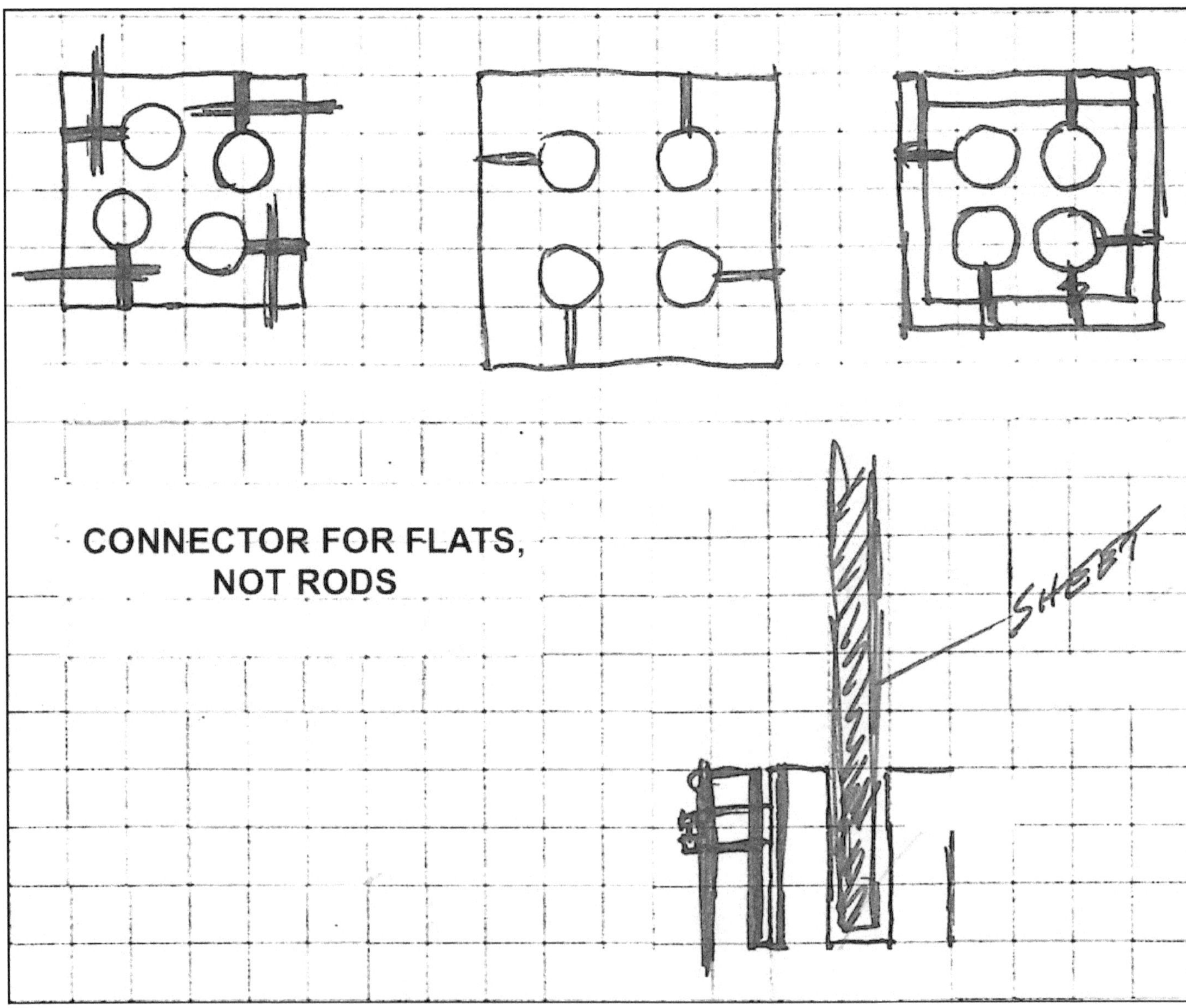

CONNECTOR FOR FLATS, NOT RODS

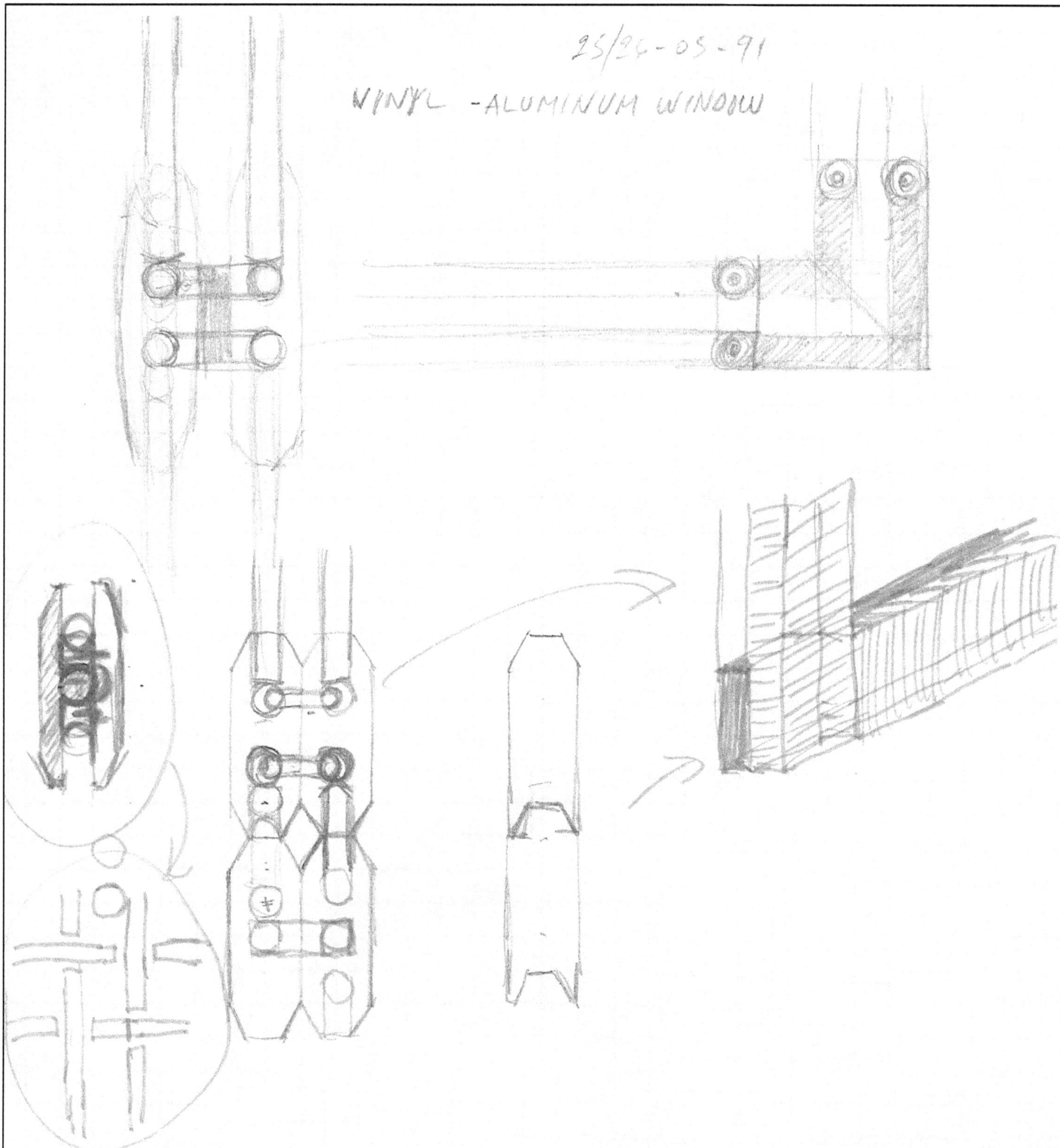

25/26-05-91
VINYL -ALUMINUM WINDOW

28-05-91

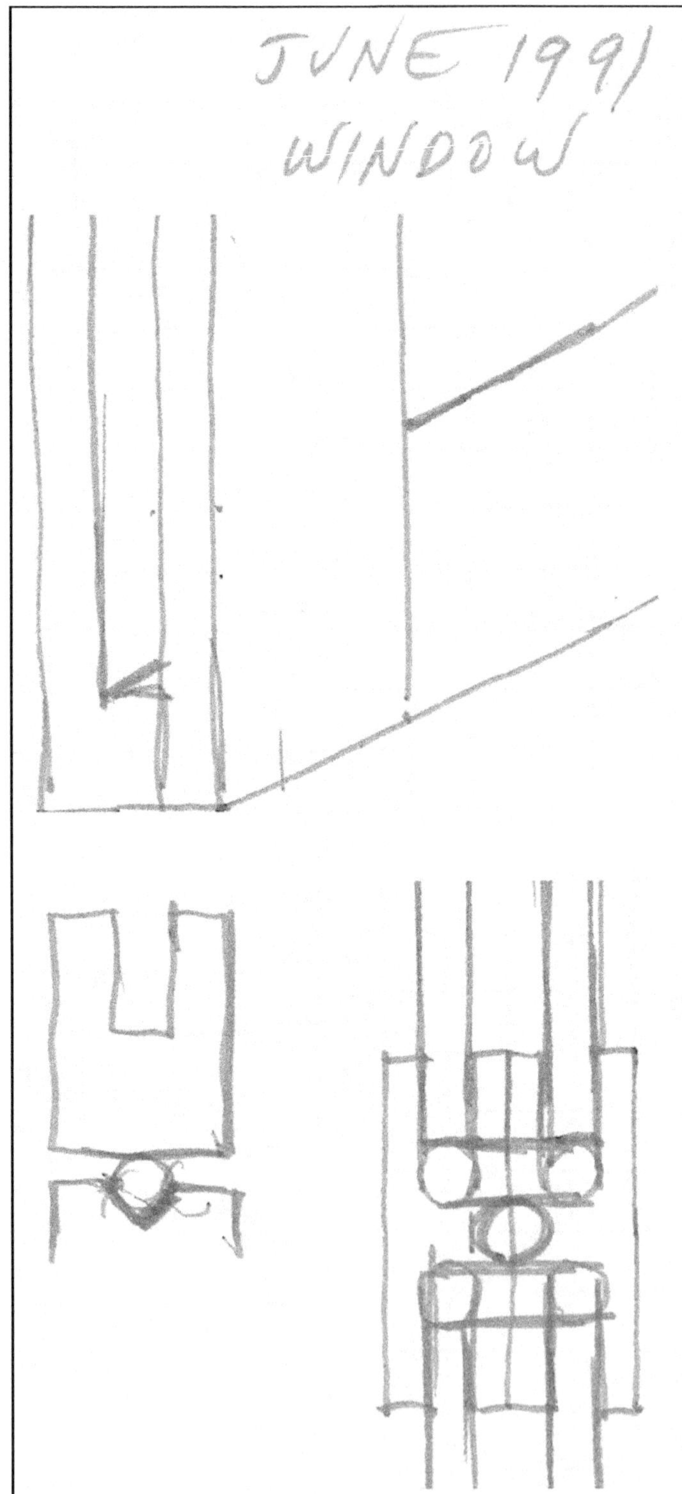
JUNE 1991
WINDOW

HOUSE WALLS

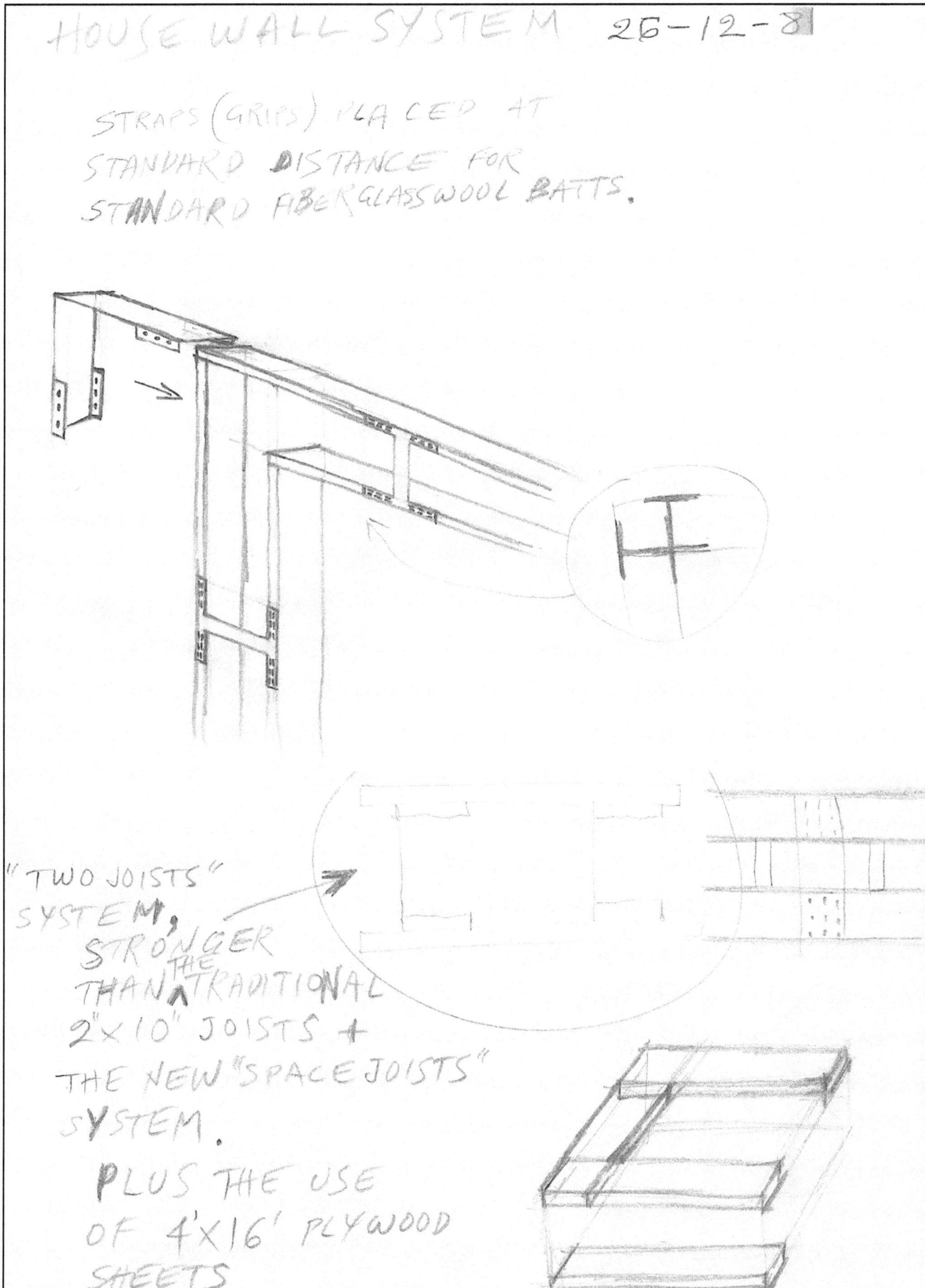
HOUSE WALL SYSTEM 26-12-8
STRAPS (GRIPS) PLACED AT
STANDARD DISTANCE FOR
STANDARD FIBERGLASSWOOL BATTS.
"TWO JOISTS"
SYSTEM,
STRONGER
THAN THE TRADITIONAL
2"x10" JOISTS +
THE NEW "SPACE JOISTS"
SYSTEM.
PLUS THE USE
OF 4'x16' PLYWOOD
SHEETS

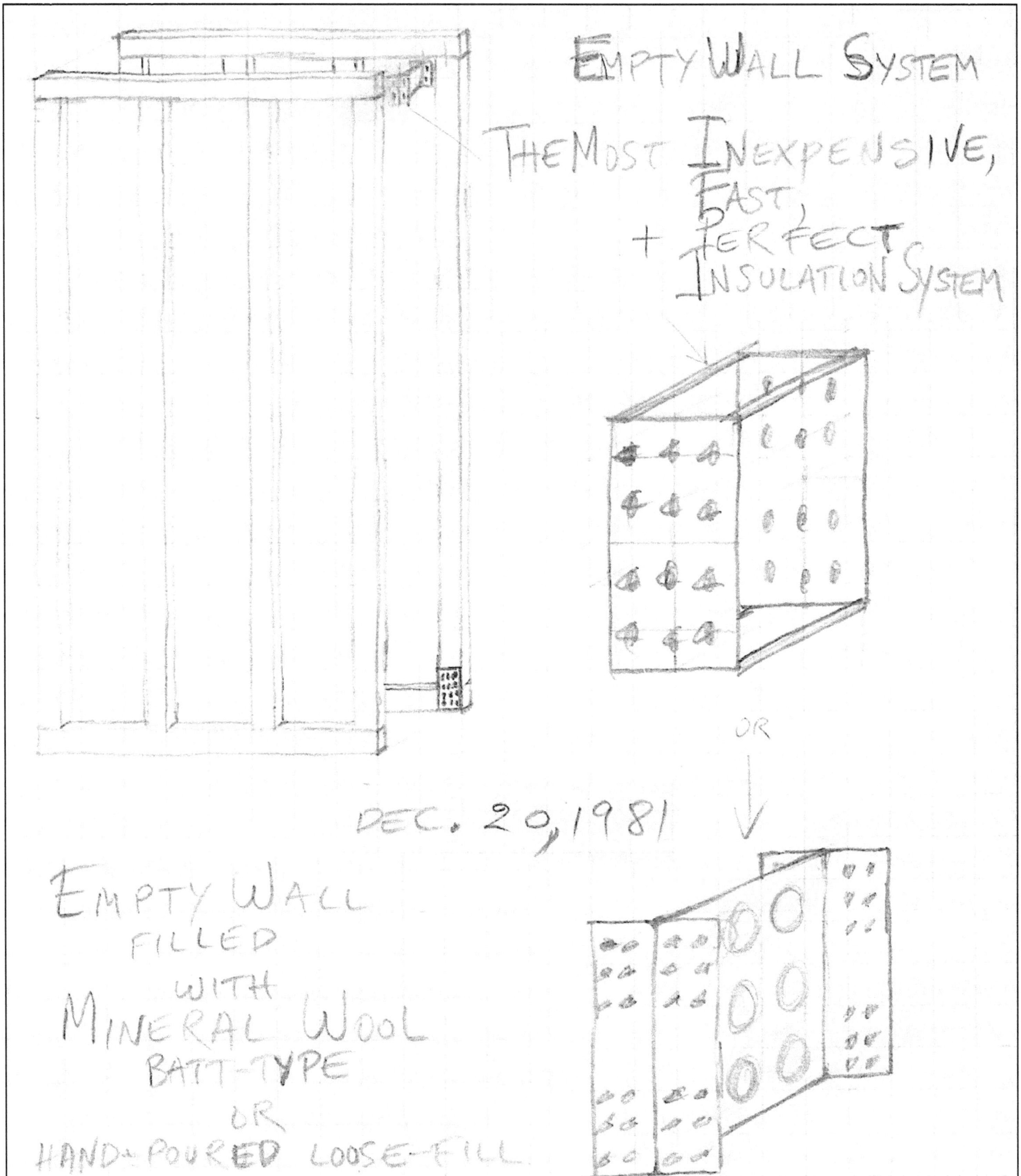
EMPTY WALL SYSTEM
THE MOST INEXPENSIVE,
FAST,
+ PERFECT
INSULATION SYSTEM
OR
DEC. 20, 1981
EMPTY WALL
FILLED
WITH
MINERAL WOOL
BATT-TYPE
OR
HAND-POURED LOOSE-FILL

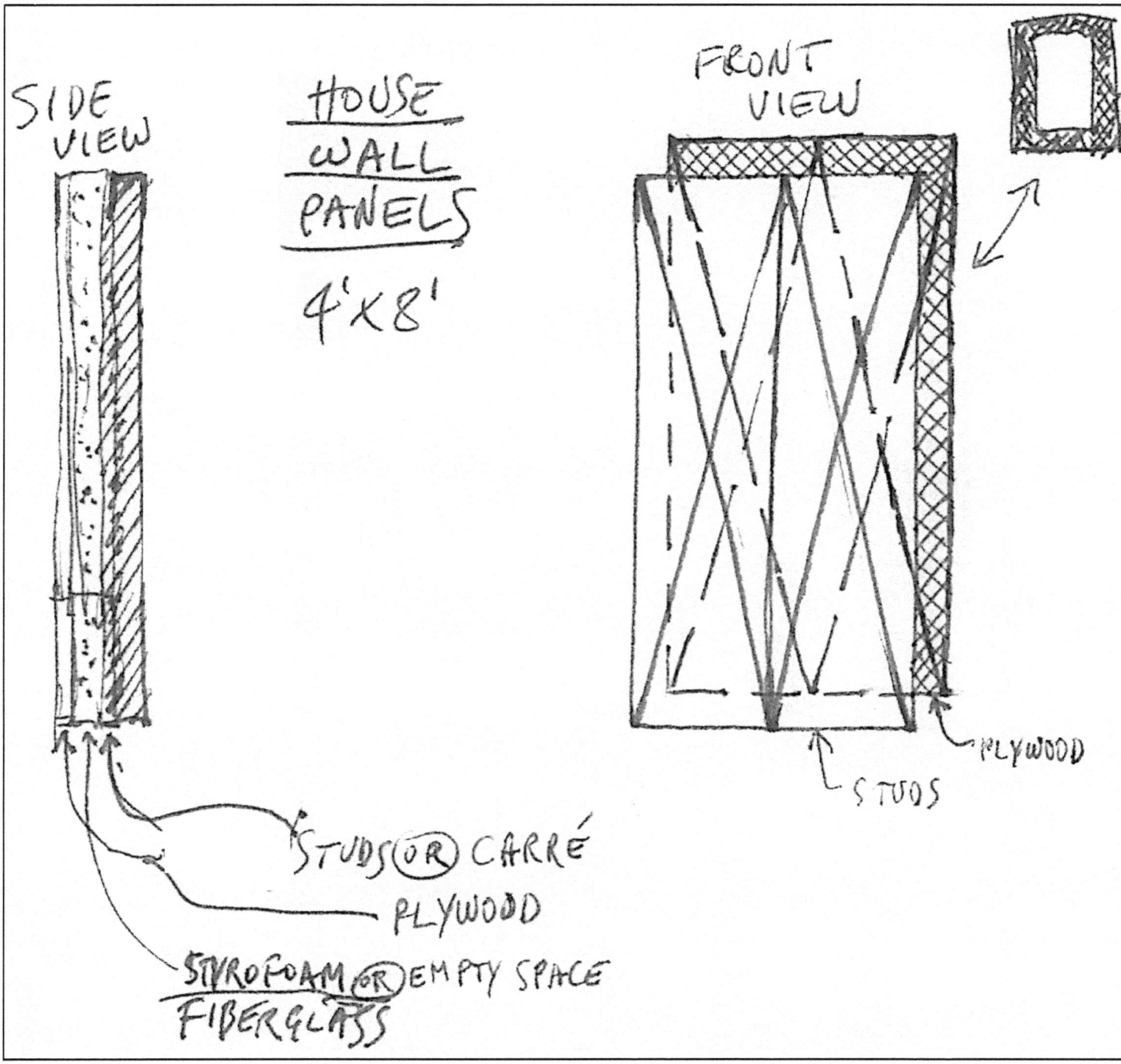
SIDE VIEW
HOUSE WALL PANELS
4'X8'
FRONT VIEW
PLYWOOD
STUDS
STUDS OR CARRÉ
PLYWOOD
STYROFOAM OR EMPTY SPACE
FIBERGLASS

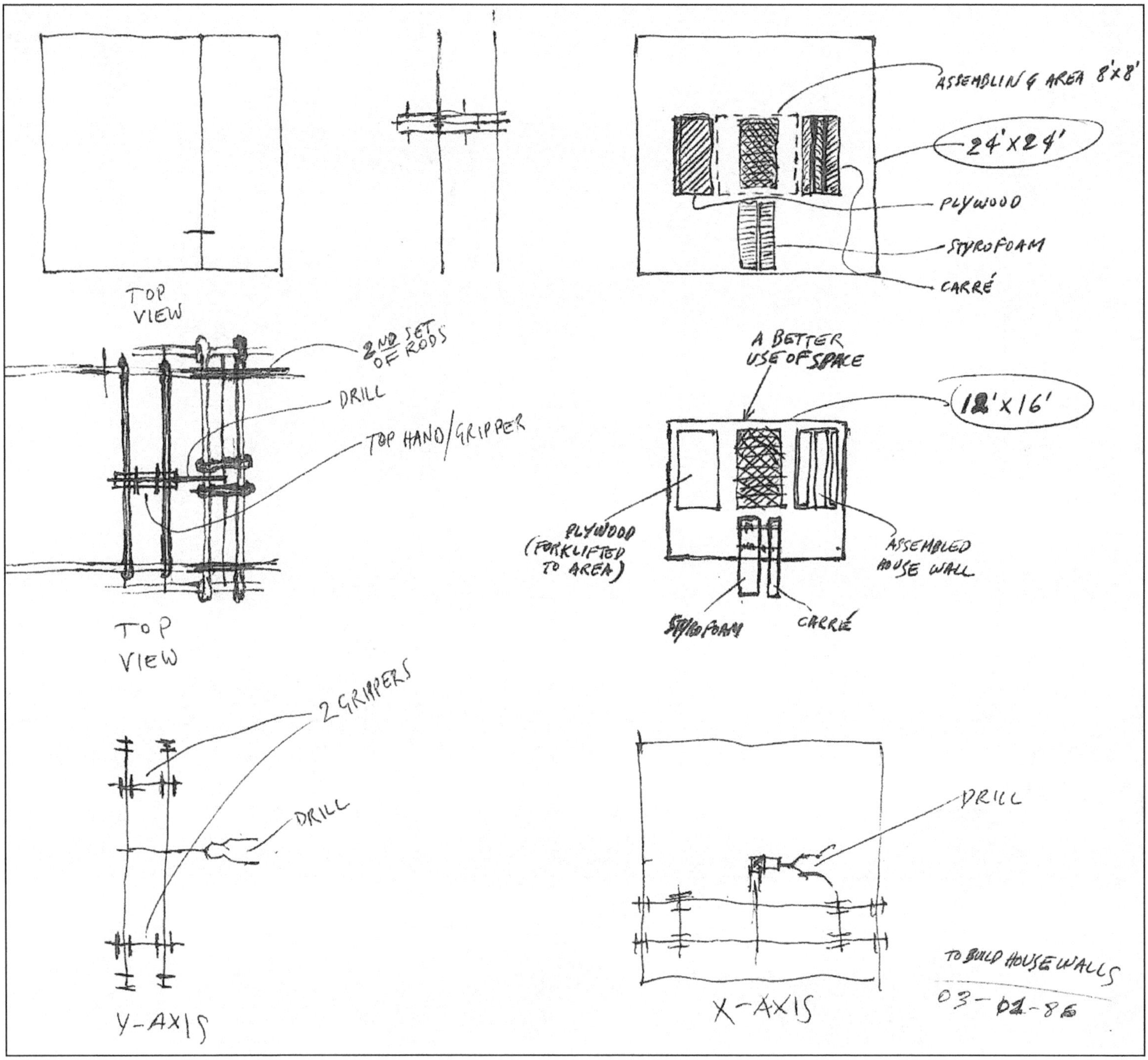
ASSEMBLING AREA 8'x8'
24'x24'
PLYWOOD
STYROFOAM
CARRÉ
TOP VIEW
2ND SET OF RODS
DRILL
TOP HAND/GRIPPER
A BETTER USE OF SPACE
12'x16'
PLYWOOD (FORKLIFTED TO AREA)
ASSEMBLED HOUSE WALL
STYROFOAM
CARRÉ
TOP VIEW
2 GRIPPERS
DRILL
Y-AXIS
DRILL
X-AXIS
TO BUILD HOUSE WALLS
03-01-86

2B. TILE SETTER (SANDCOATER)

SANDCOATER Ver1

Post with No Base: 5 parts:

1. Vertical Post: to lock against the ceiling
2. Pivot Leg: to be attached to vertical post
3. Horizontal Arm: to pivot from pivot leg
4. Spreader: to spread sandcoat
5. Handle: for the tilesetter to control the rake

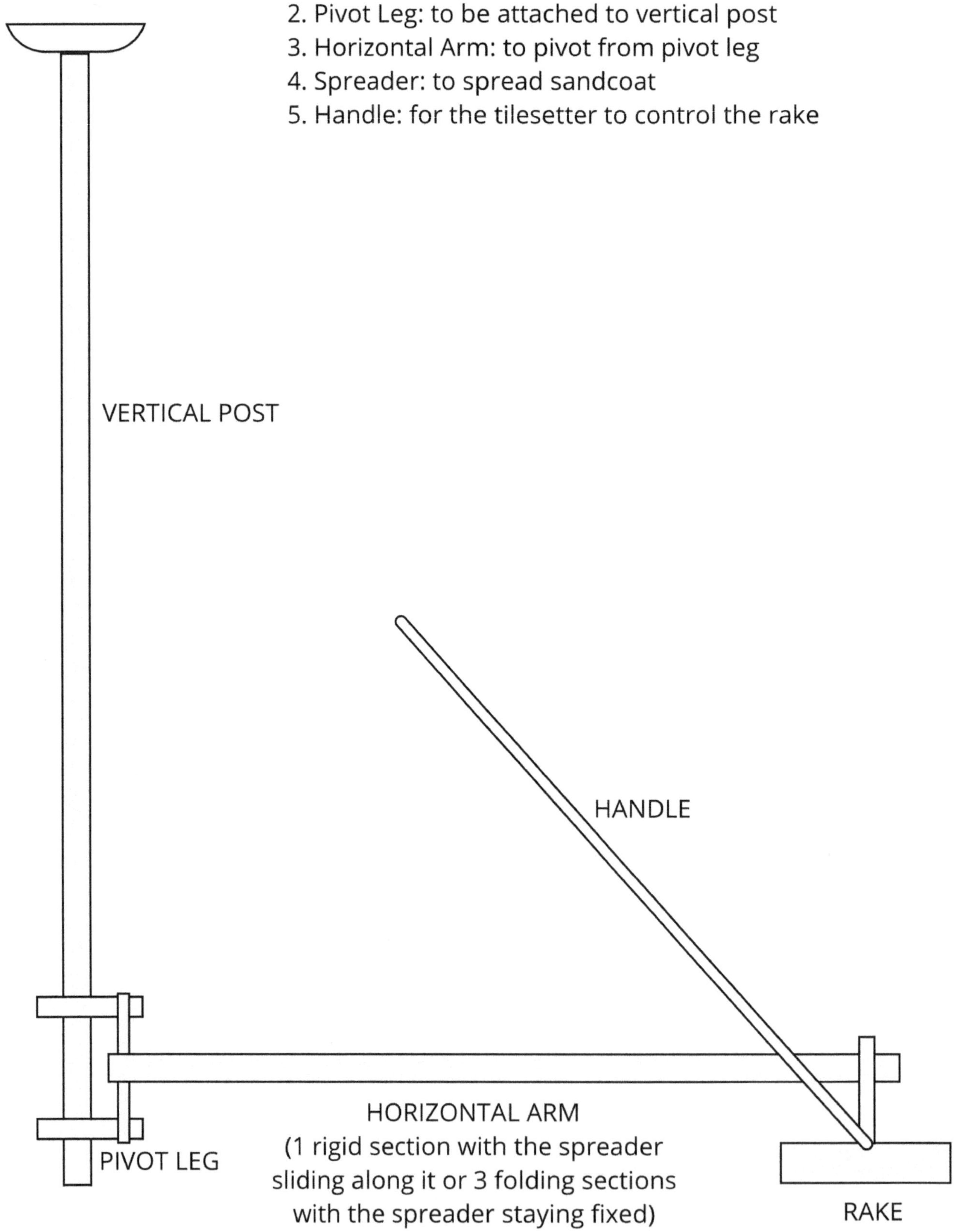

SANDCOATER Ver2

Heavy Base with No Post: 5 parts:

1. Heavy Base: to act as anchor
2. Pivot Leg: to set level
3. Horizontal Arm: to pivot from pivot leg
4. Spreader: to spread sandcoat
5. Handle: for the tilesetter to handle the rake

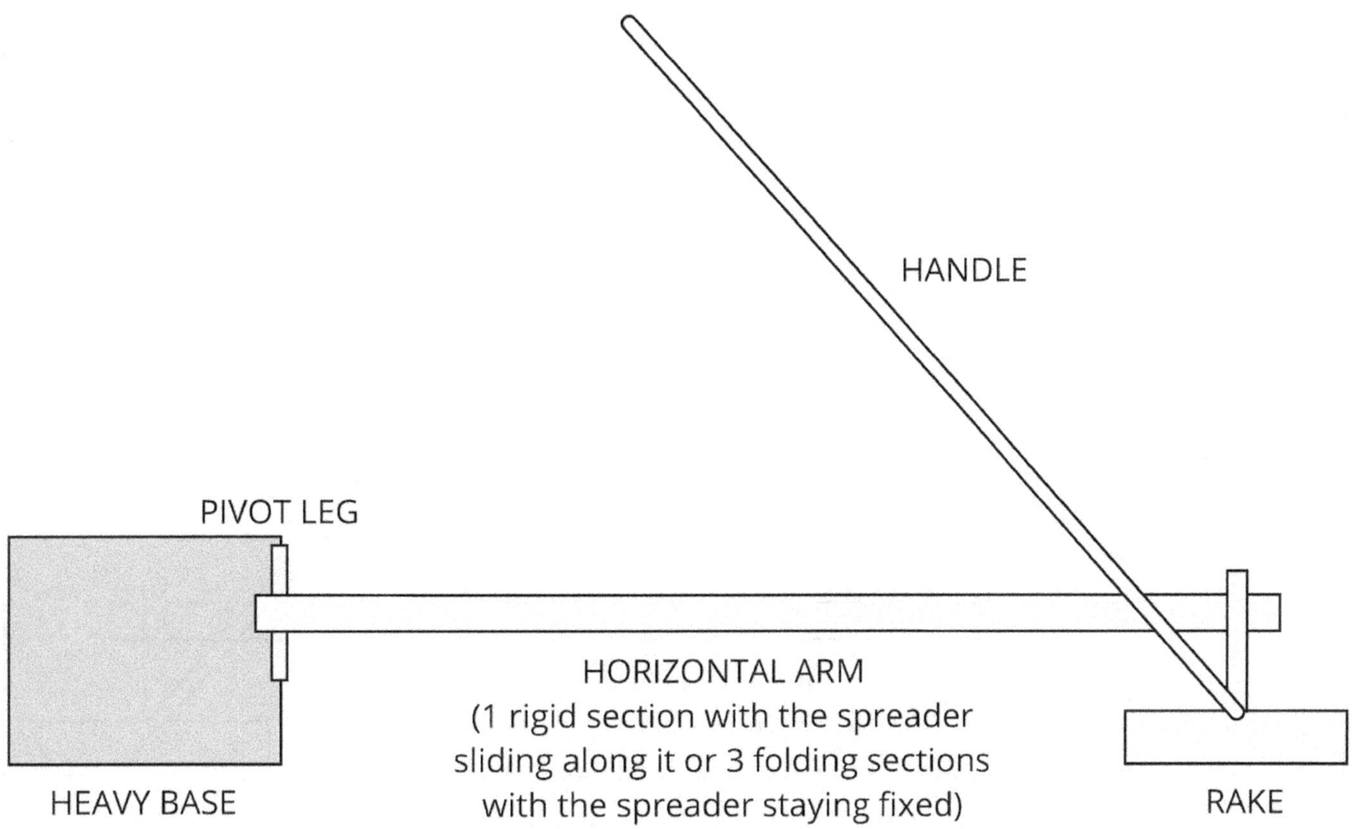

SANDCOATER Ver3

Light Base with Post: 6 parts:

1. Base: to act as anchor
2. Pivot Leg: to be set level
3. Vertical Post: to lock the base against the ceiling
4. Horizontal Arm: to pivot from pivot leg
5. Spreader: to spread sandcoat
6. Handle: for the tilesetter to control the rake

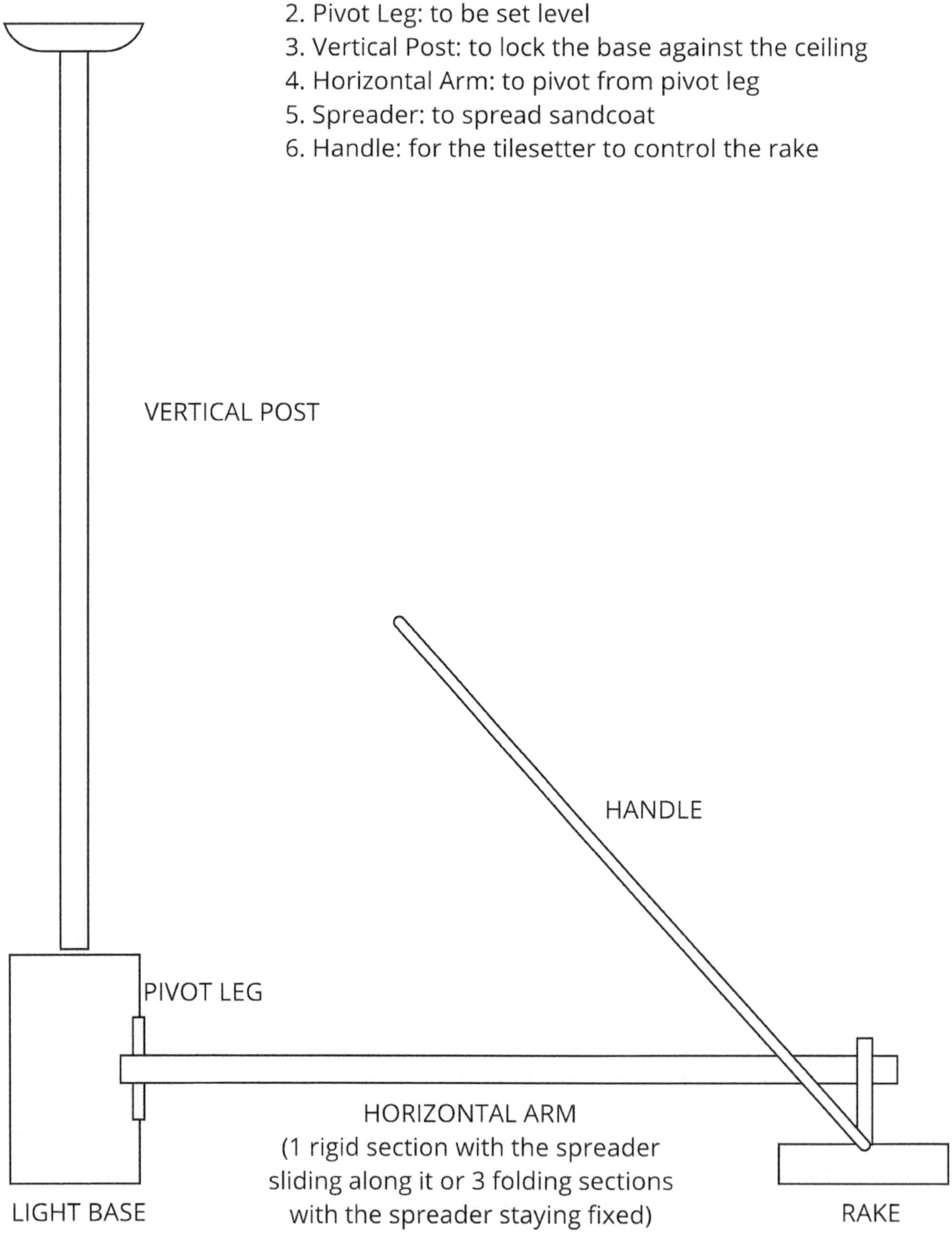

This is a variation of SANDCOATER Ver3

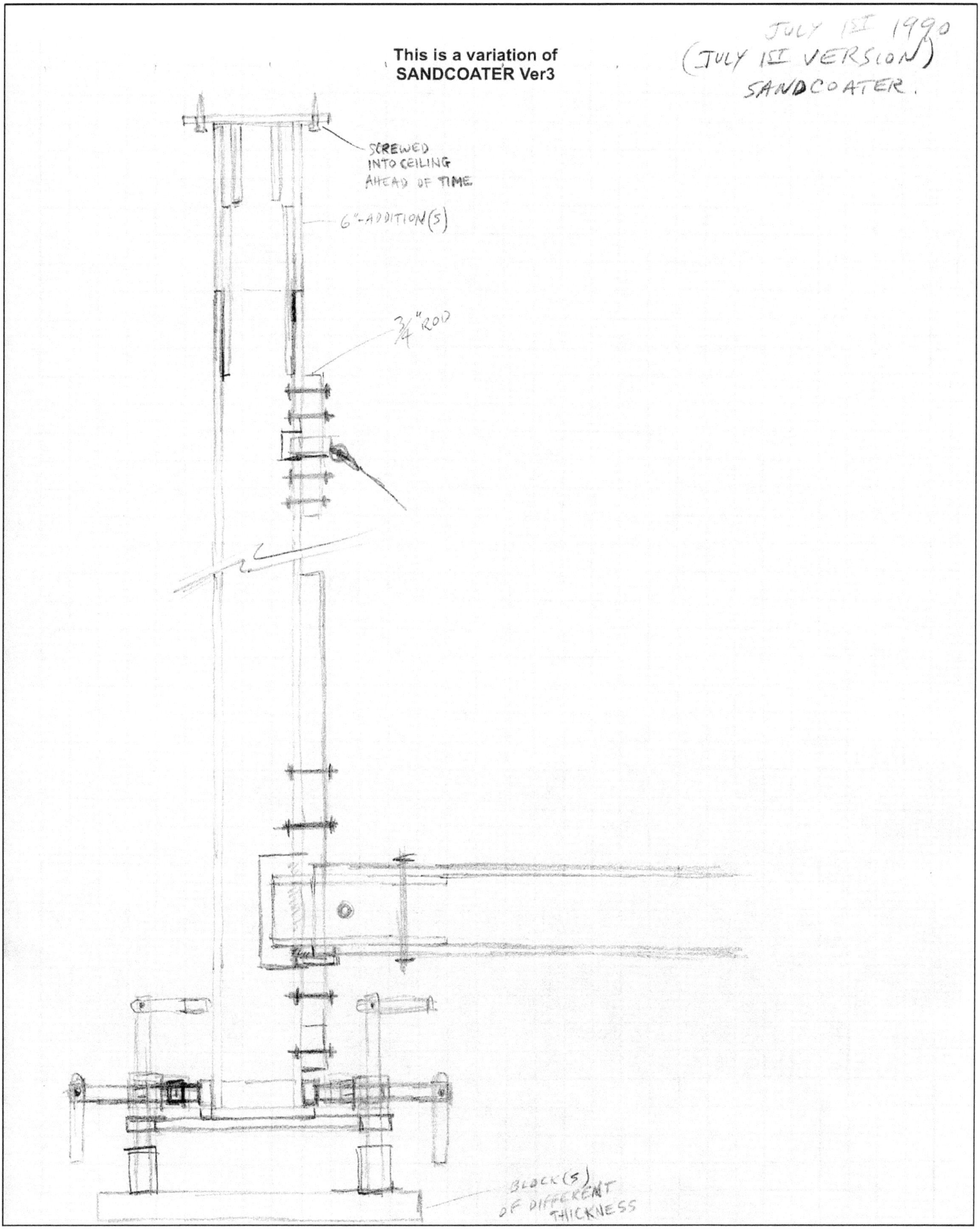

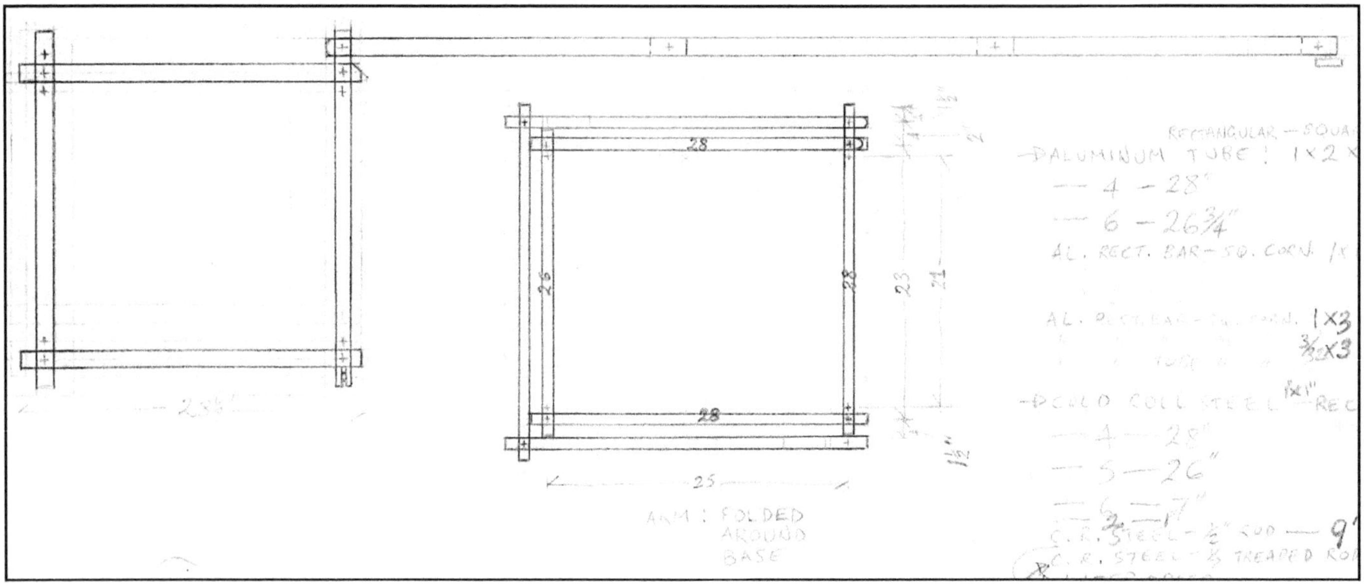

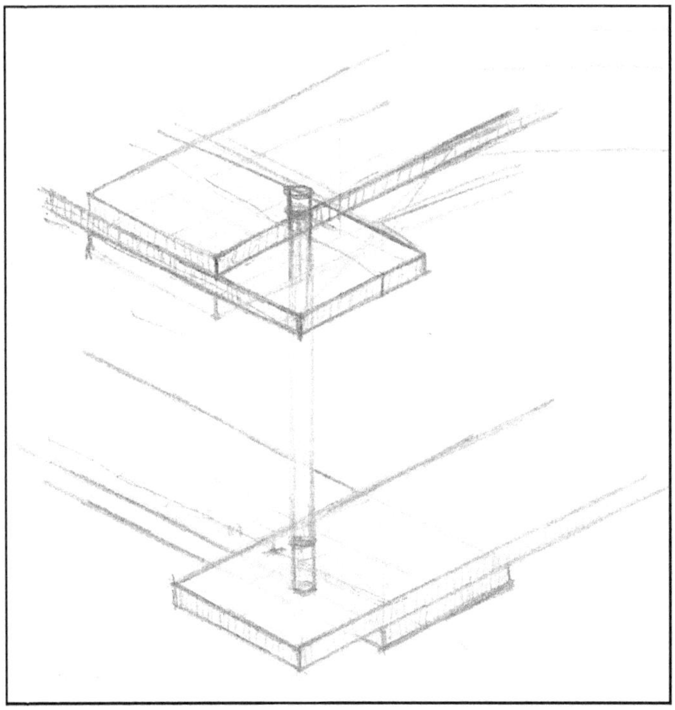

SANDCOATER'S HEAVY BASE

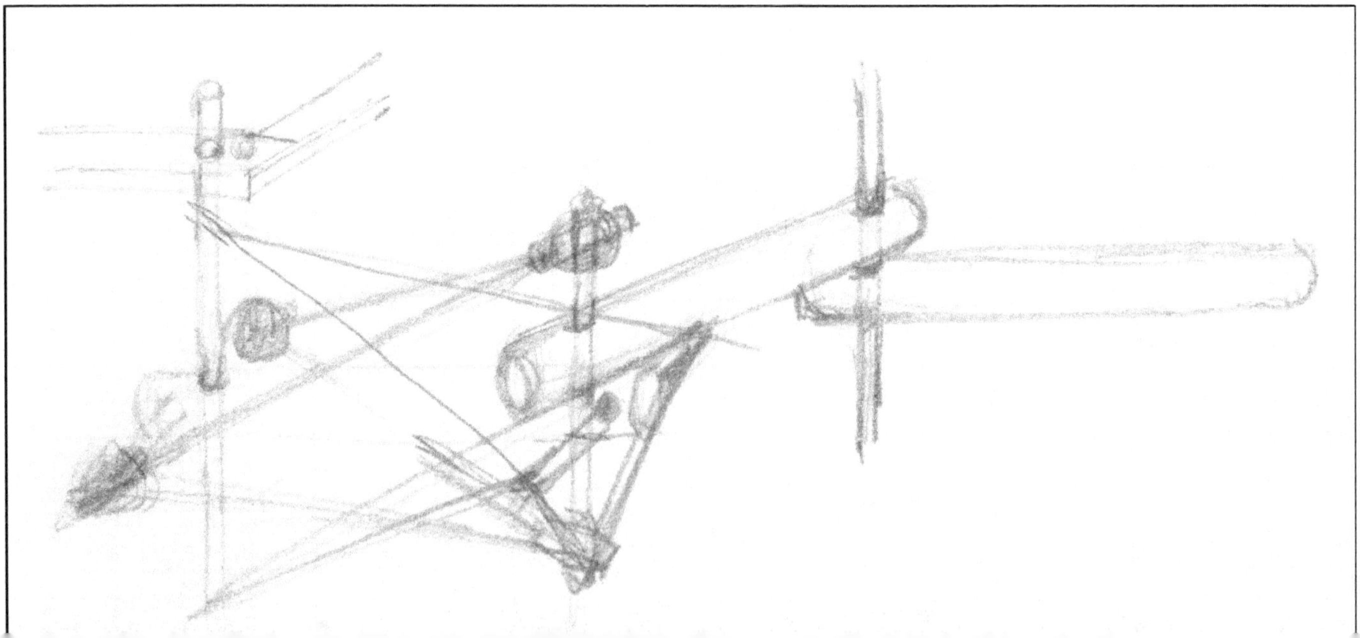

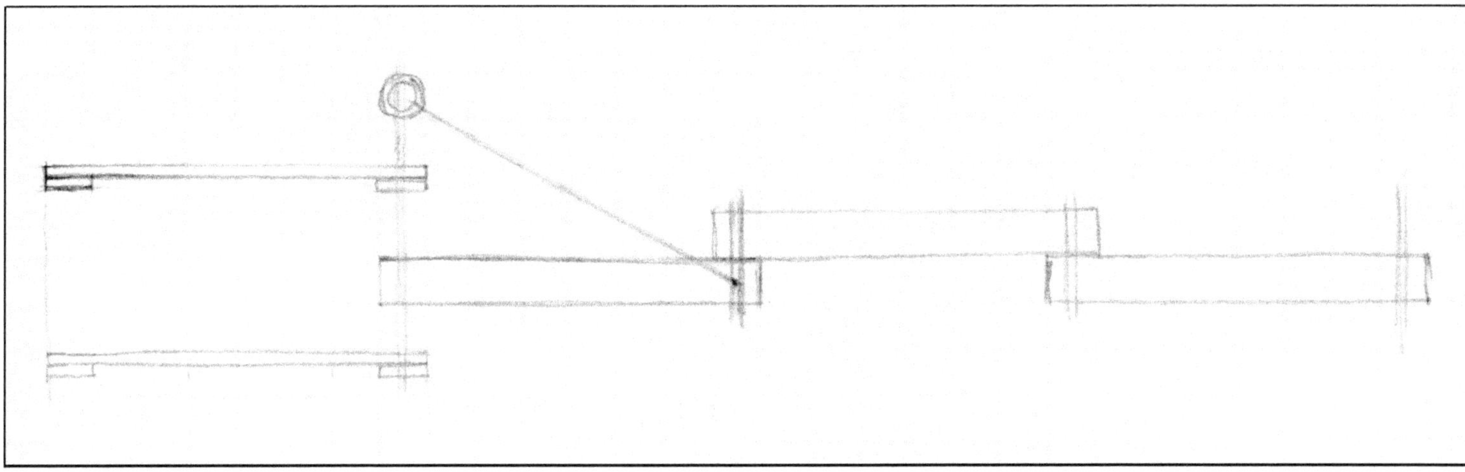

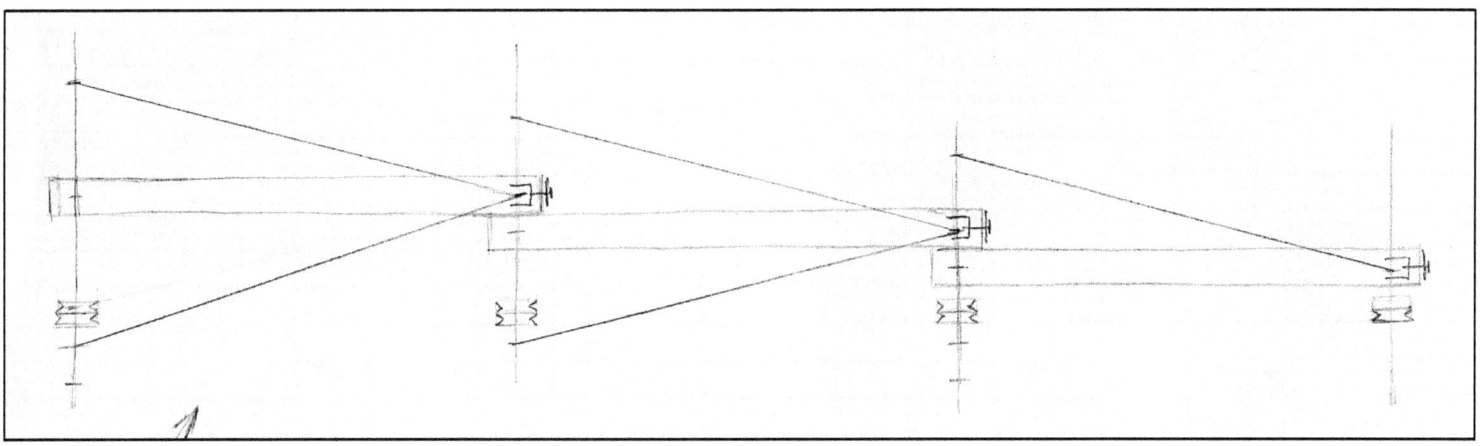

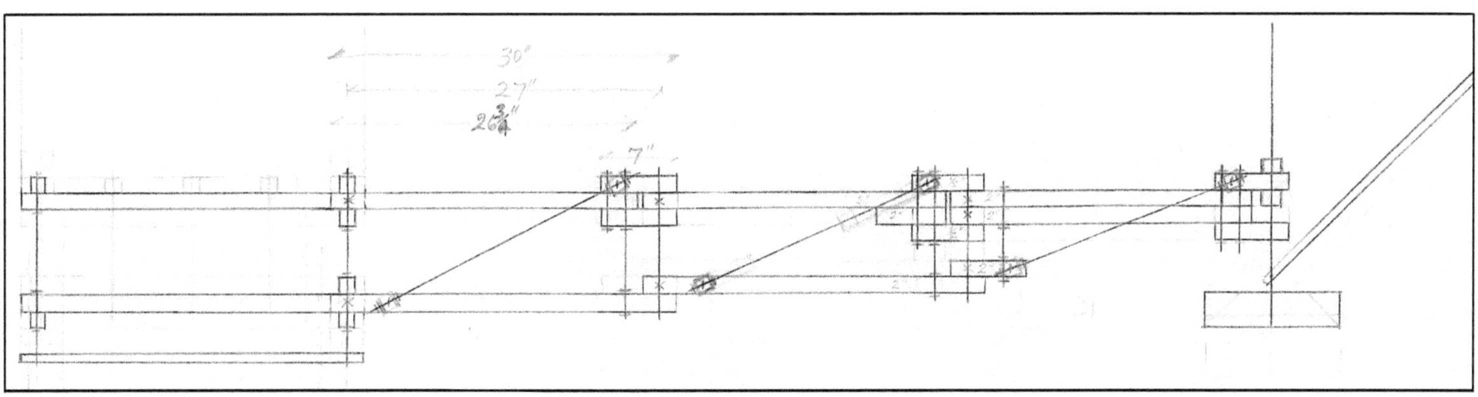
30"
27"
7"

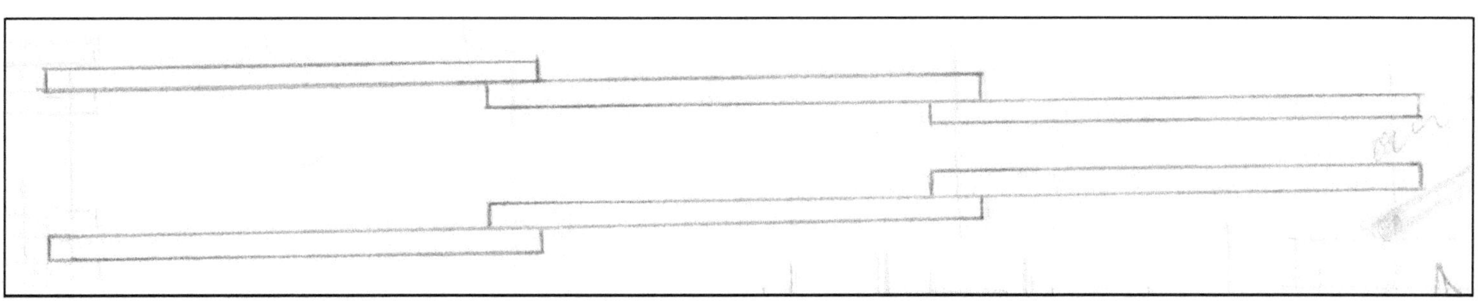

2C. LEVEL

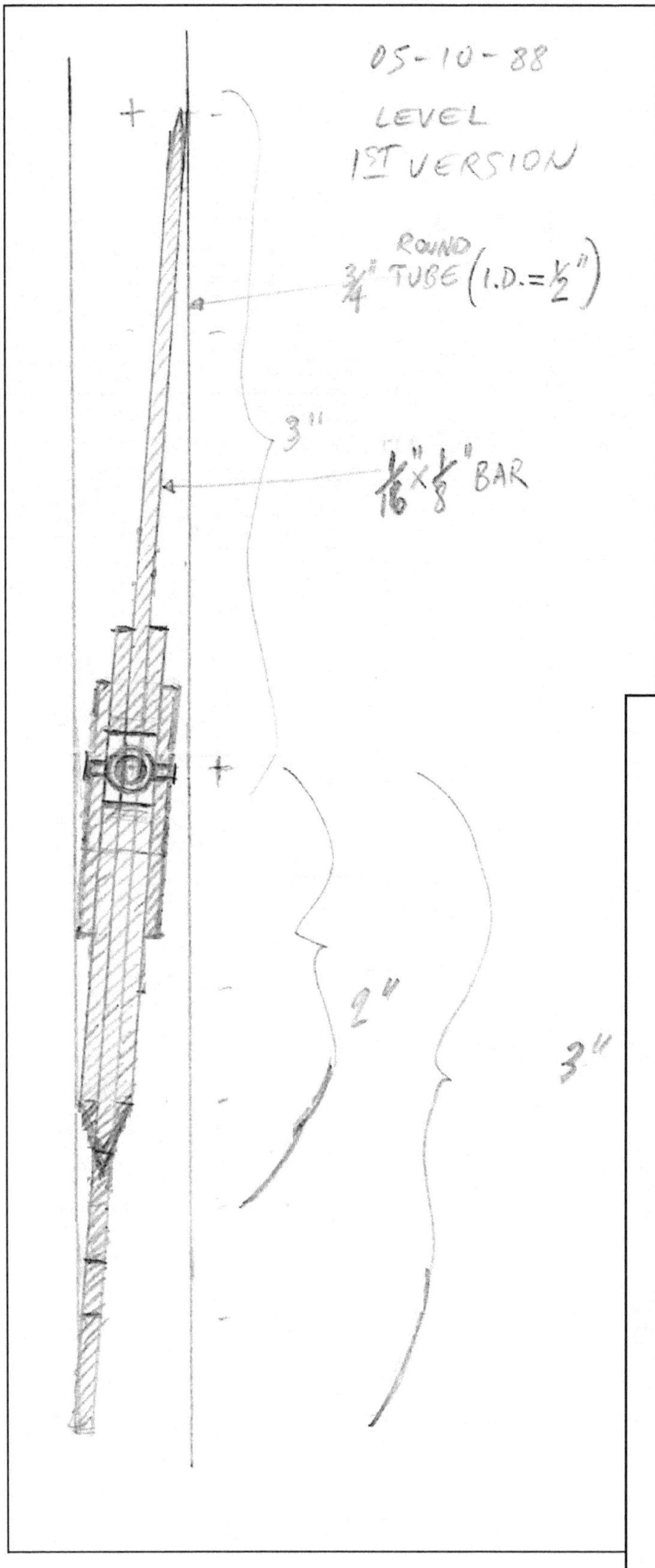

Level
(Using Pendulum Instead Of Water Bubble)

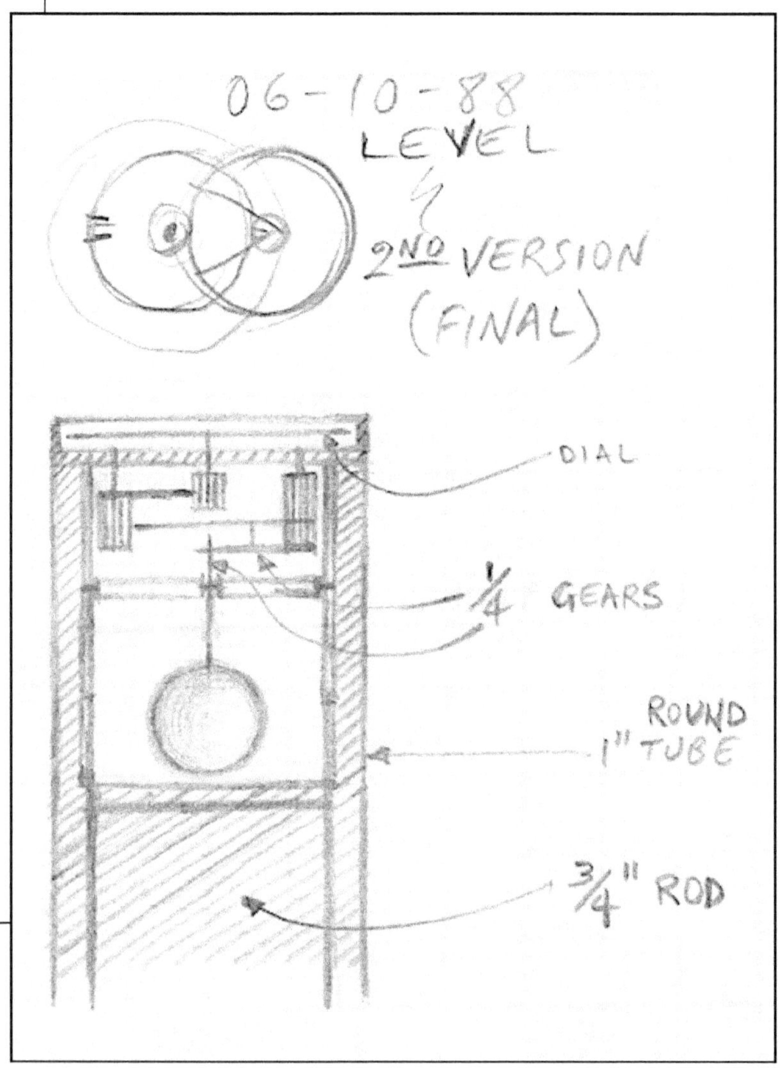

10-10-88

LEVEL

(FINAL VERSION)

ADDENTUM

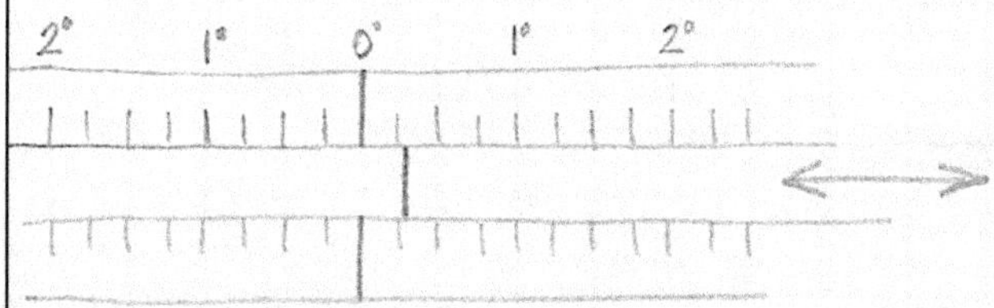

CENTRE SECTION SLIDES
RIGHT OR LEFT
SAME AS SPIRIT LEVEL

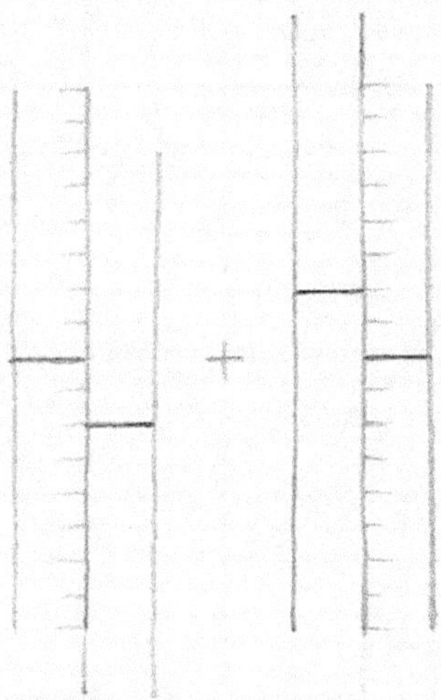

THIS ARRANGEMENT
WOULD MAKE IT LOOK
MORE LIKE A SEE-SAW
MOVEMENT — WHAT A LEVEL IS LIKE.

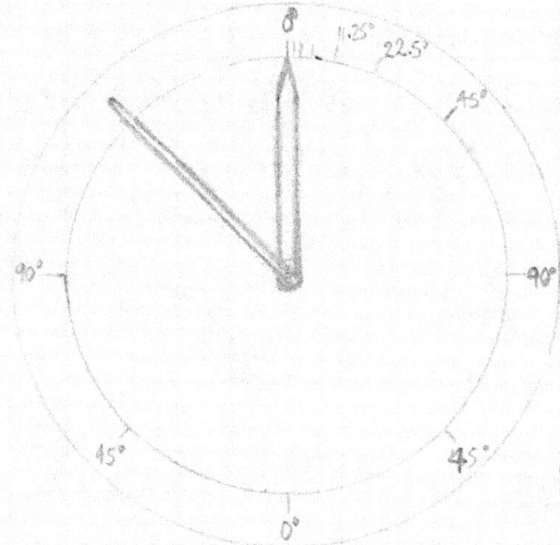

LEVEL BAR

(3 HANDS FOR EVEN GREATER PRECISION)

—TWO HANDS; THE LONG HAND HAS TO MAKE A FULL TURN (360°) FOR THE SHORT HAND TO MOVE ONE DEGREE FOR VERY PRECISE MEASUREMENT.

—LEVEL BAR ~~COULD~~ CAN BE TWIRLDED AROUND TO EITHER MEASURE HORIZONTALLY OR MEASURE VERTICALLY.

OFFSHOOTS:

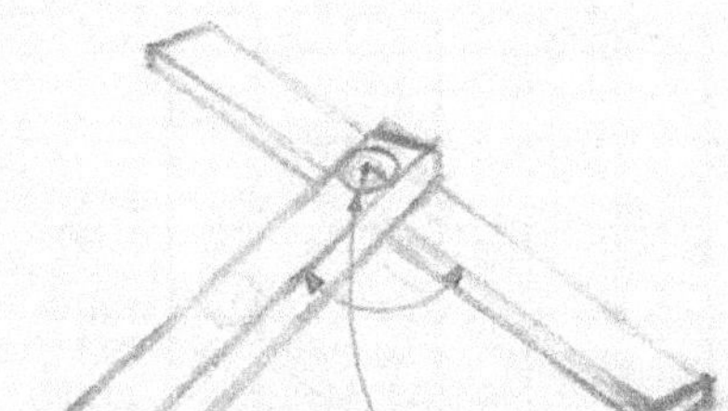

DIAL
SHOWING DEGREES—
POWERED BY MUSCLE POWER
BY OPENING OR CLOSING
"ANGLE-GAGER"
("ANGLER")

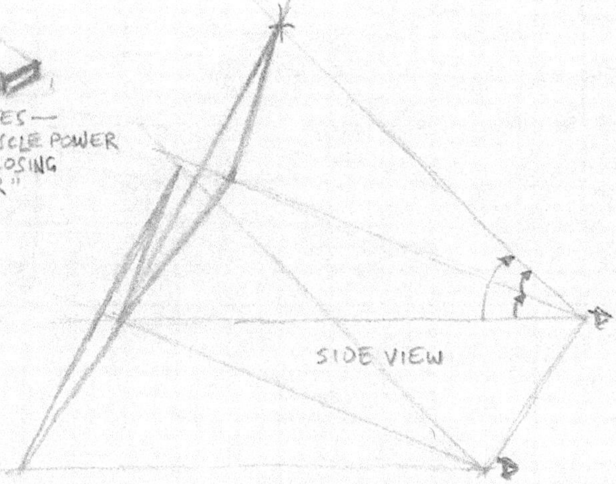

SIDE VIEW

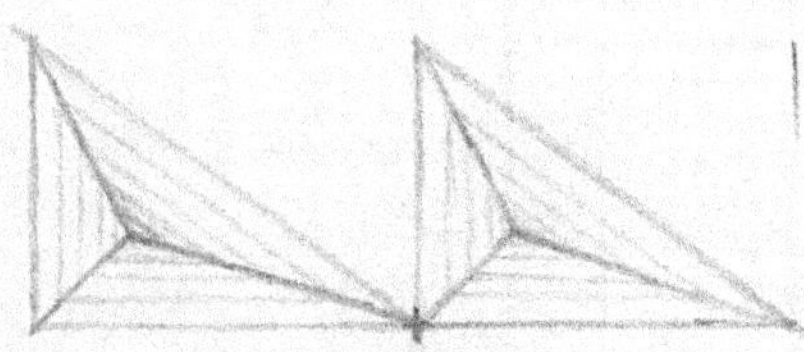

FRONT VIEW

MORE PRECISE ANGLE-GAGERS
ALLOW FOR BETTER
MEASUREMENT OF
DISTANCES

3. SEWING MACHINE

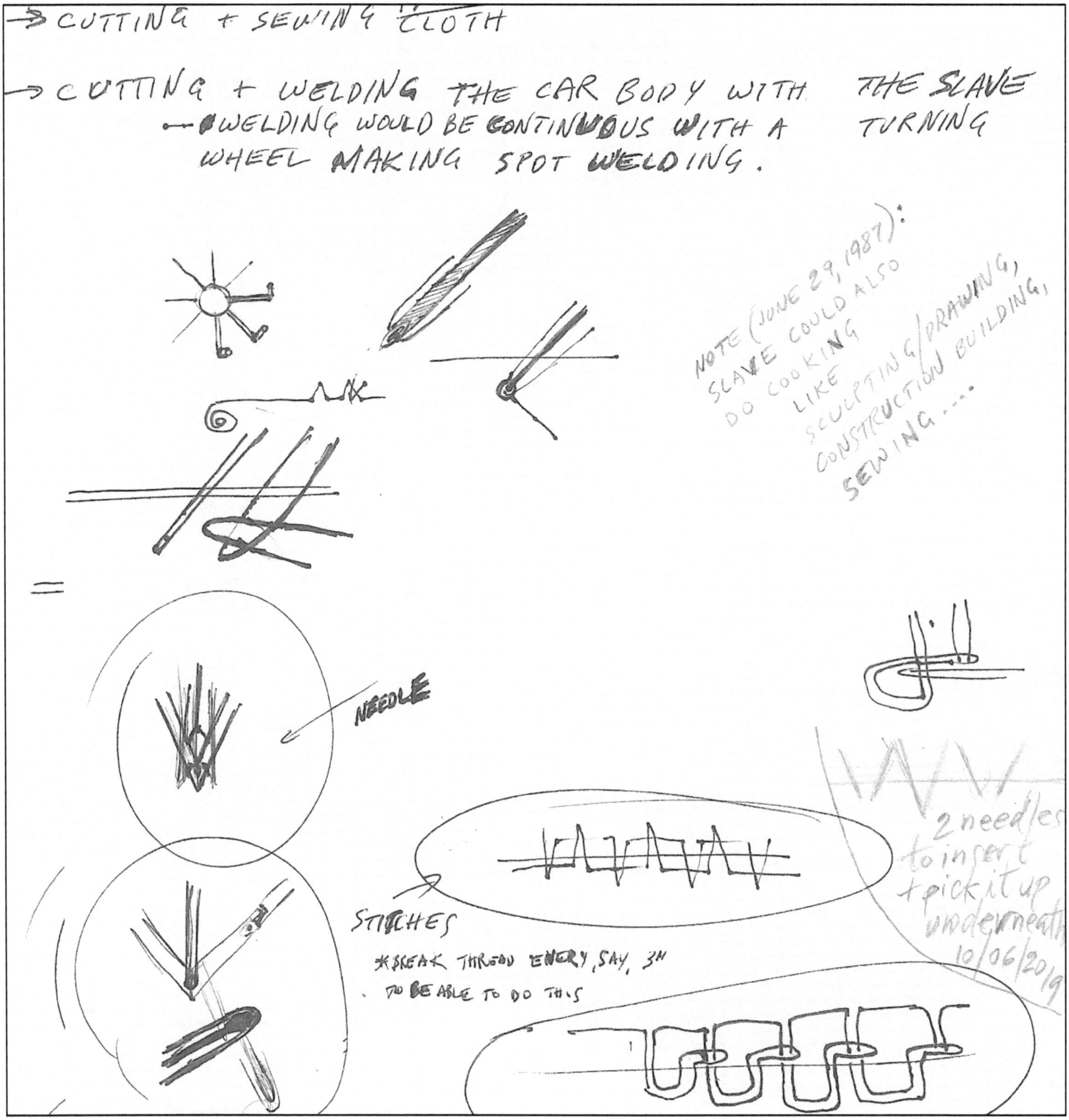

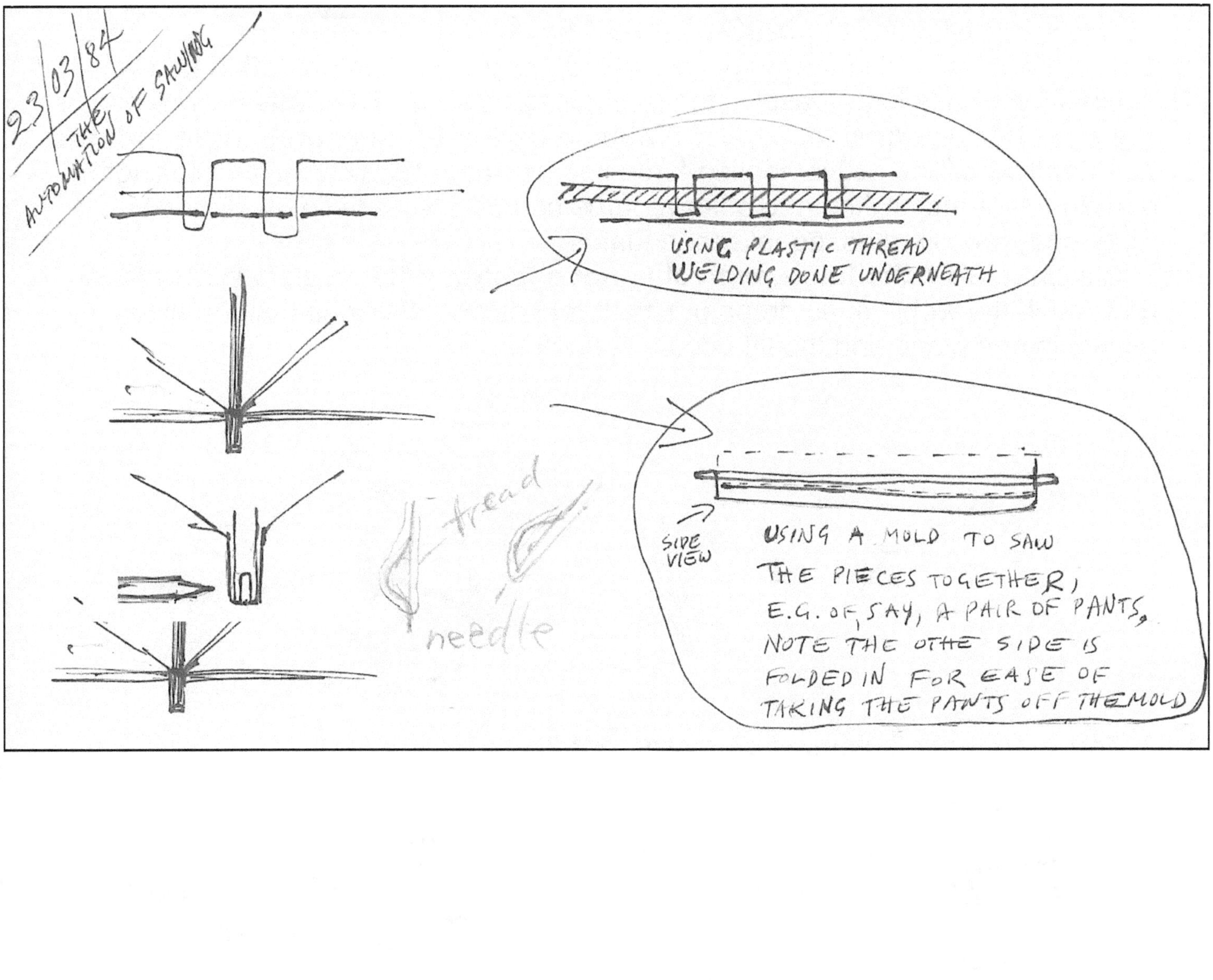
23 | 03 | 84
THE AUTOMATION OF SAWING
USING PLASTIC THREAD
WELDING DONE UNDERNEATH
thread
needle
SIDE VIEW
USING A MOLD TO SAW
THE PIECES TOGETHER,
E.G. OF, SAY, A PAIR OF PANTS,
NOTE THE OTHE SIDE IS
FOLDED IN FOR EASE OF
TAKING THE PANTS OFF THEMOLD

4A. LEVERS

LEVERS

There are 3 types of levers: Rigid, Foldable, and Flexible (by use of rope). A Rigid Lever can be **straight**, **angular**, and **circular**. A Foldable Lever can be **fixed**, **sliding**, or **free-floating**. And, a Flexible Lever can work within **1**, **2**, or **3** **planes**. So, we have …

straight, angular, circular Rigid Lever
fixed, sliding, free-floating Foldable Lever
1-, 2-, 3- plane Flexible Lever

LEVER PRINCIPLES (1993-07-19)

1. The lever, inclined plane, or pulley, all it does, it balances* the effort (active/applied force) and load (passive force) in different ways. **The active force has to travel a shorter, equal, or longer distance/period to find equilibrium with the passive force.** The active/applied force can be natural gravity, animal strength, or man-made motor. And, the passive force can be a superfluid, fluid, or solid.
* balances: equilibrates, equalizes, equates
2. **Distance from the fulcrum/pivot is not a factor.** What matters is how far or how much the weight/load drops or lifts. It makes no difference that the drop occurs in one plane and the lift occurs in another.

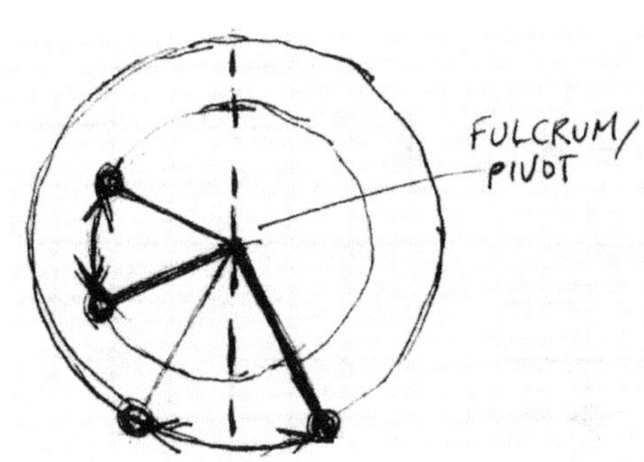

3. **There are only 2 classes of levers, not 3.**

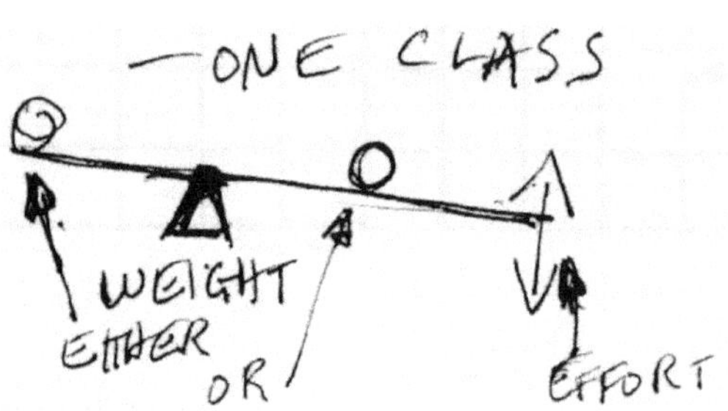

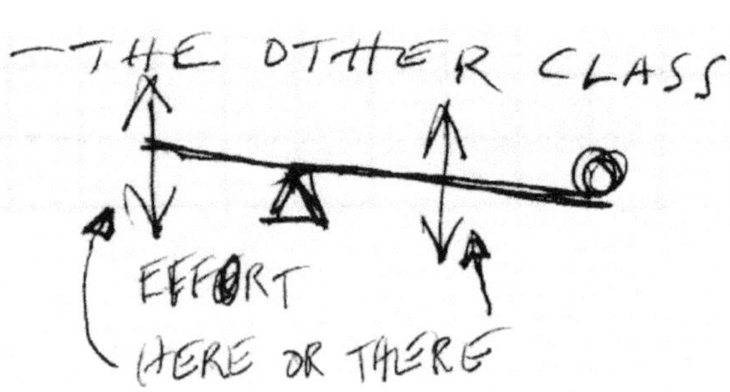

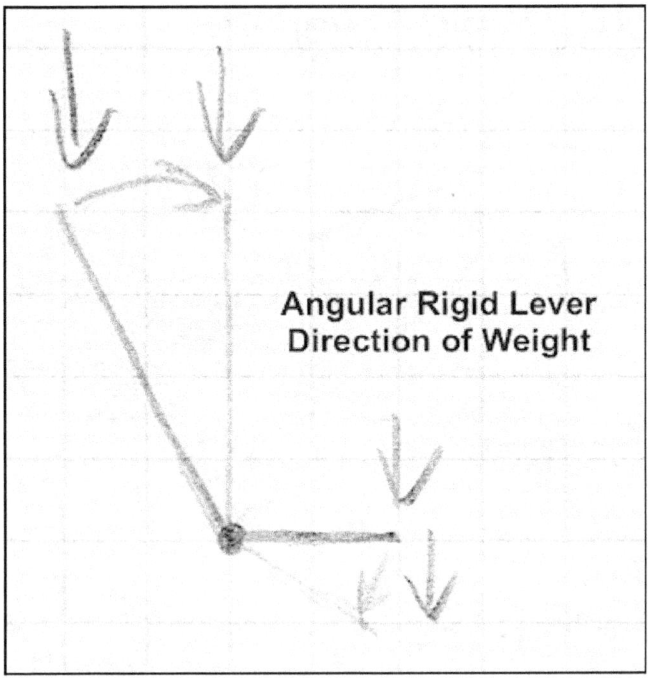
Angular Rigid Lever
Direction of Weight

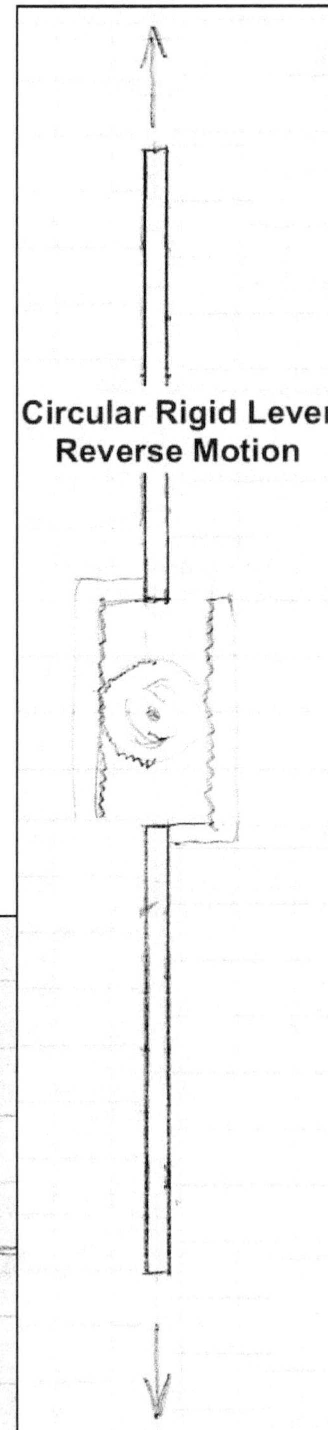
Circular Rigid Lever
Reverse Motion

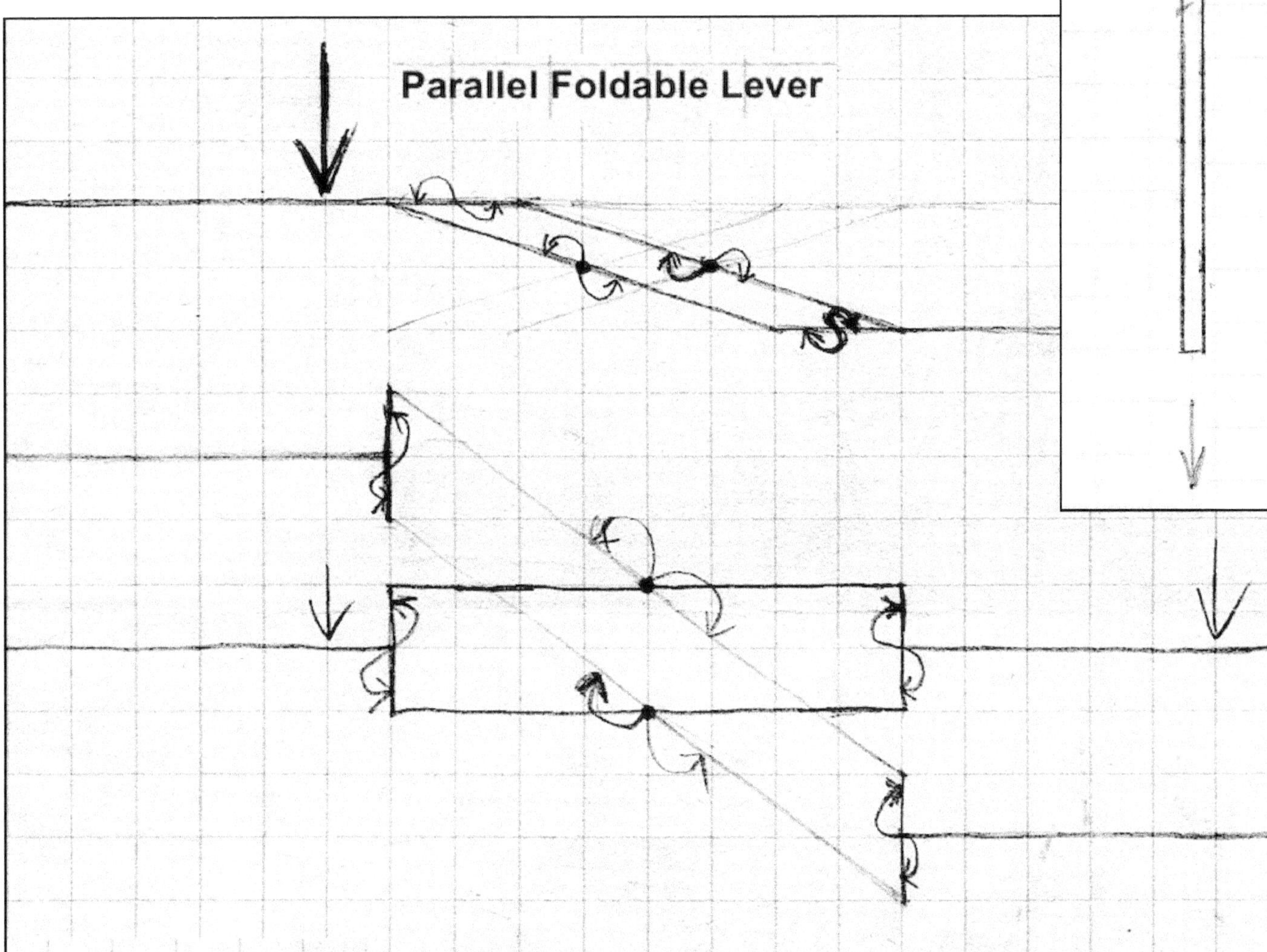
Parallel Foldable Lever

Circular Rigid Lever

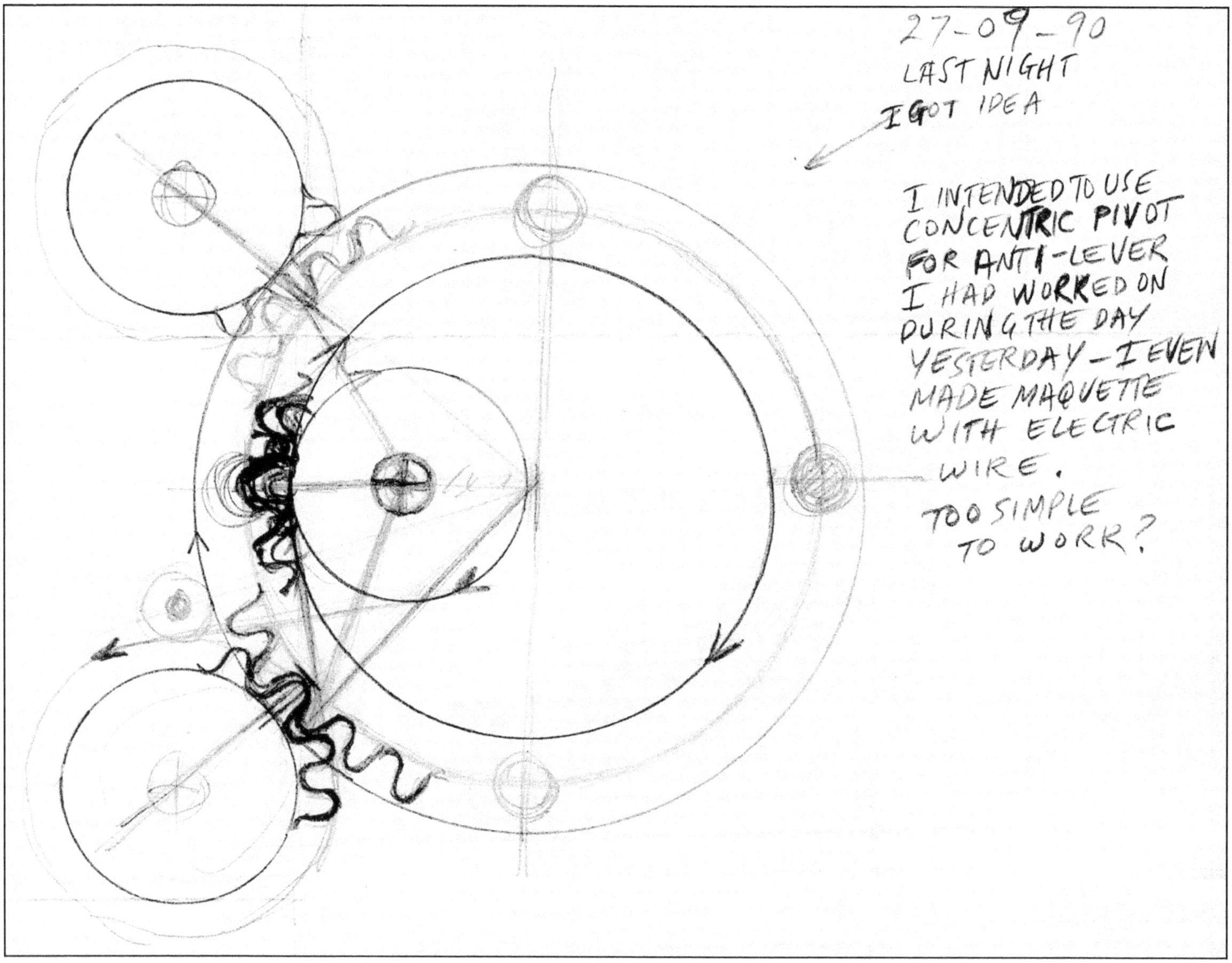

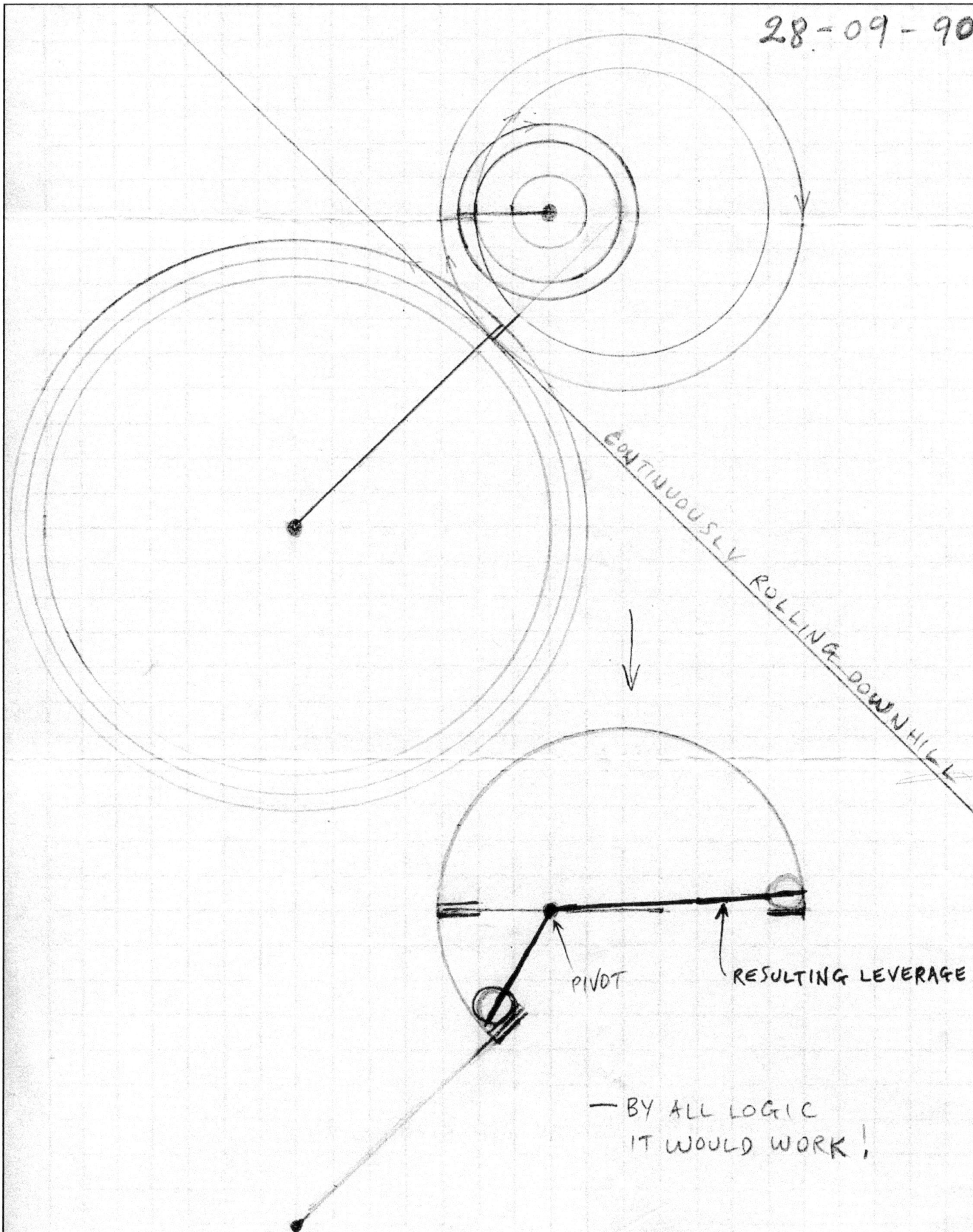
28-09-90
CONTINUOUSLY ROLLING DOWNHILL
PIVOT
RESULTING LEVERAGE
— BY ALL LOGIC
IT WOULD WORK!

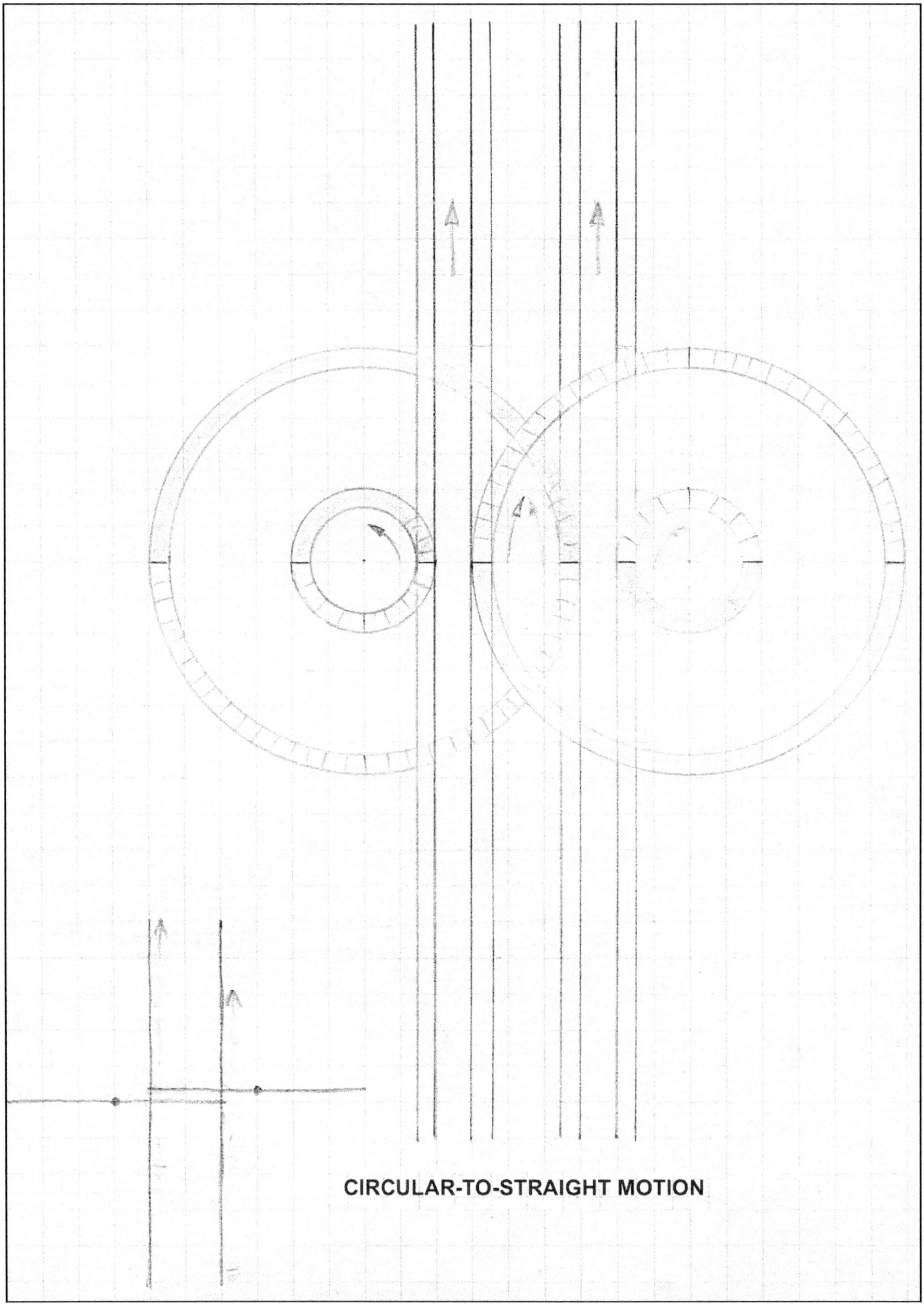

CIRCULAR-TO-STRAIGHT MOTION

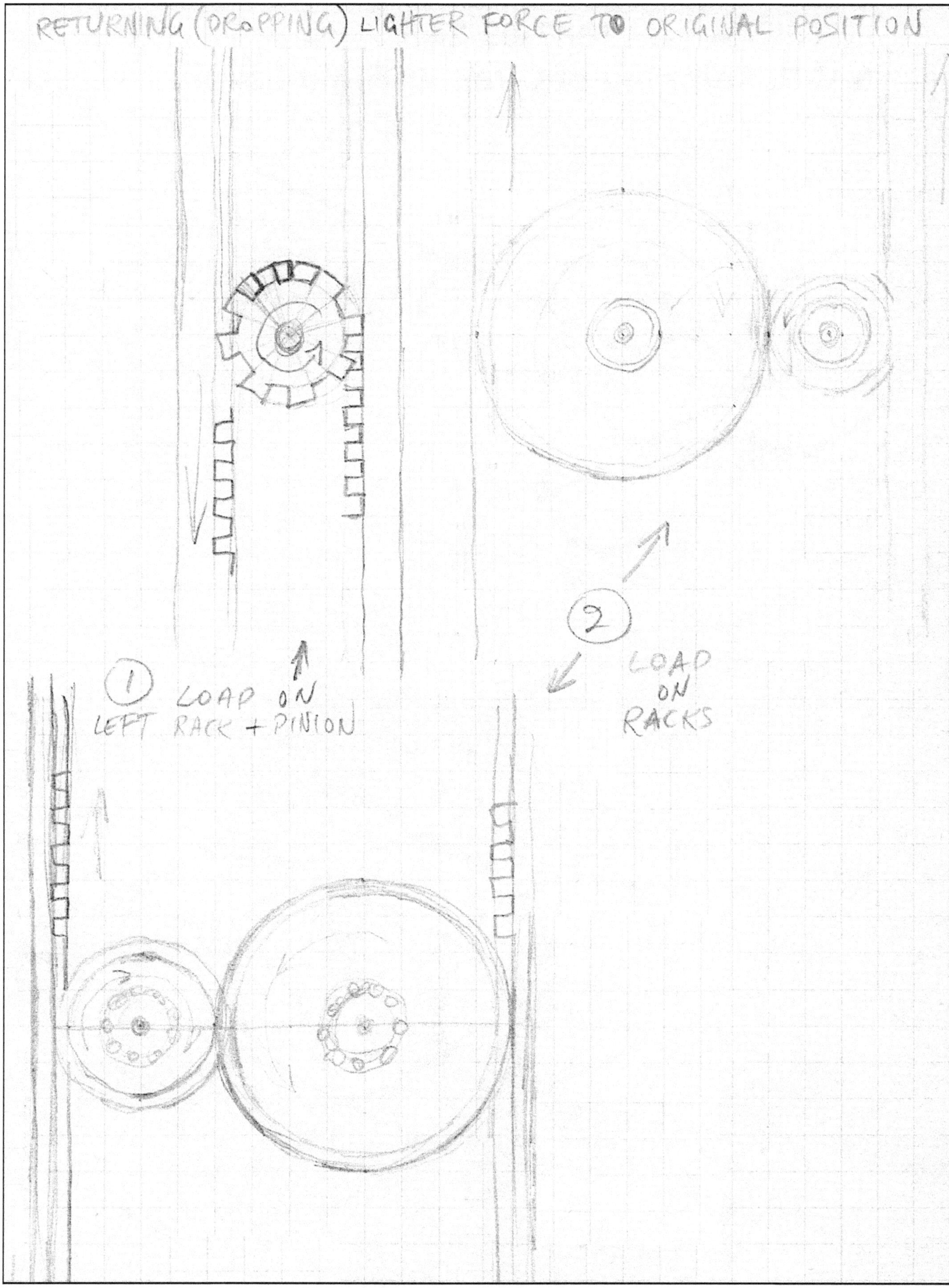
RETURNING (DROPPING) LIGHTER FORCE TO ORIGINAL POSITION
① LOAD ON LEFT RACK + PINION
② LOAD ON RACKS

1987-09-24
3-D Positioning of Circular Rigid Levers

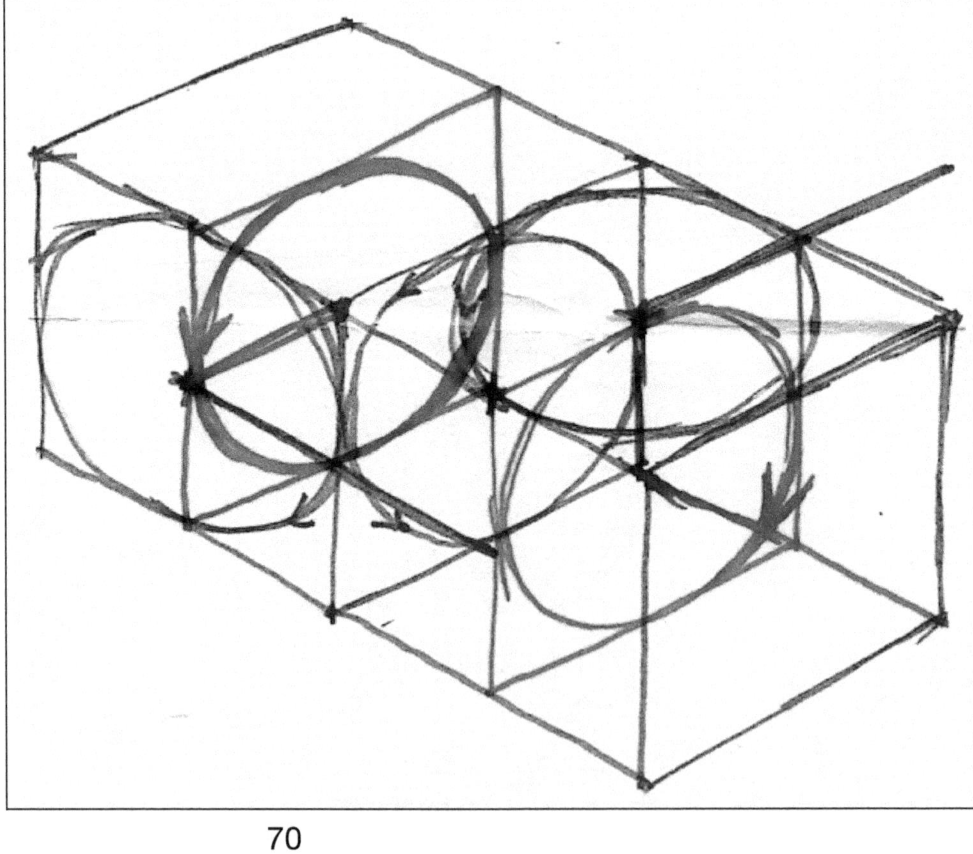

Foldable Lever

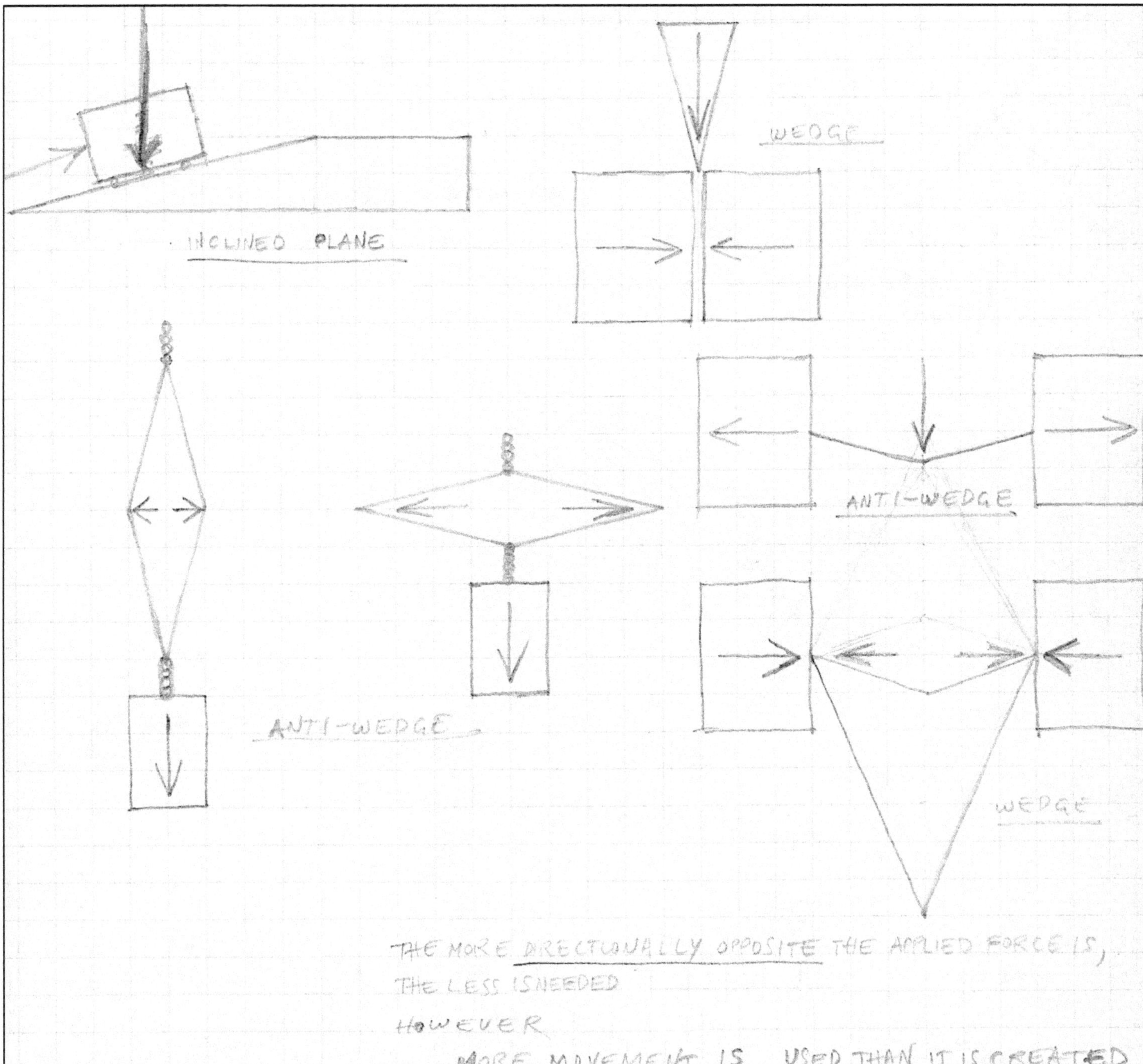

Sliding Foldable Lever

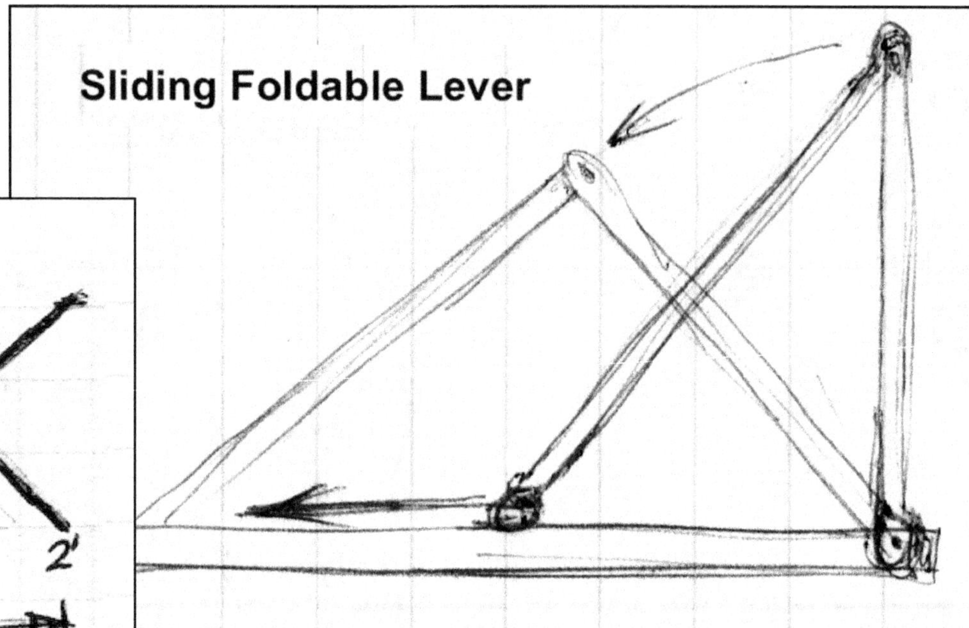

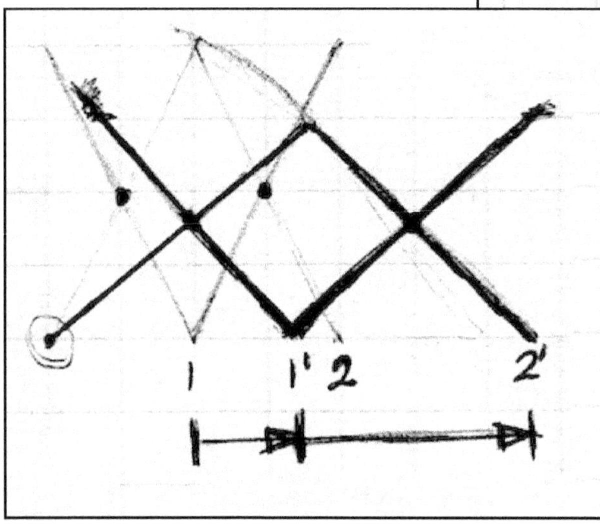

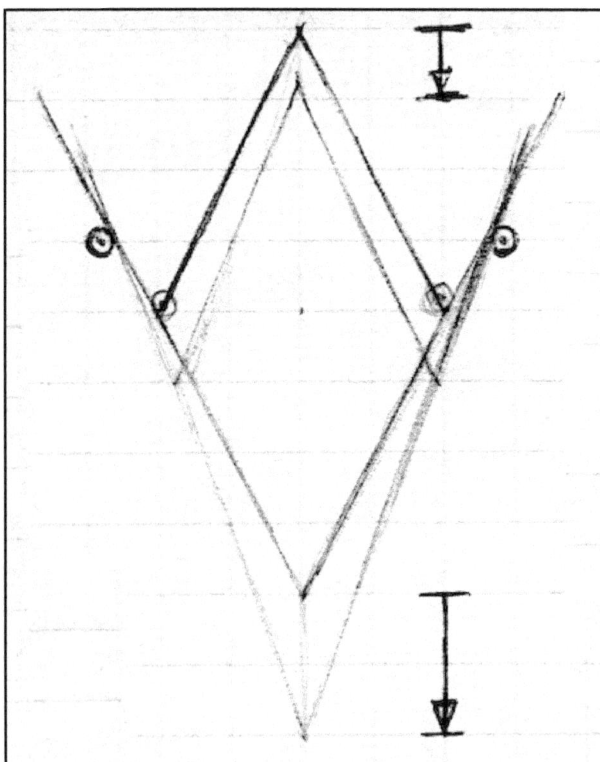

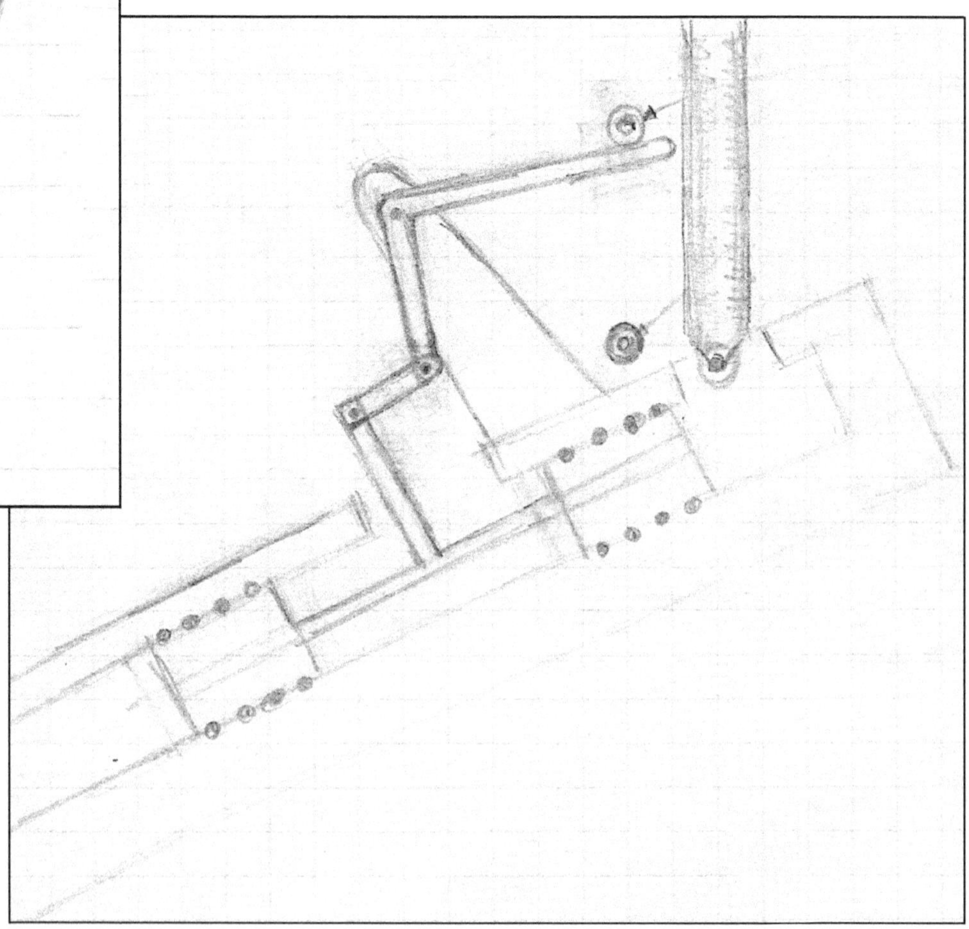

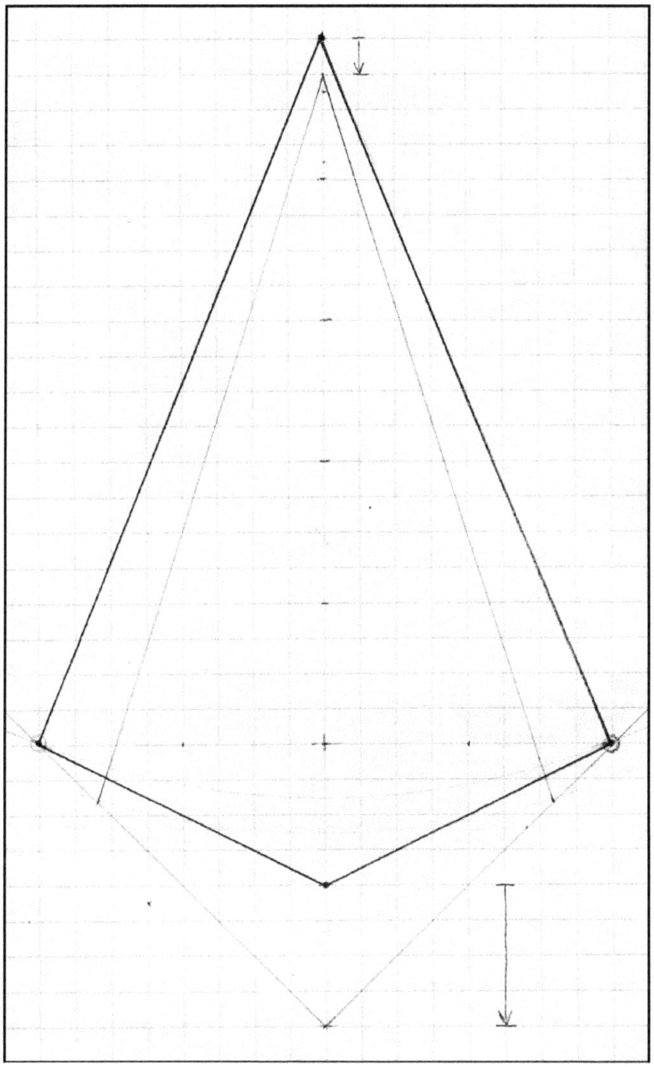

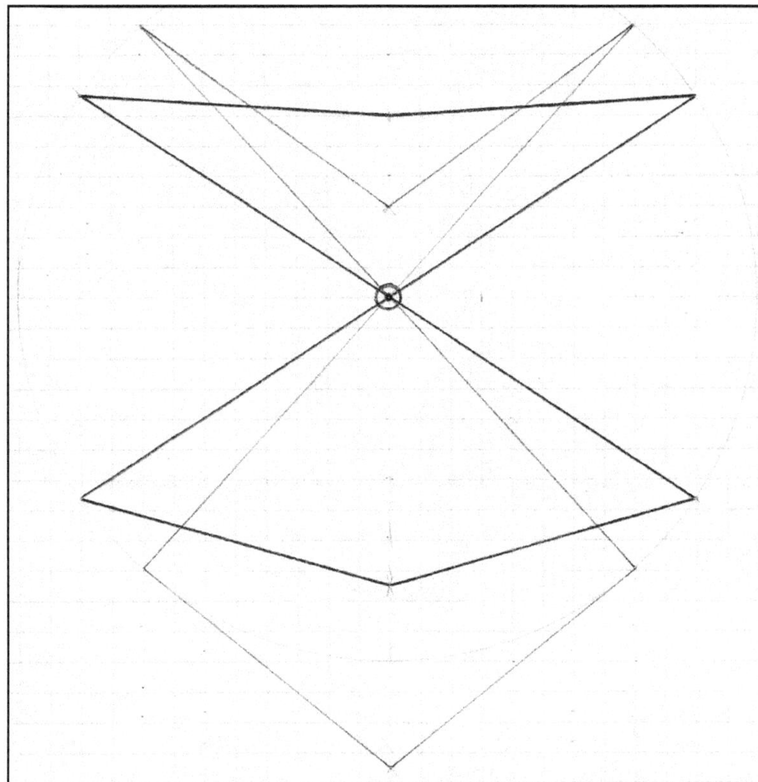

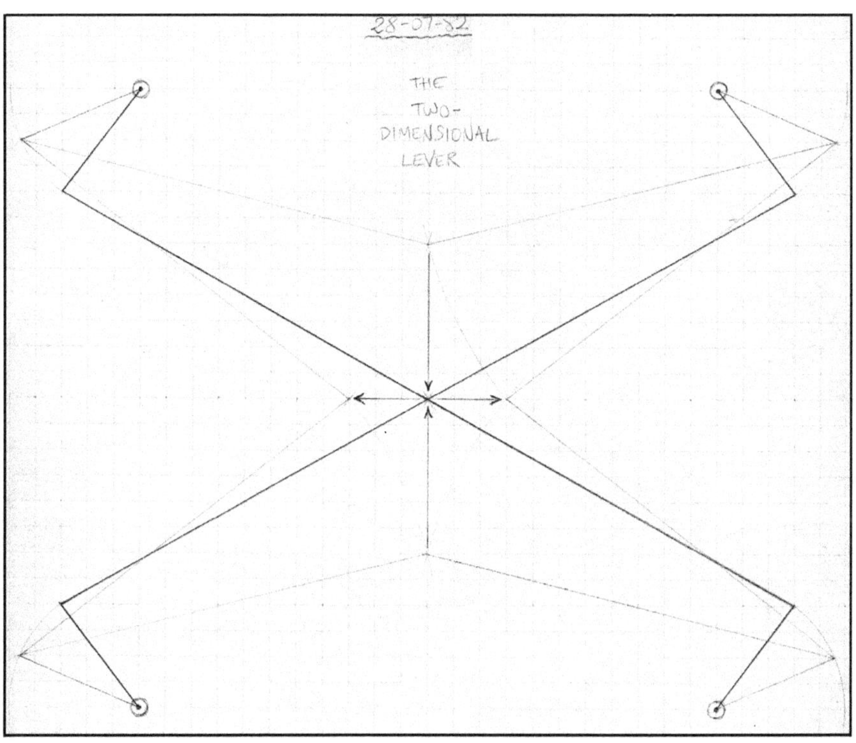

Both Straight and Angular Motion

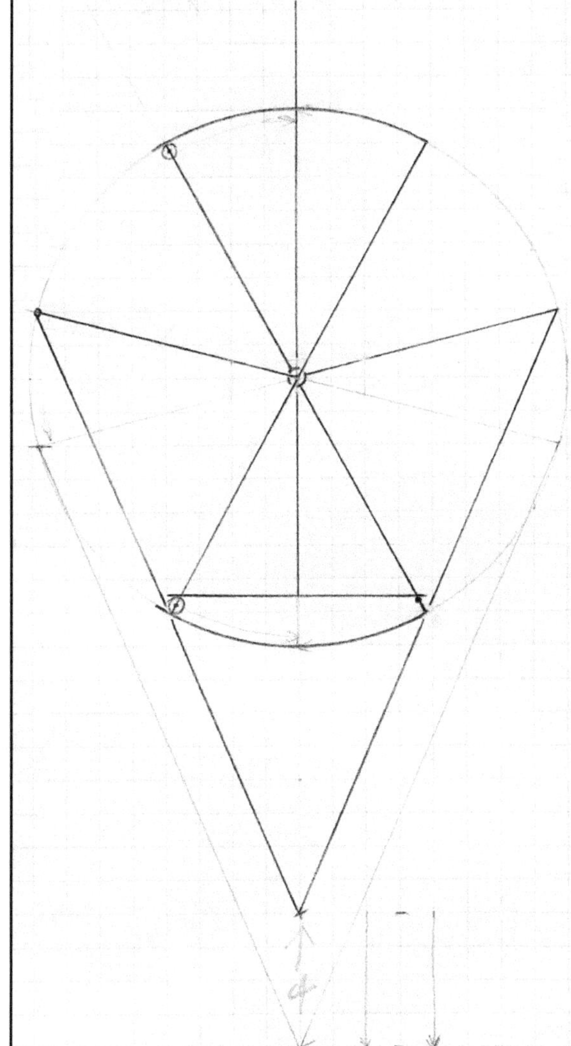

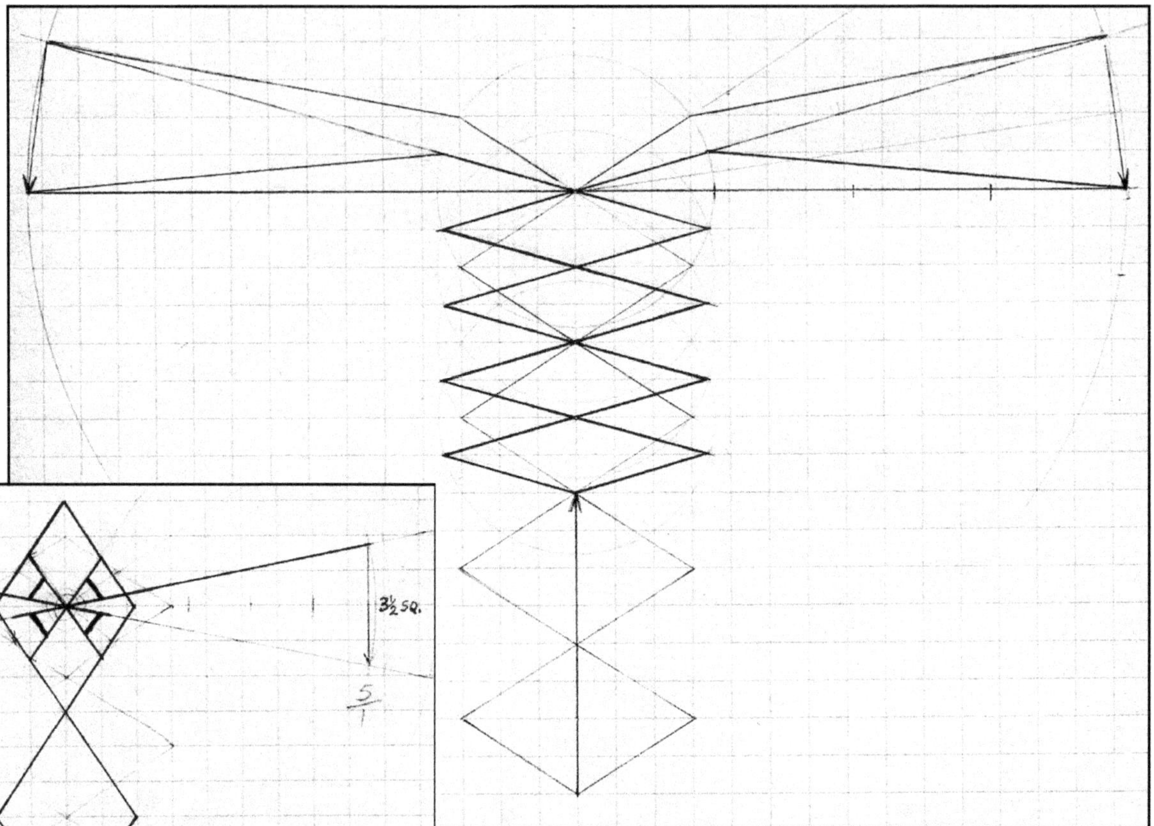

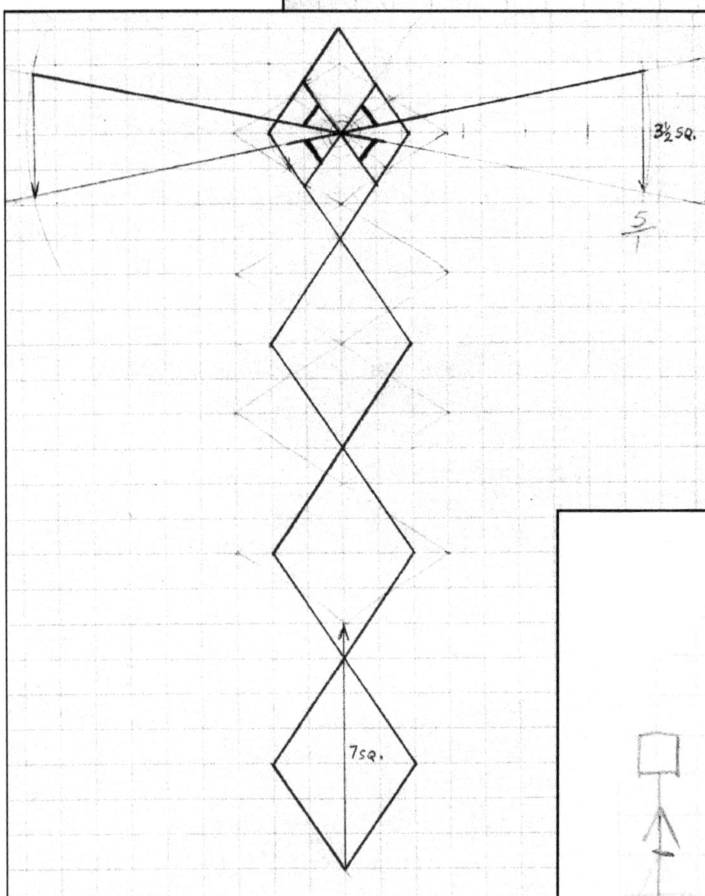
3½ SQ.
7SQ.

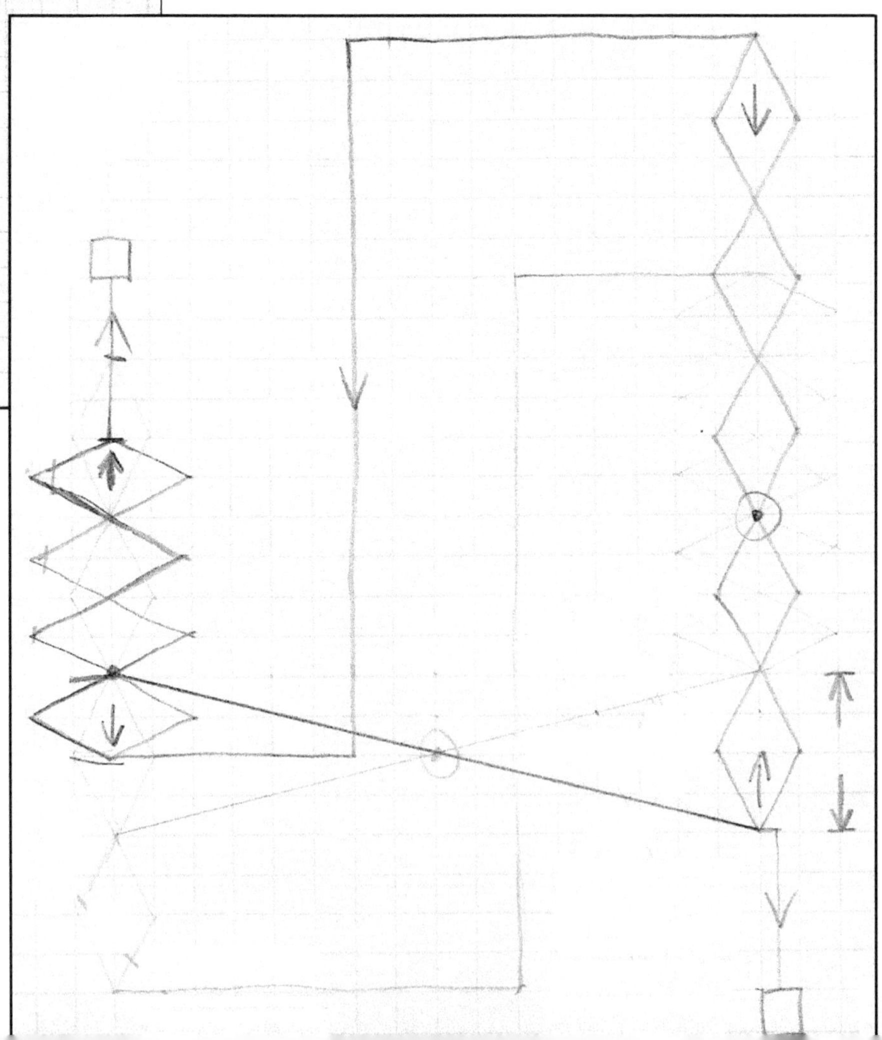

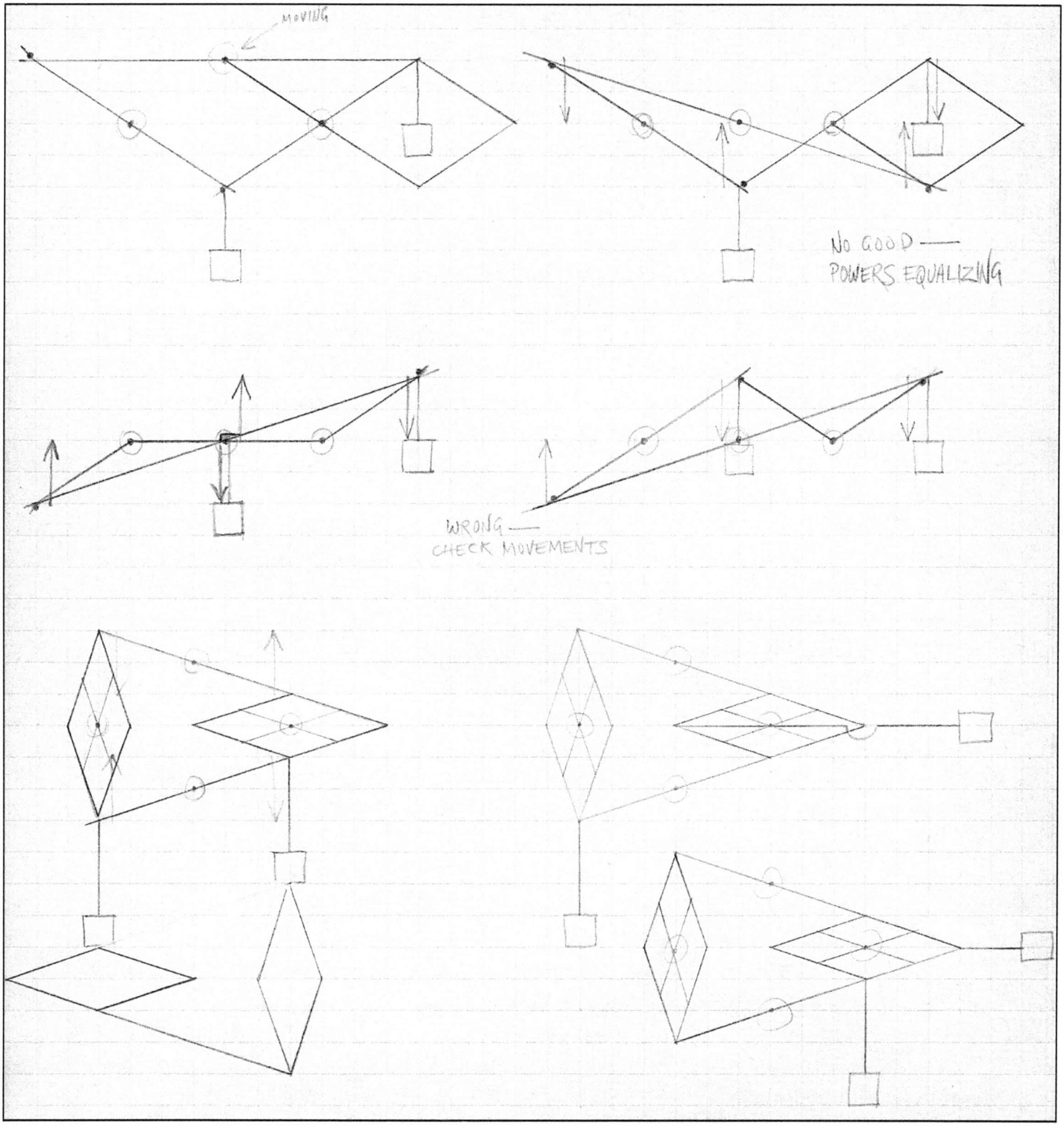
MOVING
NO GOOD —
POWERS EQUALIZING
WRONG —
CHECK MOVEMENTS

INDEX-OUTLINE —— EVEN-ANGLE WEDGE-LEVERING

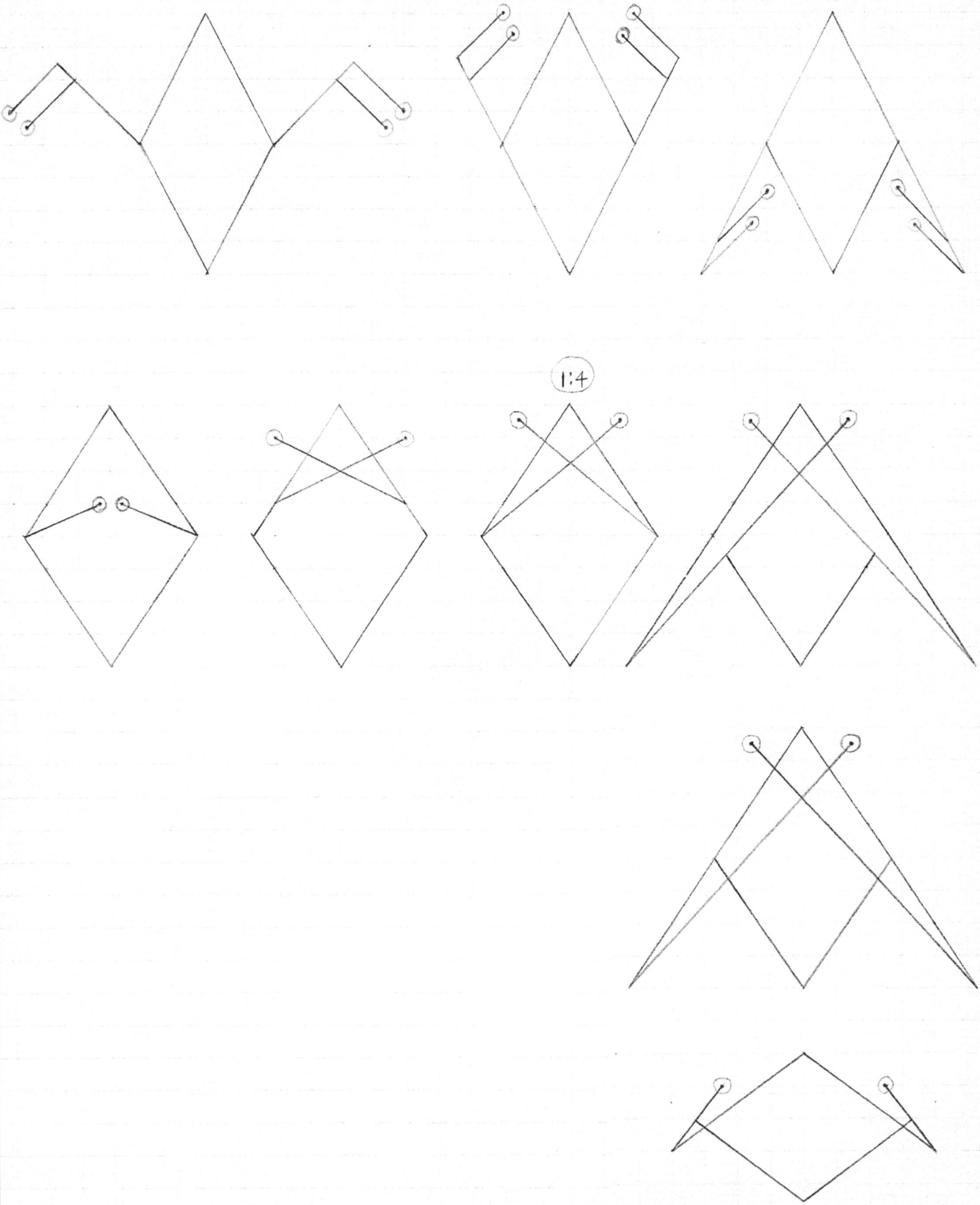

Flexible Lever

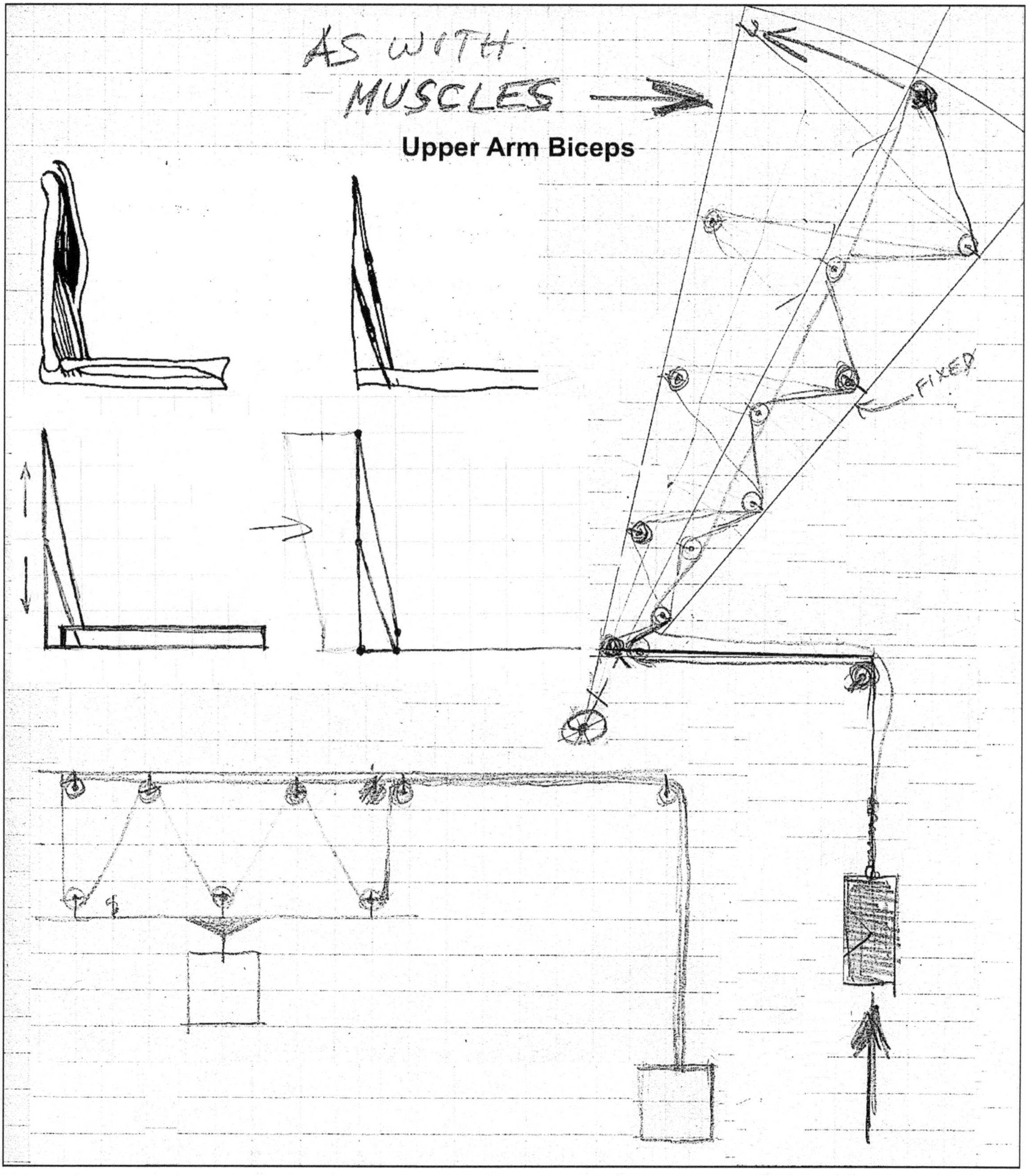

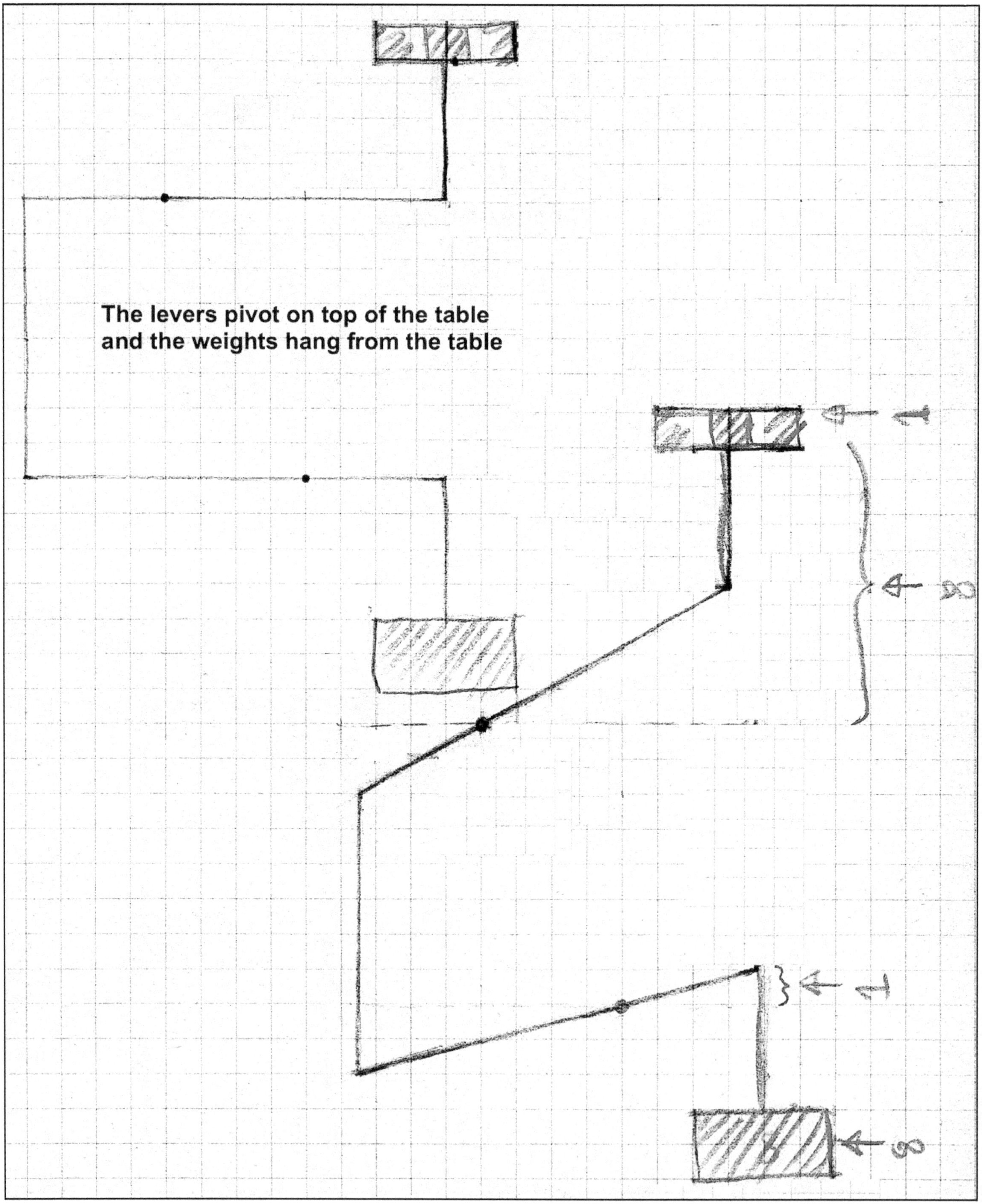
The levers pivot on top of the table
and the weights hang from the table
1
8
1
8

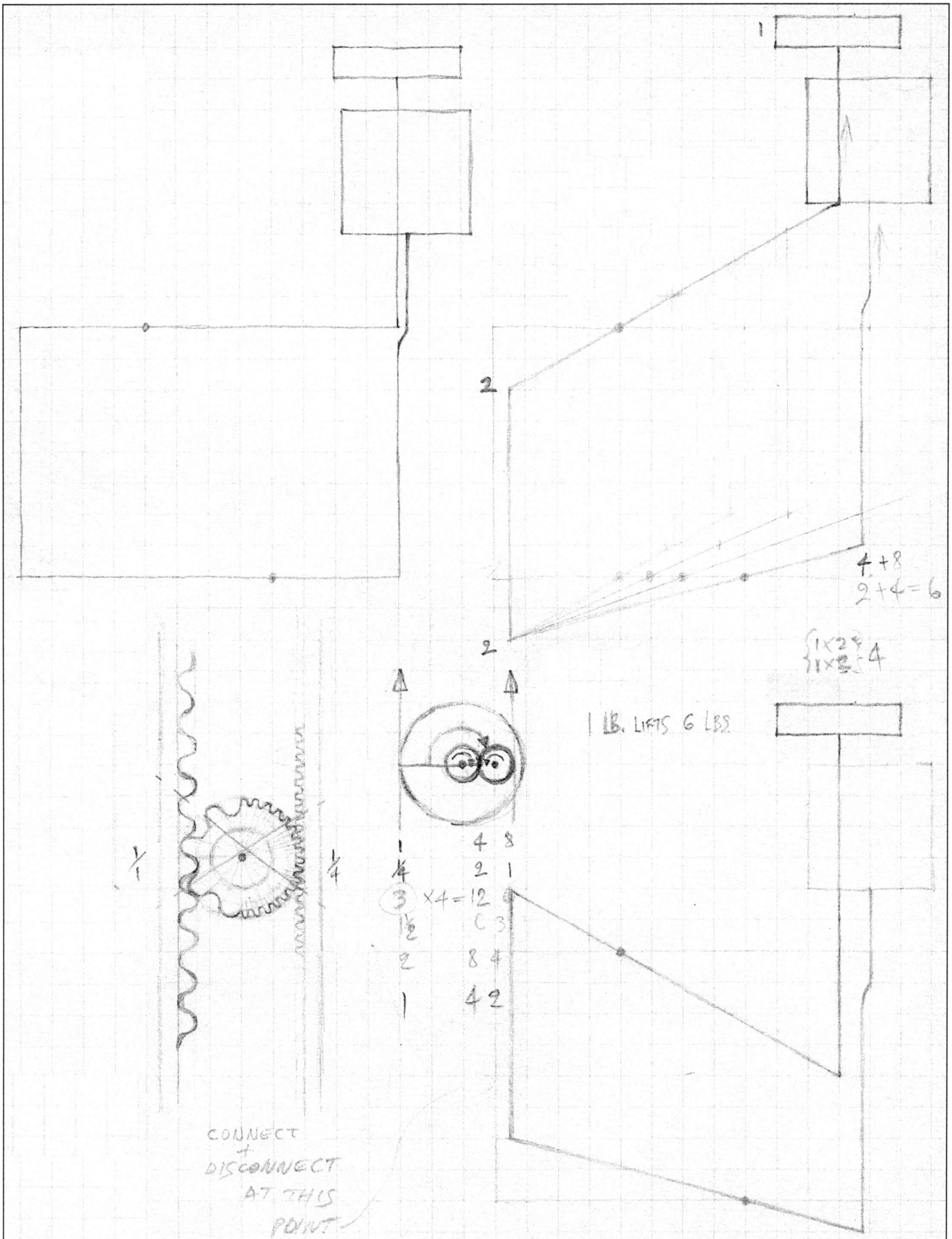
1
2
2
4+8
2+4=6
1 LB. LIFTS 6 LBS
1/1
1/4
1 4 8
1/4 2 1
3 X4=12
1/2
2 8 4
1 4 2
CONNECT
+
DISCONNECT
AT THIS
POINT

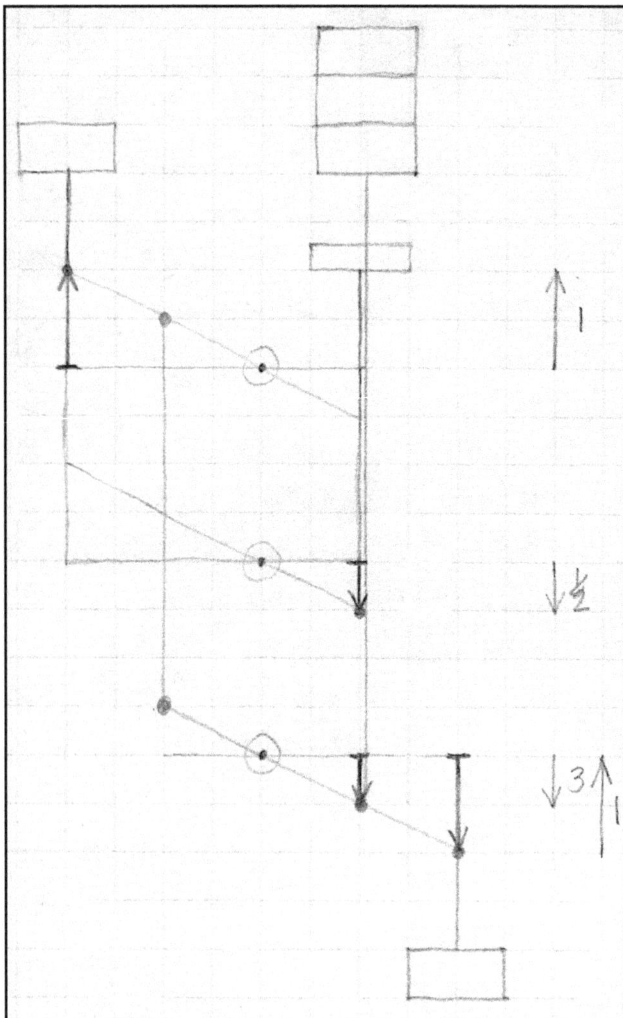
1
½
3
1

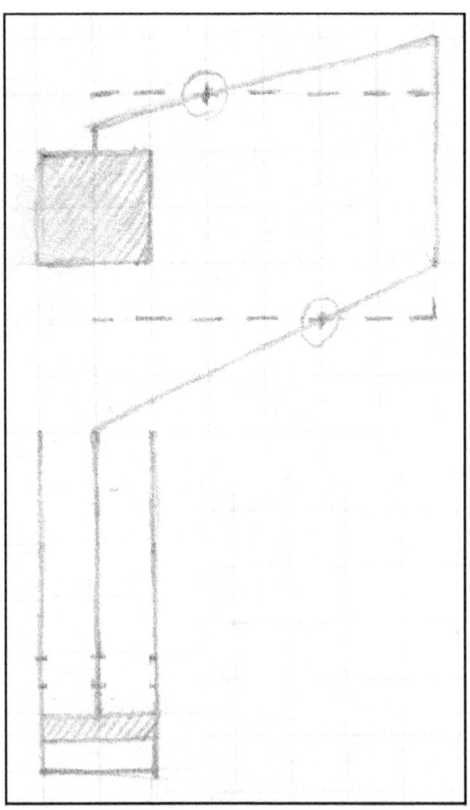

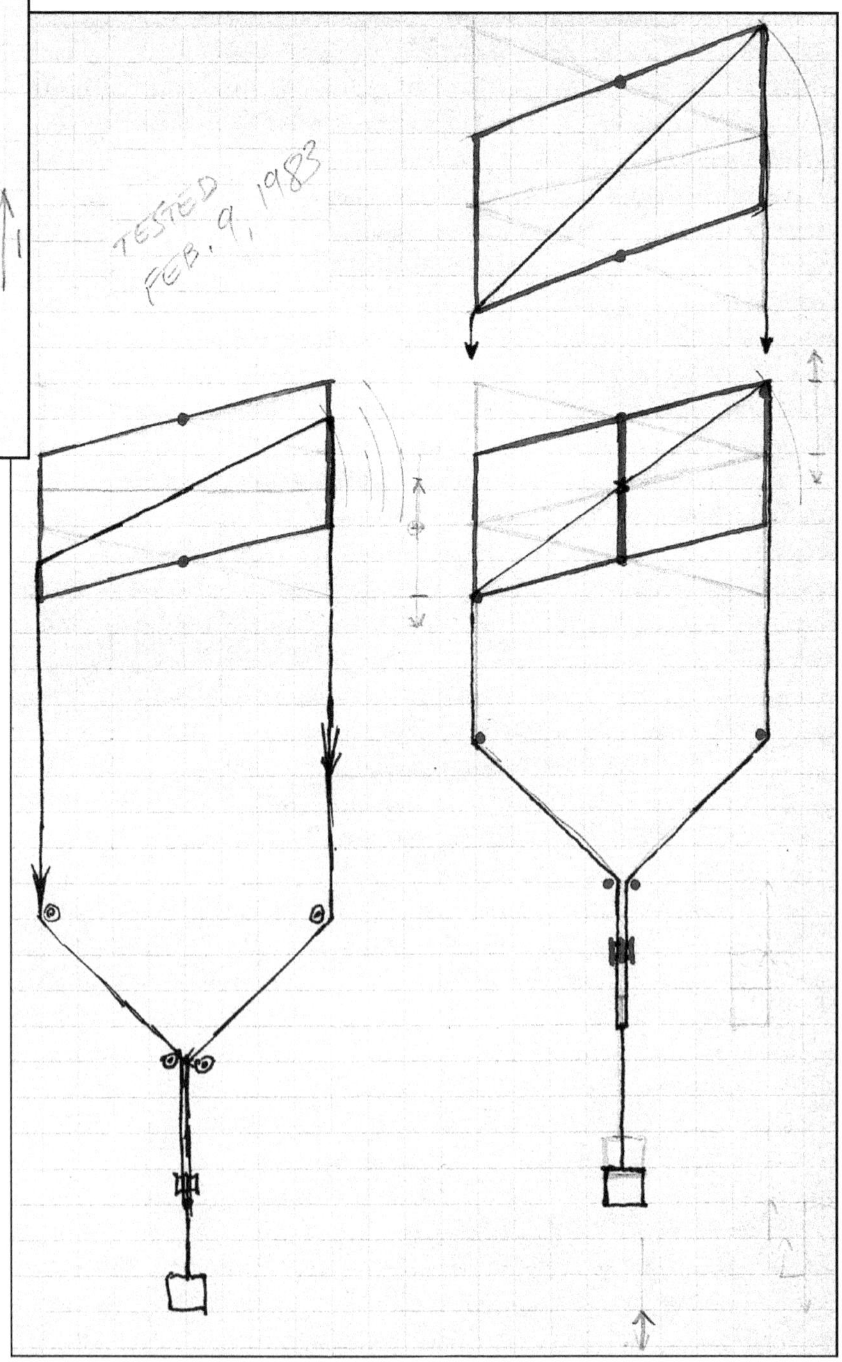
TESTED
FEB. 9, 1983

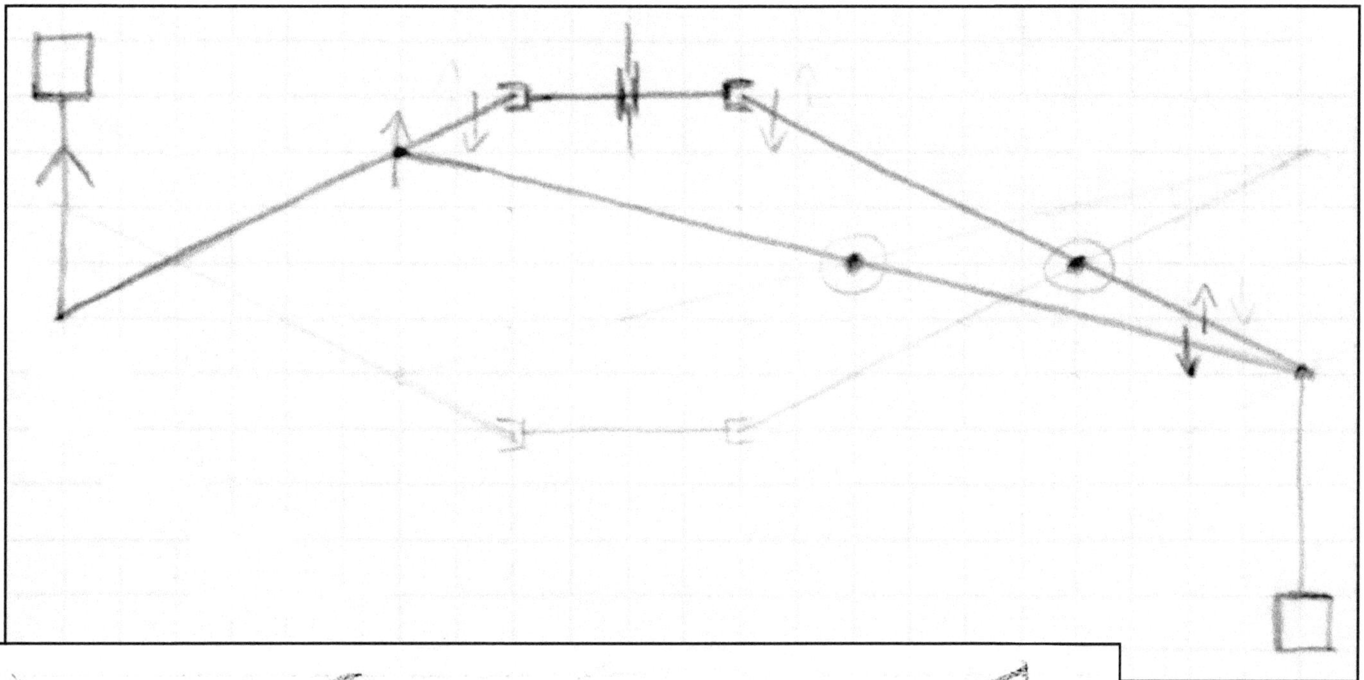

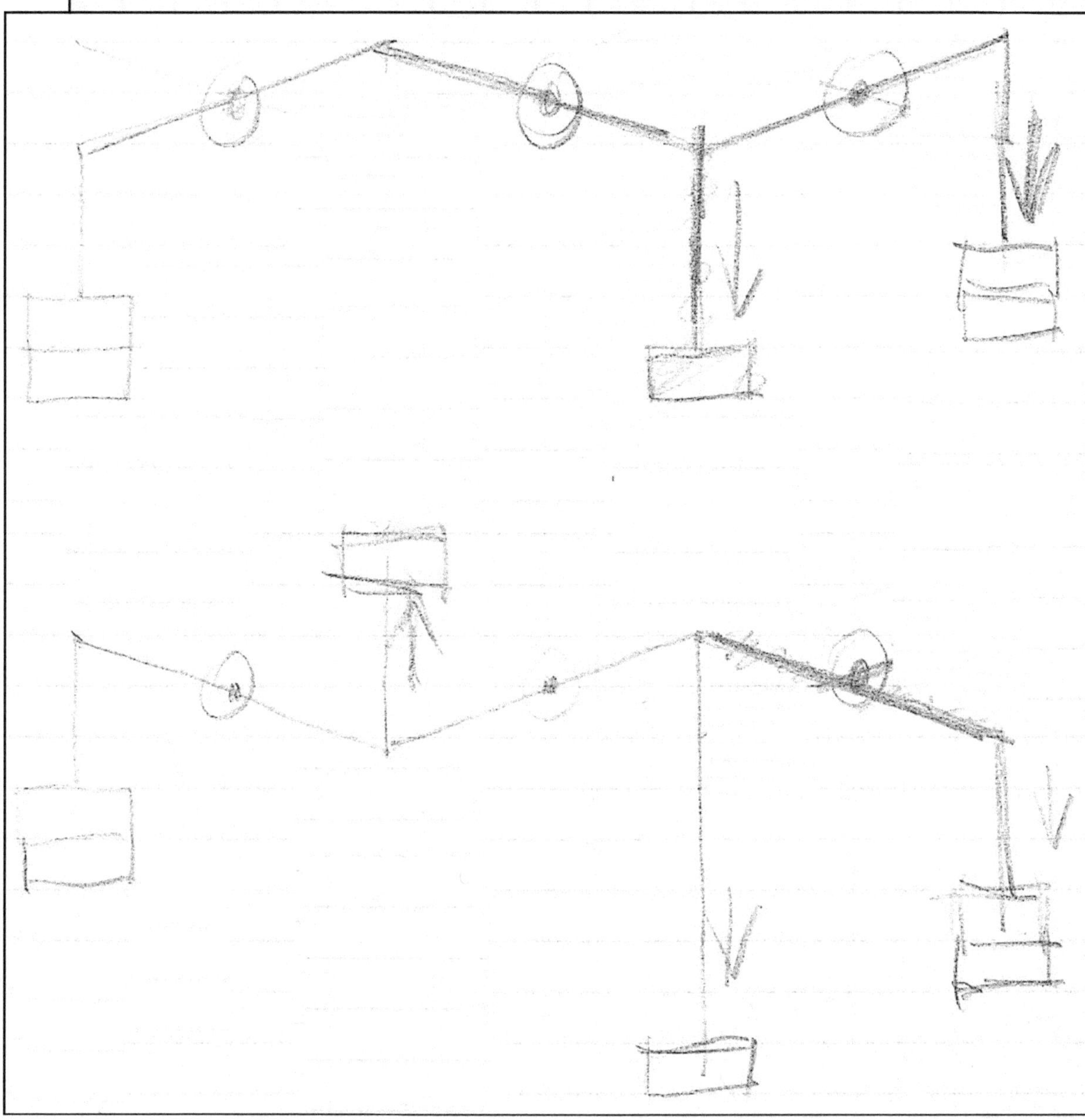

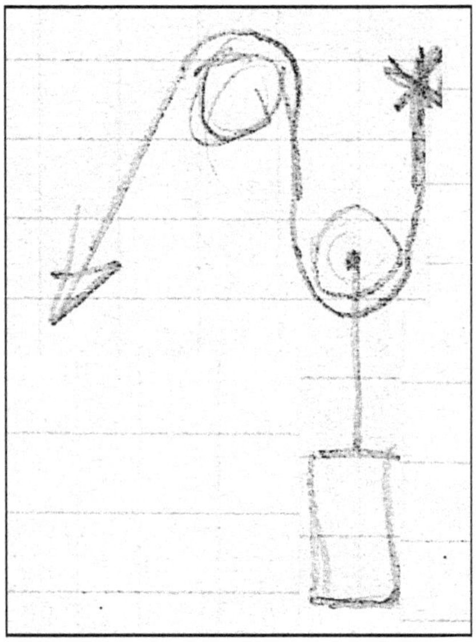

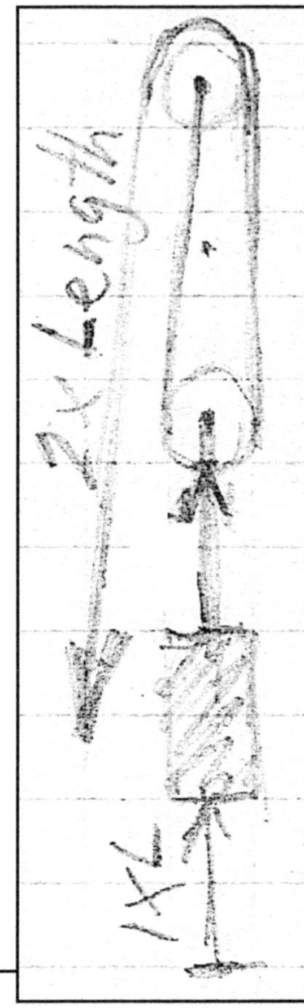
2x Length
1xL

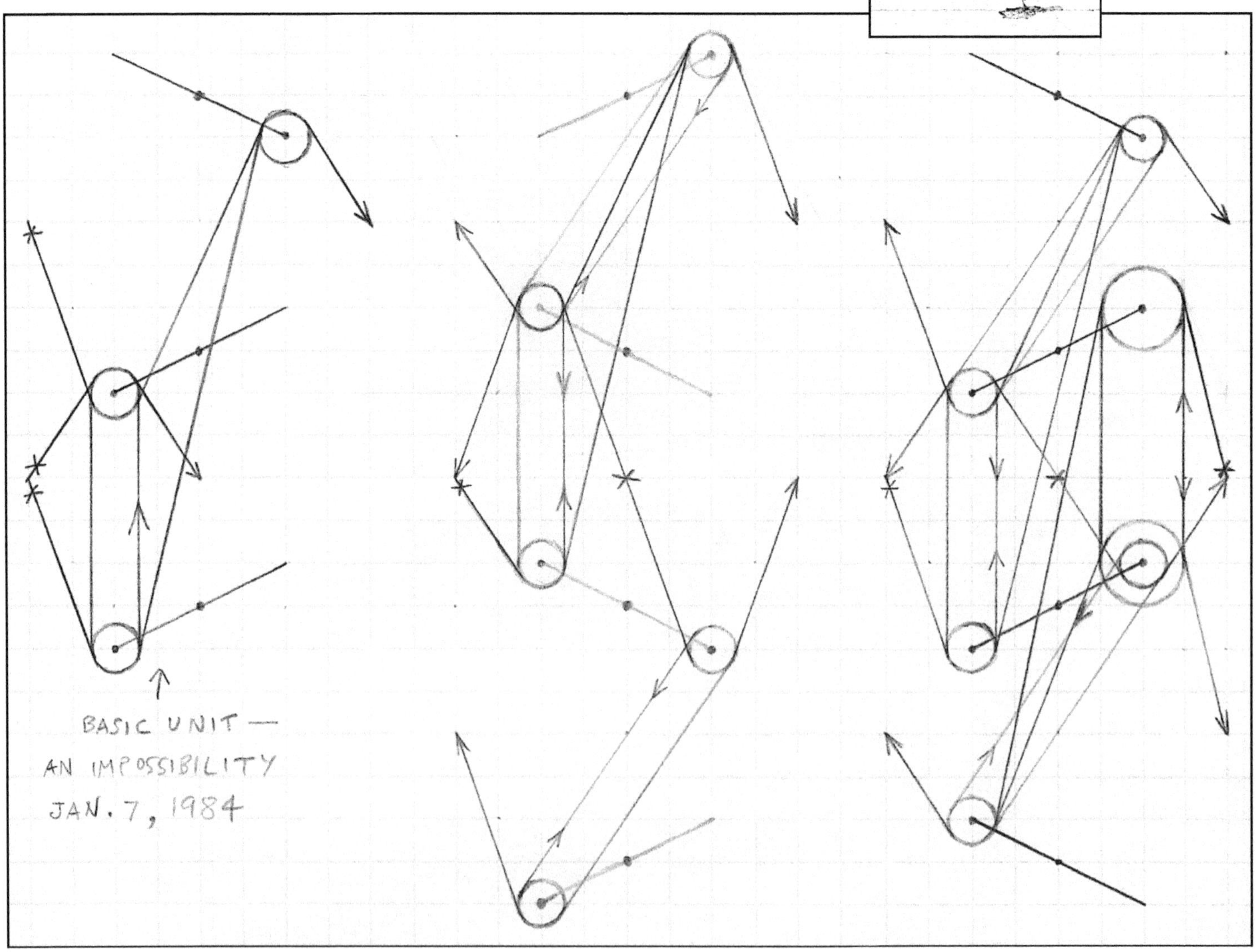
BASIC UNIT —
AN IMPOSSIBILITY
JAN. 7, 1984

4B. INCLINED PLANE

27-05-93 (3)

INEQUILIBRIUM

- MOTION WOULD BE DUE TO CONSTANT/CONTINUOUS INEQUILIBRIUM. THINK OF AN OBJECT ROLLING DOWN AN INCLINED PLANE. WHAT, THEN, GIVES INEQUILIBRIUM TO THE EARTH? DOES THE EARTH ACT AROUND THE SUN LIKE A CIRCULAR PENDULUM? OR MORE LIKELY A SWING + THE EARTH BEING THE SWINGER? THAT WOULD ACCOUNT FOR THE ACCELERATION + DECELERATION OF THE EARTH AROUND THE SUN.

- THE GREATER THE INEQUILIBRIUM, AS WITH A PENDULUM SWUNG HIGHER UP OR WITH AN OBJECT ON A MORE INCLINED PLANE, THE HEAVIER THE PENDULUM/OBJECT. THAT HAS NOTHING TO DO WITH THE FACT THAT THE PENDULUM/OBJECT IS HIGHER OR HOW HIGH THE PENDULUM/OBJECT IS. AS ANYONE CAN TELL IF HE HAS TRIED TO LIFT EVER HIGHER A PENDULUM OF A CONSIDERABLE SIZE OR A SWING WITH A CHILD SITTING ON IT.

- MOTION IS DUE TO INEQUILIBRIUM. THINK OF A BALL. IT IS IN EQUILIBRIUM. HOW DO THEY HAPPEN TO SPIN + ROTATE? — PLANETS, ELECTRONS? THINK OF A TOY BALLOON OR A ROCKET (ITS ESCAPING FROM ONE END/WALL). WE HAVE MOTION BECAUSE OF INEQUILIBRIUM.

- IF WE USE A SQUARE BLOCK (SEEN AS SQUARE) INSTEAD OF A SPHERE (SEEN A DISK) ON AN INCLINED PLANE IT IS EASY TO SEE EFFORT NEEDED TO PUSH UP WEIGHT. CHECK:

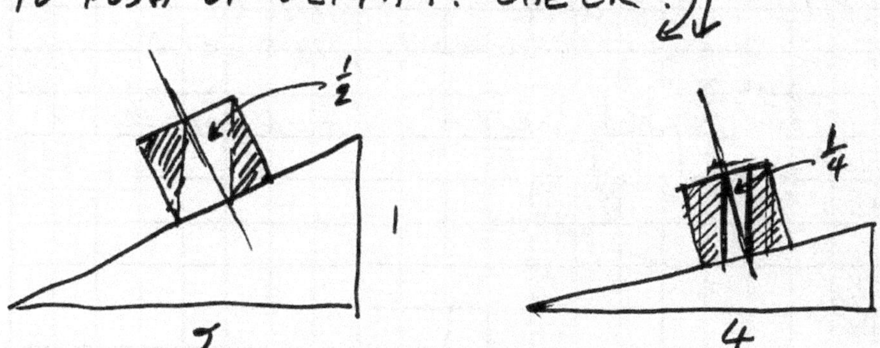

27-05-93.

QUANTIFYING INEQUILIBRIUM

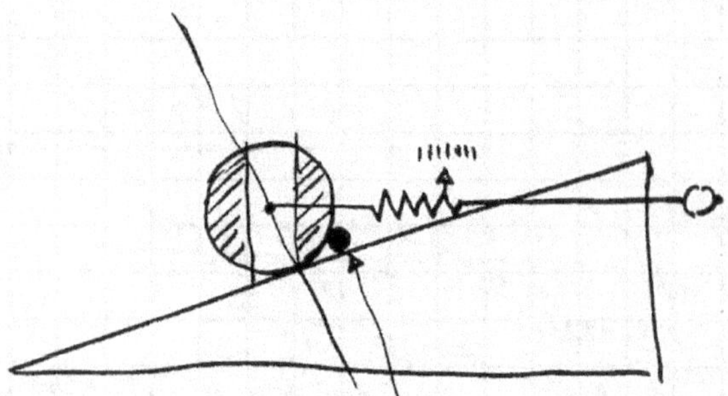

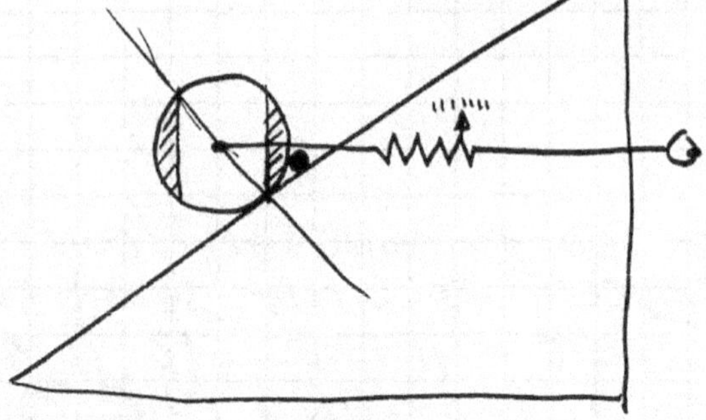

STOPPER

HOW MUCH OF A HORIZONTAL PULL IS NECESSARY TO BRING WEIGHT TO A STOP AGAINST THE STOPPER ?

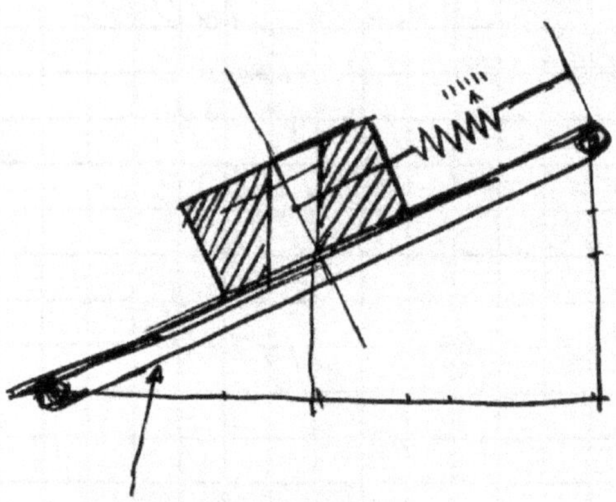

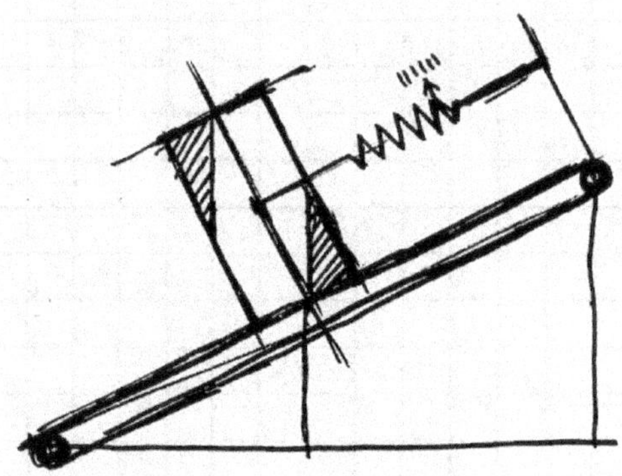

USING A ROLLING PLATFORM TO ELIMINATE THE DISCREPANCY OF FRICTION BETWEEN THE BLOCK LYING HORIZONTALLY AND THE BLOCK STANDING VERTICALLY.

TO DEMONSTRATE DIFFERENCE IN WEIGHT.
PLACEMENT OF SPRINGS (TO MEASURE WEIGHT) HERE IS ALLOWED BECAUSE THE ANGLE OF PULL/RESISTANCE IS THE SAME.

29-05-93

EQUILIBRIUM

- EQUILIBRIUM → EQUATION(S)
- TO ALTER EQUILIBRIUM (INEQUILIBRIUM) WE USE THE LEVER → A LEVER ALWAYS IN EQUILIBRIUM
- TOTAL EQUILIBRIUM : THE WHEEL, THE SPHERE
- PARTIAL EQUILIBRIUM → THE INCLINED PLANE, THE PENDULUM
- POTENTIAL ENERGY MAGNITUTE QUANTITY (HOW MUCH) = HOW FAR (DISTANCE) FROM EQUILIBRIUM LOCATION, EVEN IF OF THE SAME AMOUNT OF INEQUILIBRIUM.

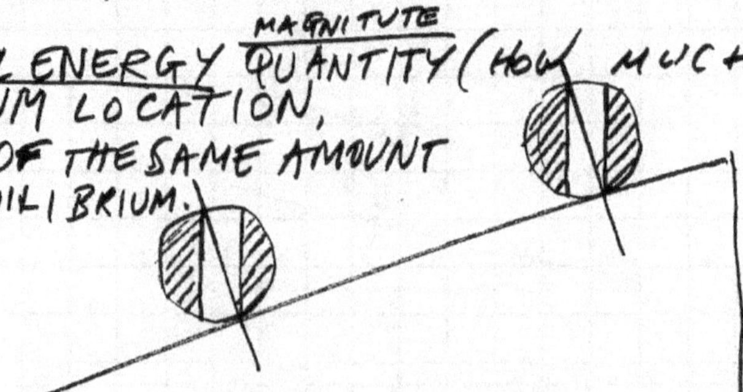

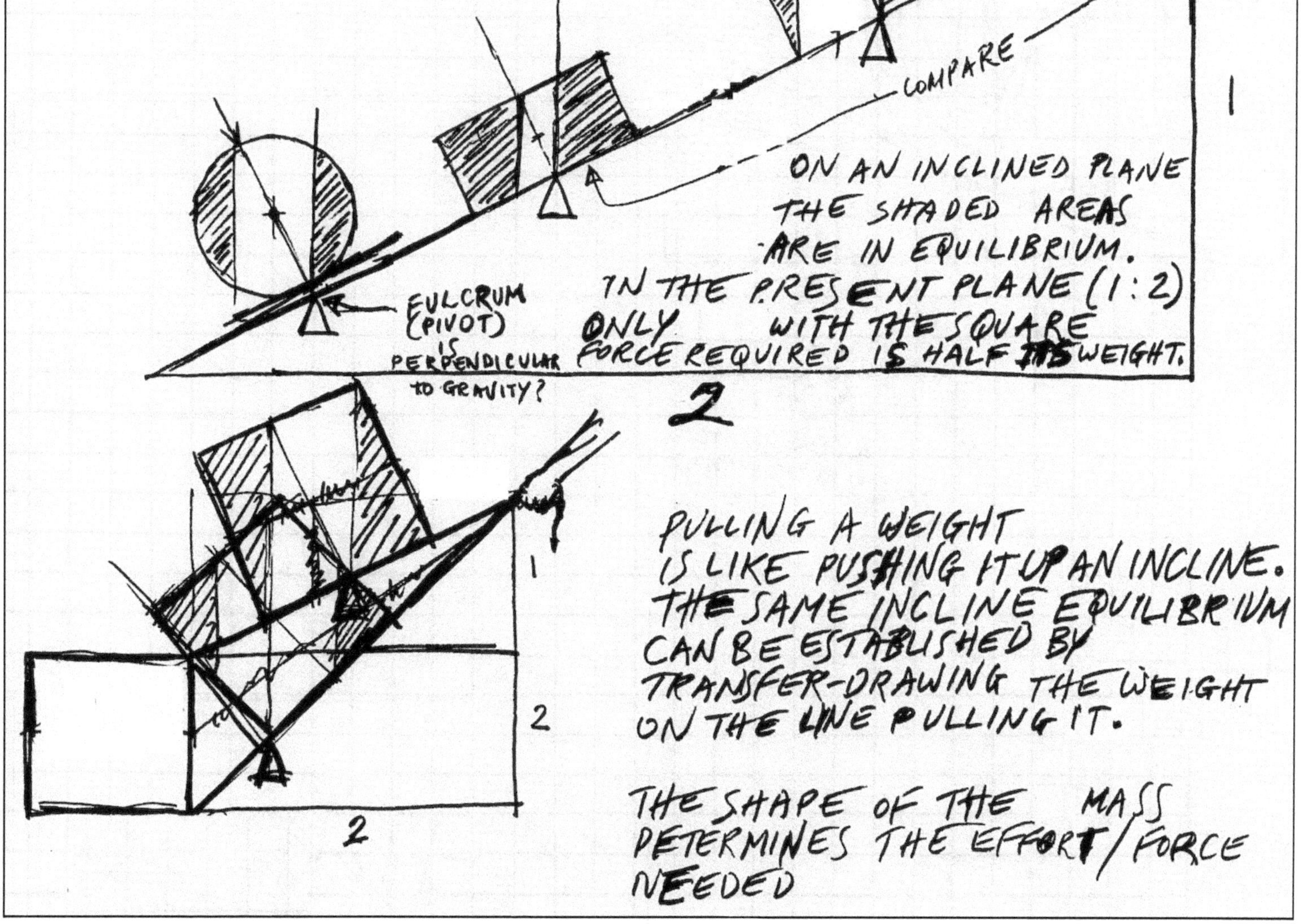
27-05-93
①
P.P. + ASS.
(MECHANICS)
A WHEEL IS A ROLLING LEVER
FULCRUM
COMPARE
1
ON AN INCLINED PLANE
THE SHADED AREAS
ARE IN EQUILIBRIUM.
IN THE PRESENT PLANE (1:2)
ONLY WITH THE SQUARE
FORCE REQUIRED IS HALF ITS WEIGHT.
FULCRUM
(PIVOT)
IS
PERPENDICULAR
TO GRAVITY?
2
1
2
2
PULLING A WEIGHT
IS LIKE PUSHING IT UP AN INCLINE.
THE SAME INCLINE EQUILIBRIUM
CAN BE ESTABLISHED BY
TRANSFER-DRAWING THE WEIGHT
ON THE LINE PULLING IT.
THE SHAPE OF THE MASS
DETERMINES THE EFFORT/FORCE
NEEDED

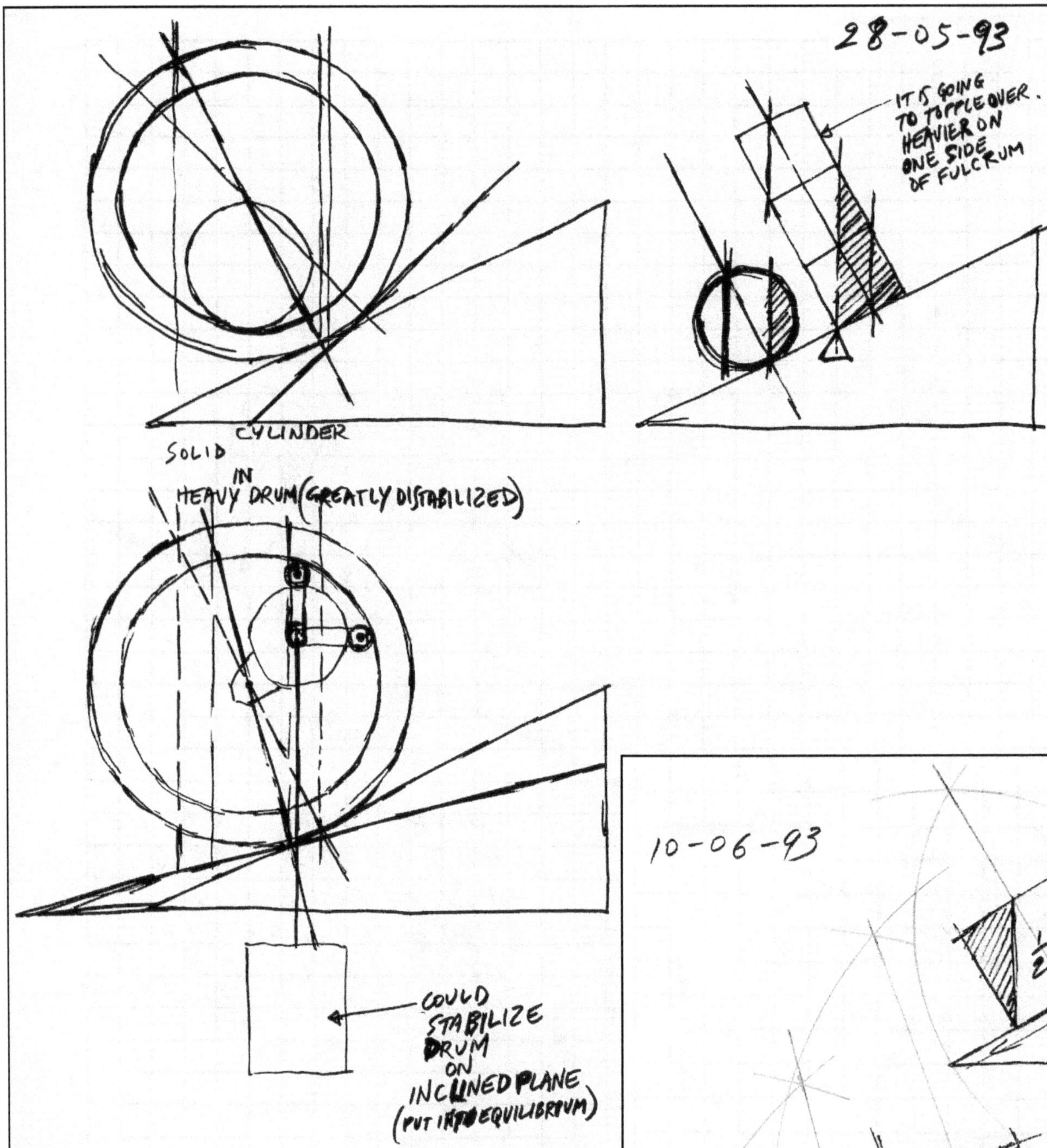
28-05-93
IT IS GOING TO TOPPLE OVER. HEAVIER ON ONE SIDE OF FULCRUM
CYLINDER
SOLID
IN
HEAVY DRUM (GREATLY DISTABILIZED)
COULD STABILIZE DRUM ON INCLINED PLANE
(PUT INTO EQUILIBRIUM)

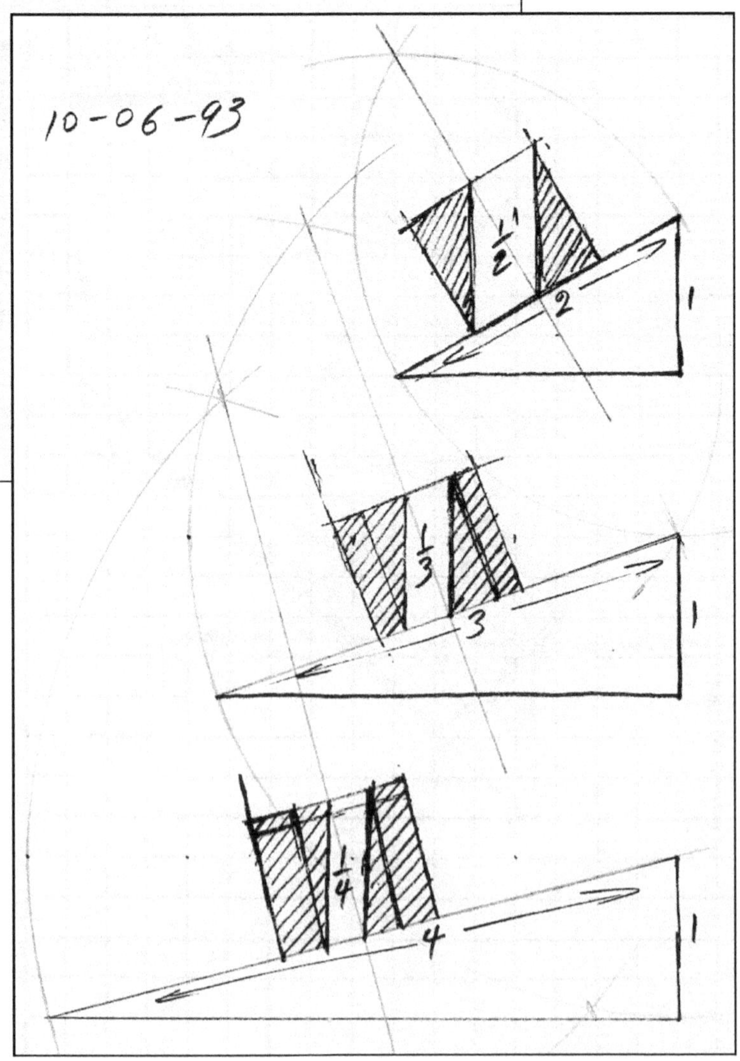
10-06-93
1/2
2
1
1/3
3
1
1/4
4
1

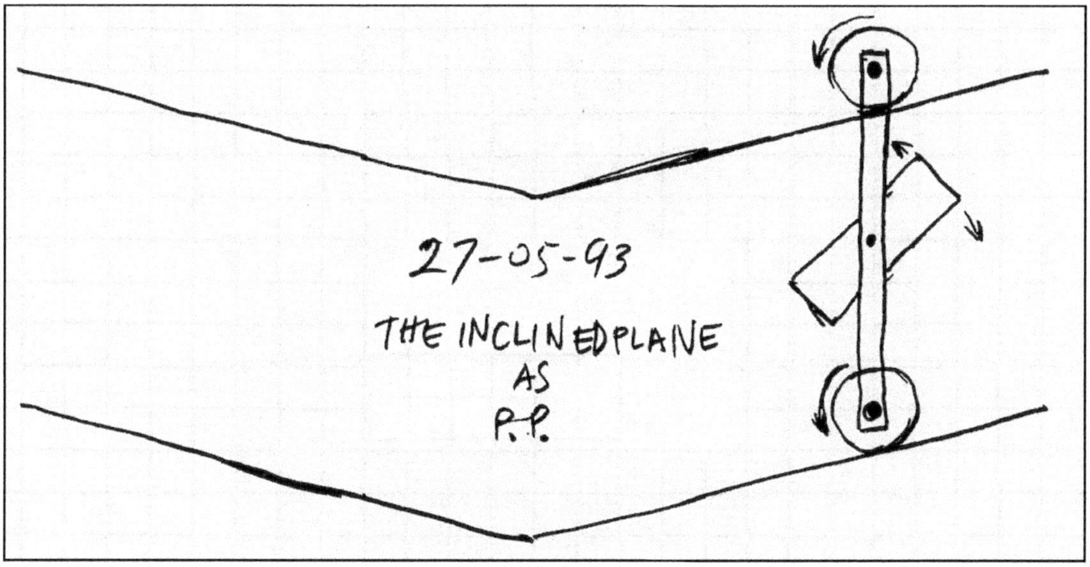

13-07-93

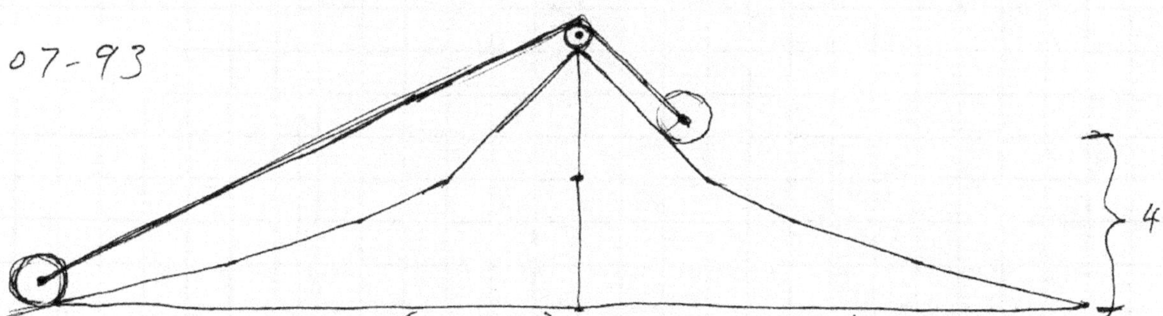

DEVICE (MACHINE) TO TEST "NEW HOPE 3"

– BY MAKING THE LEVER ARM
(DISTANCE FROM FULCRUM)
A FACTOR IN DETERMINING
LIFT, IT WAS A BIG MISTAKE
(BY DA VINCI)

WHEN THINKING
TO MAKE A
PERPETUAL MACHINE

LIKE
A LEVER

NOTE: IT A MISTAKE
TO THINK THAT A MACHINE,
ANY MACHINE,
PRODUCING A CERTAIN
FORCE WILL LIFT
A SIMILAR (EQUAL
MAGNITUDE) RESISTING
FORCE.

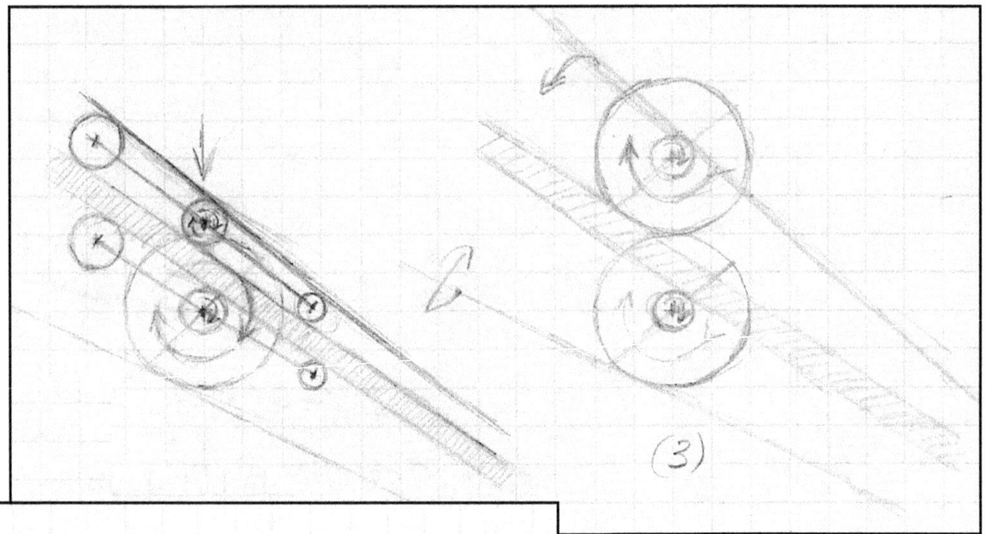
(3)

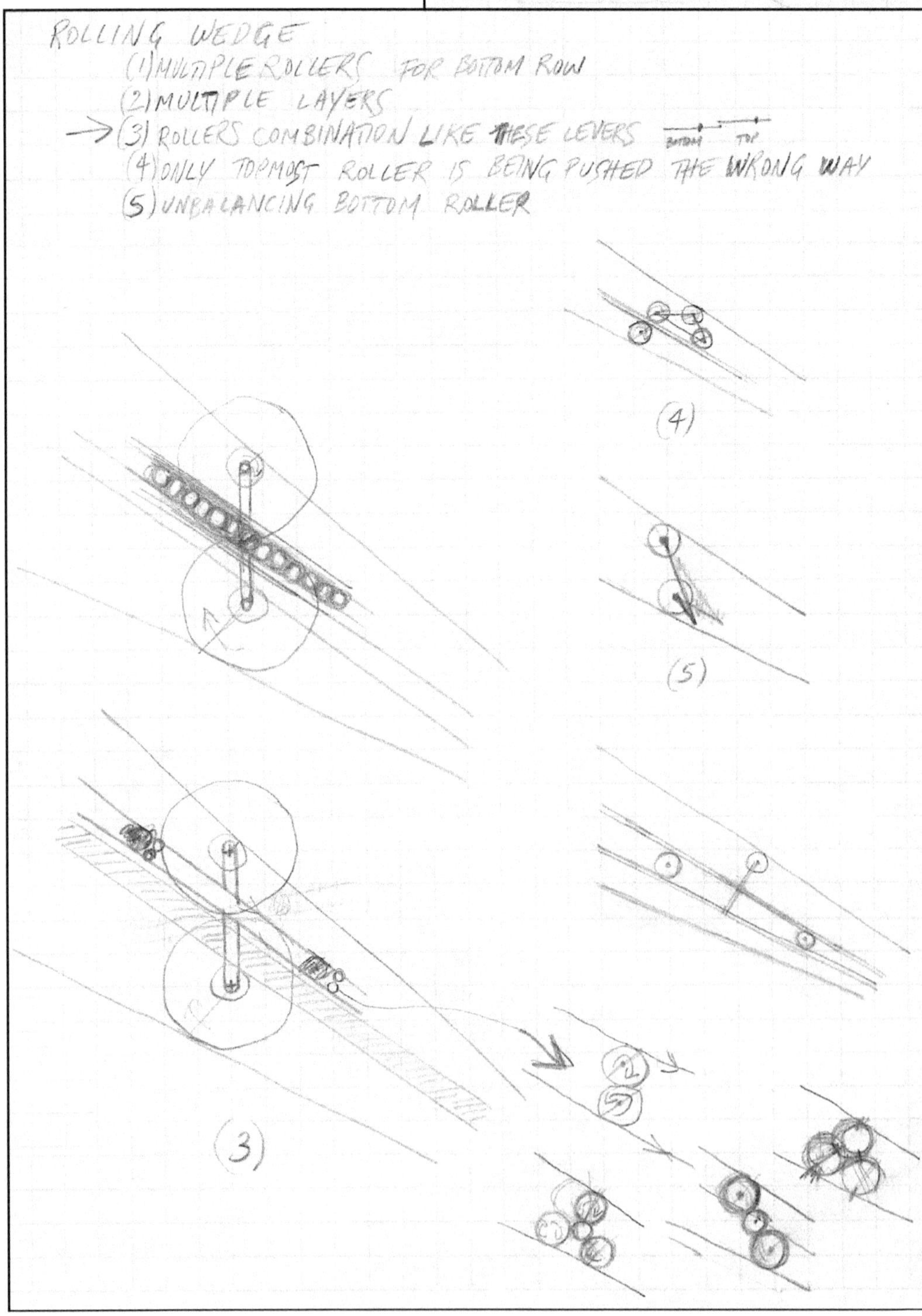
ROLLING WEDGE
(1) MULTIPLE ROLLERS FOR BOTTOM ROW
(2) MULTIPLE LAYERS
→ (3) ROLLERS COMBINATION LIKE THESE LEVERS BOTTOM TOP
(4) ONLY TOPMOST ROLLER IS BEING PUSHED THE WRONG WAY
(5) UNBALANCING BOTTOM ROLLER
(4)
(5)
(3)

4C. PENDULUM

1993 July - PP (Perpetual Pendulum)

Is it possible for a pendulum to swing continuously? The boy at the park shows us how. At the end of each half cycle he swings his body towards the outside to increase the angle of the swing. That gives him momentum and pushes him higher.

Is it possible to make a perpetual motion pendulum/swing? The weight of the pendulum would have to shift at the end of each half cycle like the boy on the swing. At the end of each half cycle, the pendulum would hit a switch that would make the weight swing towards the outside. (LATER NOTE: Even simple swinging takes effort. A dead weight with a live boy is a poor comparison.)

The following images show my analysis for a PP.

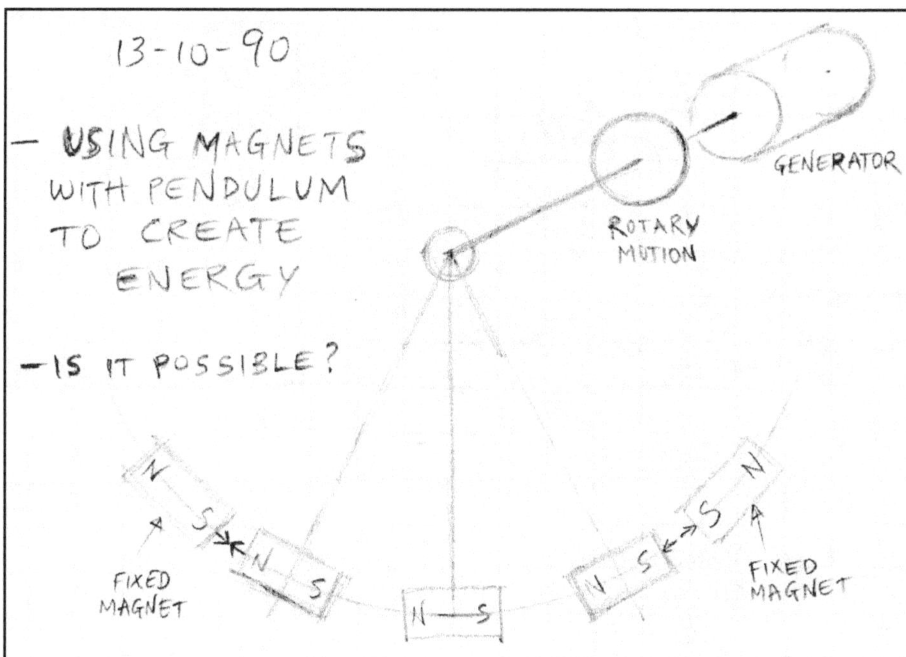

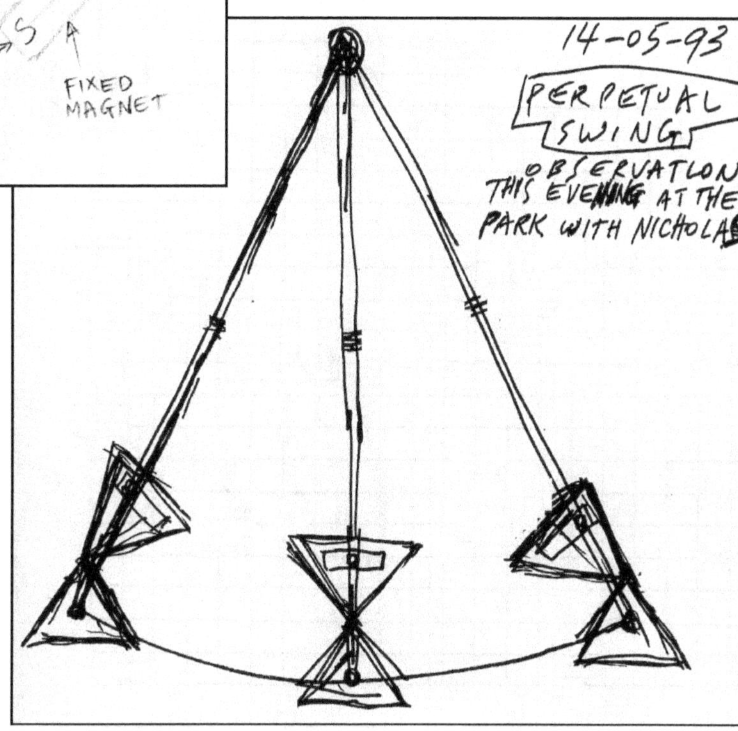

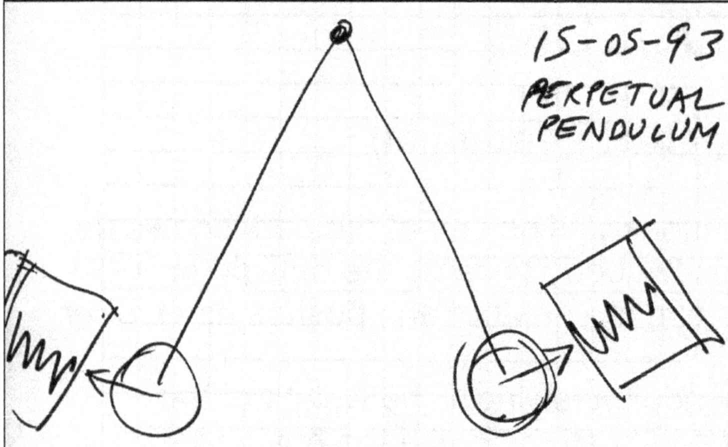

Magnet + Spring:
Because of momentum, weight attracts to spring-magnet; but, afterwards, gravity on weight is stronger + thus, weight "falls" again.
16-06-2019

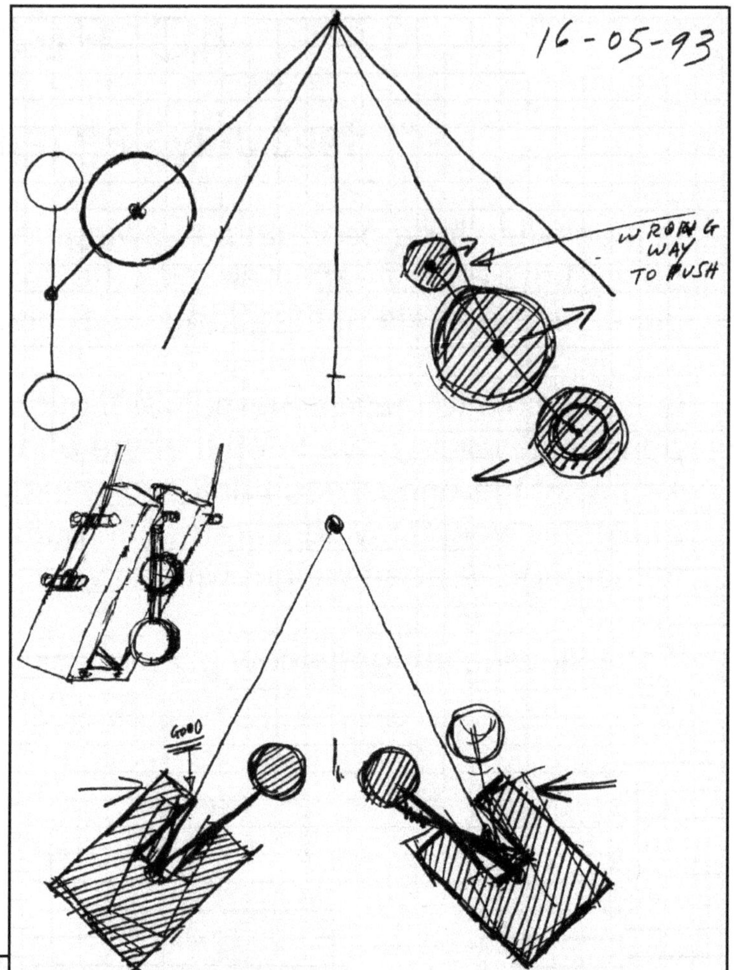

17-05-93

THE PENDULUM. TO NOTICE HOW, WE STOP A SWING WITH SOMEONE ELSE ON IT (AT THE PARK).
WE PULL (PULLING) + HANGING ON AT THE CHAINS. THAT HAS THE EFFECT OF STOPPING IT IMMEDIATELY OR OF SLOWING IT DOWN IMMESUREABLY.
FOR THE SAME REASON WE CANNOT CREATE A SECONDARY PENDULUM BY STOPPING (PUSHING) THE PRIMARY PENDULUM SOMEWHERE HALFWAY.
THERE IS NO KICK BACK (WHIPPING BACK) AND THAT IS THE REASON IN THE TWO INSTANCES WHY THE PENDULUM STOPS INSTEAD OF GAINING MOMENTUM.

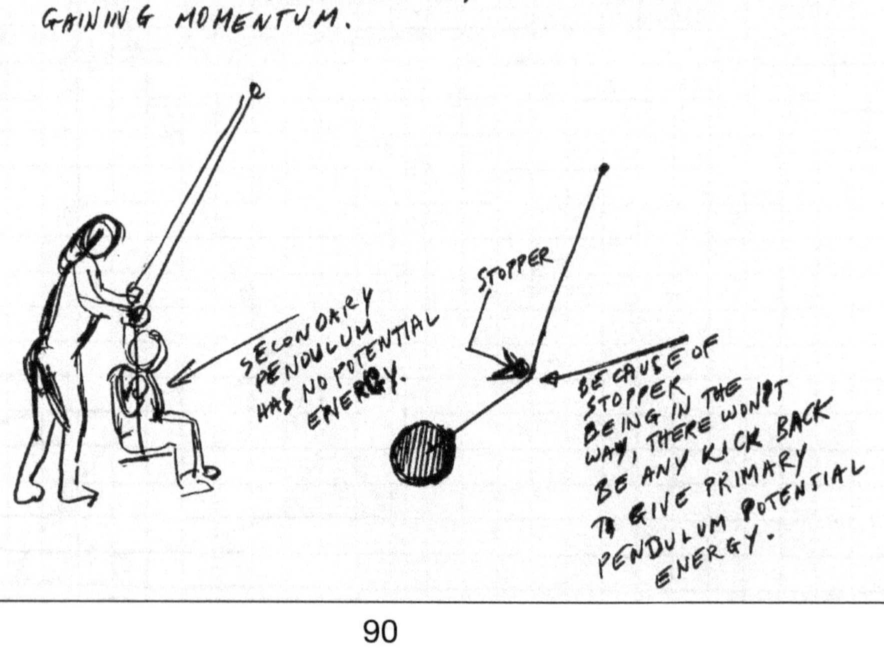

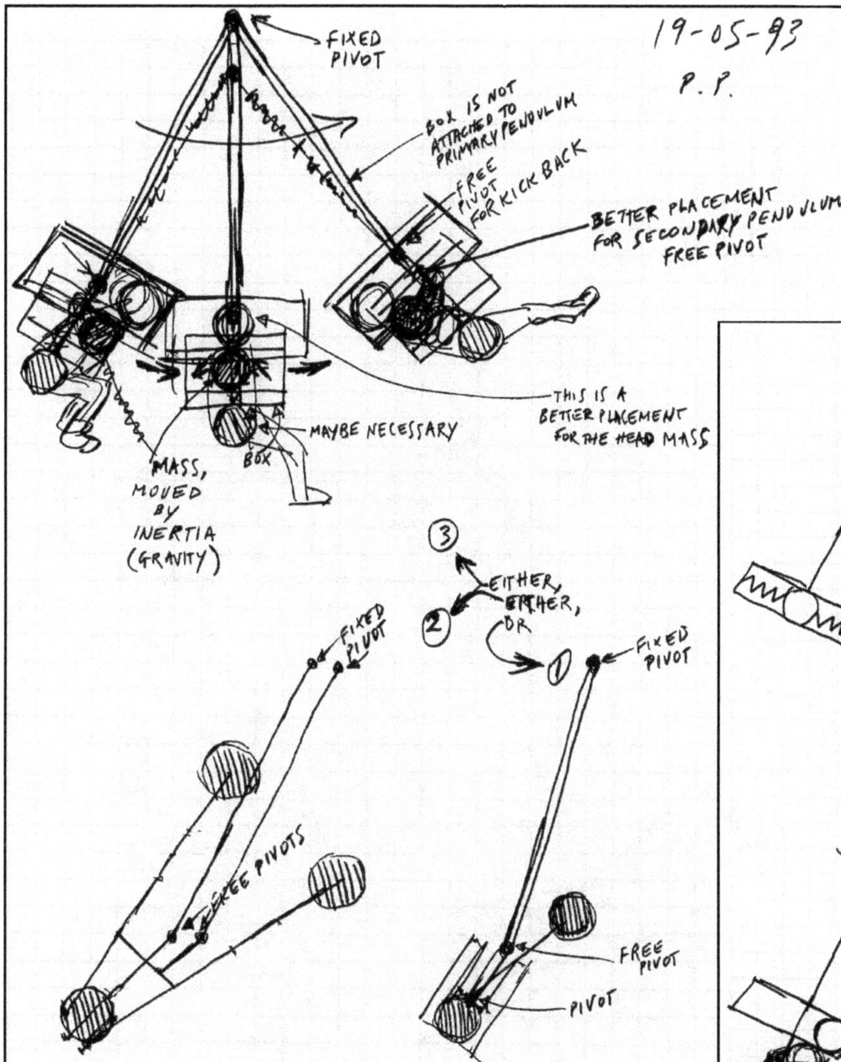

19-05-93
P. P.
FIXED PIVOT
BOX IS NOT ATTACHED TO PRIMARY PENDULUM
FREE PIVOT FOR KICK BACK
BETTER PLACEMENT FOR SECONDARY PENDULUM FREE PIVOT
THIS IS A BETTER PLACEMENT FOR THE HEAD MASS
MAYBE NECESSARY
BOX
MASS, MOVED BY INERTIA (GRAVITY)
3
EITHER,
2
EITHER, OR
1
FIXED PIVOT
FIXED PIVOT
FREE PIVOTS
FREE PIVOT
PIVOT

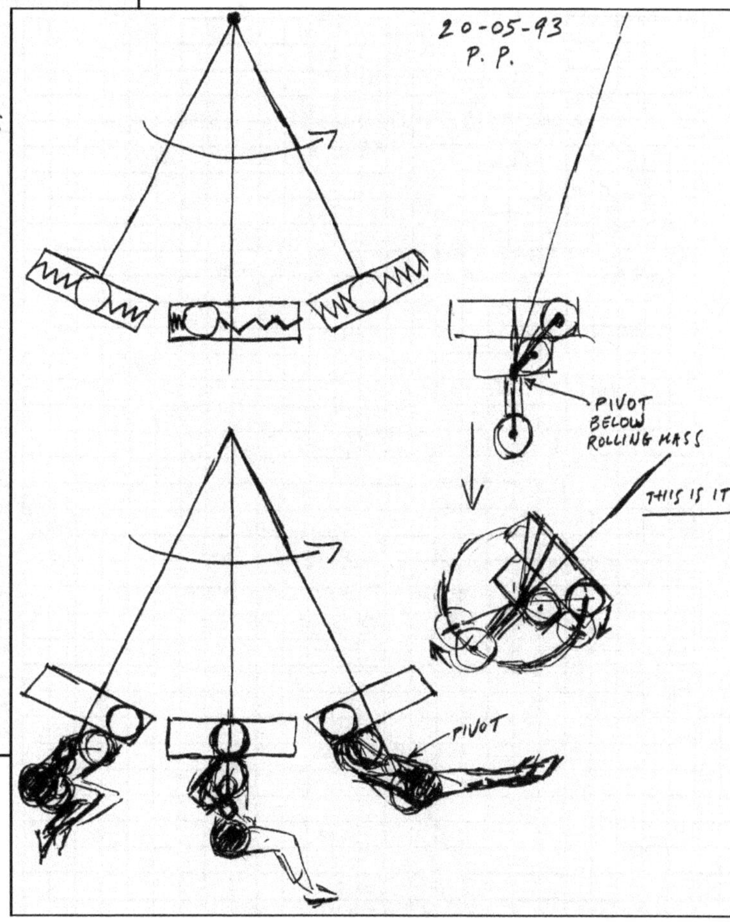

20-05-93
P. P.
PIVOT BELOW ROLLING MASS
THIS IS IT!
PIVOT

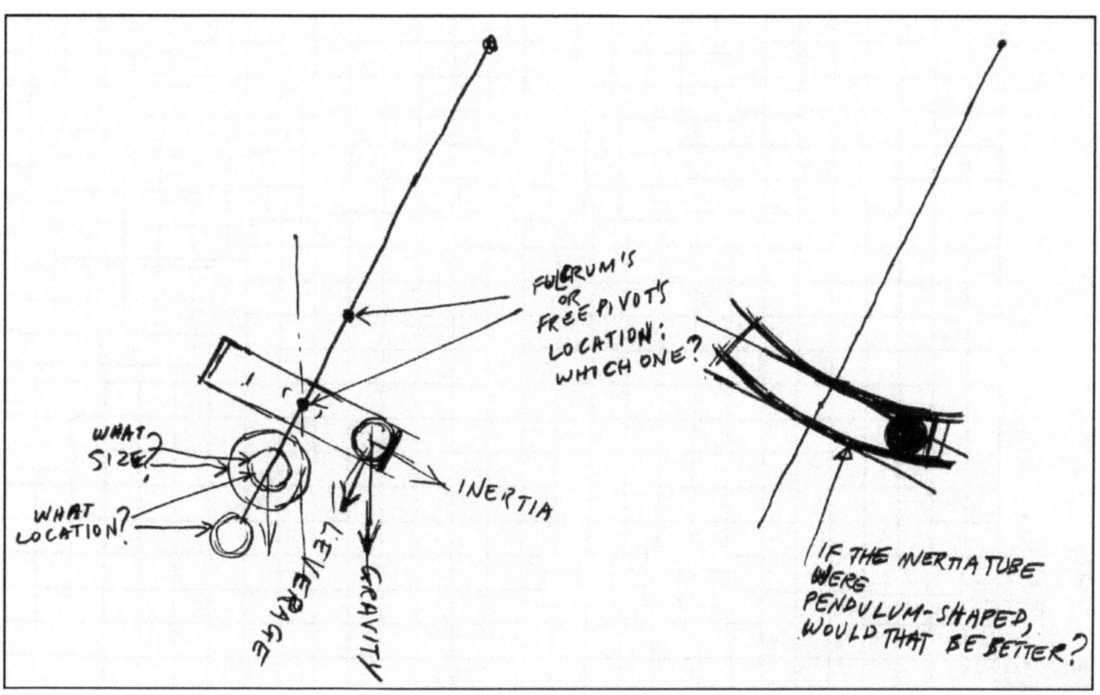

FULCRUM'S OR FREE PIVOT'S LOCATION: WHICH ONE?
WHAT SIZE?
WHAT LOCATION?
INERTIA
LEVERAGE
GRAVITY
IF THE INERTIA TUBE WERE PENDULUM-SHAPED, WOULD THAT BE BETTER?

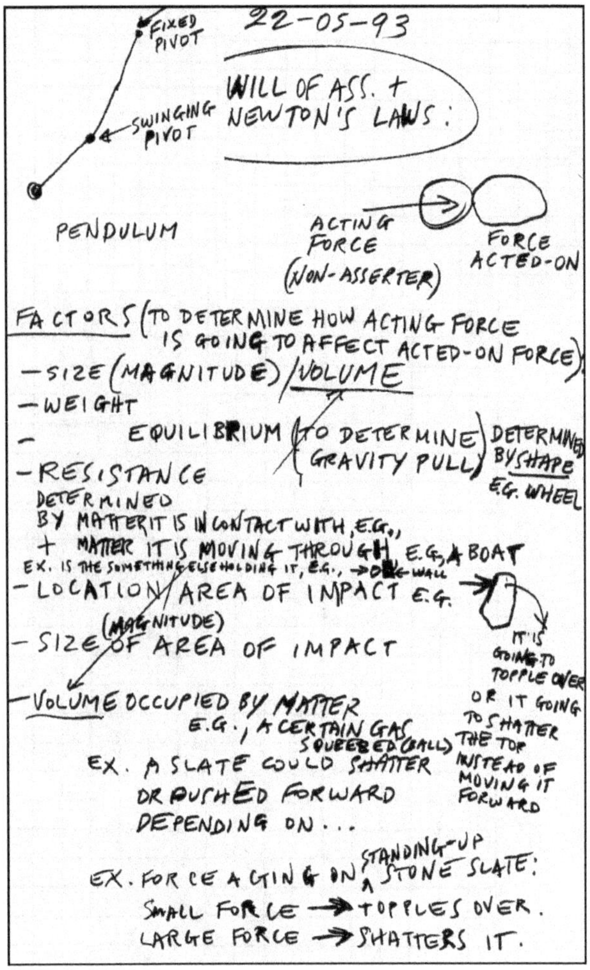
22-05-93
FIXED PIVOT
SWINGING PIVOT
PENDULUM
WILL OF ASS. + NEWTON'S LAWS:
ACTING FORCE (NON-ASSERTER)
FORCE ACTED-ON
FACTORS (TO DETERMINE HOW ACTING FORCE IS GOING TO AFFECT ACTED-ON FORCE)
- SIZE (MAGNITUDE) / VOLUME
- WEIGHT
- EQUILIBRIUM (TO DETERMINE GRAVITY PULL) DETERMINED BY SHAPE E.G. WHEEL
- RESISTANCE DETERMINED BY MATTER IT IS IN CONTACT WITH, E.G.,
+ MATTER IT IS MOVING THROUGH E.G. A BOAT
EX. IS THE SOMETHING ELSE HOLDING IT, E.G., WALL
- LOCATION / AREA OF IMPACT E.G.
IT'S GOING TO TOPPLE OVER OR IT GOING TO SHATTER THE TOP INSTEAD OF MOVING IT FORWARD
(MAGNITUDE)
- SIZE OF AREA OF IMPACT
- VOLUME OCCUPIED BY MATTER E.G., A CERTAIN GAS SQUEEZED (BALL)
EX. A SLATE COULD SHATTER OR PUSHED FORWARD DEPENDING ON...
EX. FORCE ACTING ON STANDING-UP STONE SLATE:
SMALL FORCE → TOPPLES OVER.
LARGE FORCE → SHATTERS IT.

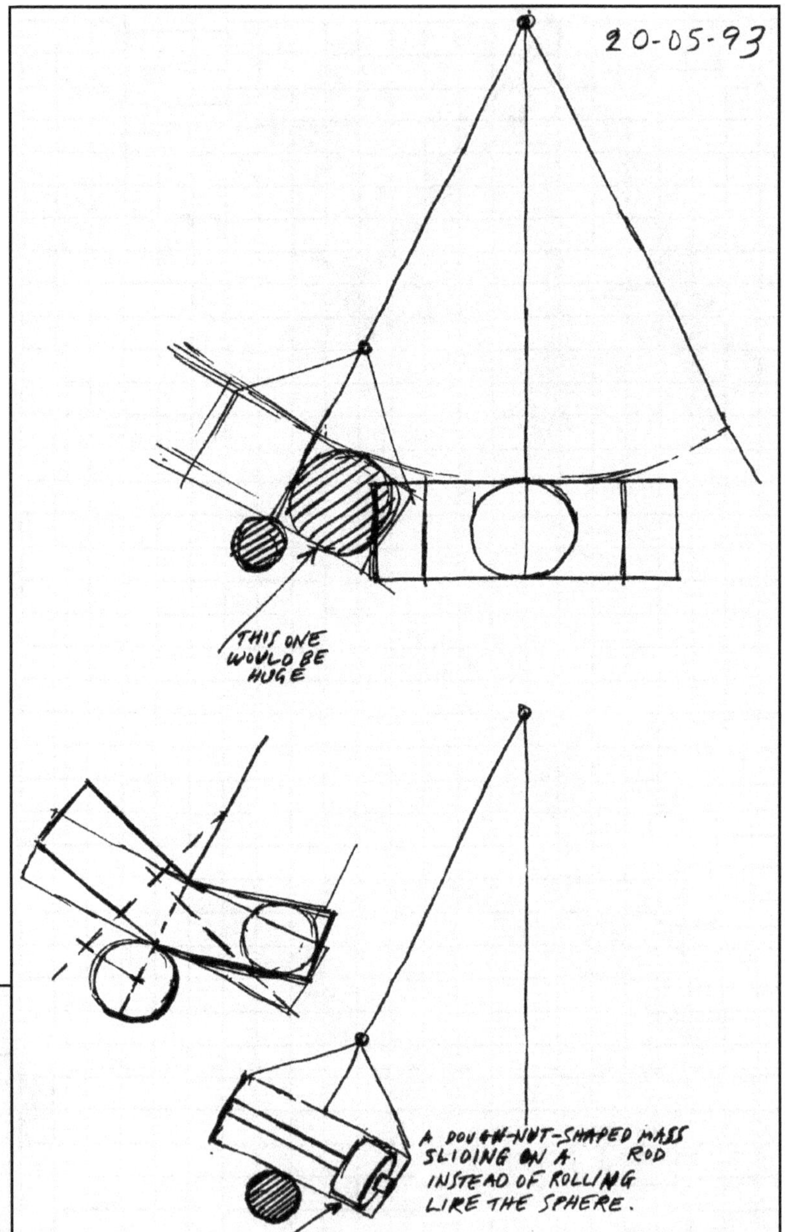
20-05-93
THIS ONE WOULD BE HUGE
A DOUGH-NUT-SHAPED MASS SLIDING ON A ROD INSTEAD OF ROLLING LIKE THE SPHERE.

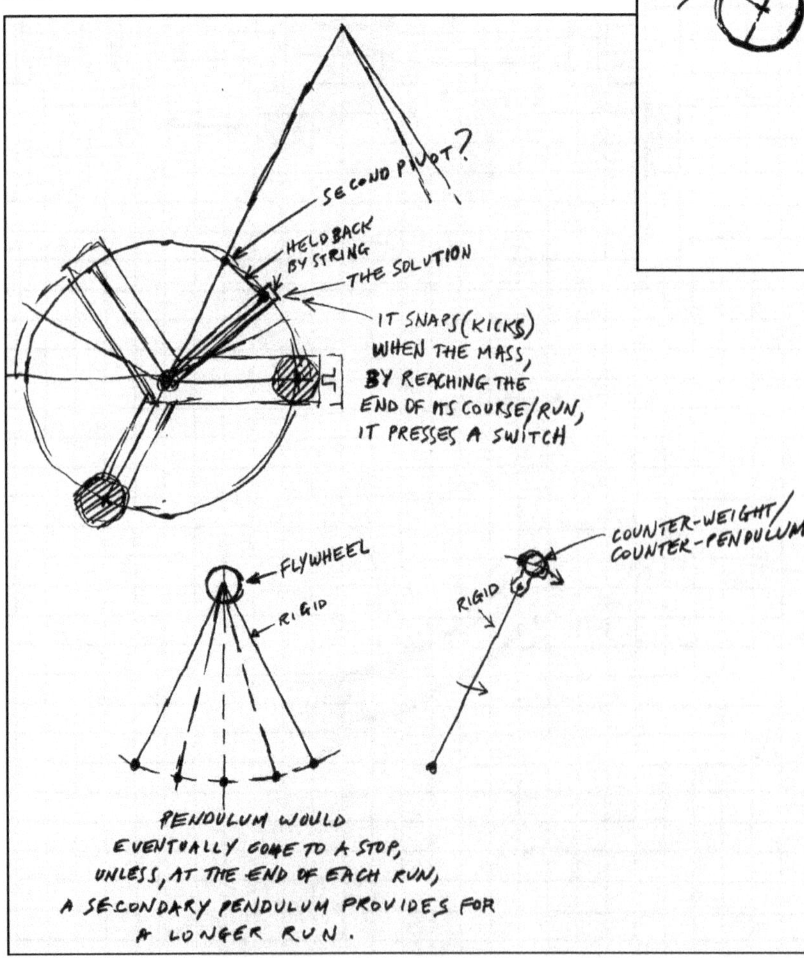
SECOND PIVOT?
HELD BACK BY STRING
THE SOLUTION
IT SNAPS (KICKS) WHEN THE MASS, BY REACHING THE END OF ITS COURSE/RUN, IT PRESSES A SWITCH
FLYWHEEL
RIGID
COUNTER-WEIGHT/ COUNTER-PENDULUM
RIGID
PENDULUM WOULD EVENTUALLY COME TO A STOP, UNLESS, AT THE END OF EACH RUN, A SECONDARY PENDULUM PROVIDES FOR A LONGER RUN.

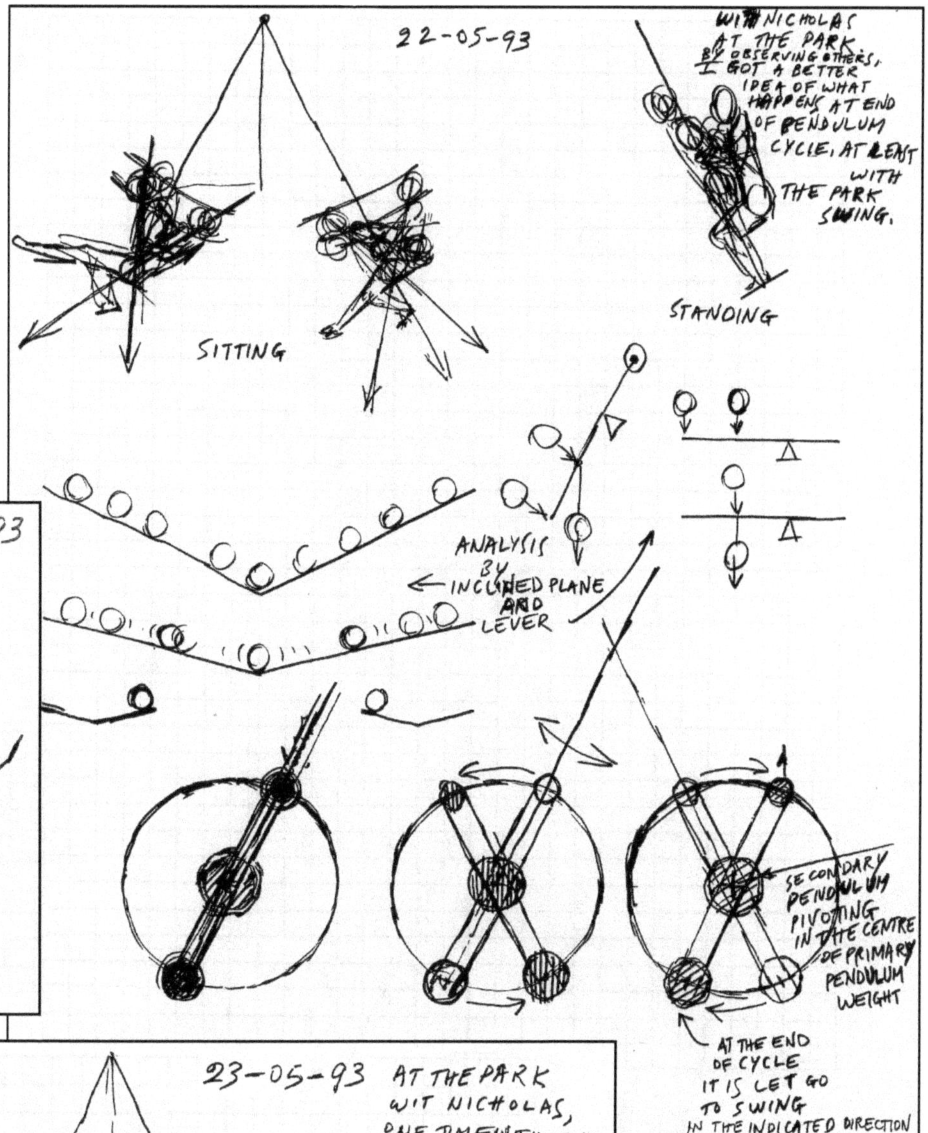
22-05-93
WITH NICHOLAS AT THE PARK
BY OBSERVING OTHERS, I GOT A BETTER IDEA OF WHAT HAPPENS AT END OF PENDULUM CYCLE, AT LEAST WITH THE PARK SWING.
SITTING
STANDING
ANALYSIS BY INCLINED PLANE AND LEVER
SECONDARY PENDULUM PIVOTING IN THE CENTRE OF PRIMARY PENDULUM WEIGHT
AT THE END OF CYCLE IT IS LET GO TO SWING IN THE INDICATED DIRECTION

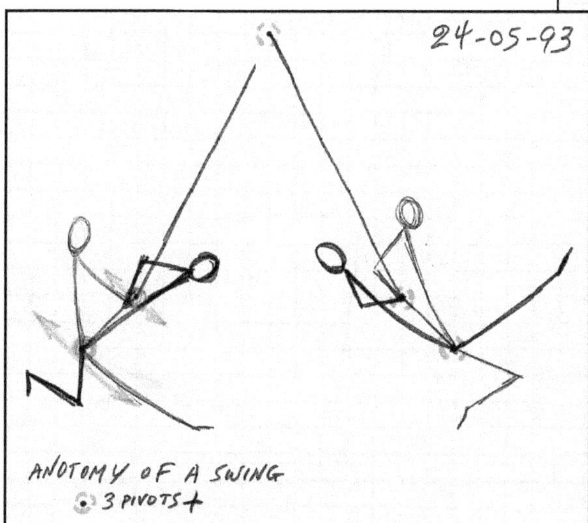
24-05-93
ANOTOMY OF A SWING
3 PIVOTS +

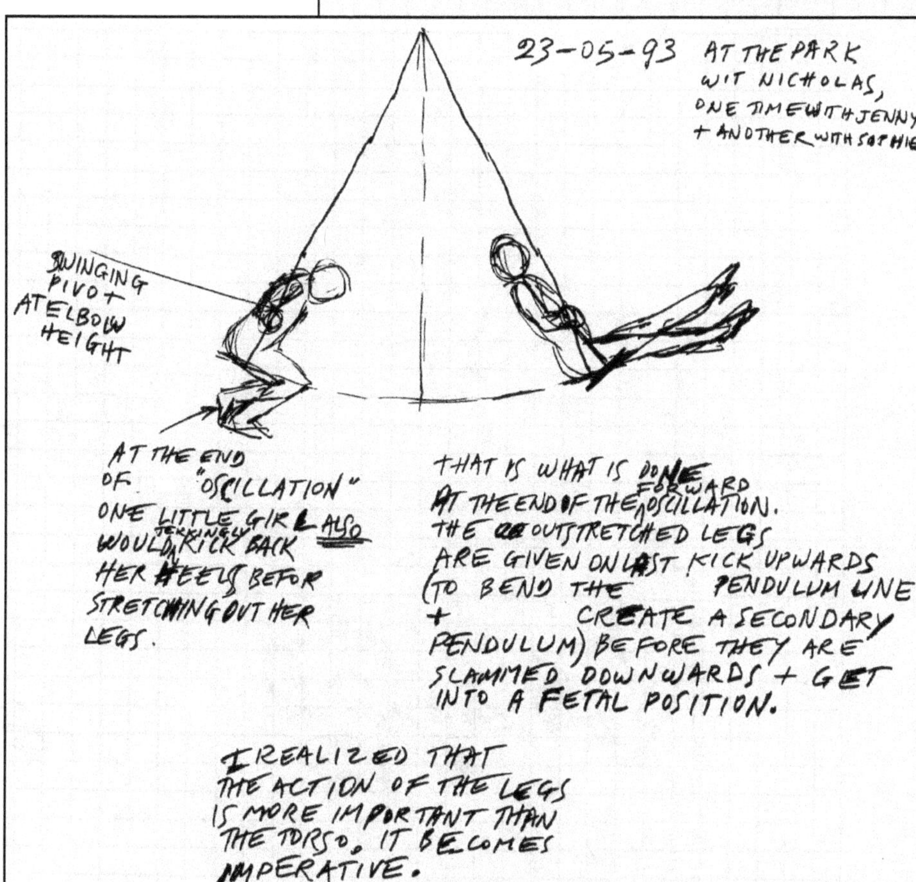
23-05-93 AT THE PARK WIT NICHOLAS, ONE TIME WITH JENNY + ANOTHER WITH SOPHIE.
SWINGING PIVOT AT ELBOW HEIGHT
AT THE END OF "OSCILLATION" ONE LITTLE GIRL ALSO WOULD JERKINGLY KICK BACK HER HEELS BEFOR STRETCHING OUT HER LEGS.
THAT IS WHAT IS DONE AT THE END OF THE FORWARD OSCILLATION. THE OUTSTRETCHED LEGS ARE GIVEN ON LAST KICK UPWARDS (TO BEND THE PENDULUM LINE + CREATE A SECONDARY PENDULUM) BEFORE THEY ARE SLAMMED DOWNWARDS + GET INTO A FETAL POSITION.
I REALIZED THAT THE ACTION OF THE LEGS IS MORE IMPORTANT THAN THE TORSO. IT BECOMES IMPERATIVE.

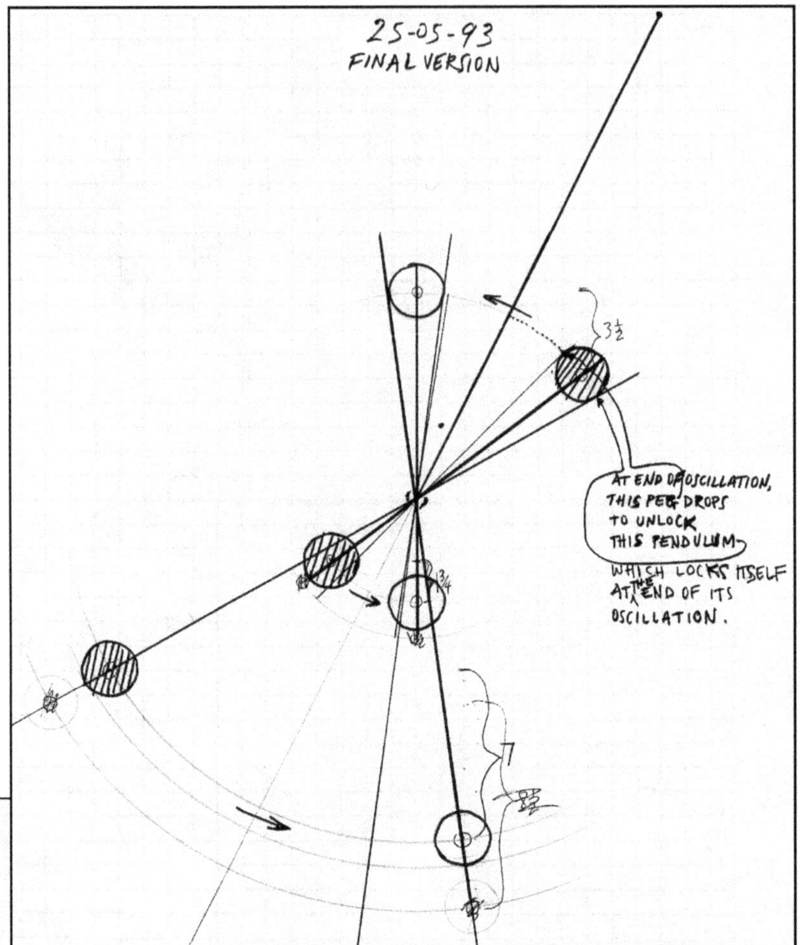
25-05-93
FINAL VERSION
3½
AT END OF OSCILLATION, THIS PEG DROPS TO UNLOCK THIS PENDULUM
WHICH LOCKS ITSELF AT THE END OF ITS OSCILLATION.
1¾
7

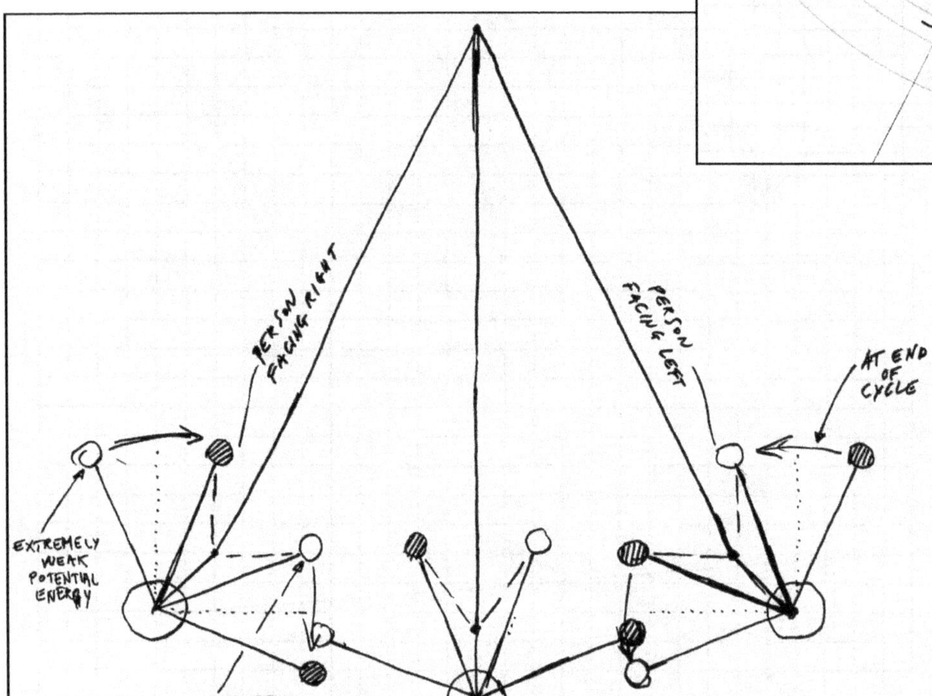
PERSON FACING RIGHT
PERSON FACING LEFT
AT END OF CYCLE
EXTREMELY WEAK POTENTIAL ENERGY
EXTREMELY STRONG POTENTIAL ENERGY

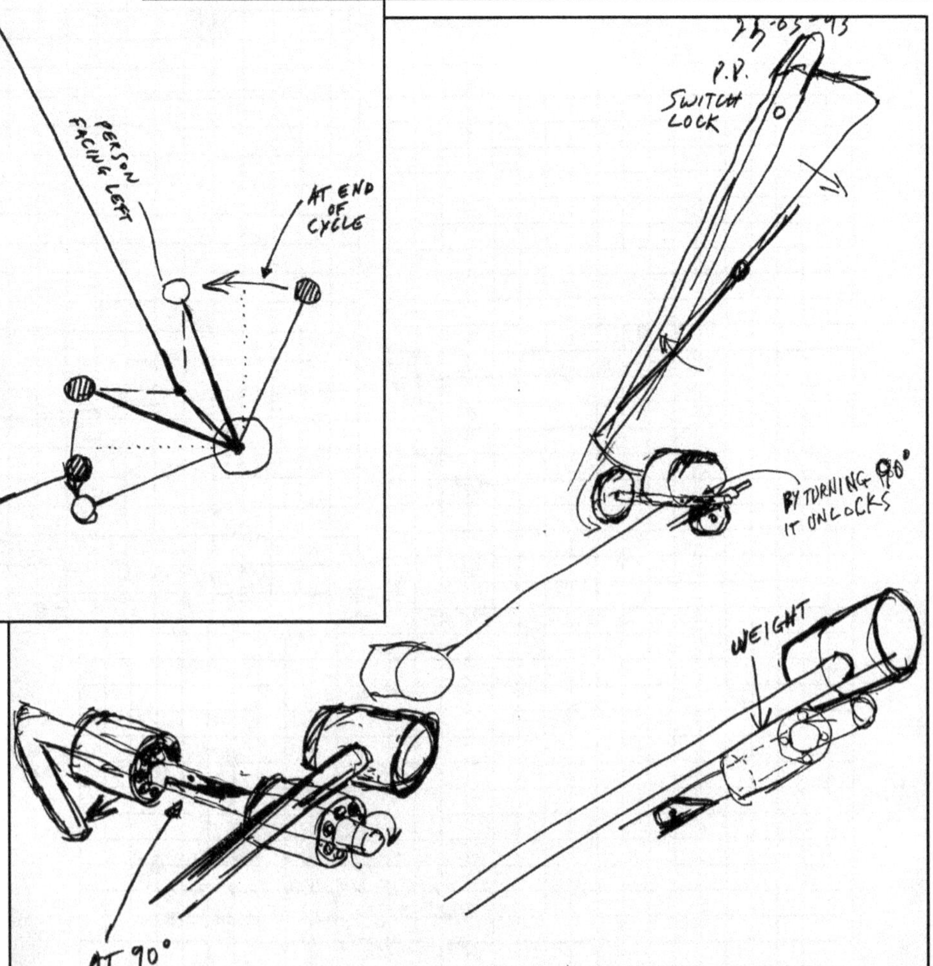
P.P.
SWITCH LOCK
BY TURNING 90° IT UNLOCKS
WEIGHT
AT 90°

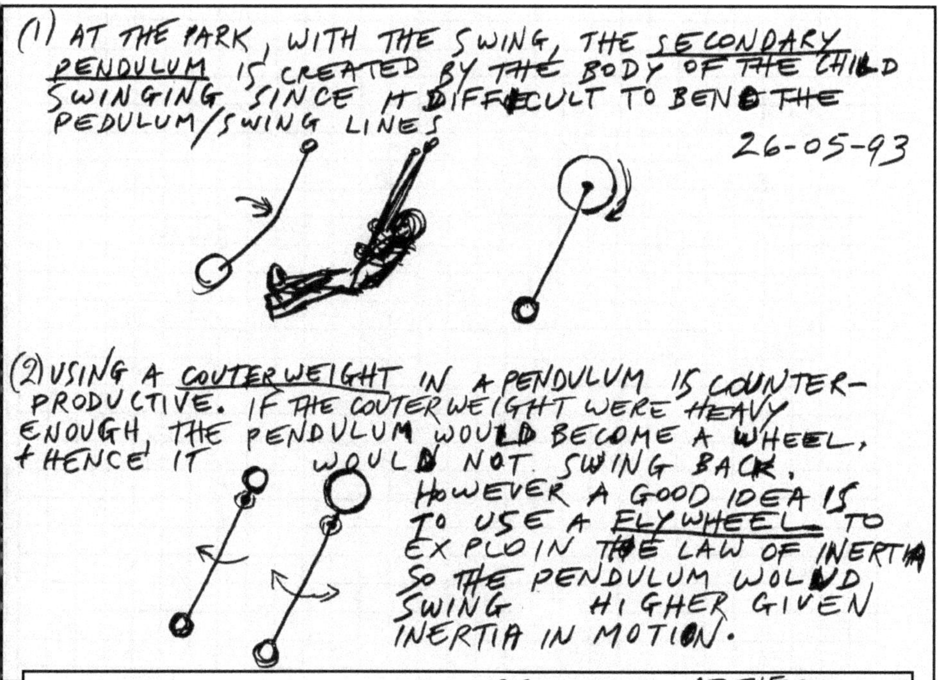
(1) AT THE PARK, WITH THE SWING, THE SECONDARY PENDULUM IS CREATED BY THE BODY OF THE CHILD SWINGING SINCE IT DIFFICULT TO BEND THE PEDULUM/SWING LINES
26-05-93
(2) USING A COUTERWEIGHT IN A PENDULUM IS COUNTER-PRODUCTIVE. IF THE COUTERWEIGHT WERE HEAVY ENOUGH, THE PENDULUM WOULD BECOME A WHEEL, +HENCE IT WOULD NOT SWING BACK.
HOWEVER A GOOD IDEA IS TO USE A FLYWHEEL TO EXPLOIN THE LAW OF INERTIA SO THE PENDULUM WOULD SWING HIGHER GIVEN INERTIA IN MOTION.

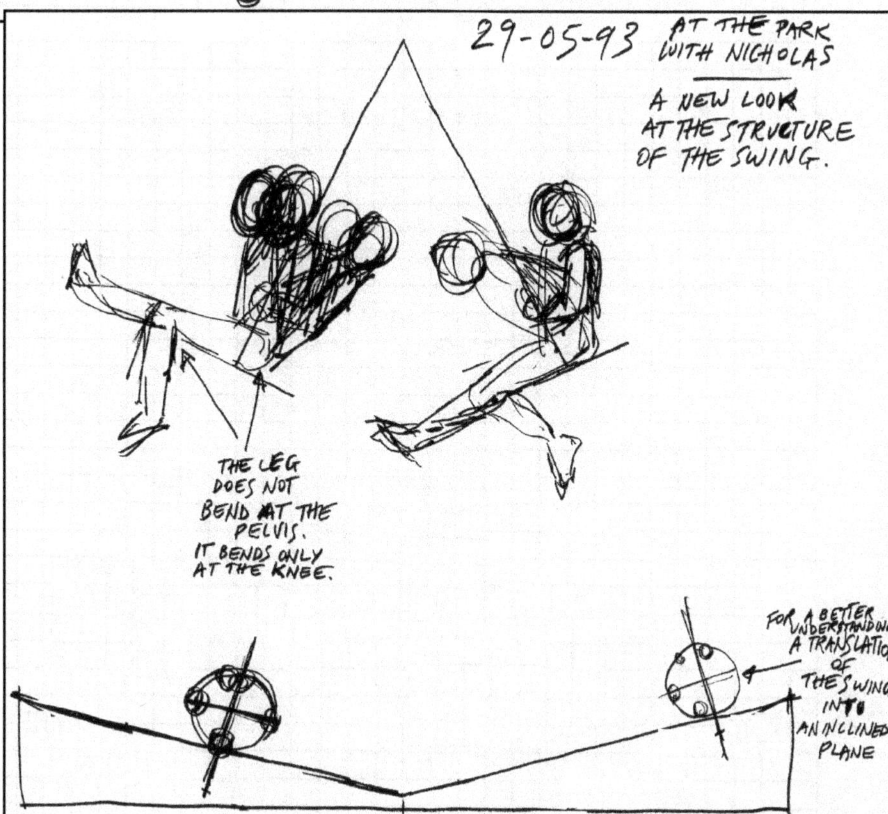
29-05-93
AT THE PARK WITH NICHOLAS
A NEW LOOK AT THE STRUCTURE OF THE SWING.
THE LEG DOES NOT BEND AT THE PELVIS. IT BENDS ONLY AT THE KNEE.
FOR A BETTER UNDERSTANDING A TRANSLATION OF THE SWING INTO AN INCLINED PLANE

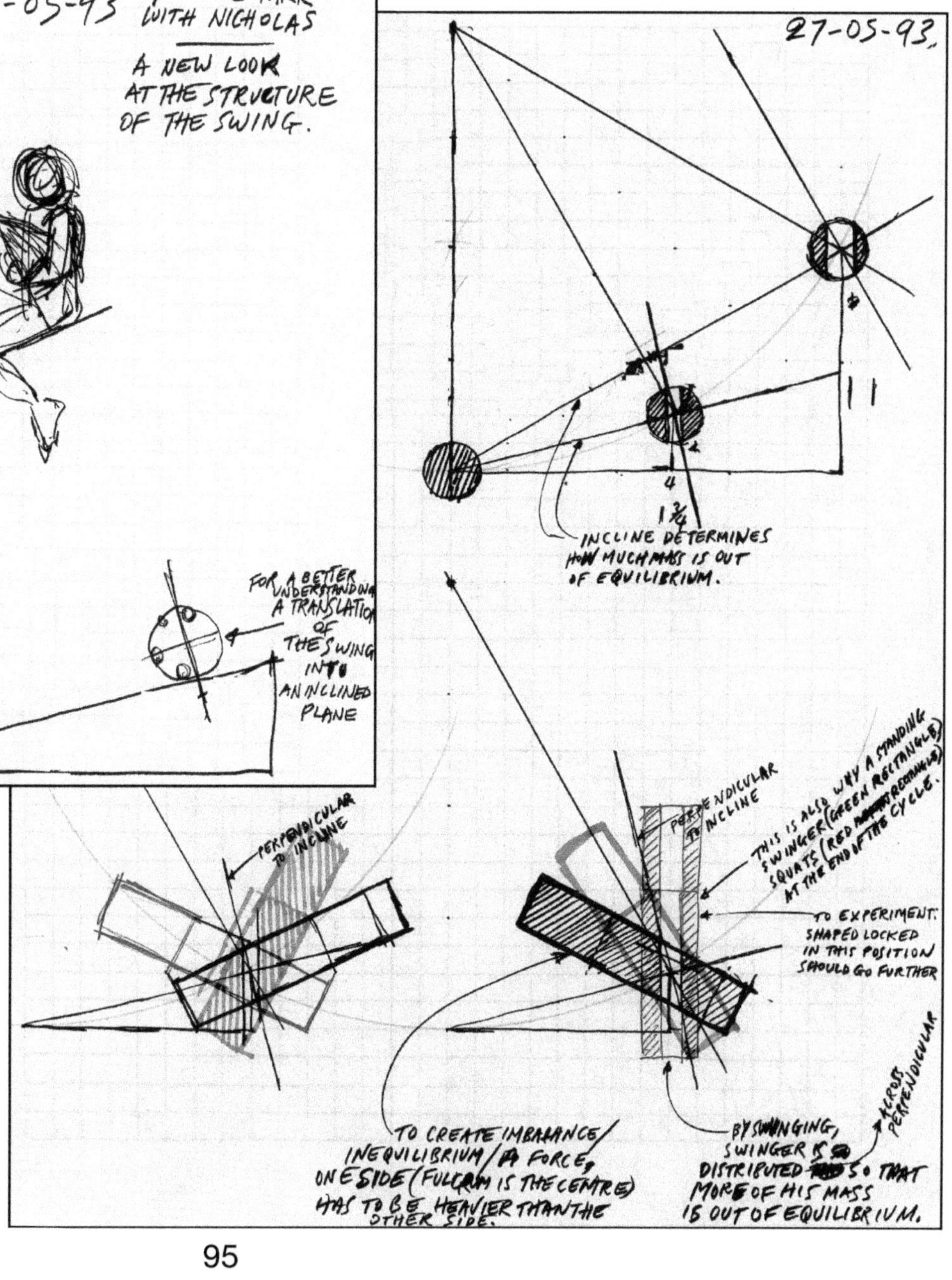
27-05-93
11
4
1 3/4
INCLINE DETERMINES HOW MUCH MASS IS OUT OF EQUILIBRIUM.
PERPENDICULAR TO INCLINE
PERPENDICULAR TO INCLINE
THIS IS ALSO WHY A STANDING SWINGER (GREEN RECTANGLE) SQUATS (RED RECTANGLE) AT THE END OF THE CYCLE.
TO EXPERIMENT: SHAPED LOCKED IN THIS POSITION SHOULD GO FURTHER
TO CREATE IMBALANCE/INEQUILIBRIUM/A FORCE, ONE SIDE (FULCRUM IS THE CENTRE) HAS TO BE HEAVIER THAN THE OTHER SIDE.
BY SWINGING, SWINGER IS DISTRIBUTED SO THAT MORE OF HIS MASS IS OUT OF EQUILIBRIUM.
ACROSS PERPENDICULAR

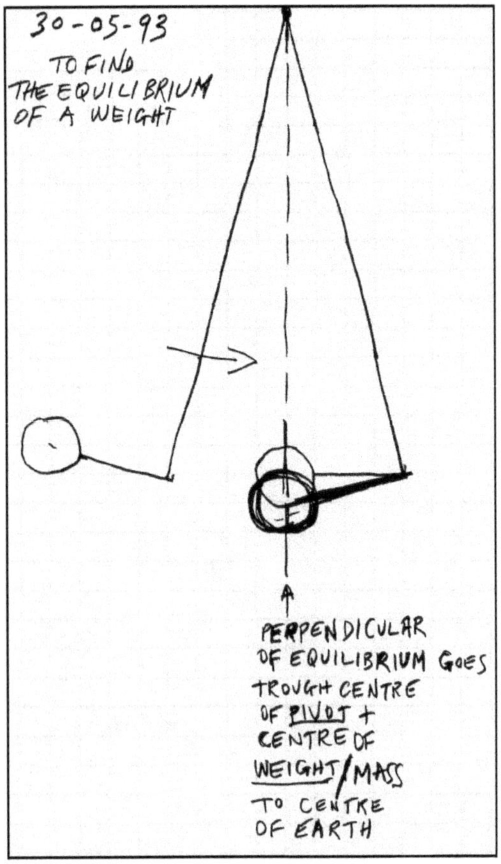
30-05-93
TO FIND THE EQUILIBRIUM OF A WEIGHT
PERPENDICULAR OF EQUILIBRIUM GOES TROUGH CENTRE OF PIVOT + CENTRE OF WEIGHT/MASS TO CENTRE OF EARTH

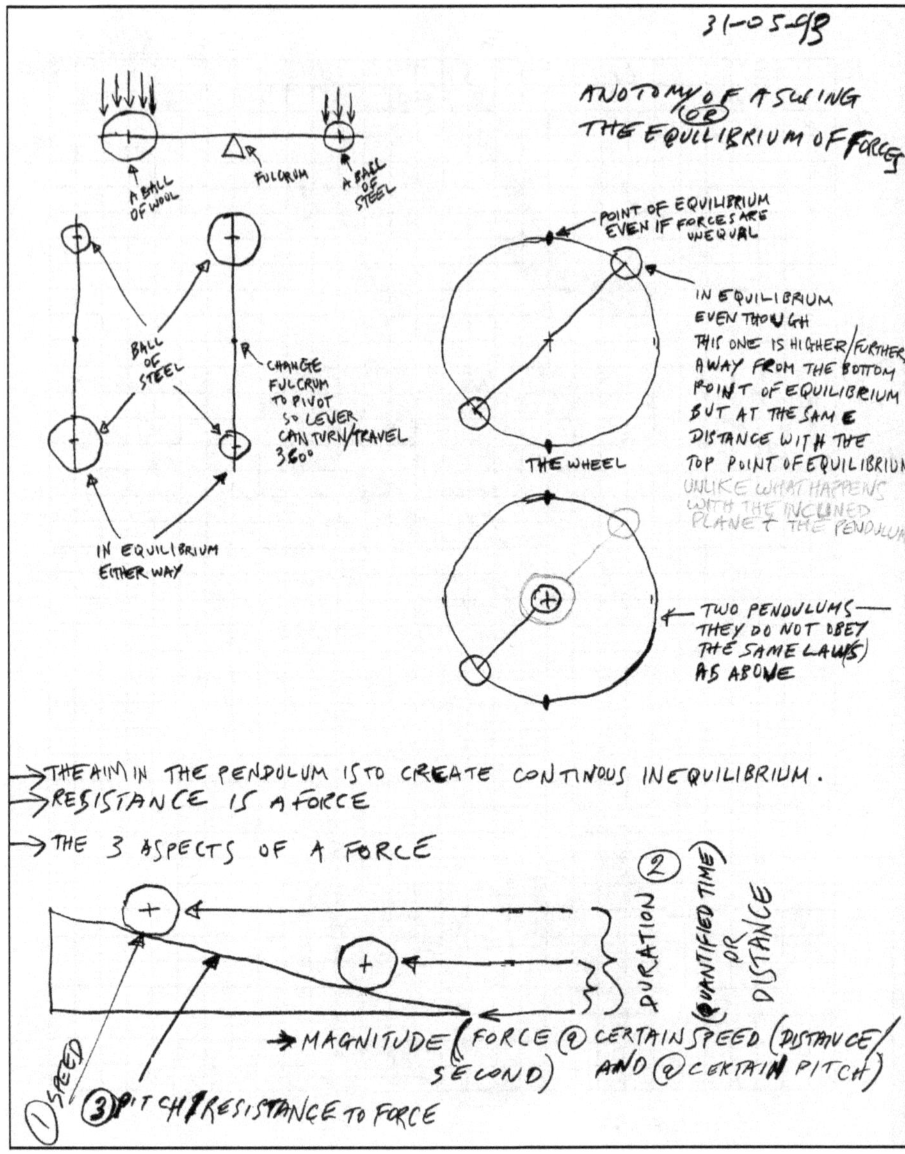
31-05-93
ANOTOMY OF A SWING OR THE EQUILIBRIUM OF FORCES
A BALL OF WOOL
FULCRUM
A BALL OF STEEL
BALL OF STEEL
CHANGE FULCRUM TO PIVOT SO LEVER CAN TURN/TRAVEL 360°
IN EQUILIBRIUM EITHER WAY
POINT OF EQUILIBRIUM EVEN IF FORCES ARE UNEQUAL
IN EQUILIBRIUM EVEN THOUGH THIS ONE IS HIGHER/FURTHER AWAY FROM THE BOTTOM POINT OF EQUILIBRIUM BUT AT THE SAME DISTANCE WITH THE TOP POINT OF EQUILIBRIUM
UNLIKE WHAT HAPPENS WITH THE INCLINED PLANE + THE PENDULUM
THE WHEEL
TWO PENDULUMS — THEY DO NOT OBEY THE SAME LAWS AS ABOVE
THE AIM IN THE PENDULUM IS TO CREATE CONTINOUS INEQUILIBRIUM.
RESISTANCE IS A FORCE
THE 3 ASPECTS OF A FORCE
② DURATION (QUANTIFIED TIME) OR DISTANCE
MAGNITUDE (FORCE @ CERTAIN SPEED (DISTANCE/SECOND) AND @ CERTAIN PITCH)
① SPEED
③ PITCH/RESISTANCE TO FORCE

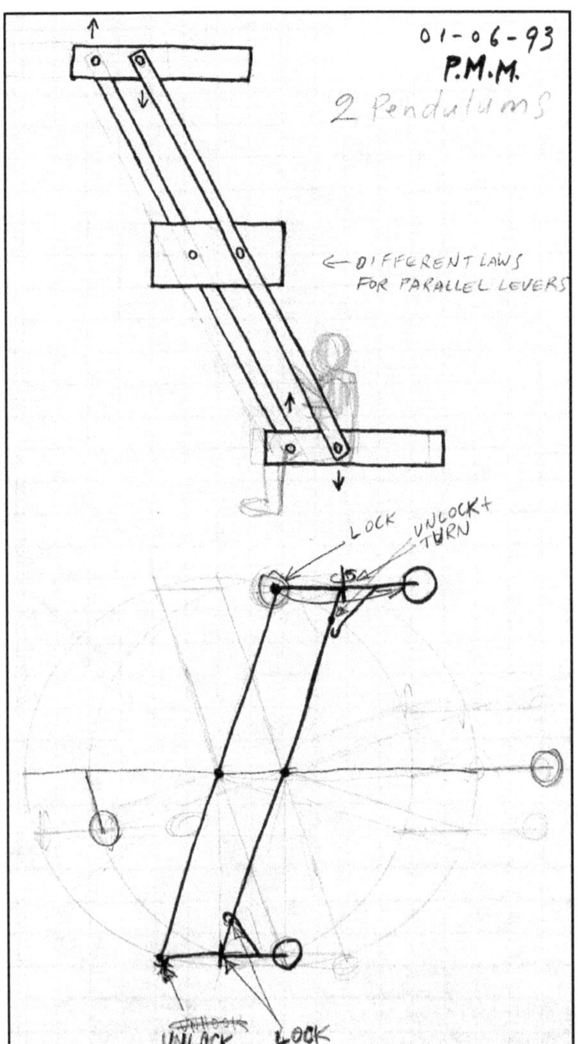
01-06-93
P.M.M.
2 Pendulums
DIFFERENT LAWS FOR PARALLEL LEVERS
LOCK
UNLOCK + TURN
UNLOCK
LOCK

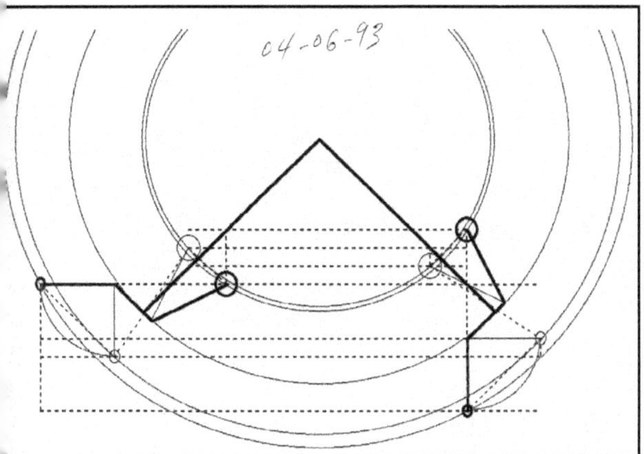
04-06-93

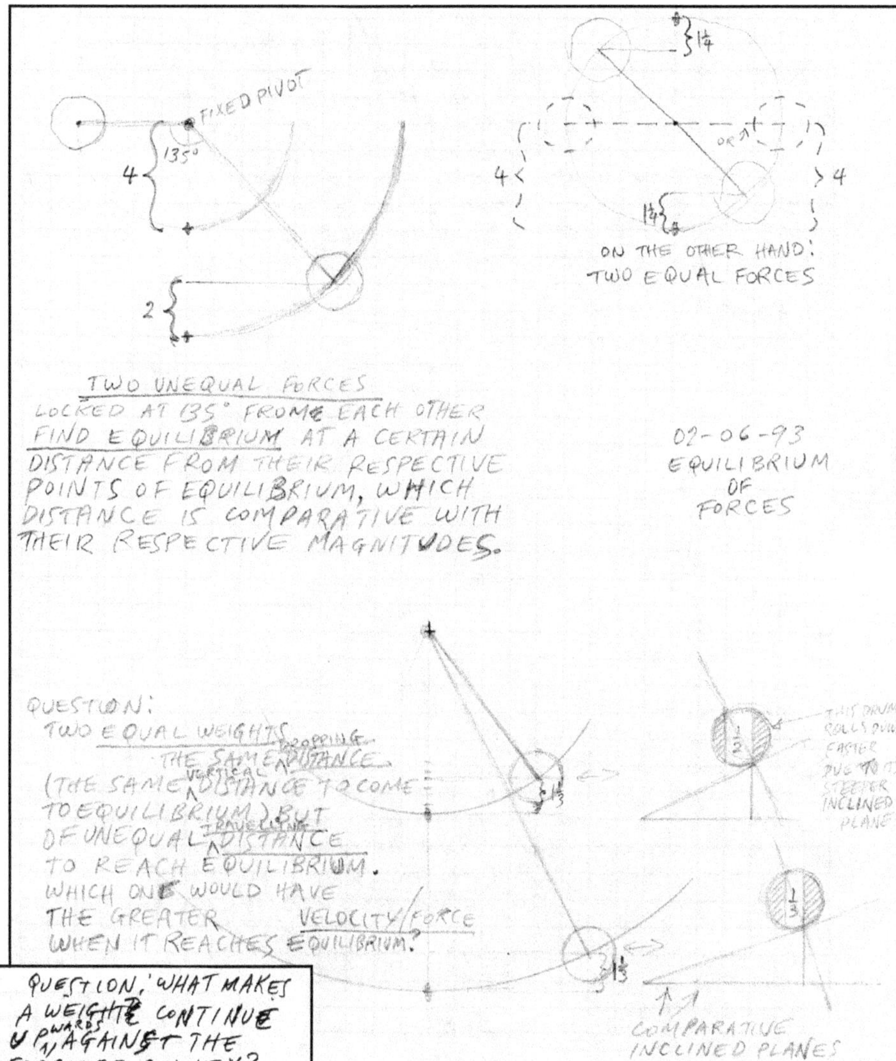
FIXED PIVOT
135°
4
2
1¼
4
4
1¼
ON THE OTHER HAND:
TWO EQUAL FORCES
TWO UNEQUAL FORCES
LOCKED AT 135° FROM EACH OTHER
FIND EQUILIBRIUM AT A CERTAIN
DISTANCE FROM THEIR RESPECTIVE
POINTS OF EQUILIBRIUM, WHICH
DISTANCE IS COMPARATIVE WITH
THEIR RESPECTIVE MAGNITUDES.
02-06-93
EQUILIBRIUM
OF
FORCES
QUESTION:
TWO EQUAL WEIGHTS DROPPING
THE SAME DISTANCE
(THE SAME VERTICAL DISTANCE TO COME
TO EQUILIBRIUM) BUT
OF UNEQUAL TRAVELLING DISTANCE
TO REACH EQUILIBRIUM.
WHICH ONE WOULD HAVE
THE GREATER VELOCITY/FORCE
WHEN IT REACHES EQUILIBRIUM?
COMPARATIVE
INCLINED PLANES

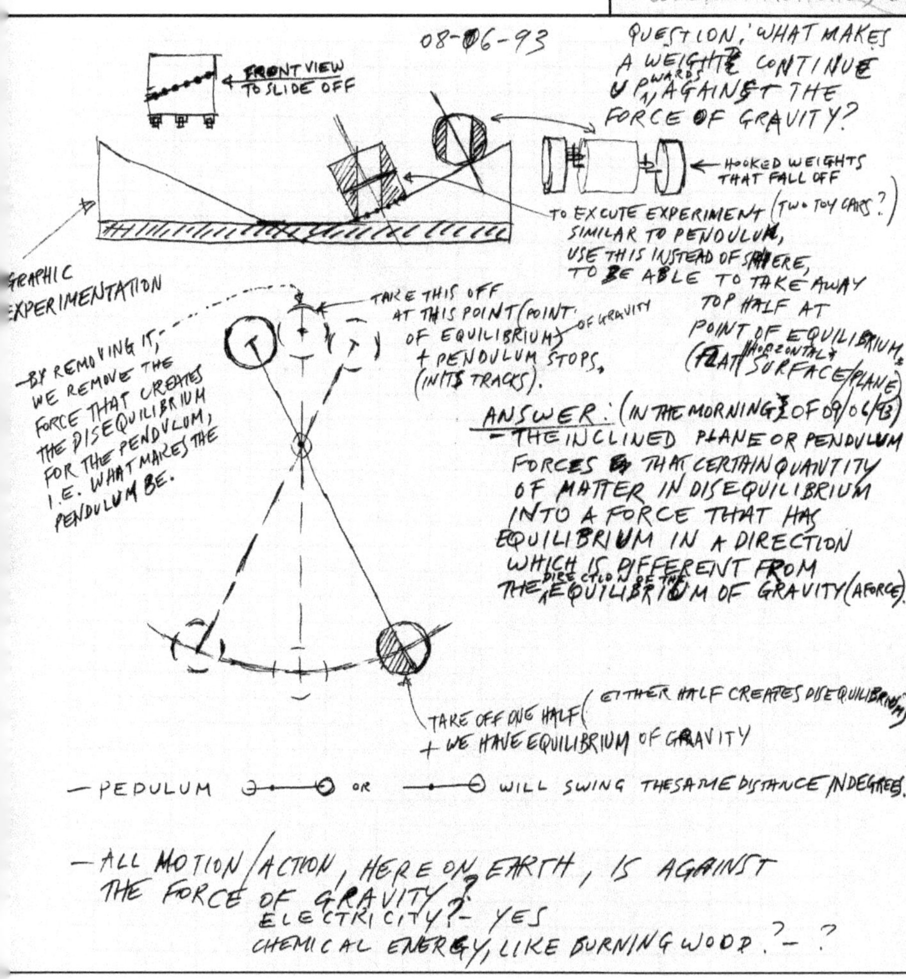
08-06-93
QUESTION: WHAT MAKES
A WEIGHT CONTINUE
UPWARDS AGAINST THE
FORCE OF GRAVITY?
FRONT VIEW
TO SLIDE OFF
HOOKED WEIGHTS
THAT FALL OFF
TO EXCUTE EXPERIMENT (TWO TOY CARS?)
SIMILAR TO PENDULUM,
USE THIS INSTEAD OF SPHERE,
TO BE ABLE TO TAKE AWAY
TOP HALF AT
POINT OF EQUILIBRIUM,
(FLAT HORIZONTAL SURFACE/PLANE)
GRAPHIC
EXPERIMENTATION
TAKE THIS OFF
AT THIS POINT (POINT OF GRAVITY
OF EQUILIBRIUM)
+ PENDULUM STOPS,
(IN ITS TRACKS).
-BY REMOVING IT,
WE REMOVE THE
FORCE THAT CREATES
THE DISEQUILIBRIUM
FOR THE PENDULUM,
I.E. WHAT MAKES THE
PENDULUM BE.
ANSWER: (IN THE MORNING OF 09/06/93)
-THE INCLINED PLANE OR PENDULUM
FORCES THAT CERTAIN QUANTITY
OF MATTER IN DISEQUILIBRIUM
INTO A FORCE THAT HAS
EQUILIBRIUM IN A DIRECTION
WHICH IS DIFFERENT FROM
THE DIRECTION OF THE EQUILIBRIUM OF GRAVITY (A FORCE).
TAKE OFF ONE HALF (EITHER HALF CREATES DISEQUILIBRIUM)
+ WE HAVE EQUILIBRIUM OF GRAVITY
-PEDULUM OR WILL SWING THE SAME DISTANCE IN DEGREES.
-ALL MOTION/ACTION, HERE ON EARTH, IS AGAINST
THE FORCE OF GRAVITY?
ELECTRICITY?- YES
CHEMICAL ENERGY, LIKE BURNING WOOD?- ?

– A PROJECTILE'S (E.G. A STONE) (OR A SPACE SHUTTLE) EQUILIBRIUM IS IN THE DIRECTION OF THE FORCE THAT PUT IT IN THAT TRAJECTORY UNTIL IT MEETS WITH RESISTANCE (A FORCE) i.e.) AIR, A WALL, GRAVITY.

– IF A DROPPED WEIGHT WOULD MEET NO RESISTANCE, COULD GO THROUGH + REACH THE CENTRE OF THE EARTH. IT WOULD GO PAST IT (THE CENTRE) + GET TO THE OTHER SIDE + THEN BOUNCE BACK, LIKE A BALL, TO THE POINT OF ORIGIN.

– THE EARTH ROTATES AROUND THE SUN IN THE SAME WAY A PENDULUM WORKS/DOES.

09-06-93

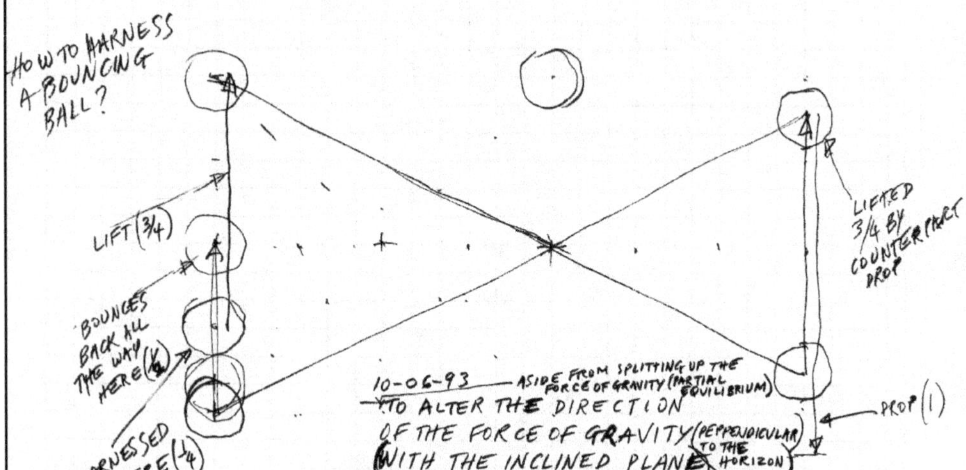

10-06-93 – ASIDE FROM SPLITTING UP THE FORCE OF GRAVITY (PARTIAL EQUILIBRIUM)

– TO ALTER THE DIRECTION OF THE FORCE OF GRAVITY (PERPENDICULAR TO THE HORIZON) WITH THE INCLINED PLANE INTO AN INCLINED DIRECTION, SO WE CAN MAKE IT COME BACK (BOUNCE IT BACK). (+ SO DOES THE PENDULUM)

– WITH PIVOT (LEVER OR WHEEL) WE ARTIFICIALLY CREATE A HIGHER REST PLANE (POINT OF EQUILIBRIUM)

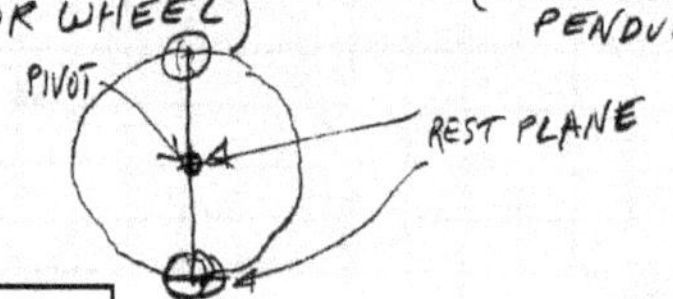

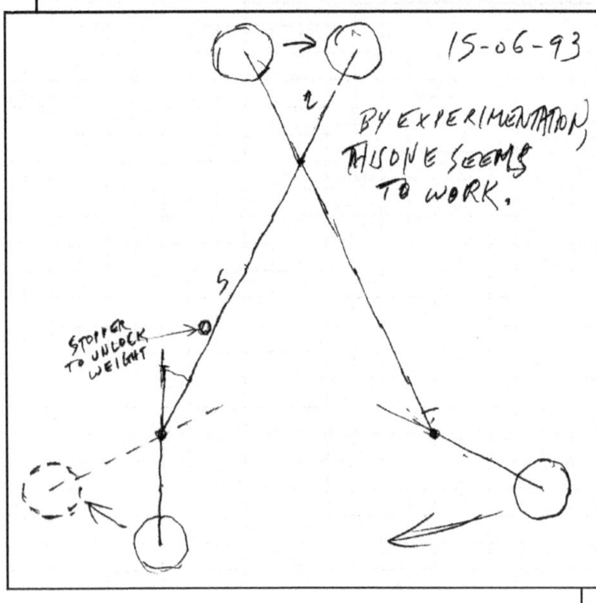

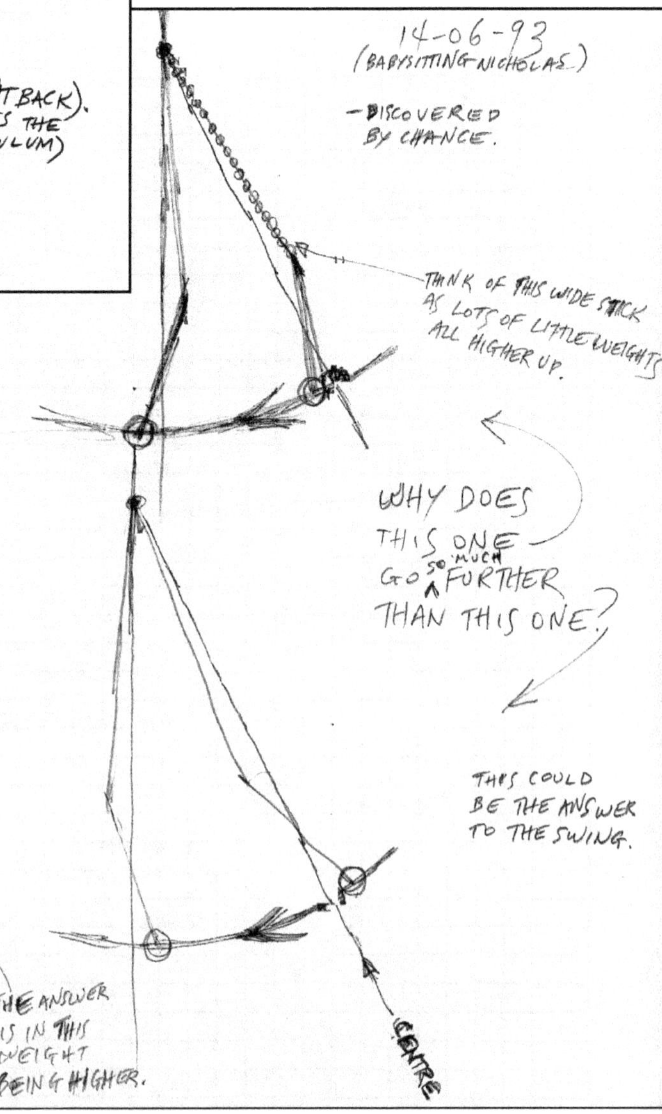

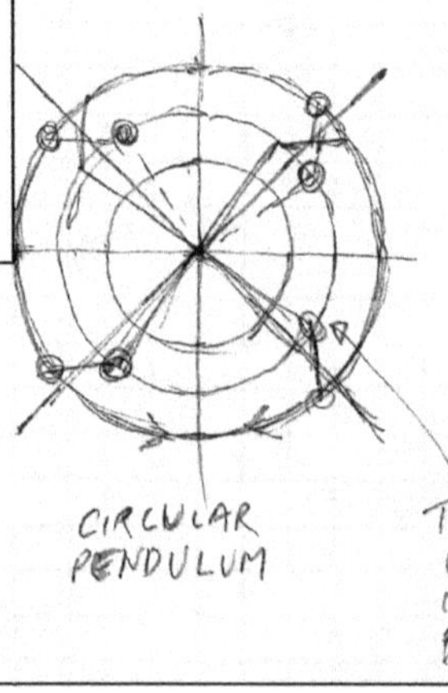

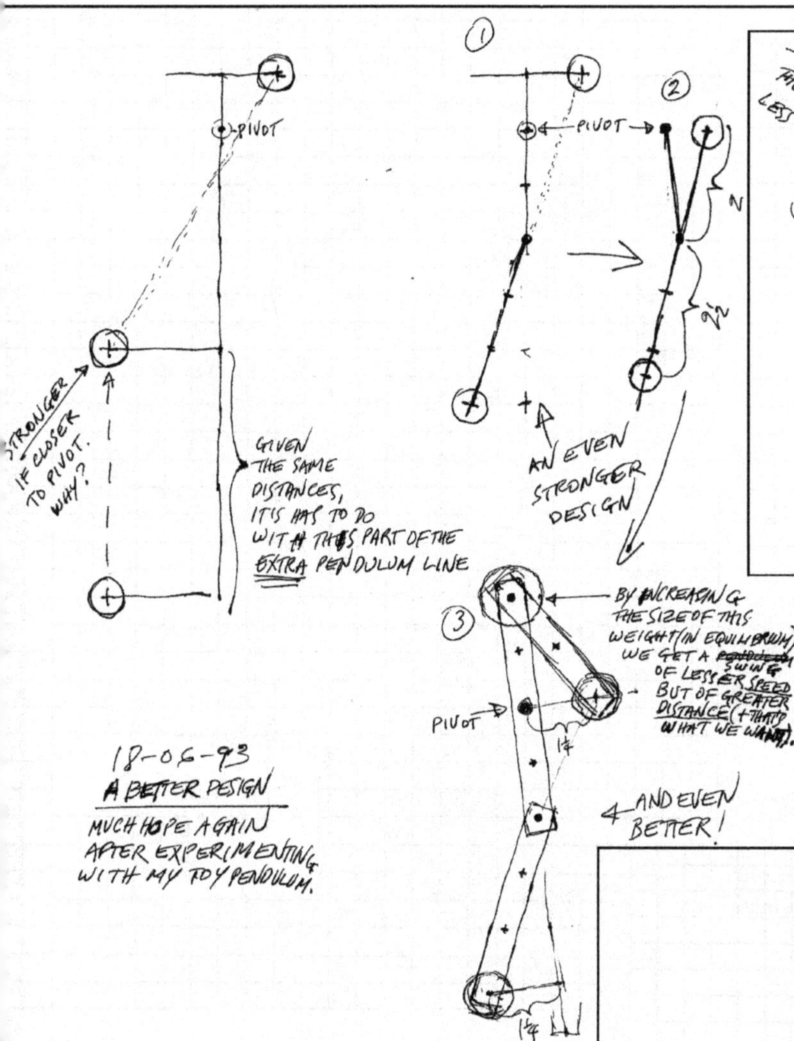
PIVOT
STRONGER IF CLOSER TO PIVOT. WHY?
GIVEN THE SAME DISTANCES, IT'S HAS TO DO WITH THIS PART OF THE EXTRA PENDULUM LINE
①
②
AN EVEN STRONGER DESIGN
③
BY INCREASING THE SIZE OF THIS WEIGHT (IN EQUILIBRIUM) WE GET A SWING OF LESSER SPEED BUT OF GREATER DISTANCE (+ THAT'S WHAT WE WANT).
AND EVEN BETTER!
18-06-93
A BETTER DESIGN
MUCH HOPE AGAIN AFTER EXPERIMENTING WITH MY TOY PENDULUM.

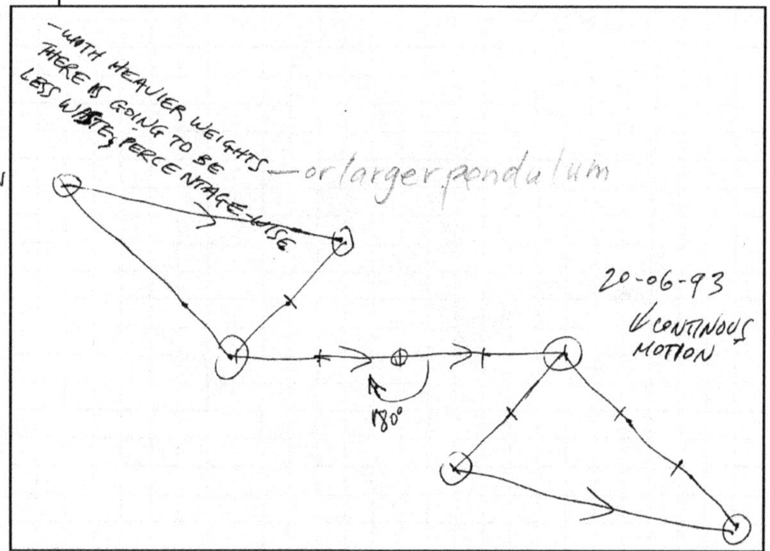
–WITH HEAVIER WEIGHTS THERE IS GOING TO BE LESS WASTE, PERCENTAGE-WISE —or larger pendulum
20-06-93
CONTINOUS MOTION
180°

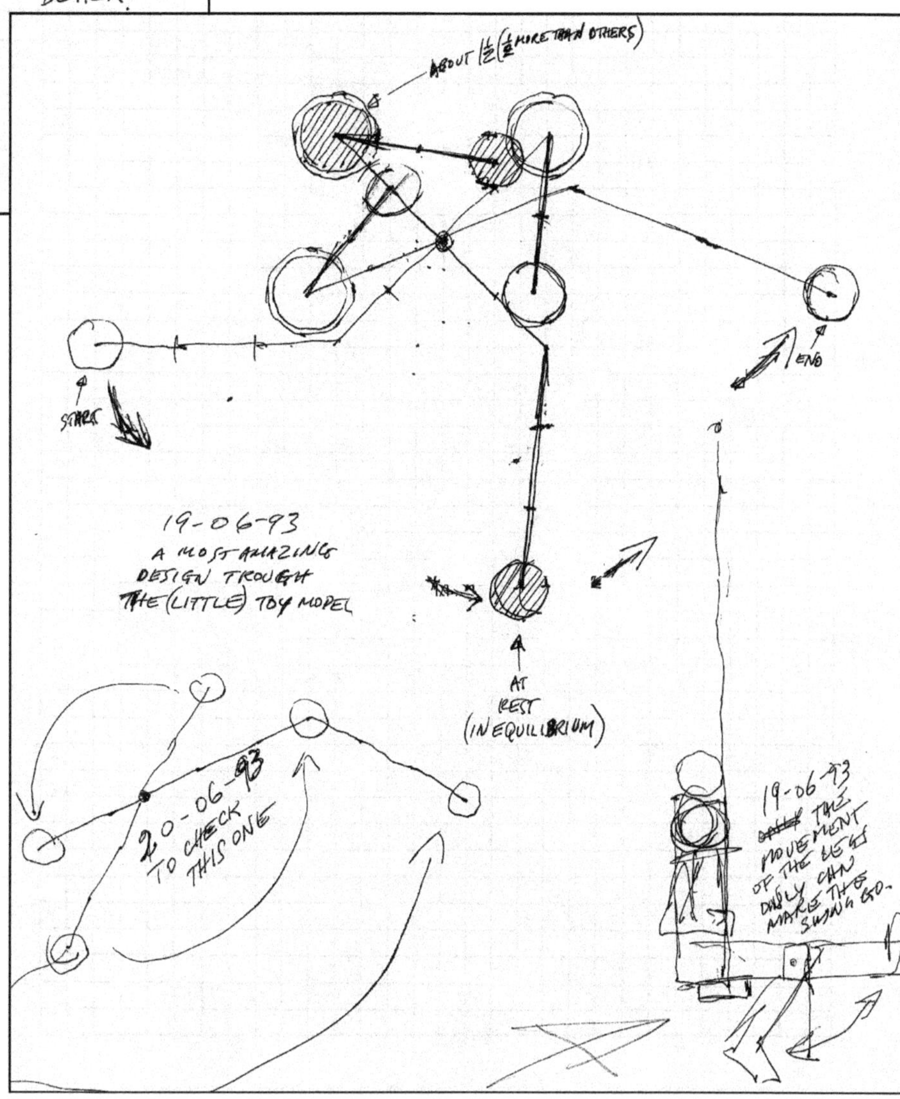
ABOUT 1½ (½ MORE THAN OTHERS)
START
END
19-06-93
A MOST AMAZING DESIGN TROUGH THE (LITTLE) TOY MODEL
AT REST (IN EQUILIBRIUM)
20-06-93 TO CHECK THIS ONE
19-06-93 THE MOVEMENT OF THE LEGS ONLY CAN MAKE THE SWING GO.

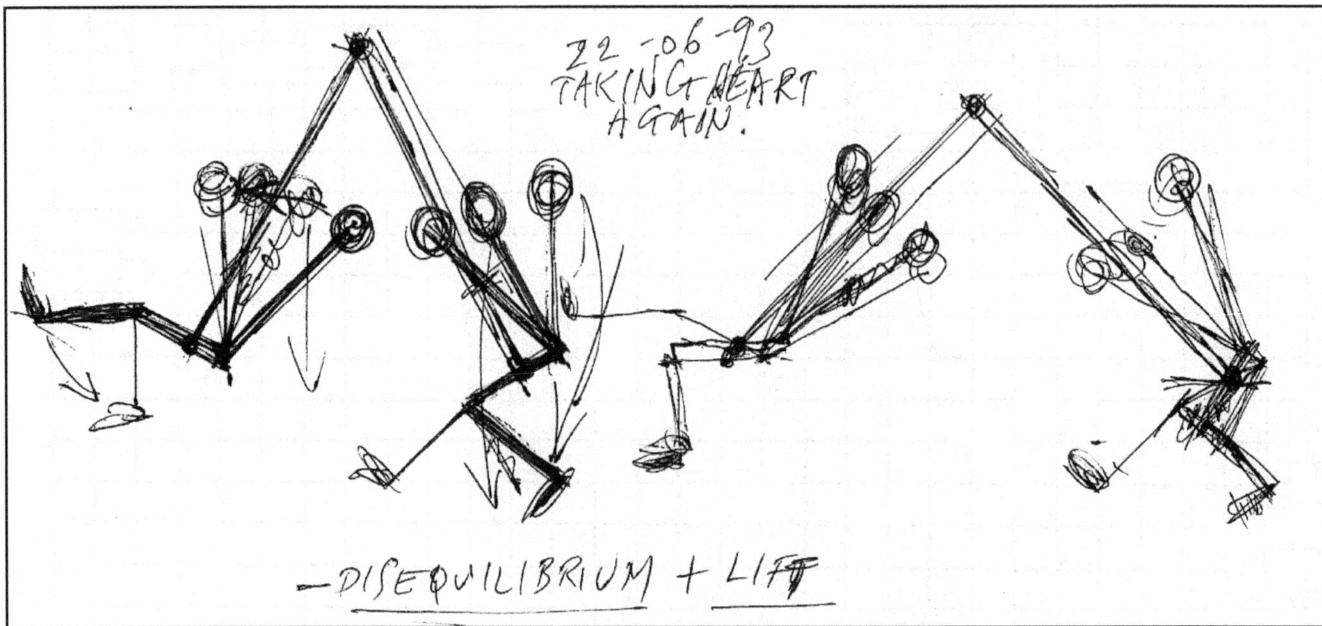
22-06-93
TAKING HEART
AGAIN.
—DISEQUILIBRIUM + LIFT

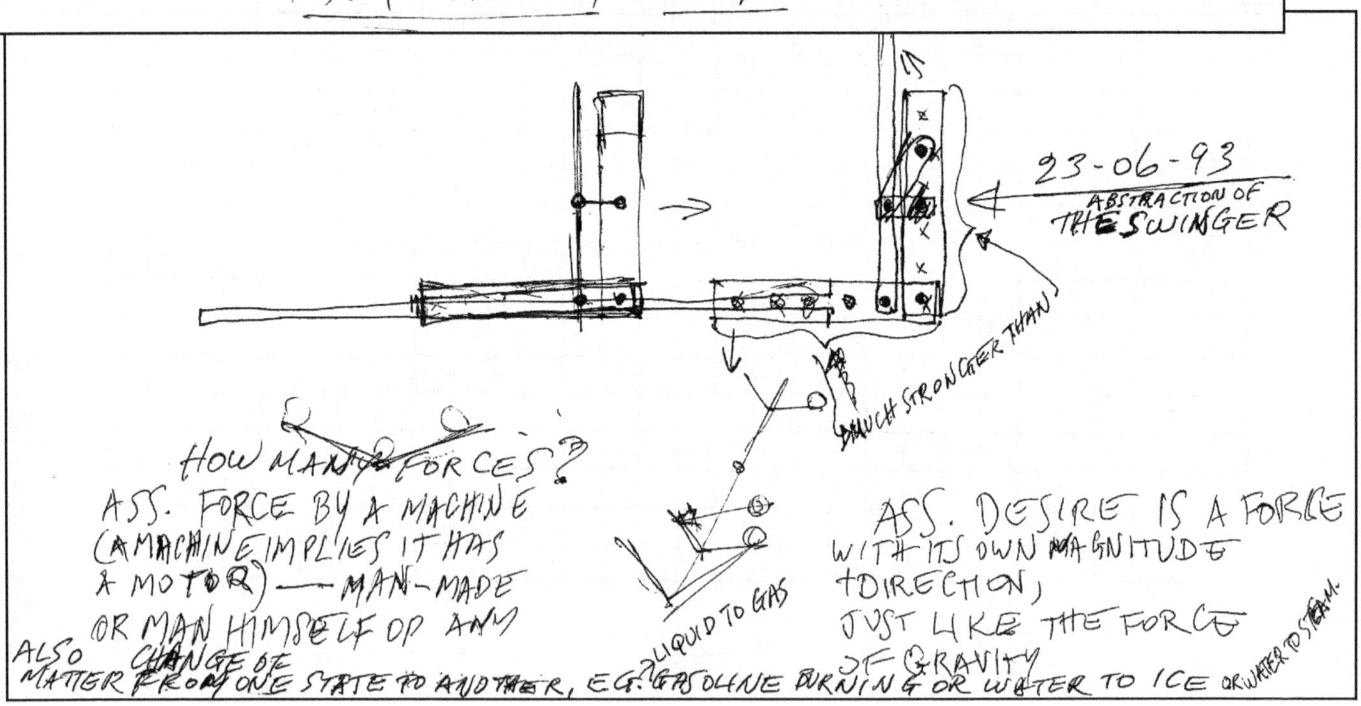
23-06-93
ABSTRACTION OF
THE SWINGER
MUCH STRONGER THAN
HOW MANY FORCES?
ASS. FORCE BY A MACHINE
(A MACHINE IMPLIES IT HAS
A MOTOR) — MAN-MADE
OR MAN HIMSELF OR ANY
ALSO CHANGE OF
MATTER FROM ONE STATE TO ANOTHER, E.G. GASOLINE BURNING OR WATER TO ICE OR WATER TO STEAM.
LIQUID TO GAS
ASS. DESIRE IS A FORCE
WITH ITS OWN MAGNITUDE
+ DIRECTION,
JUST LIKE THE FORCE
OF GRAVITY

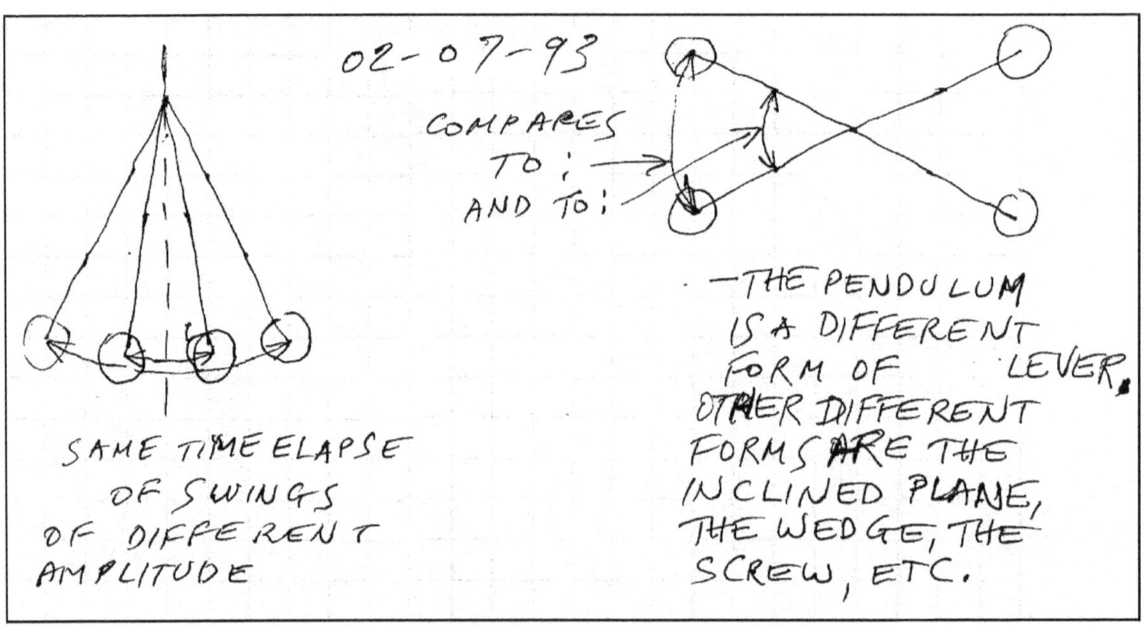
02-07-93
COMPARES
TO:
AND TO:
SAME TIME ELAPSE
OF SWINGS
OF DIFFERENT
AMPLITUDE
—THE PENDULUM
IS A DIFFERENT
FORM OF LEVER.
OTHER DIFFERENT
FORMS ARE THE
INCLINED PLANE,
THE WEDGE, THE
SCREW, ETC.

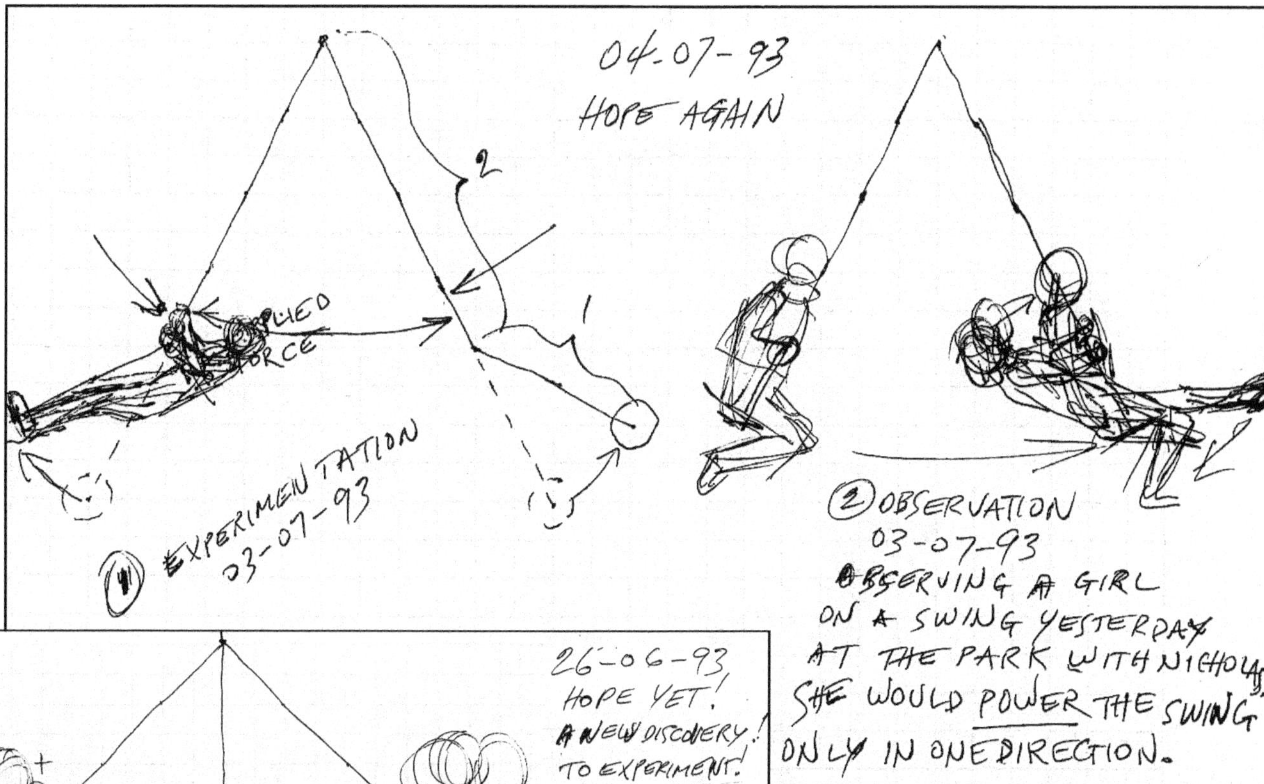
04-07-93
HOPE AGAIN
2
1
APPLIED FORCE
① EXPERIMENTATION 03-07-93
② OBSERVATION 03-07-93
OBSERVING A GIRL ON A SWING YESTERDAY AT THE PARK WITH NICHOLAS. SHE WOULD POWER THE SWING ONLY IN ONE DIRECTION.

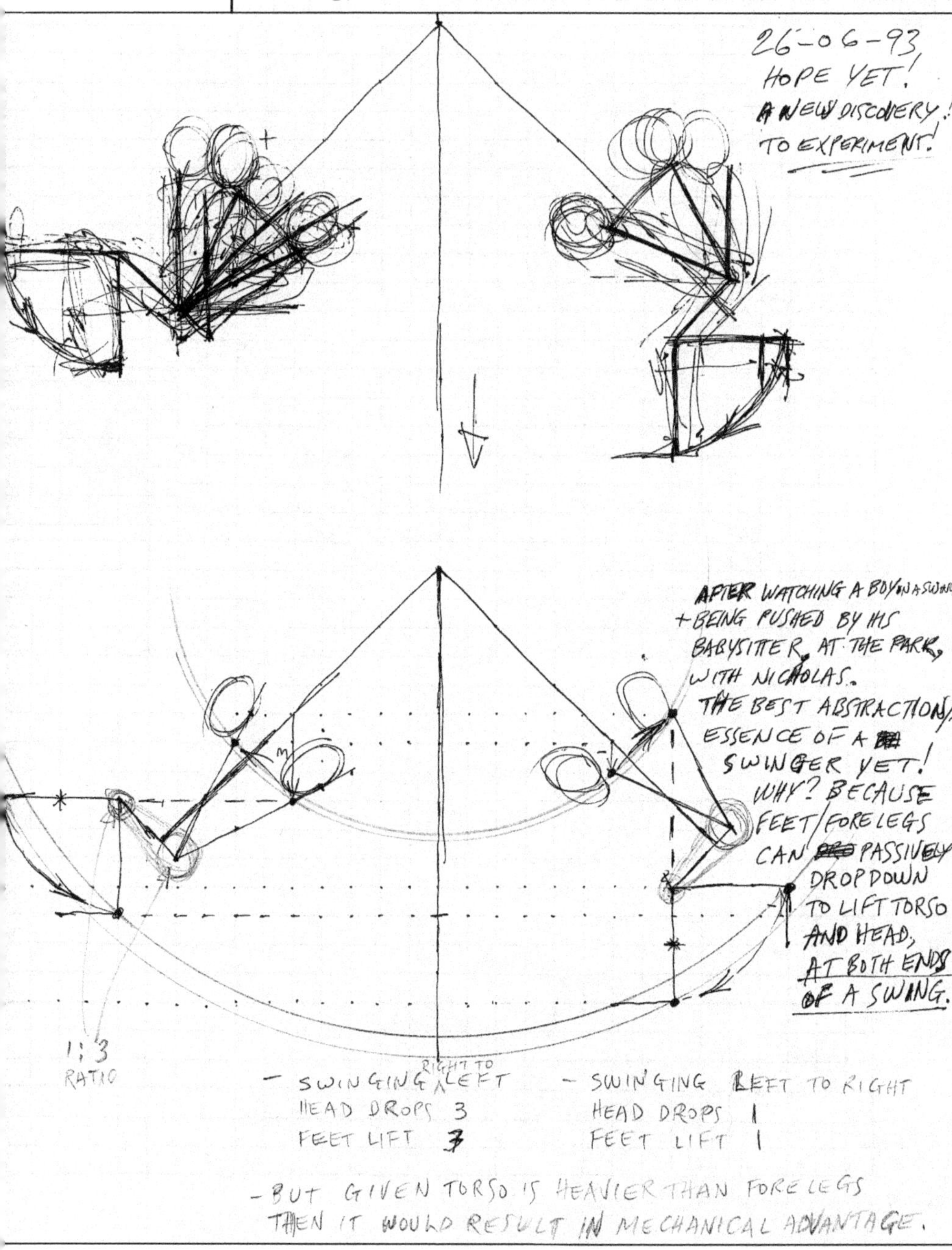
26-06-93
HOPE YET!
A NEW DISCOVERY! TO EXPERIMENT!
AFTER WATCHING A BOY ON A SWING + BEING PUSHED BY HIS BABYSITTER, AT THE PARK, WITH NICHOLAS.
THE BEST ABSTRACTION/ESSENCE OF A SWINGER YET!
WHY? BECAUSE FEET/FORELEGS CAN PASSIVELY DROP DOWN TO LIFT TORSO AND HEAD, AT BOTH ENDS OF A SWING.
1:3 RATIO
- SWINGING RIGHT TO LEFT
HEAD DROPS 3
FEET LIFT 3
- SWINGING LEFT TO RIGHT
HEAD DROPS 1
FEET LIFT 1
- BUT GIVEN TORSO IS HEAVIER THAN FORELEGS THEN IT WOULD RESULT IN MECHANICAL ADVANTAGE.

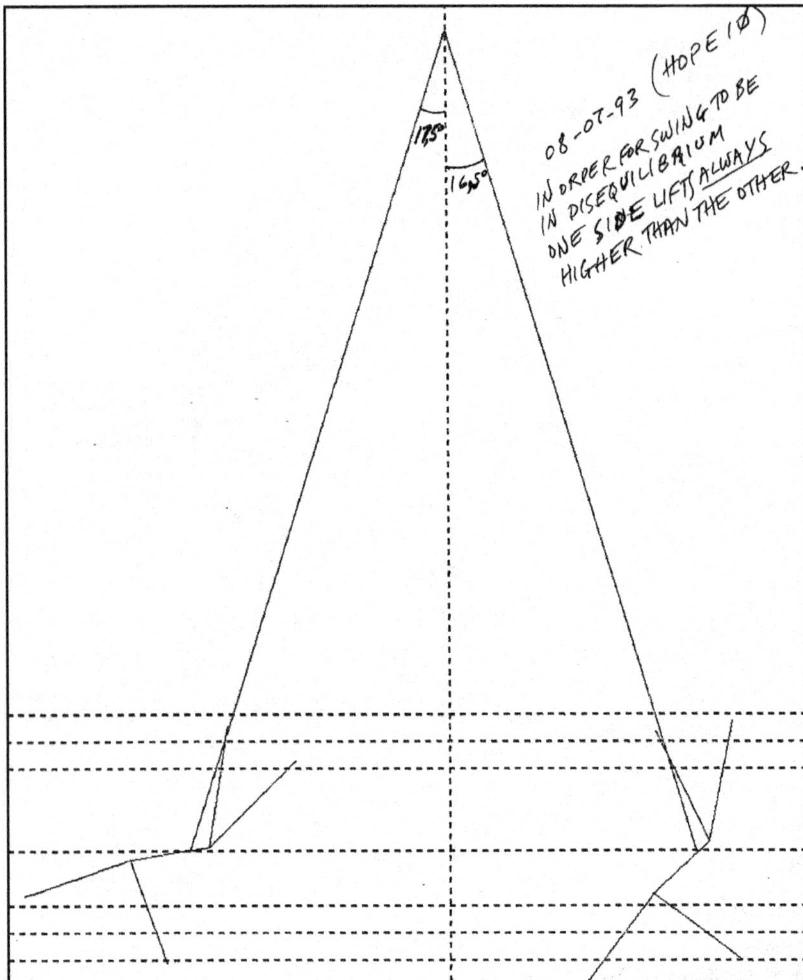
08-07-93 (HOPE 10)
IN ORDER FOR SWING TO BE IN DISEQUILIBRIUM ONE SIDE LIFTS ALWAYS HIGHER THAN THE OTHER.

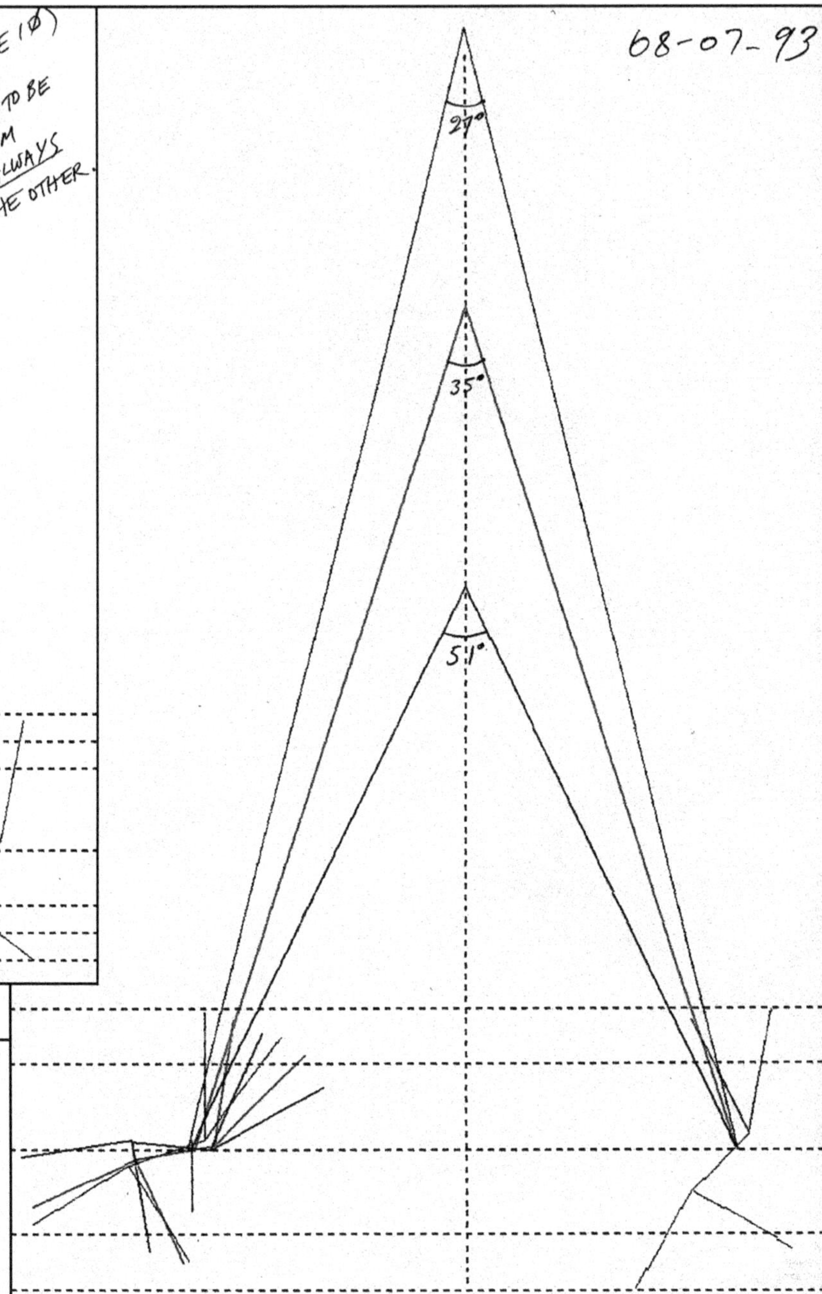
08-07-93
27°
35°
51°

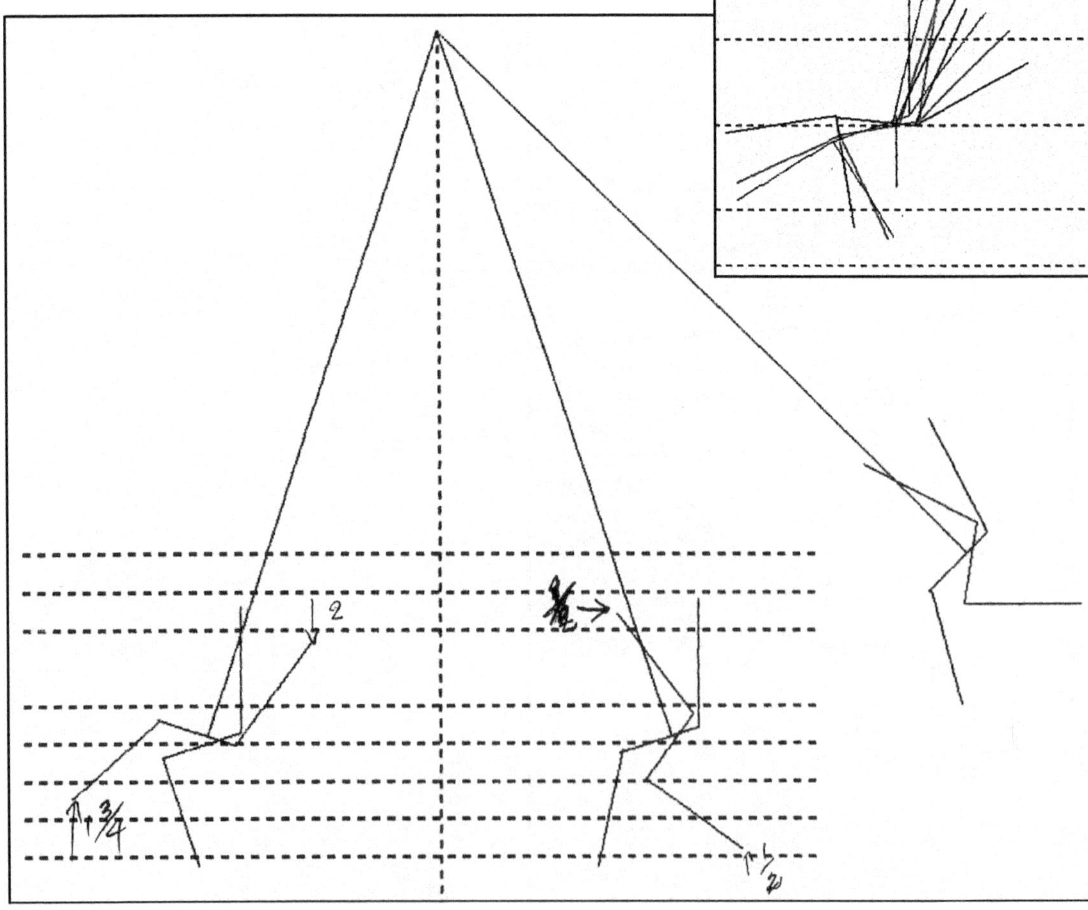
2
3/4

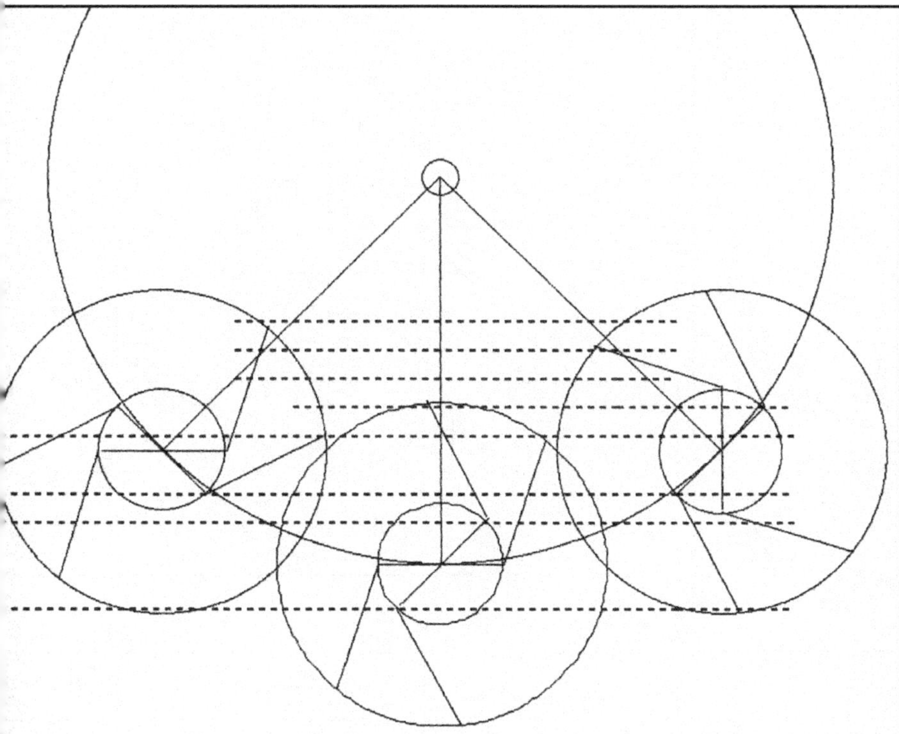

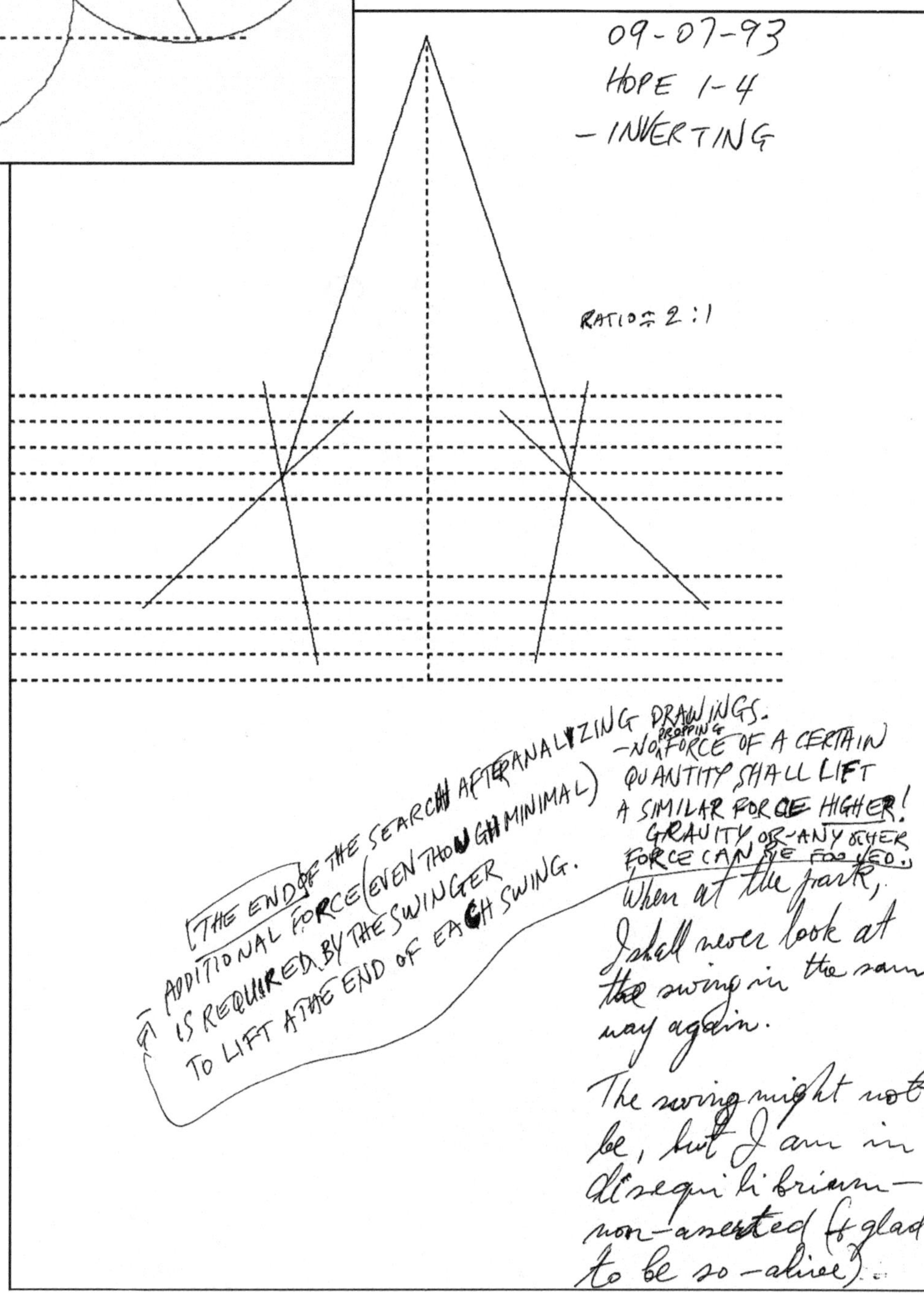
09-07-93
HOPE 1-4
-INVERTING
RATIO ≑ 2:1
THE END OF THE SEARCH AFTER ANALYZING DRAWINGS.
- ADDITIONAL FORCE (EVEN THOUGH MINIMAL)
IS REQUIRED BY THE SWINGER
TO LIFT AT THE END OF EACH SWING.
-NO PROPPING FORCE OF A CERTAIN
QUANTITY SHALL LIFT
A SIMILAR FORCE HIGHER!
GRAVITY OR ANY OTHER
FORCE CAN BE FOOLED.
When at the park,
I shall never look at
the swing in the same
way again.
The swing might not
be, but I am in
disequilibrium—
non-asserted (& glad
to be so—alive).

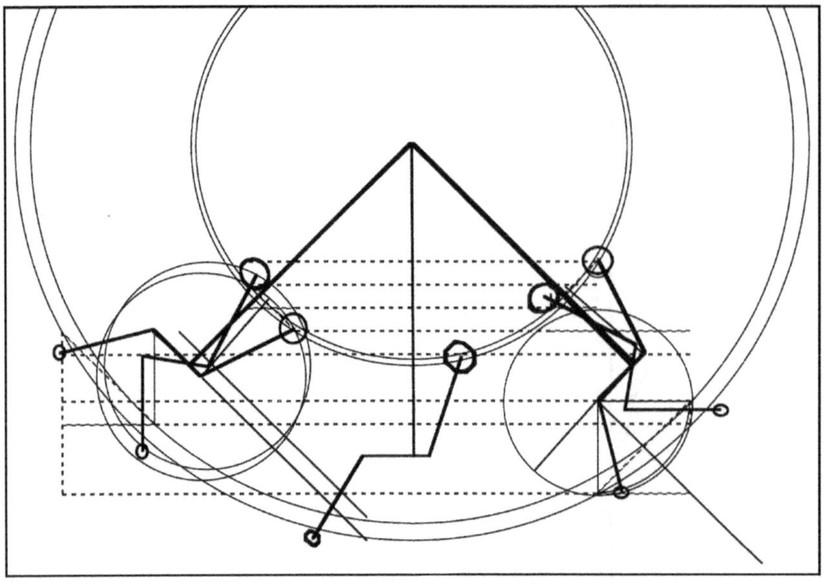

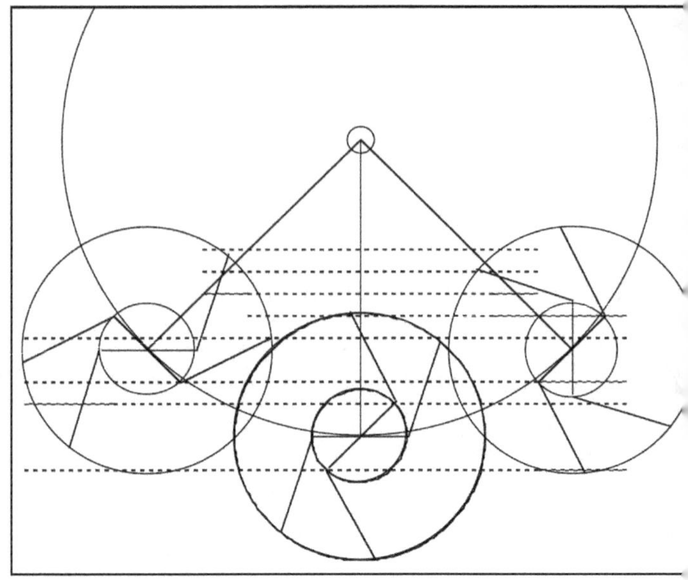

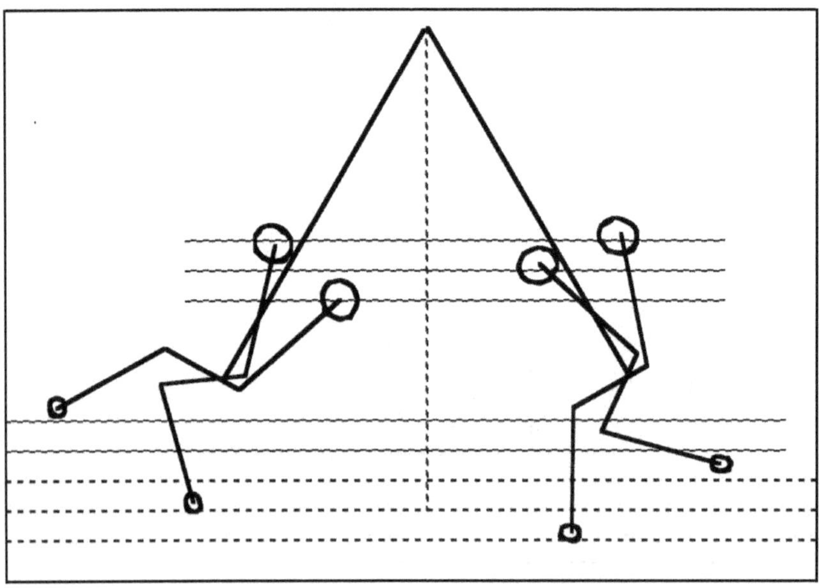

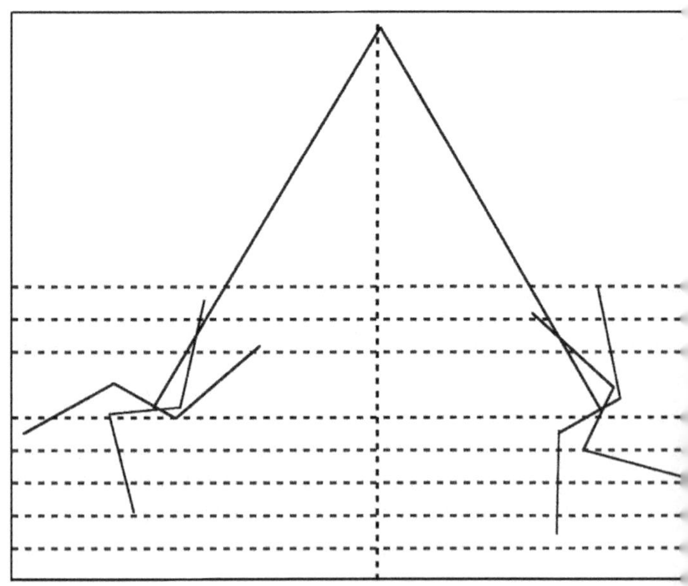

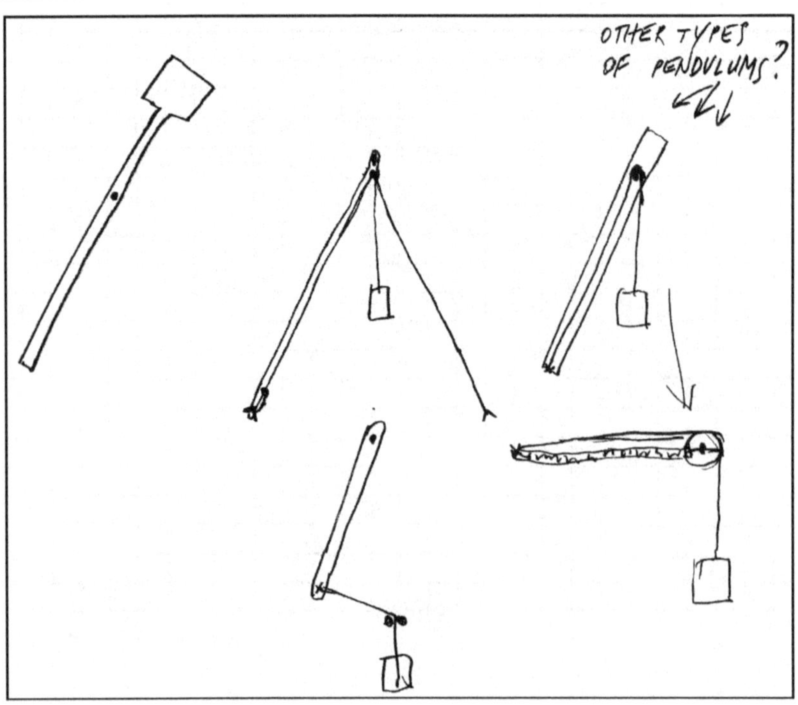
OTHER TYPES
OF PENDULUMS?

5. GAS ENGINE

GAS ENGINE or ICE (Internal Combustion Engine)

The problem with the STRAIGHT PISTON Internal Combustion Engine: There is an enormous amount of wasted energy. When the explosion occurs, the piston is allowed to travel a very short distance. The potential energy of the piston is cut short. But, what if the piston is allowed to do its work until it no longer can. In this case, there would be the least possible waste of energy. All of this can be demonstrated with a piston turning a wheel: as you can see, a **straight piston** would have a short run; but, as you can imagine, a **circular piston** would turn the wheel completely and, probably, more than once. Of course, the force would not be even; but that can be overcome 1) by changing the number of ignitions in relation to the number of rotations, 2) by synchronizing with several pistons, and 3) by regulating with a transmission (this would have a similar effect as number 1).

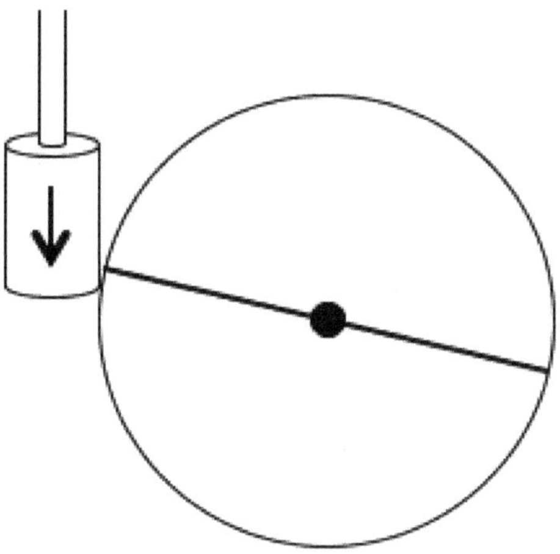

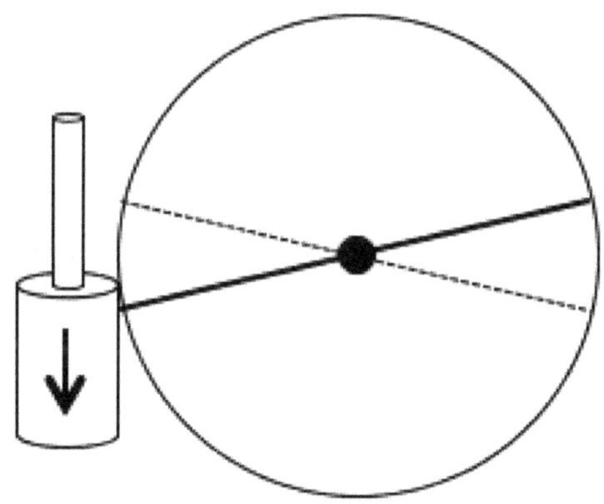

The fundamental aim of this idea is to invent a CIRCULAR PISTON Internal Combustion

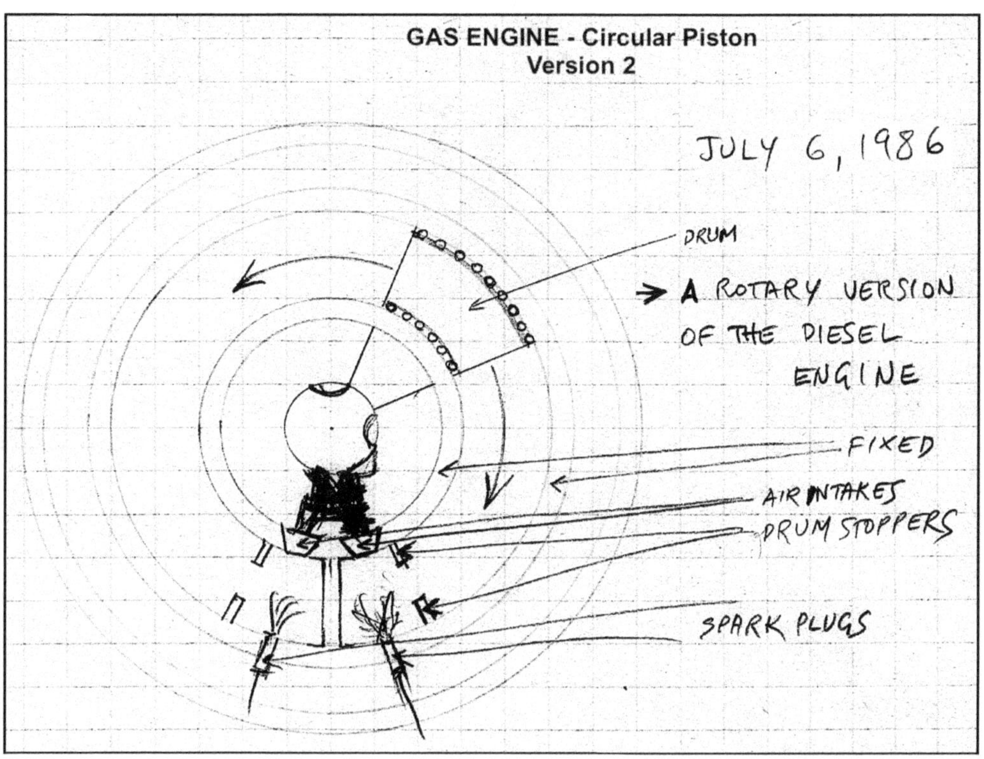

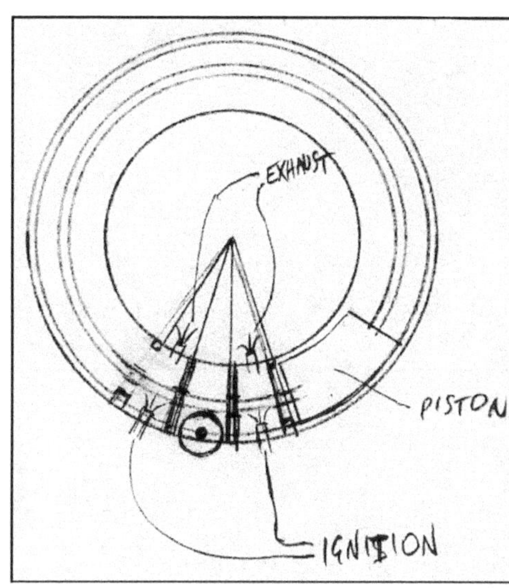

GAS ENGINE - Circular Piston
Version 1

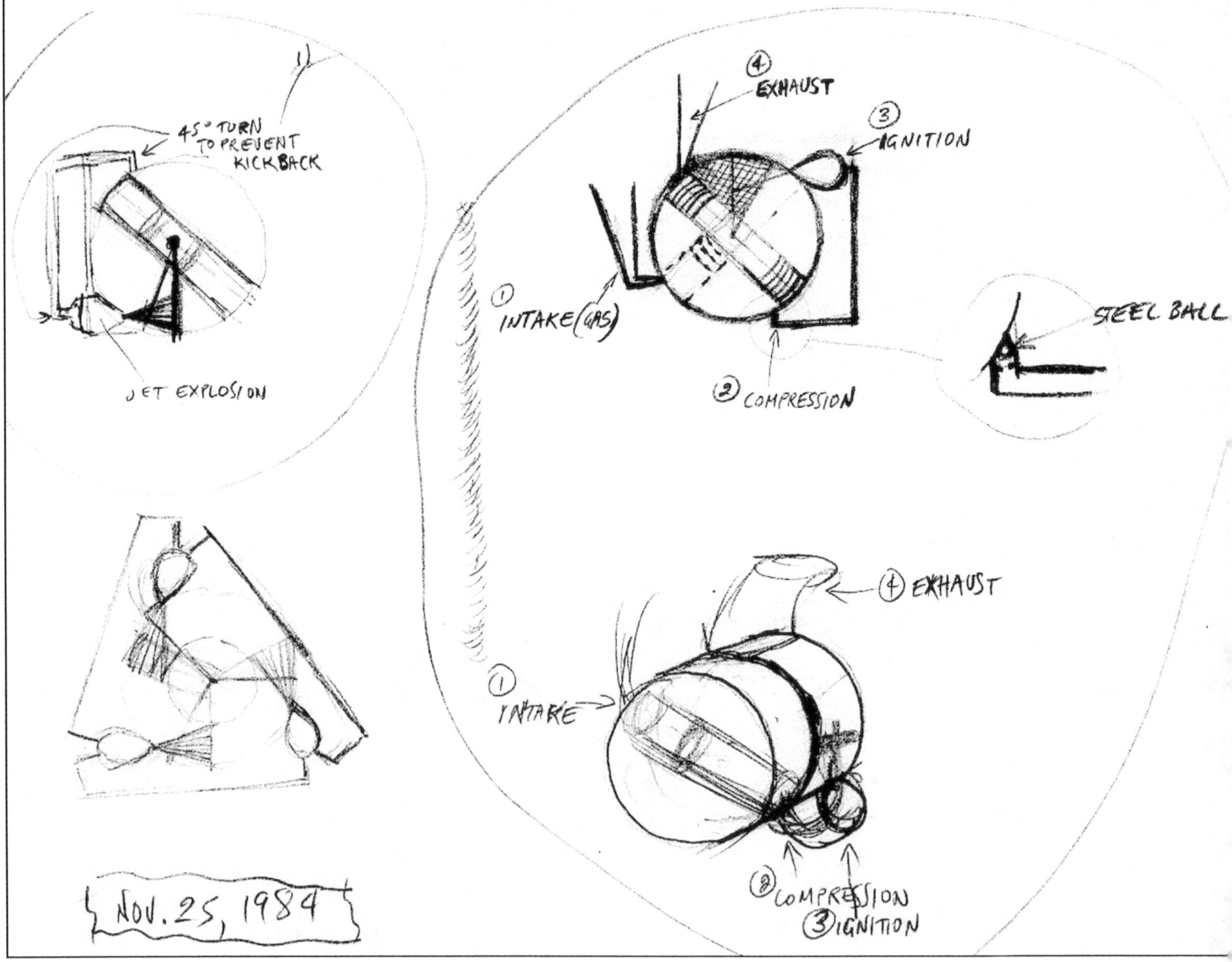

GAS ENGINE - Circular Piston
Version 3

JUNE 24, 1987

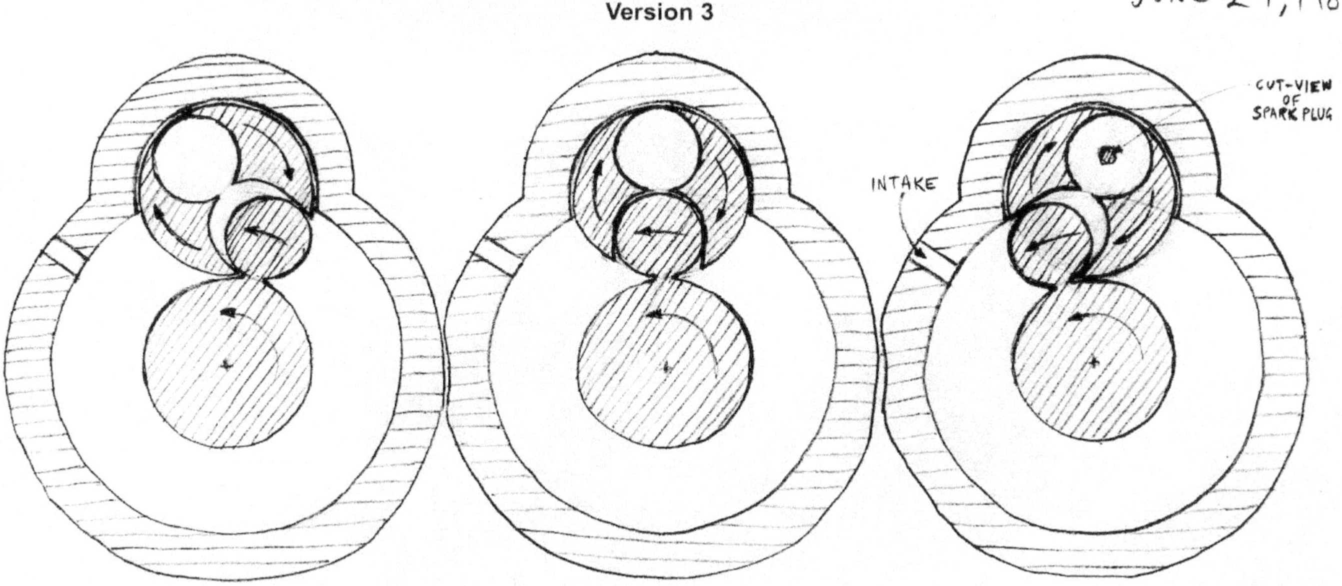

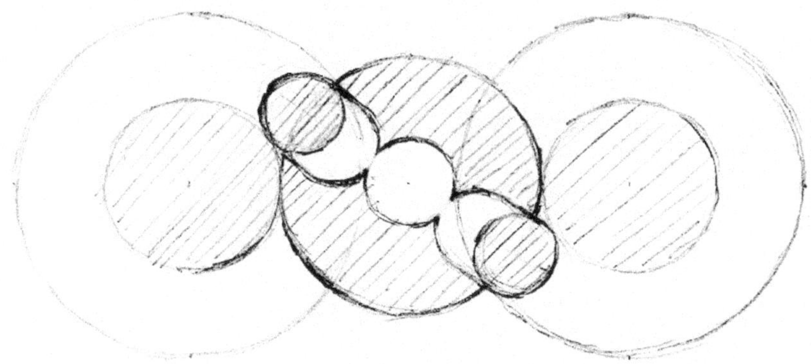

SUNDAY, JUNE 28, 1987
(MARRIAGE COURSE AT 3:00 AT POMPEI'S)
(DA RUCCI ENGINE
(USING TWO ENGINES
TO PREVENT RECOIL)

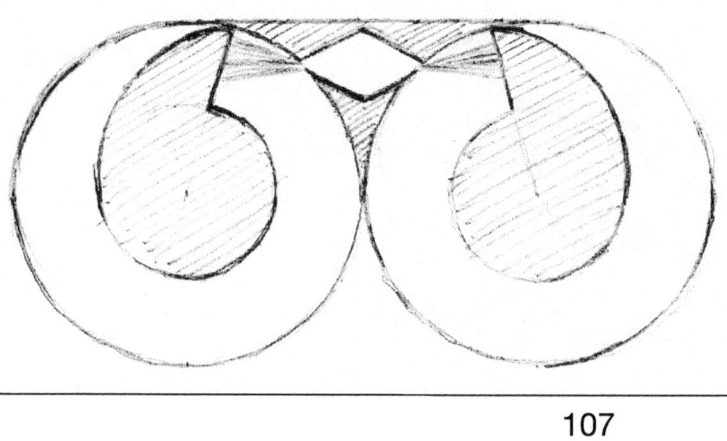

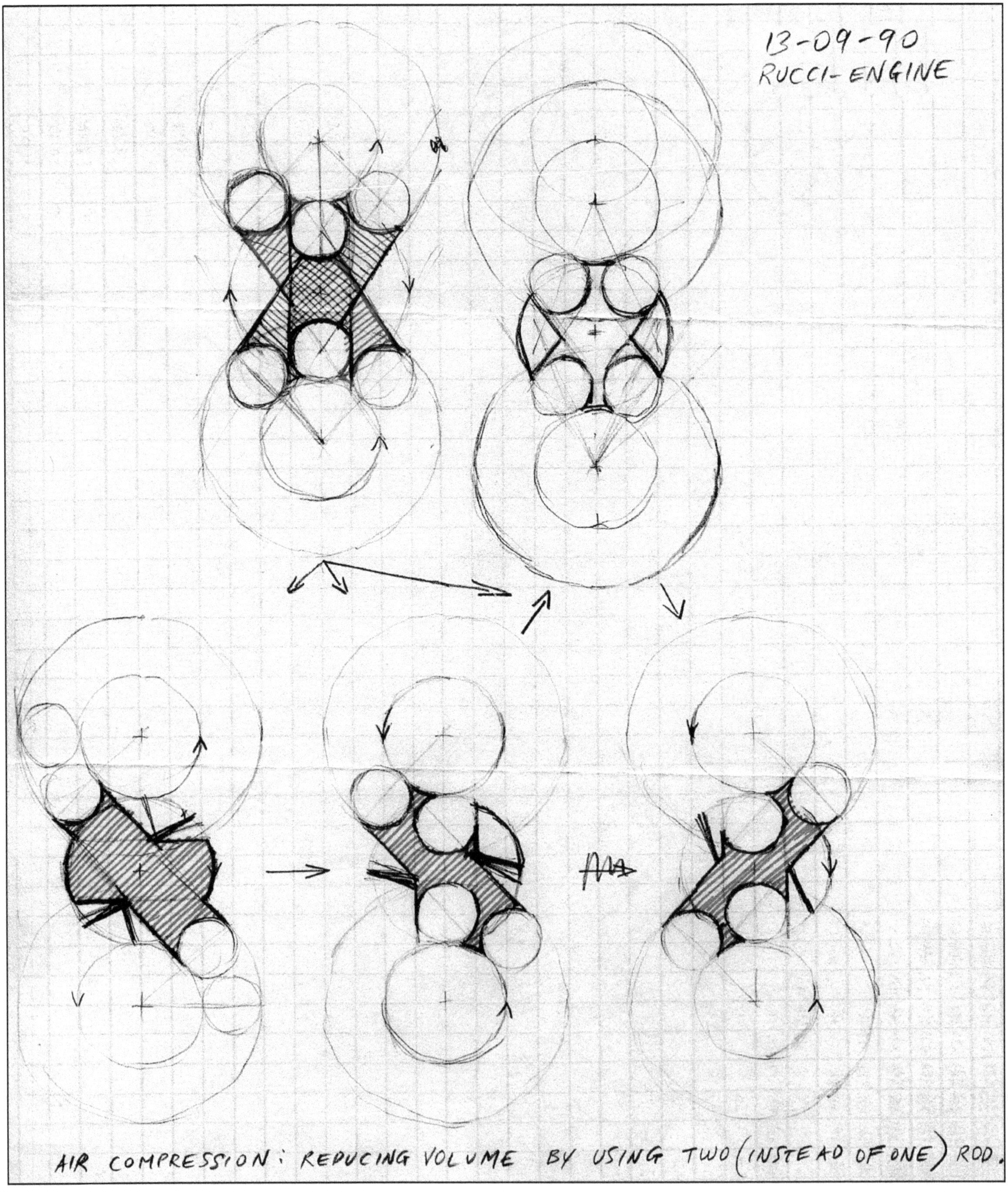

13-09-90
RUCCI-ENGINE
AIR COMPRESSION: REDUCING VOLUME BY USING TWO (INSTEAD OF ONE) ROD.

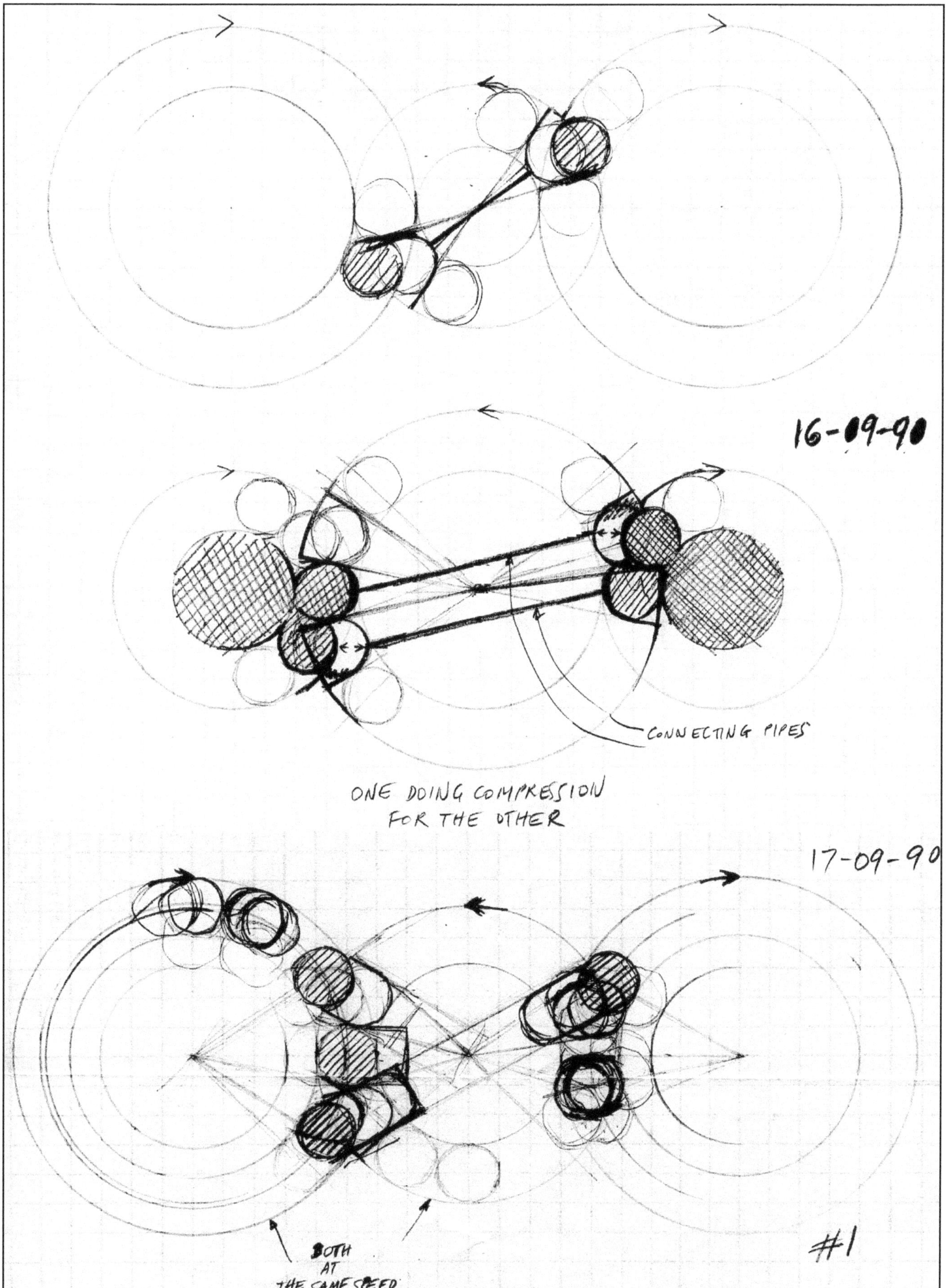
16-09-90
CONNECTING PIPES
ONE DOING COMPRESSION
FOR THE OTHER
17-09-90
BOTH
AT
THE SAME SPEED
#1

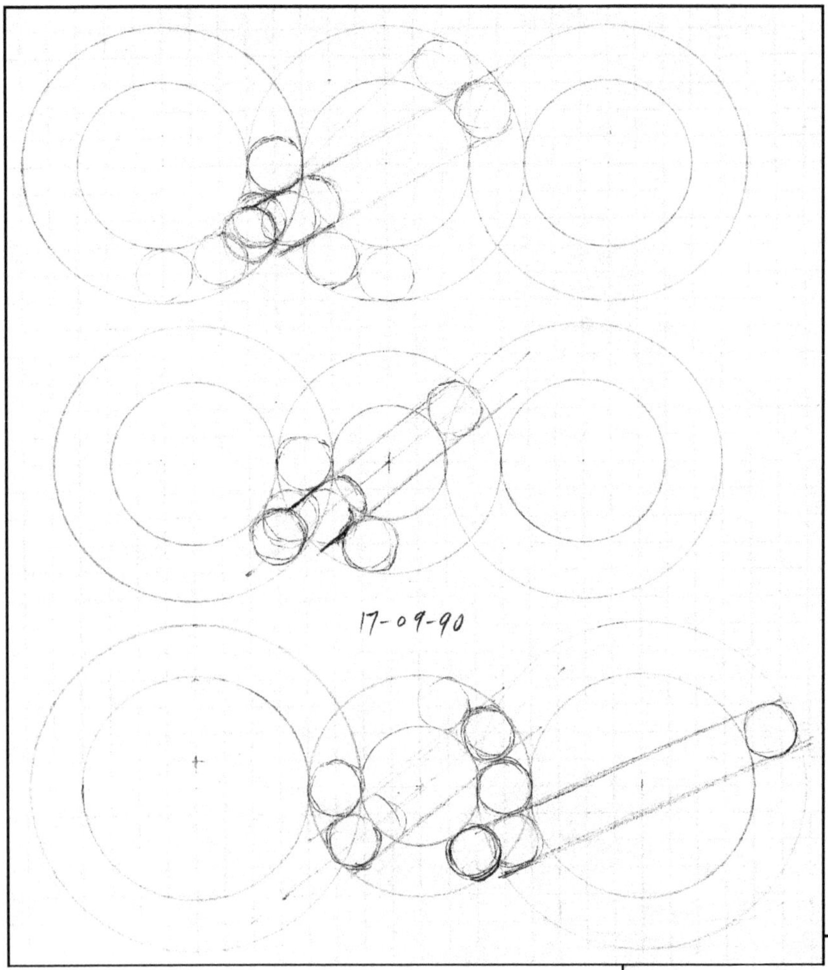
17-09-90

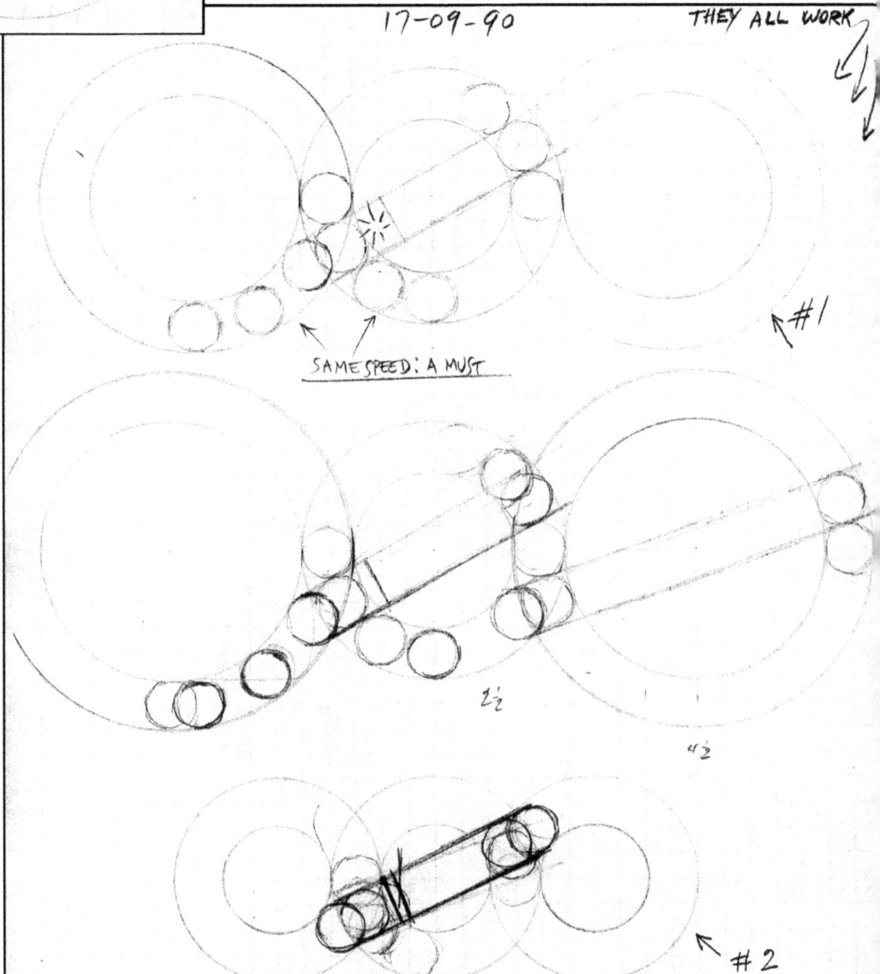
17-09-90
THEY ALL WORK
SAME SPEED: A MUST
#1
#2

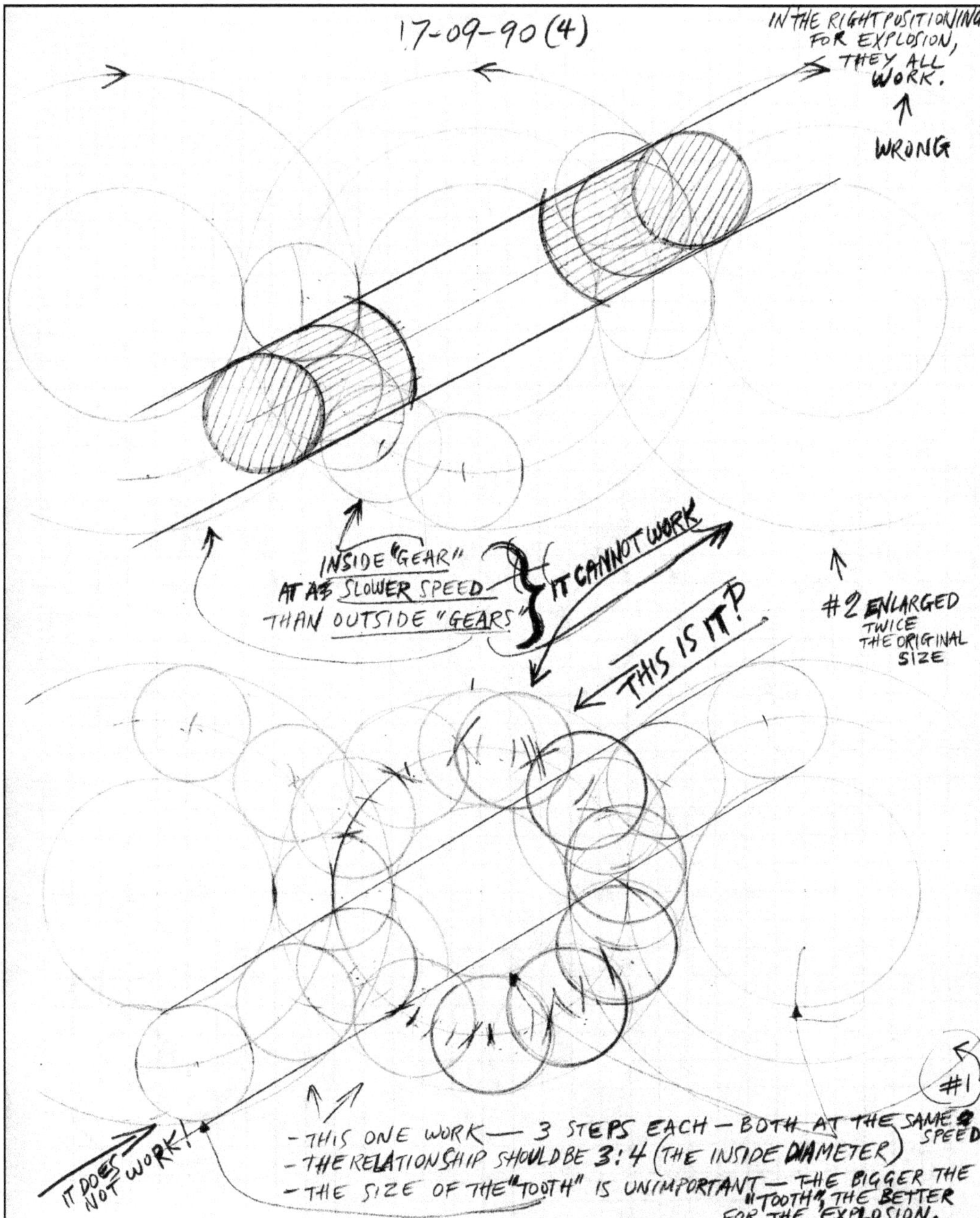
17-09-90 (4)
IN THE RIGHT POSITIONING FOR EXPLOSION, THEY ALL WORK.
WRONG
INSIDE "GEAR" AT A SLOWER SPEED THAN OUTSIDE "GEARS"
IT CANNOT WORK
THIS IS IT?
#2 ENLARGED TWICE THE ORIGINAL SIZE
#1
IT DOES NOT WORK!
- THIS ONE WORK — 3 STEPS EACH — BOTH AT THE SAME SPEED.
- THE RELATIONSHIP SHOULD BE 3:4 (THE INSIDE DIAMETER)
- THE SIZE OF THE "TOOTH" IS UNIMPORTANT — THE BIGGER THE "TOOTH", THE BETTER FOR THE EXPLOSION.

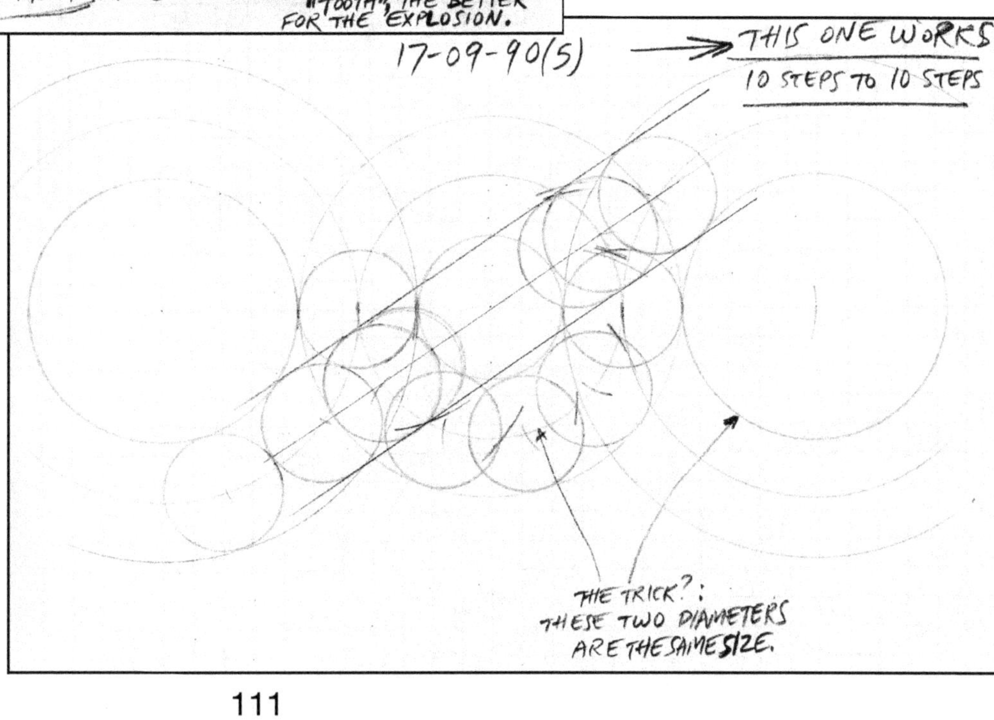
17-09-90 (5)
THIS ONE WORKS!
10 STEPS TO 10 STEPS
THE TRICK?: THESE TWO DIAMETERS ARE THE SAME SIZE.

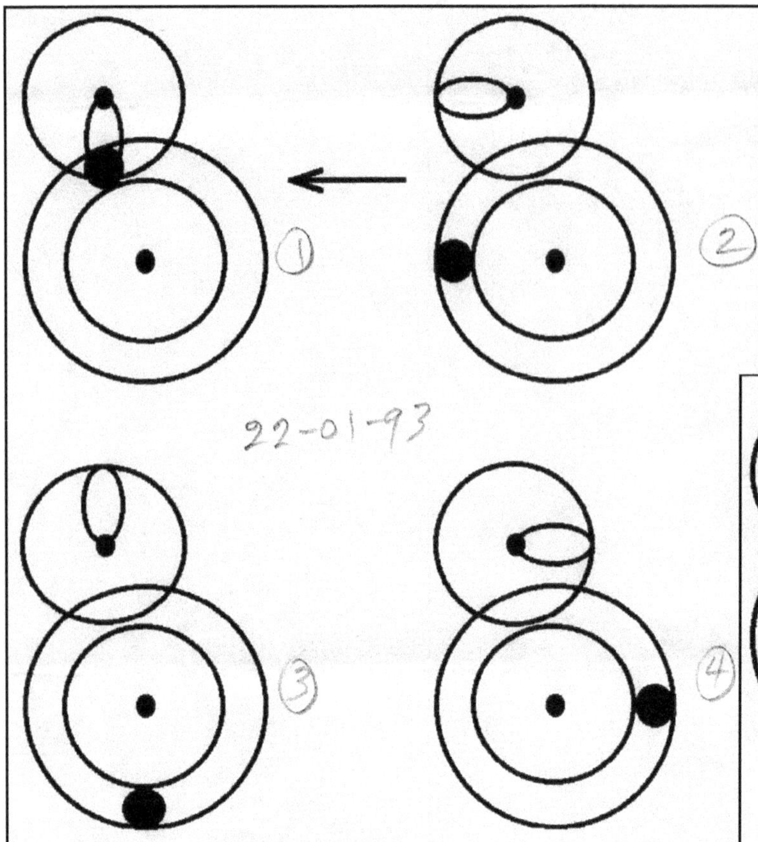
1
2
22-01-93
3
4

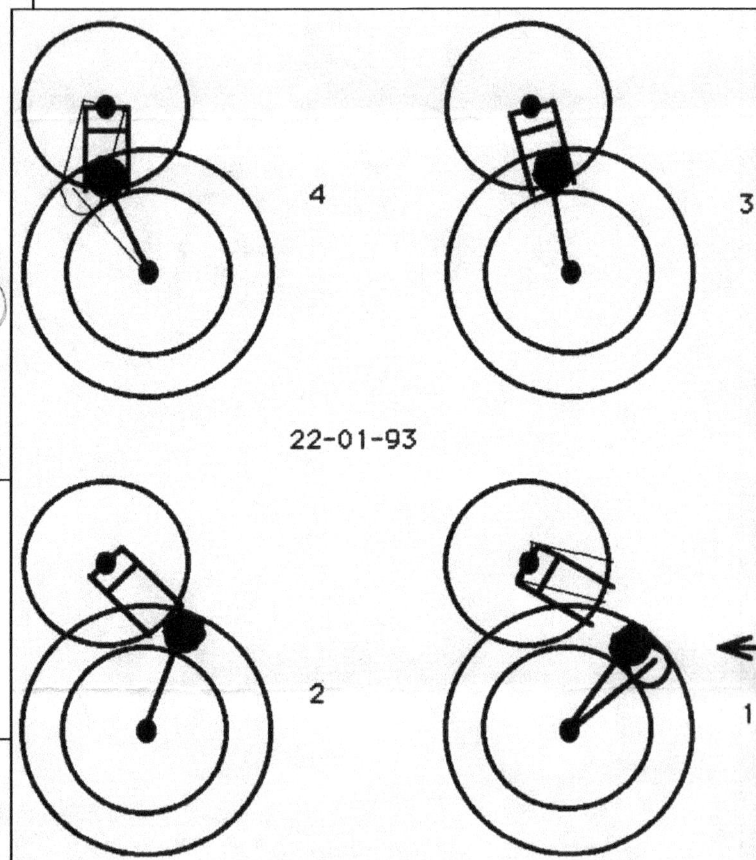
4
3
22-01-93
2
1

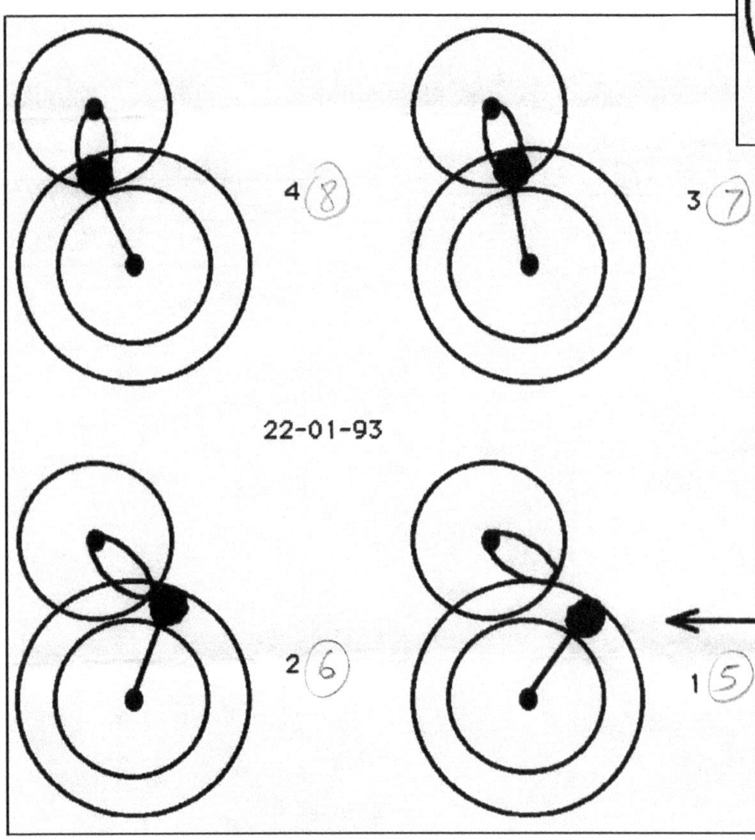
4 8
3 7
22-01-93
2 6
1 5

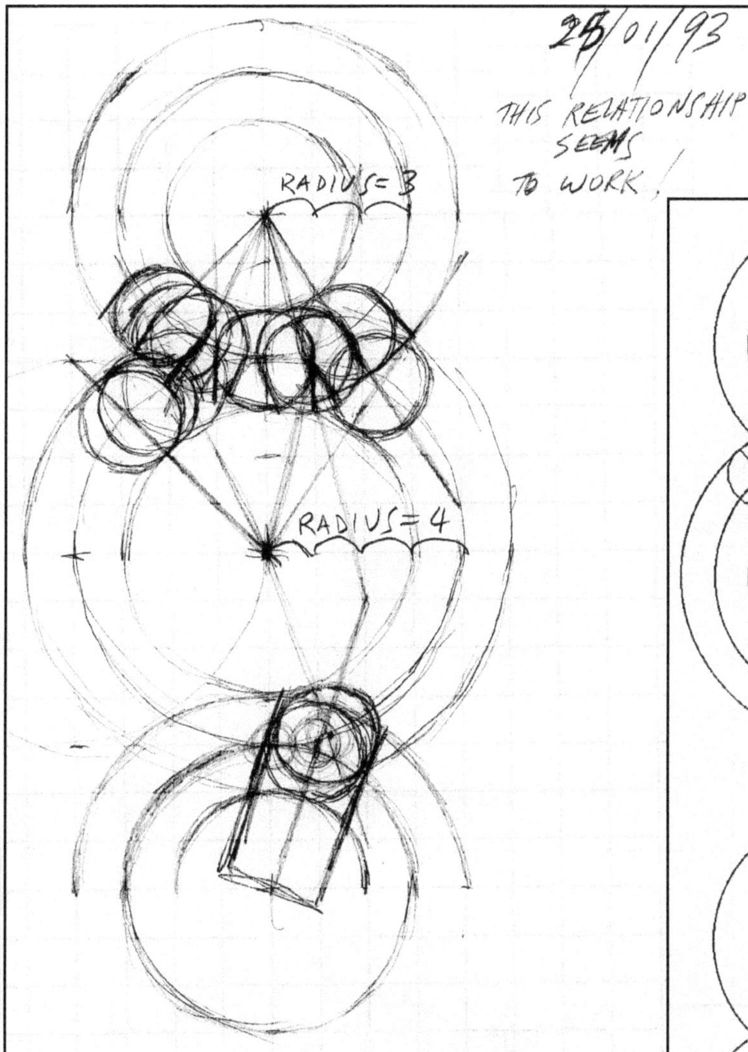
25/01/93
THIS RELATIONSHIP
SEEMS
TO WORK!
RADIUS=3
RADIUS=4

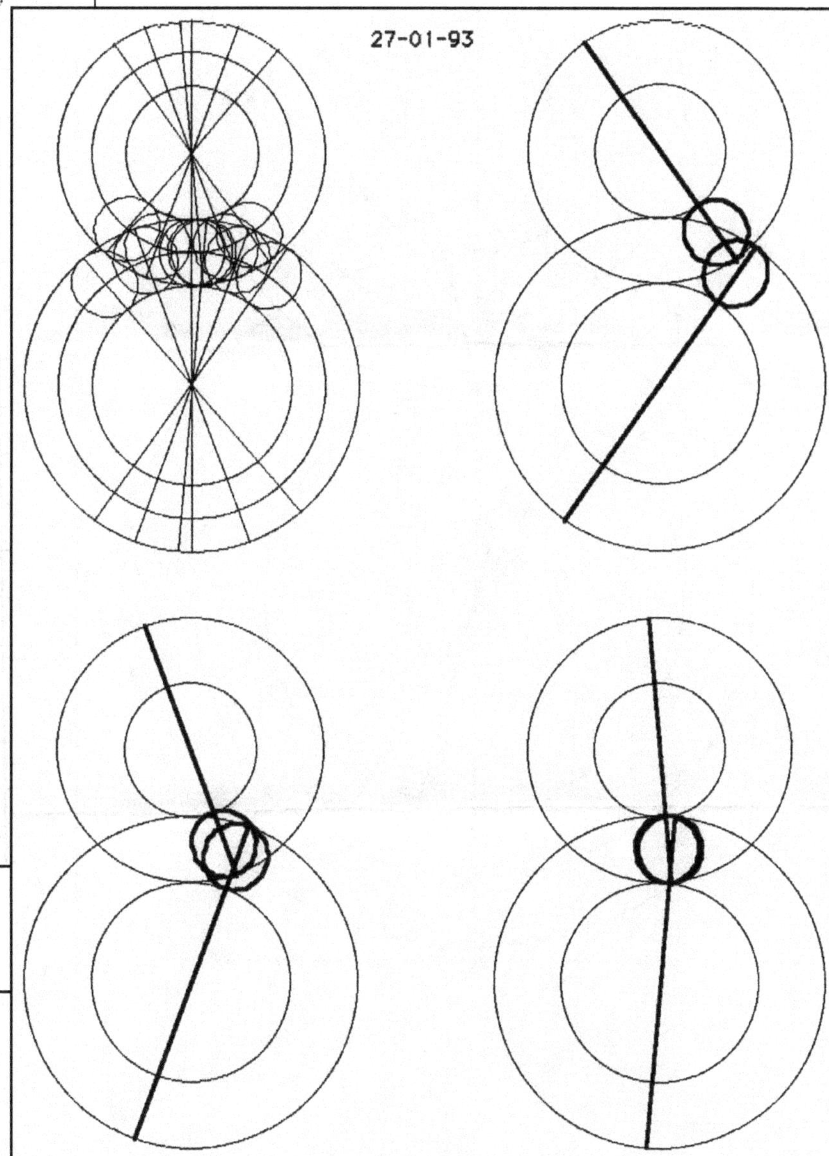
27-01-93

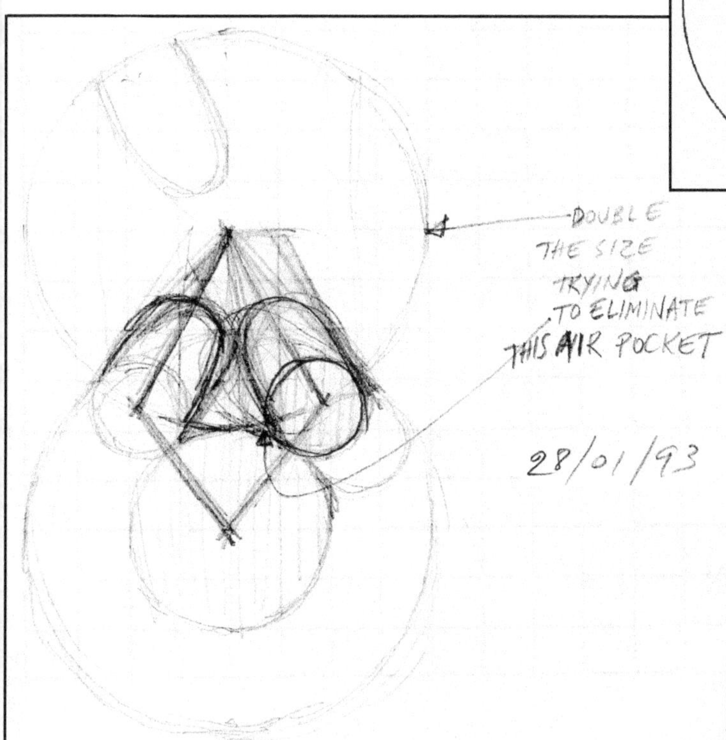
DOUBLE
THE SIZE
TRYING
TO ELIMINATE
THIS AIR POCKET
28/01/93

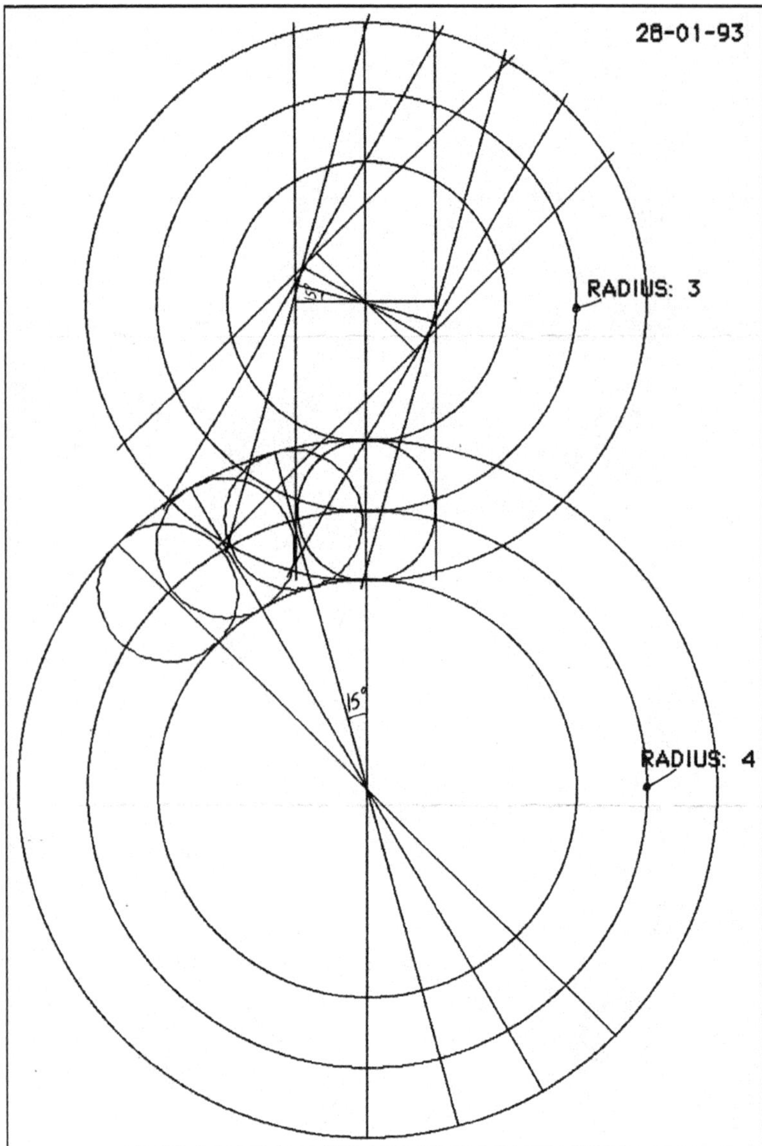
28-01-93
RADIUS: 3
15°
RADIUS: 4

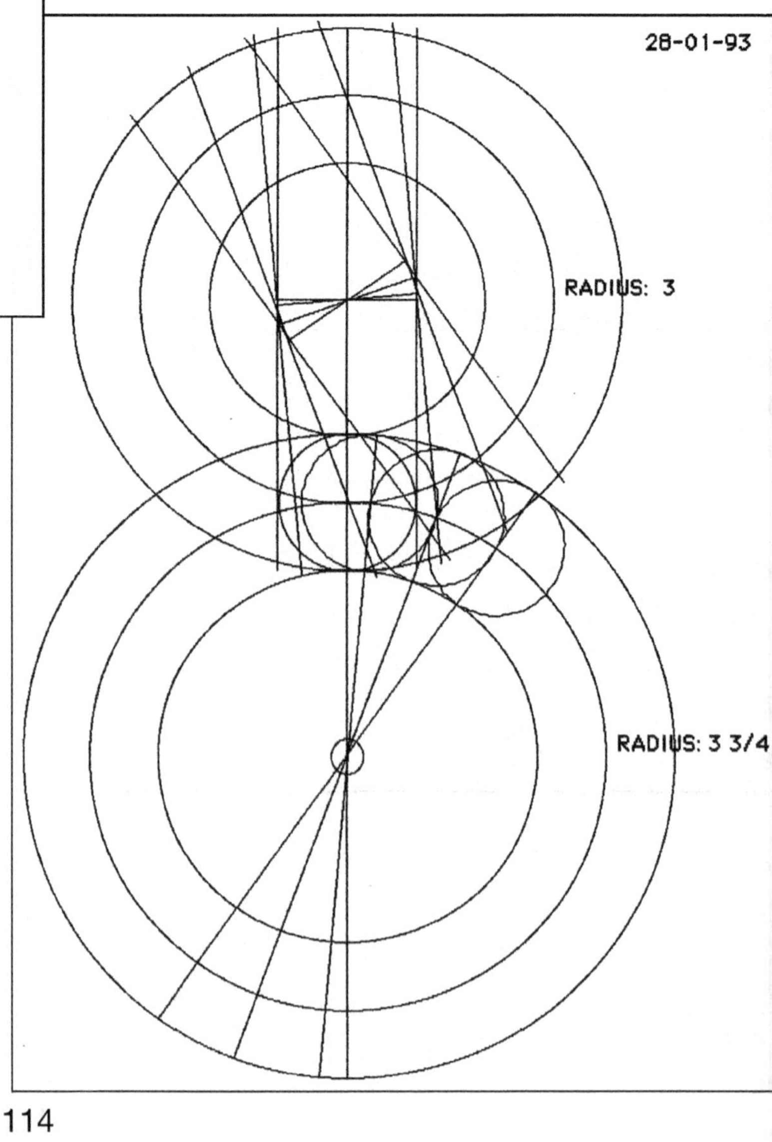
28-01-93
RADIUS: 3
RADIUS: 3 3/4

6A. VEHICLE

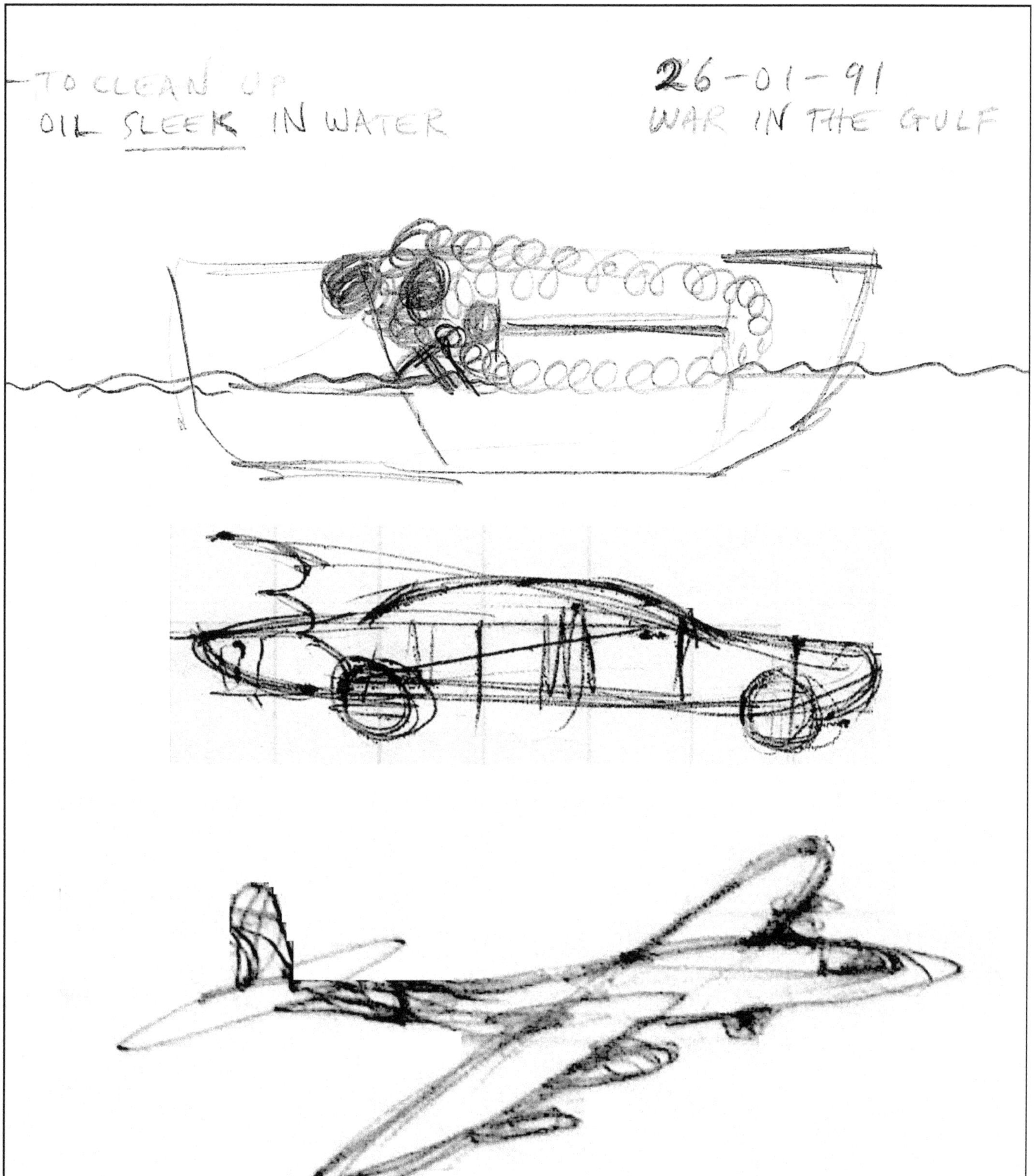

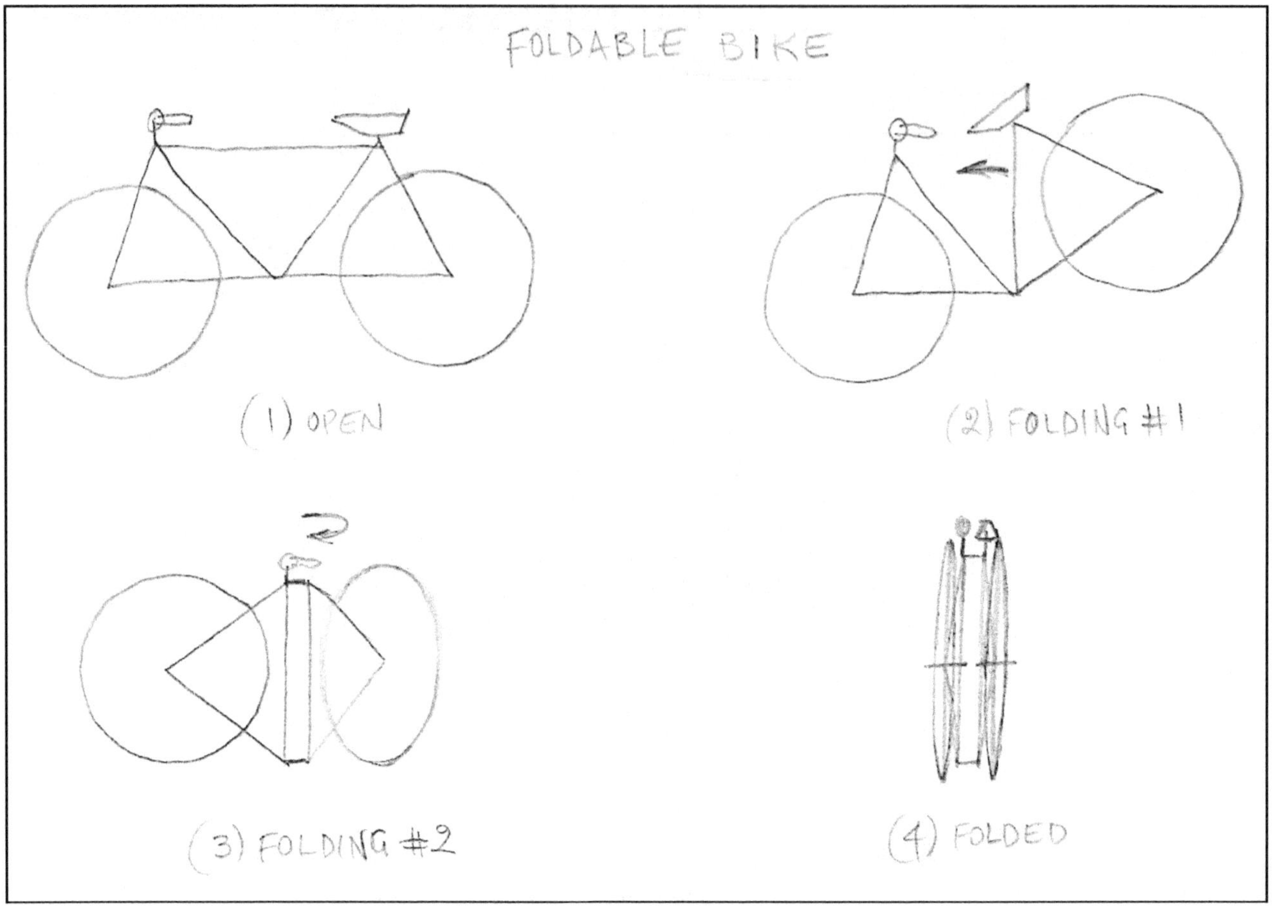
FOLDABLE BIKE
(1) OPEN
(2) FOLDING #1
(3) FOLDING #2
(4) FOLDED

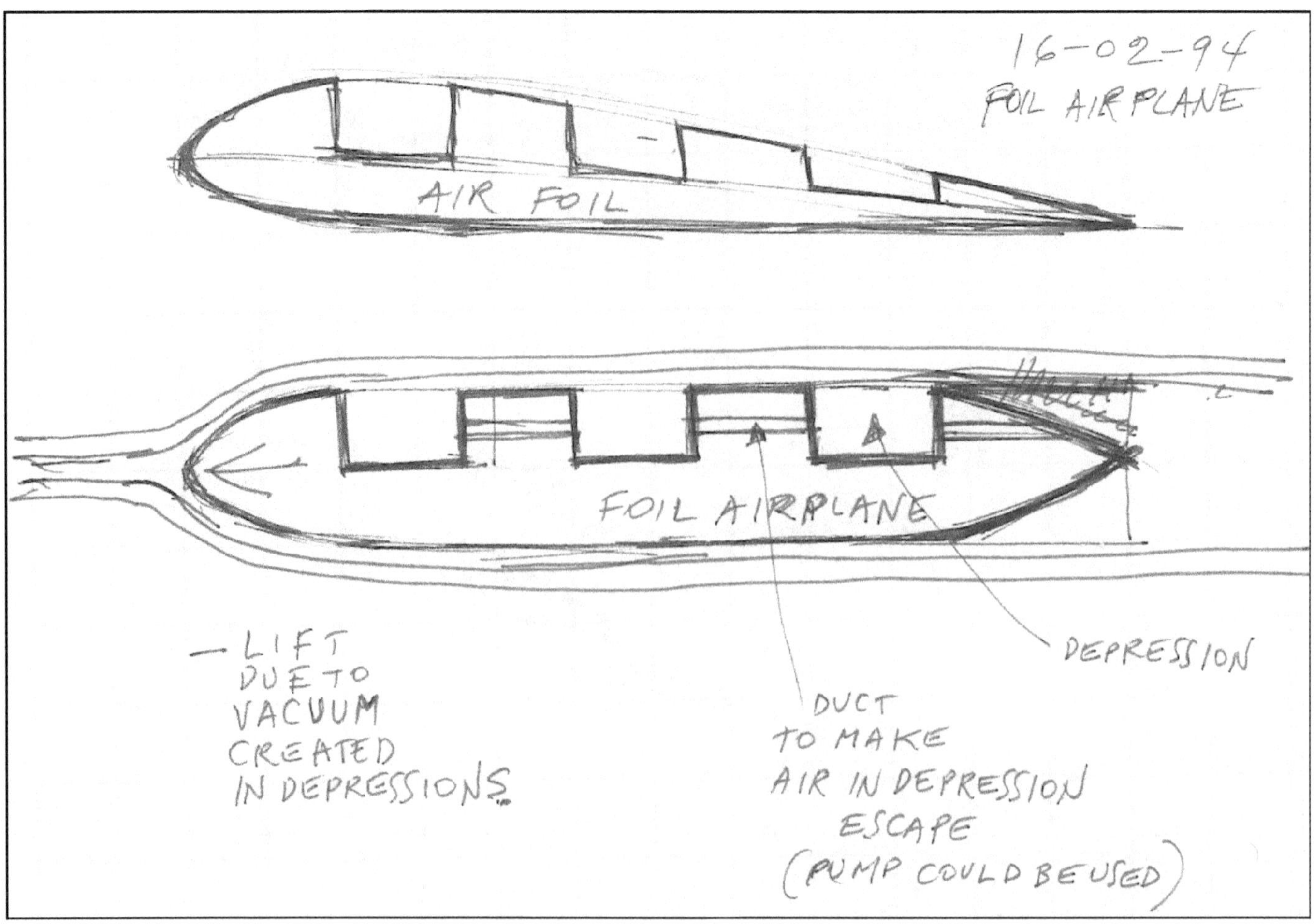
16-02-94
FOIL AIRPLANE
AIR FOIL
FOIL AIRPLANE
DEPRESSION
– LIFT DUE TO VACUUM CREATED IN DEPRESSIONS
DUCT TO MAKE AIR IN DEPRESSION ESCAPE (PUMP COULD BE USED)

14-02-94

BERNOULLI'S PRINCIPLE

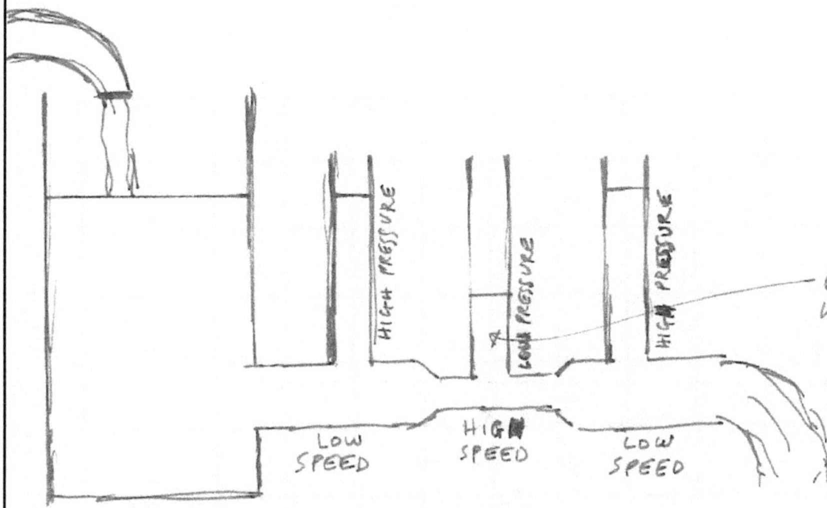

WHAT IF THIS TUBE WERE PROPORTIONALLY-SIZED, THAT IS, SMALLER, LIKE THE HIGH SPEED TUBE? WOULD, THEN, PRESSURE BE COMPARATIVELY/PROPORTIONALLY JUST AS HIGH AS THE OTHERS? COULD THIS, BERNOULLI'S EXPERIMENT, BE A WRONG EXPERIMENT?

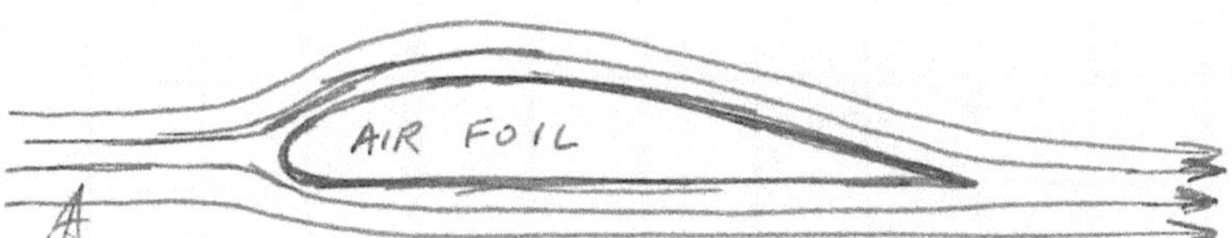

- TEST CONDUCTED ON AIR FOIL BY SHOOTING SMOKE AT IT IS WRONG BECAUSE FOIL/WING MOVES INTO IT, NOT VICEVERSA, AIR MOVING TO FOIL/WING.

- IT IS NOT AIR SPEEDING PAST FOIL, BUT FOIL SPEEDING THROUGH PAST AIR.

- LIFT IS DUE TO VACUUM CREATED BY FOIL AS IT MOVES INTO/THROUGH AIR.

- WHY NOT USE WATER CASCADING IN THE SHAPE OF A SHEET ONTO A FOIL?

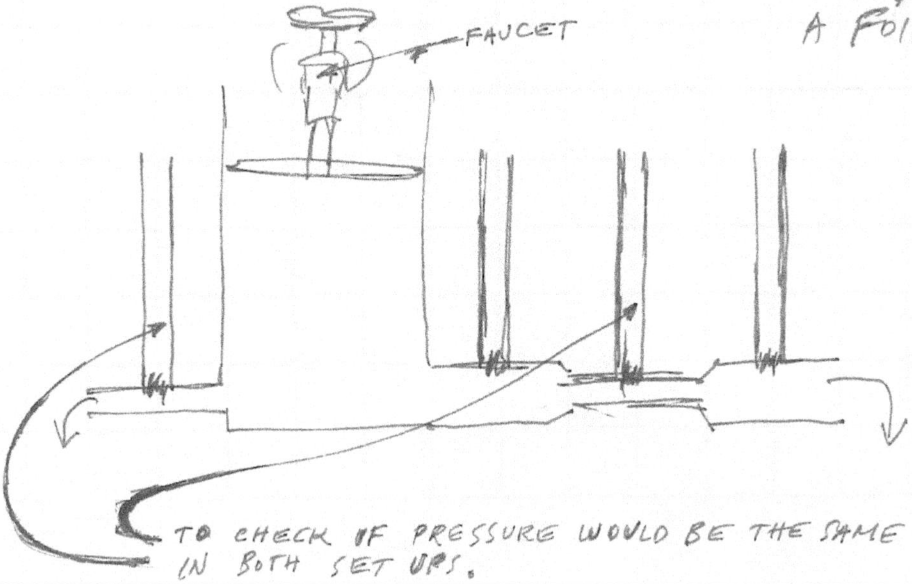

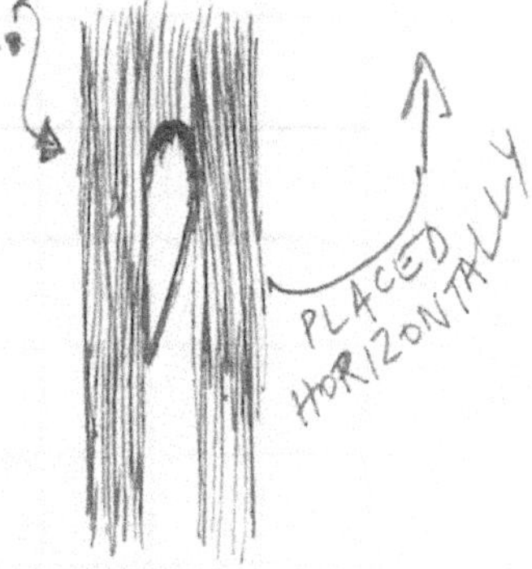

TO CHECK IF PRESSURE WOULD BE THE SAME IN BOTH SET UPS.

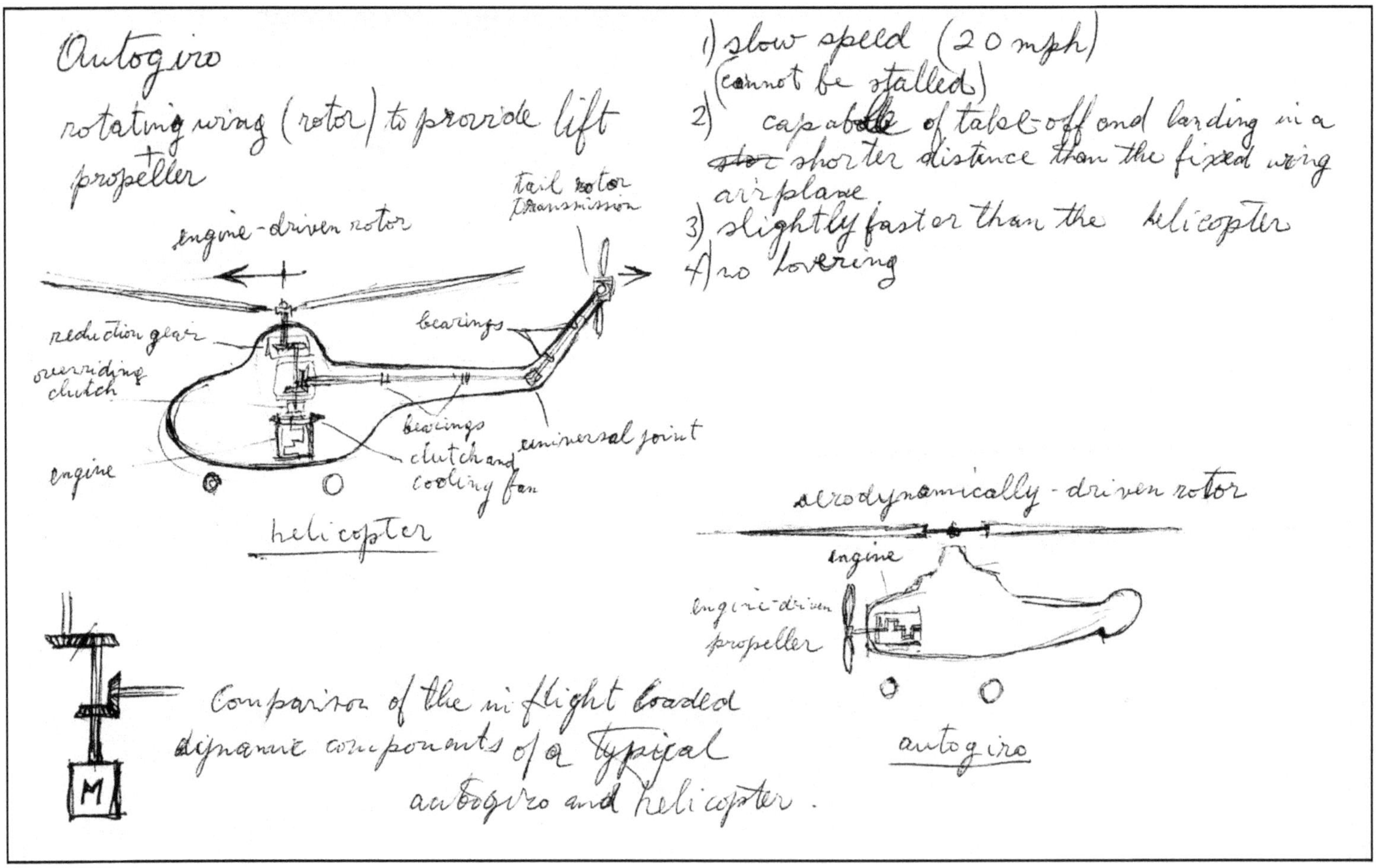
Autogiro
rotating wing (rotor) to provide lift
+
propeller
1) slow speed (20 mph)
(cannot be stalled)
2) capable of take-off and landing in a shorter distance than the fixed wing airplane
3) slightly faster than the helicopter
4) no hovering
engine-driven rotor
tail rotor
transmission
reduction gear
overriding clutch
bearings
bearings
universal joint
engine
clutch and cooling fan
helicopter
aerodynamically-driven rotor
engine
engine-driven propeller
autogiro
M
Comparison of the in flight loaded dynamic components of a typical autogiro and helicopter.

Cycloidal propellers

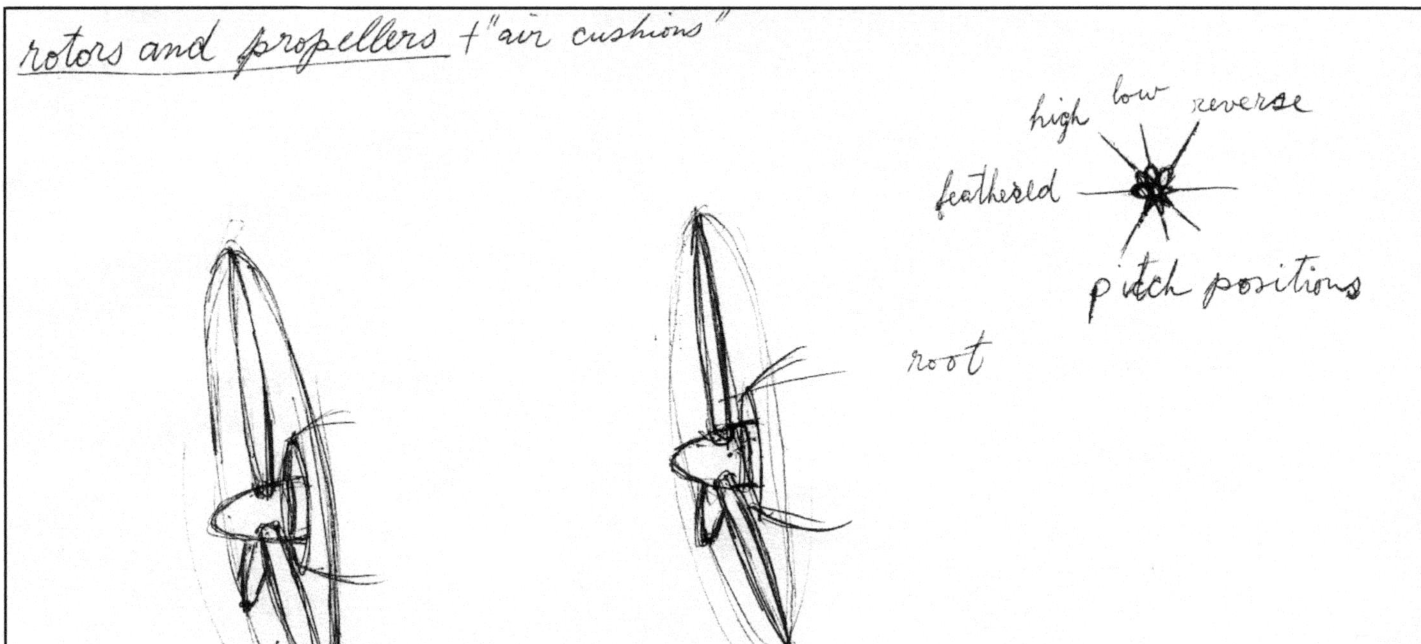
rotors and propellers + "air cushions"
high
low
reverse
feathered
pitch positions
root
Principal parts of aircraft propeller and typical pitch positions.

6B. TRANSMISSION

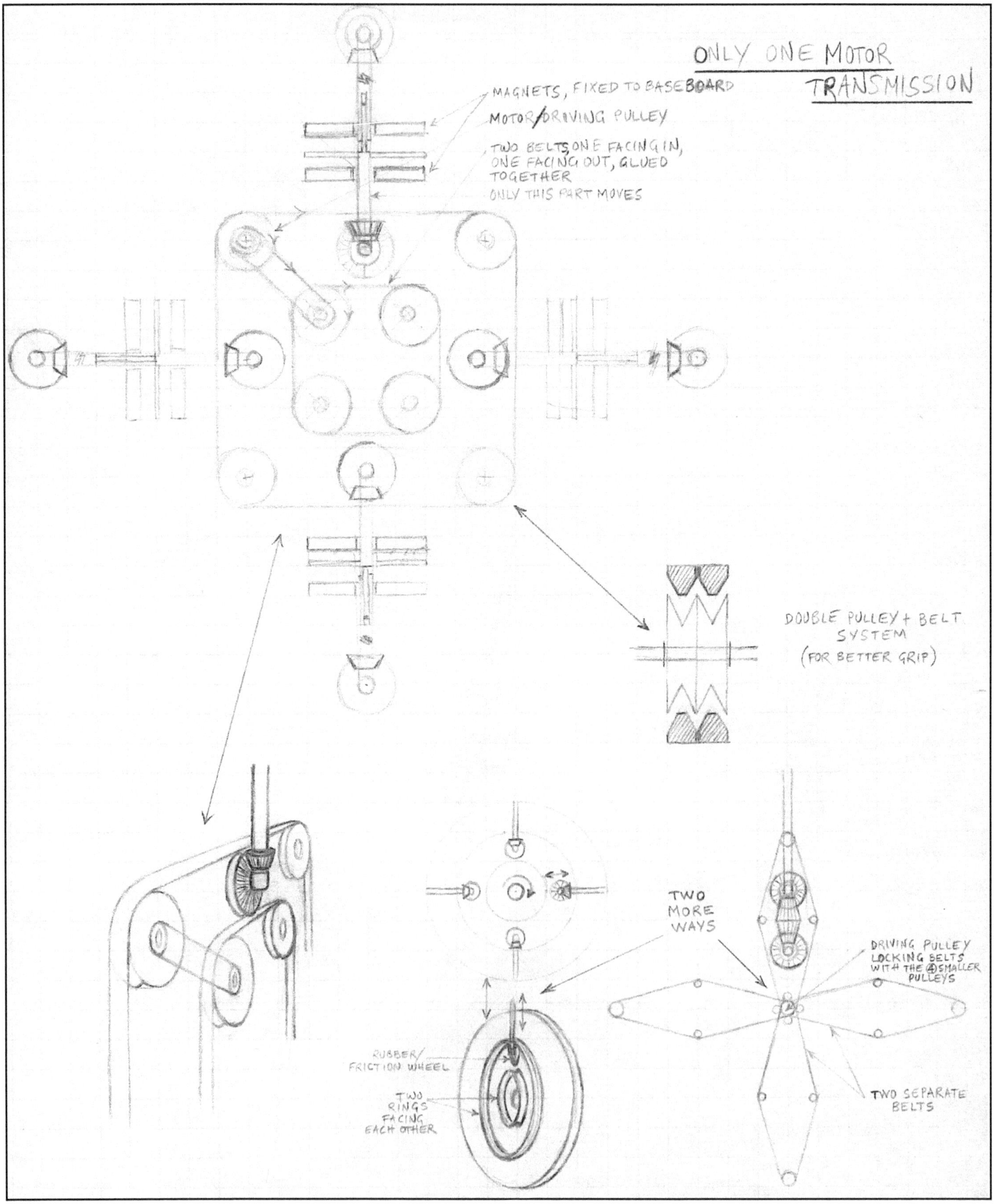

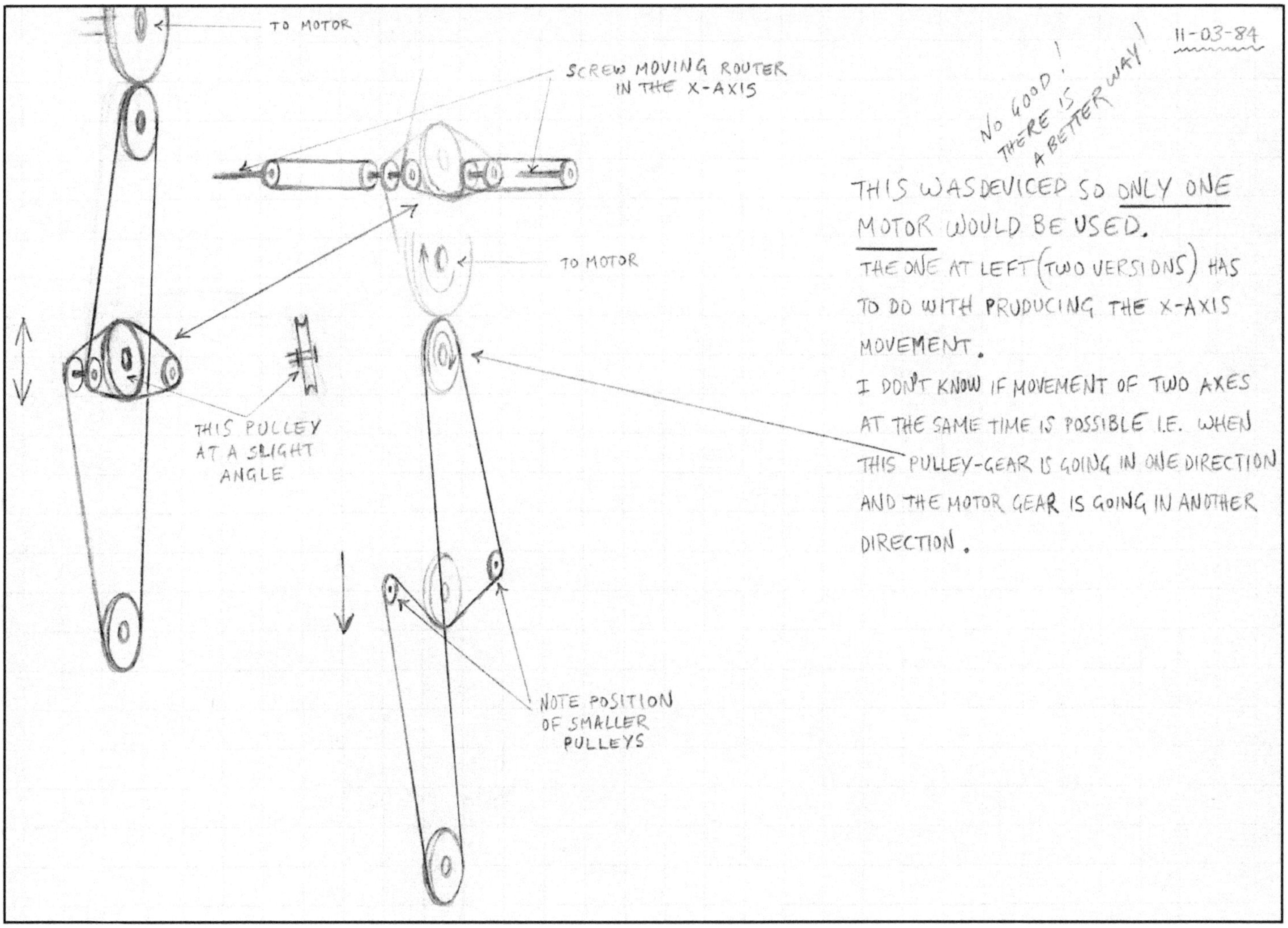
TO MOTOR
SCREW MOVING ROUTER IN THE X-AXIS
11-03-84
NO GOOD! THERE IS A BETTER WAY!
THIS WAS DEVICED SO ONLY ONE MOTOR WOULD BE USED.
THE ONE AT LEFT (TWO VERSIONS) HAS TO DO WITH PRUDUCING THE X-AXIS MOVEMENT.
I DON'T KNOW IF MOVEMENT OF TWO AXES AT THE SAME TIME IS POSSIBLE I.E. WHEN THIS PULLEY-GEAR IS GOING IN ONE DIRECTION AND THE MOTOR GEAR IS GOING IN ANOTHER DIRECTION.
TO MOTOR
THIS PULLEY AT A SLIGHT ANGLE
NOTE POSITION OF SMALLER PULLEYS

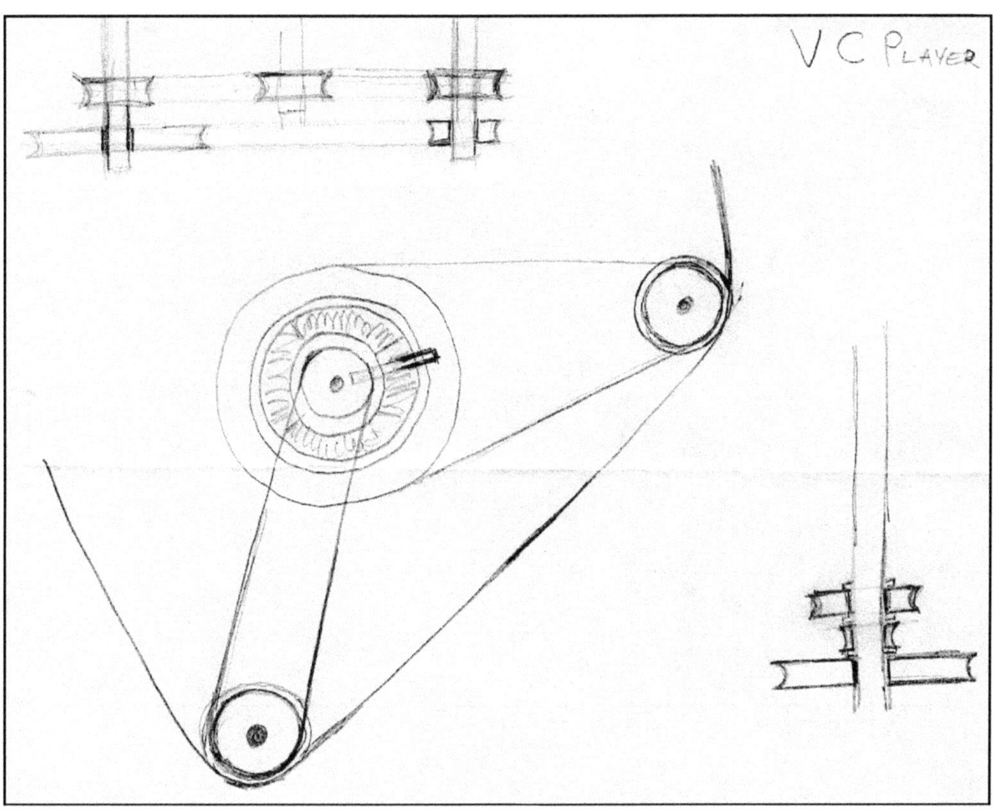
VC PLAYER

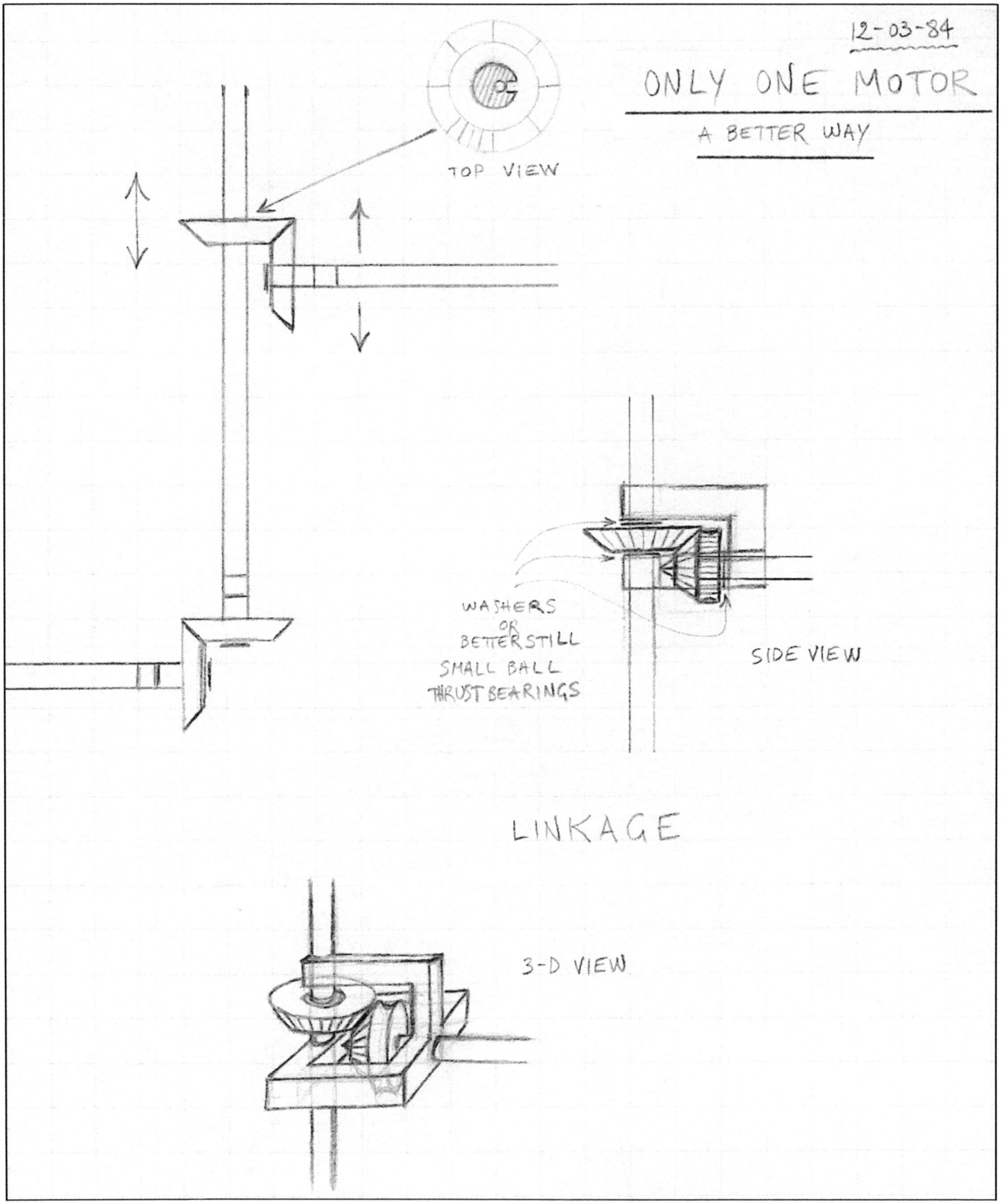
12-03-84
ONLY ONE MOTOR
A BETTER WAY
TOP VIEW
WASHERS
OR
BETTER STILL
SMALL BALL
THRUST BEARINGS
SIDE VIEW
LINKAGE
3-D VIEW

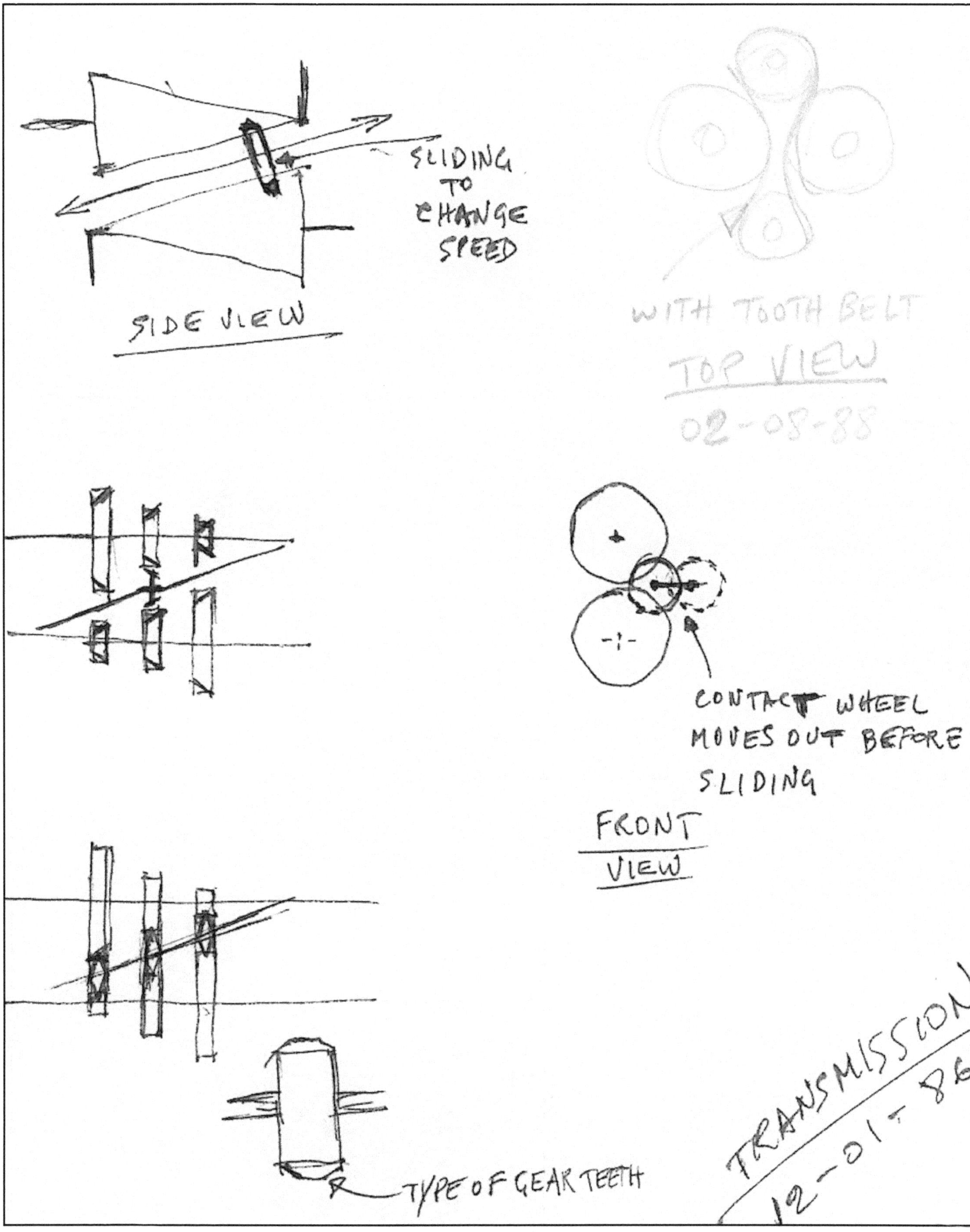
SLIDING TO CHANGE SPEED
SIDE VIEW
WITH TOOTH BELT
TOP VIEW
02-05-88
CONTACT WHEEL MOVES OUT BEFORE SLIDING
FRONT VIEW
TYPE OF GEAR TEETH
TRANSMISSION
12-01-86

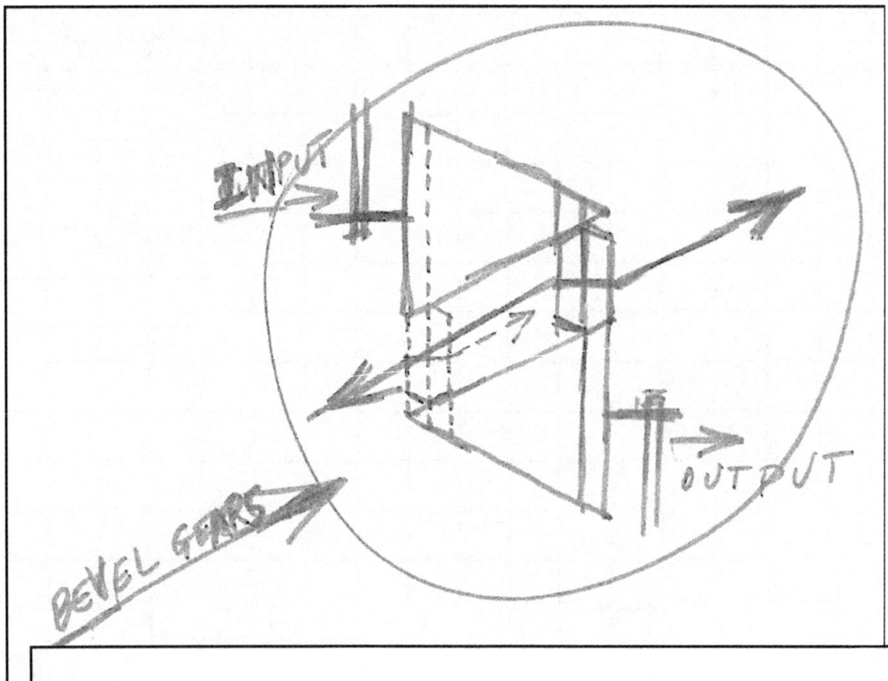
INPUT
OUTPUT
BEVEL GEARS

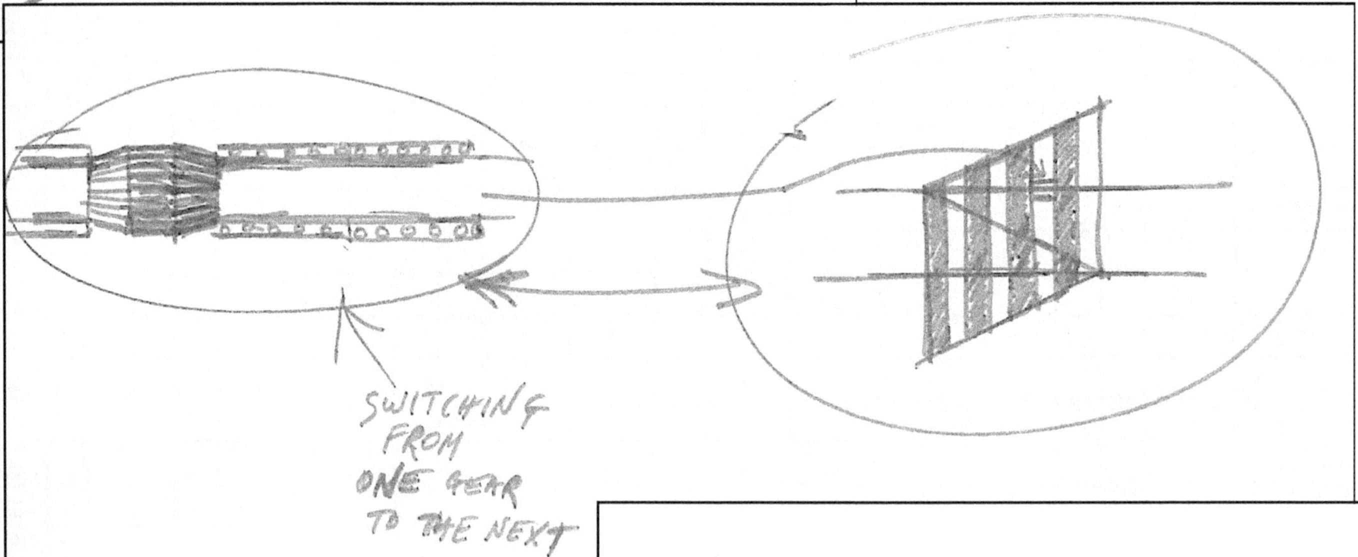
SWITCHING
FROM
ONE GEAR
TO THE NEXT

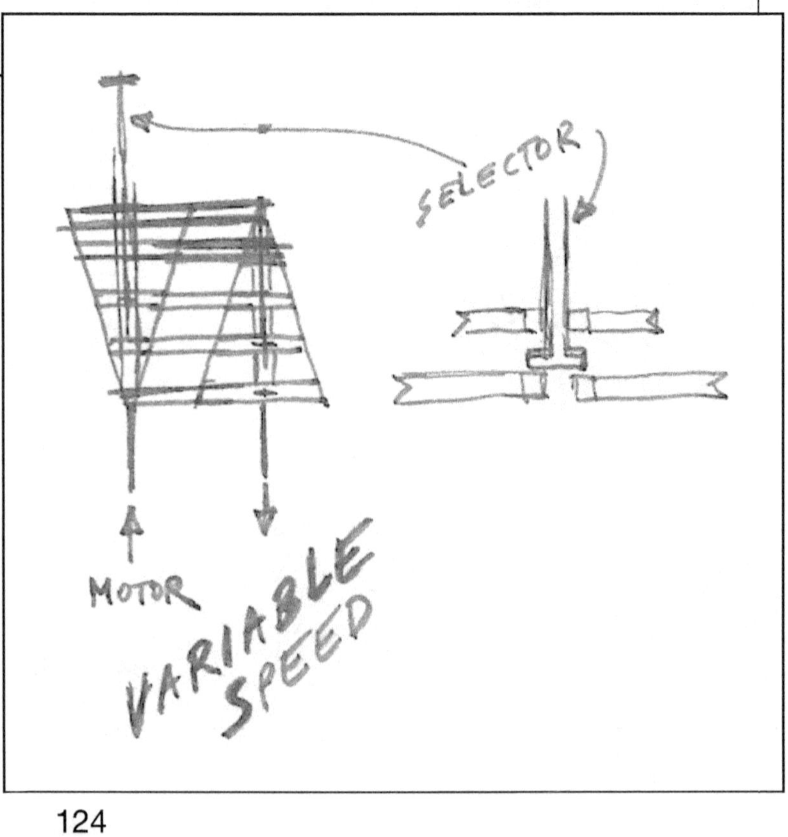
SELECTOR
MOTOR
VARIABLE
SPEED

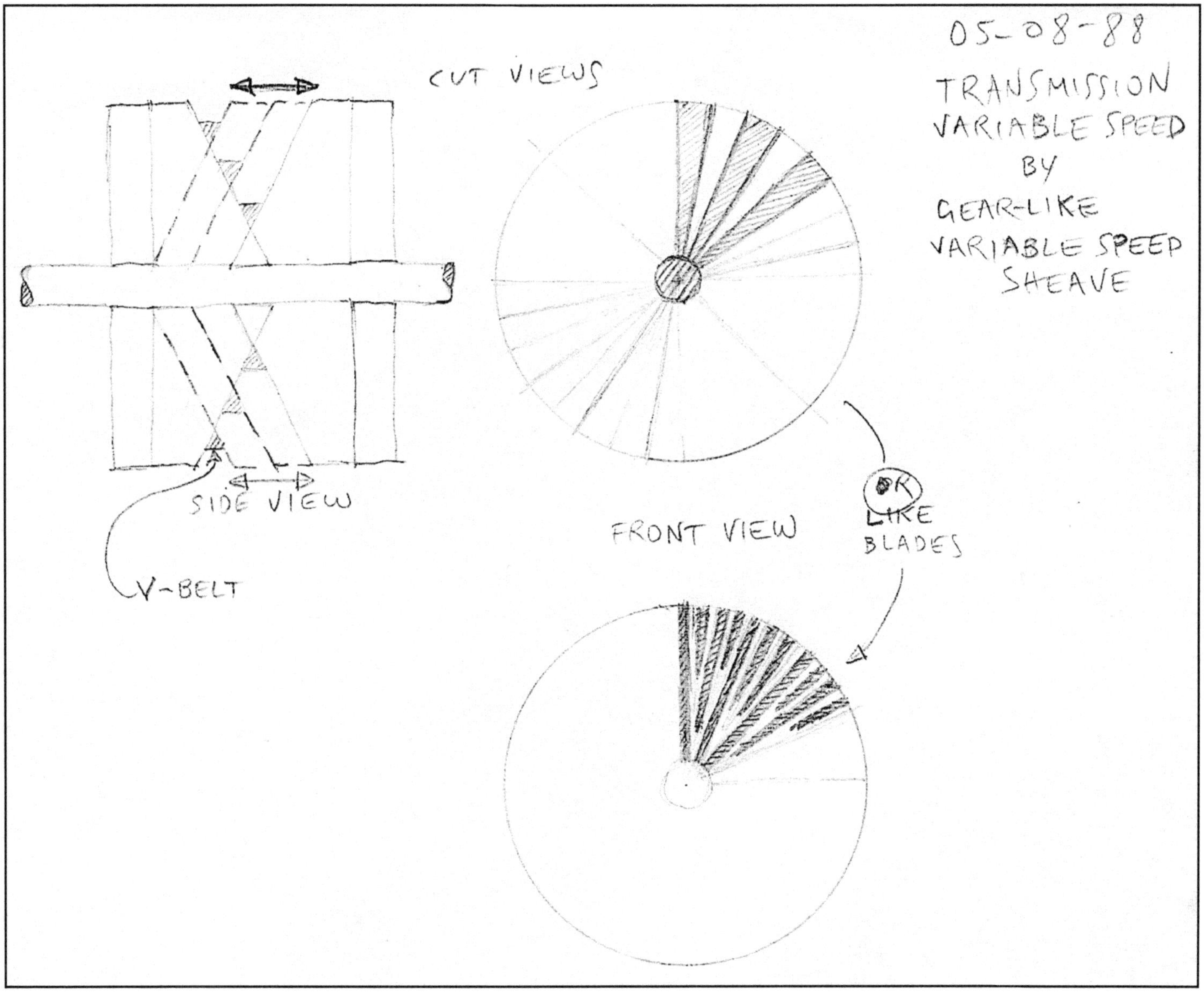
05-08-88
TRANSMISSION
VARIABLE SPEED
BY
GEAR-LIKE
VARIABLE SPEED
SHEAVE
CUT VIEWS
SIDE VIEW
V-BELT
FRONT VIEW
OK
LIKE
BLADES

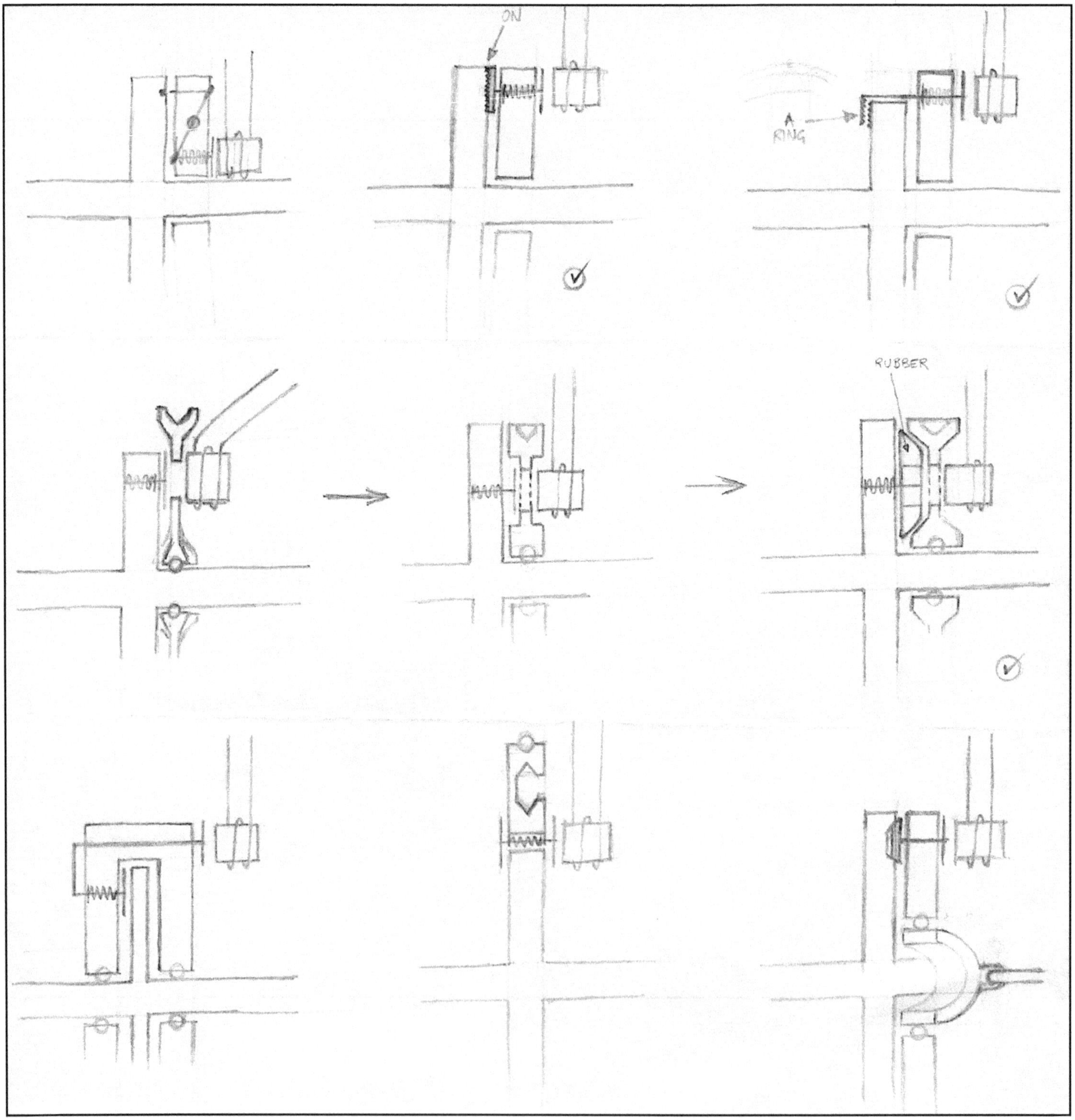
ON
A RING
RUBBER

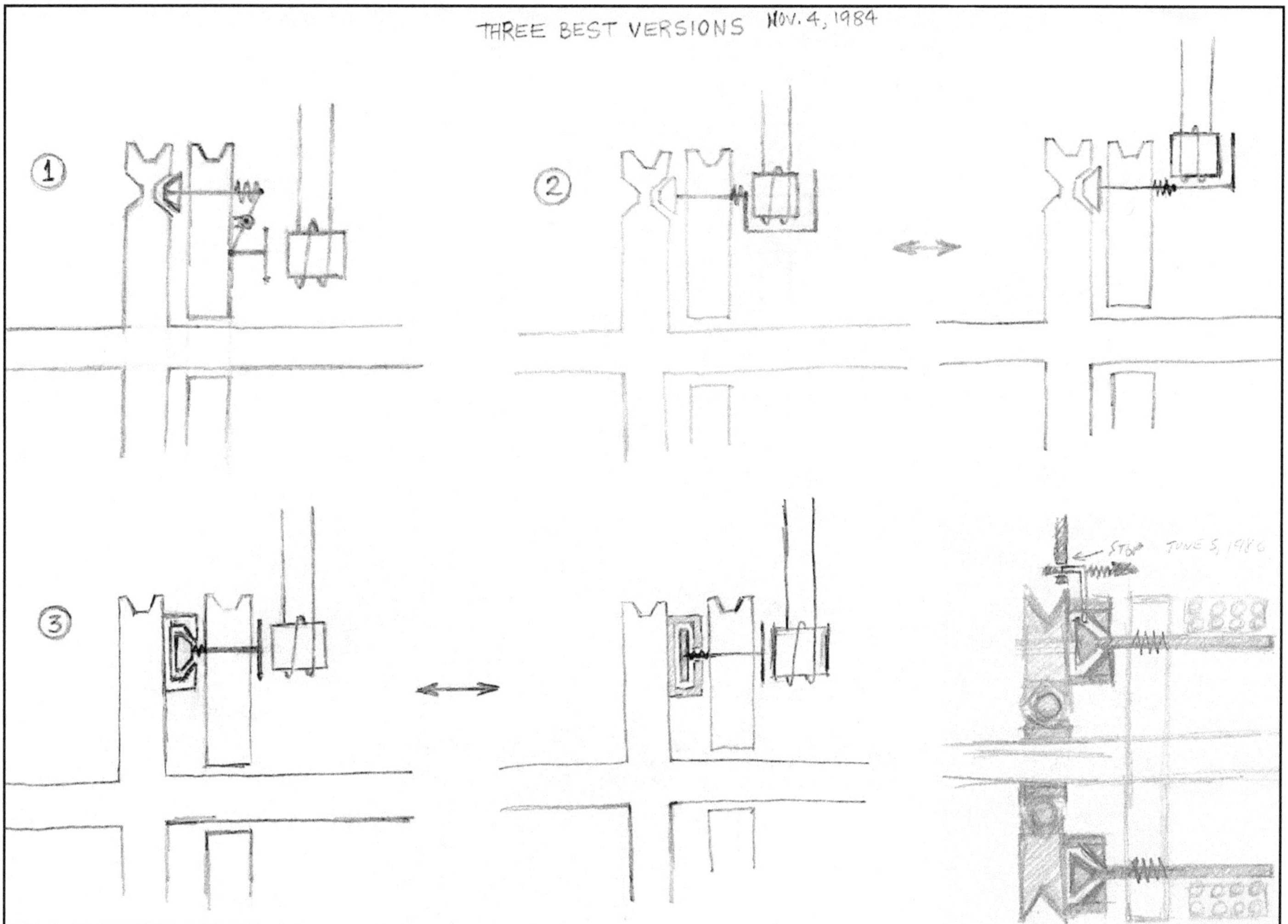
THREE BEST VERSIONS NOV. 4, 1984
1
2
3
STOP
JUNE 5, 1986

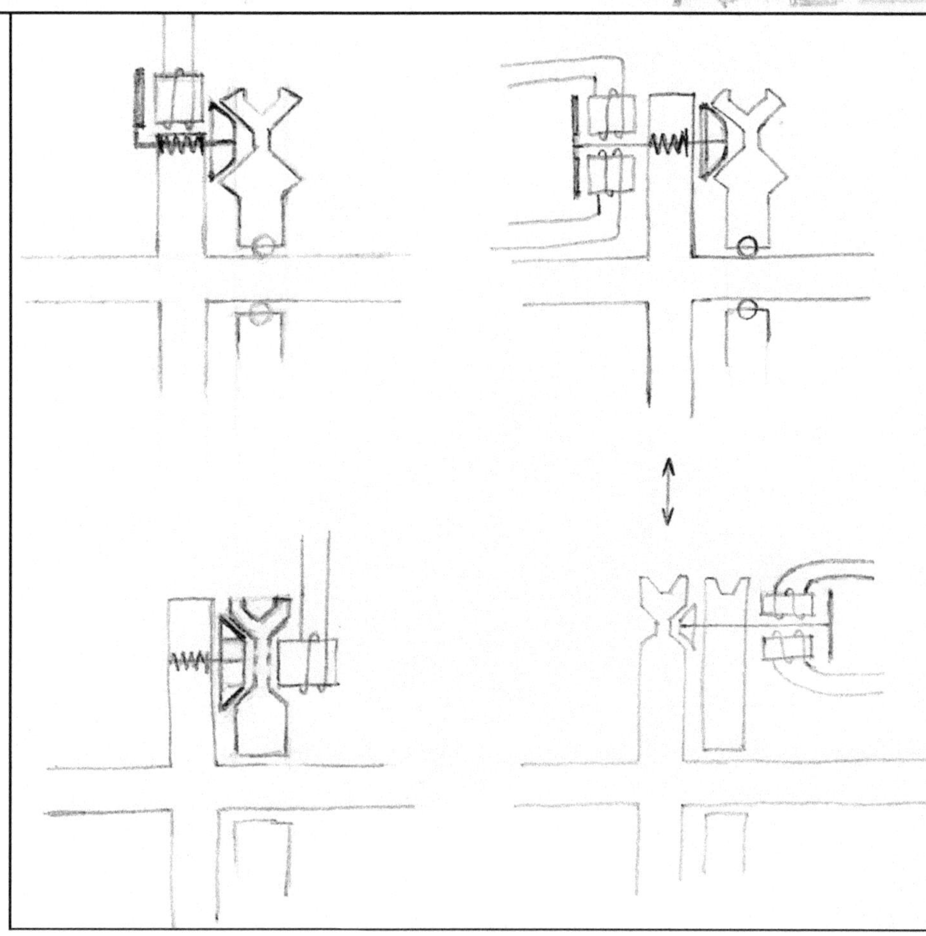

6C. IFO (Identified Flying Object)

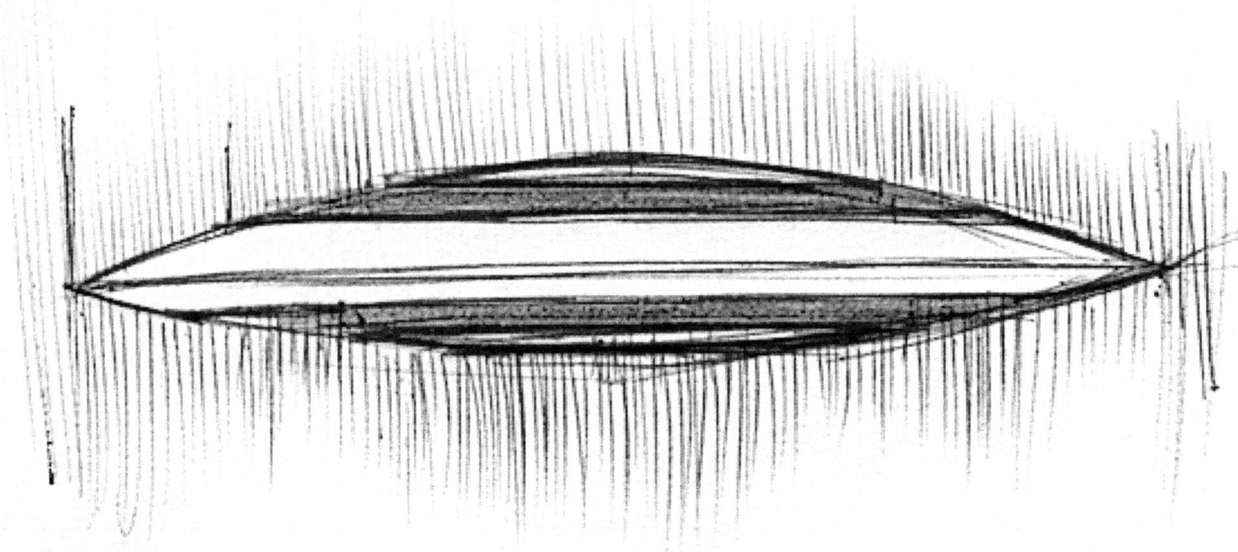

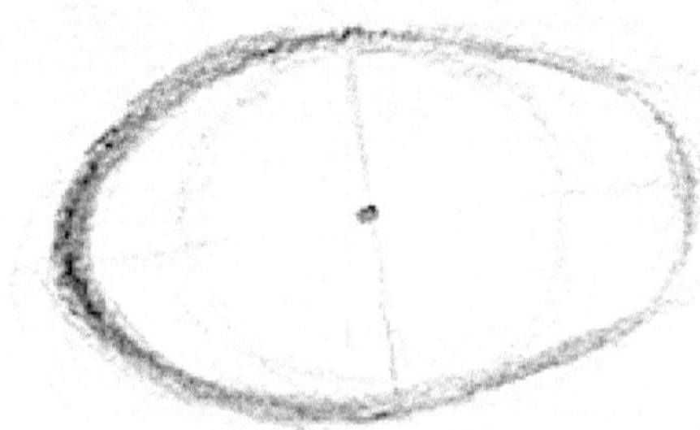

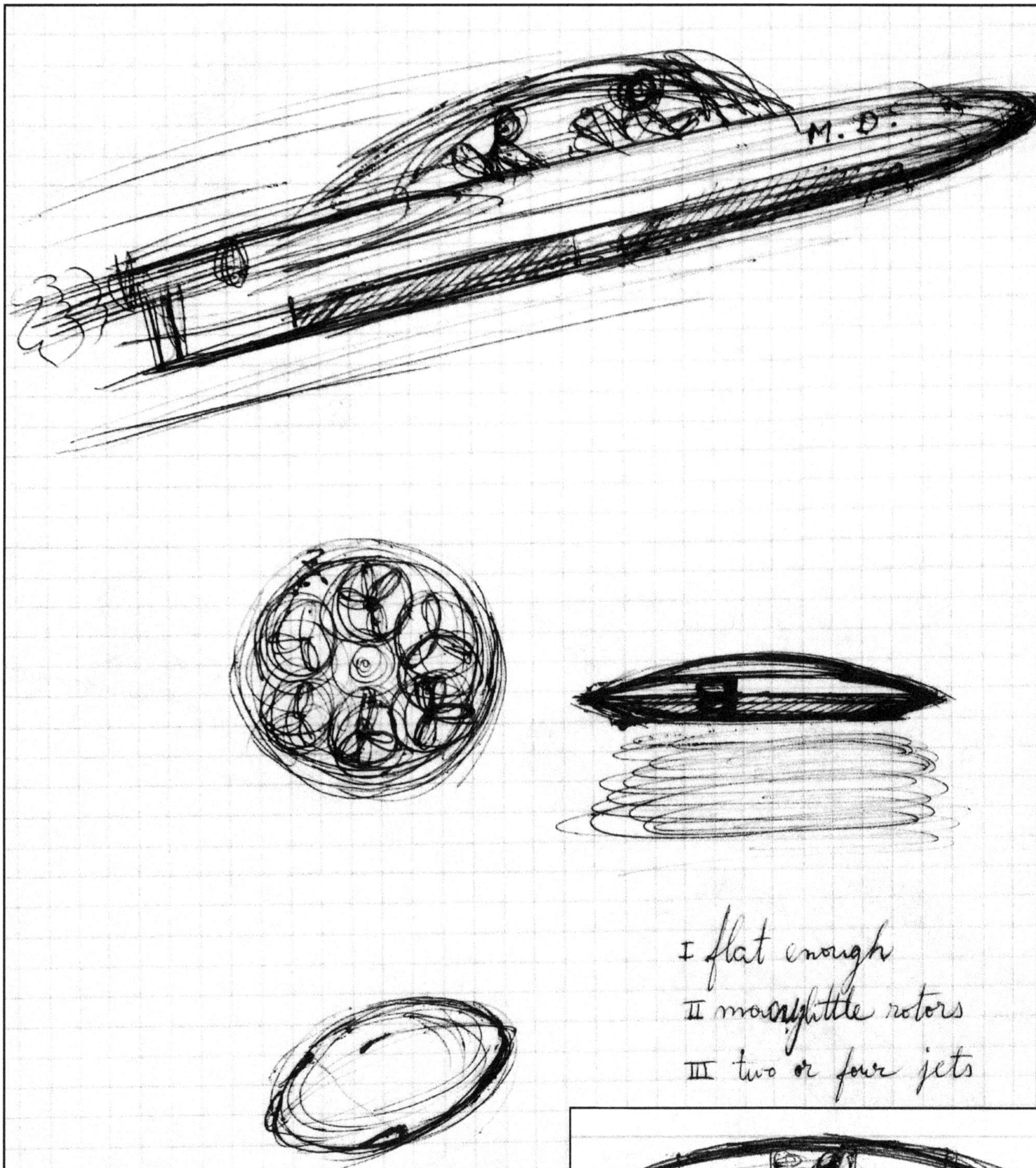
M.D.
I flat enough
II many little rotors
III two or four jets

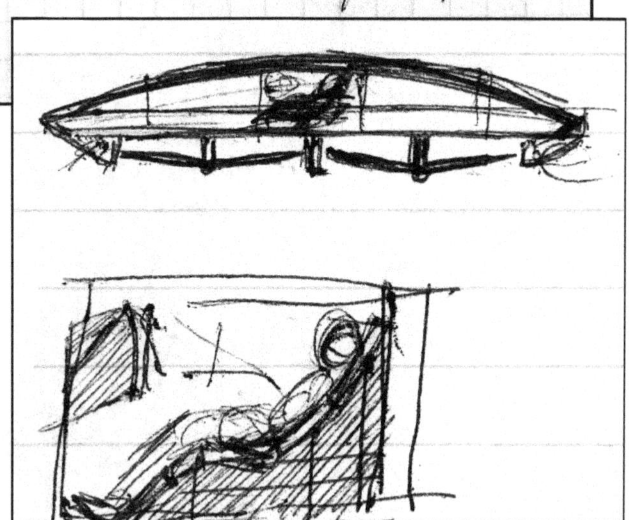

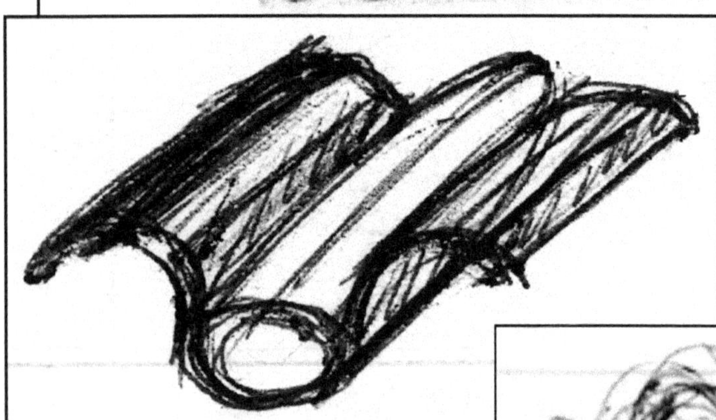

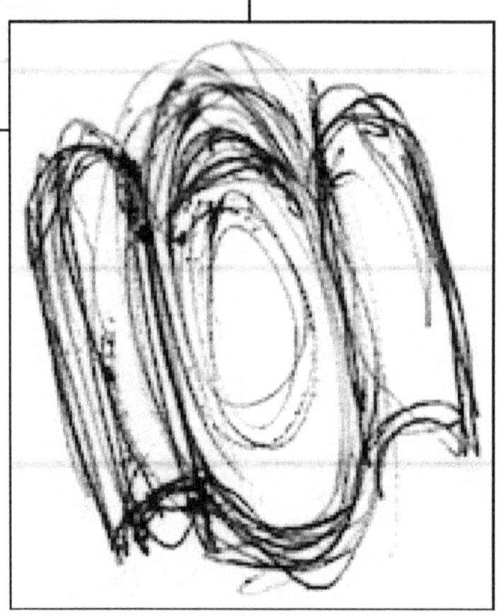

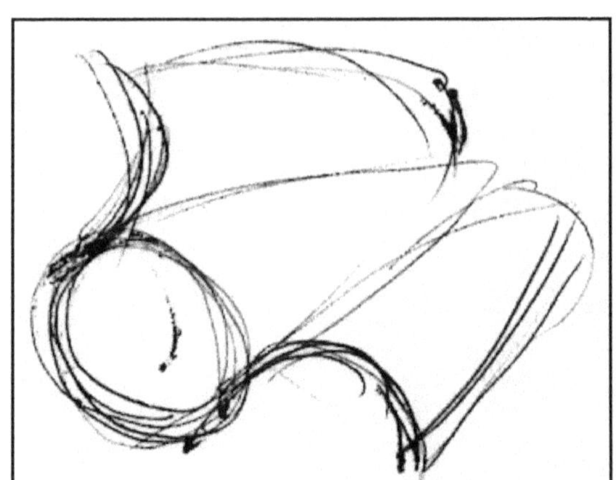

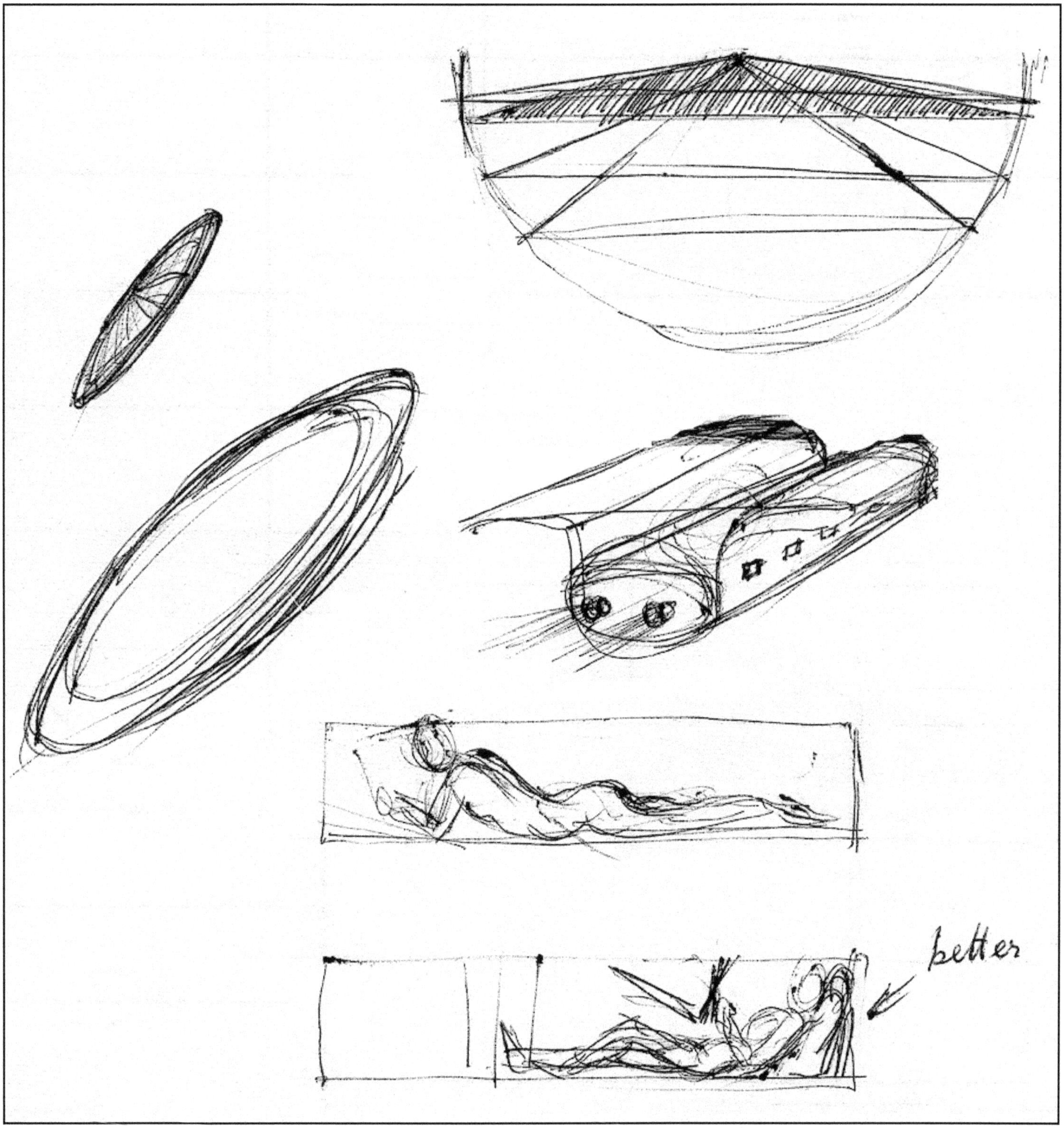
better

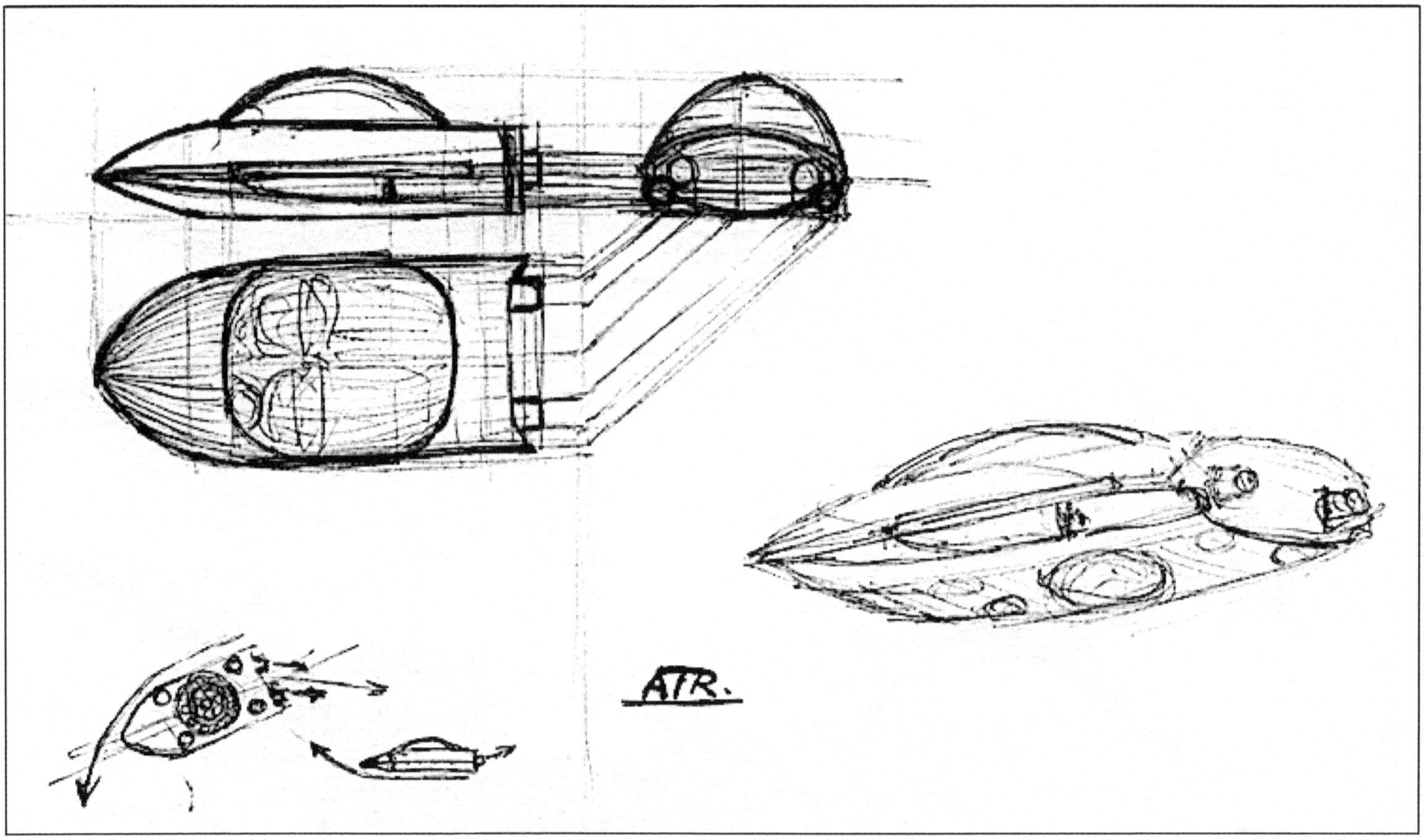
ATR.

ATR.

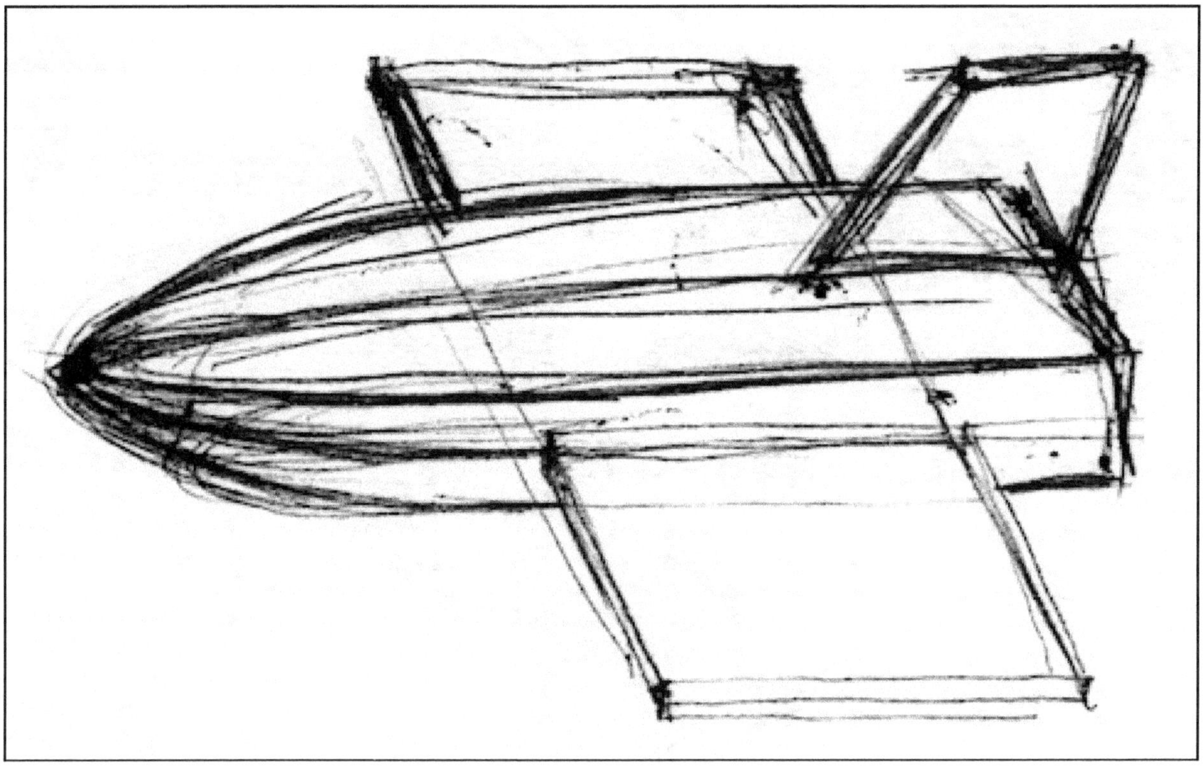

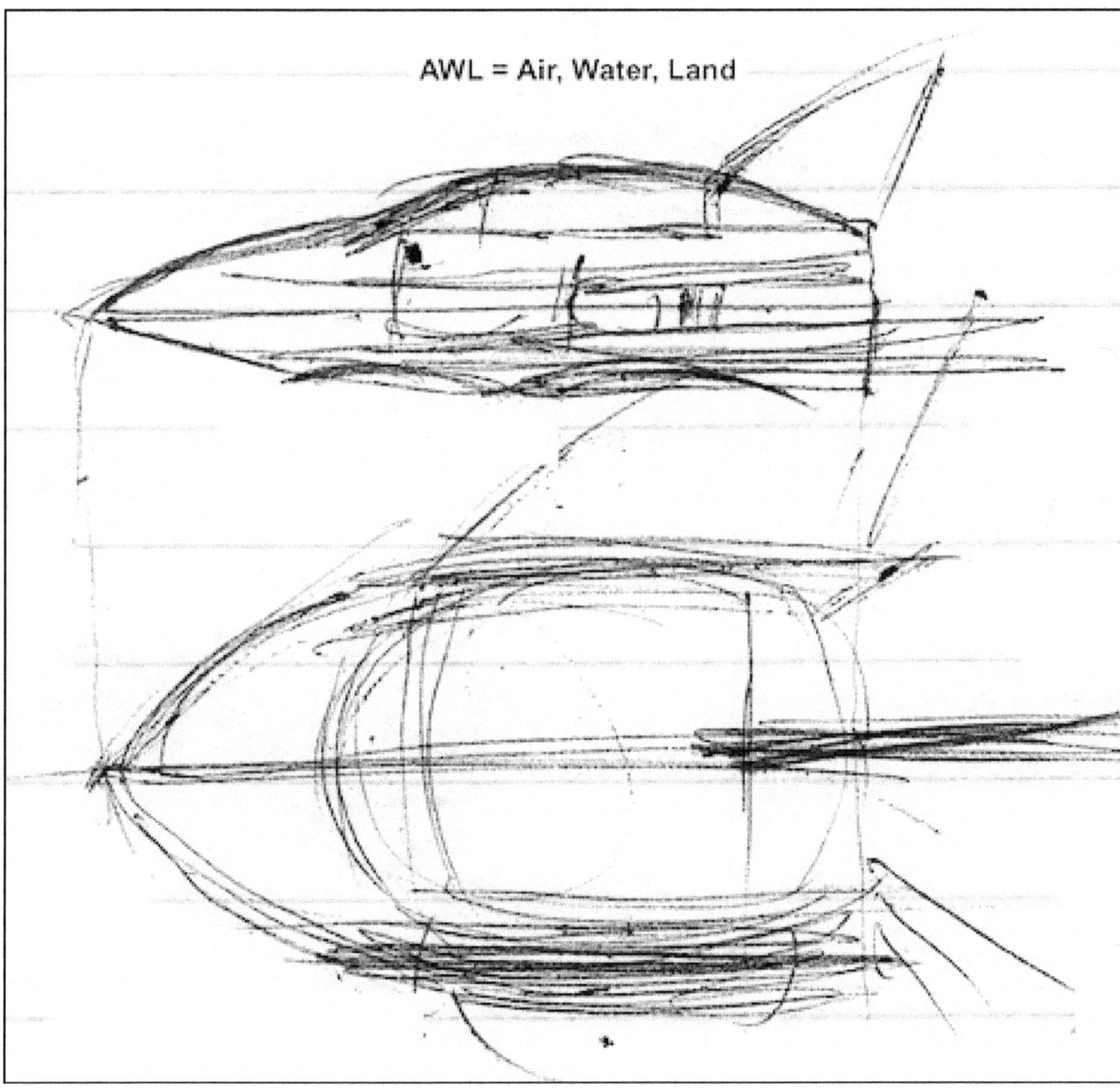
AWL = Air, Water, Land

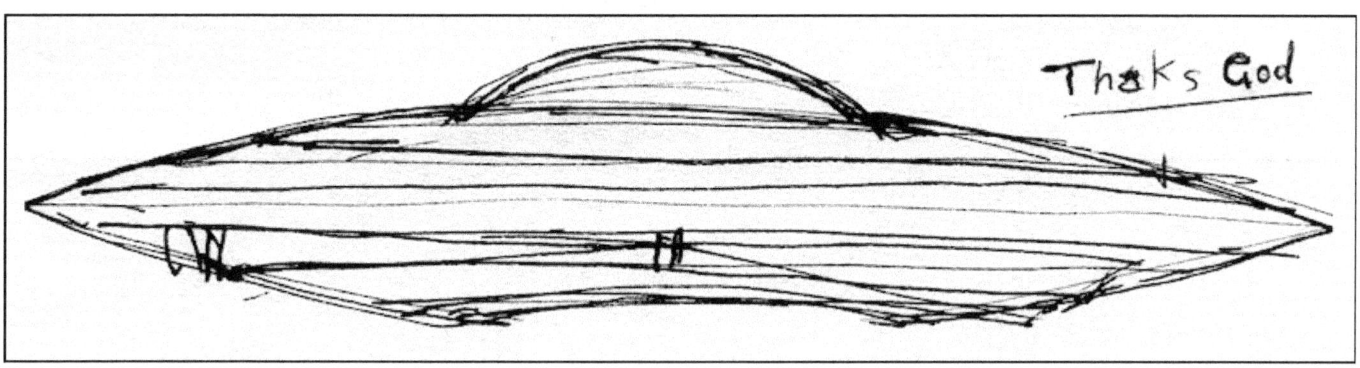
Thaks God

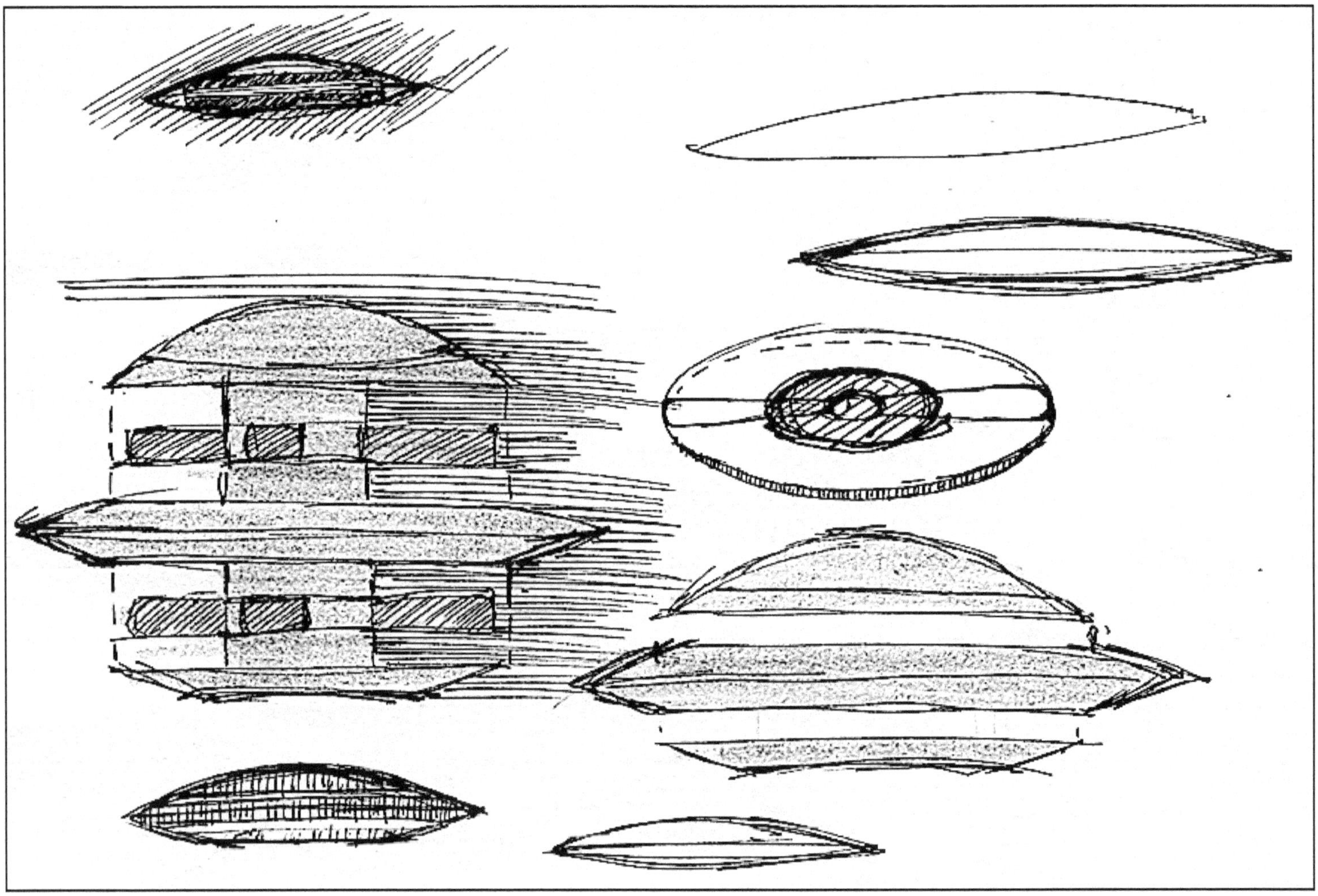

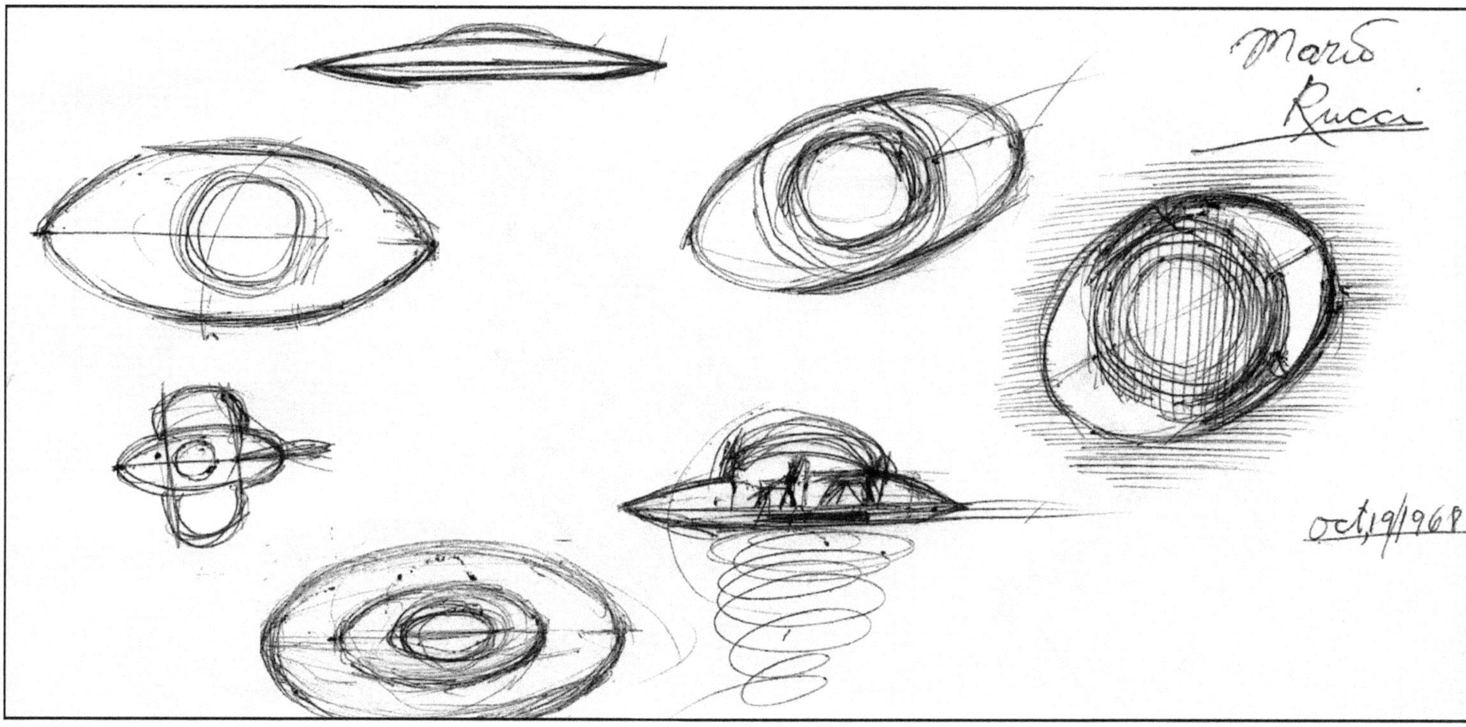
Mario
Rucci
Oct,19/1968

7. GUN

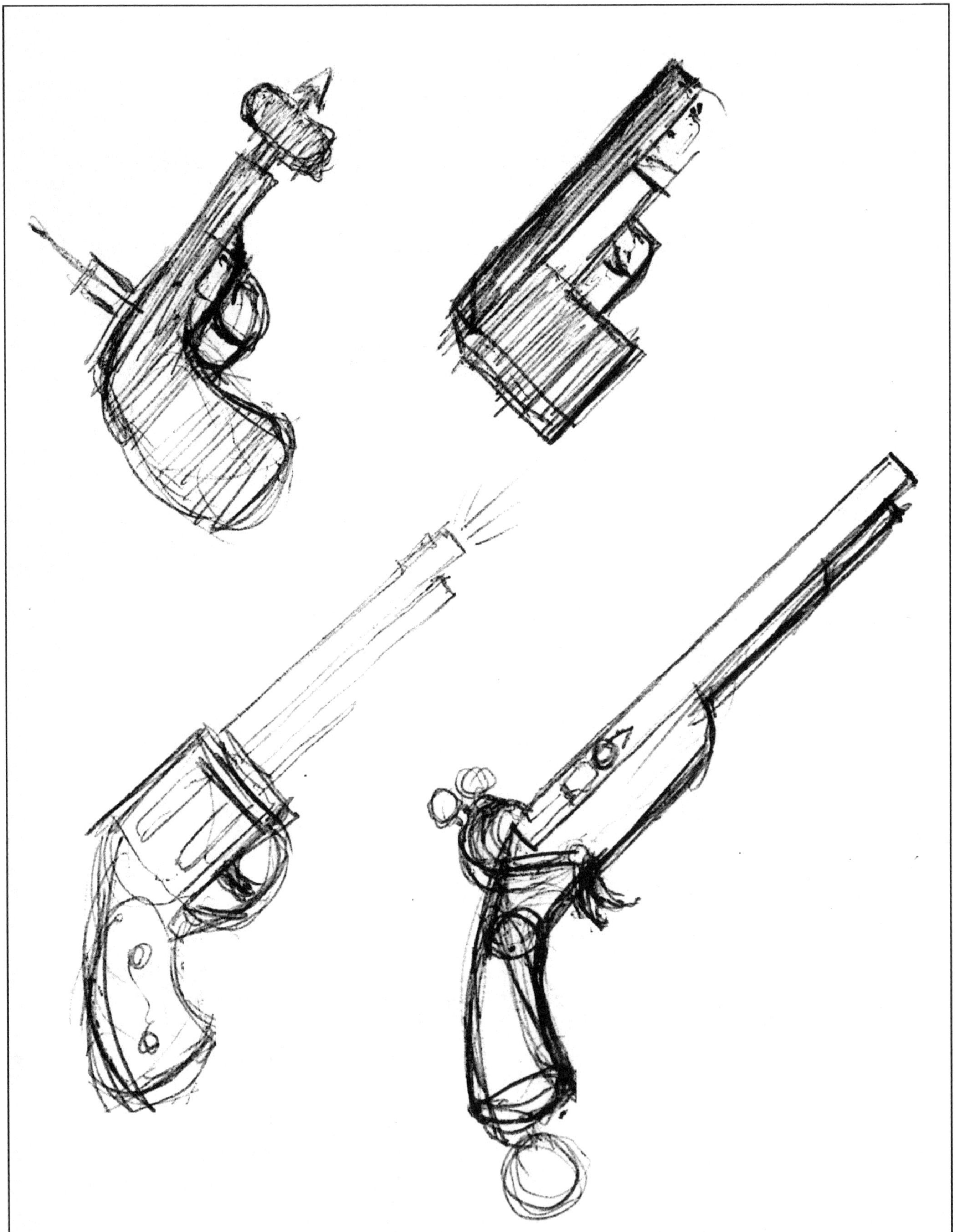

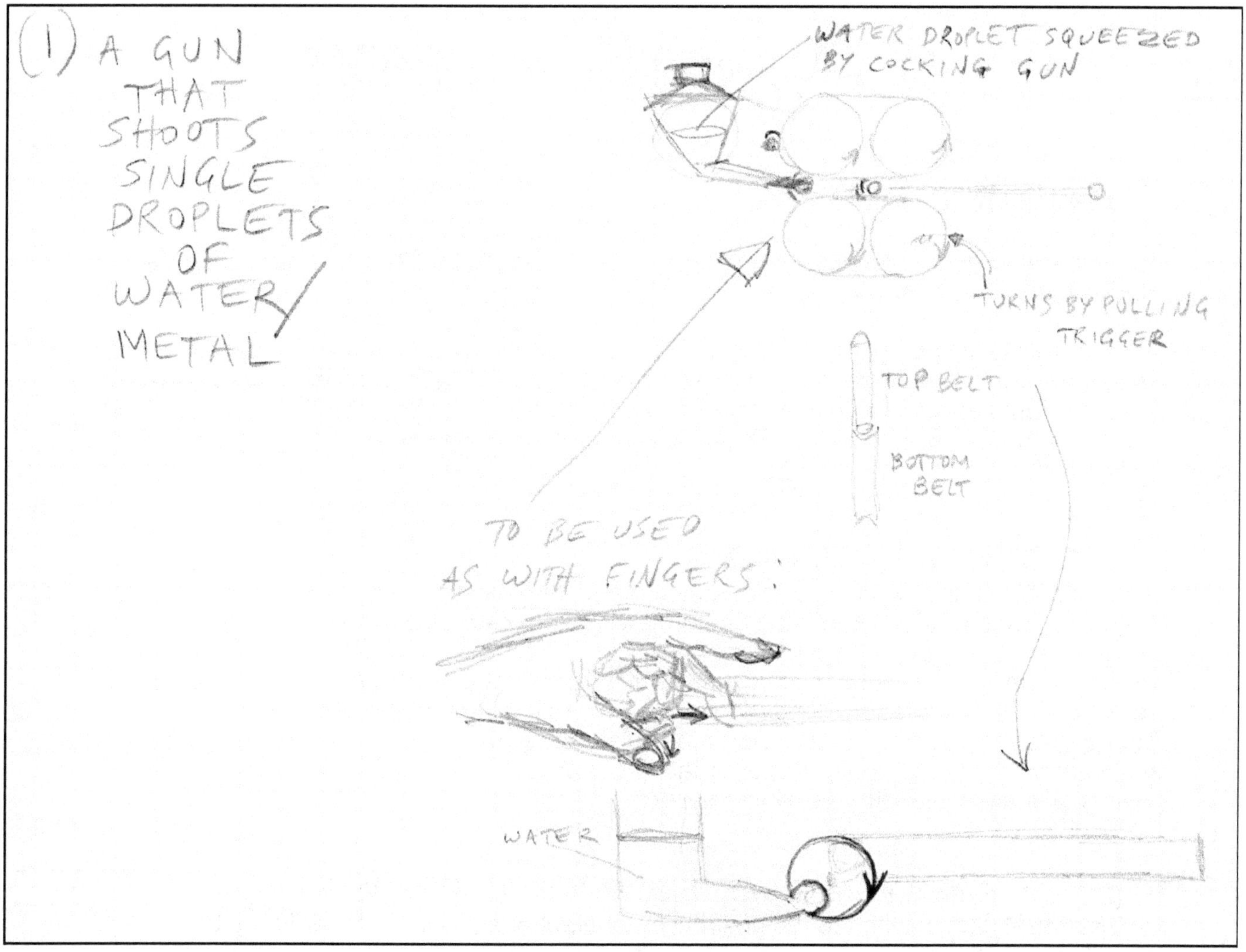
(1) A GUN THAT SHOOTS SINGLE DROPLETS OF WATER/ METAL
WATER DROPLET SQUEEZED BY COCKING GUN
TURNS BY PULLING TRIGGER
TOP BELT
BOTTOM BELT
TO BE USED AS WITH FINGERS:
WATER

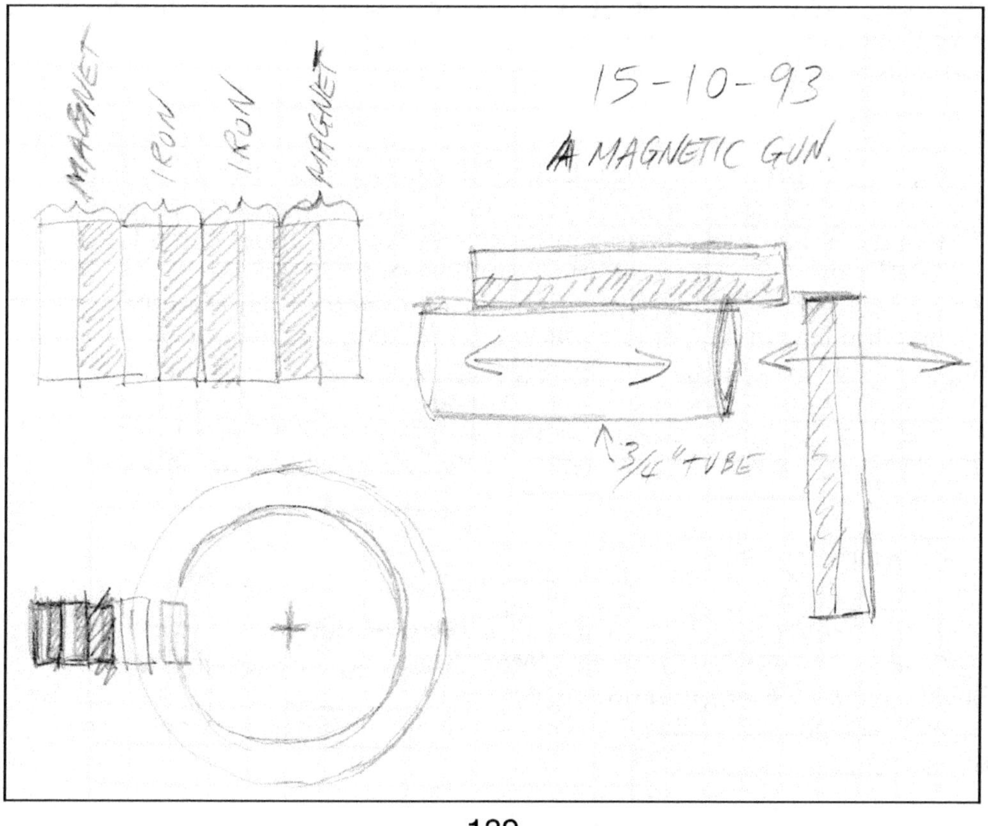
15-10-93
A MAGNETIC GUN.
MAGNET
IRON
IRON
MAGNET
3/4" TUBE

8.MAGNETIC ACTIVITY

MAGNETIC ACTIVITY

Matter is P/magnetic, S/organic, and M/synthetic. Magnetic matter is **superfluid** (subatomic particles), **fluid**, and **solid**.

We refer to activity as energy, motion, and force. Motion implies transfer. Electronic transfer = **transmission**, Atomic transfer = **transformation**, and Cellular transfer = **transportation**. Transmission/delivery/handing over/passing on has to do with SPACE/INFO; transformation/conversion has to do with MATTER; and, transportation/movement has to do with MOTION/ENERGY.

ELECTRONIC TRANSMISSION

Electronic Transmission is magnetic, atmospheric, and electric.
Magnetic Transmission (Radiation or strings vibration) causes earth's or celestial*/cosmic rotation, revolution, and galactic spiral
Atmospheric Transmission (Convection or fluid temperature change or rise and fall of atoms) causes evaporation, condensation, winds, rain, rivers, waterfalls, tides.
Electric Transmission (Conduction or molecules undulation / wavelike motion) causes heat, light, lightning, auroras, static and alternating current
* celestial/cosmic: planetary, stellar, galactic

Magnetism, Atmosphere/Gravity, Electricity OCCURRENCE

MAGNETISM occurs with the radial direction of electrons around a solid due to the concentration/ amassing of **Superfluid* Matter** (from electronic to atomic elements).
ATMOSPHERE/GRAVITY occurs with the weight/pressure of atoms due to the concentration/amassing of **Fluid Matter** (from gas to liquid molecules).
ELECTRICITY occurs with the alternation of electrons spin rate due to the concentration/amassing of **Solid Matter** (from particulate to solid aggregates/masses).
* superfluid: non-resistant

Magnetism, Atmosphere/Gravity, Electricity ON AND OFF SWITCH

In order for us to make a Motor, we need to control Electricity, Gravity, Magnetism.
We turn MAGNETISM on and off by **switching* the flow of Superfluid Matter**.
We turn GRAVITY on and off by **switching the flow of Fluid Matter**.
We turn ELECTRICITY on and off by **switching the flow of Solid Matter**.
* fluid: stopping the flow of a fluid, like water/air, by cutting off the supply
* solid: stopping the flow of a solid by resting (supporting/hanging) weight
NOTE: We inventors are pleased to see that we can move ferric objects with natural magnets (vs. artificial magnets or solenoids). However, only if we are able to *switch on and off* Superfluid Matter, we will produce a motor powered by natural magnets.

Mechanical APPLICATION

MAGNETISM leads to **Magnetic Motor (MM)**
ATMOSPHERE/GRAVITY leads to **Atmospheric Motor (AM)**
MAGNETISM & GRAVITY leads to **Mag-Gra** (**Mag**netic-**Gra**vity) **Motor**
ELECTRICITY leads to **Electric Motor (EM)**
Sometimes a machine is a combination of 2 activities, like when a machine uses gravity and magnets.

8A. MAGNETIC MOTOR (GIRA-GIRA)

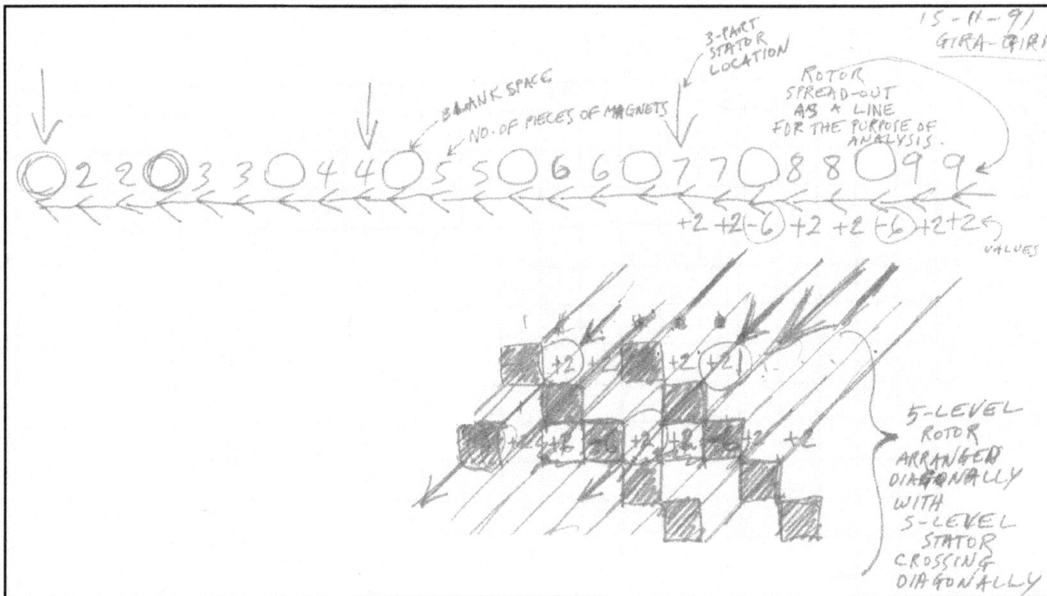

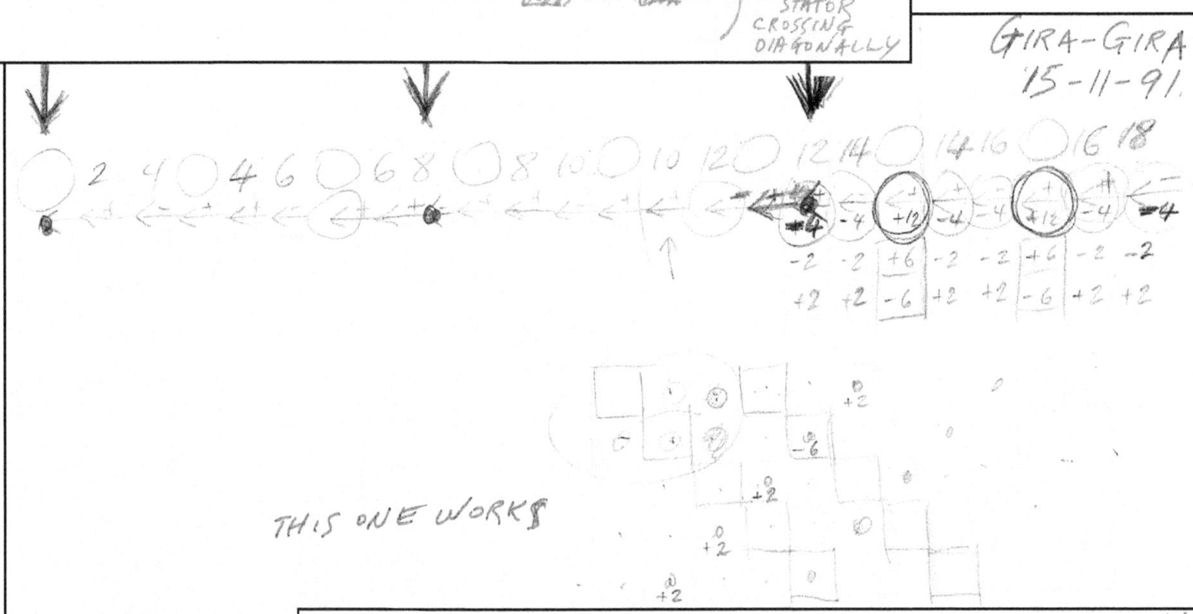

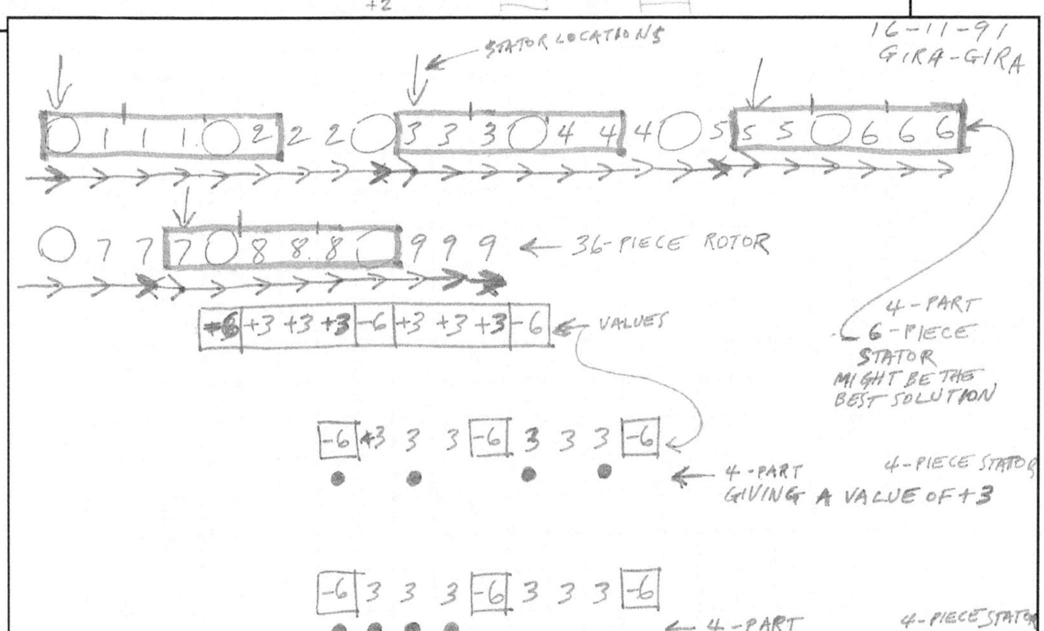

Rotor (top) and Stator (bottom), arranged in a straight line, to better see how they affect each other

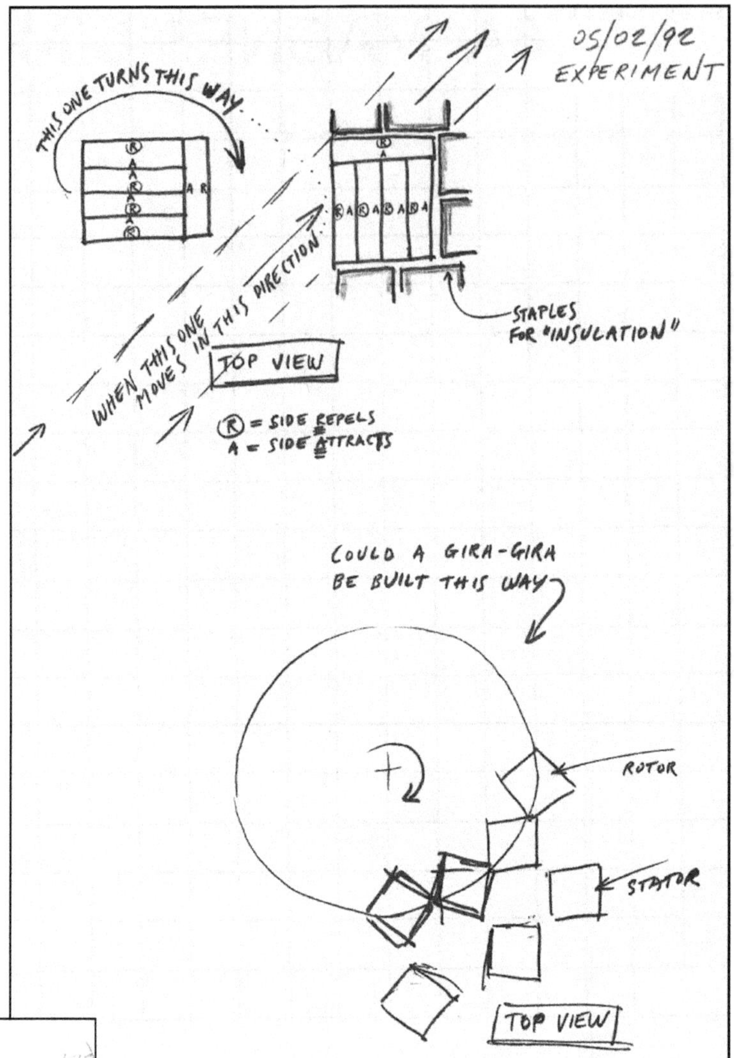

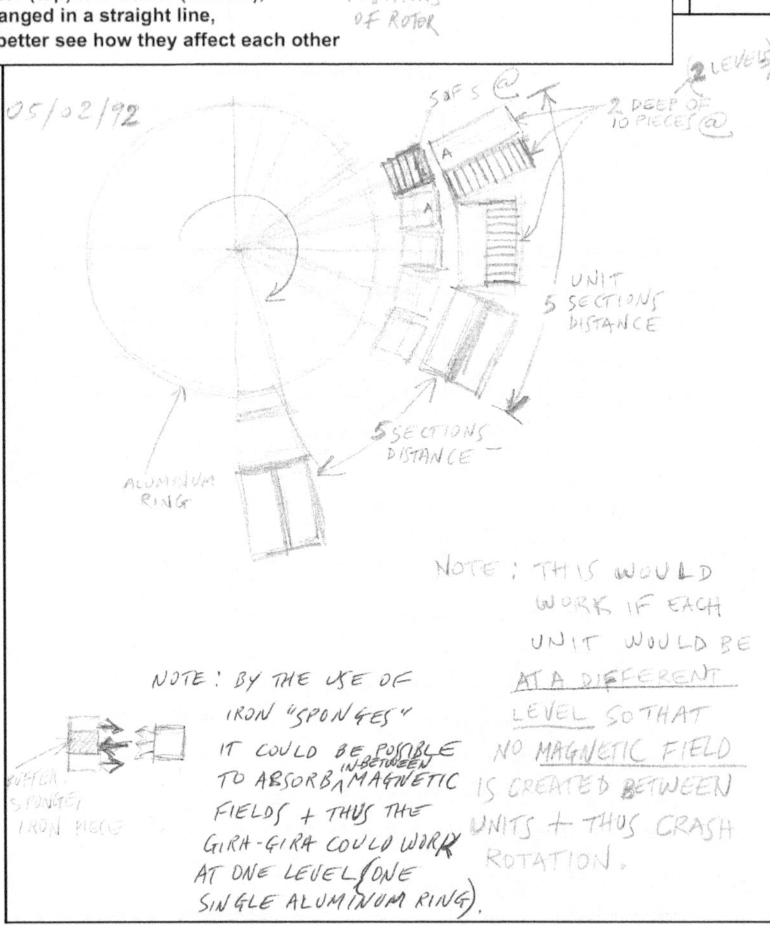

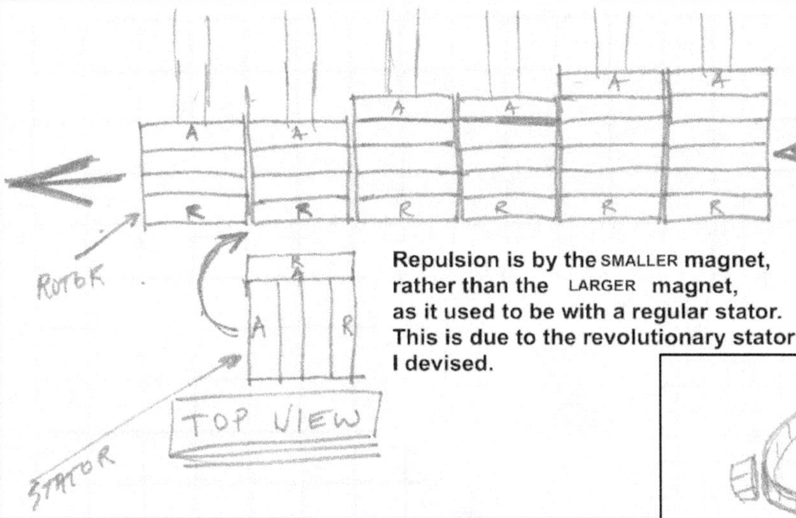

08-02-92
Linear Arrangement
A = Attraction
R = Repulsion

Repulsion is by the SMALLER magnet, rather than the LARGER magnet, as it used to be with a regular stator. This is due to the revolutionary stator I devised.

11/02/92
EVOLUTION OF IDEAS

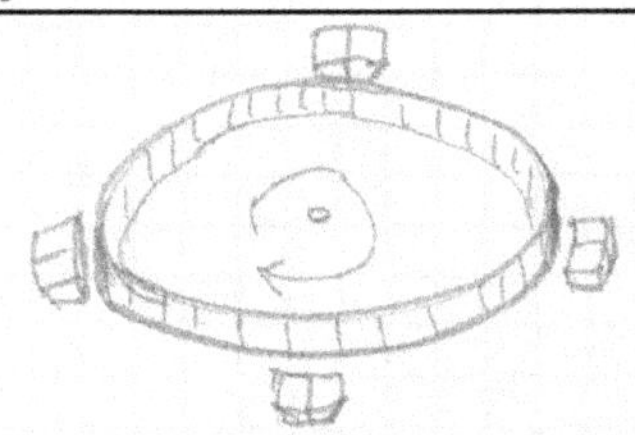

1) 4-SECTION GIRA-GIRA USING 3-SECTION STATOR TO BE ABLE TO "MAGNETIZE/REPEL" ALL 1/2 DIFFERENT PIECES OF ROTOR ALL THE TIME.

2) 9 2-PIECE-WIDE STATORS SPACED AT A DISTANCE OF 2 PIECES.

3) 4 2-PIECE-WIDE STATORS, ONE FOR EACH OF THE 4-SECTION ROTOR, TO THE STONGEST PUSH WHICH IS NEEDED WHEN THE ∅ POINT (PIECE#8) OF THE GIRA-GIRA PREVENTS THE ROTOR FROM TURNING.

4) ROTOR'S SECTIONS SHOULD BE MODIFIED TO MAKE IT EASIER FOR STATORS TO GO OVER THE ∅ POINT.

→2 DISCOVERIES TONIGHT:

1) THE ROTOR THAT WORKS BEST IS:

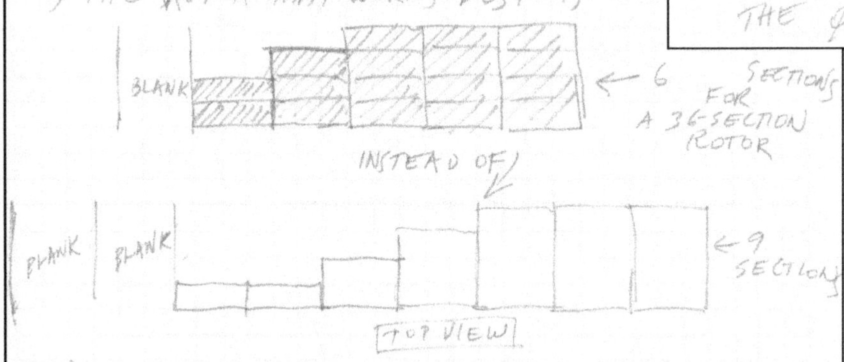

2) A METAL BAND RAPPED AROUND A FEW SECTIONS OF THE ROTOR CREATES ITS OWN POLARITIES, AS IF THERE WERE A BLANK SECTION WHERE THE BAND IS:

1) TO TEST 6-SECTION ROTOR

2) TO USE BOTH REPULSION + ATTRATION OF THE ROTOR: 12/02/92

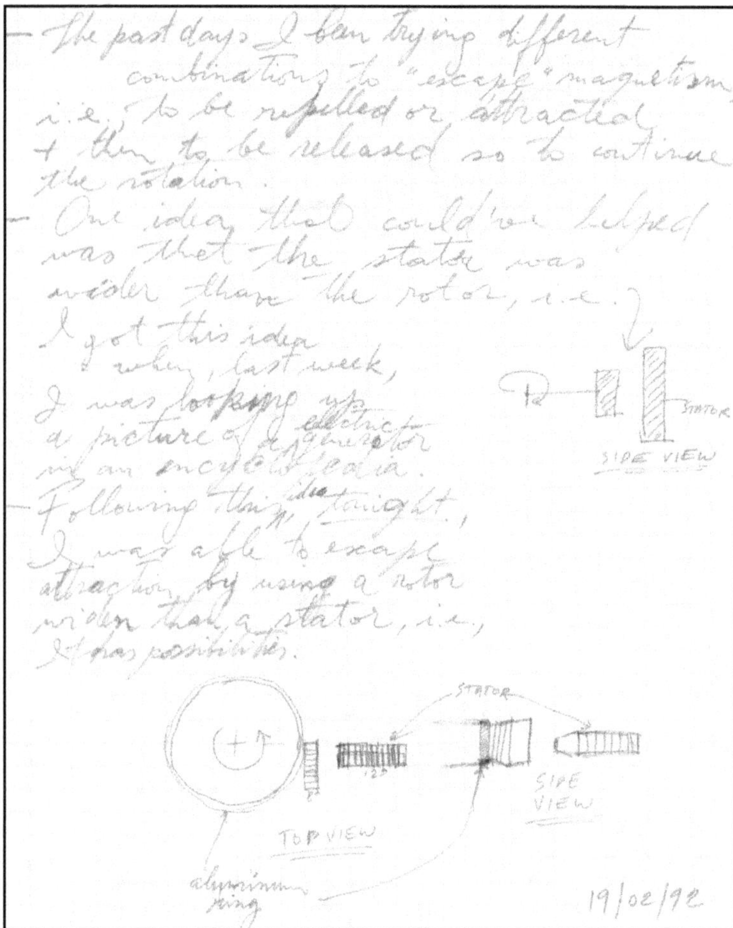
- The past days I been trying different combinations to "escape" magnetism, i.e., to be repelled or attracted + then to be released so to continue the rotation.
- One idea that could've helped was that the stator was wider than the rotor, i.e.
I got this idea when, last week, I was looking up a picture of a electric generator in an encyclopedia.
Following this idea tonight, I was able to escape attraction by using a rotor wider than a stator, i.e.,
It has possibilities.
R
STATOR
SIDE VIEW
STATOR
SIDE VIEW
TOP VIEW
aluminum ring
19/02/92

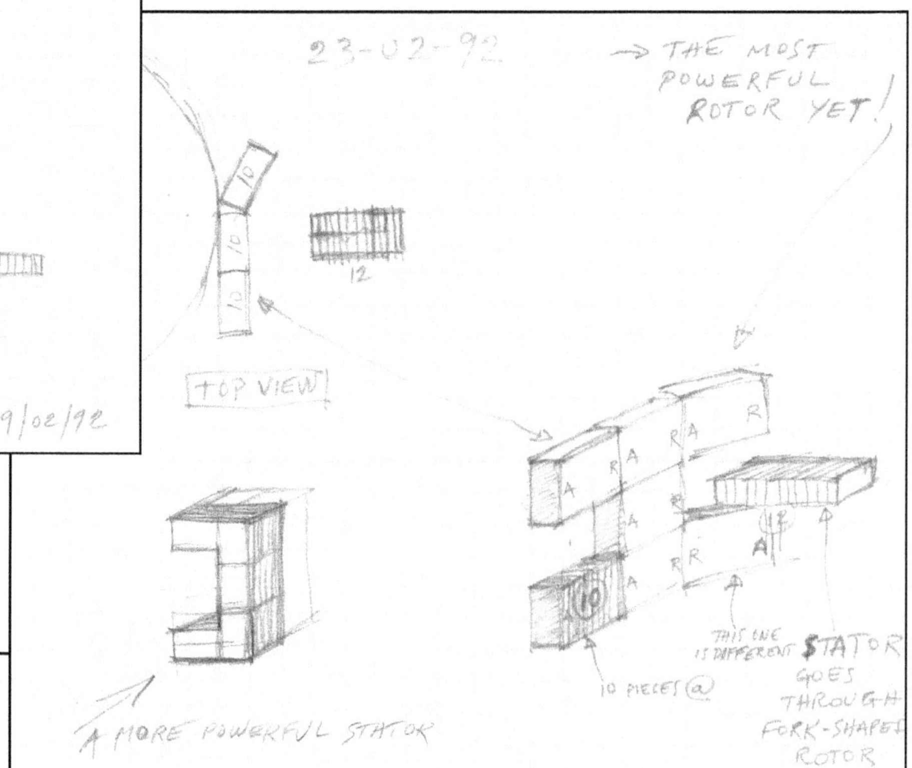
23-02-92
→ THE MOST POWERFUL ROTOR YET!
10
12
TOP VIEW
A
R
10
THIS ONE IS DIFFERENT
STATOR GOES THROUGH FORK-SHAPED ROTOR
10 PIECES @
A MORE POWERFUL STATOR

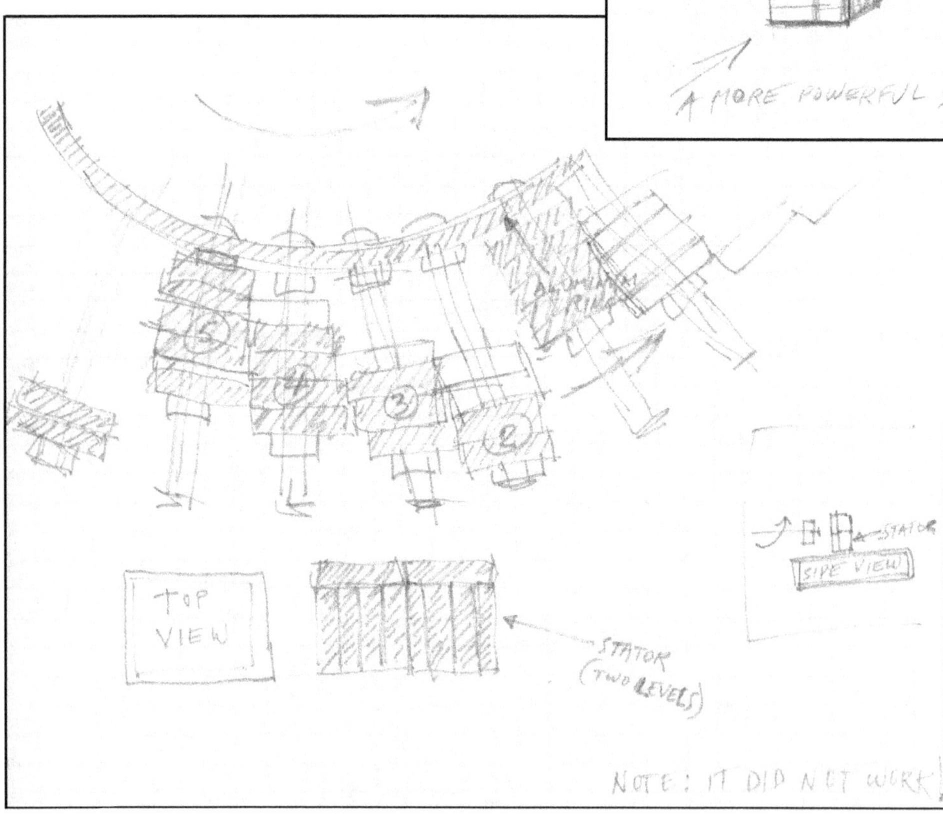
Aluminum Ring
5
4
3
2
STATOR
SIDE VIEW
TOP VIEW
STATOR (TWO LEVELS)
NOTE: IT DID NOT WORK!

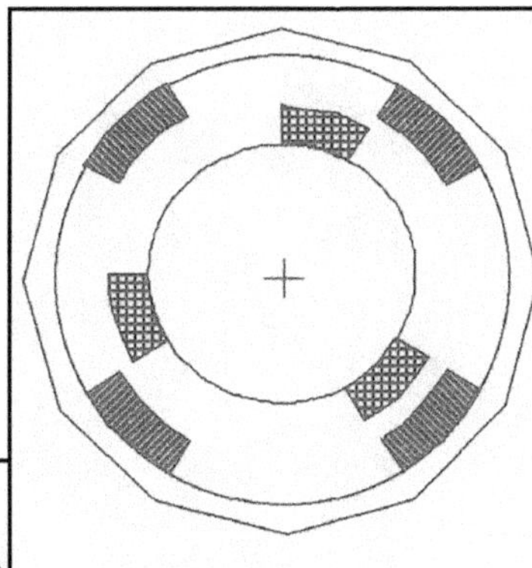

28-02-92
12-section Gira-Gira
rotating clockwise

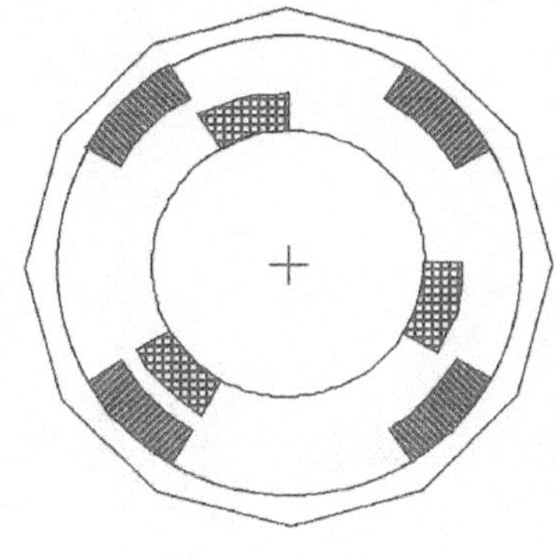

28-02-92
12-section Gira-Gira
rotating clockwise

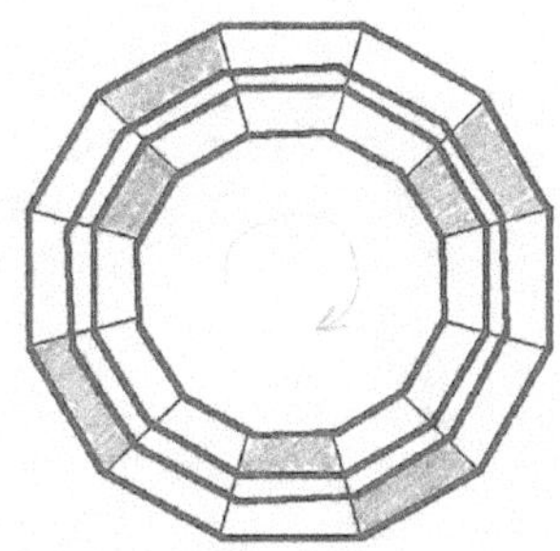

25-02-92
12-sided Gira-Gira
3-section rotor
4-section stator

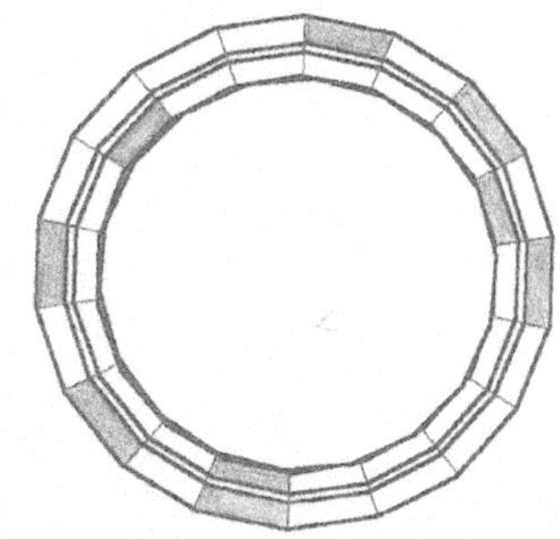

25-02-92
18-sided Gira-Gira
3-section rotor
6-section stator

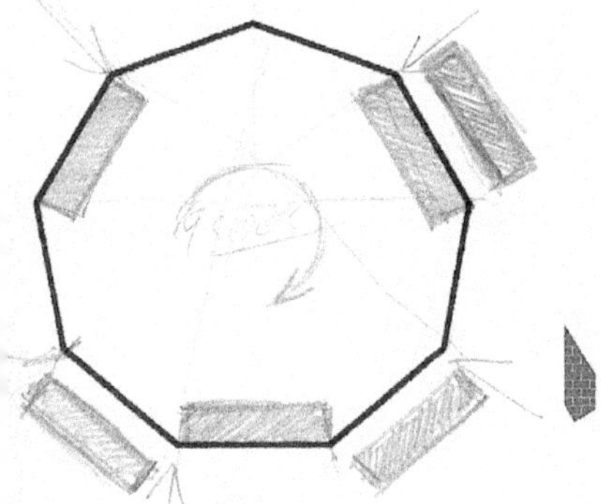

26-02-92
9-sided Gira-Gira
rotating clockwise
3-section rotor
3-section stator

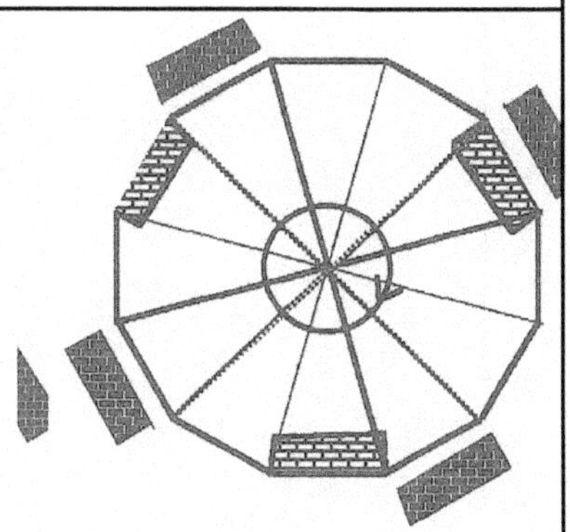

26-02-92, Wed
12-sided Gira-Gira
rotating clockwise
3-section rotor
4-section stator
Nichols wanted this design, when he saw it.

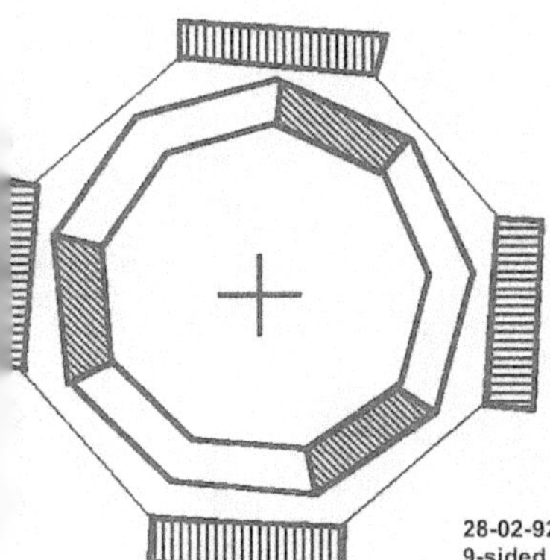

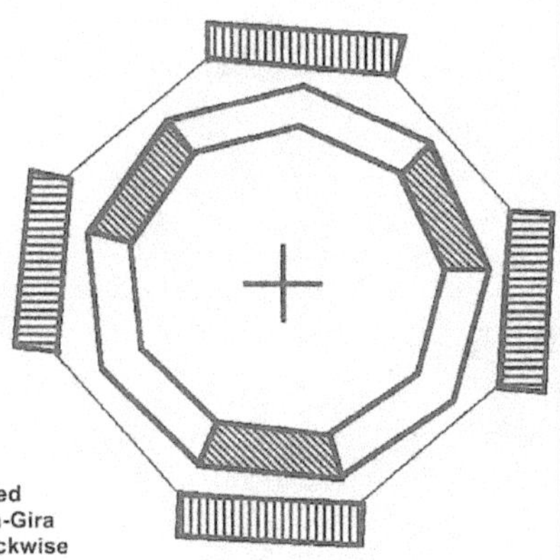

28-02-92, Wed
9-sided Gira-Gira
rotating clockwise
3-section rotor
4-section stator

version 2

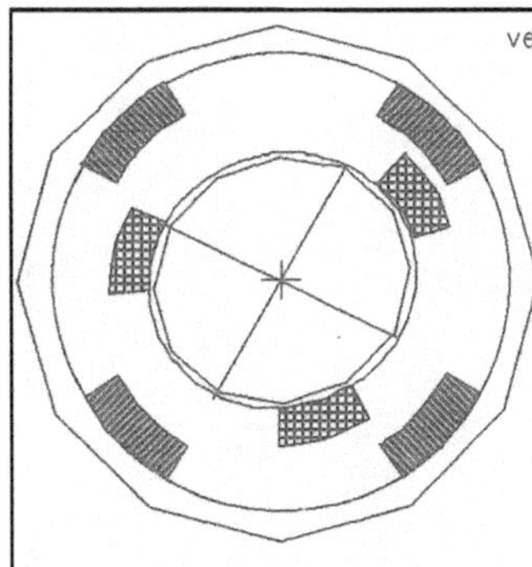

28-02-92
12-section Gira-Gira
rotating clockwise

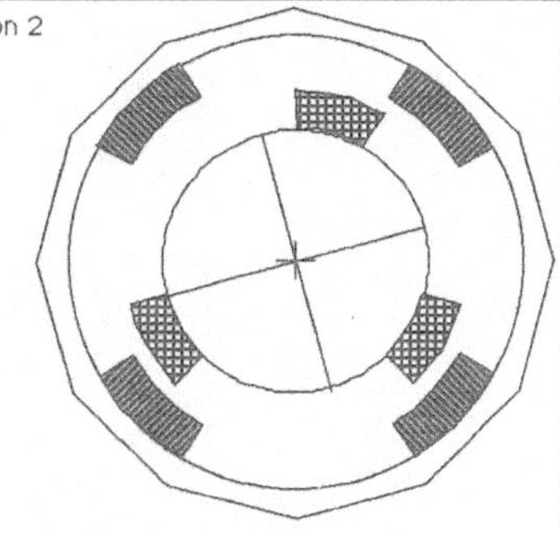

28-02-92
12-section Gira-Gira
rotating clockwise

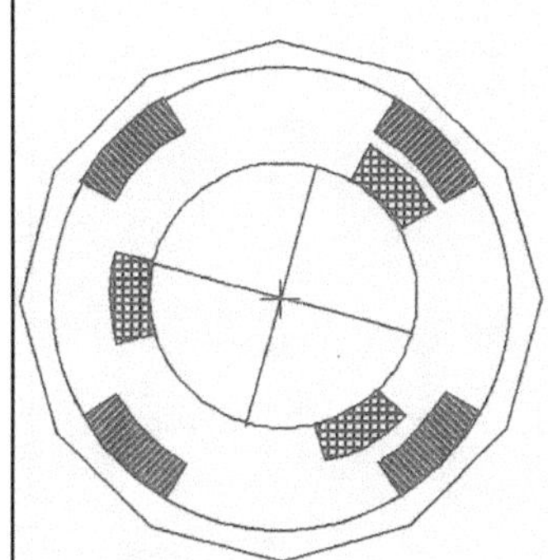

28-02-92
12-section Gira-Gira
rotating clockwise

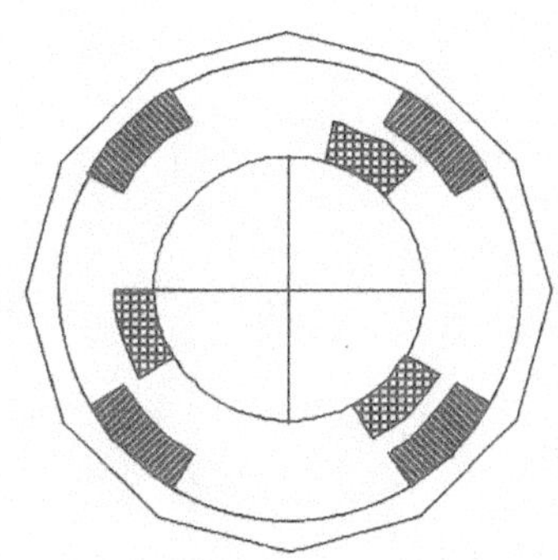

28-02-92
12-section Gira-Gira
rotating clockwise

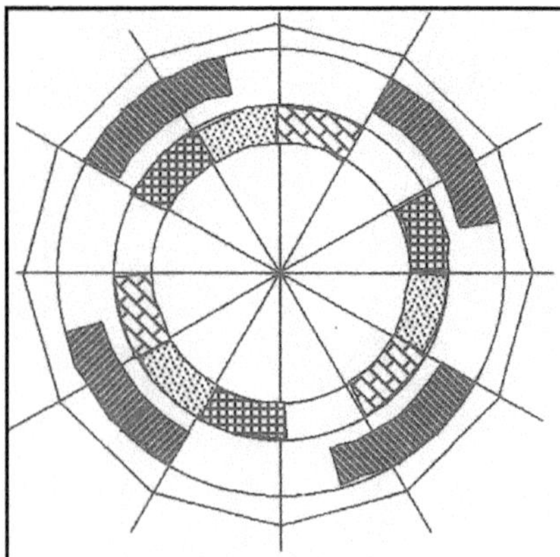

29-02-92
12-section Gira-Gira
rotating clockwise

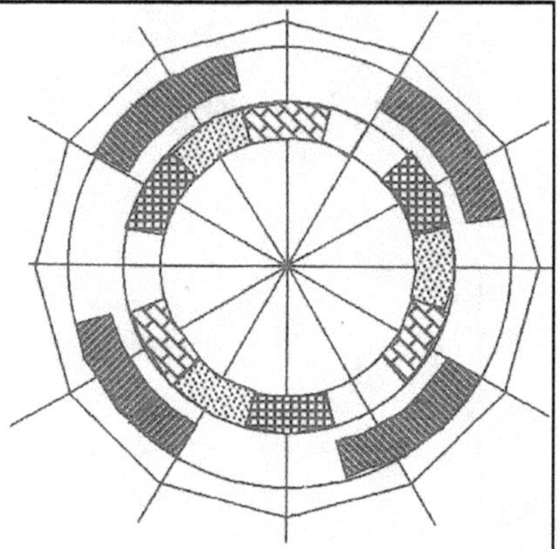

29-02-92
12-section Gira-Gira
rotating clockwise

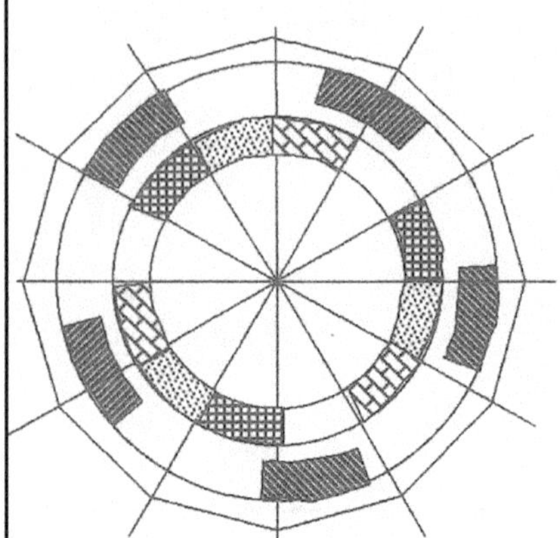

01-03-92
12-section Gira-Gira
rotating clockwise

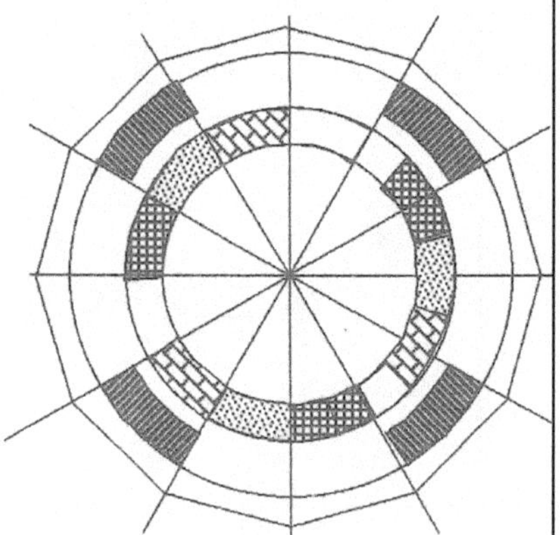

28-02-92
12-section Gira-Gira
rotating clockwise

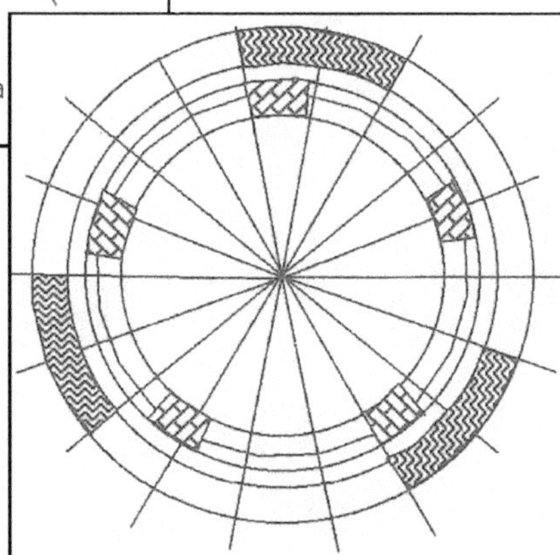

01-03-92 Gira-Gira
9-section stator
20-section rotor
rotating clockwise

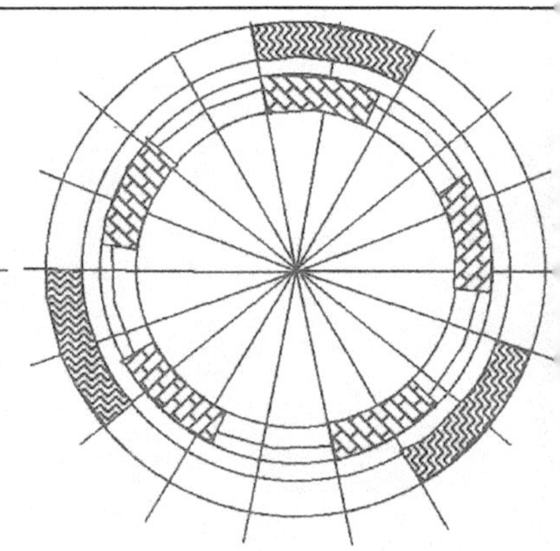

01-03-92 Gira-Gira
9-section stator
10-section rotor
rotating clockwise

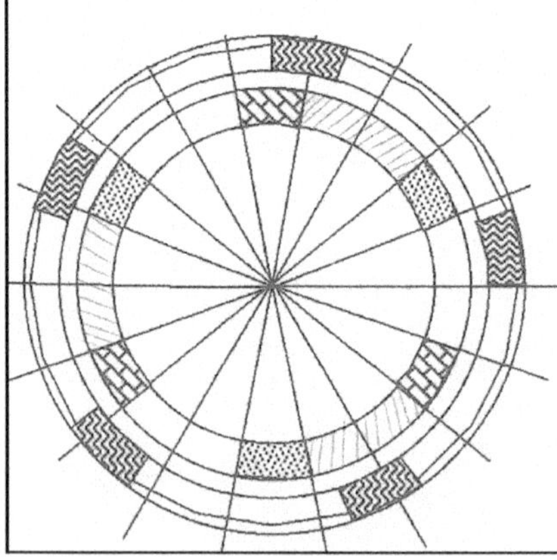

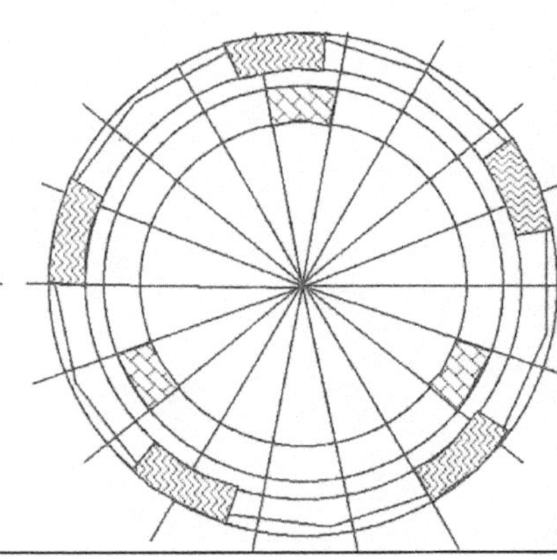

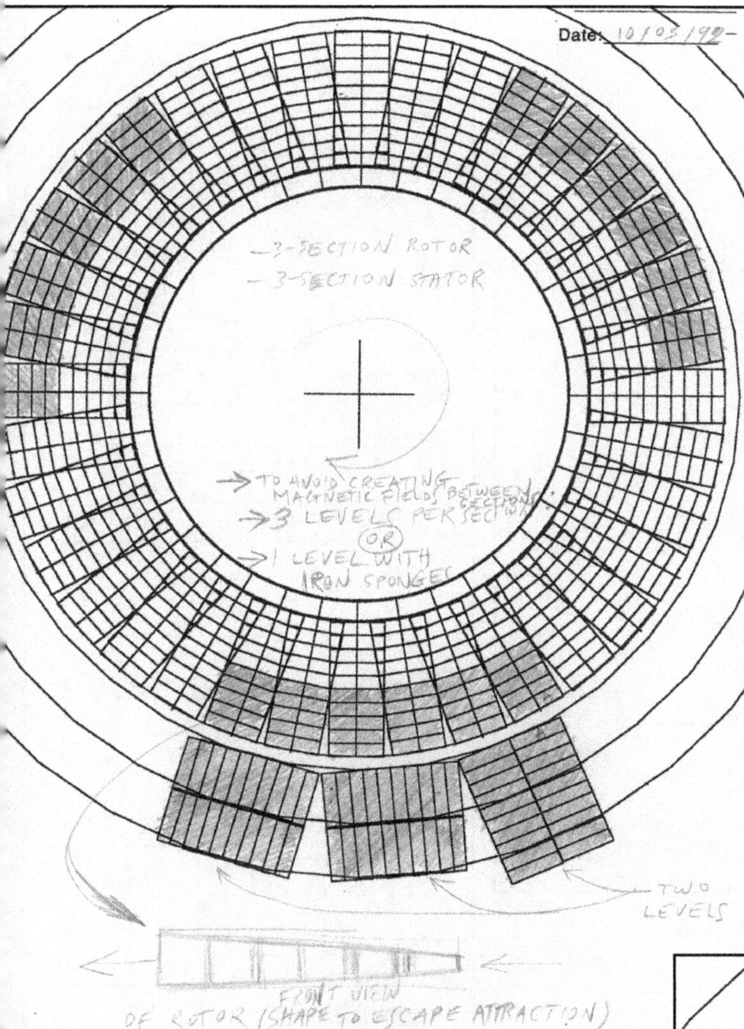
Date: 10/03/92
-3-SECTION ROTOR
-3-SECTION STATOR
→TO AVOID CREATING MAGNETIC FIELDS BETWEEN SECTIONS
→3 LEVELS PER SECTION
OR
→1 LEVEL WITH IRON SPONGES
TWO LEVELS
FRONT VIEW OF ROTOR (SHAPE TO ESCAPE ATTRACTION)

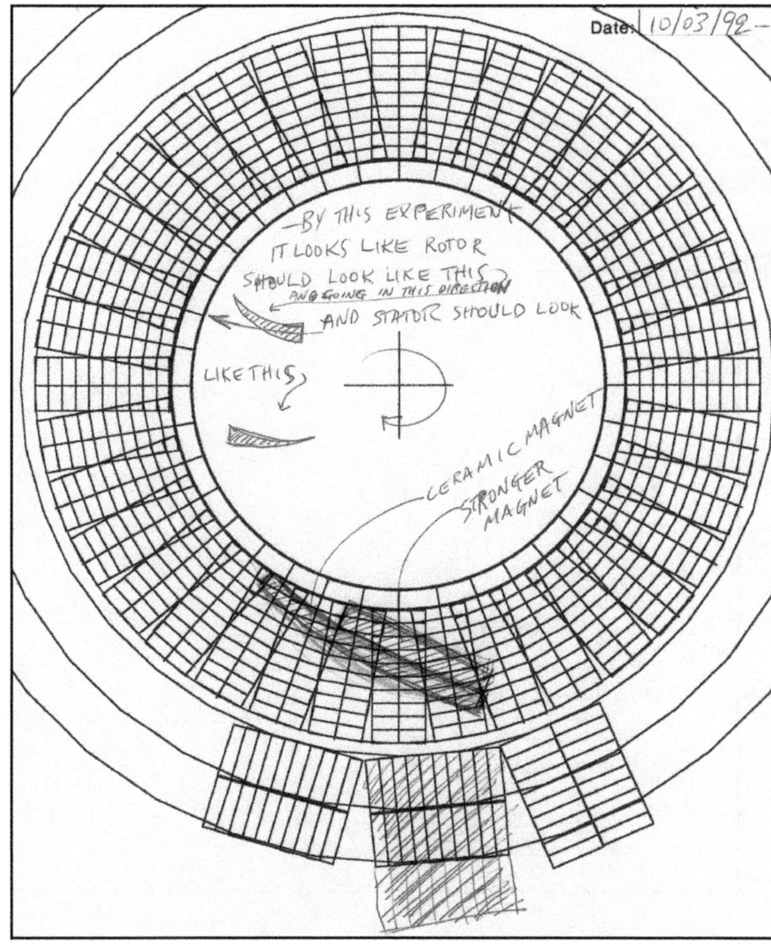
Date: 10/03/92
-BY THIS EXPERIMENT IT LOOKS LIKE ROTOR SHOULD LOOK LIKE THIS
AND GOING IN THIS DIRECTION
AND STATOR SHOULD LOOK
LIKE THIS
CERAMIC MAGNET
STRONGER MAGNET

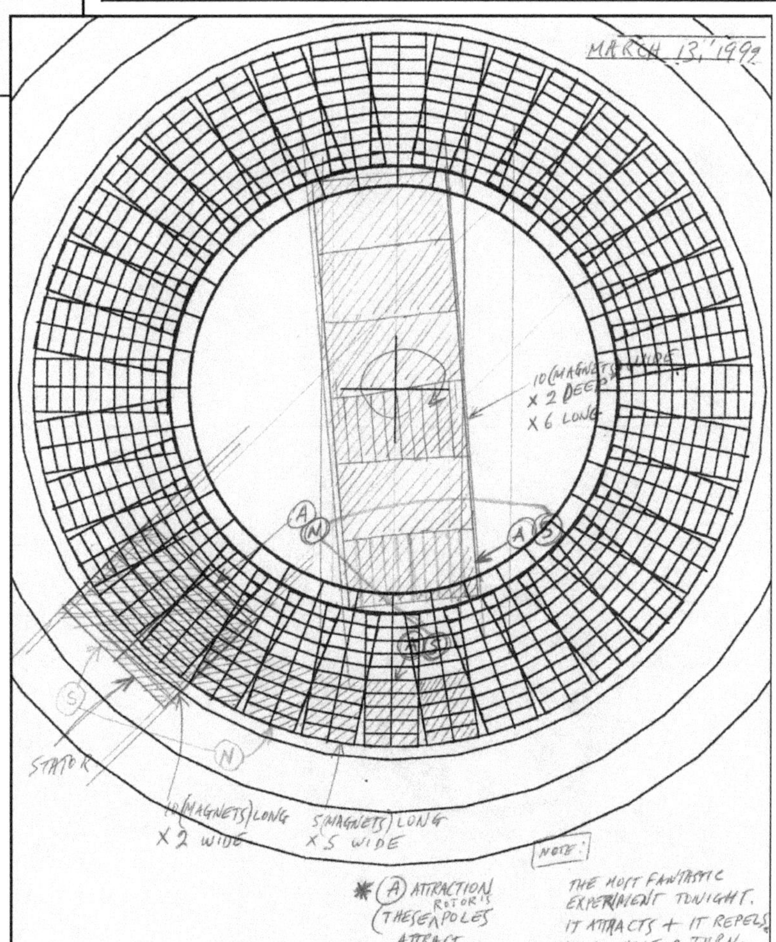
MARCH 13, 1999
10 (MAGNETS) WIDE
X 2 DEEP
X 6 LONG
A
N
A
S
S
N
STATOR
10 (MAGNETS) LONG
X 2 WIDE
5 (MAGNETS) LONG
X 5 WIDE
NOTE:
* (A) ATTRACTION
(THESE ROTOR'S POLES ATTRACT THIS STATOR'S POLE)
THE MOST FANTASTIC EXPERIMENT TONIGHT.
IT ATTRACTS + IT REPELS. ABOUT HALF A TURN.

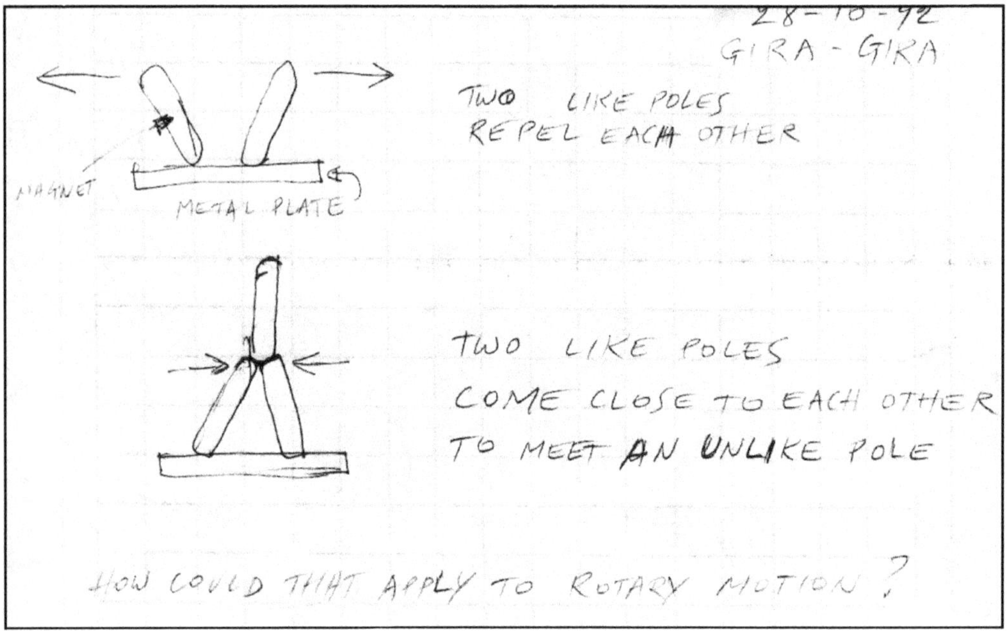
28-10-92
GIRA - GIRA
TWO LIKE POLES
REPEL EACH OTHER
MAGNET
METAL PLATE
TWO LIKE POLES
COME CLOSE TO EACH OTHER
TO MEET AN UNLIKE POLE
HOW COULD THAT APPLY TO ROTARY MOTION?

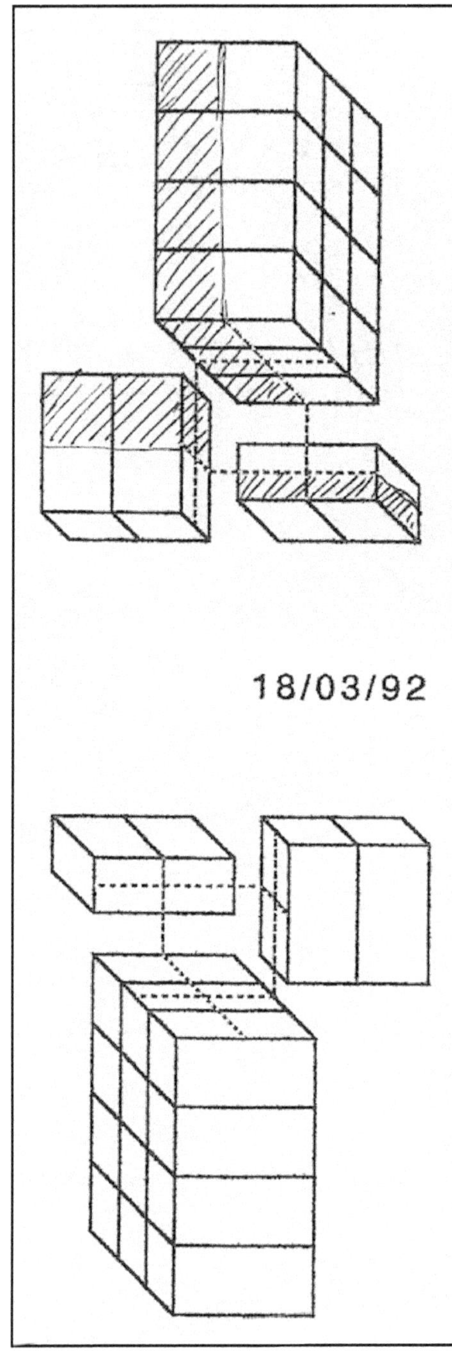
18/03/92

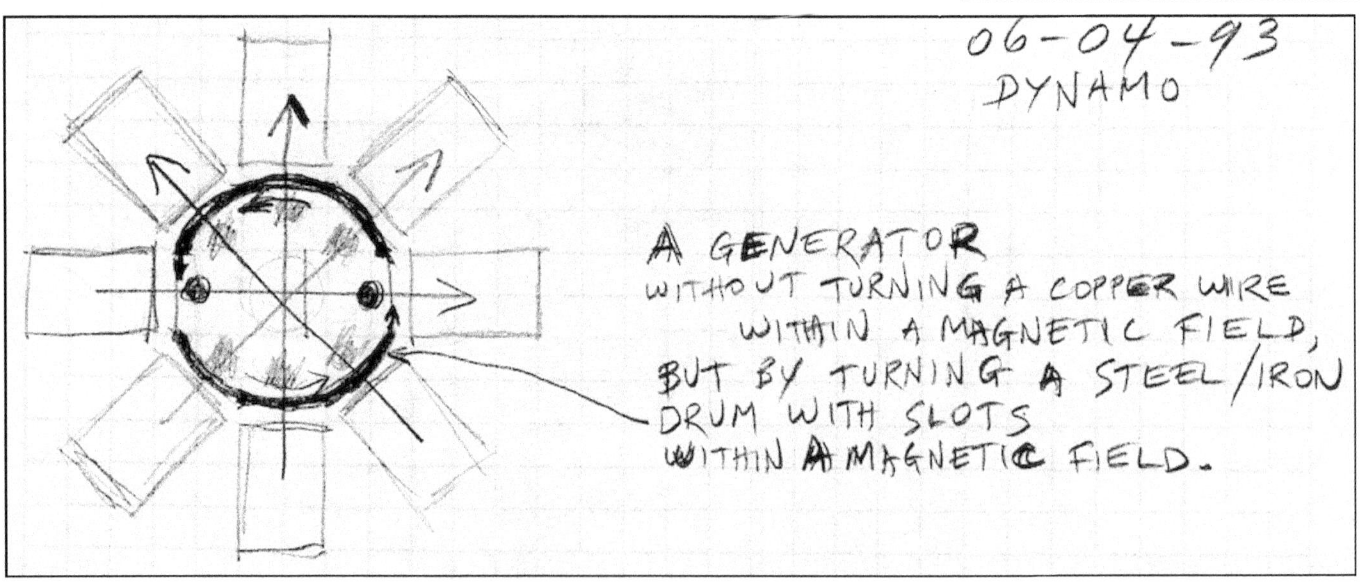
06-04-93
DYNAMO
A GENERATOR
WITHOUT TURNING A COPPER WIRE
WITHIN A MAGNETIC FIELD,
BUT BY TURNING A STEEL/IRON
DRUM WITH SLOTS
WITHIN A MAGNETIC FIELD.

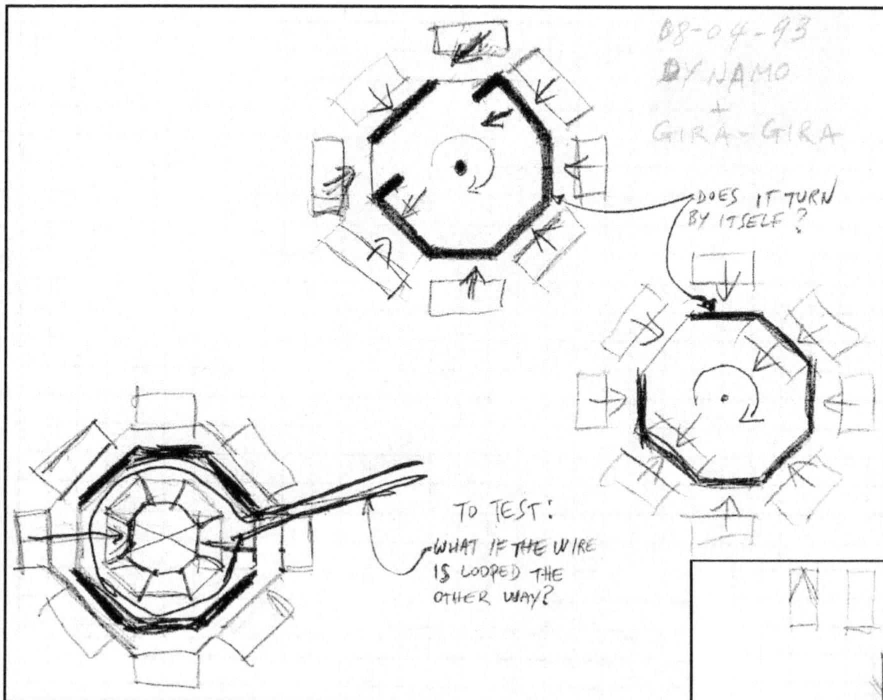
08-04-93
DYNAMO
+
GIRA-GIRA
DOES IT TURN BY ITSELF?
TO TEST:
WHAT IF THE WIRE IS LOOPED THE OTHER WAY?

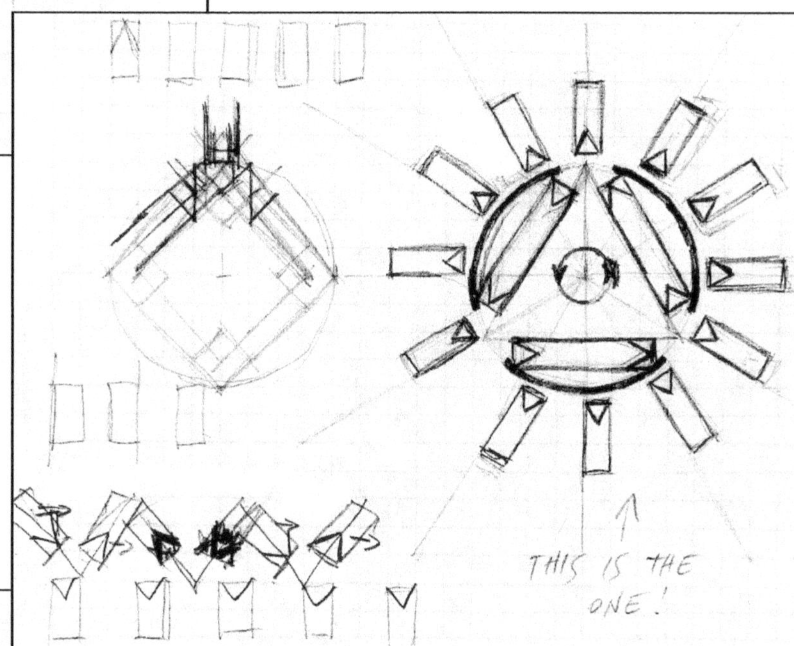
THIS IS THE ONE!

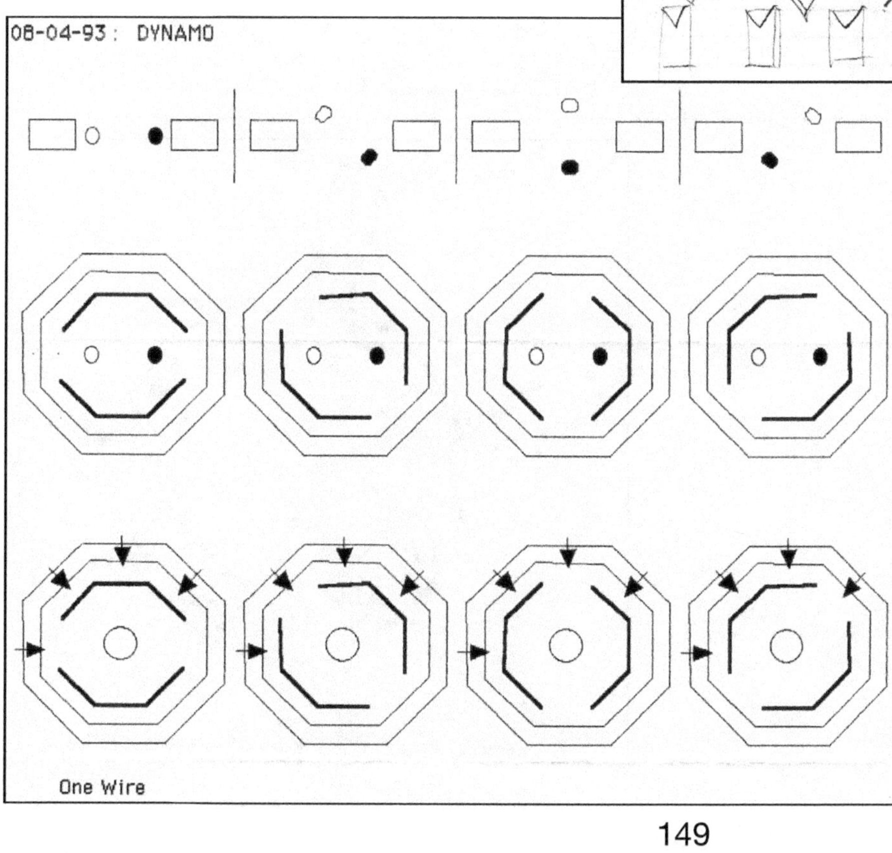
08-04-93 : DYNAMO
One Wire

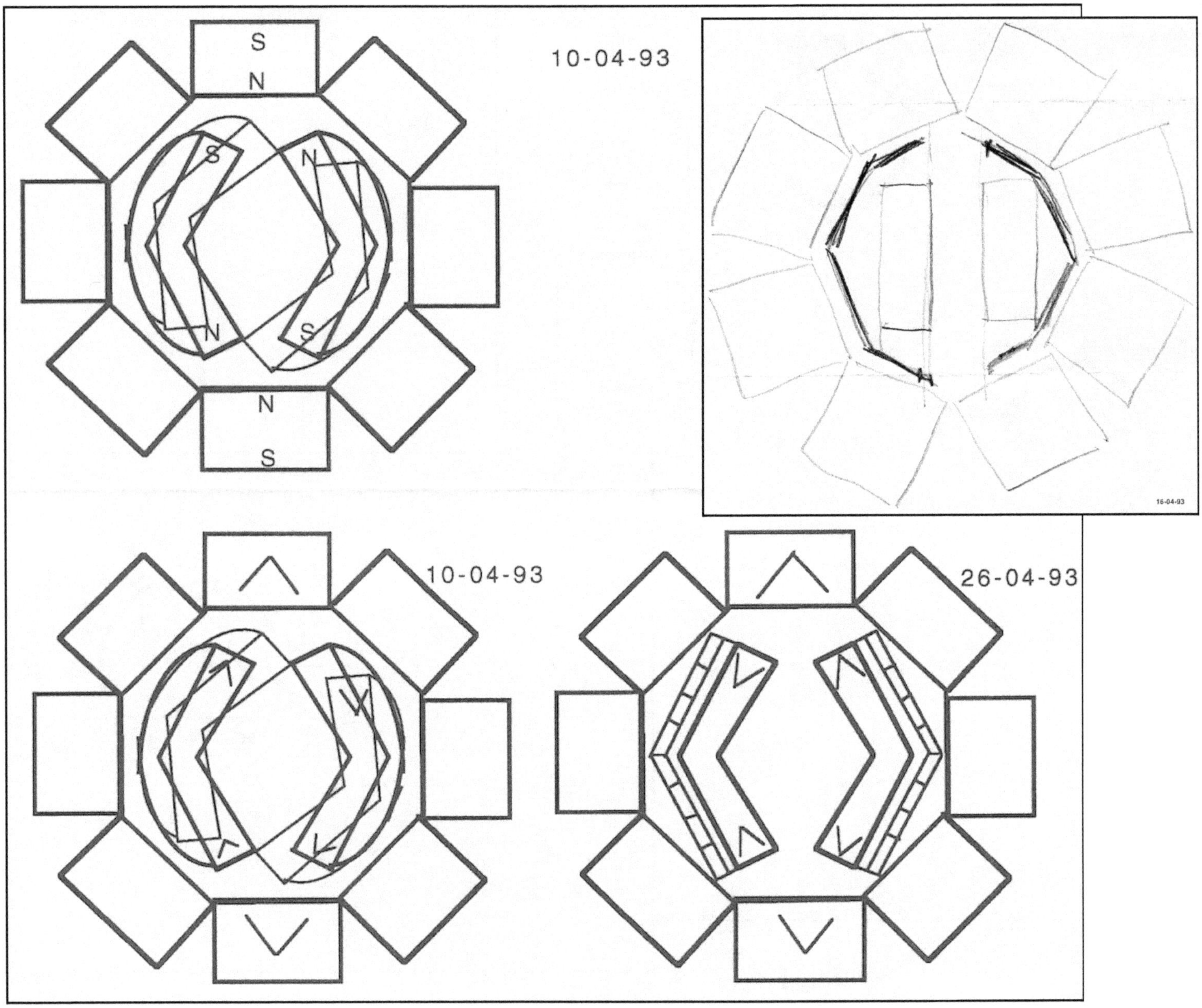
10-04-93
S
N
S
N
N
S
N
S
16-04-93
10-04-93
26-04-93

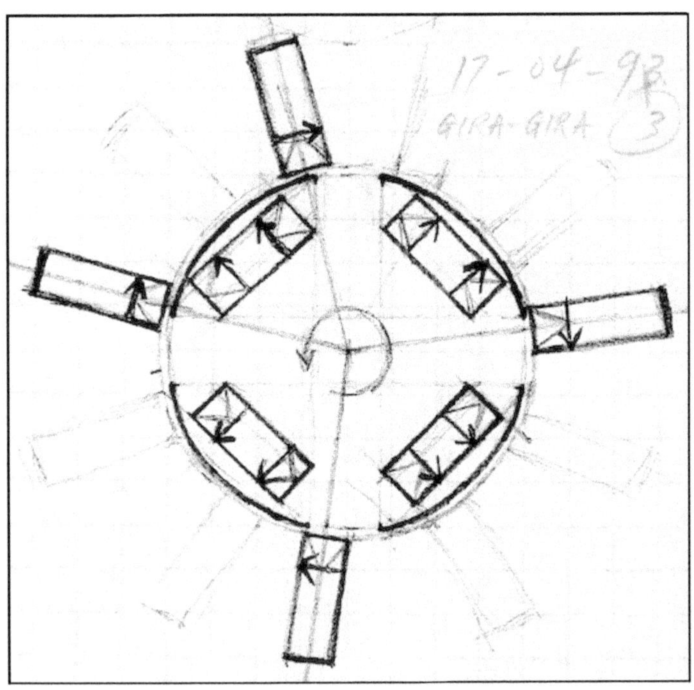
17-04-93
GIRA-GIRA
3

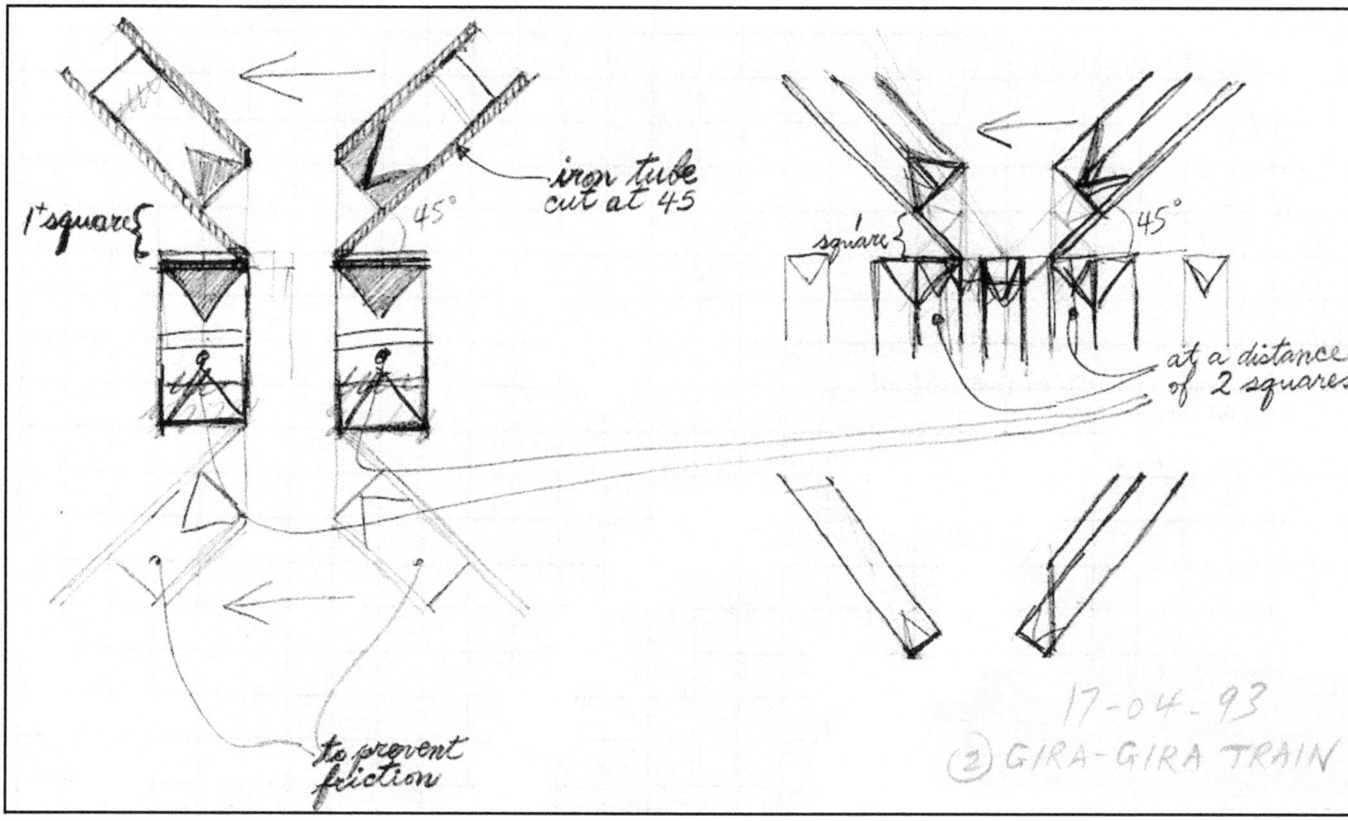

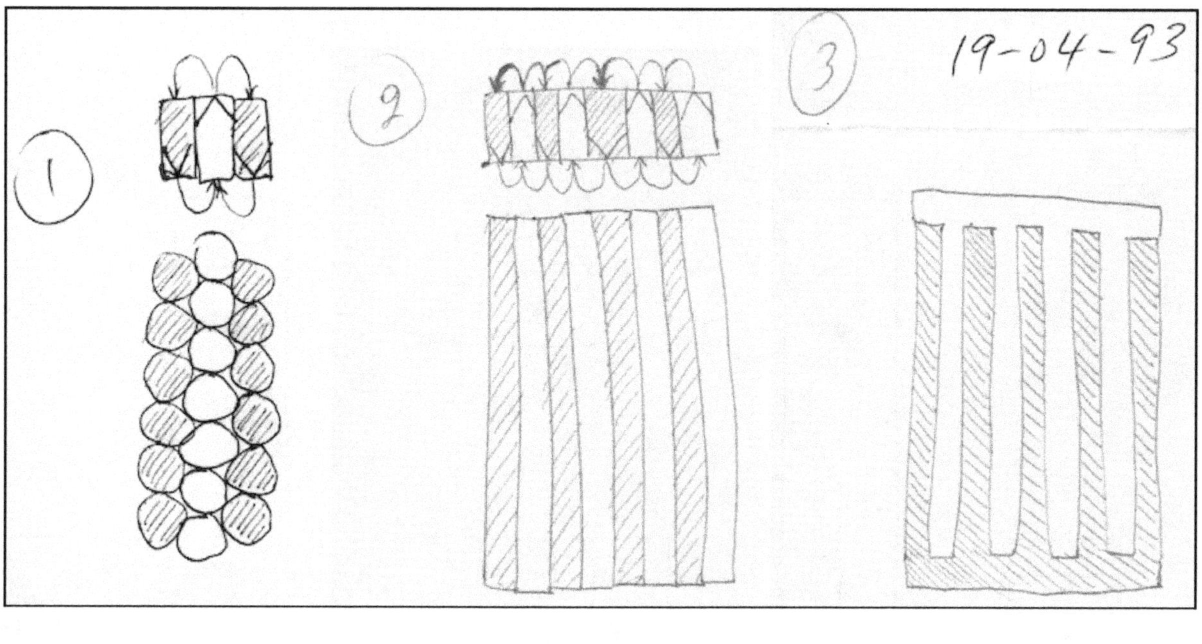

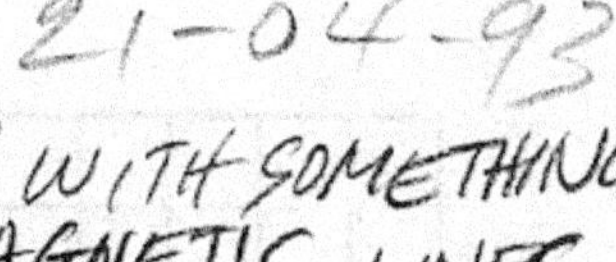

-TRYING TO COME UP WITH SOMETHING THAT WOULD ABSORB MAGNETIC LINES OF FORCE AND YET BE UNMAGNETIZED

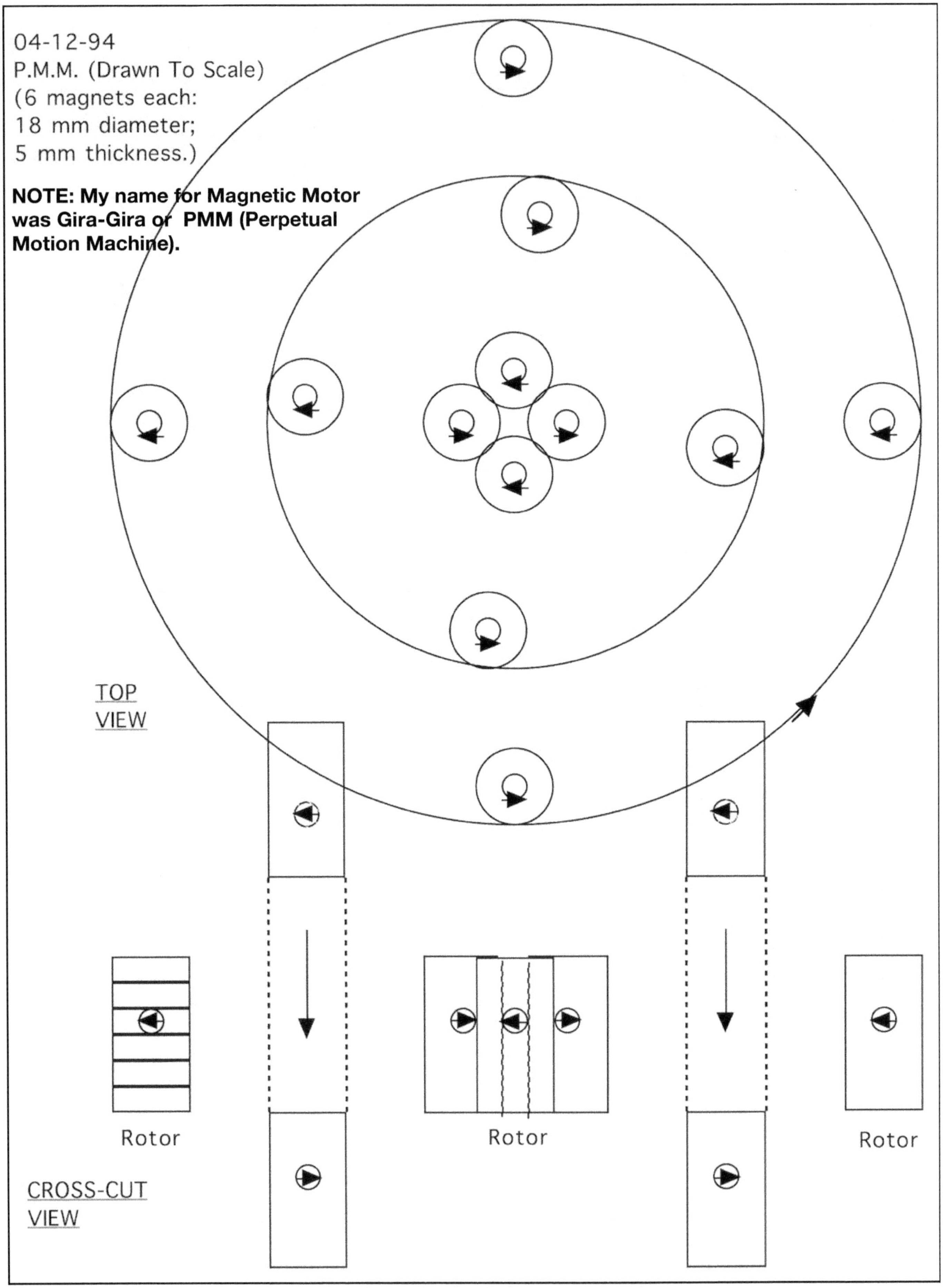

04-12-94
P.M.M. (Drawn To Scale)
(6 magnets each:
18 mm diameter;
5 mm thickness.)
NOTE: My name for Magnetic Motor was Gira-Gira or PMM (Perpetual Motion Machine).
TOP VIEW
Rotor
Rotor
Rotor
CROSS-CUT VIEW

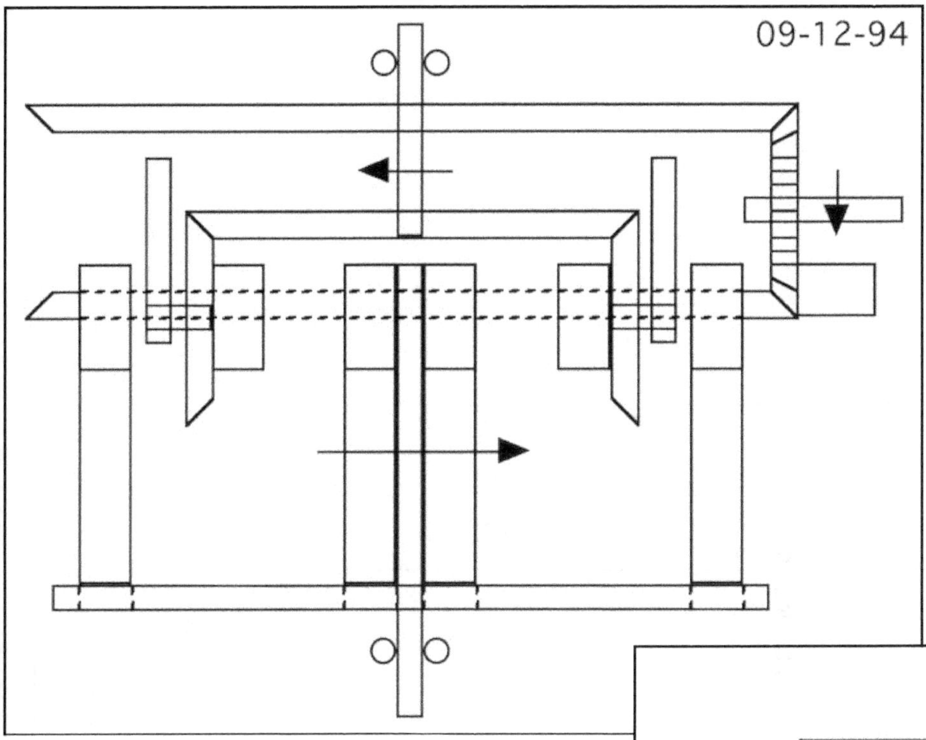
09-12-94

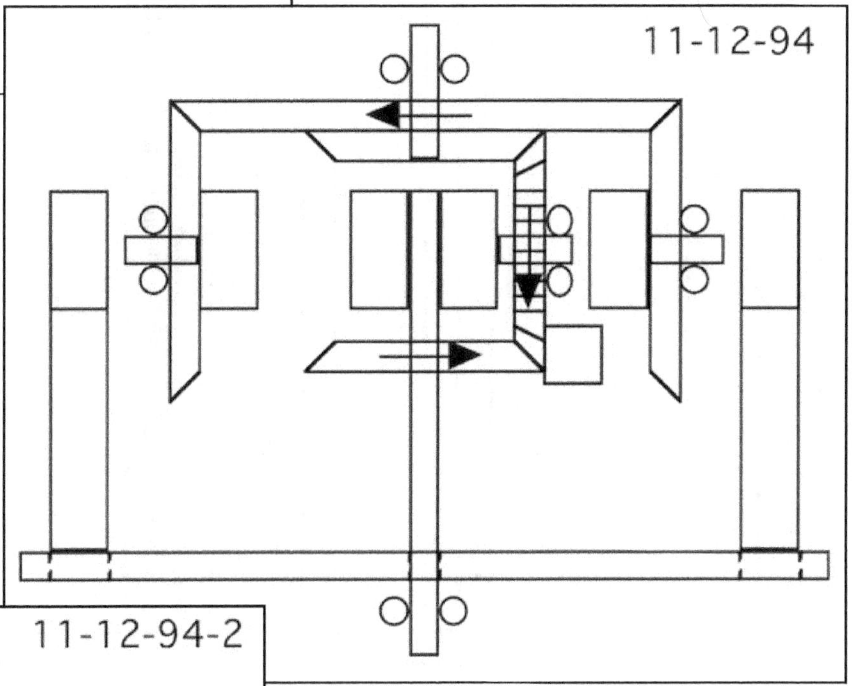
11-12-94

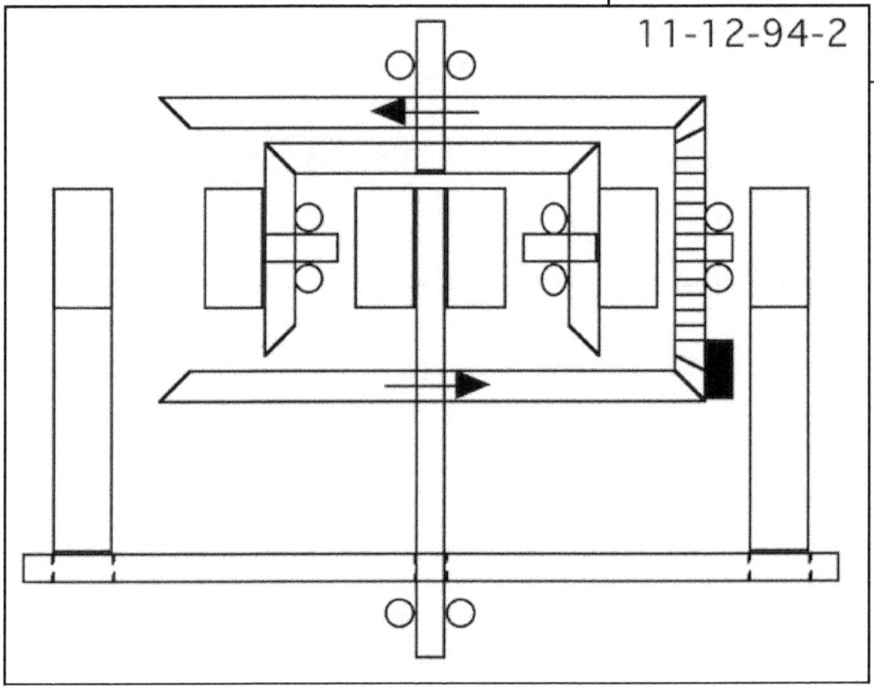
11-12-94-2

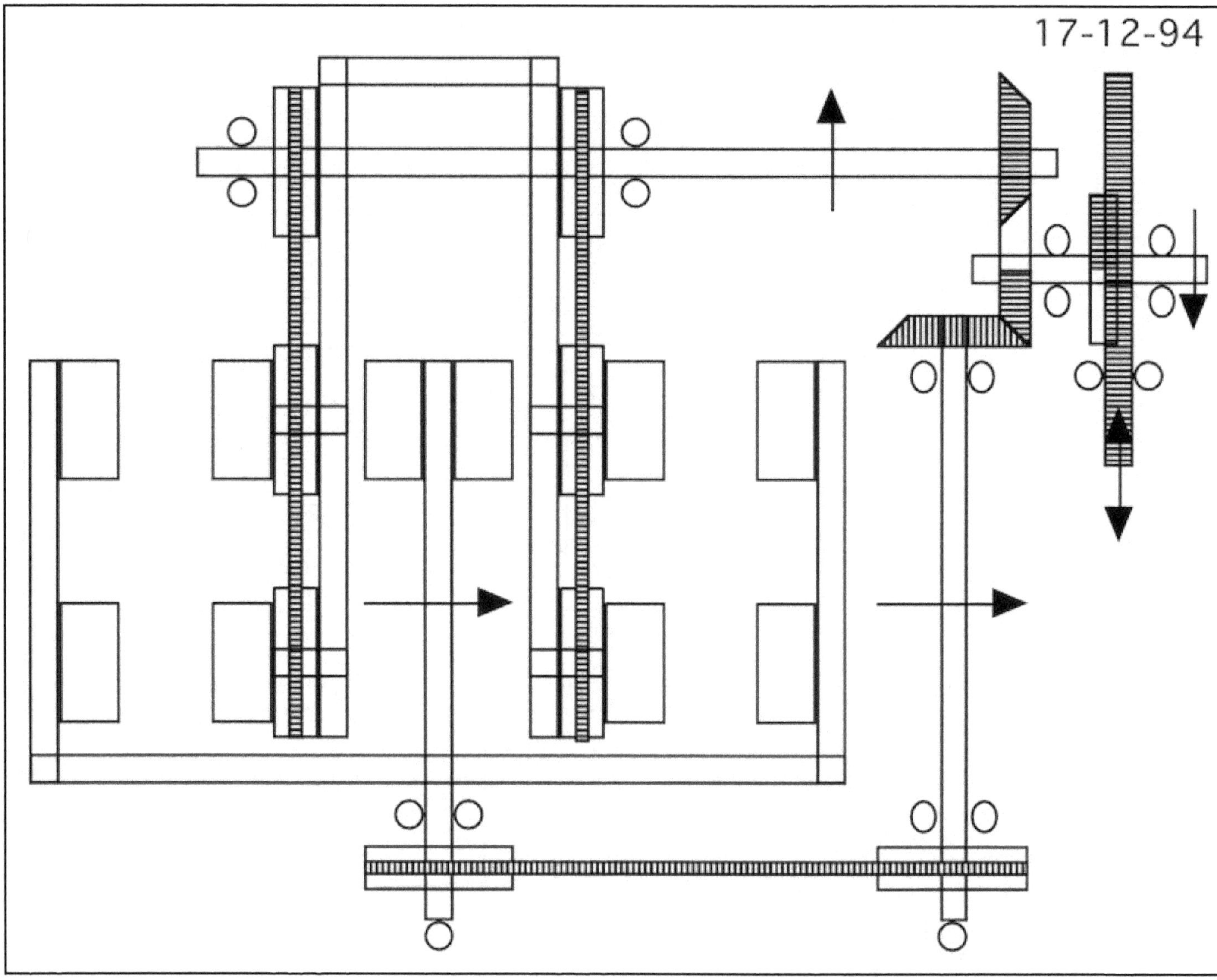
17-12-94

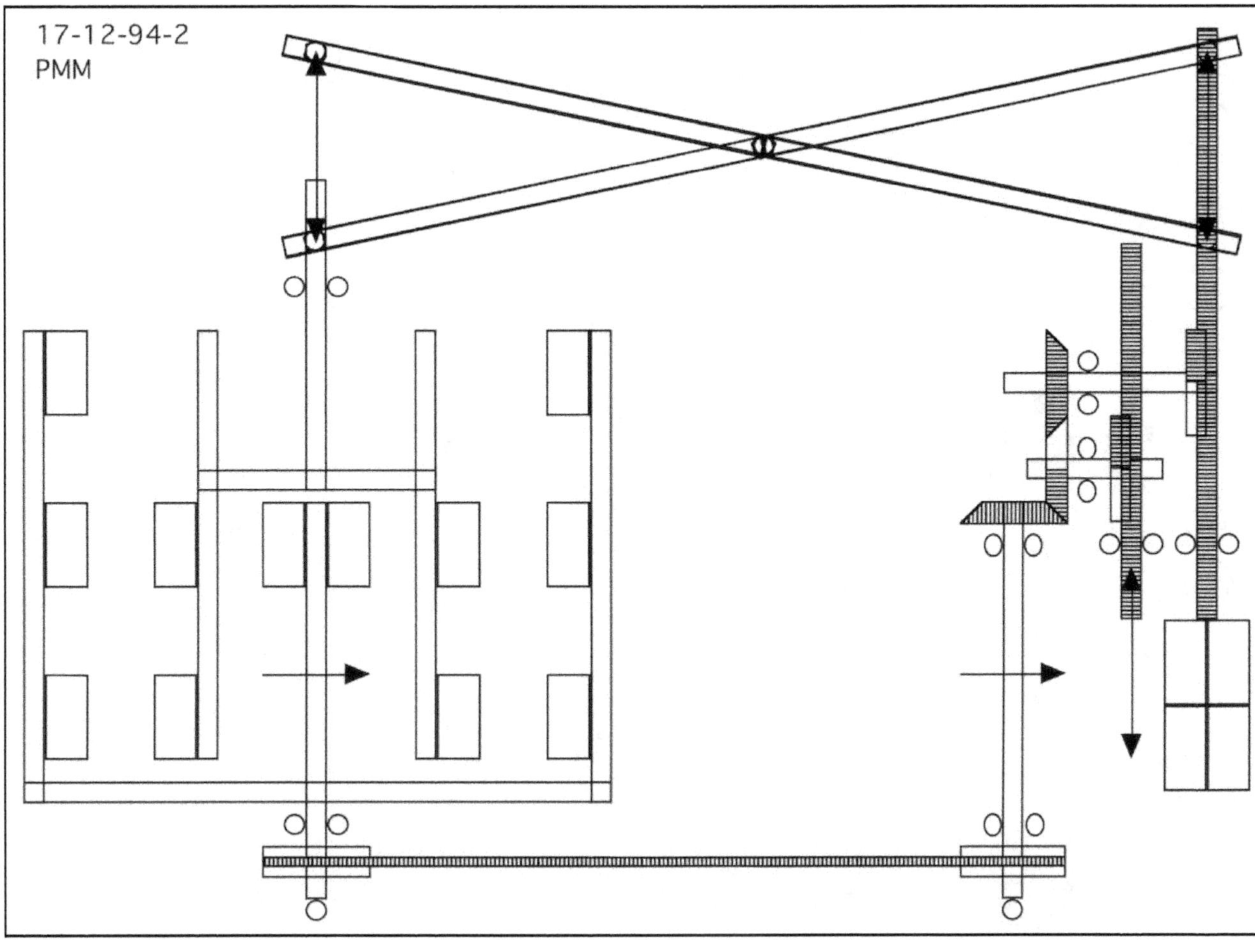
17-12-94-2
PMM

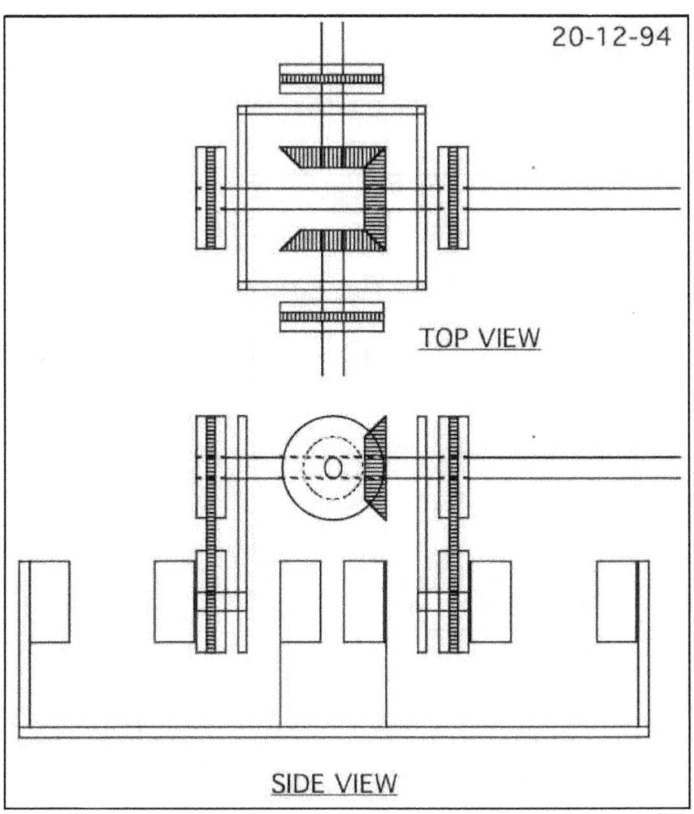
20-12-94
TOP VIEW
SIDE VIEW

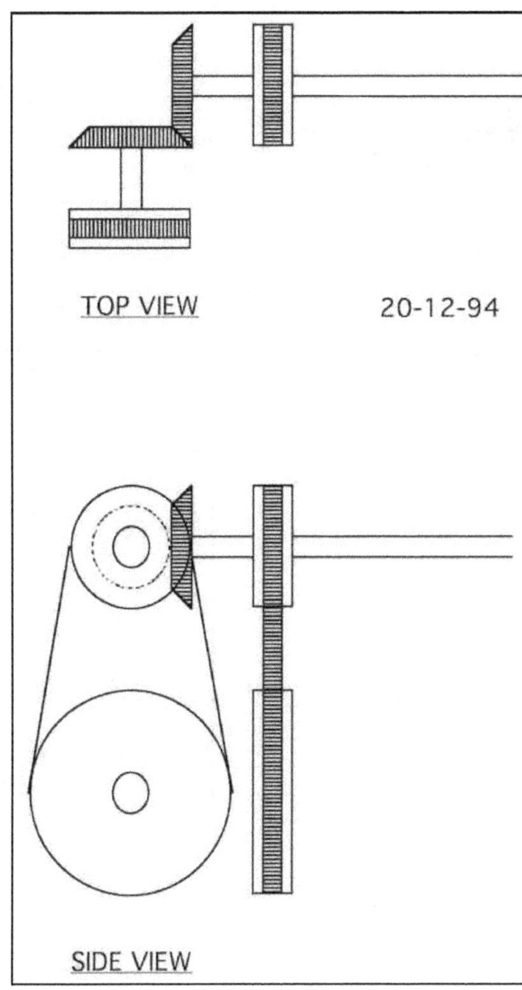
TOP VIEW
20-12-94
SIDE VIEW

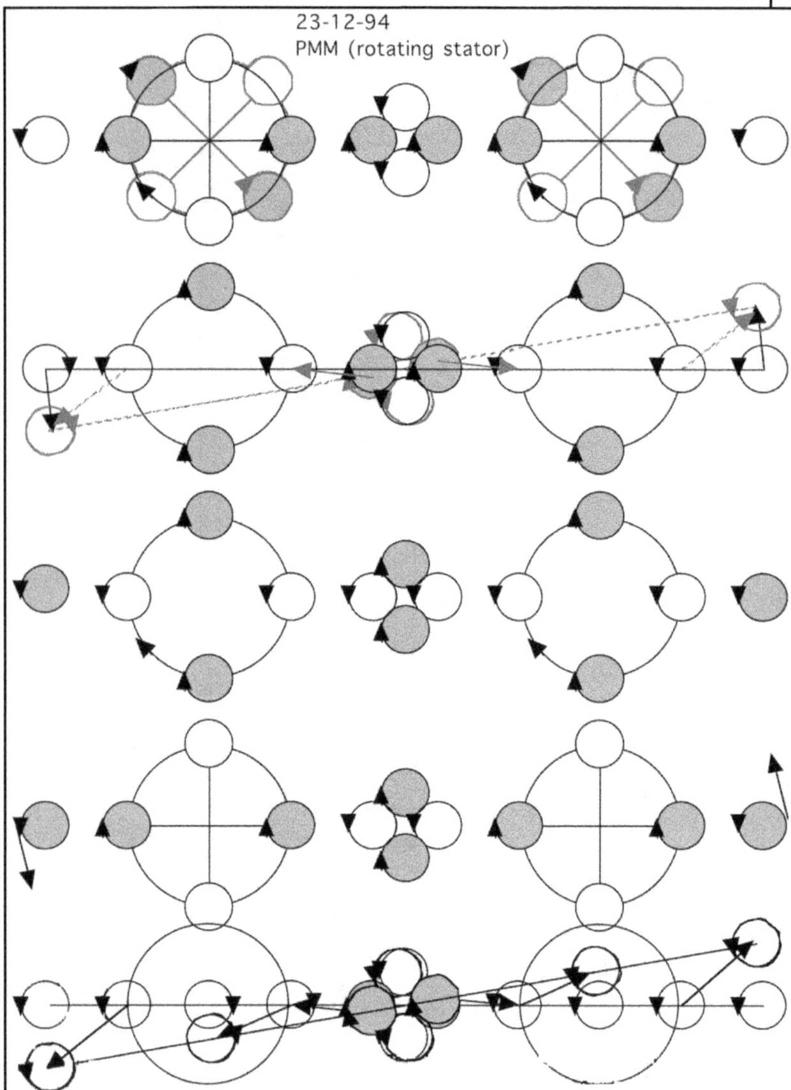

In this 10-degree rotation by rotor, we have a very significant improvement when the stator is enlarged as above, also a 10-degree rotation. The difference is in the angle of attack. The closer to a right angle, the more effective the force.

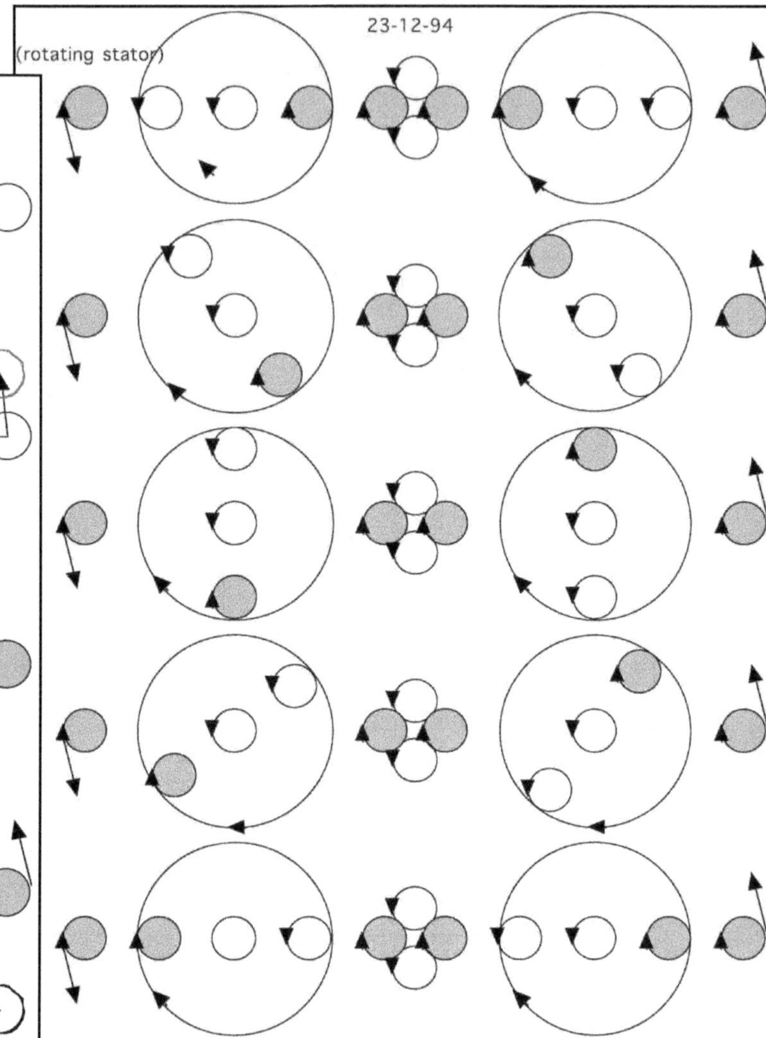

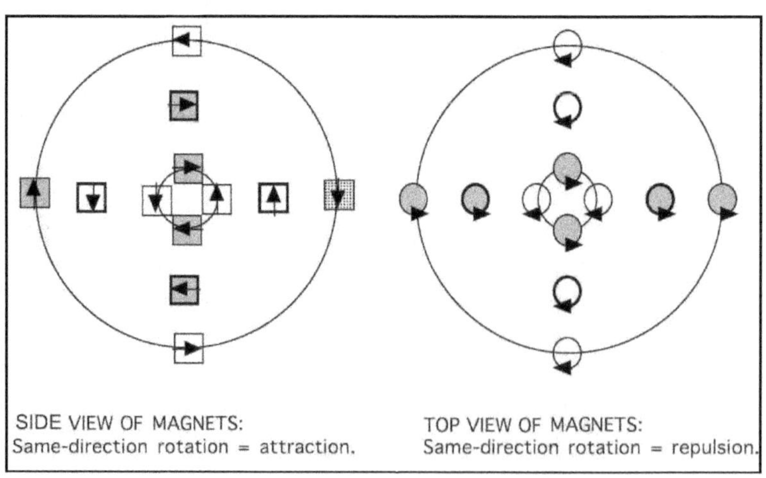

30-12-94 PMM Final Version!
(size of magnets: 3/8 x 7/8 x 1 7/8 inches)

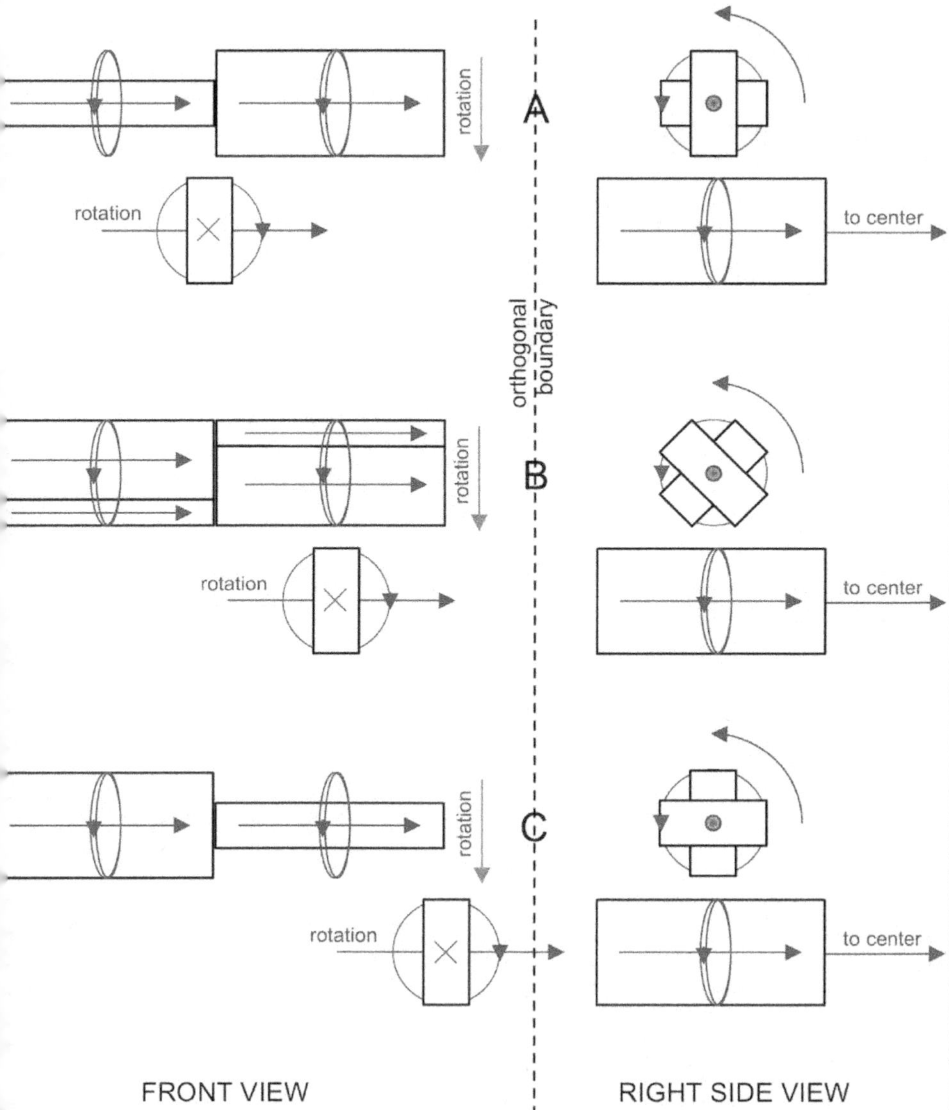

02-01-95 PMM Equilizing Forces of Rotation
(size of magnets: 3/8 x 7/8 x 1 7/8 inches)

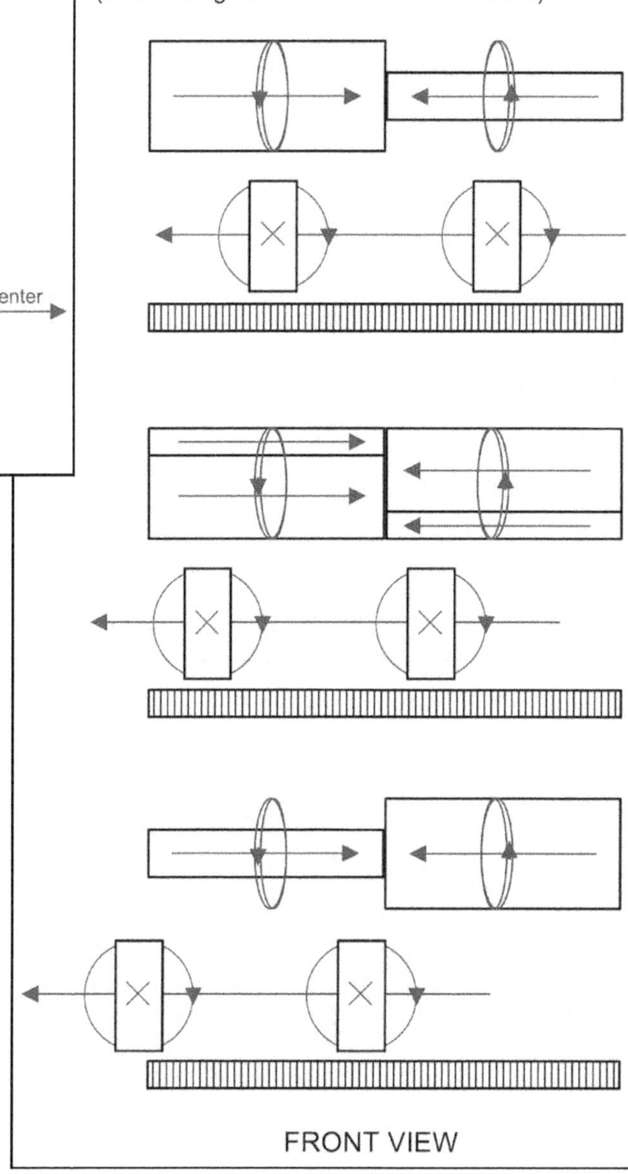

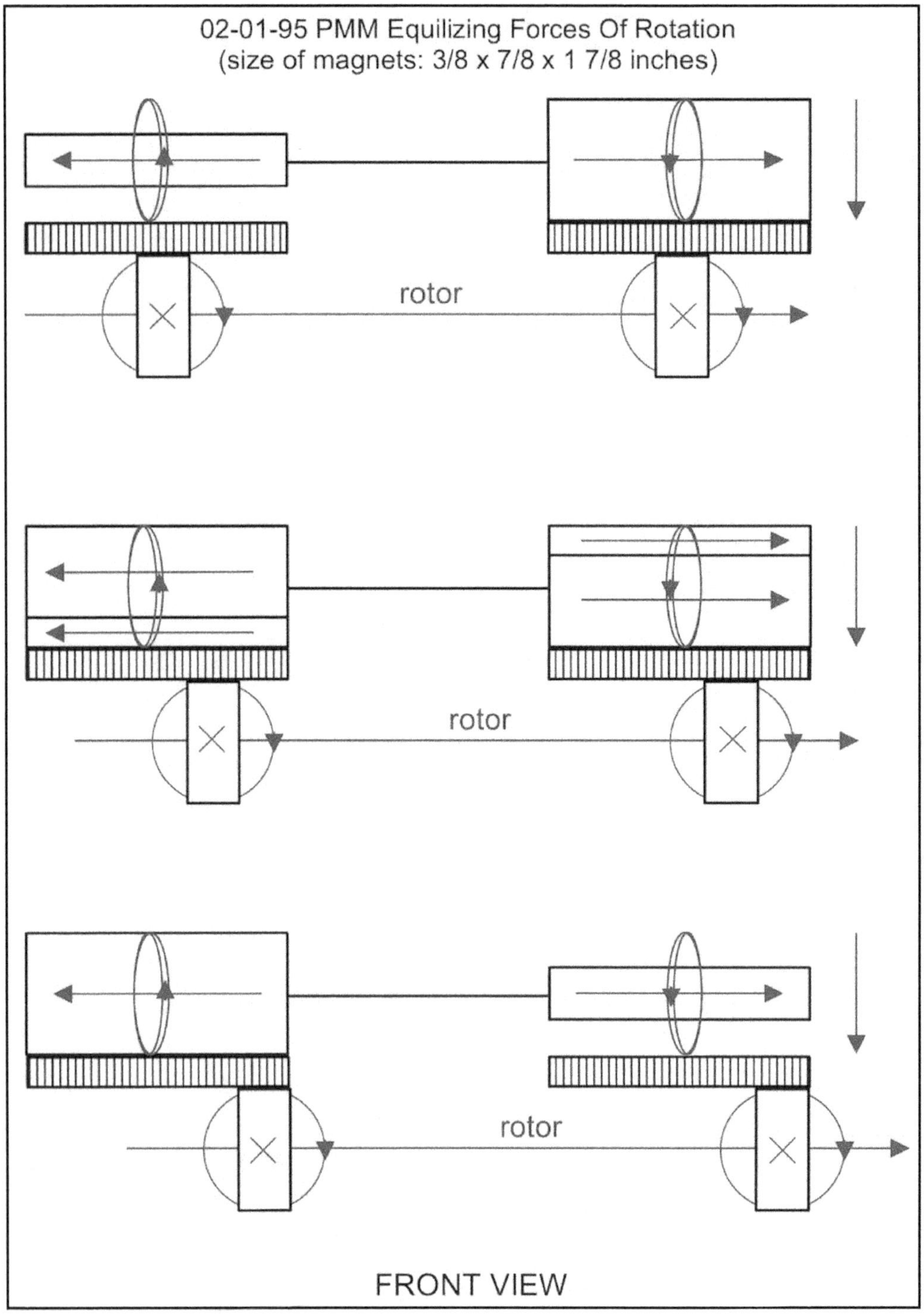
02-01-95 PMM Equilizing Forces Of Rotation
(size of magnets: 3/8 x 7/8 x 1 7/8 inches)
rotor
rotor
rotor
FRONT VIEW

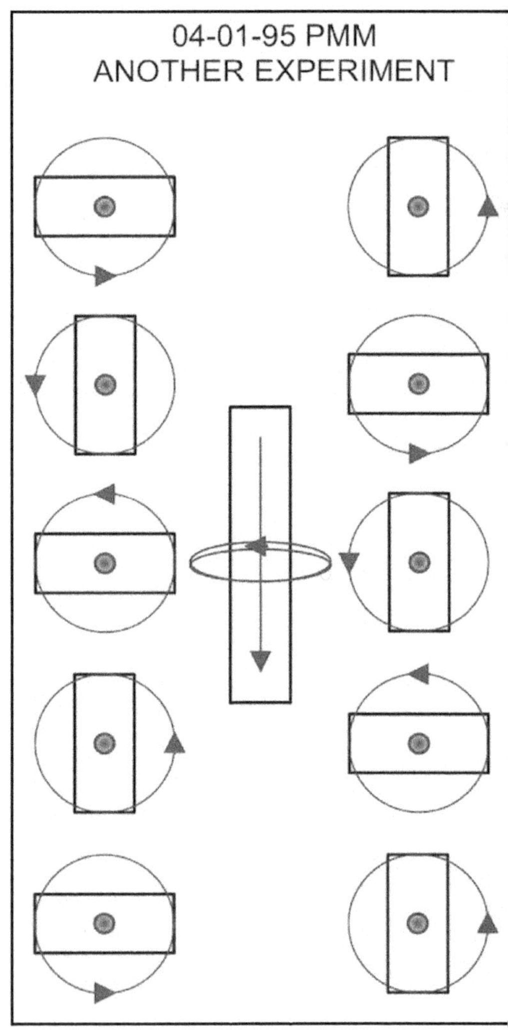
04-01-95 PMM
ANOTHER EXPERIMENT

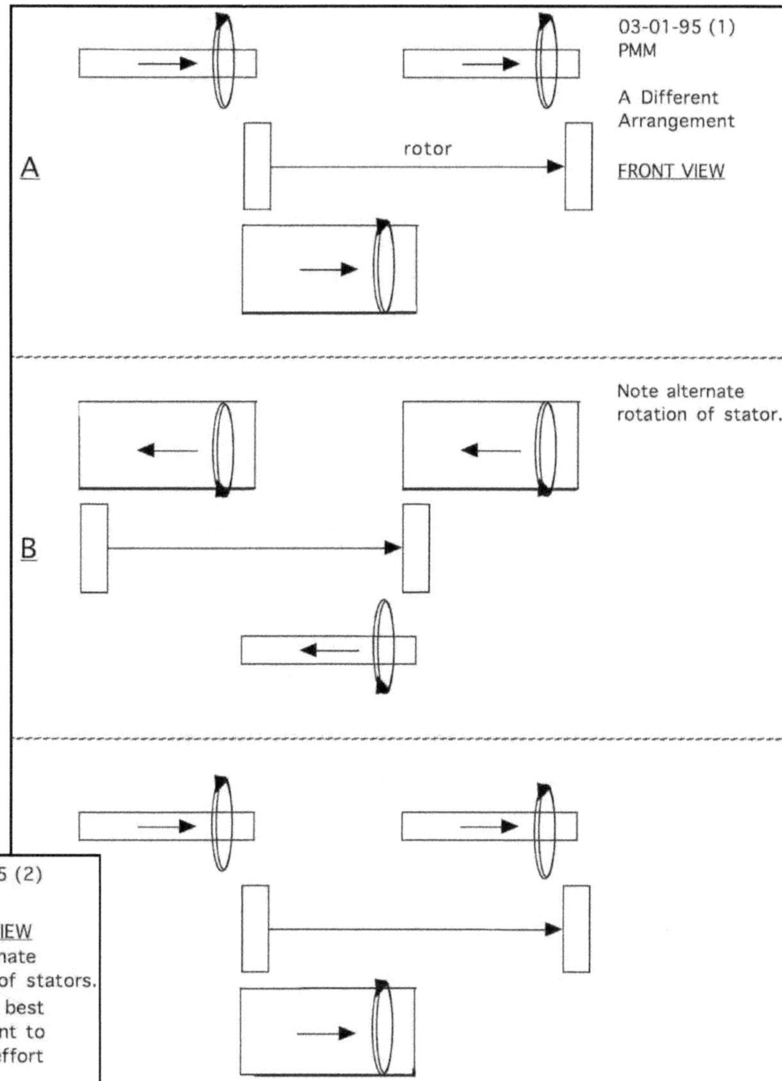
03-01-95 (1)
PMM
A Different Arrangement
FRONT VIEW
A
rotor
B
Note alternate rotation of stator.

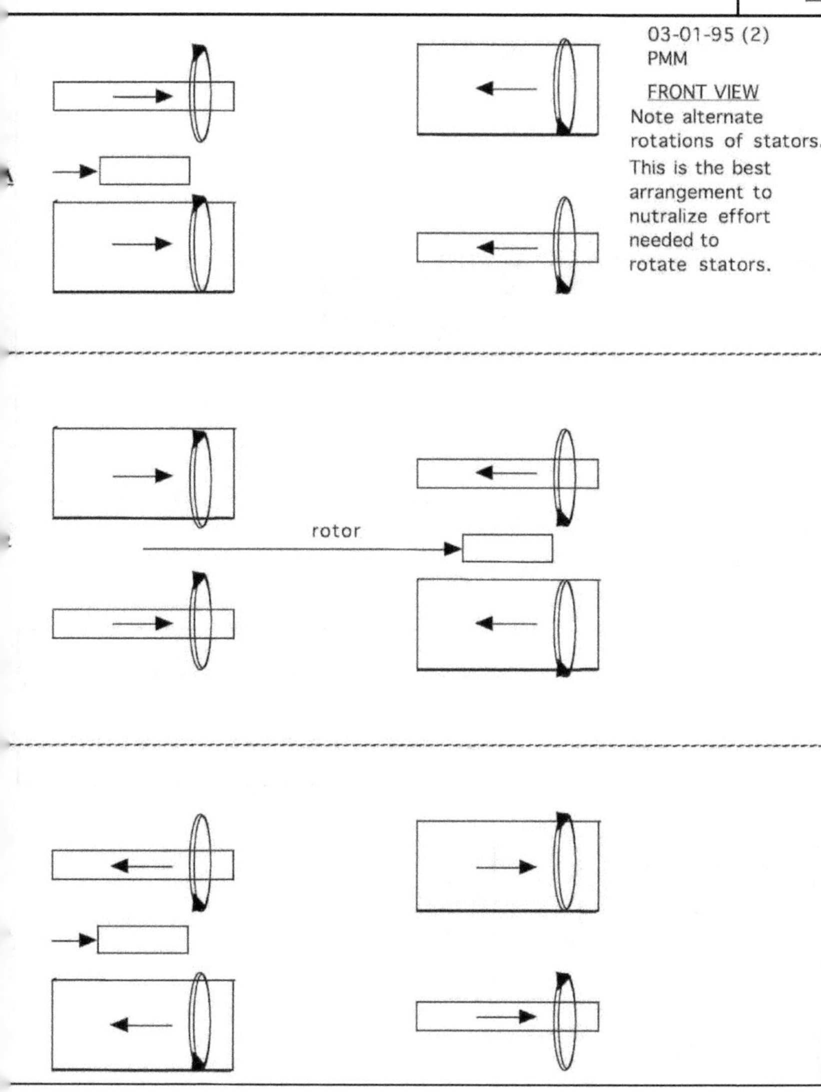
03-01-95 (2)
PMM
FRONT VIEW
Note alternate rotations of stators. This is the best arrangement to nutralize effort needed to rotate stators.
rotor

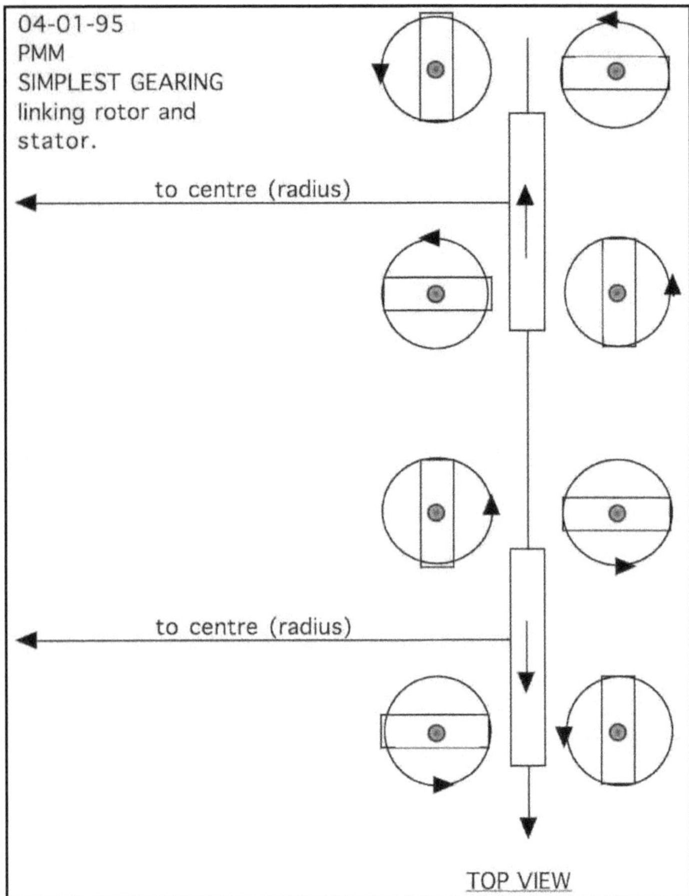
04-01-95
PMM
SIMPLEST GEARING
linking rotor and
stator.
to centre (radius)
to centre (radius)
TOP VIEW

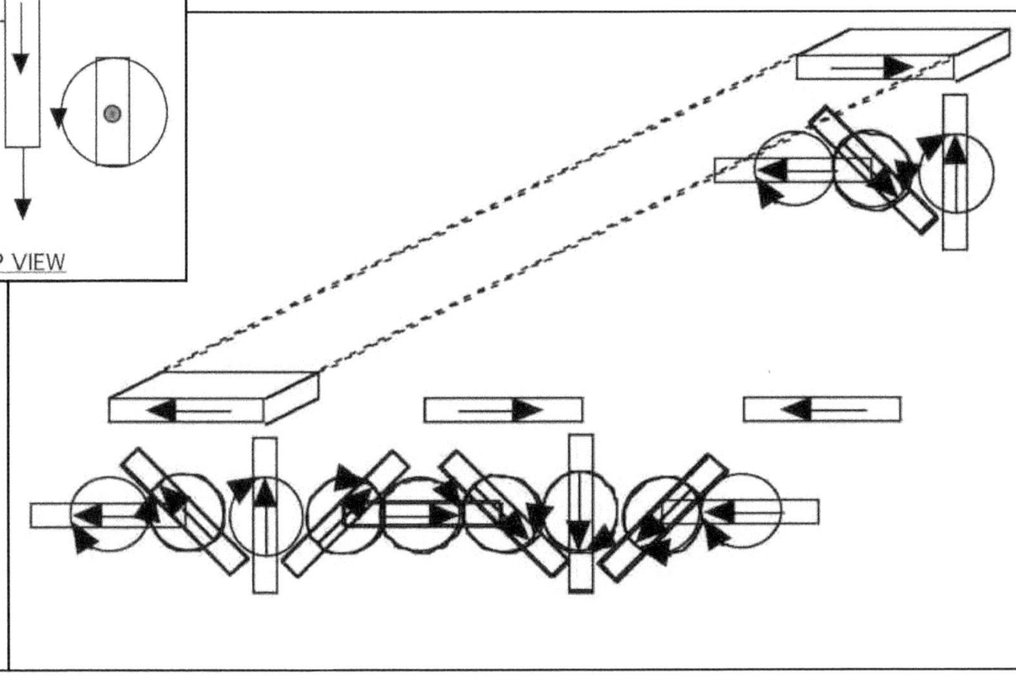

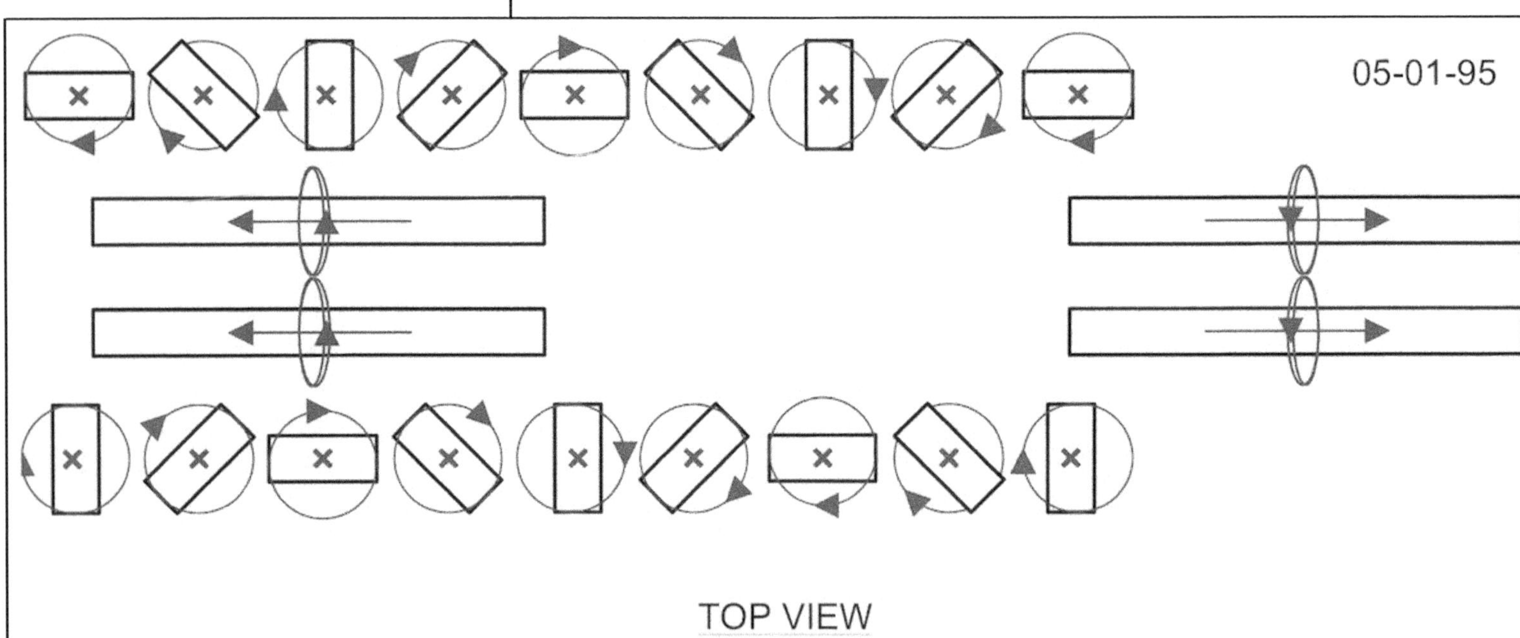
05-01-95
TOP VIEW

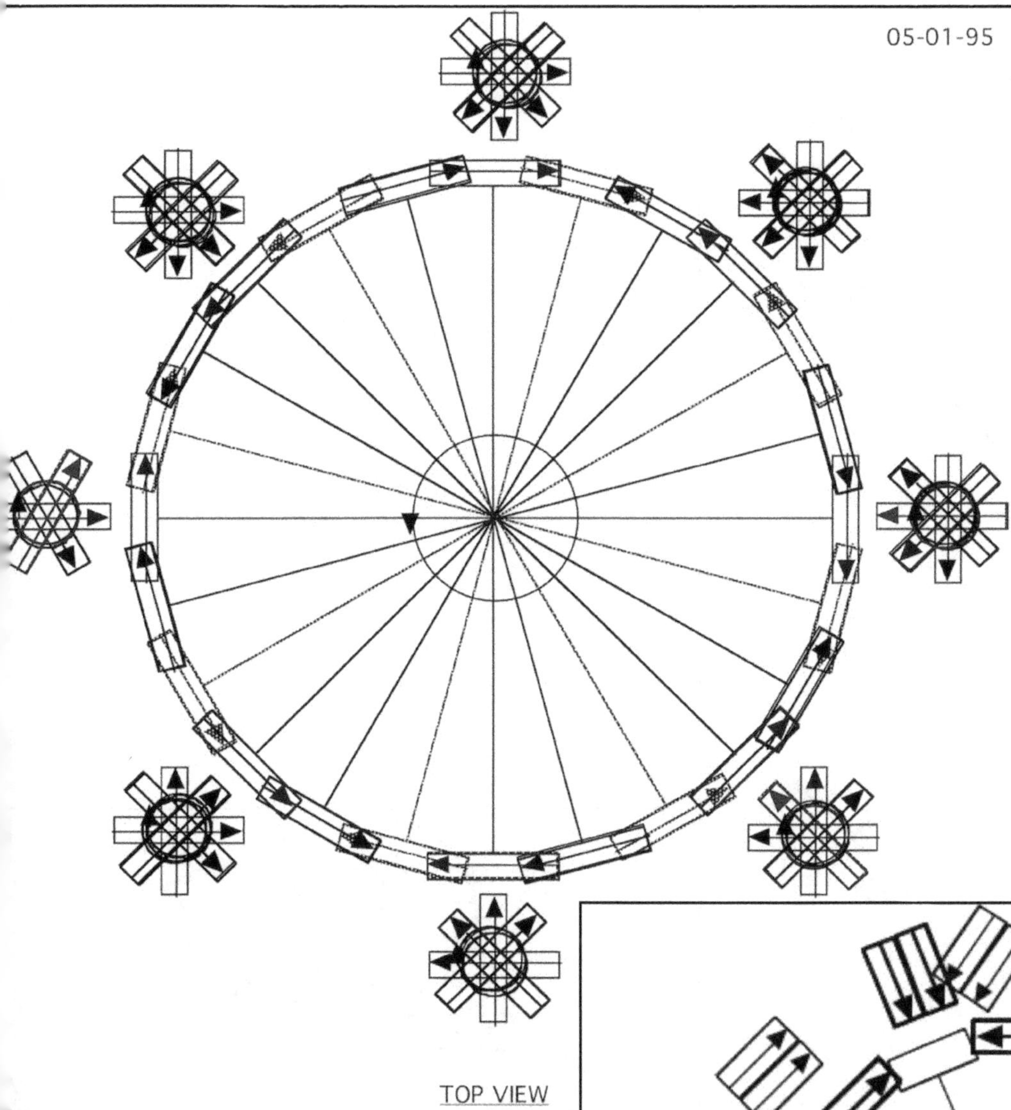
05-01-95
TOP VIEW

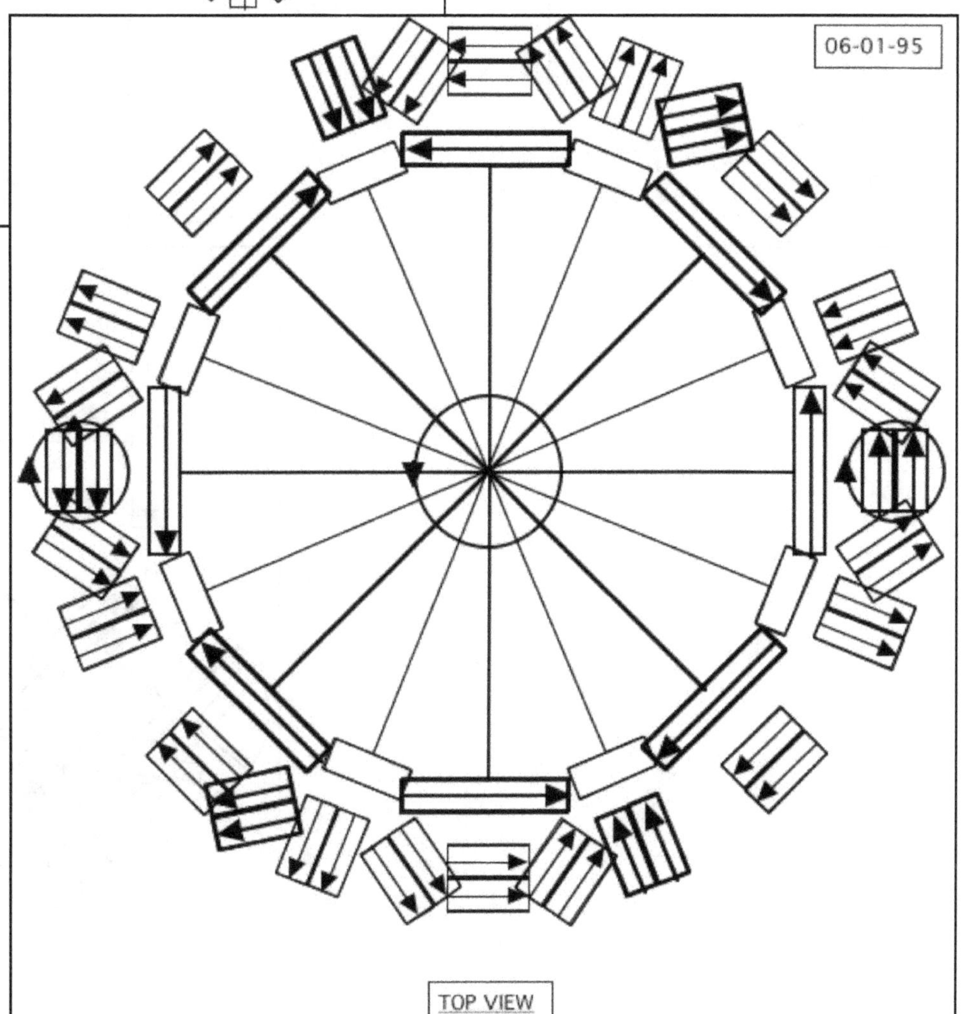
06-01-95
TOP VIEW

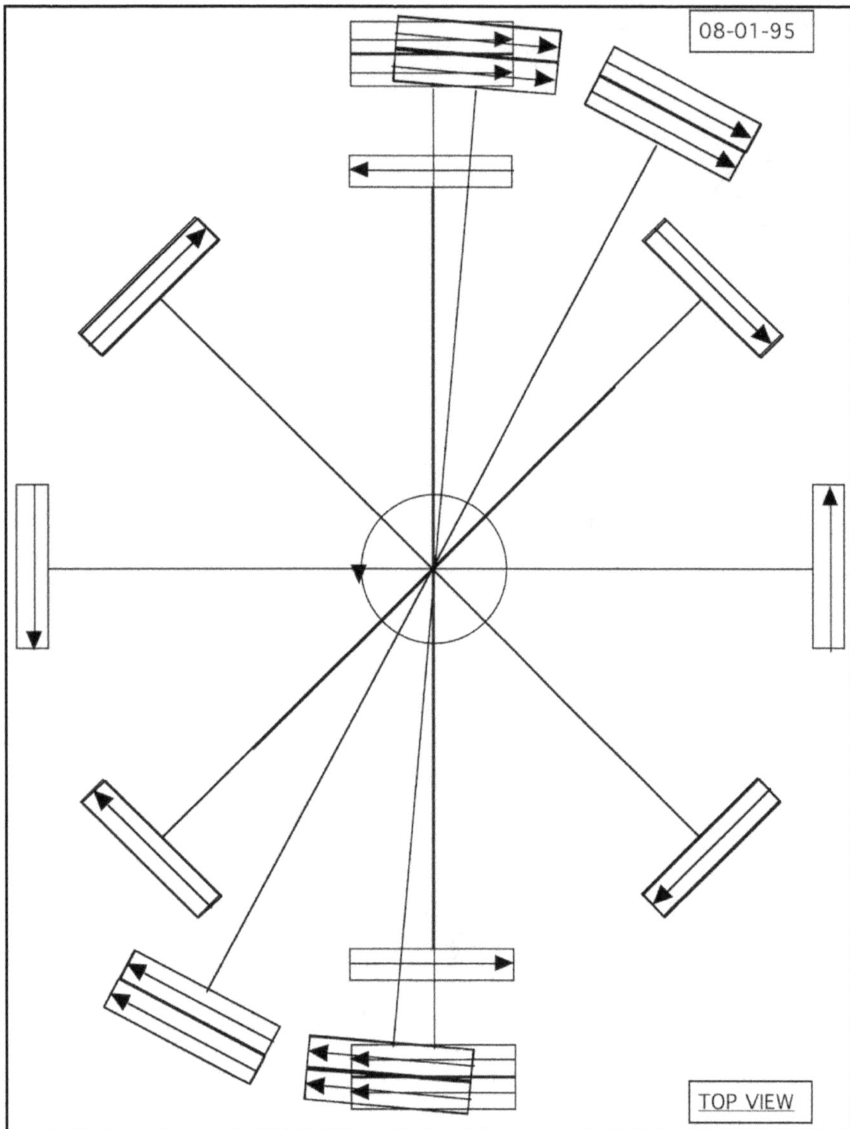
08-01-95
TOP VIEW

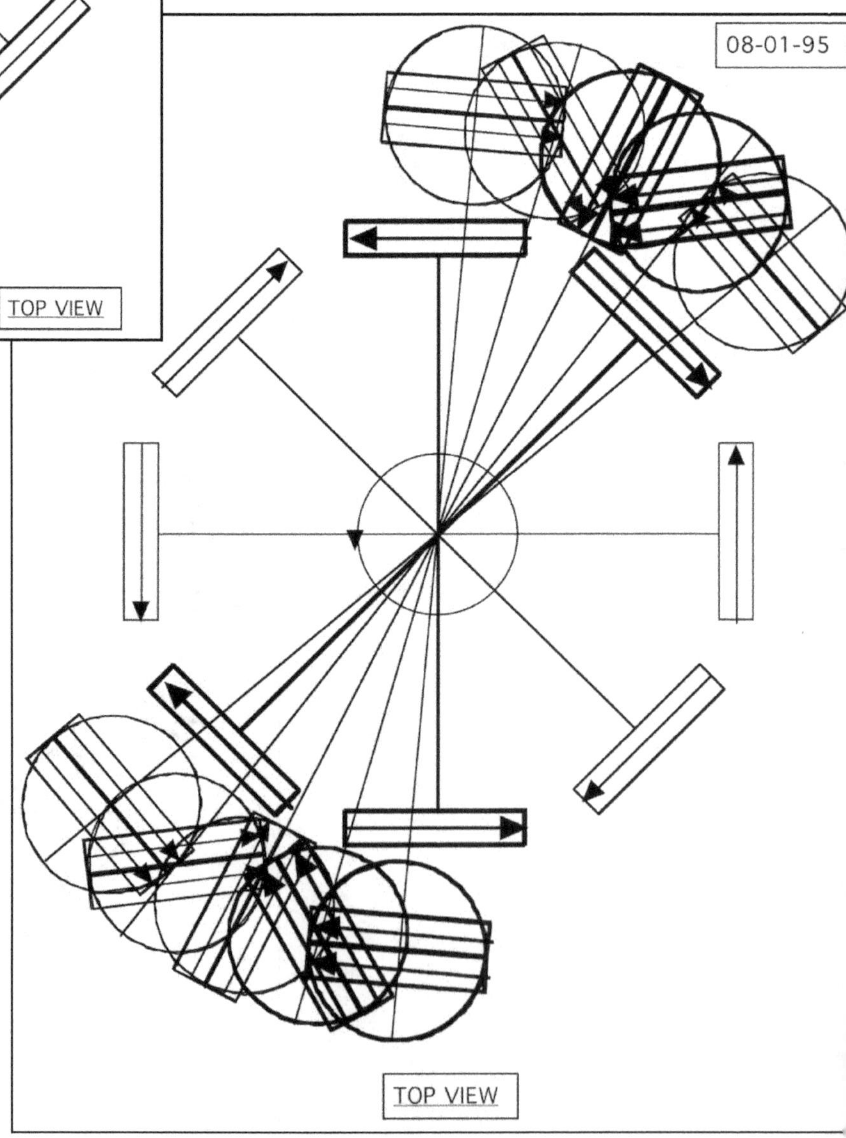
08-01-95
TOP VIEW

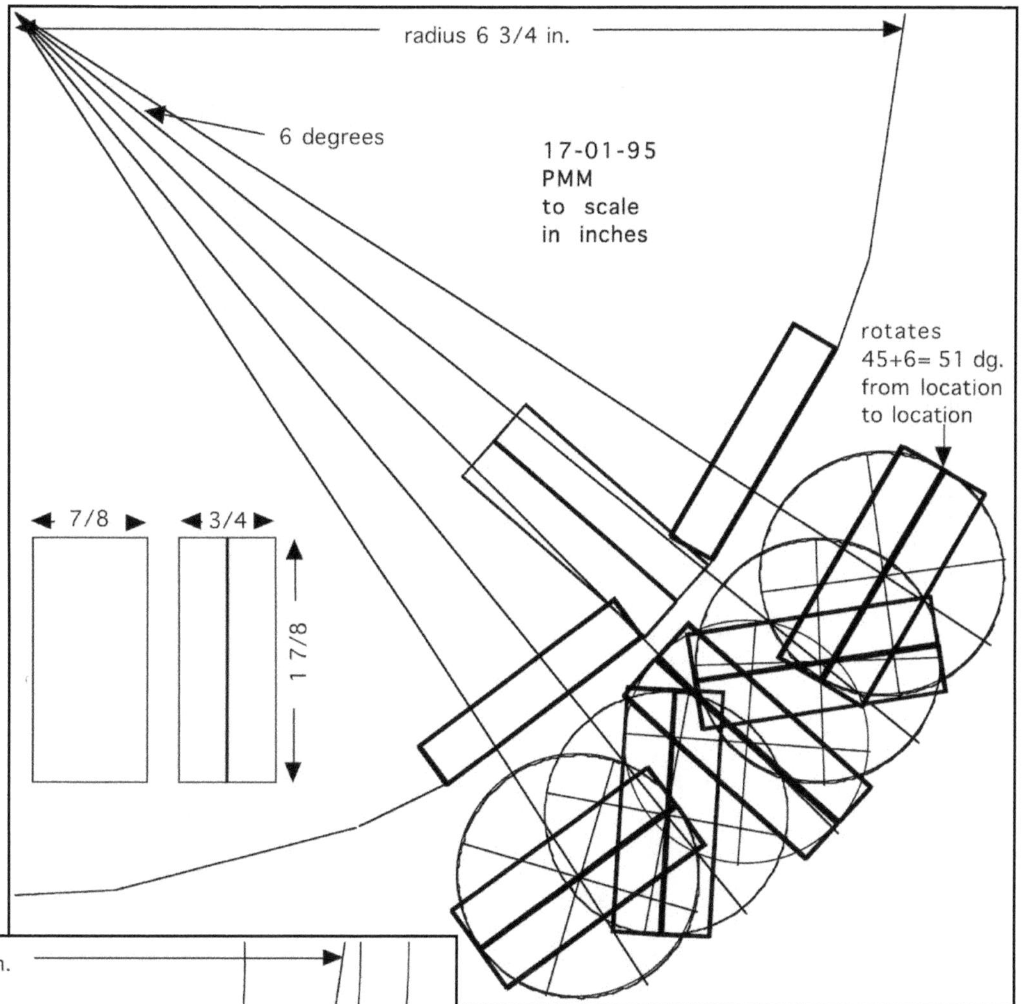
radius 6 3/4 in.
6 degrees
17-01-95
PMM
to scale
in inches
rotates
45+6= 51 dg.
from location
to location
7/8
3/4
1 7/8

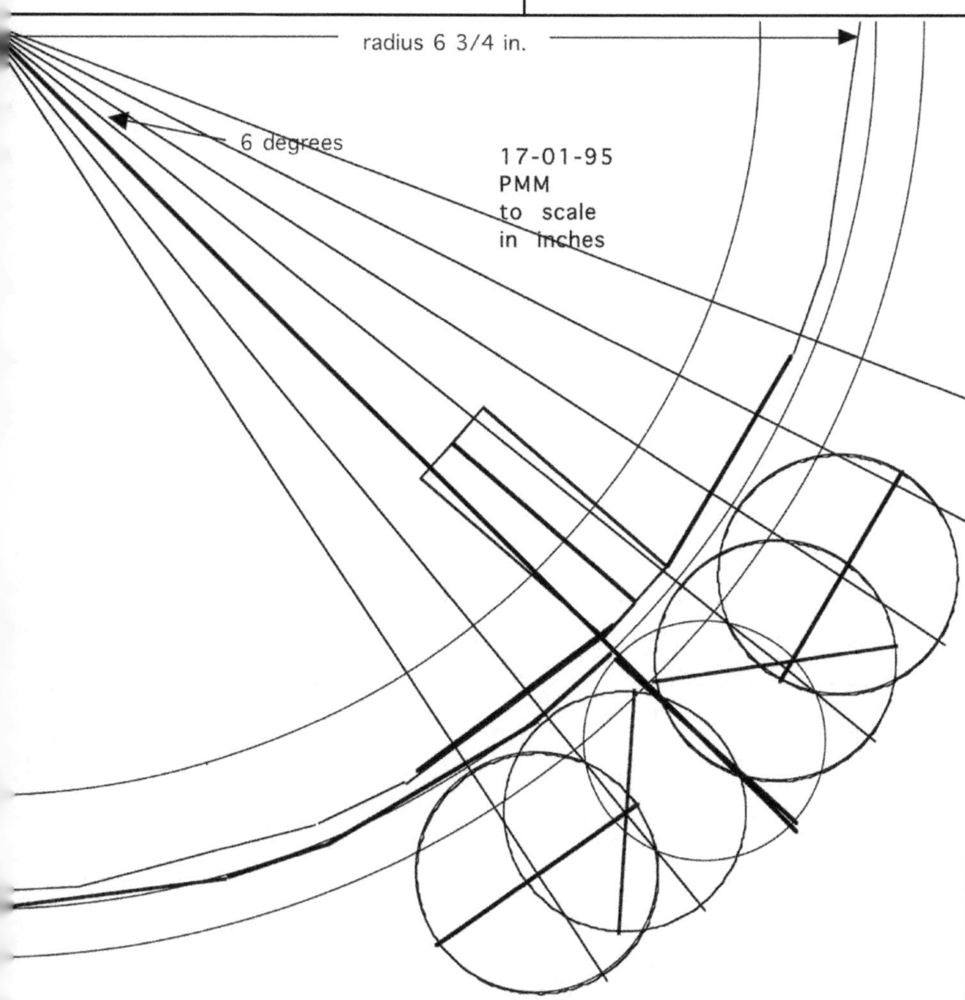
radius 6 3/4 in.
6 degrees
17-01-95
PMM
to scale
in inches

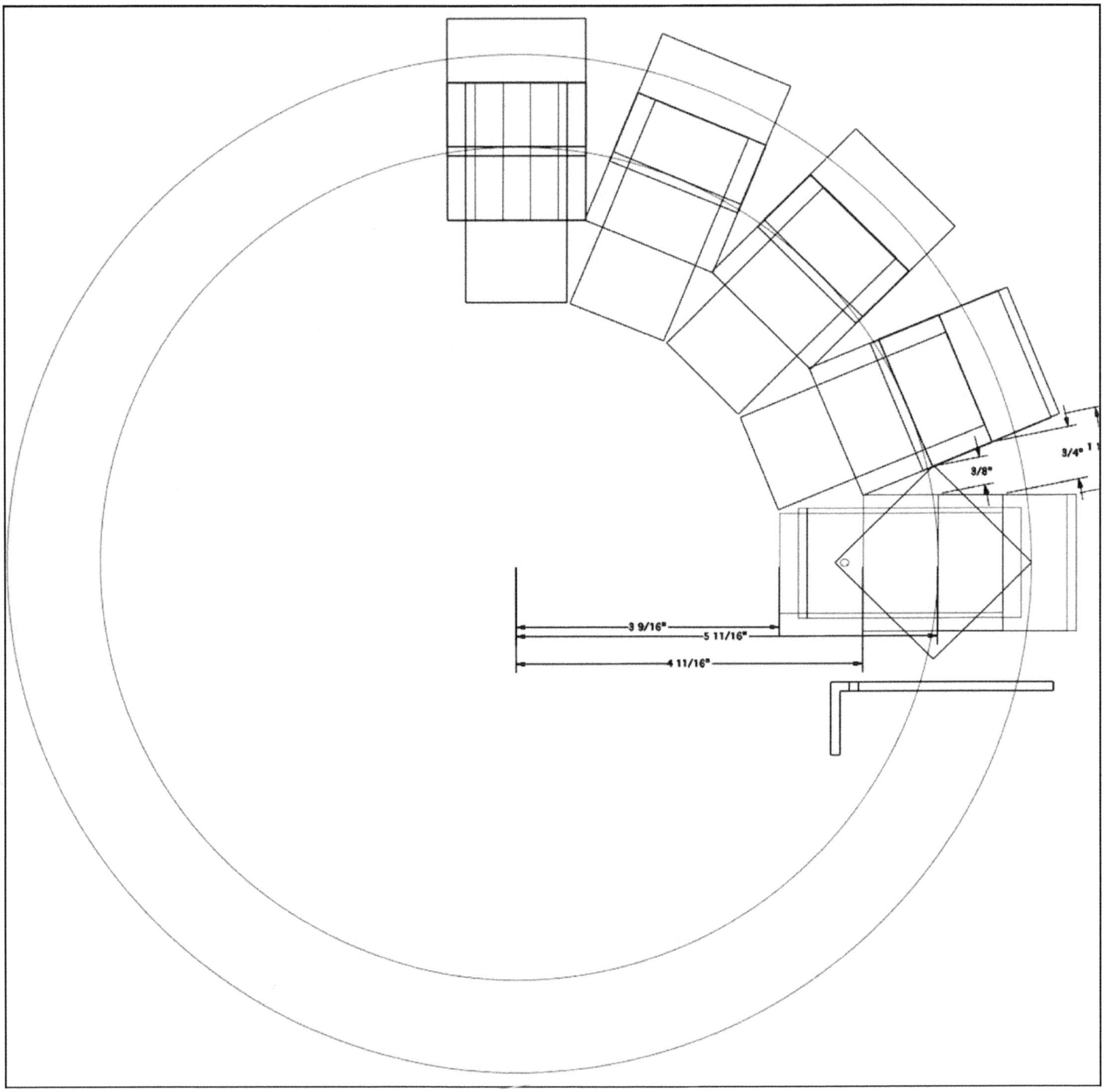
3/4"
3/8"
3 9/16"
5 11/16"
4 11/16"

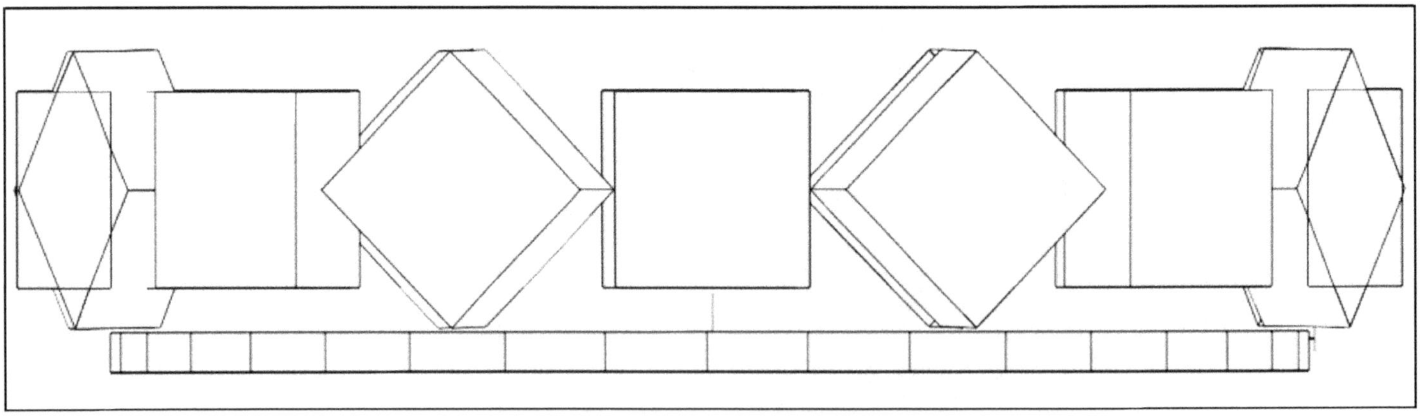

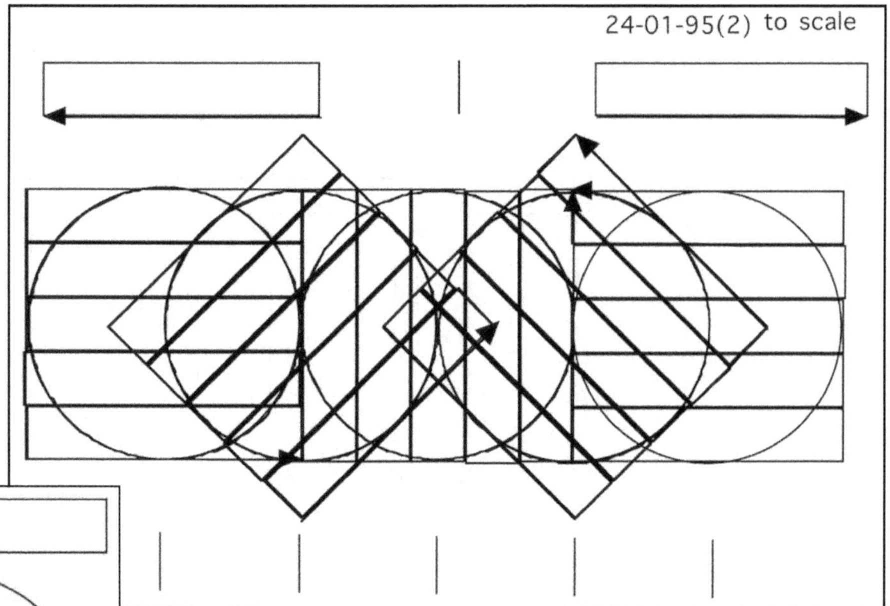
24-01-95(2) to scale

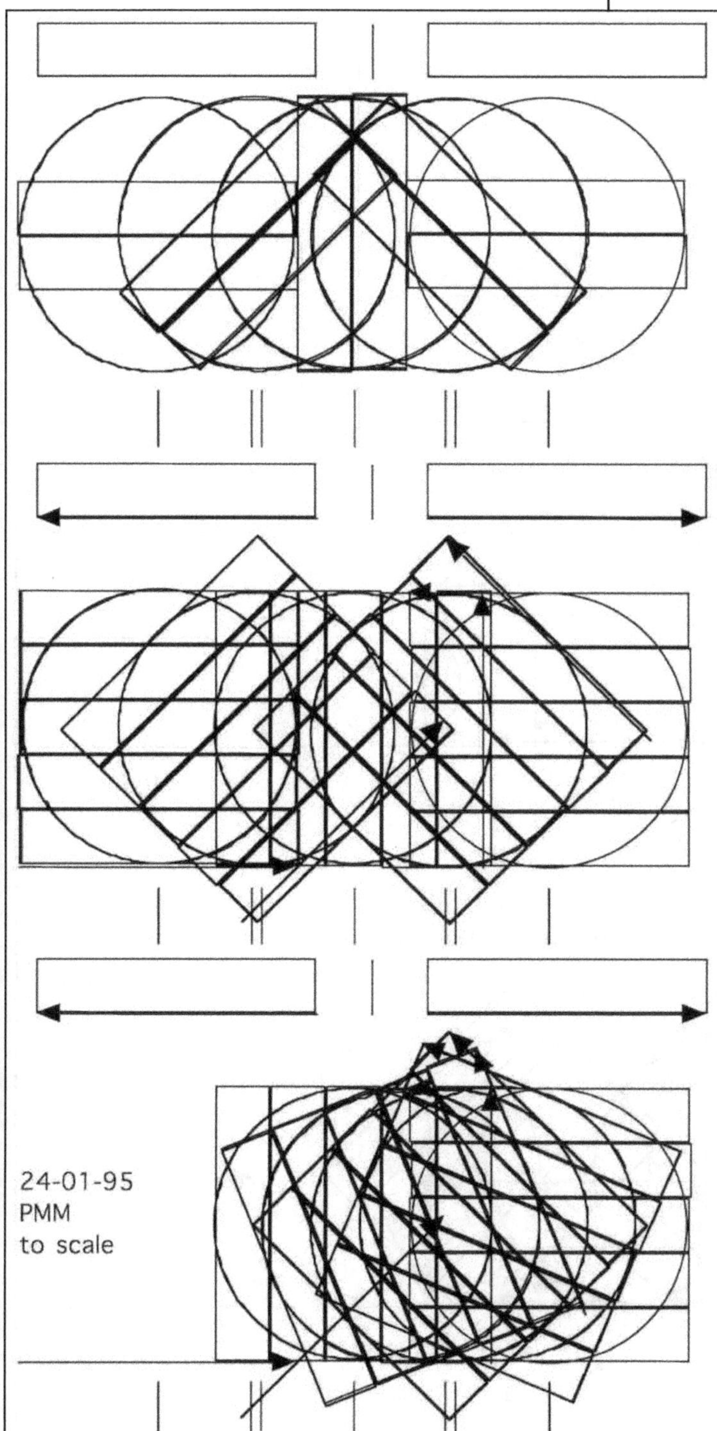
24-01-95
PMM
to scale

04-02-95 PMM to scale. This is it! Rotation of stator in perfect equilibrium. Both rotor and stator rotate counterclockwise.

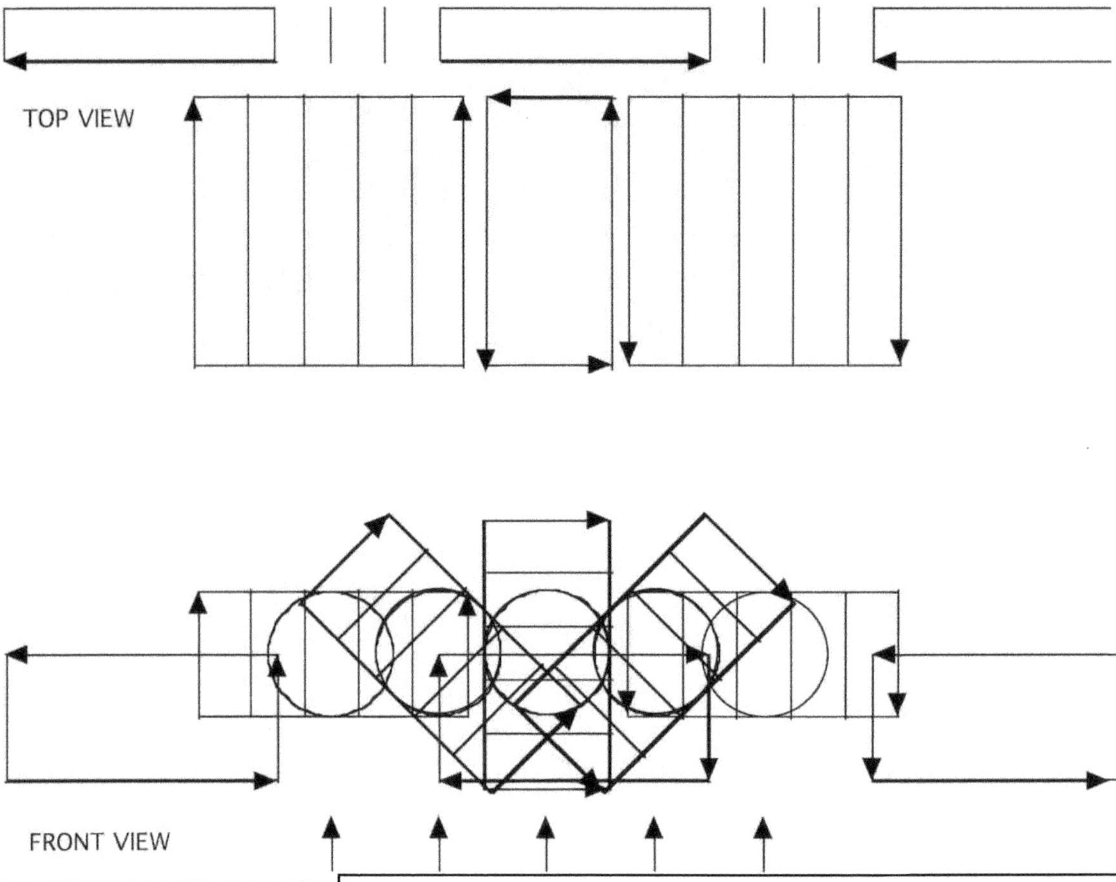

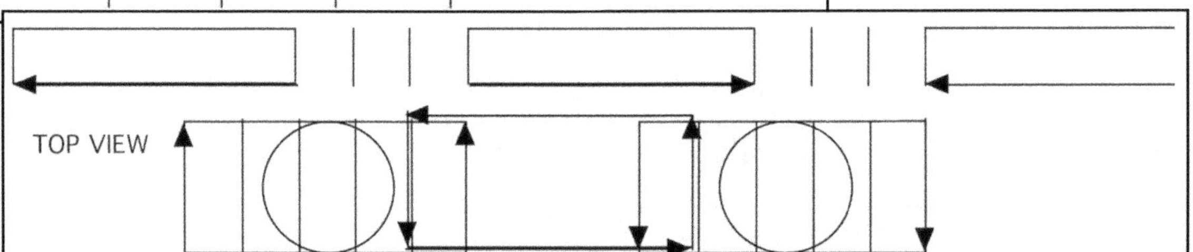

04-02-95 (2) ANOTHER ARRANGEMENT
PMM to scale.
This is it!
Rotation of stator in perfect equilibrium.
Both rotor and stator rotate counterclockwise.

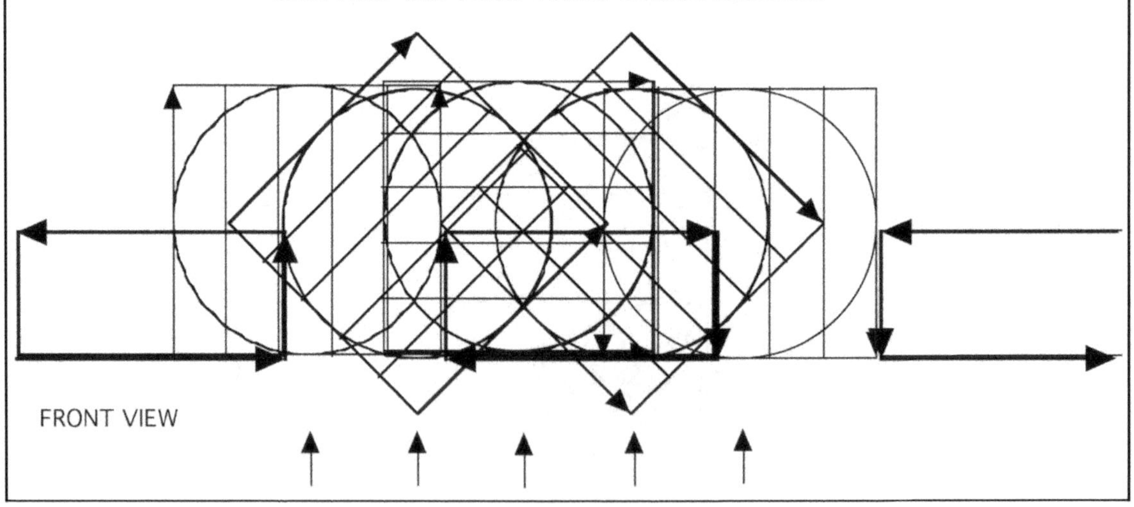

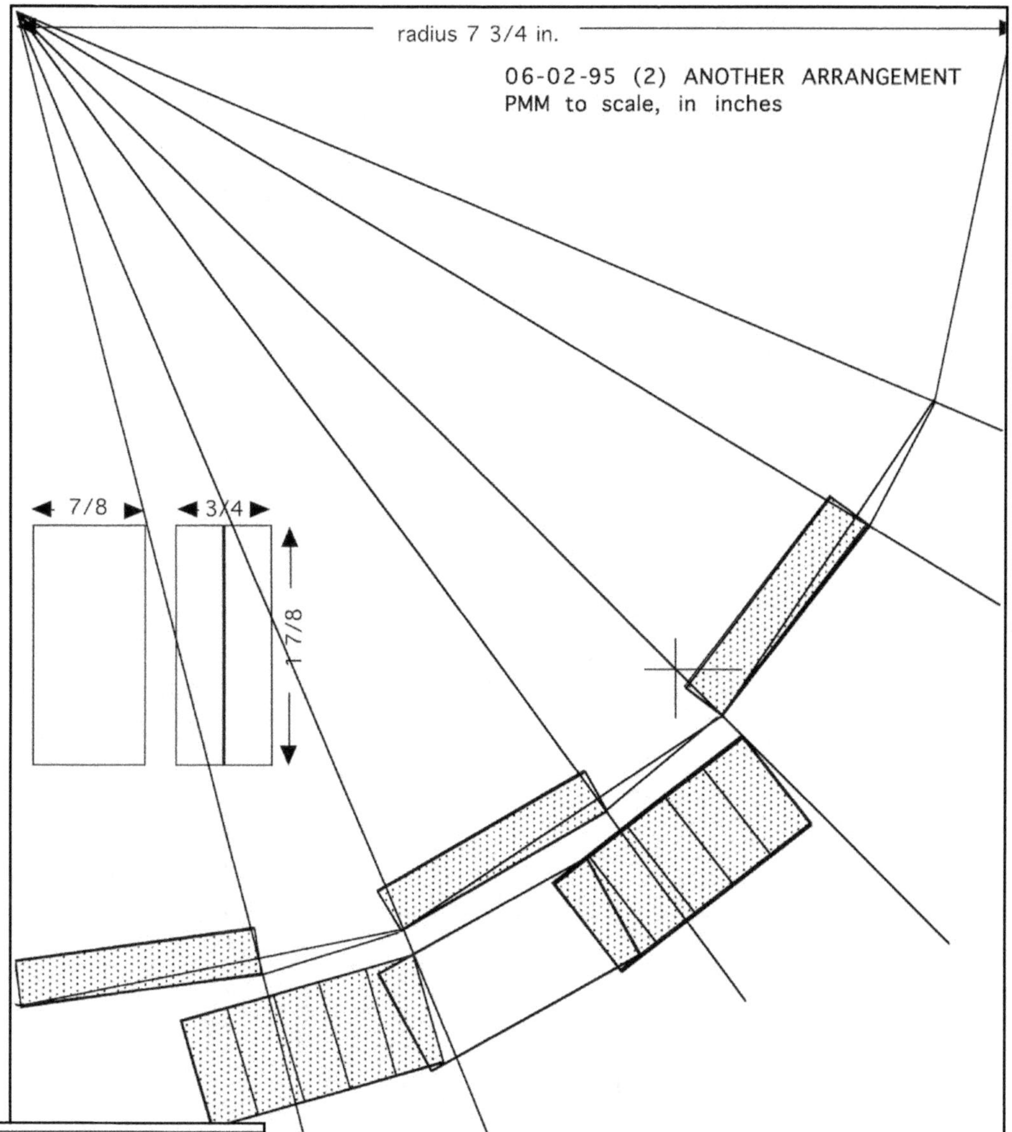
radius 7 3/4 in.
06-02-95 (2) ANOTHER ARRANGEMENT
PMM to scale, in inches
7/8
3/4
1 7/8

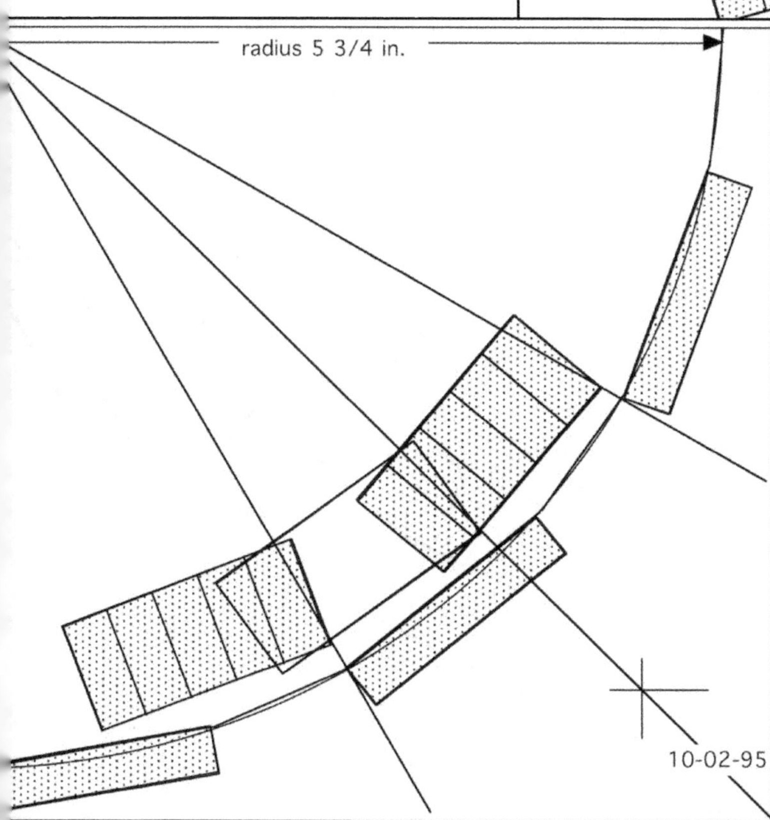
radius 5 3/4 in.
10-02-95

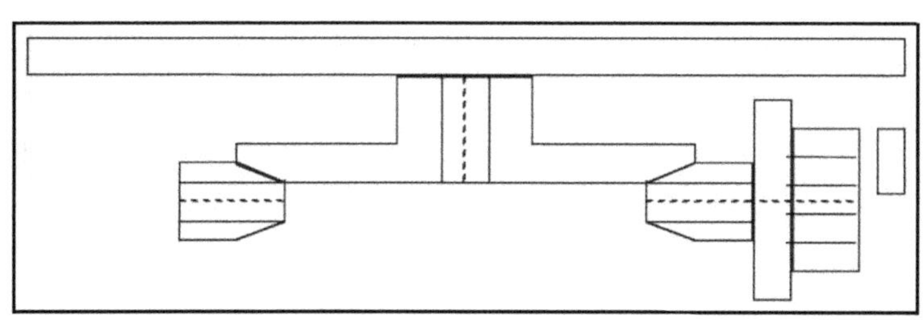

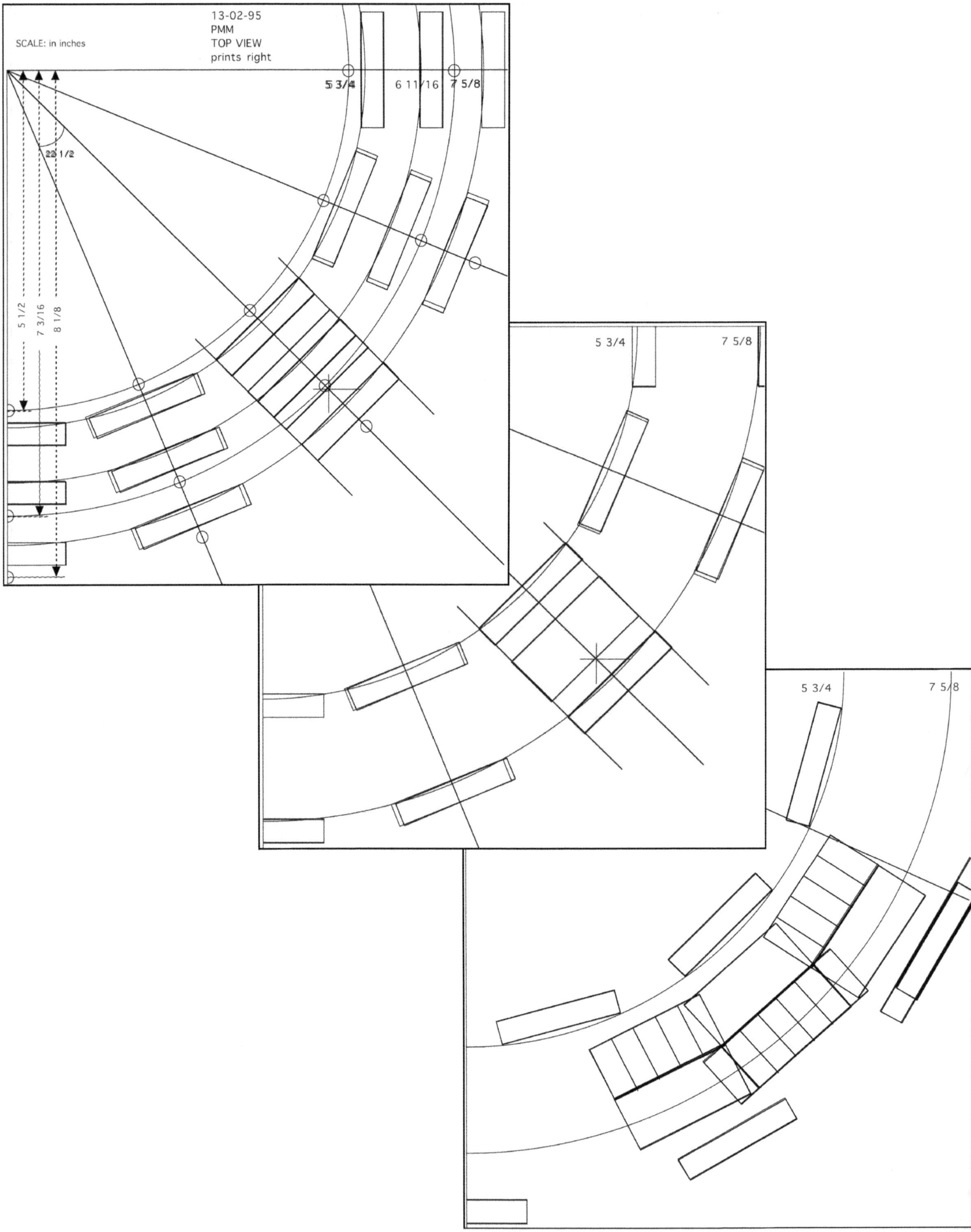
13-02-95
PMM
TOP VIEW
prints right
SCALE: in inches
5 3/4
6 11/16
7 5/8
22 1/2
5 1/2
7 3/16
8 1/8
5 3/4
7 5/8
5 3/4
7 5/8

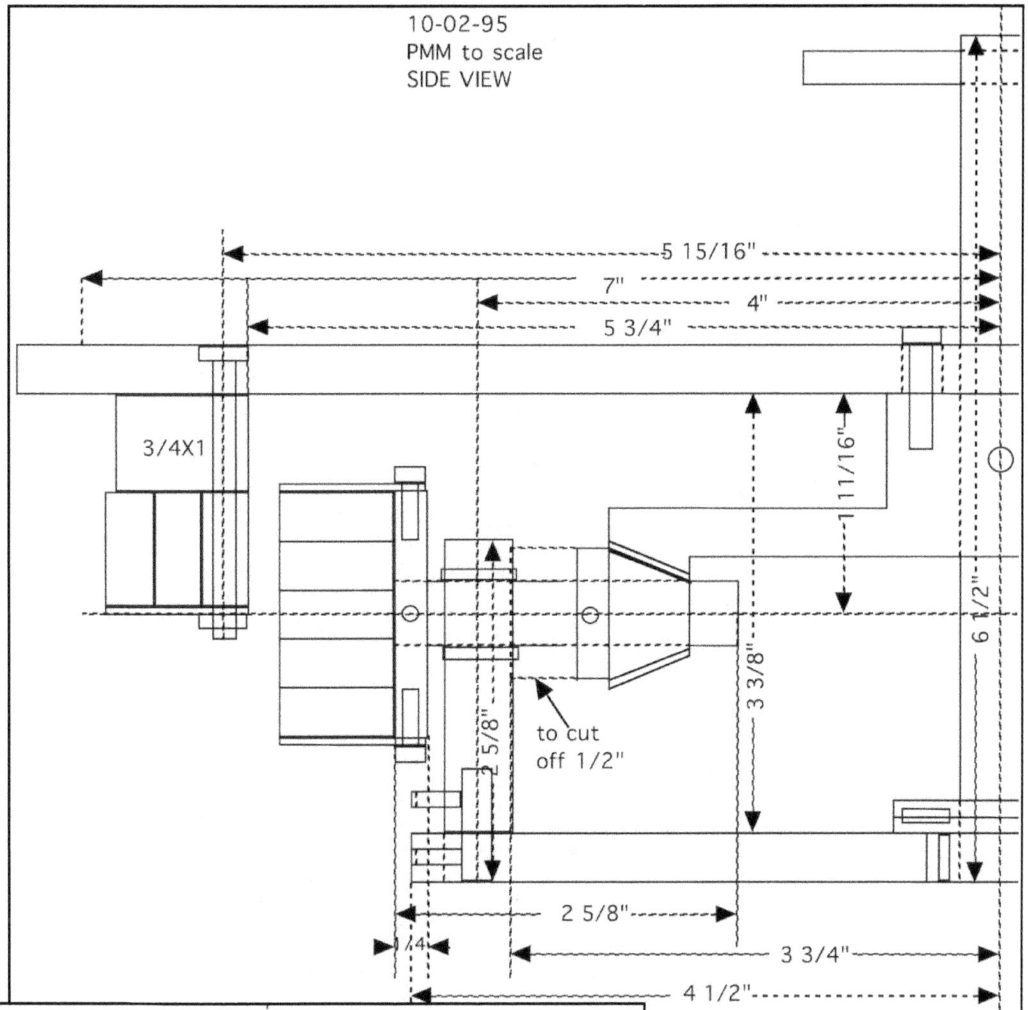
10-02-95
PMM to scale
SIDE VIEW
5 15/16"
7"
4"
5 3/4"
3/4X1
1 11/16"
6 1/2"
3 3/8"
2 5/8"
to cut
off 1/2"
2 5/8"
1/4
3 3/4"
4 1/2"

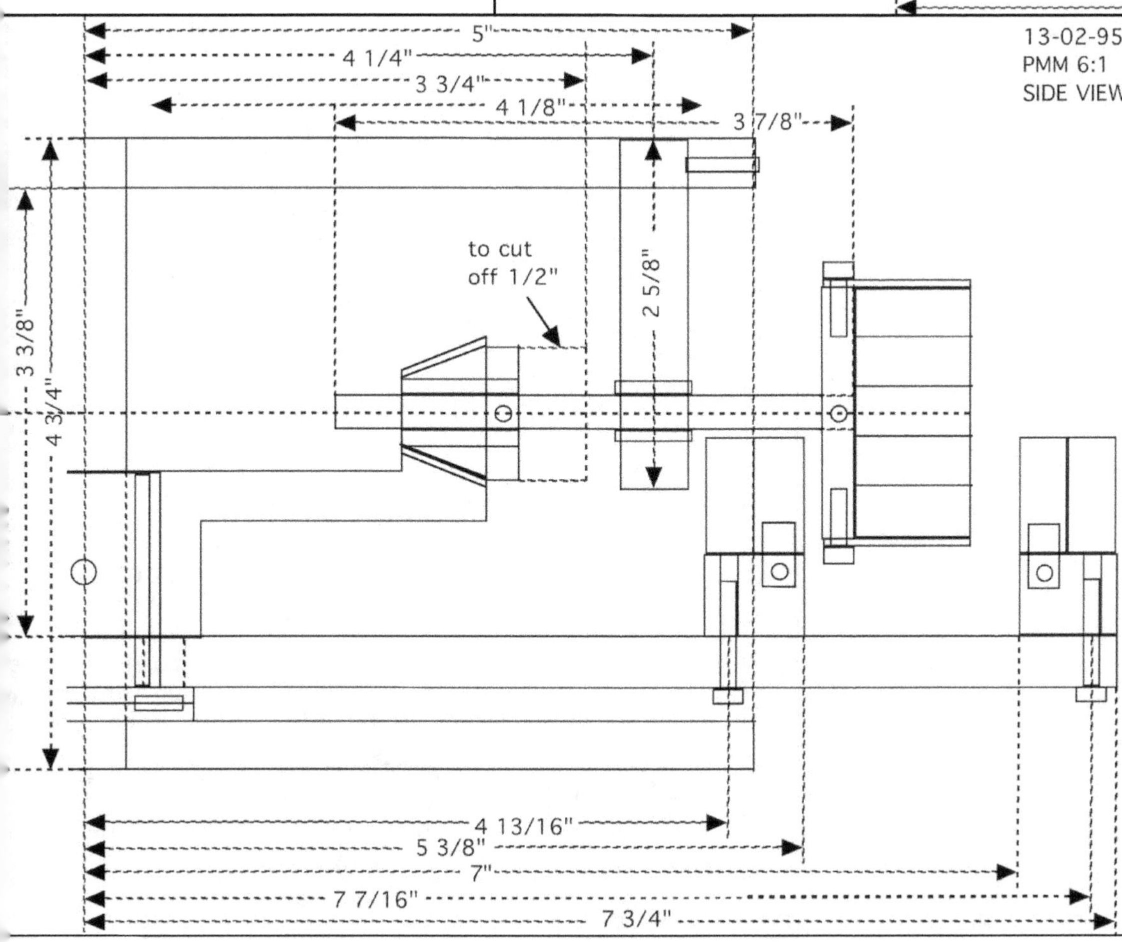
13-02-95
PMM 6:1
SIDE VIEW
5"
4 1/4"
3 3/4"
4 1/8"
3 7/8"
to cut
off 1/2"
2 5/8"
3 3/8"
4 3/4"
4 13/16"
5 3/8"
7"
7 7/16"
7 3/4"

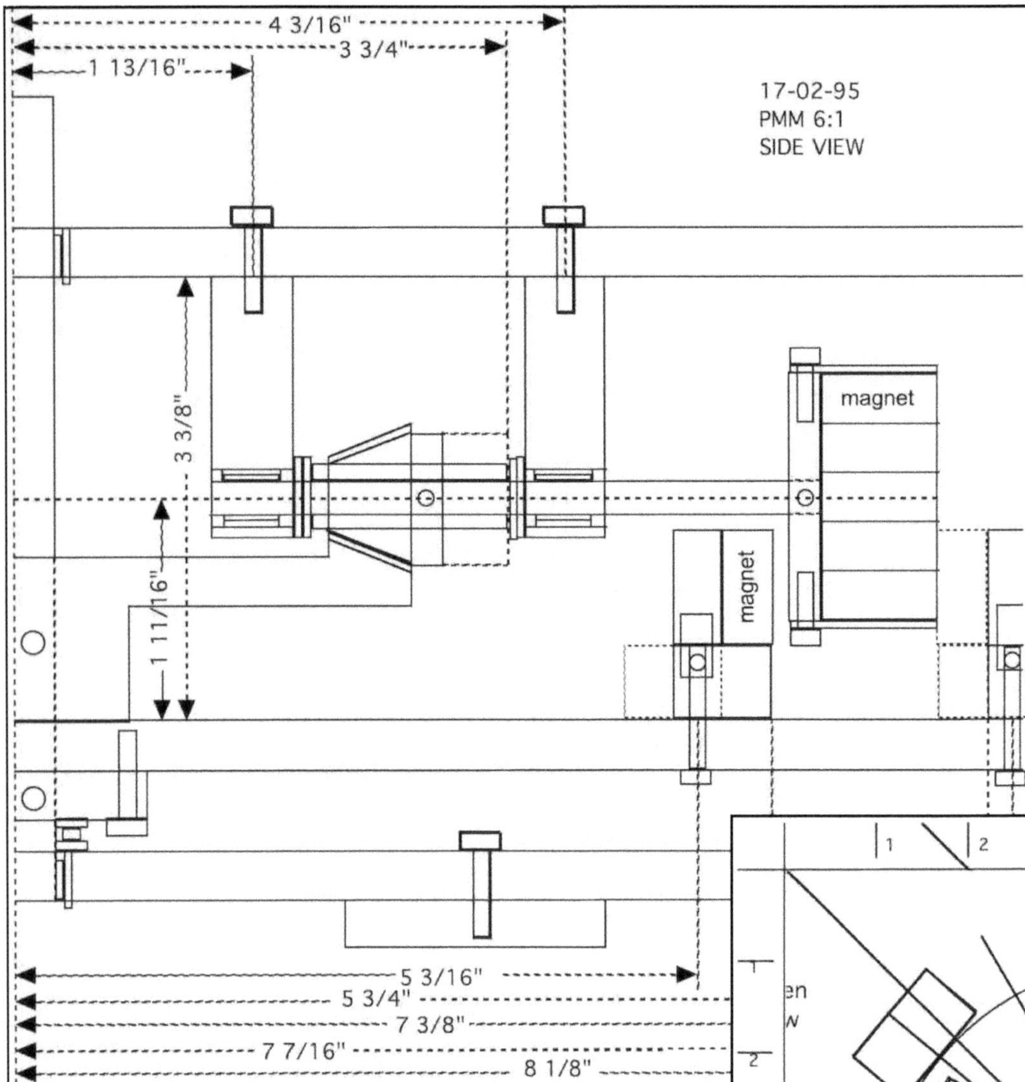
4 3/16"
3 3/4"
1 13/16"
17-02-95
PMM 6:1
SIDE VIEW
magnet
3 3/8"
1 11/16"
magnet
5 3/16"
5 3/4"
7 3/8"
7 7/16"
8 1/8"

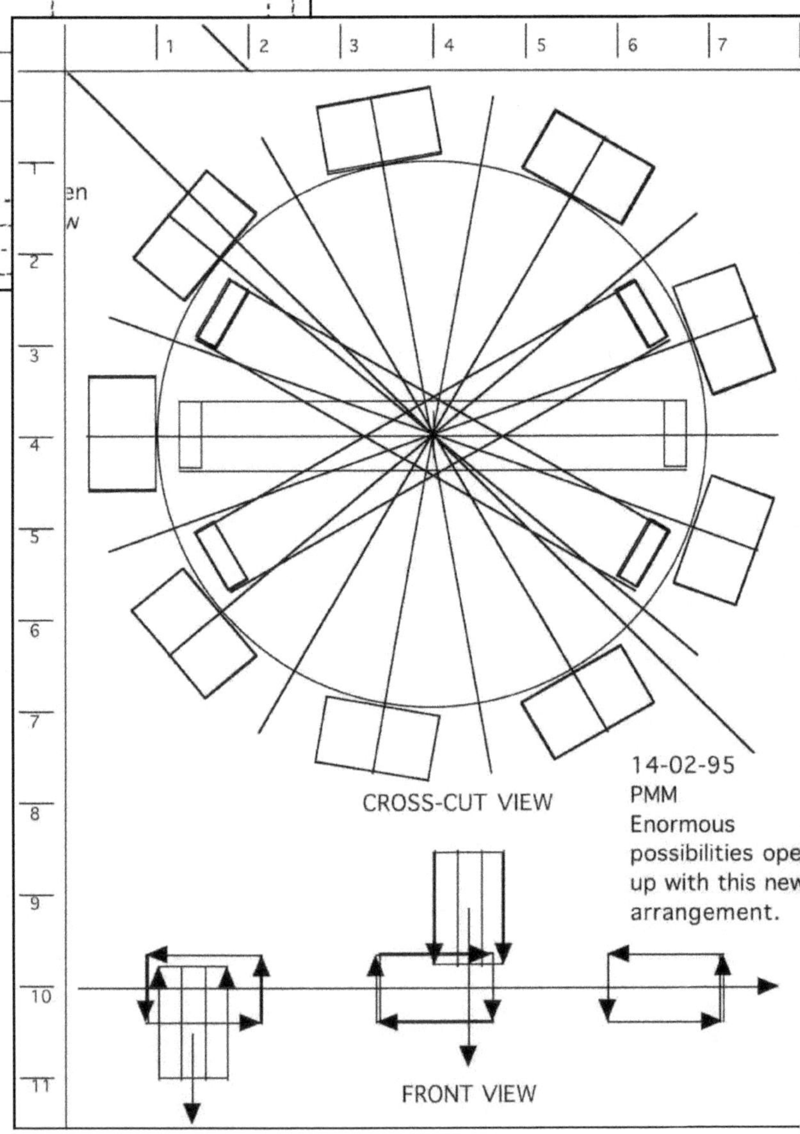
1
2
3
4
5
6
7
1
2
3
4
5
6
7
8
9
10
11
CROSS-CUT VIEW
14-02-95
PMM
Enormous possibilities ope
up with this new
arrangement.
FRONT VIEW

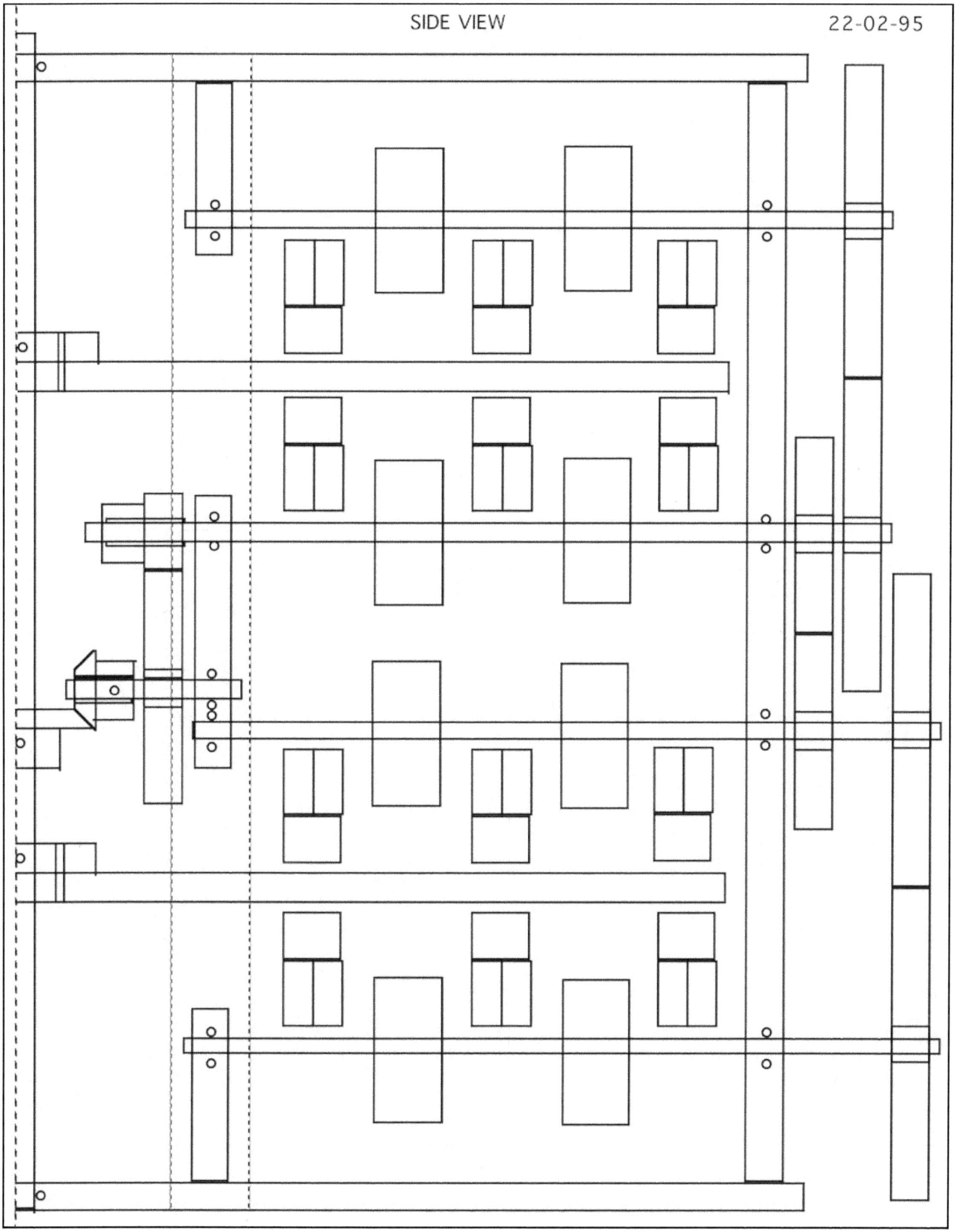
SIDE VIEW
22-02-95

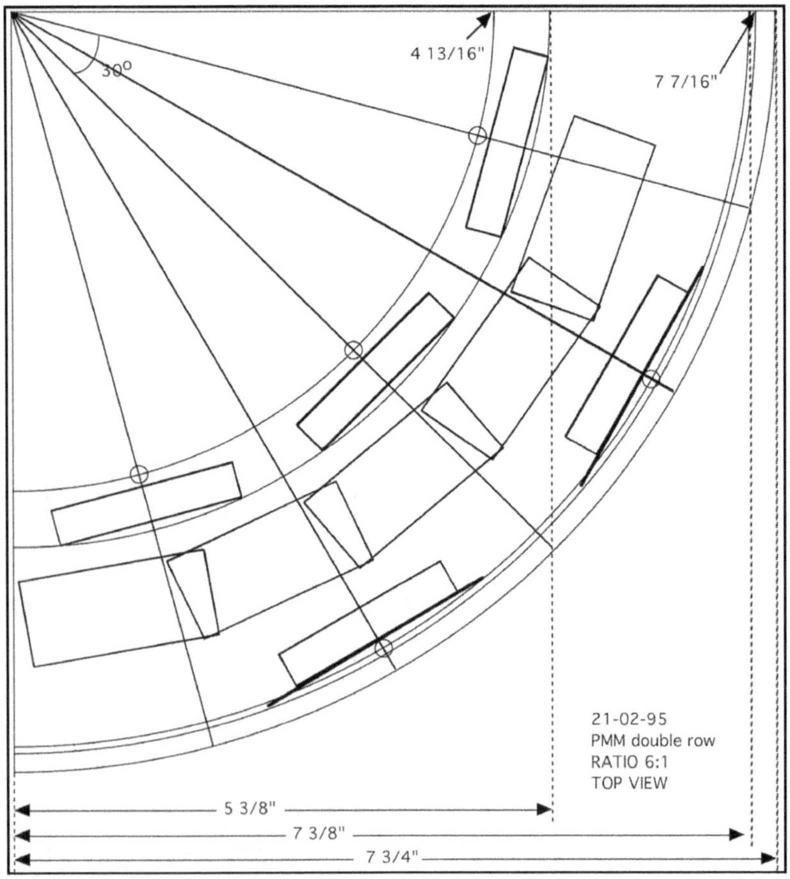
30°
4 13/16"
7 7/16"
21-02-95
PMM double row
RATIO 6:1
TOP VIEW
5 3/8"
7 3/8"
7 3/4"

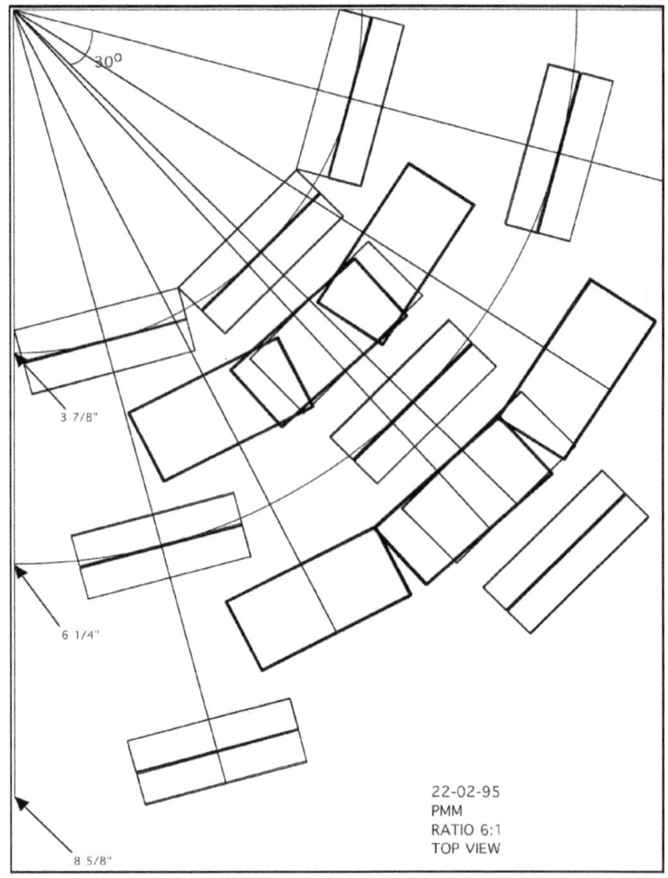
30°
3 7/8"
6 1/4"
8 5/8"
22-02-95
PMM
RATIO 6:1
TOP VIEW

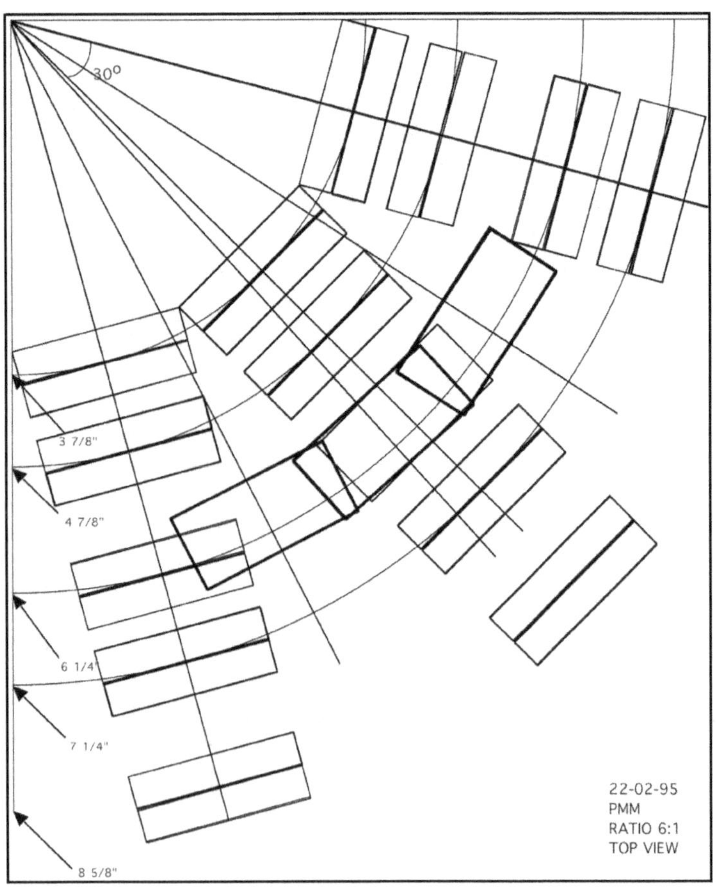
30°
3 7/8"
4 7/8"
6 1/4"
7 1/4"
8 5/8"
22-02-95
PMM
RATIO 6:1
TOP VIEW

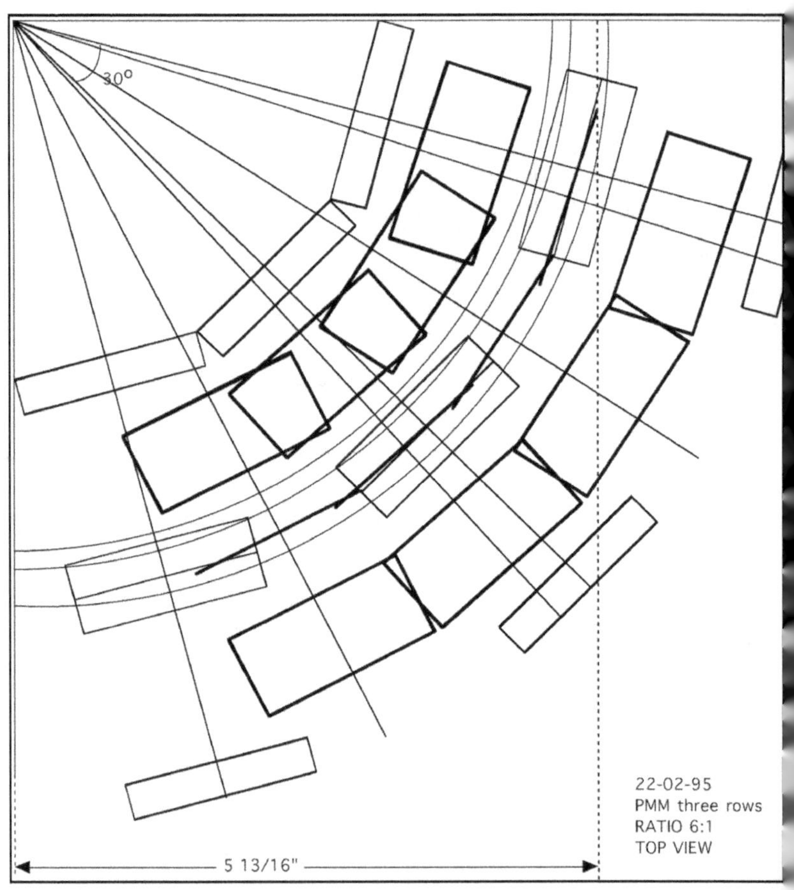
30°
22-02-95
PMM three rows
RATIO 6:1
TOP VIEW
5 13/16"

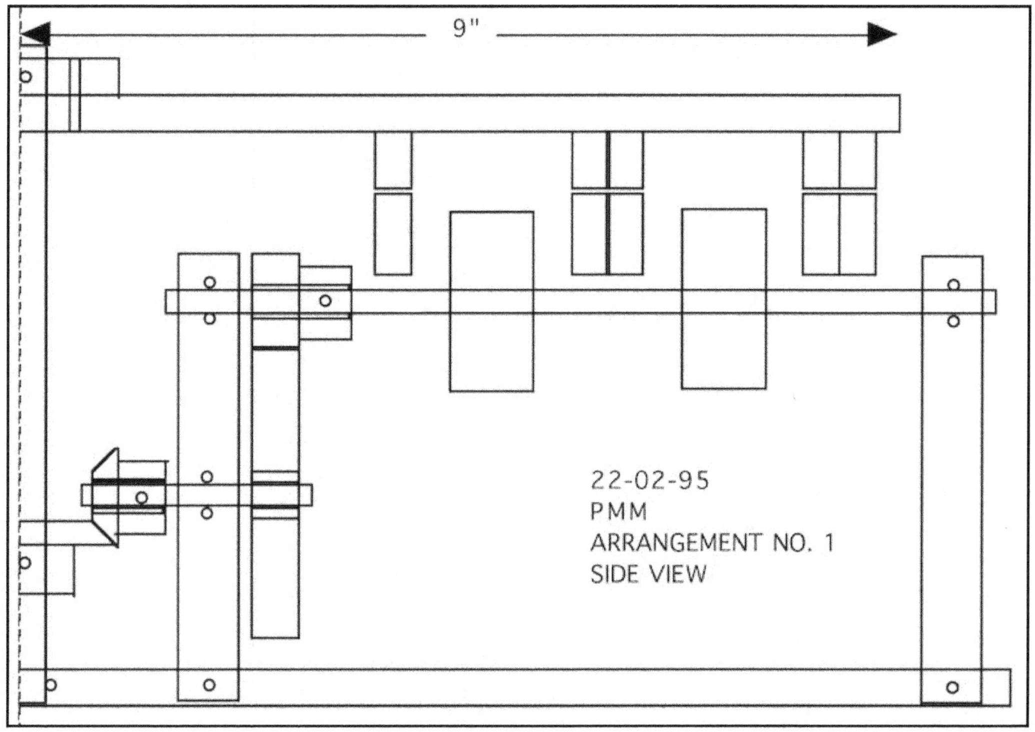
9"
22-02-95
PMM
ARRANGEMENT NO. 1
SIDE VIEW

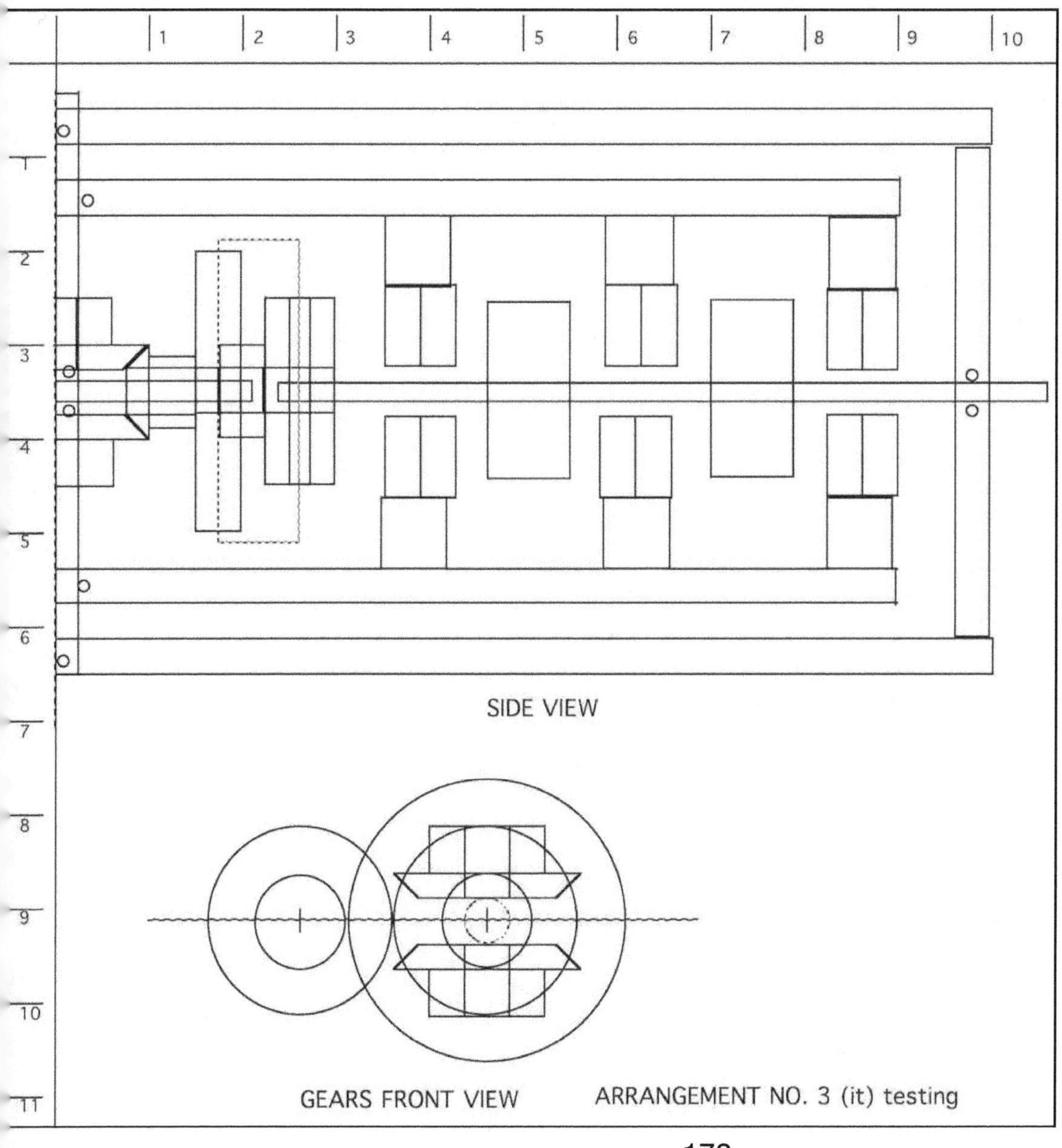
SIDE VIEW
GEARS FRONT VIEW
ARRANGEMENT NO. 3 (it) testing

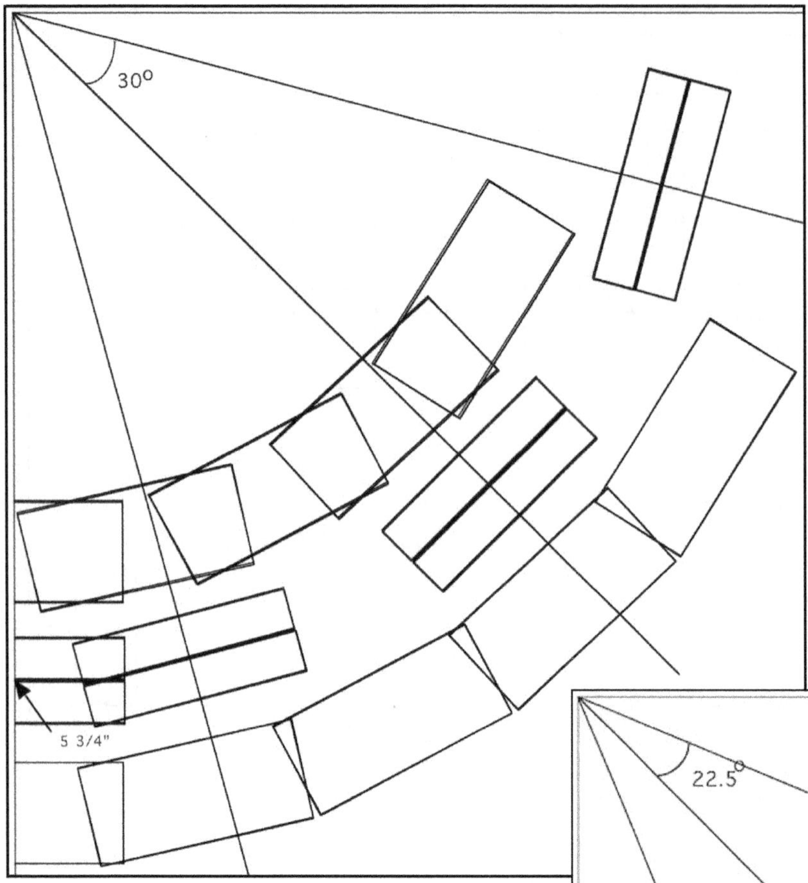
30o
5 3/4"

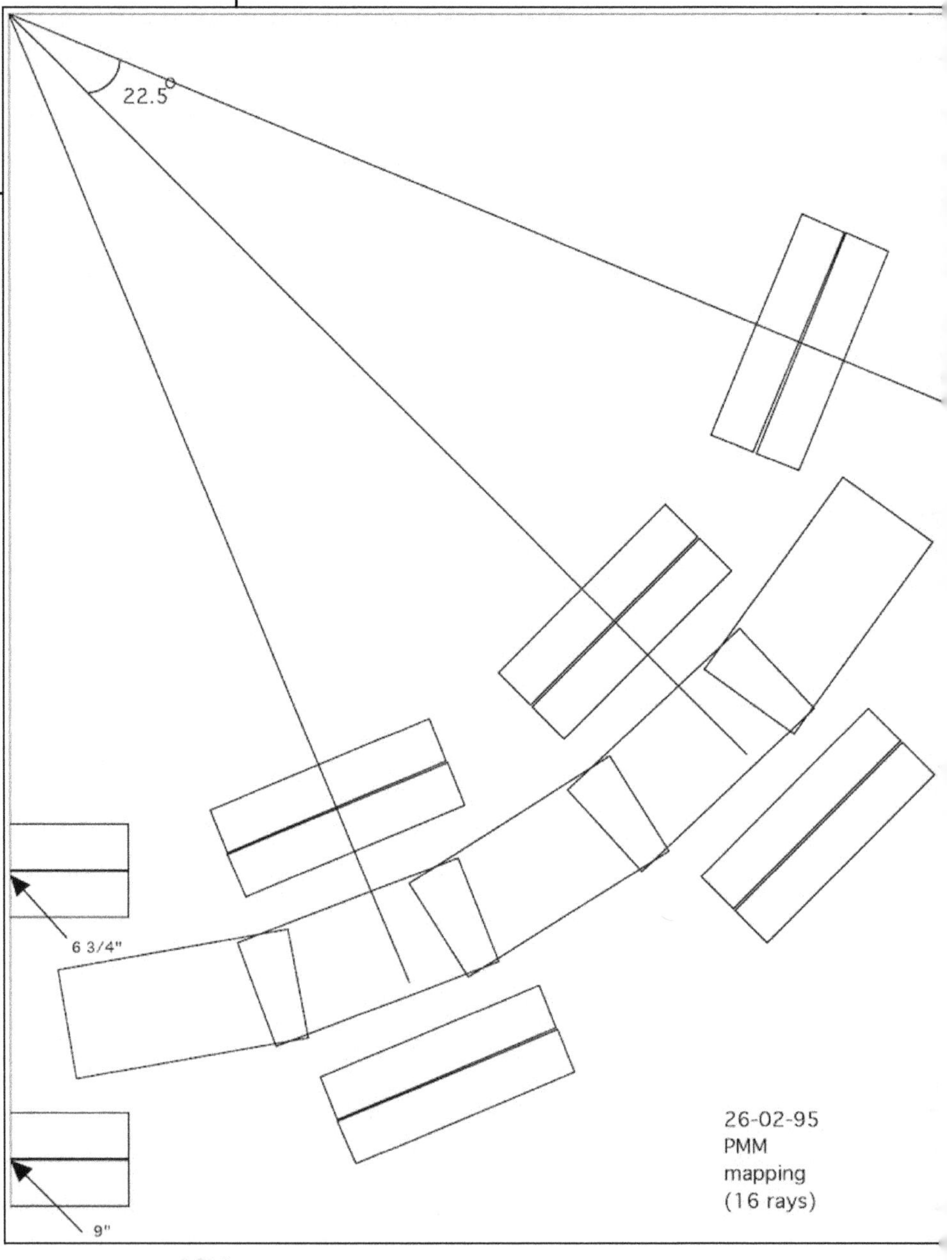
22.5o
6 3/4"
9"
26-02-95
PMM
mapping
(16 rays)

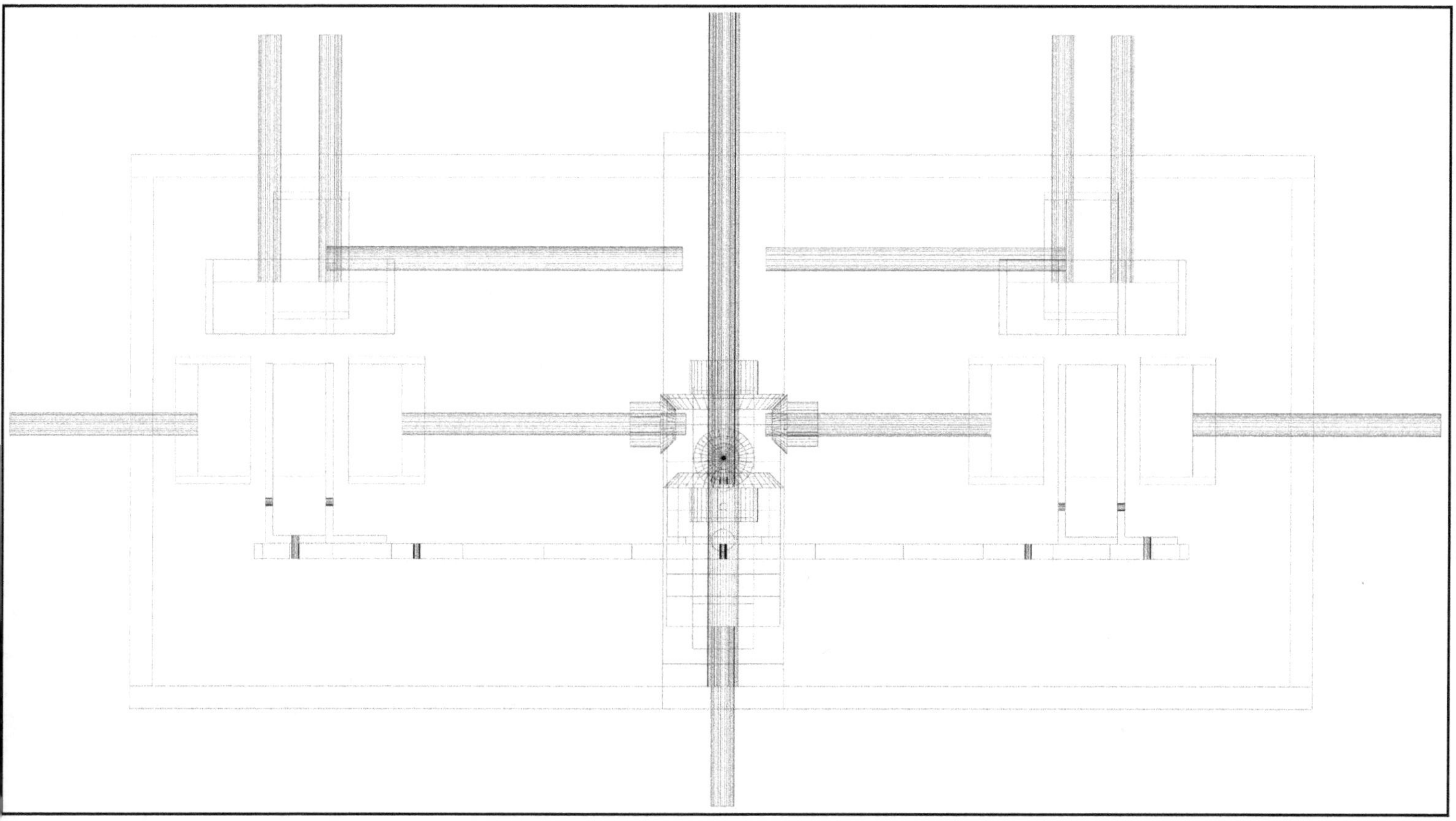

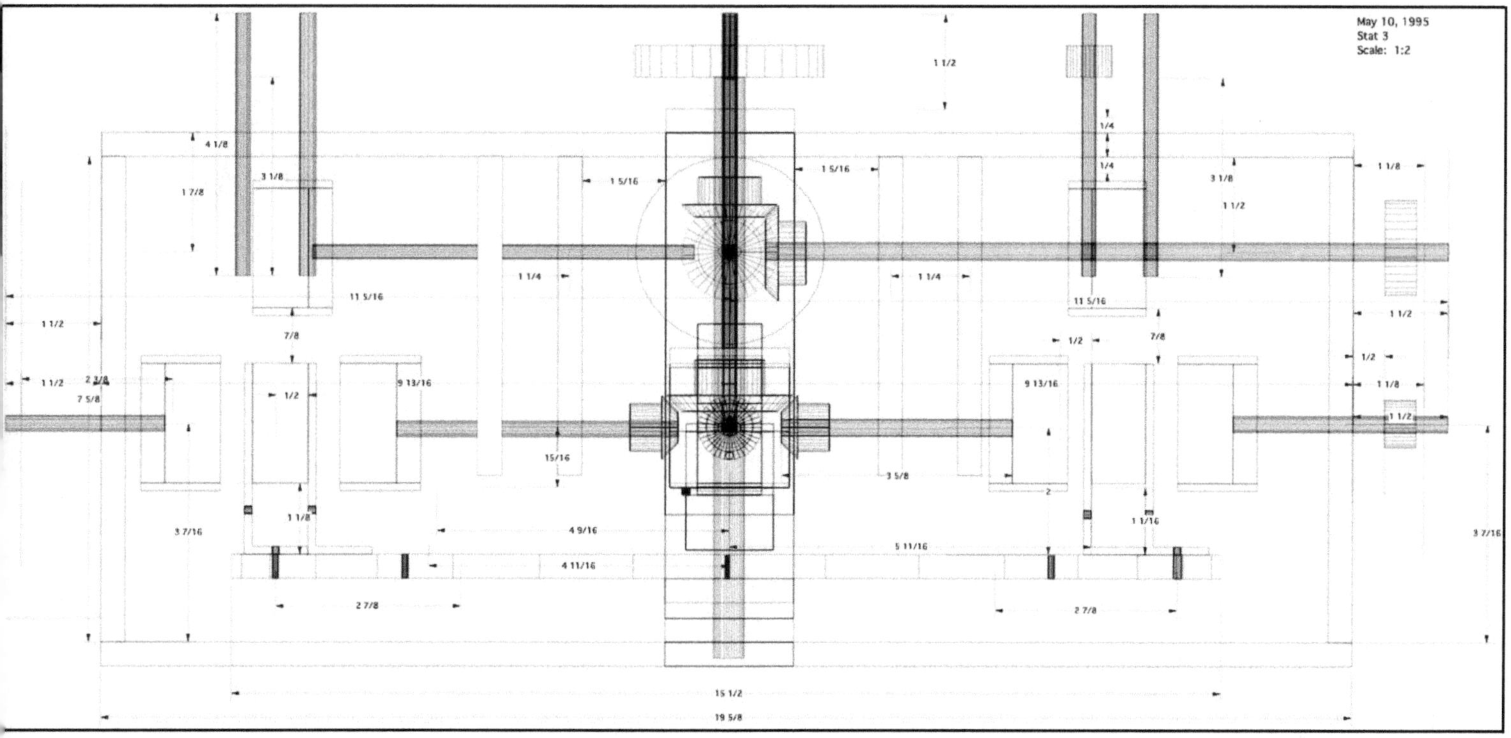
May 10, 1995
Stat 3
Scale: 1:2

8B. SPIRAL MAGNETIC MOTOR

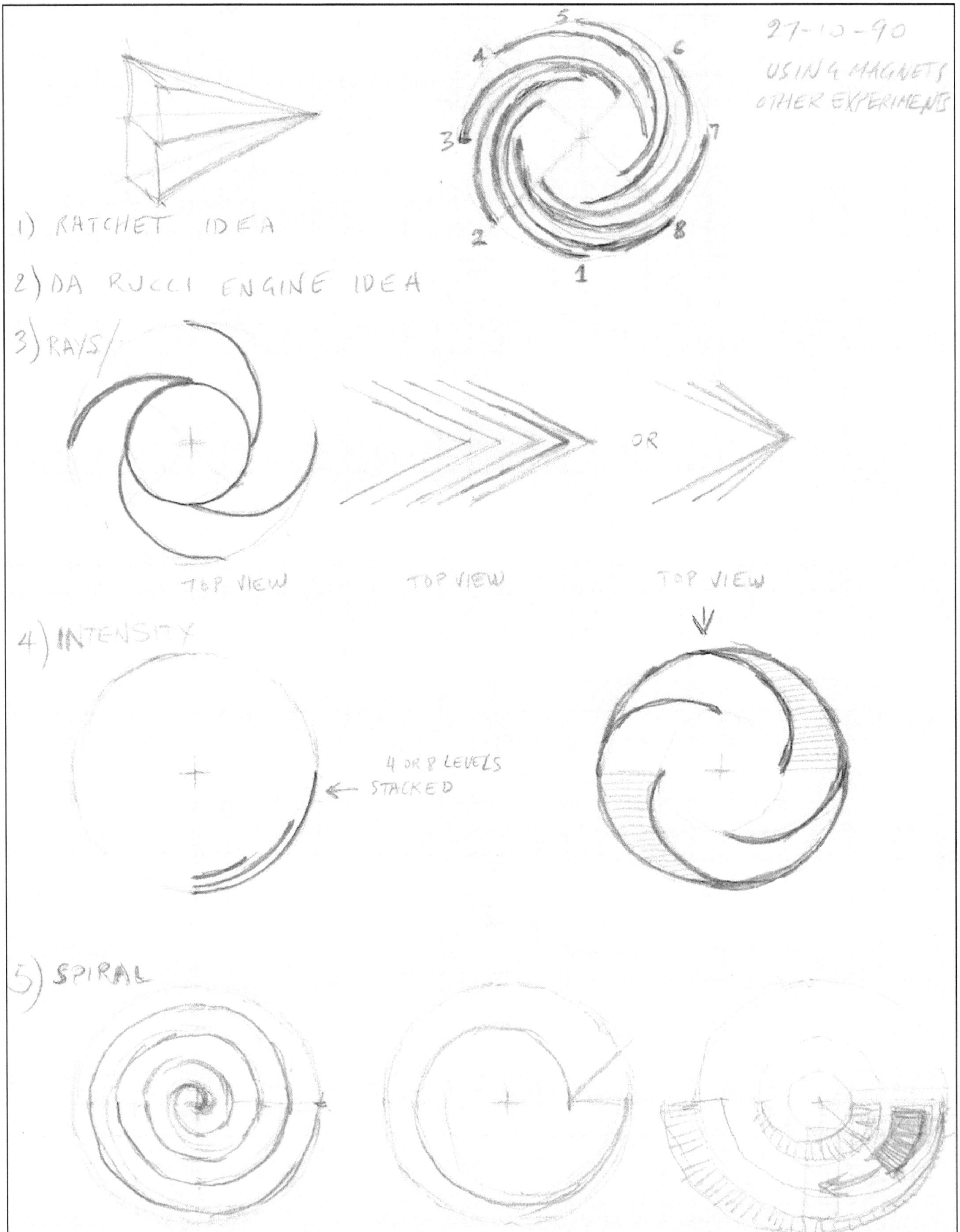

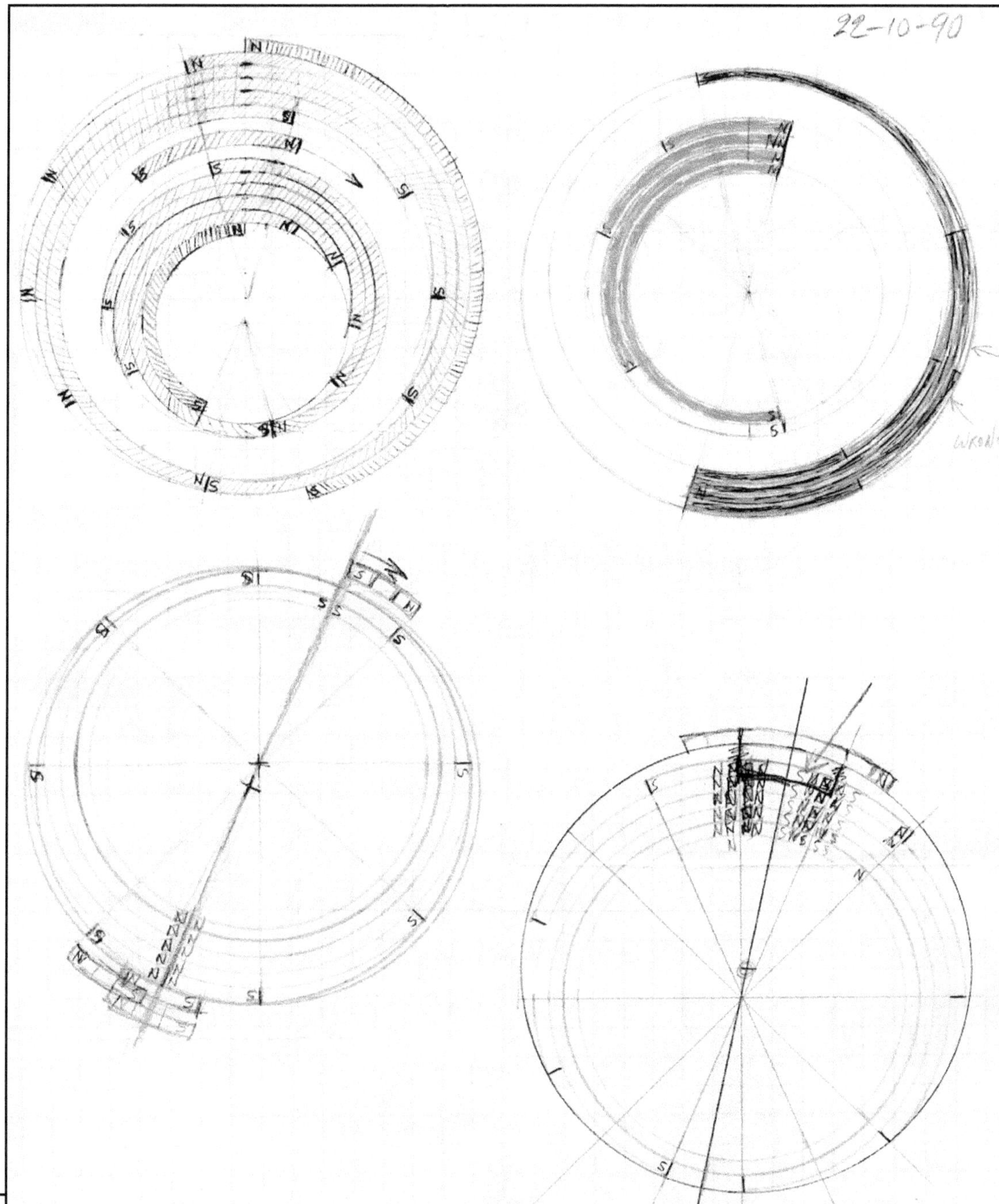
22-10-90

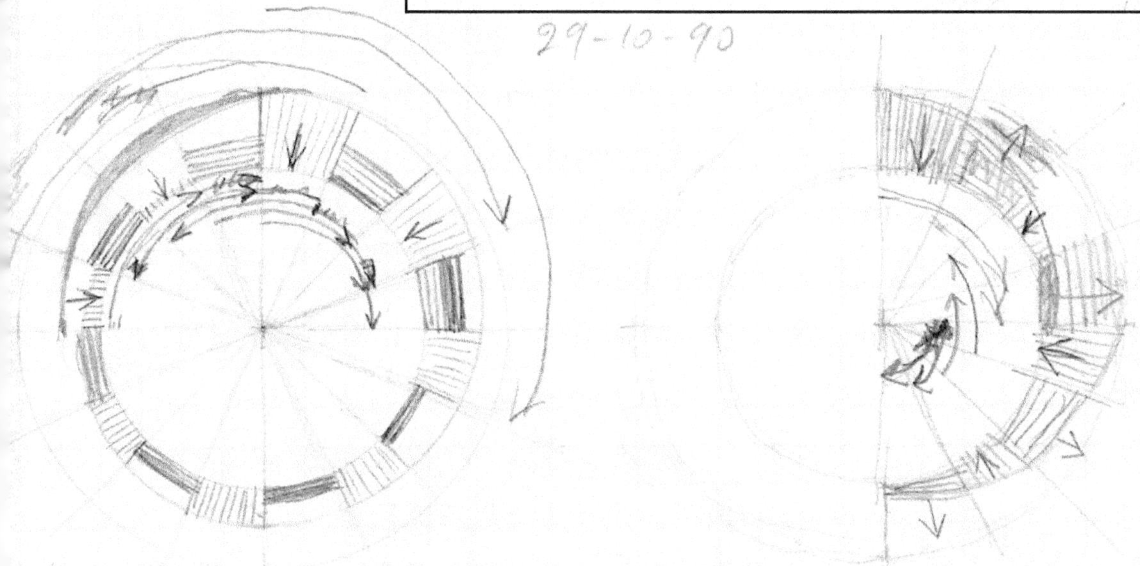
29-10-90

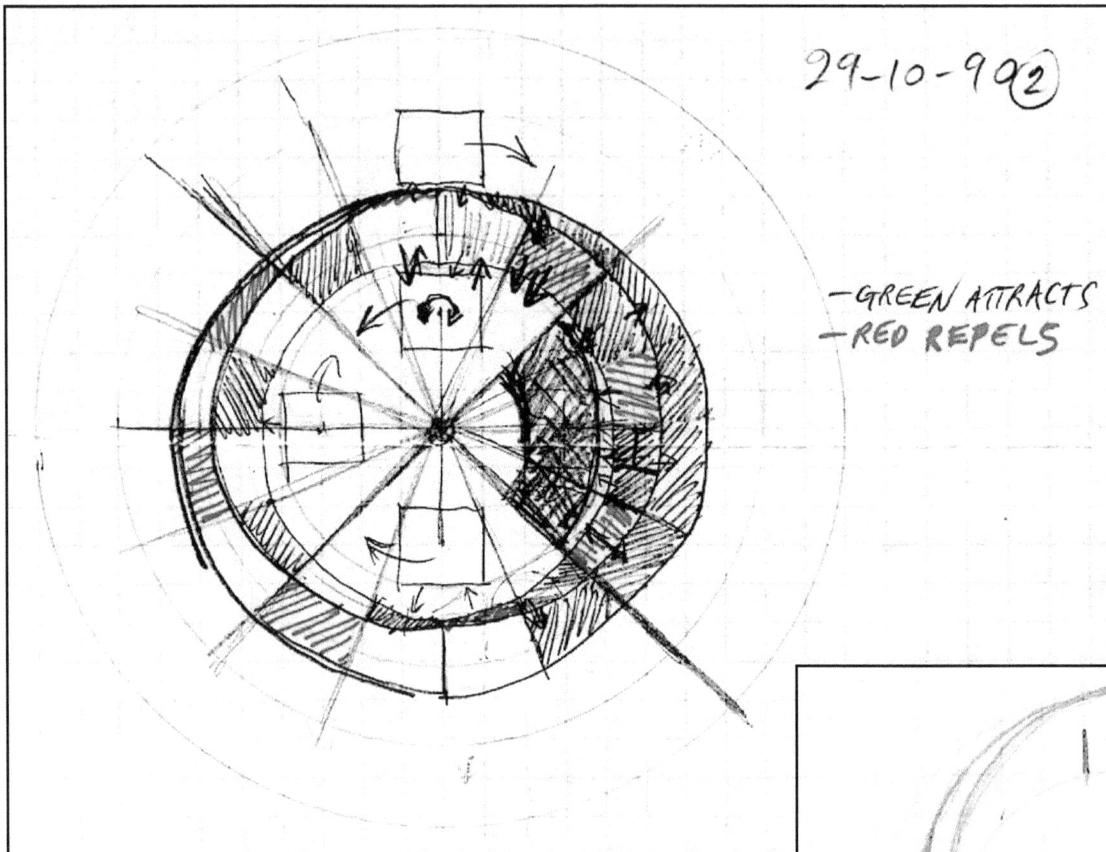
29-10-90 ②
—GREEN ATTRACTS
—RED REPELS

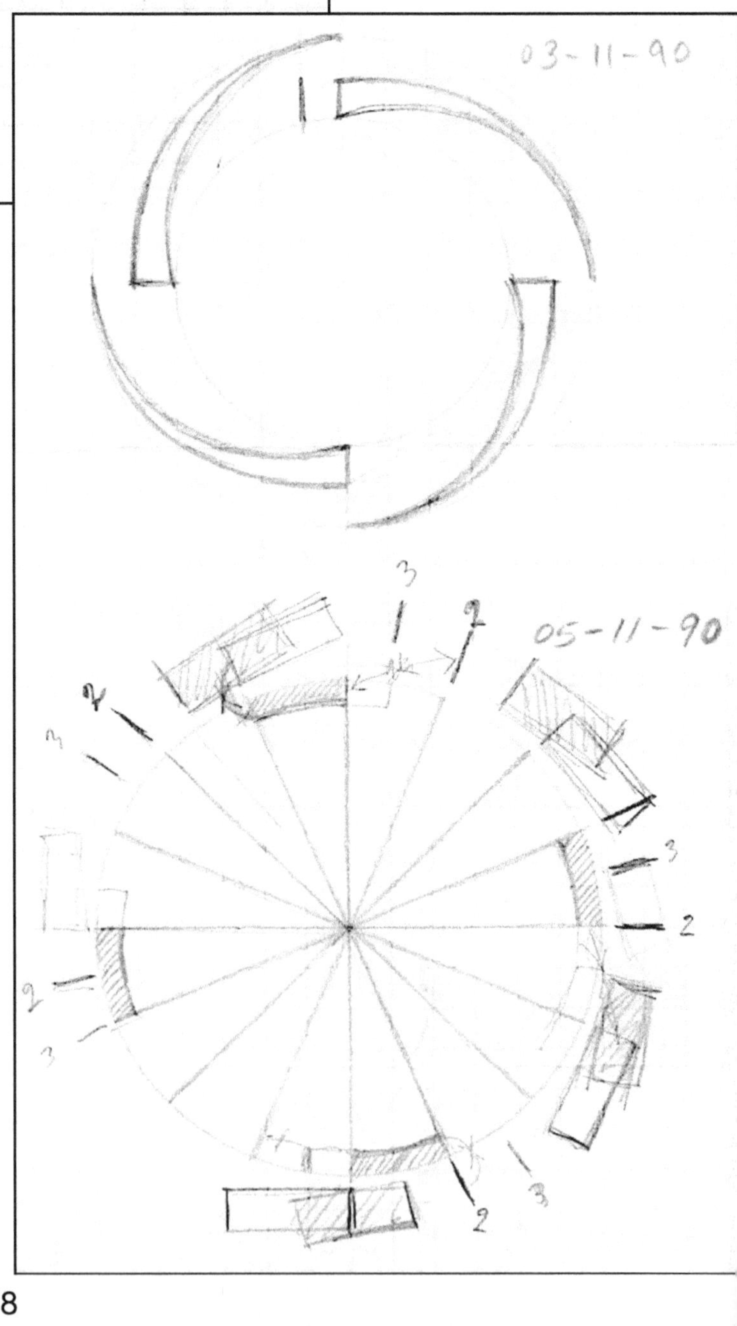
03-11-90
05-11-90

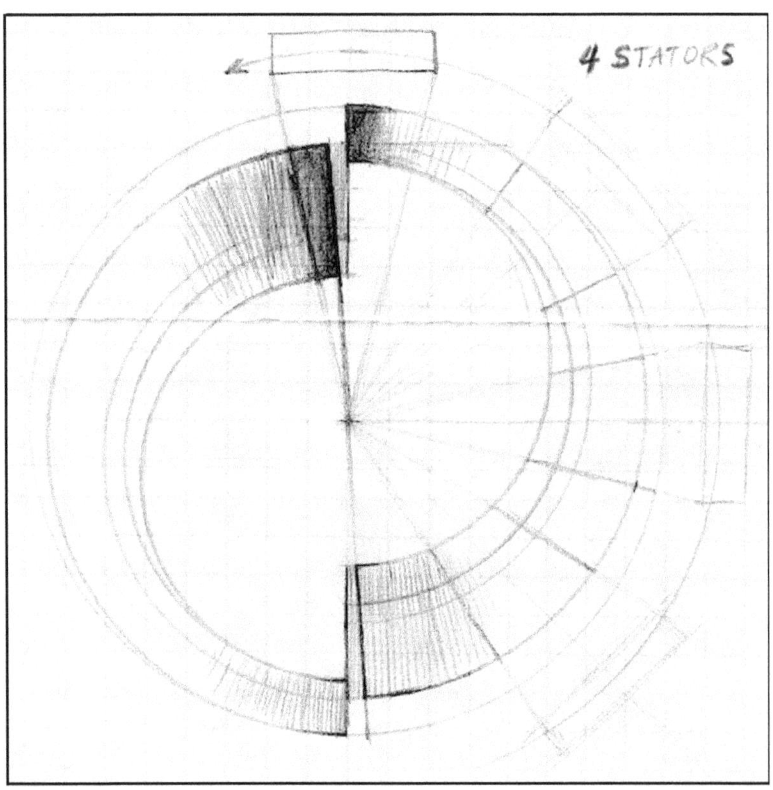
4 STATORS

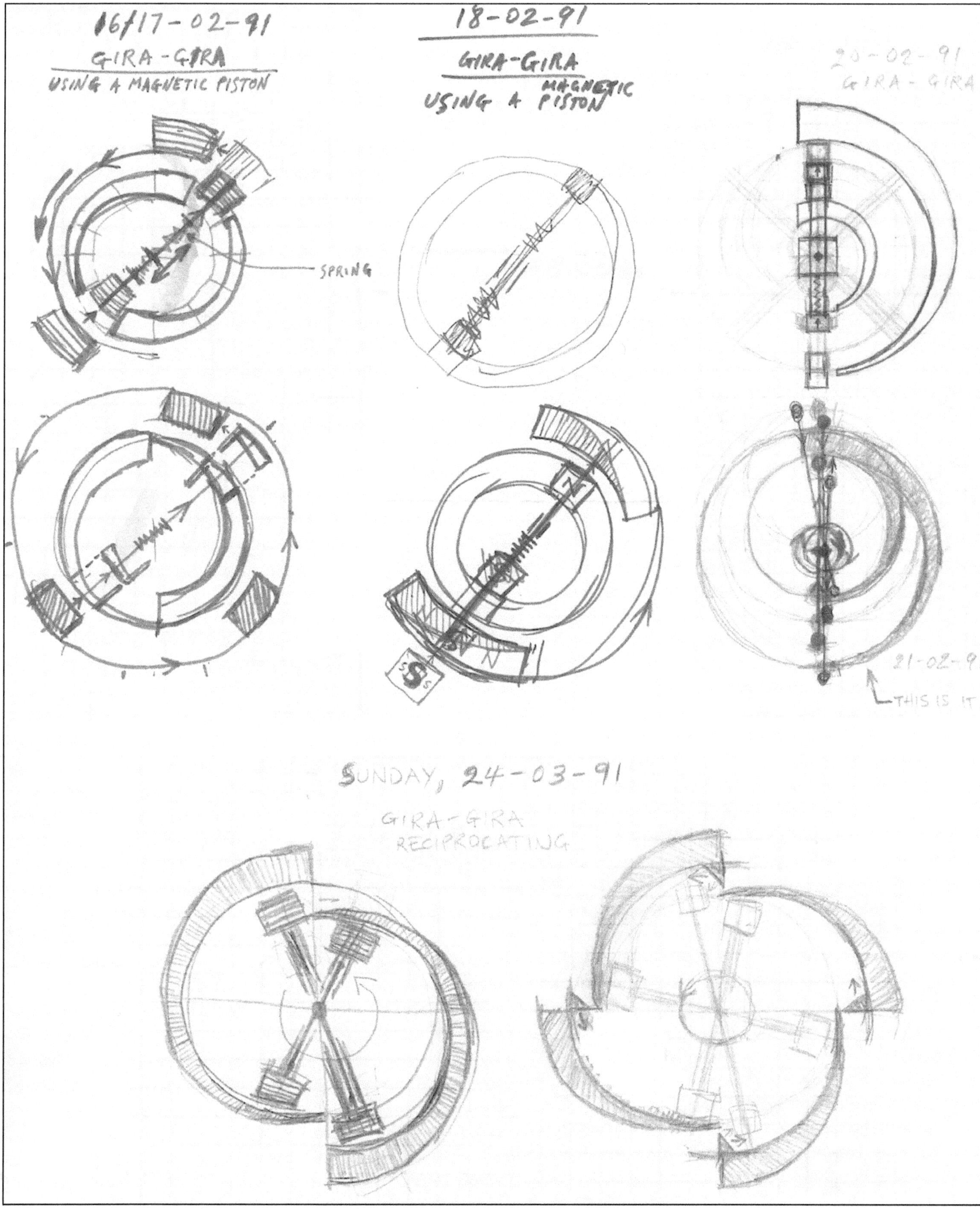
16/17-02-91
GIRA-GIRA
USING A MAGNETIC PISTON
SPRING
18-02-91
GIRA-GIRA
USING A MAGNETIC PISTON
20-02-91
GIRA-GIRA
21-02-91
THIS IS IT!
SUNDAY, 24-03-91
GIRA-GIRA
RECIPROCATING

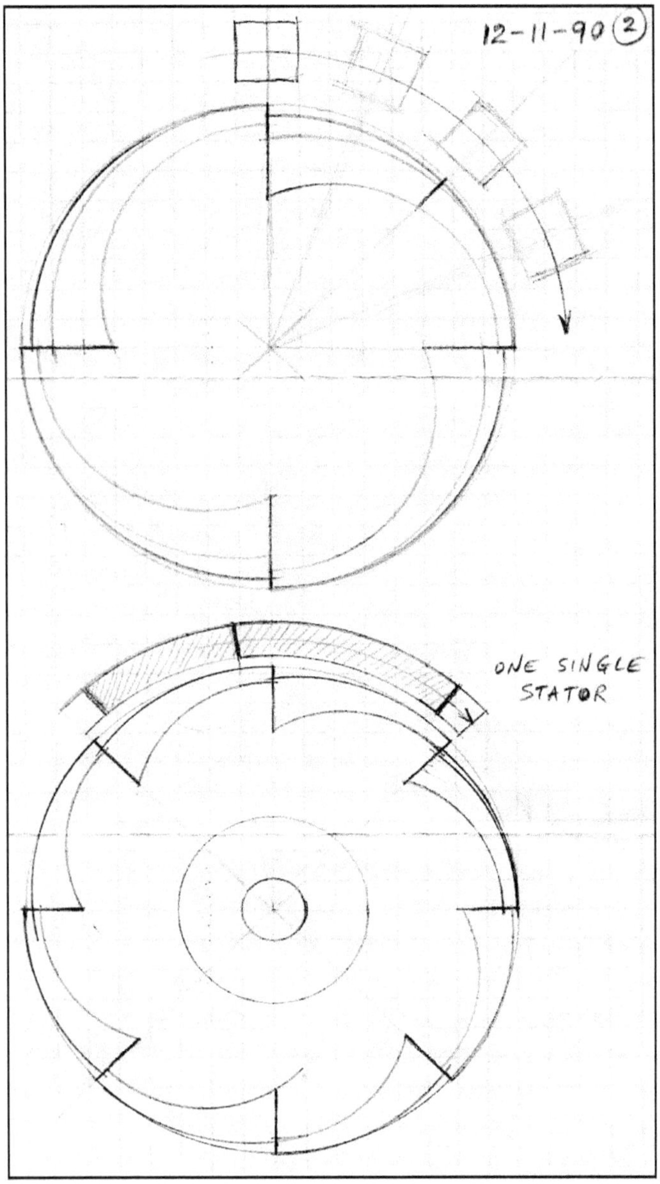
12-11-90 (2)
ONE SINGLE STATOR

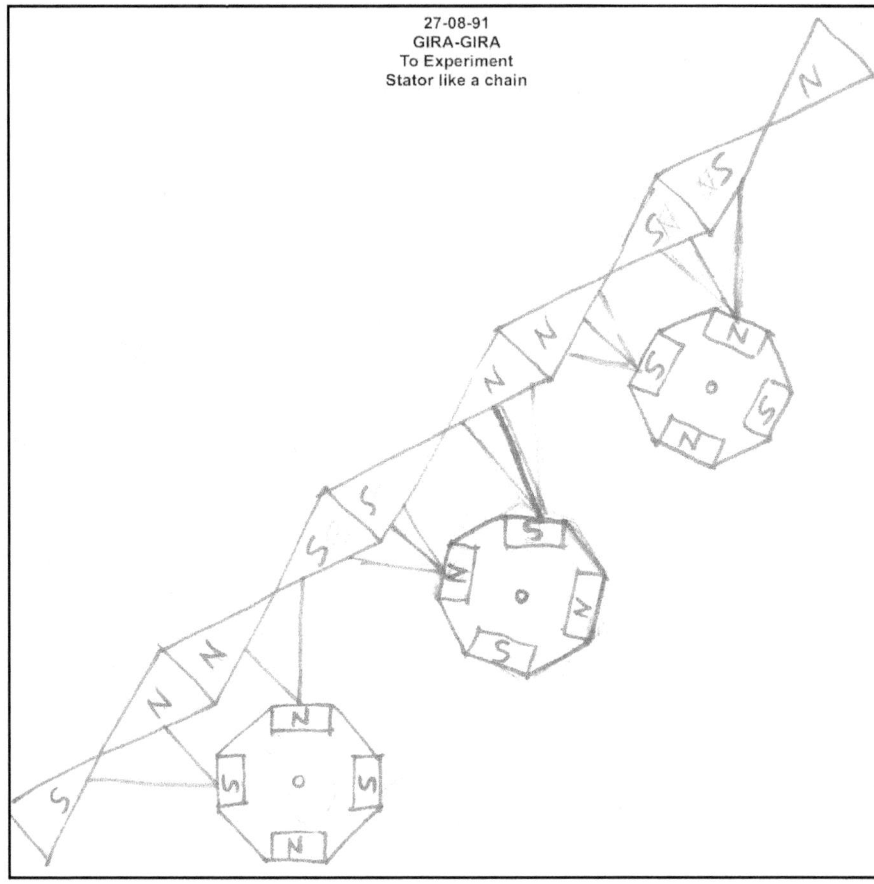
27-08-91
GIRA-GIRA
To Experiment
Stator like a chain

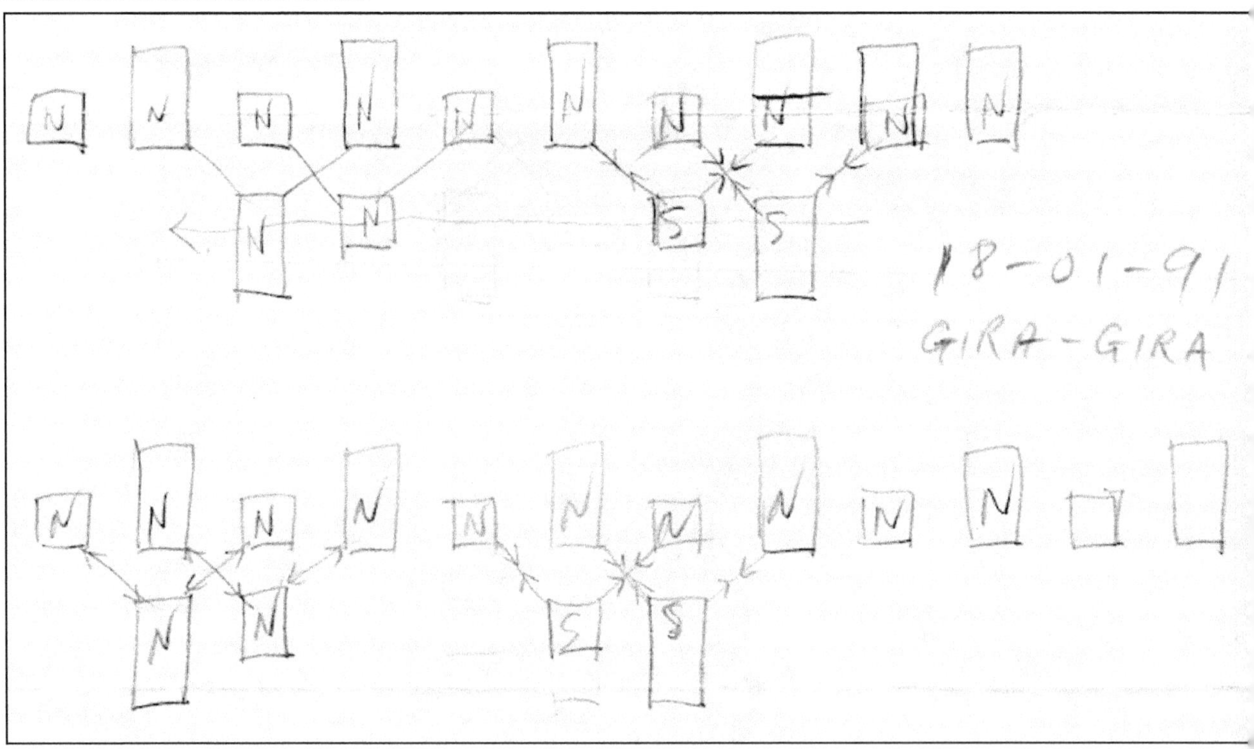
18-01-91
GIRA-GIRA

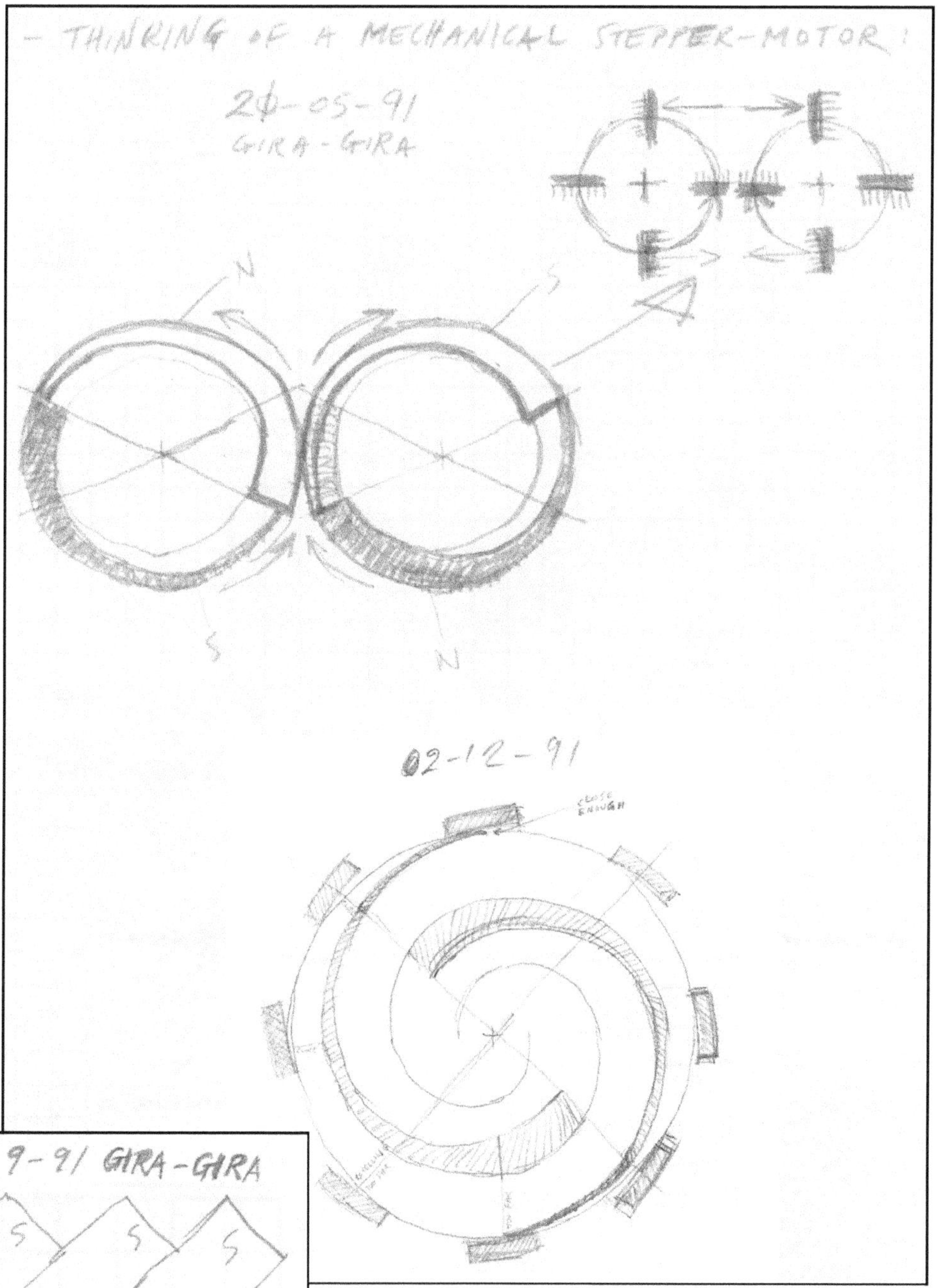
- THINKING OF A MECHANICAL STEPPER-MOTOR !
2Ø-05-91
GIRA-GIRA
N
S
S
N
02-12-91
CLOSE ENOUGH

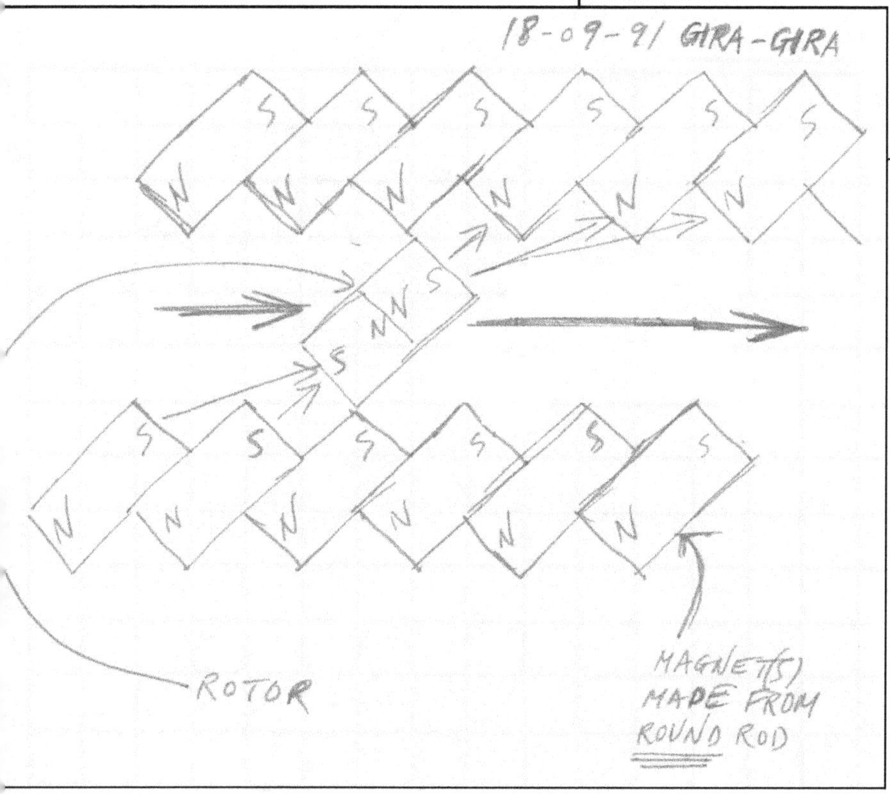
18-09-91 GIRA-GIRA
ROTOR
MAGNET(S) MADE FROM ROUND ROD

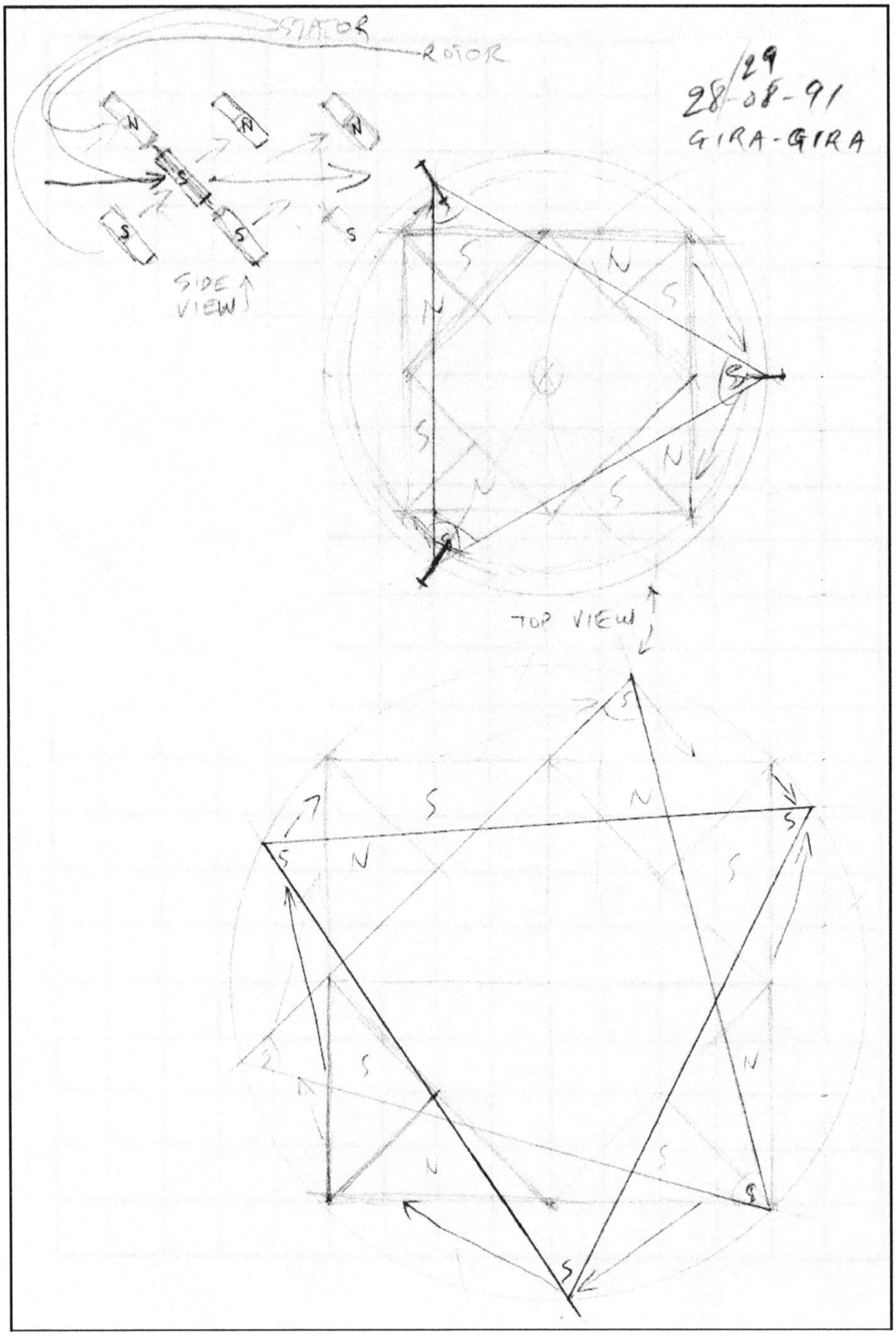
STATOR
ROTOR
29
28-08-91
GIRA-GIRA
SIDE VIEW
TOP VIEW

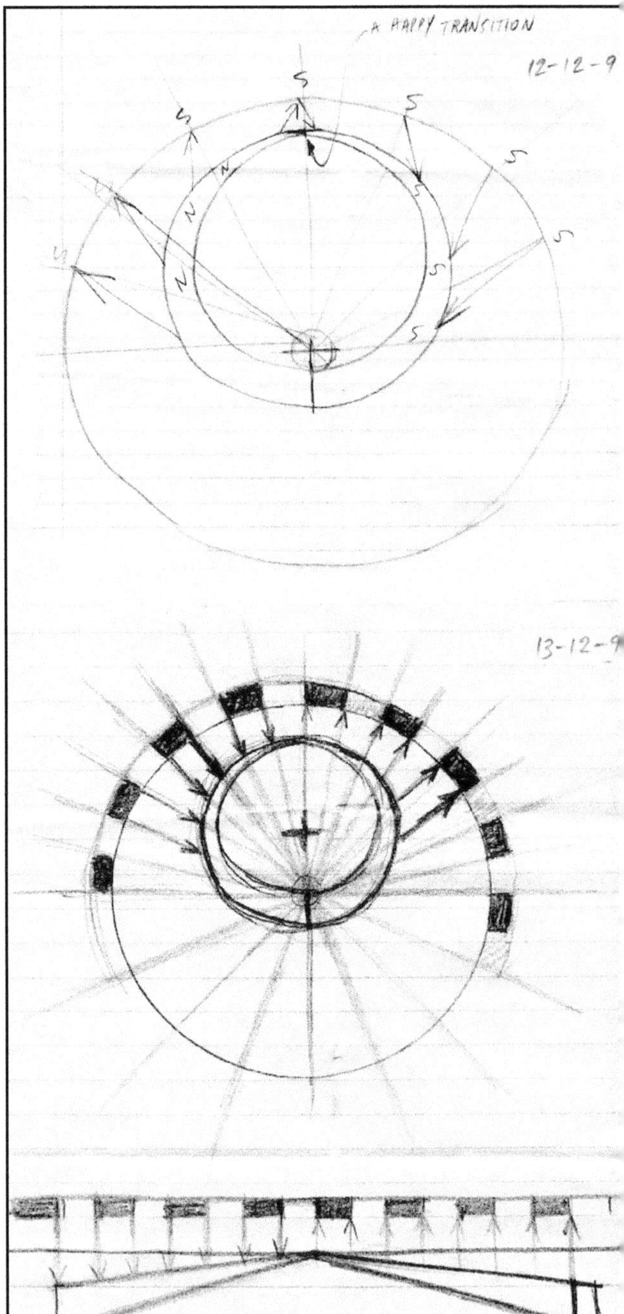
A HAPPY TRANSITION

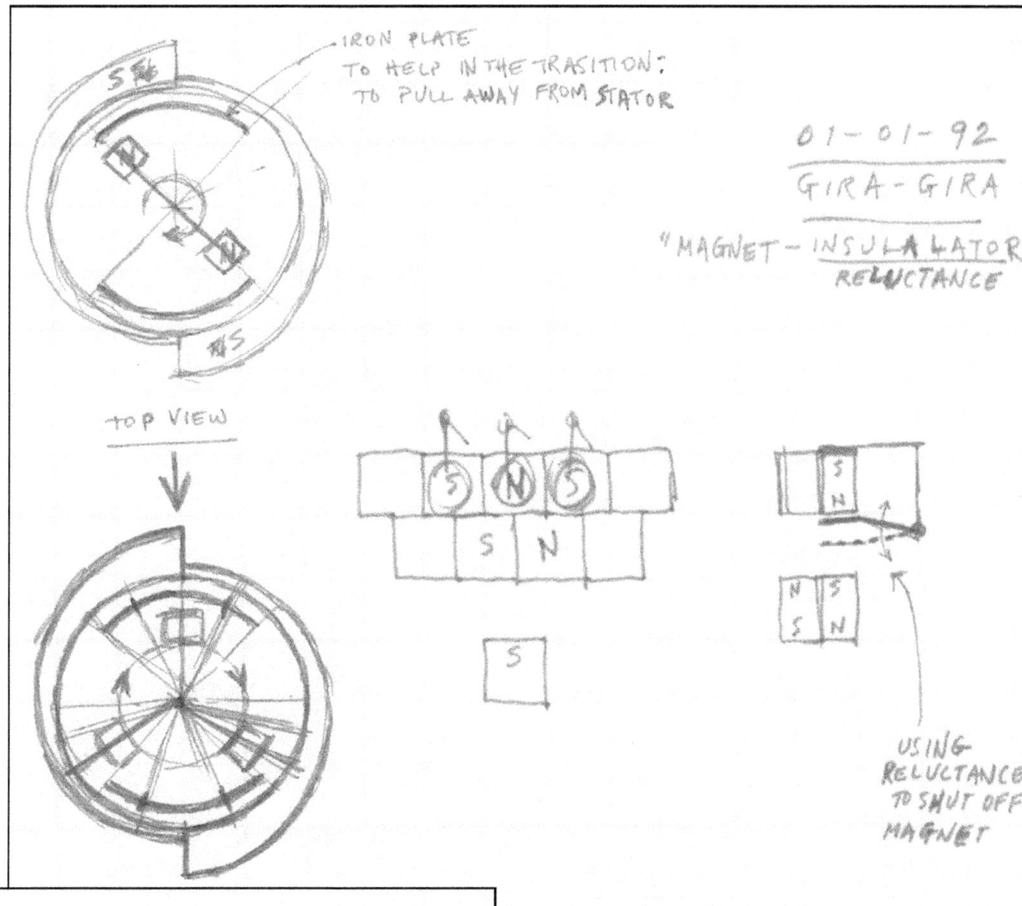
IRON PLATE
TO HELP IN THE TRASITION:
TO PULL AWAY FROM STATOR
01-01-92
GIRA-GIRA
"MAGNET-INSULALATOR
RELUCTANCE
TOP VIEW
S N S
S N
S
USING
RELUCTANCE
TO SHUT OFF
MAGNET

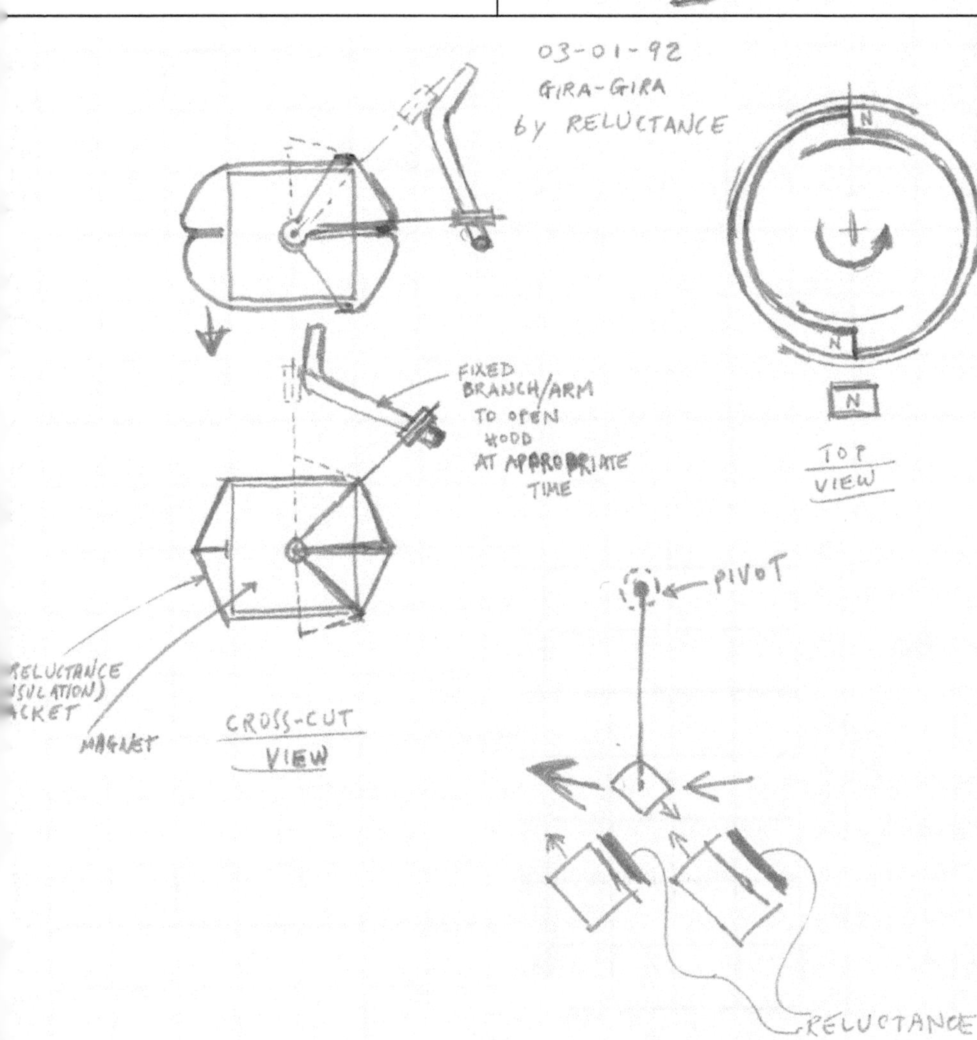
03-01-92
GIRA-GIRA
by RELUCTANCE
FIXED
BRANCH/ARM
TO OPEN
HOOD
AT APPROPRIATE
TIME
TOP
VIEW
PIVOT
CROSS-CUT
VIEW
MAGNET
RELUCTANCE

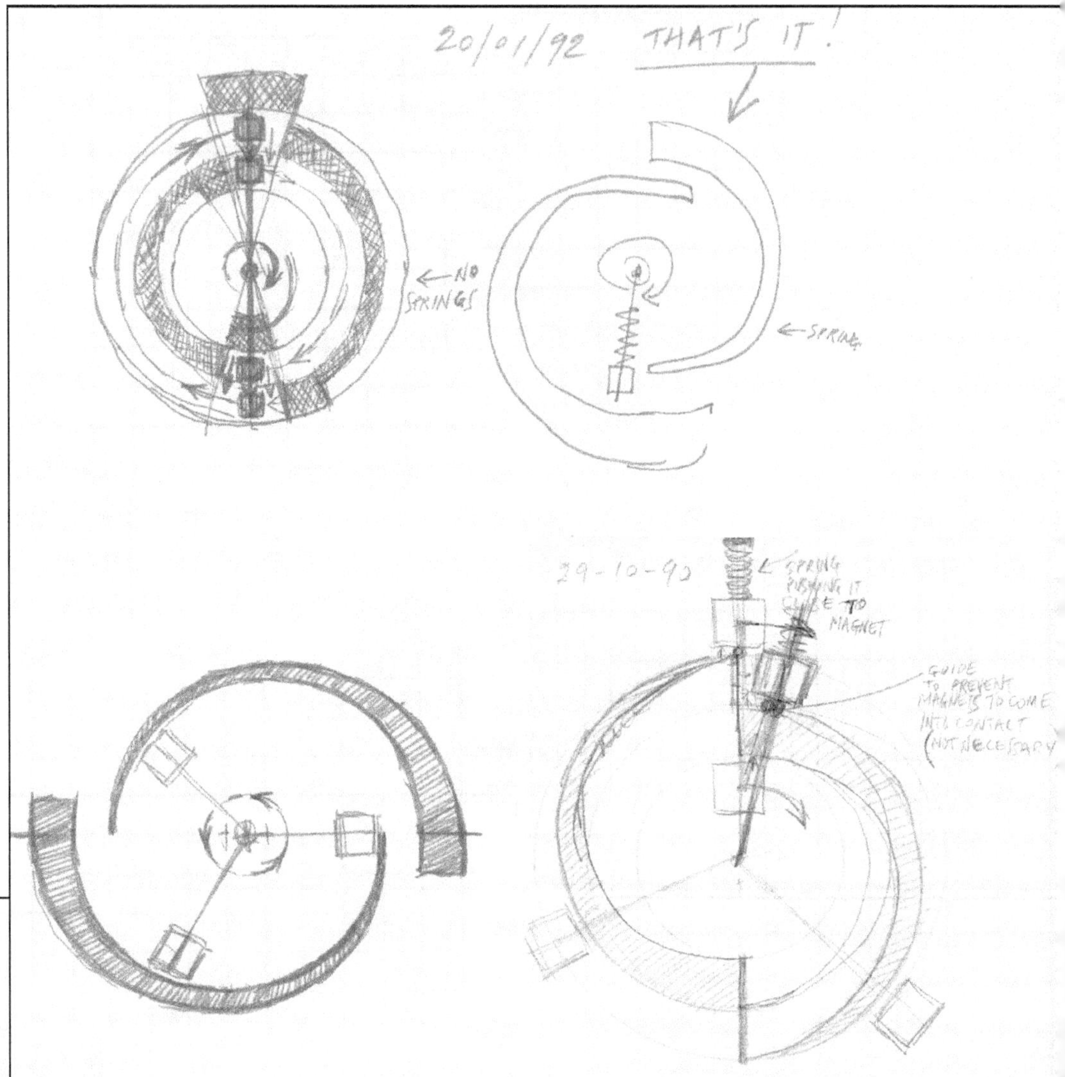
20/01/92 THAT'S IT!
←NO SPRINGS
←SPRING
29-10-92
SPRING PUSHING IT CLOSE TO MAGNET
GUIDE TO PREVENT MAGNETS TO COME INTO CONTACT (NOT NECESSARY)

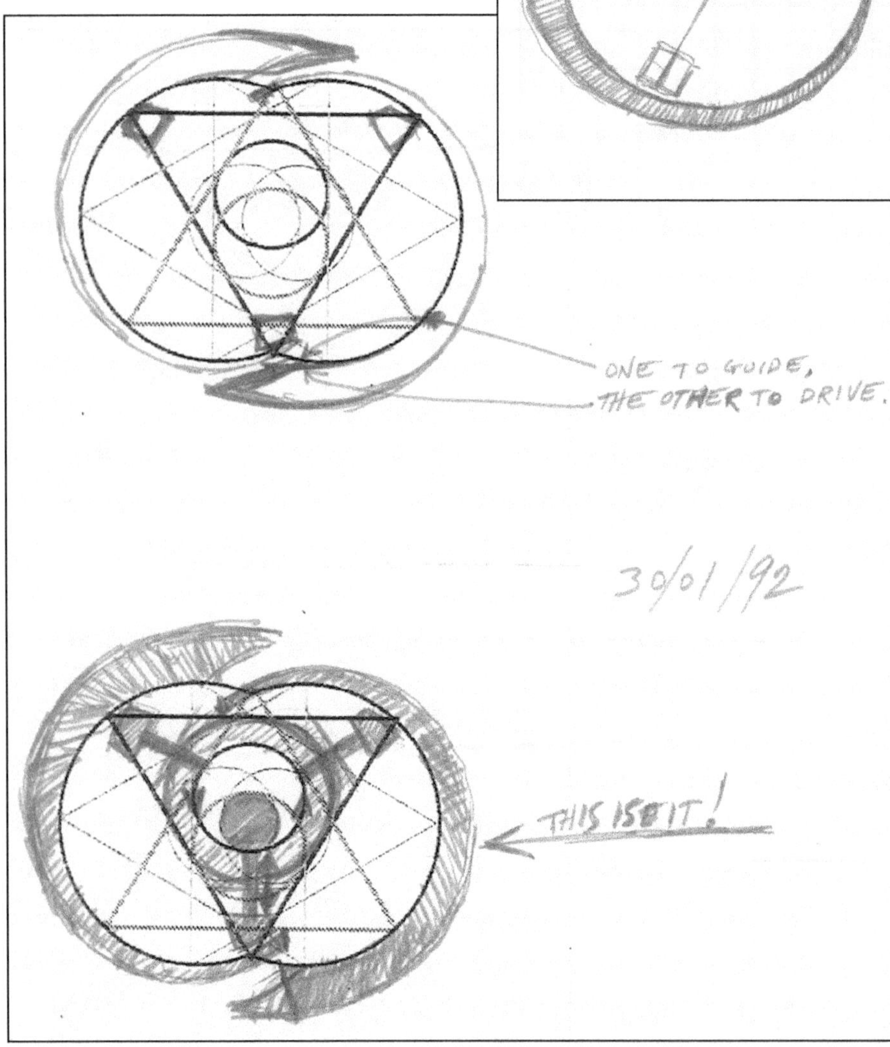
ONE TO GUIDE,
THE OTHER TO DRIVE.
30/01/92
THIS IS IT!

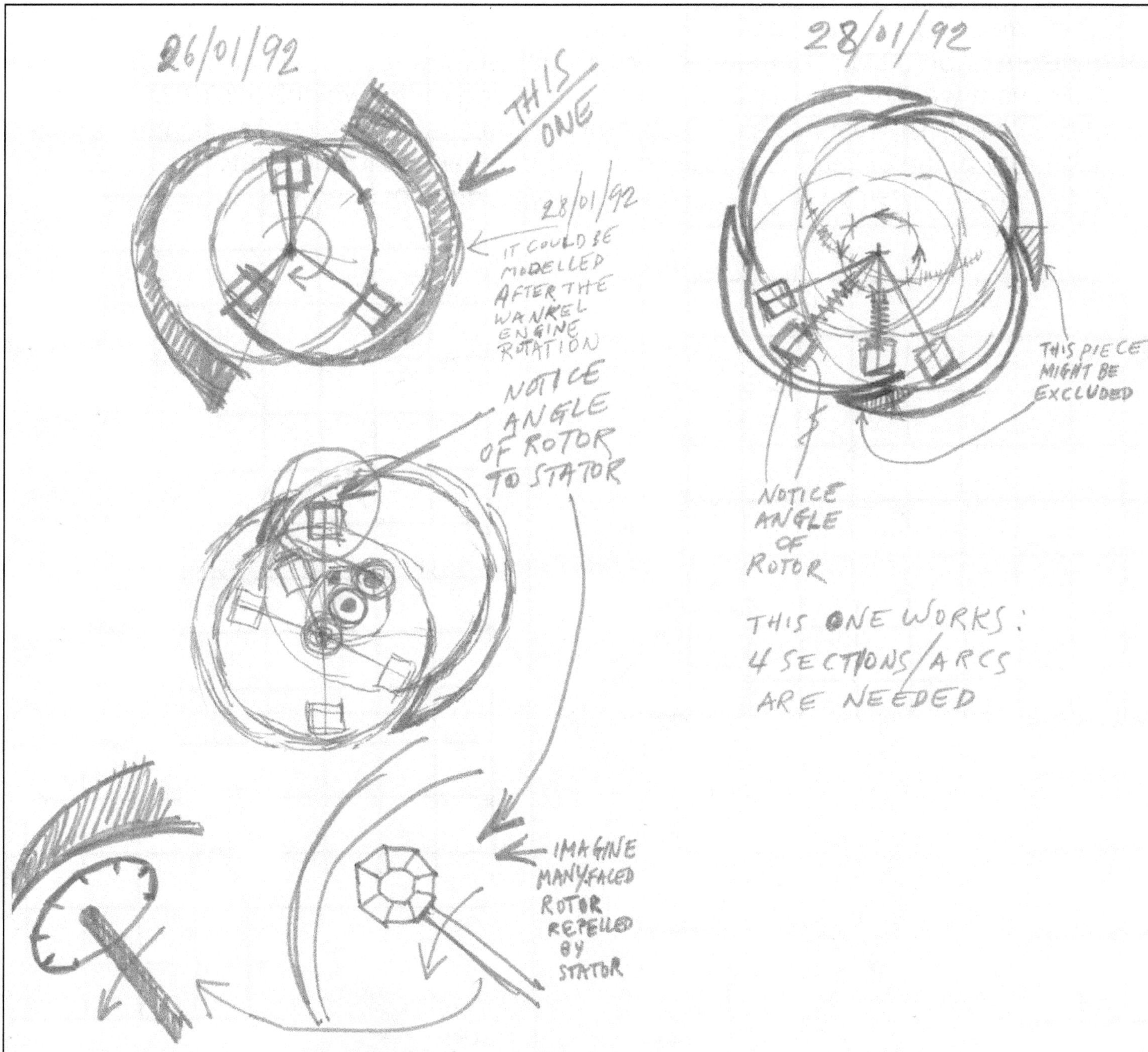
26/01/92
THIS ONE
28/01/92
IT COULD BE MODELLED AFTER THE WANKEL ENGINE ROTATION
NOTICE ANGLE OF ROTOR TO STATOR
IMAGINE MANY-FACED ROTOR REPELLED BY STATOR
28/01/92
THIS PIECE MIGHT BE EXCLUDED
NOTICE ANGLE OF ROTOR
THIS ONE WORKS: 4 SECTIONS/ARCS ARE NEEDED

8C. ATMOSPHERIC MOTOR (PMM)

There are 3 types of Transportation/Atmosphere Motors: the **wind turbine**, the **water wheel**, and the **PMM** (Perpetual Motion Machine) or Weight Wheel. Let us compare the last two. On the one hand, the water wheel works because of the disequilibrium created by the *unequal pressure* of the falling water at one side of the wheel. On the other hand, the PMM or weight wheel cannot work because, when we try to achieve continuous disequilibrium, we end up with *equal pressure* at both sides of the wheel with whatever movement we conceive. By movement we mean the flipping, sliding, or rolling of the weights inside the arms of the wheel. The result is always the same. There is no way we can beat the geometry of the circle. This problem is better understood as a matter of *equilibrium of energy* rather than one of *conservation of energy*.

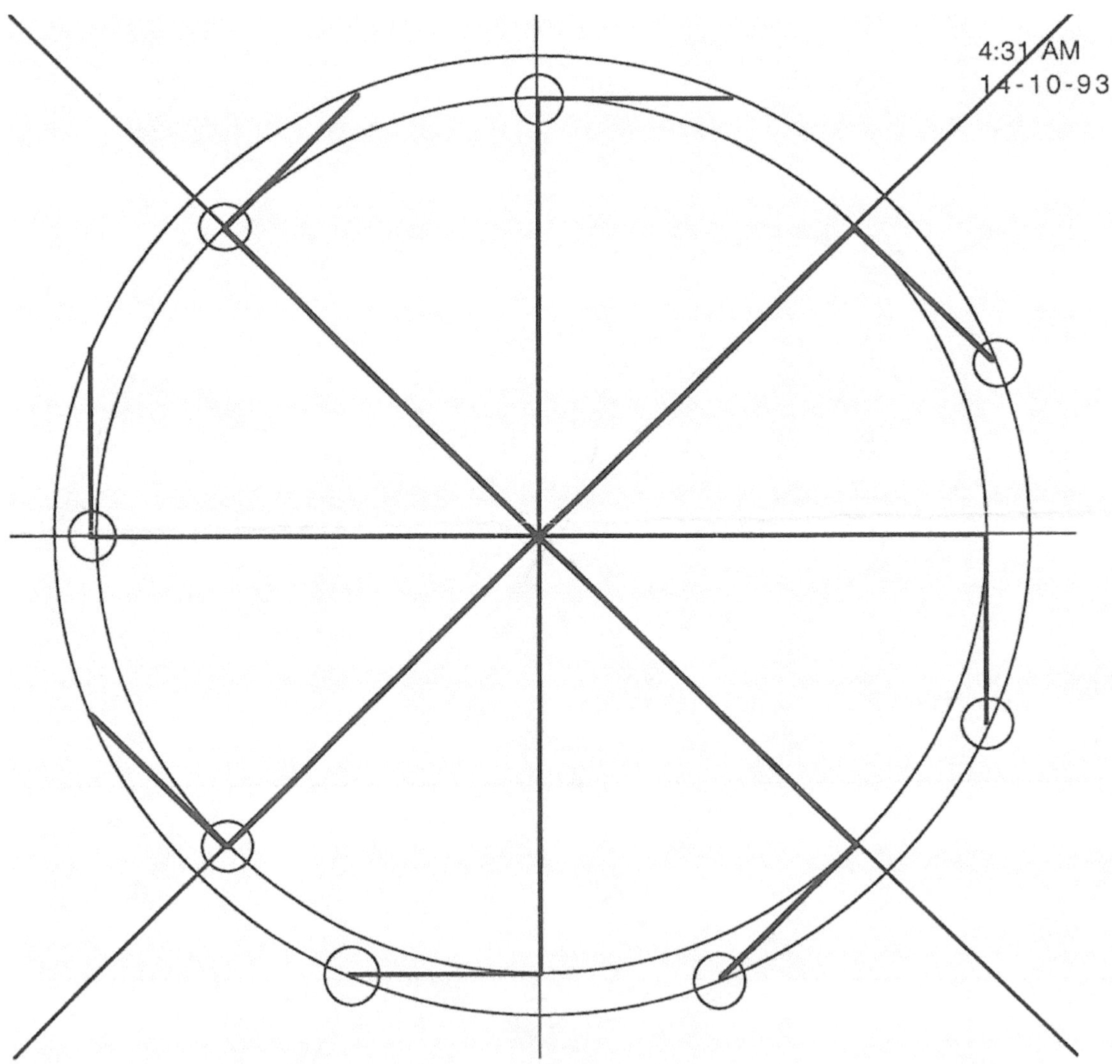

Clockwise rotation as weights slide on rods
Top and bottom weights cancels themselves out.
There is no movement because
the load on the left = the load on the right,
even though we have 3 weights vs. 2 weights.

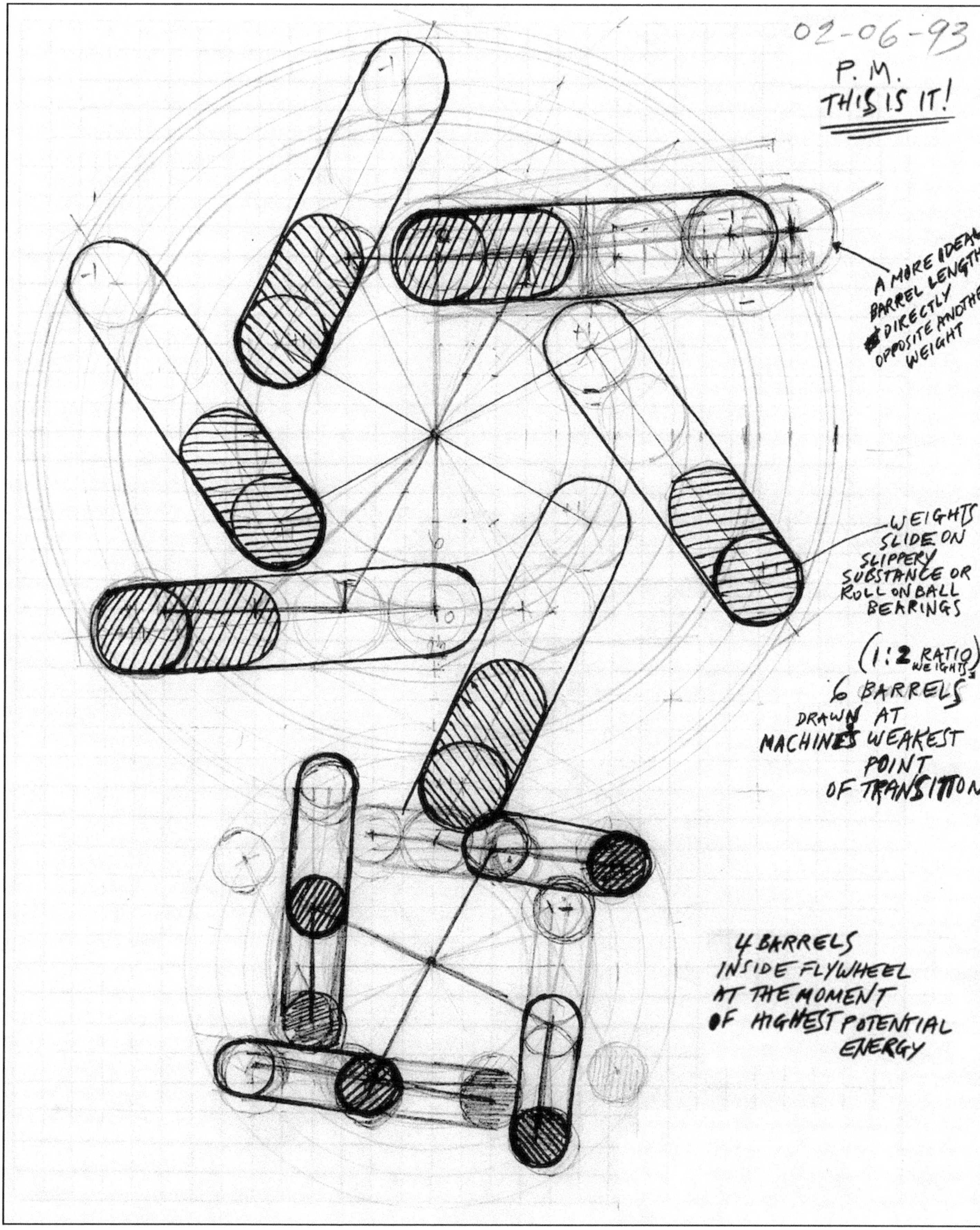

02-06-93
P.M.
THIS IS IT!
A MORE IDEAL BARREL LENGTH DIRECTLY OPPOSITE ANOTHER WEIGHT
WEIGHTS SLIDE ON SLIPPERY SUBSTANCE OR ROLL ON BALL BEARINGS
(1:2 RATIO WEIGHTS)
6 BARRELS DRAWN AT MACHINES WEAKEST POINT OF TRANSITION
4 BARRELS INSIDE FLYWHEEL AT THE MOMENT OF HIGHEST POTENTIAL ENERGY

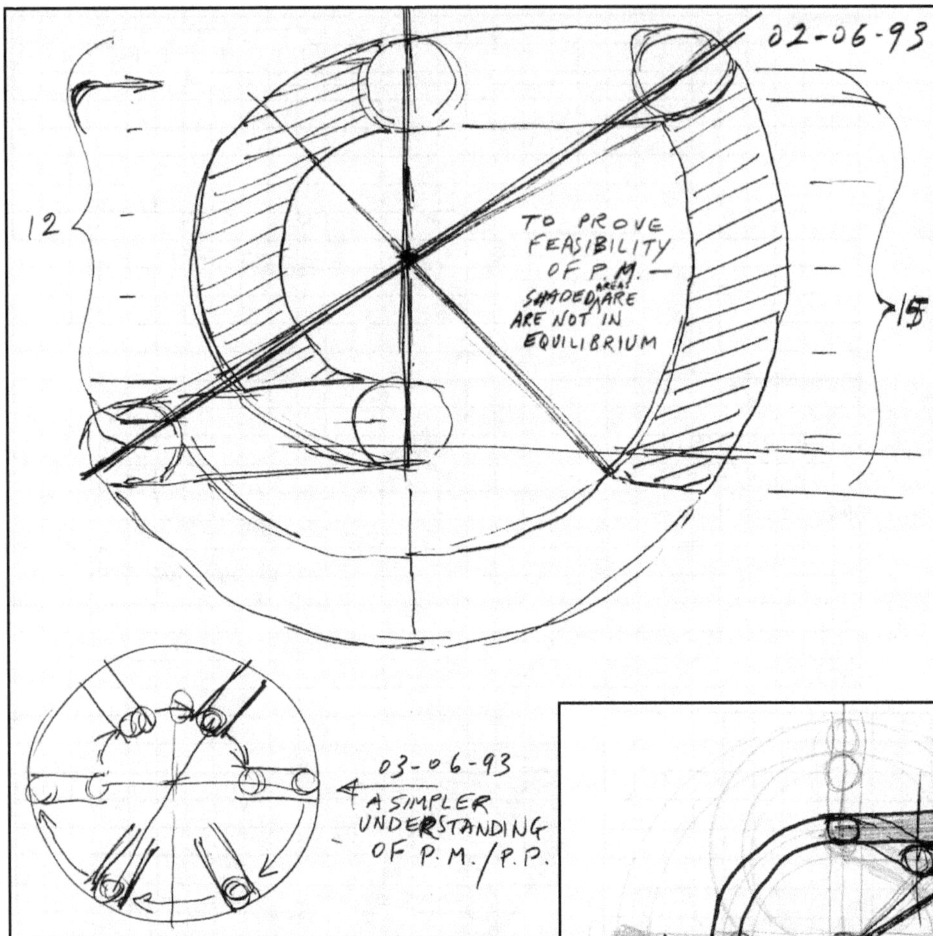
02-06-93
12
15
TO PROVE
FEASIBILITY
OF P.M. —
SHADED AREAS ARE
ARE NOT IN
EQUILIBRIUM
03-06-93
A SIMPLER
UNDERSTANDING
OF P.M./P.P.

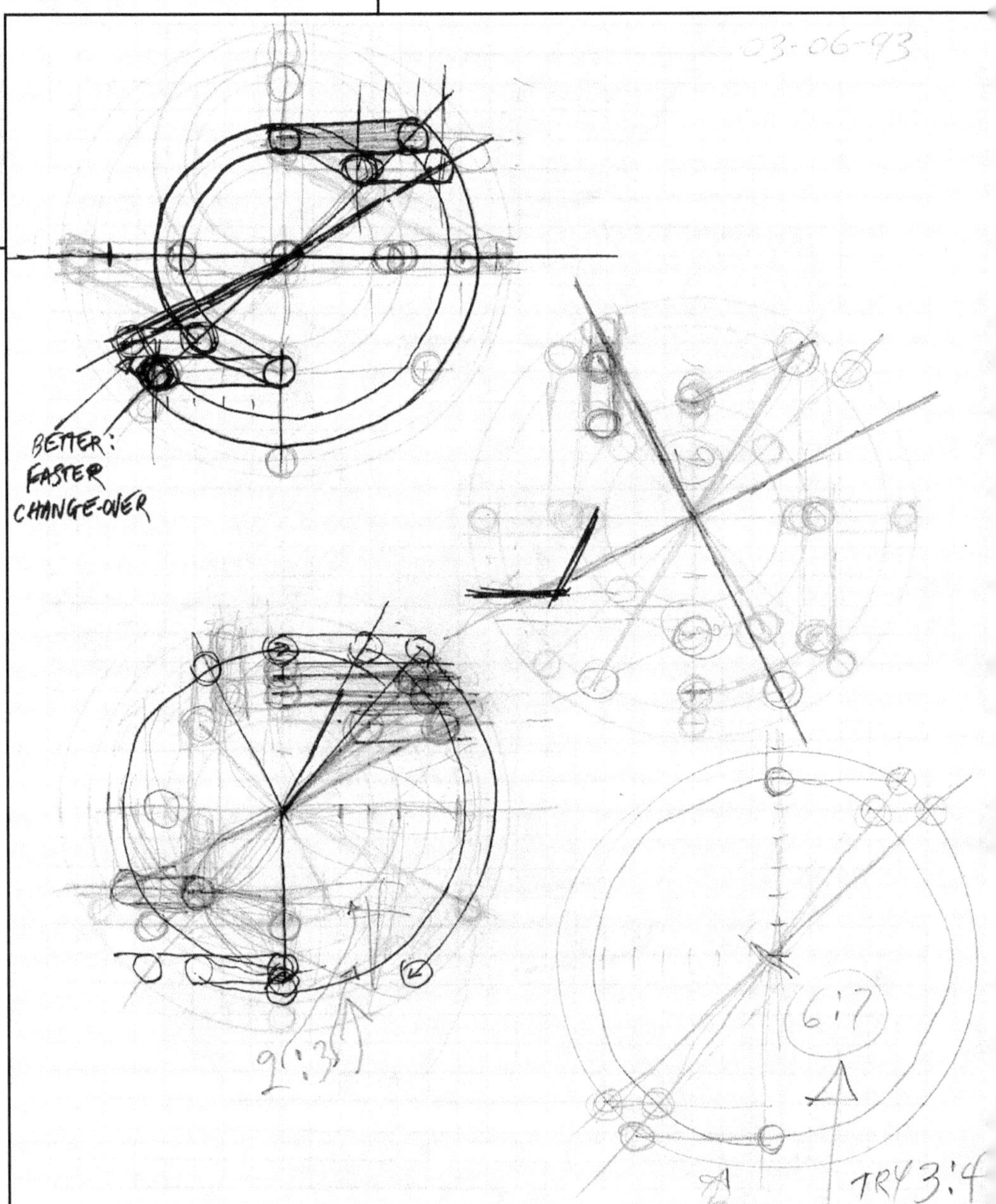
03-06-93
BETTER:
FASTER
CHANGE-OVER
6:7
TRY 3:4

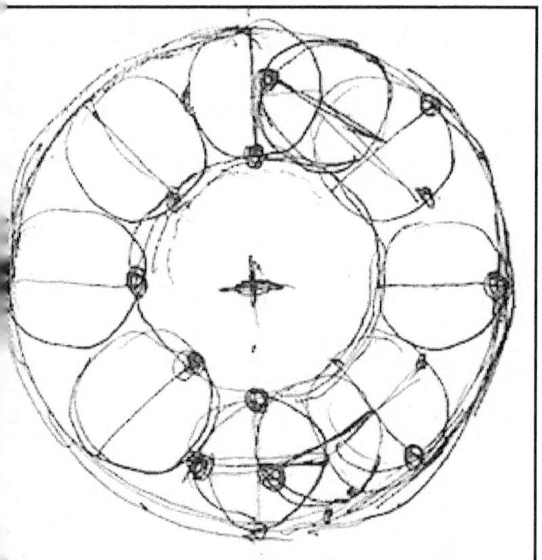

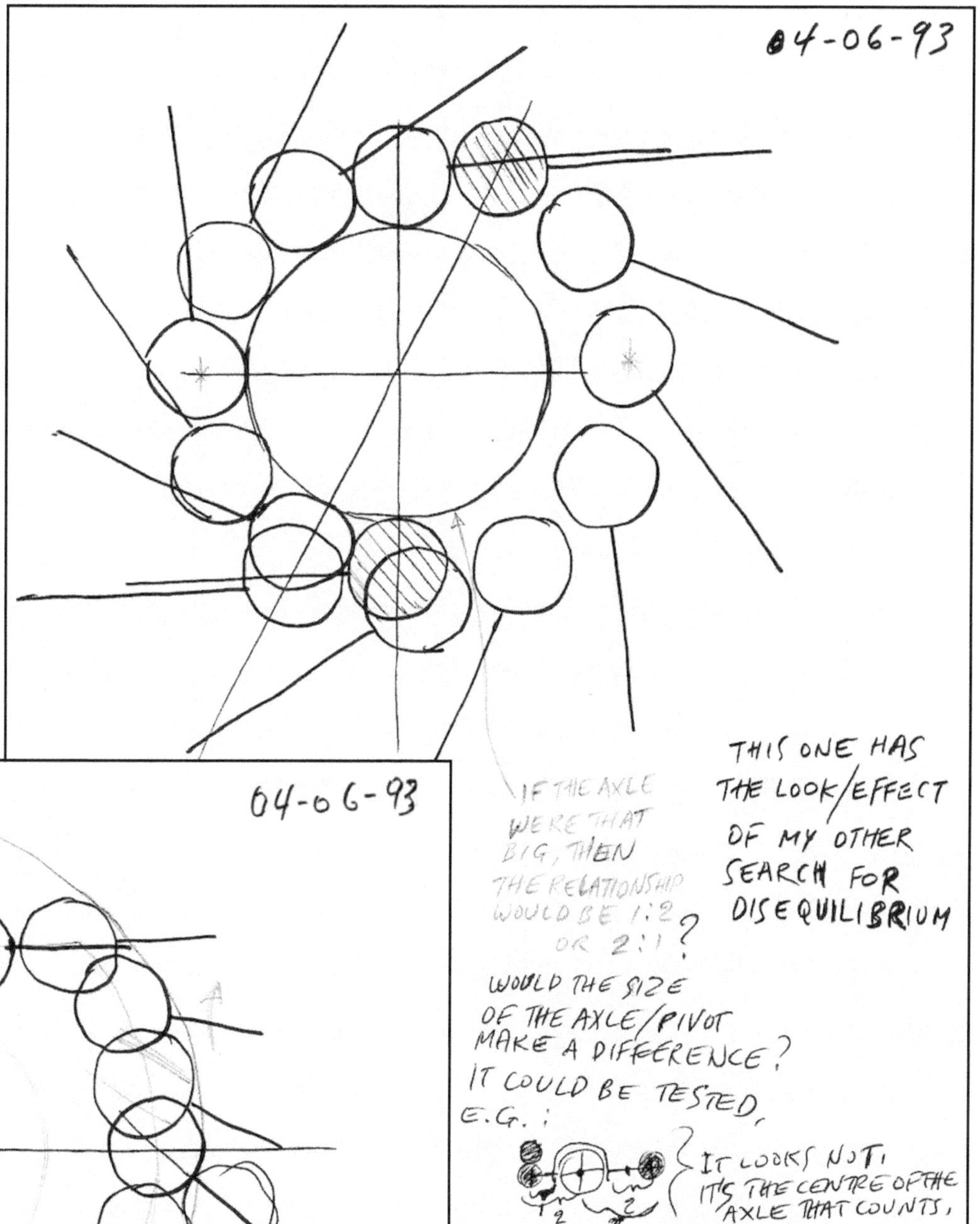
04-06-93
IF THE AXLE WERE THAT BIG, THEN THE RELATIONSHIP WOULD BE 1:2 OR 2:1?
THIS ONE HAS THE LOOK/EFFECT OF MY OTHER SEARCH FOR DISEQUILIBRIUM
WOULD THE SIZE OF THE AXLE/PIVOT MAKE A DIFFERENCE? IT COULD BE TESTED, E.G.:
IT LOOKS NOT! IT'S THE CENTRE OF THE AXLE THAT COUNTS.

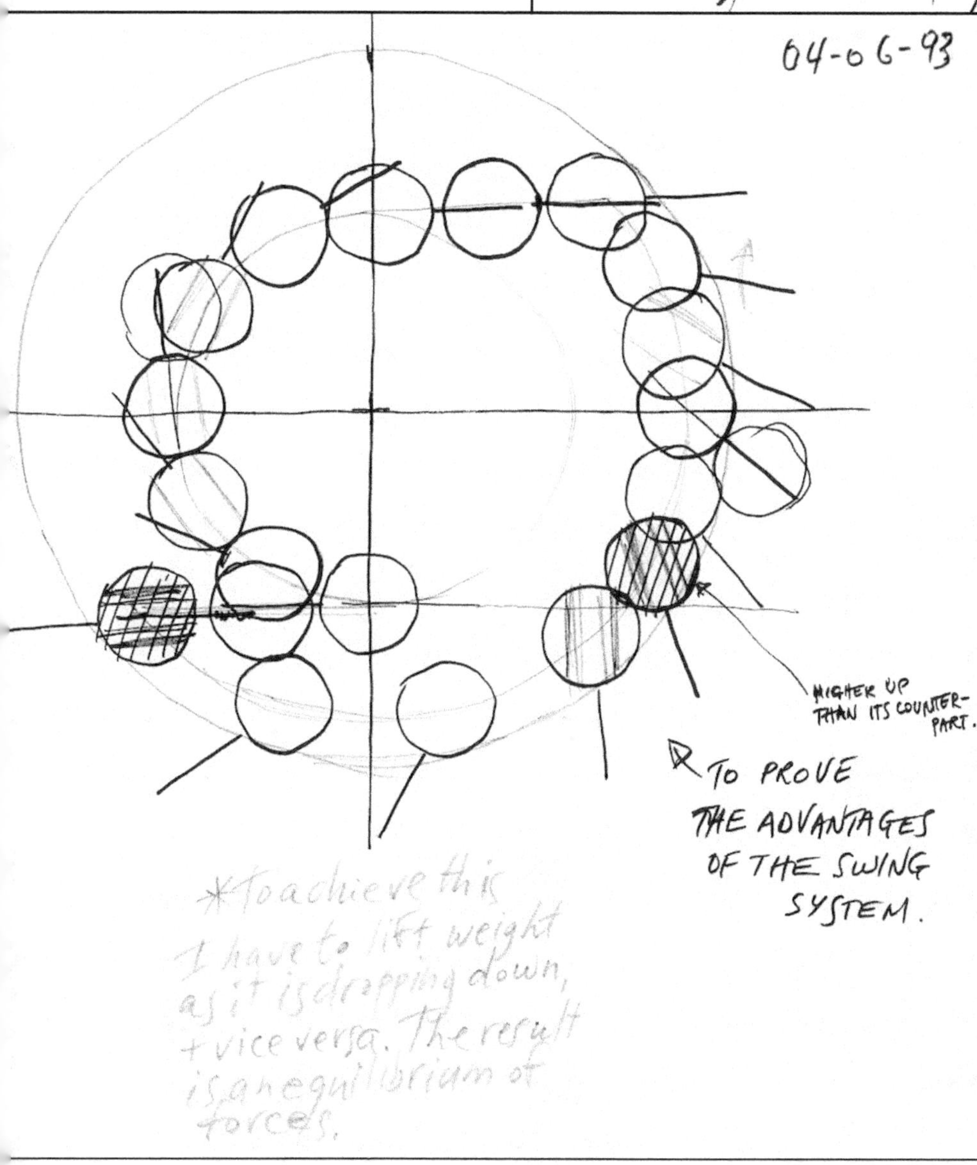
04-06-93
HIGHER UP THAN ITS COUNTER-PART.
TO PROVE THE ADVANTAGES OF THE SWING SYSTEM.
*To achieve this I have to lift weight as it is dropping down, + vice versa. The result is an equilibrium of forces.

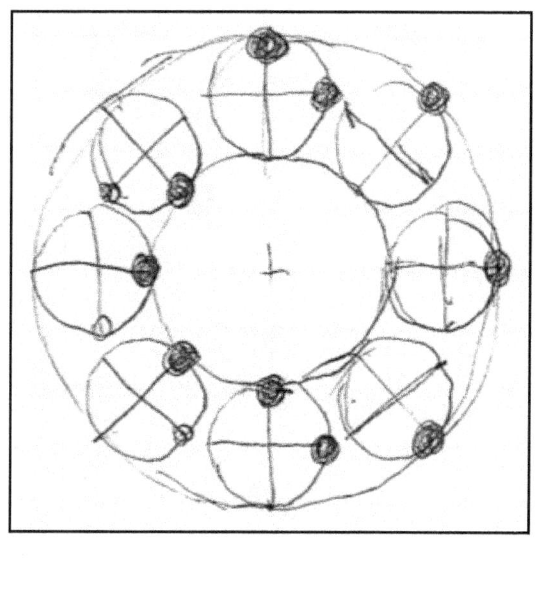

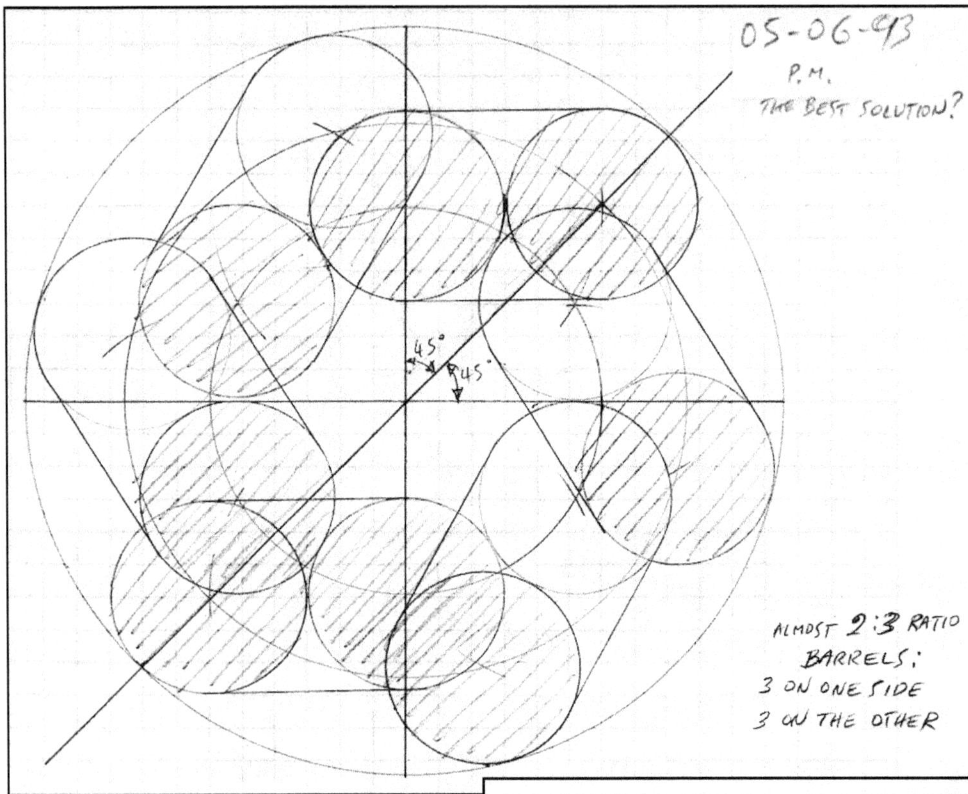
05-06-93
P.M.
THE BEST SOLUTION?
45°
45°
ALMOST 2:3 RATIO
BARRELS:
3 ON ONE SIDE
3 ON THE OTHER

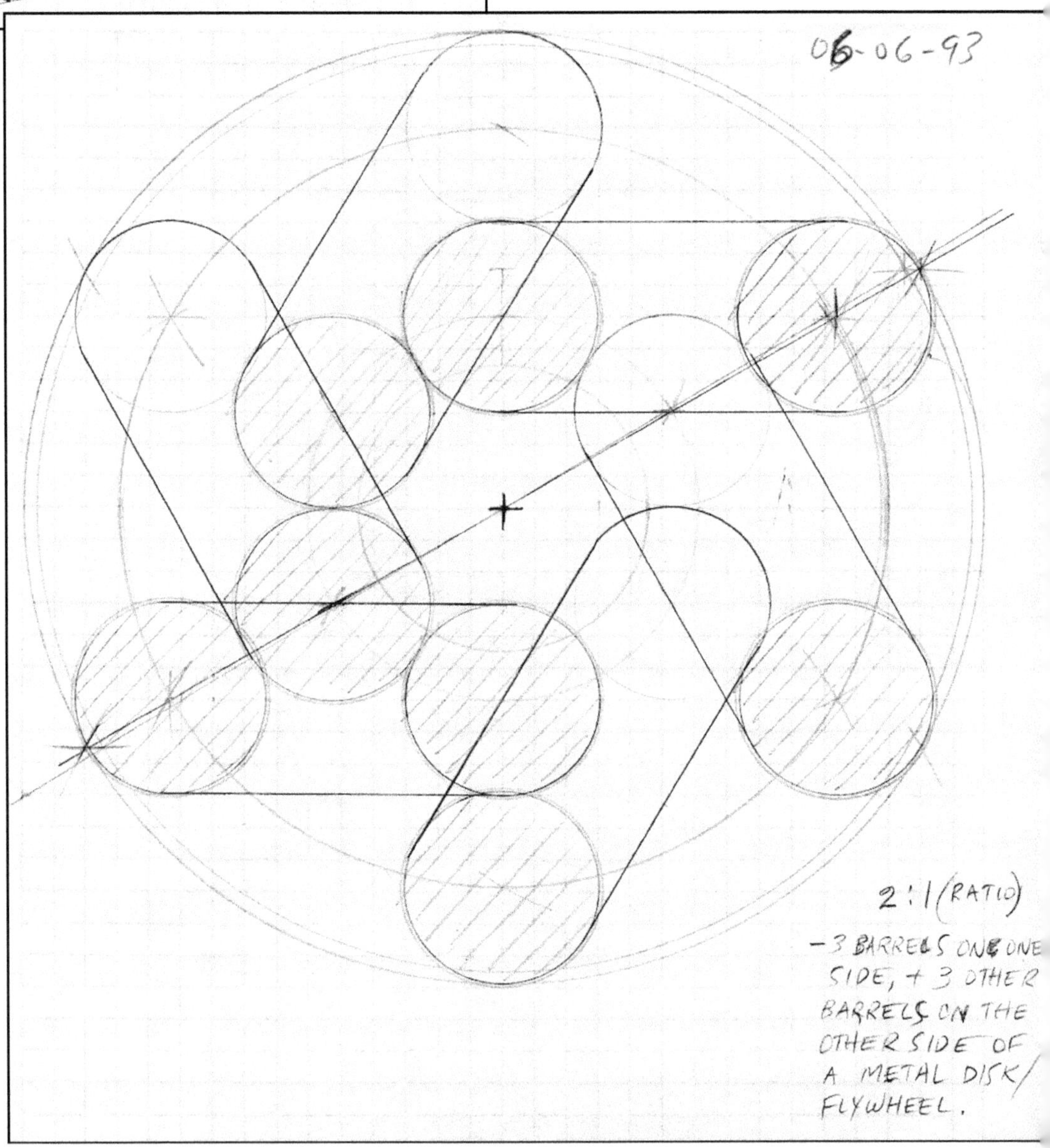
06-06-93
2:1 (RATIO)
- 3 BARRELS ON ONE SIDE, + 3 OTHER BARRELS ON THE OTHER SIDE OF A METAL DISK/FLYWHEEL.

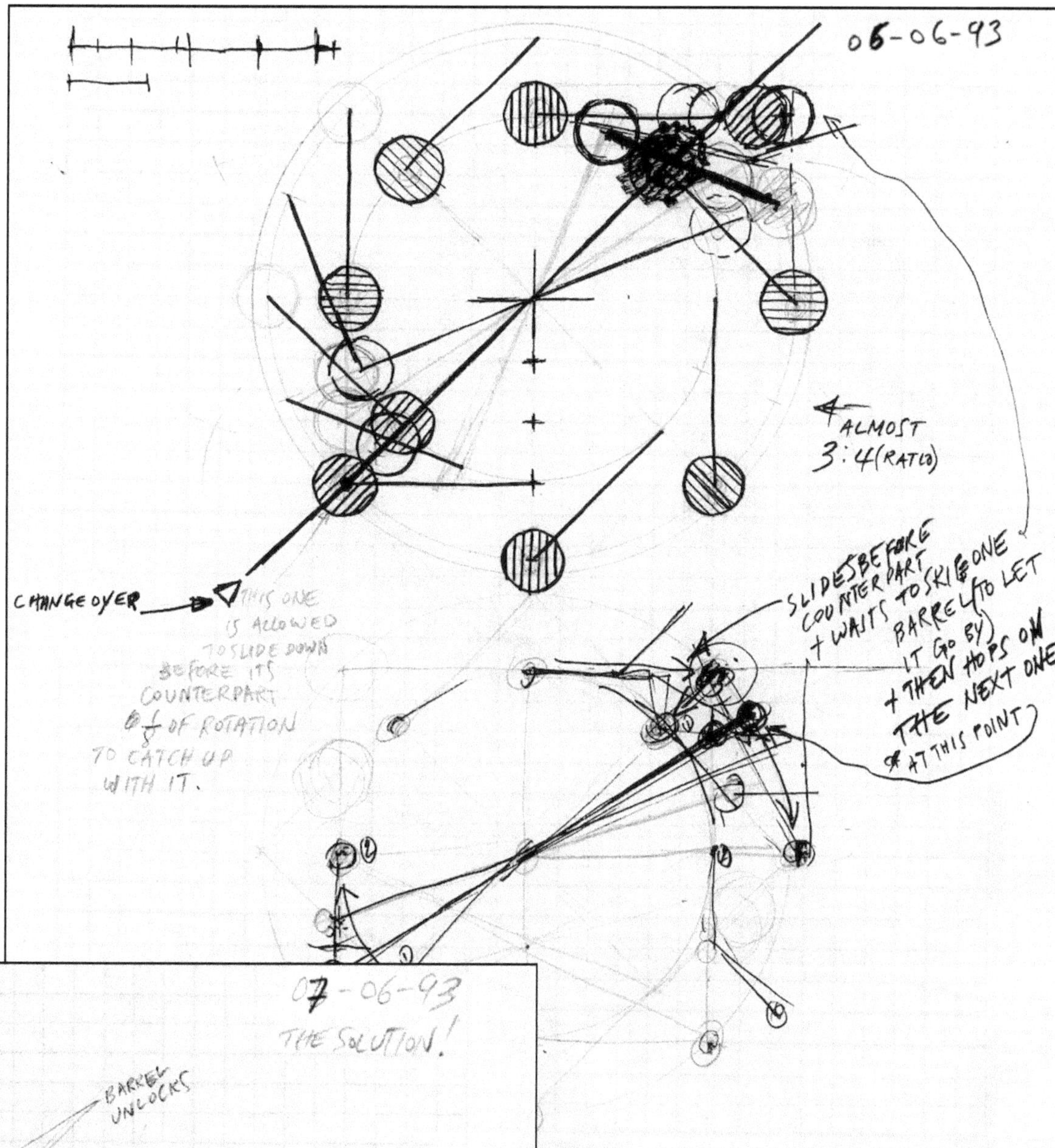
06-06-93
ALMOST 3:4 (RATIO)
CHANGEOVER
THIS ONE IS ALLOWED TO SLIDE DOWN BEFORE ITS COUNTERPART 1/5 OF ROTATION TO CATCH UP WITH IT.
SLIDES BEFORE COUNTERPART + WAITS TO SKIP ONE BARREL (TO LET IT GO BY) + THEN HOPS ON THE NEXT ONE AT THIS POINT

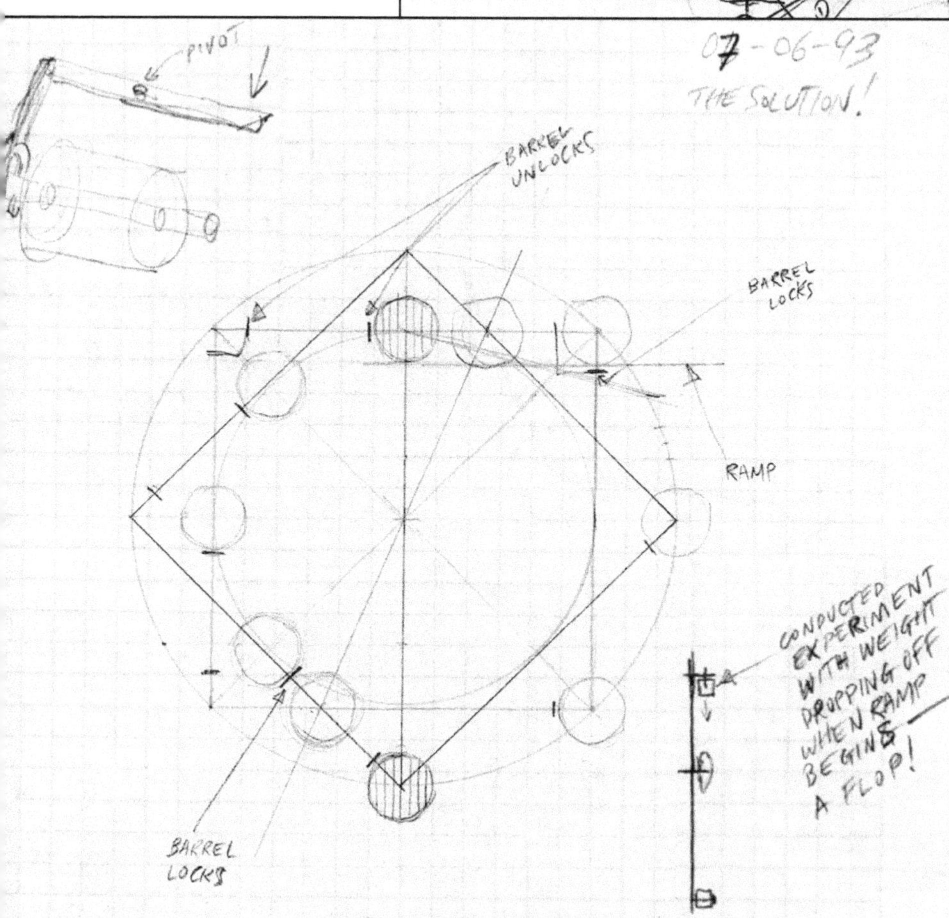
07-06-93
THE SOLUTION!
PIVOT
BARREL UNLOCKS
BARREL LOCKS
RAMP
CONDUCTED EXPERIMENT WITH WEIGHT DROPPING OFF WHEN RAMP BEGINS A FLOP!
BARREL LOCKS

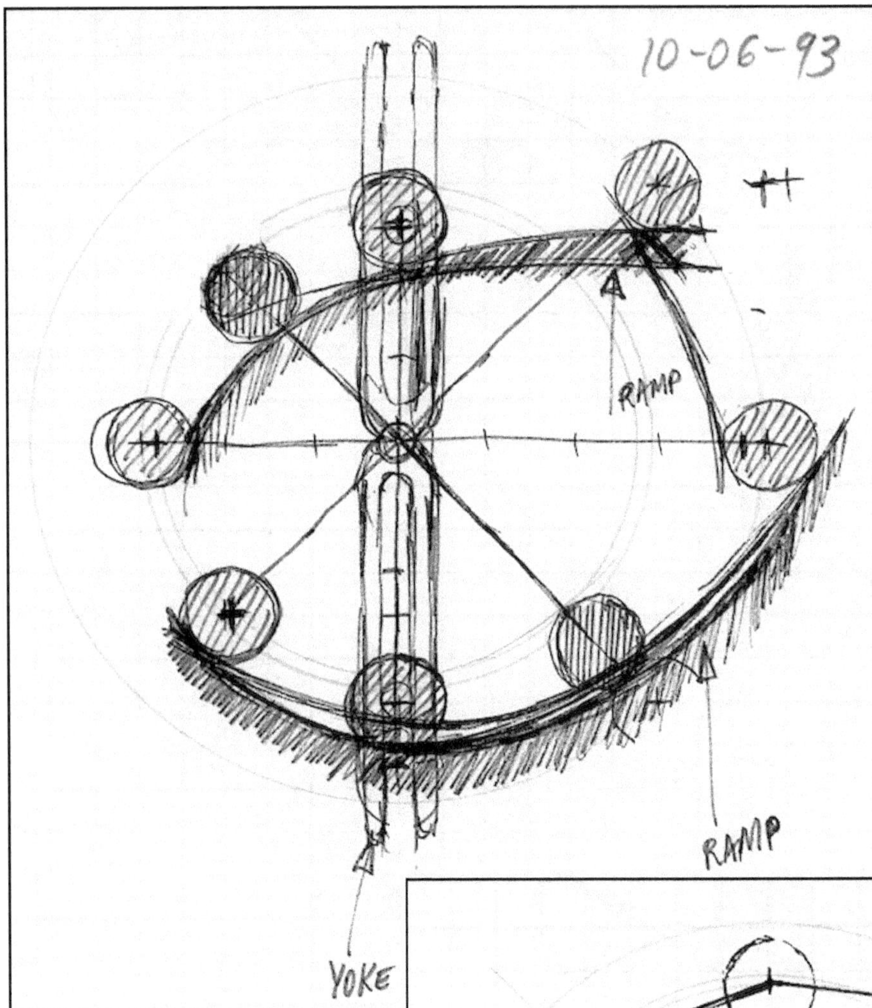
10-06-93
RAMP
RAMP
YOKE

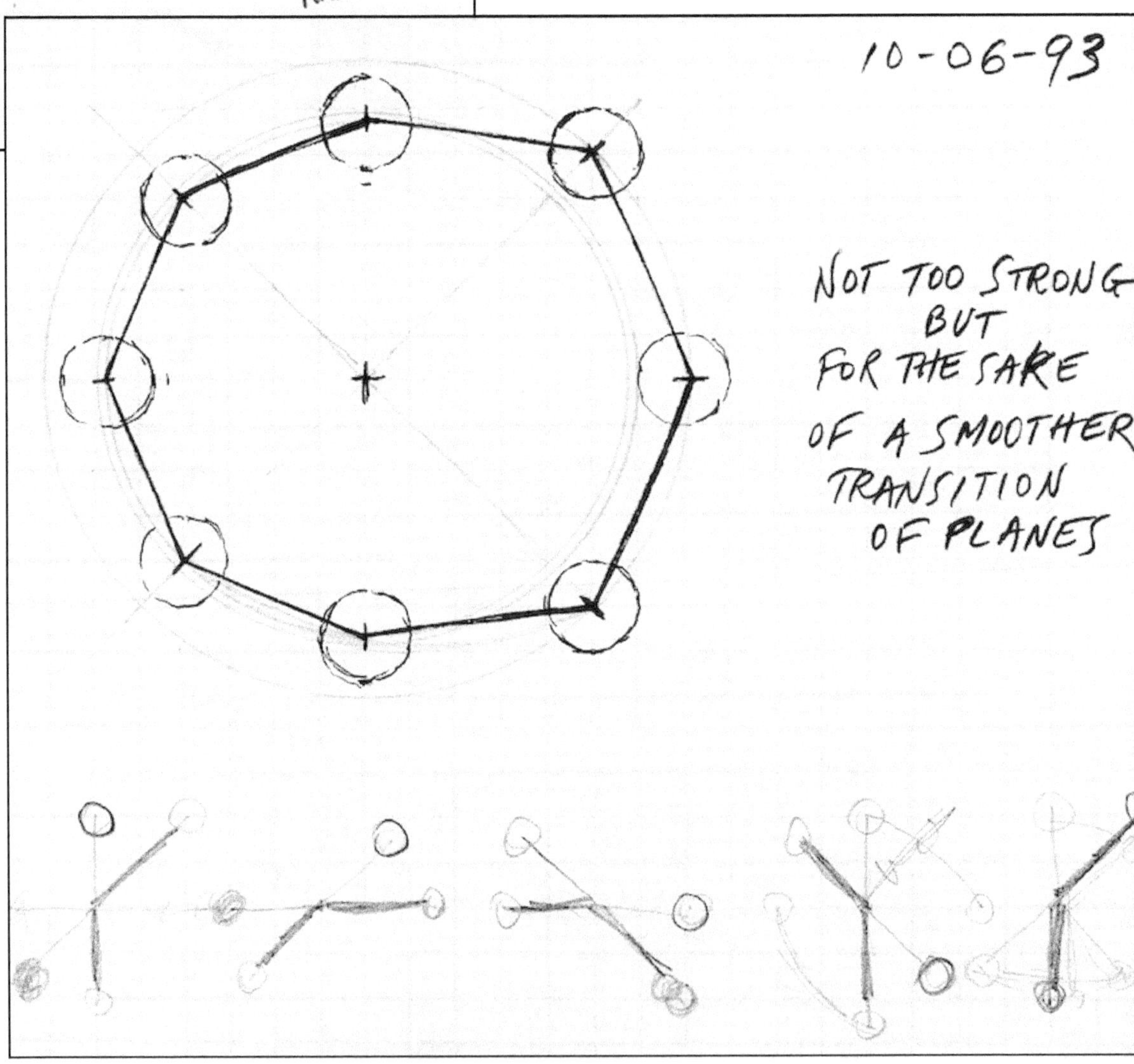
10-06-93
NOT TOO STRONG
BUT
FOR THE SAKE
OF A SMOOTHER
TRANSITION
OF PLANES

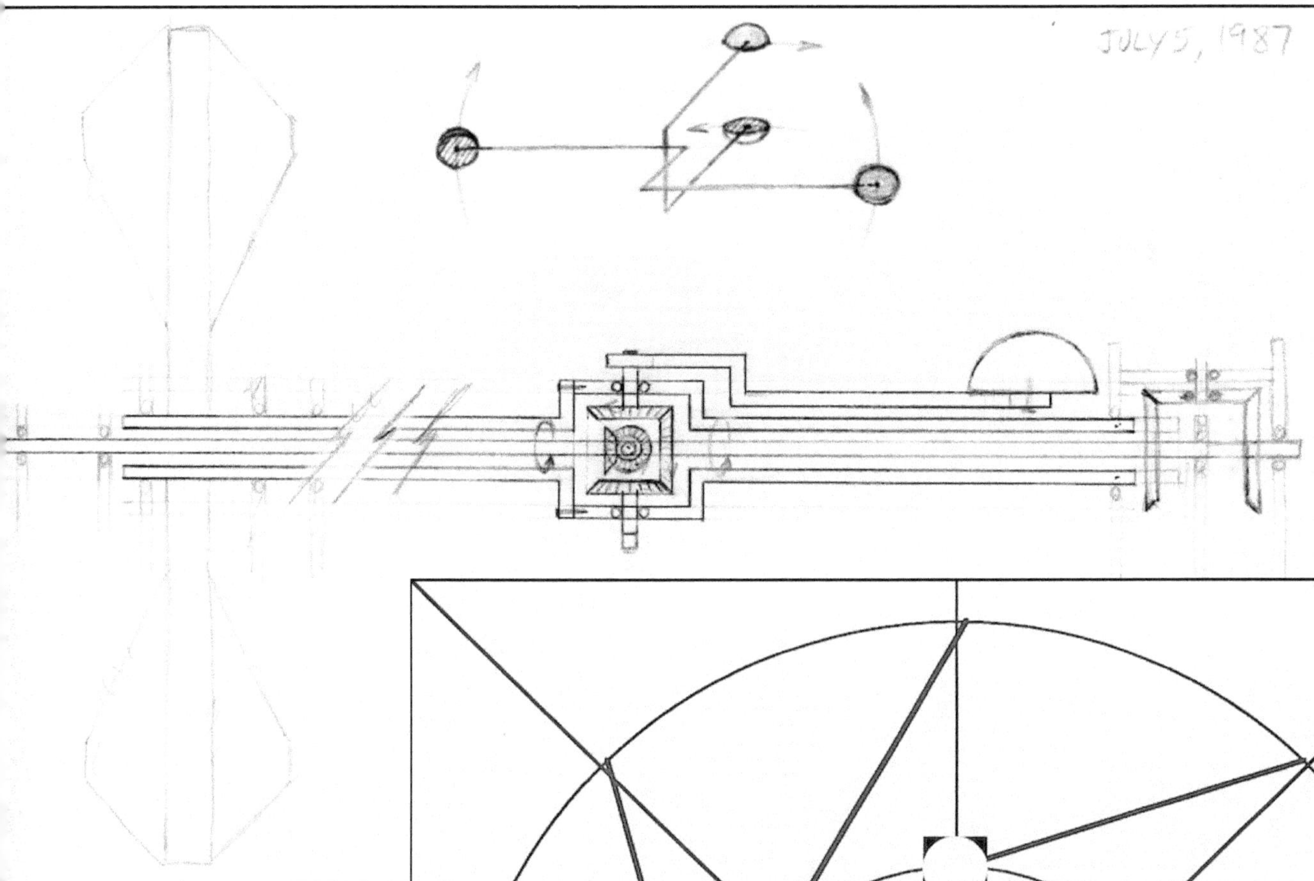

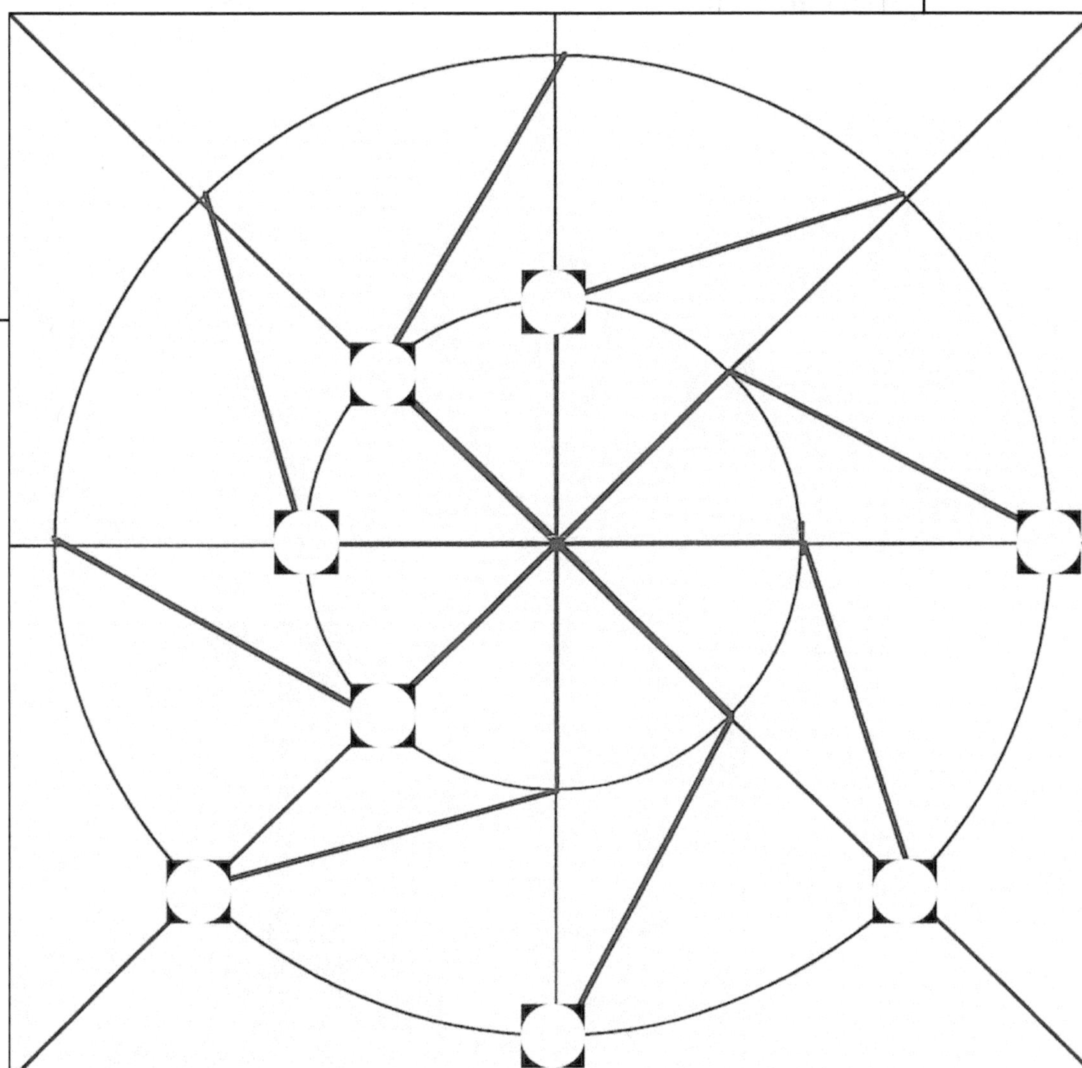

Clockwise rotation as weights slide on rods
Top and bottom weights cancels themselves out.
There is no movement because
the load on the left = the load on the right,
even though we have 3 weights vs. 1 weight.

This way of outsmarting gravity is a dead end.

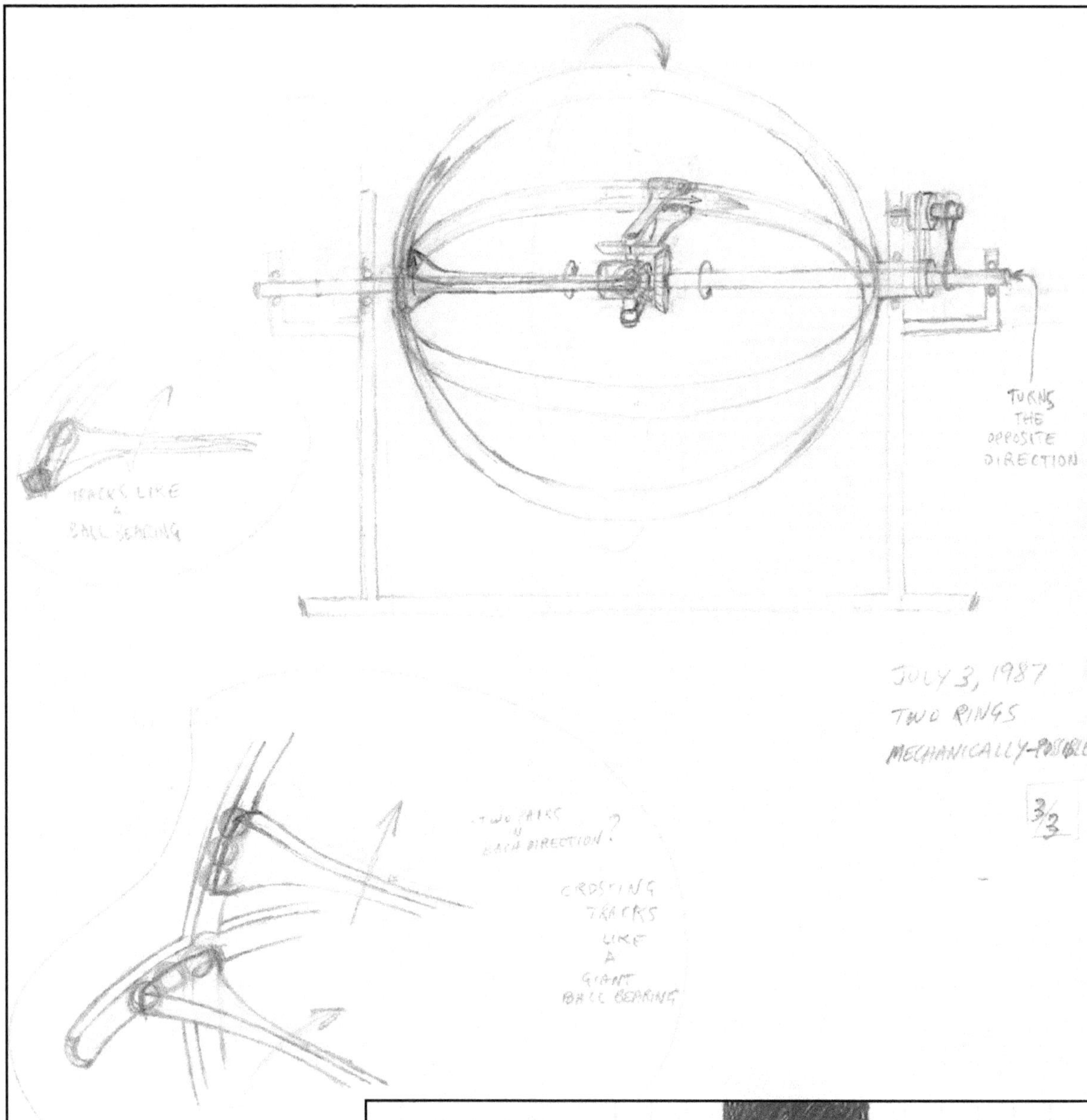
TURNS THE OPPOSITE DIRECTION
TRACKS LIKE A BALL BEARING
JULY 3, 1987
TWO RINGS
MECHANICALLY-POSSIBLE
3/3
TWO PAIRS IN EACH DIRECTION?
CROSSING TRACKS LIKE A GIANT BALL BEARING

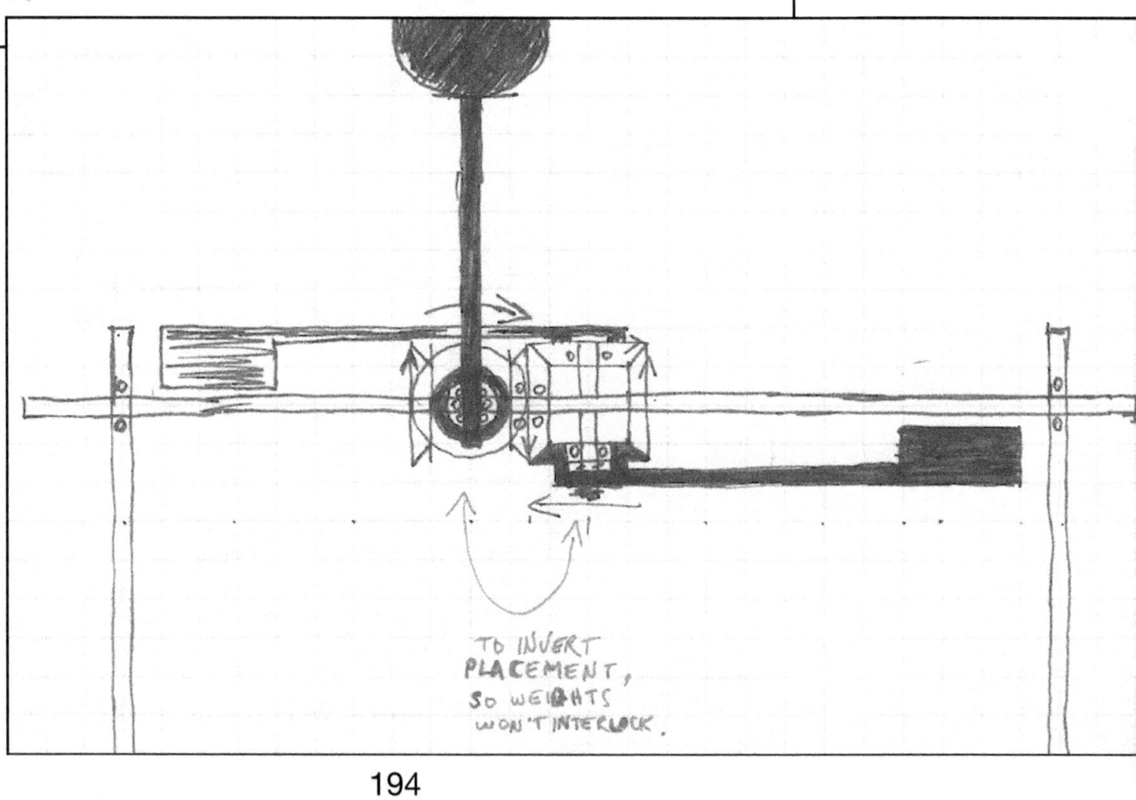
TO INVERT PLACEMENT, SO WEIGHTS WON'T INTERLOCK.

8D. MAG-GRA MOTOR

In a Maggra (**Mag**netic-**Gra**vity) Motor, the aim is to use Magnets and the Gravity/Atmosphere to power the motor. To understand the problem at hand, we have to go back to the electric motor. The way an electric motor works is by turning solenoids or stators on and off consecutively and continuously. Because of that, the stator is in disequilibrium and forces to rotate. One way we could create the same effect with a maggra motor is by making magnets switch from attraction to repulsion and vice versa. With the maggra motor I built, that is accomplished by rotating magnets so that they attract or repulse as they rotate. The work of rotating the magnets is done by gravity/atmosphere. And that landed me back into the quagmire that plagues the earth motor. We need to find a different way of switching magnets on and off. What follows is what led me to build a maggra motor.

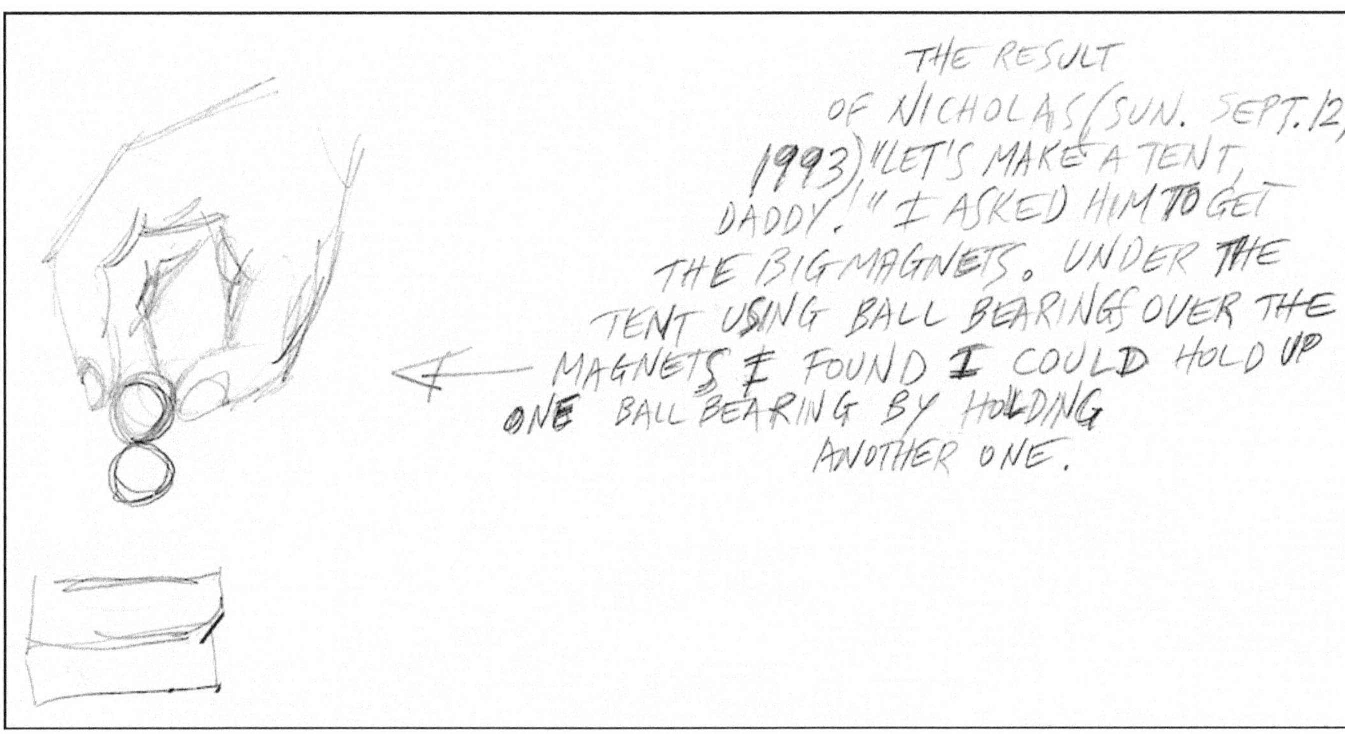

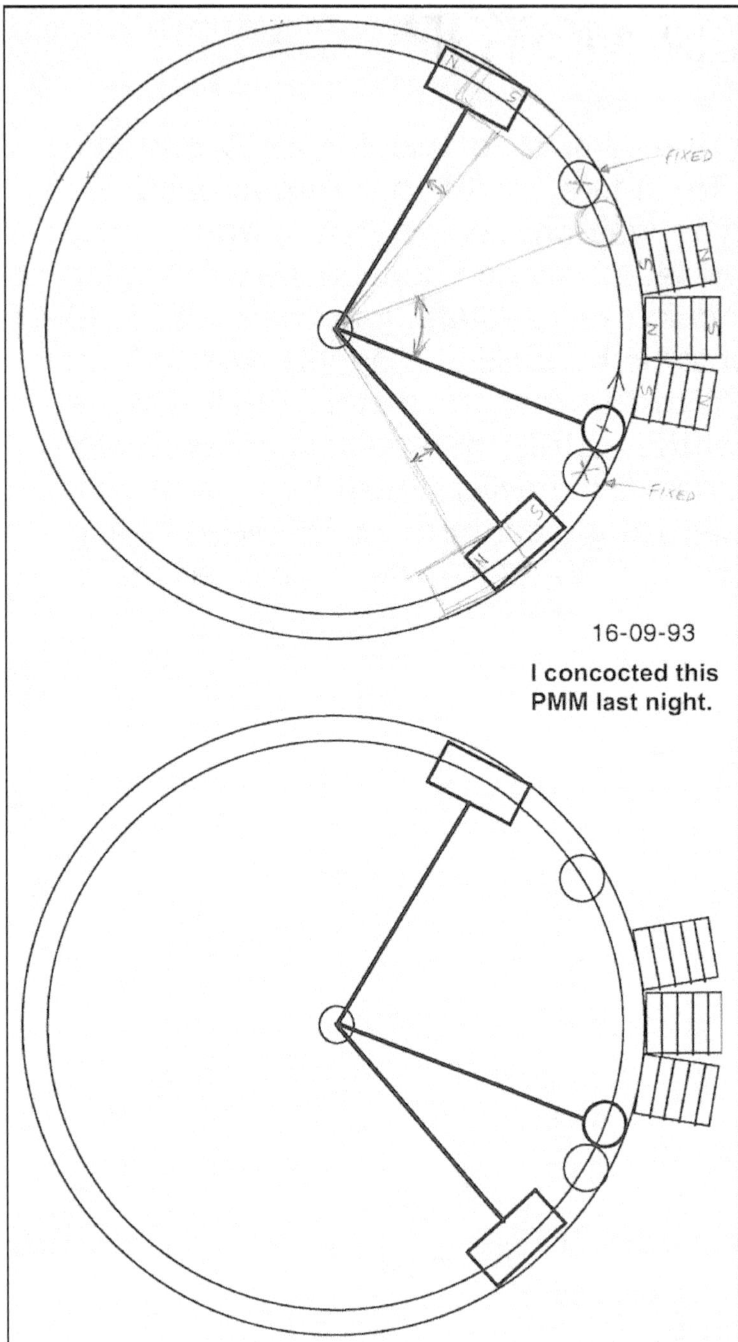
FIXED
FIXED
16-09-93
I concocted this PMM last night.

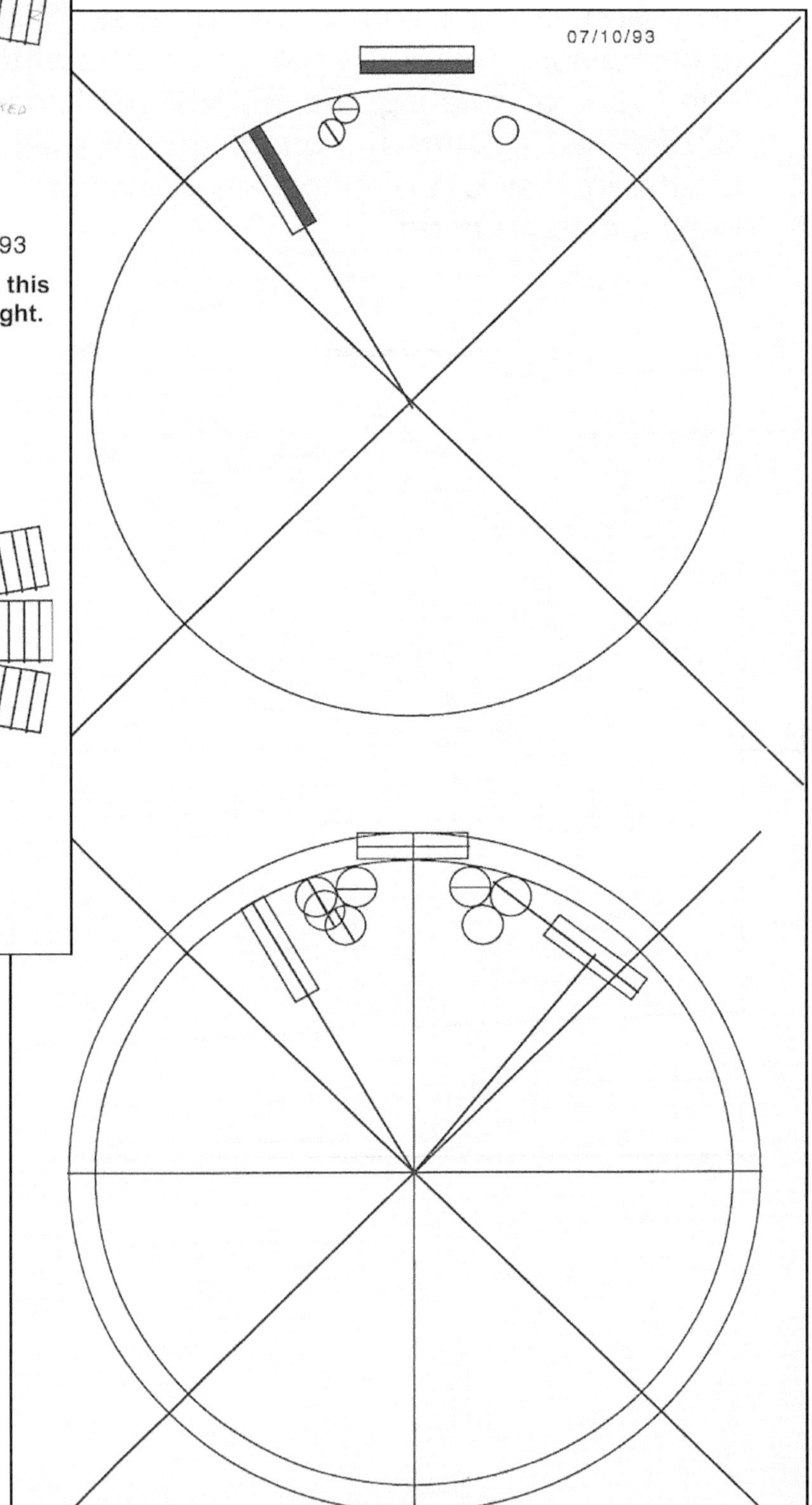
07/10/93

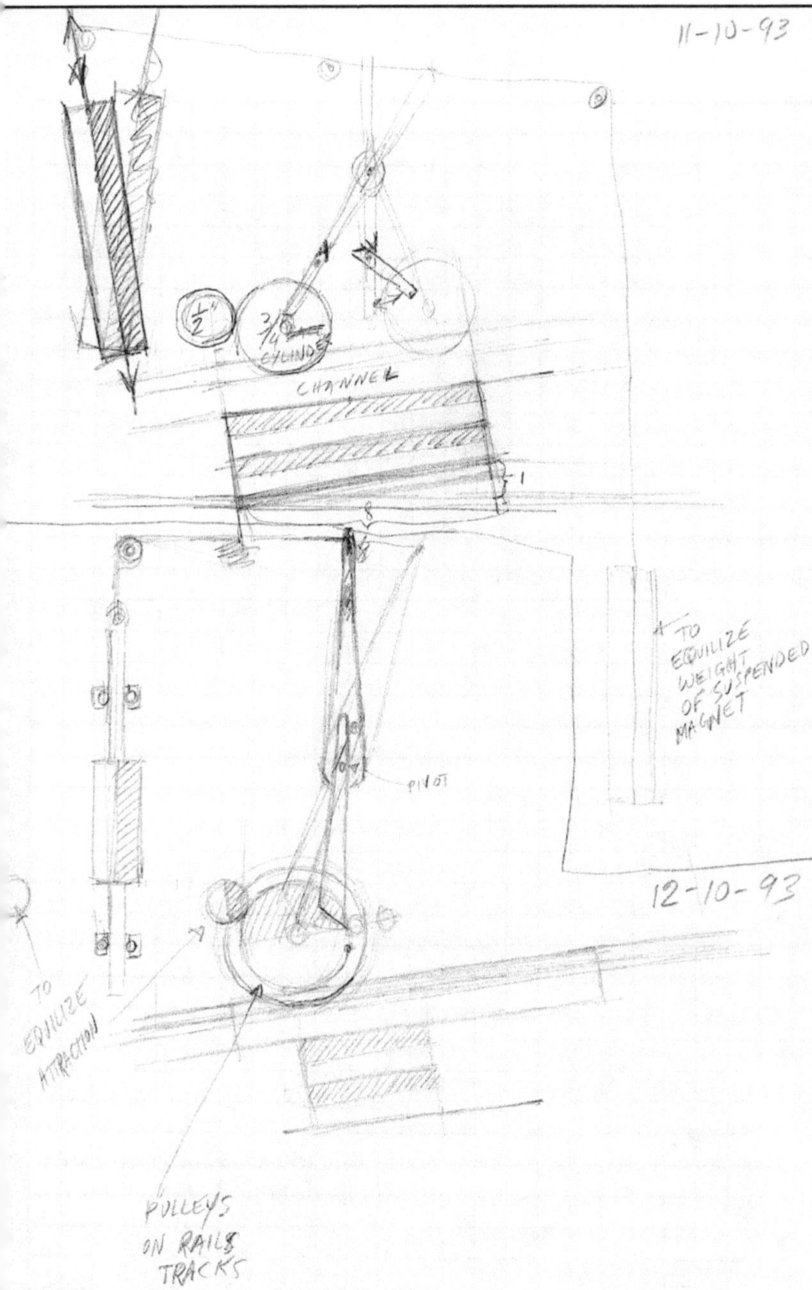
11-10-93
1/2"
3/4" CYLINDER
CHANNEL
PIVOT
TO EQUILIZE WEIGHT OF SUSPENDED MAGNET
12-10-93
TO EQUILIZE ATTRACTION
PULLEYS ON RAILS TRACKS

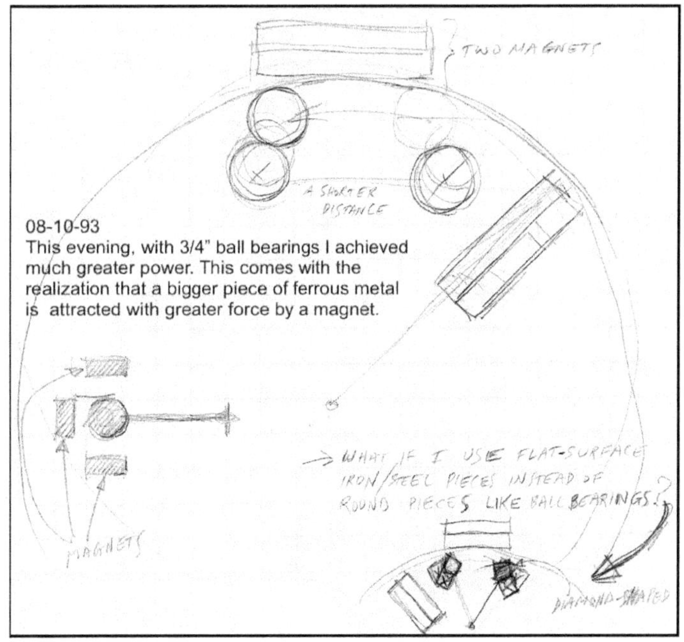
TWO MAGNETS
A SHORTER DISTANCE
08-10-93
This evening, with 3/4" ball bearings I achieved much greater power. This comes with the realization that a bigger piece of ferrous metal is attracted with greater force by a magnet.
WHAT IF I USE FLAT-SURFACE IRON/STEEL PIECES INSTEAD OF ROUND PIECES LIKE BALL BEARINGS?
MAGNETS
DIAMOND-SHAPED

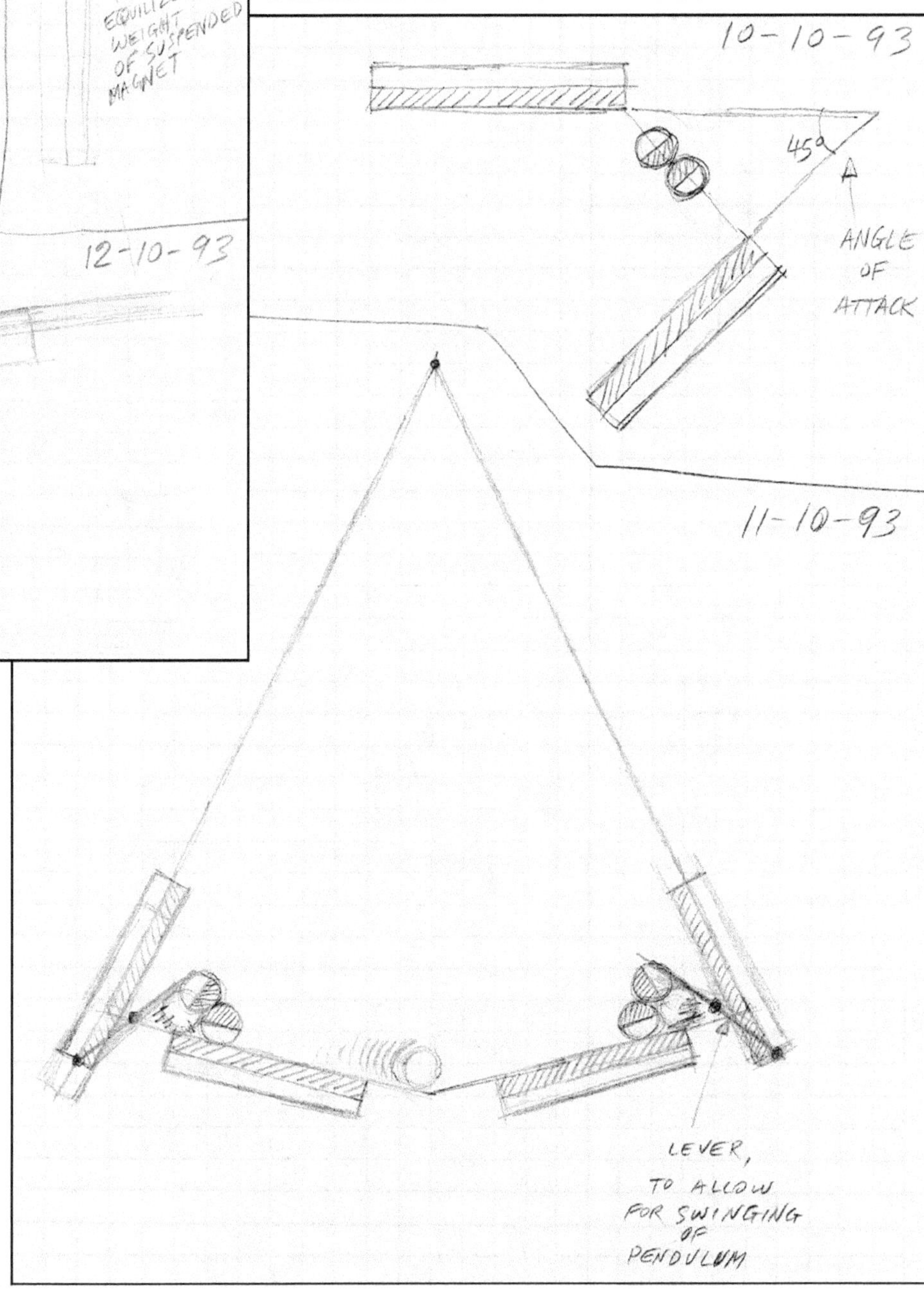
10-10-93
45°
ANGLE OF ATTACK
11-10-93
LEVER, TO ALLOW FOR SWINGING OF PENDULUM

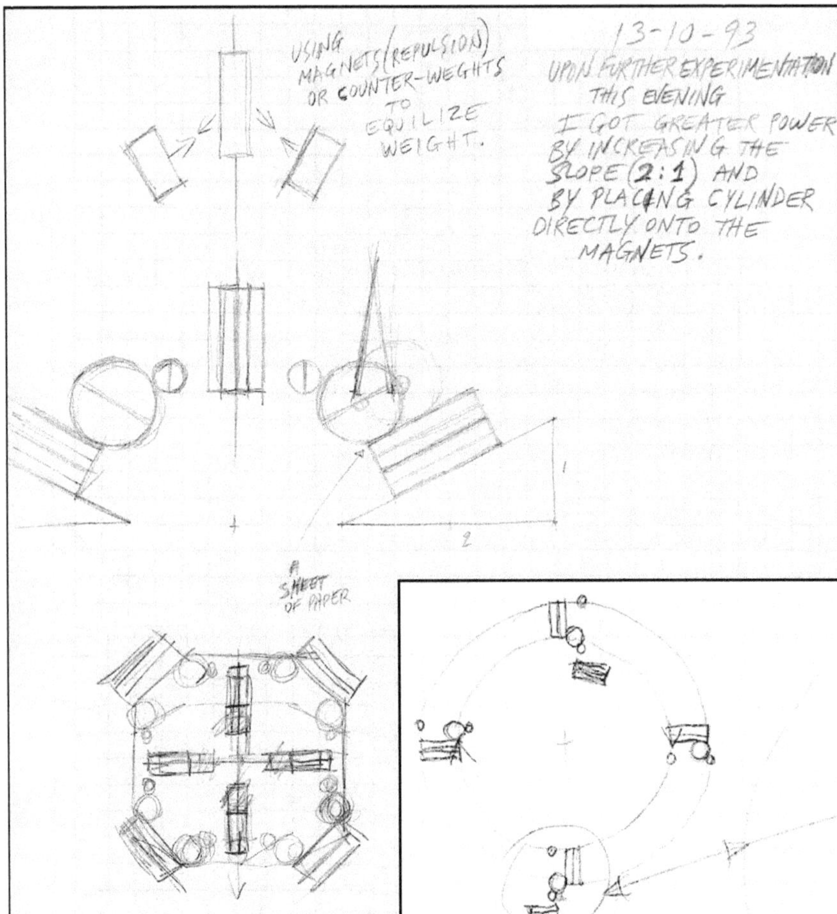

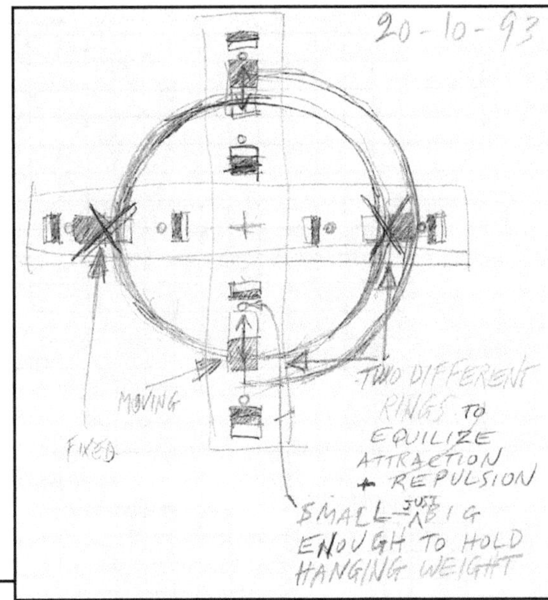

15-10-93
Attraction counteracts Repulsio

In this idea, we make use of 1 3/4" cylinder (2" long) and 2 fixed, small rods (same length). The larger cylinder rolls to the other side of the rotor-magnet when it meets the fixed, magnet stators pointing in the opposite direction.

The only disequilibrium is created by gravity's attraction.

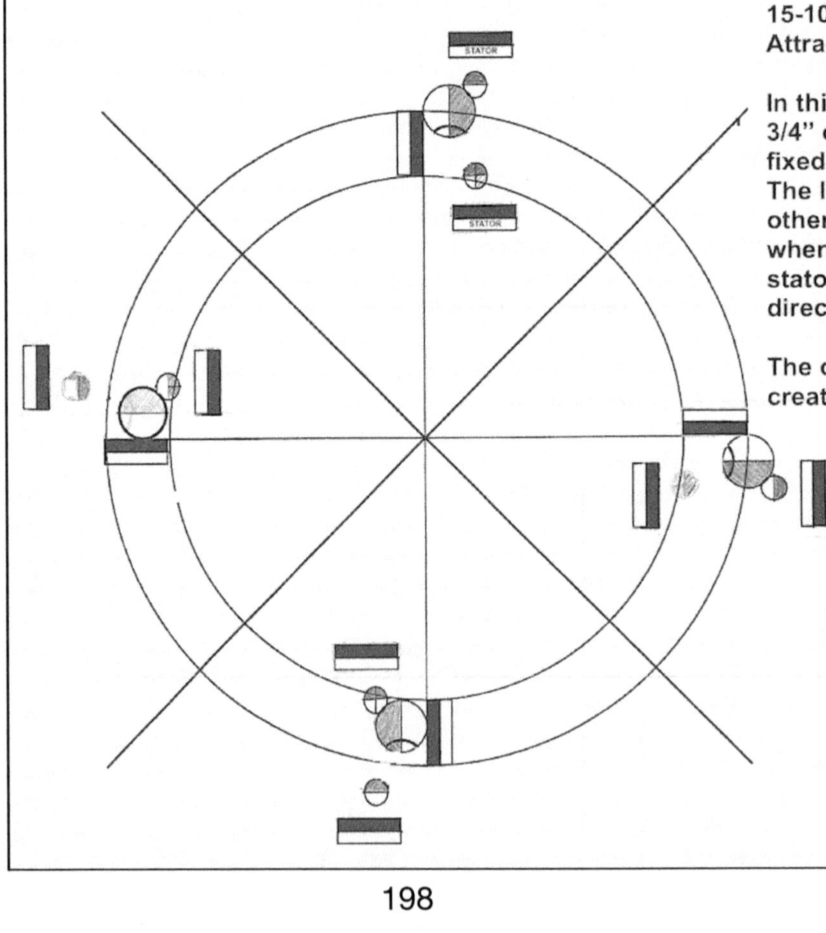

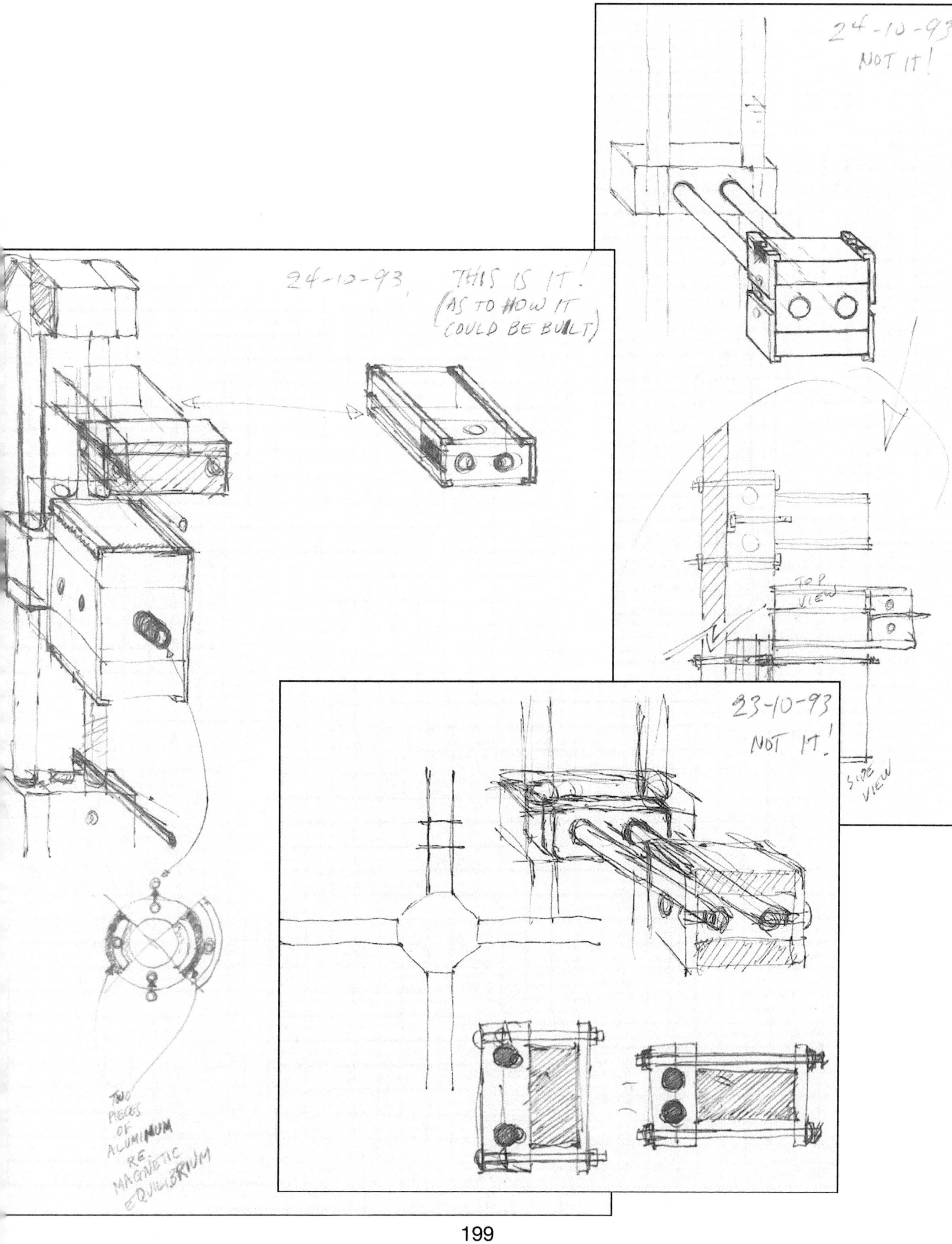
24-10-93
THIS IS IT!
(AS TO HOW IT
COULD BE BUILT)
TWO PIECES OF ALUMINUM RE. MAGNETIC EQUILIBRIUM
24-10-93
NOT IT!
TOP VIEW
SIDE VIEW
23-10-93
NOT IT!

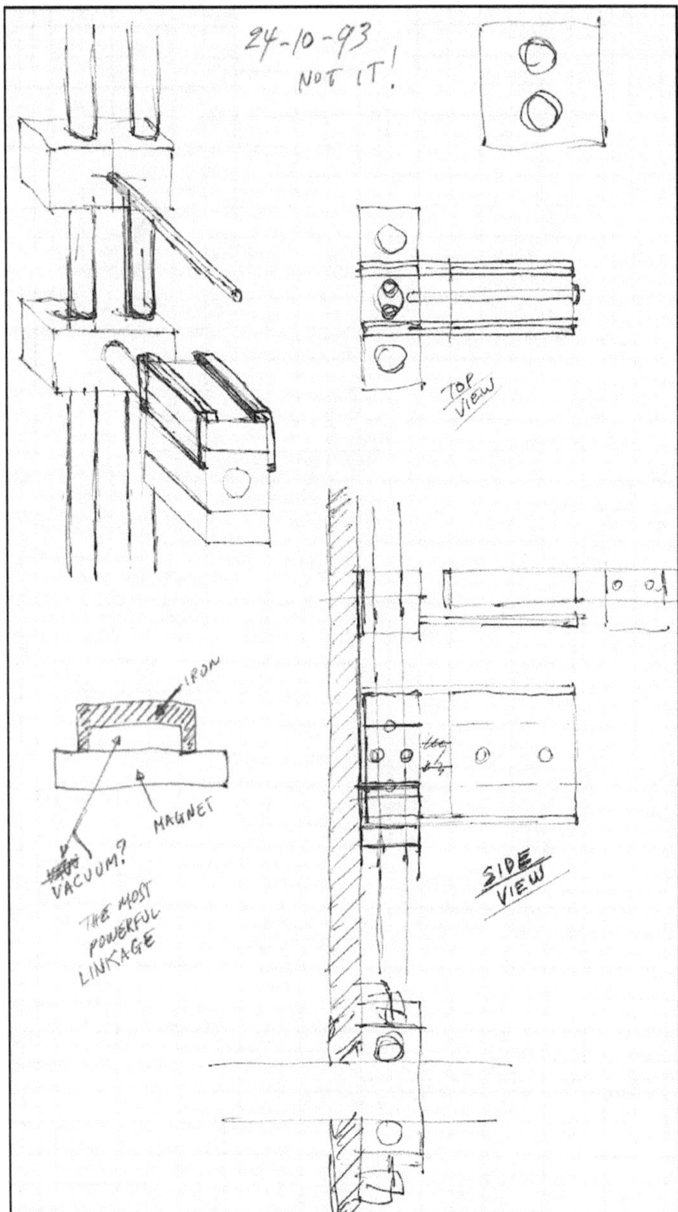
24-10-93
NOT IT!
TOP VIEW
IRON
MAGNET
VACUUM?
THE MOST POWERFUL LINKAGE
SIDE VIEW

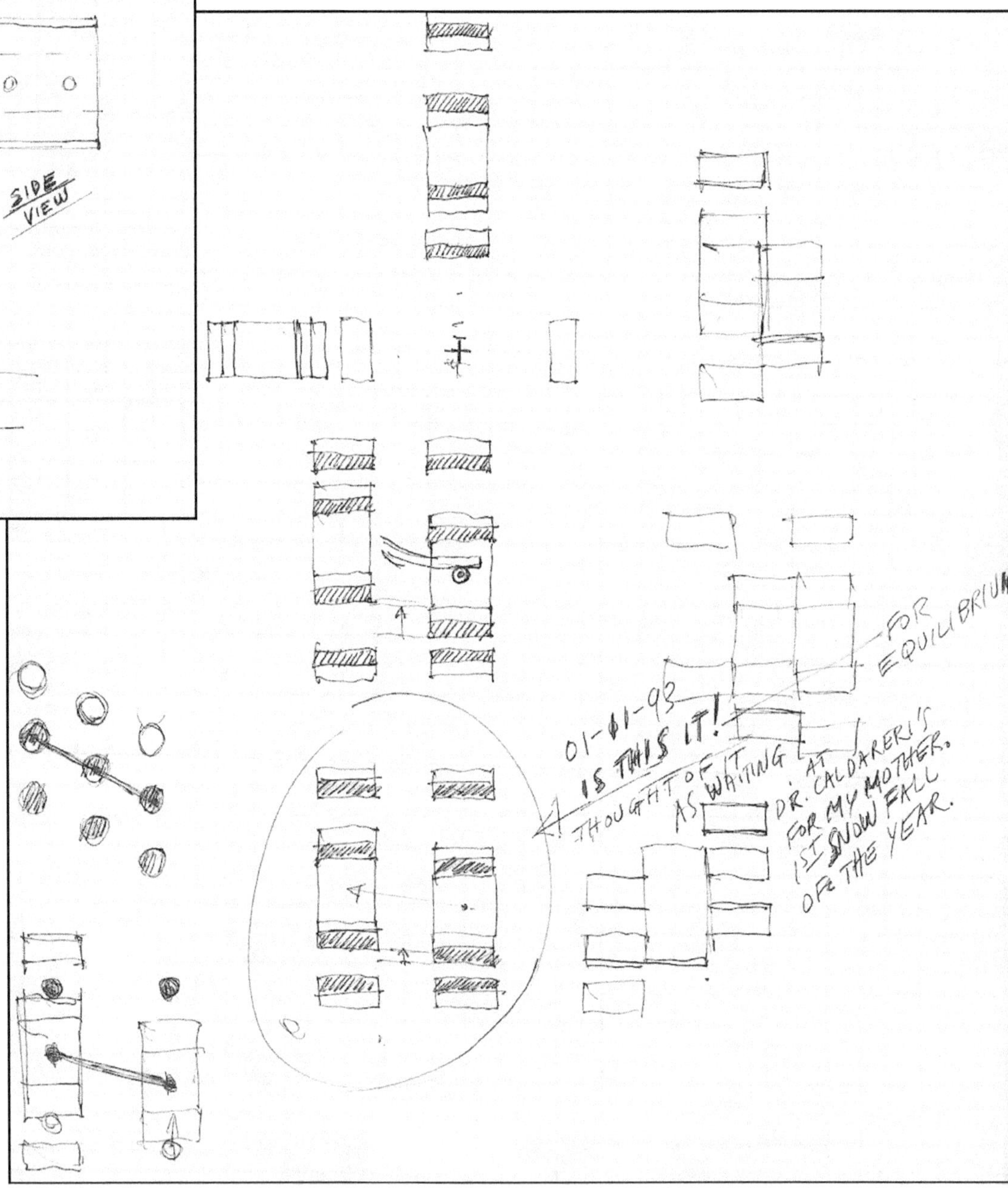
FOR EQUILIBRIUM
01-01-93
IS THIS IT!
THOUGHT OF IT AS WAITING AT DR. CALDARERI'S FOR MY MOTHER.
1ST SNOW FALL OF THE YEAR.

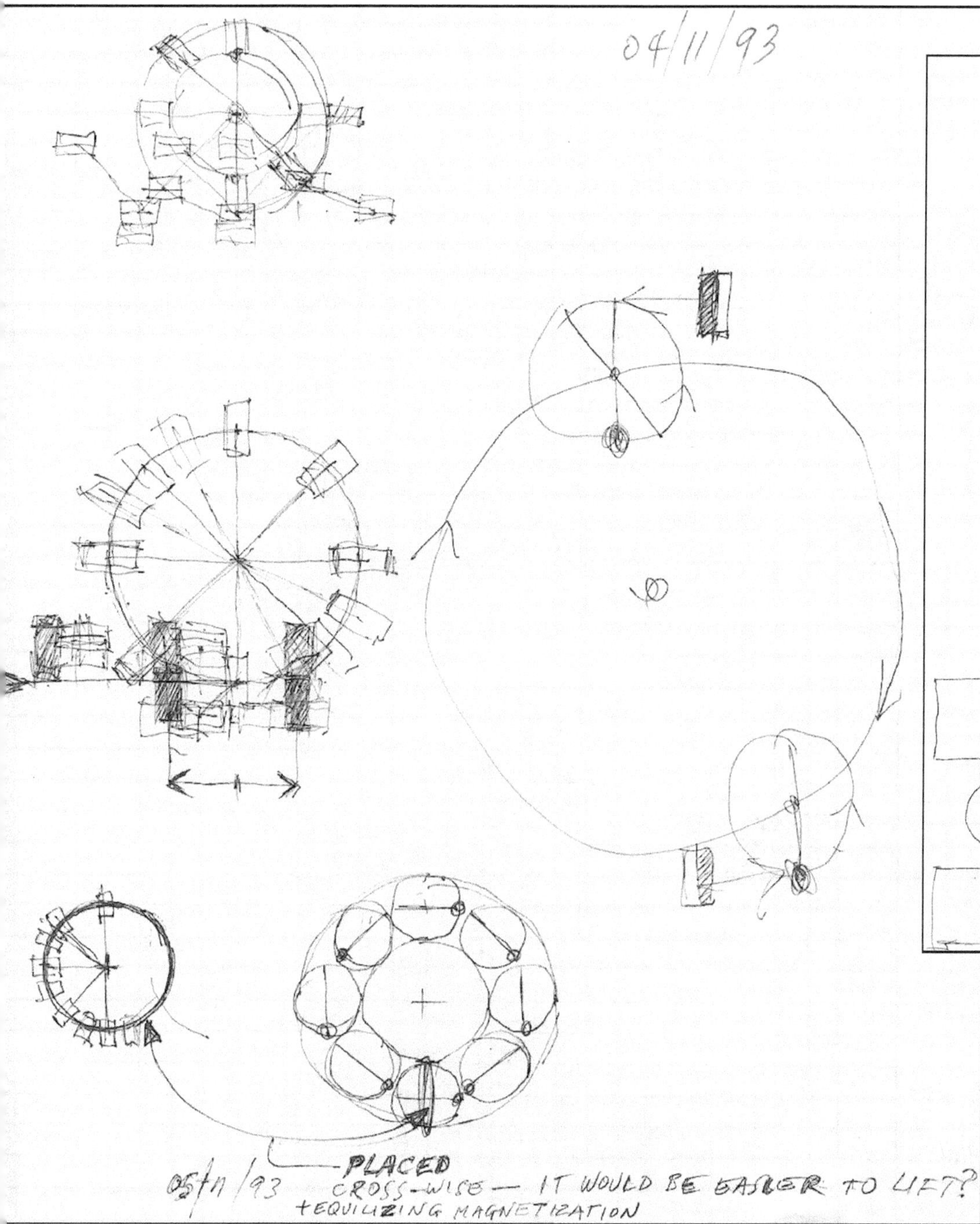
04/11/93
05/11/93
PLACED
CROSS-WISE — IT WOULD BE EASBER TO LIFT?
+EQUILIZING MAGNETIZATION

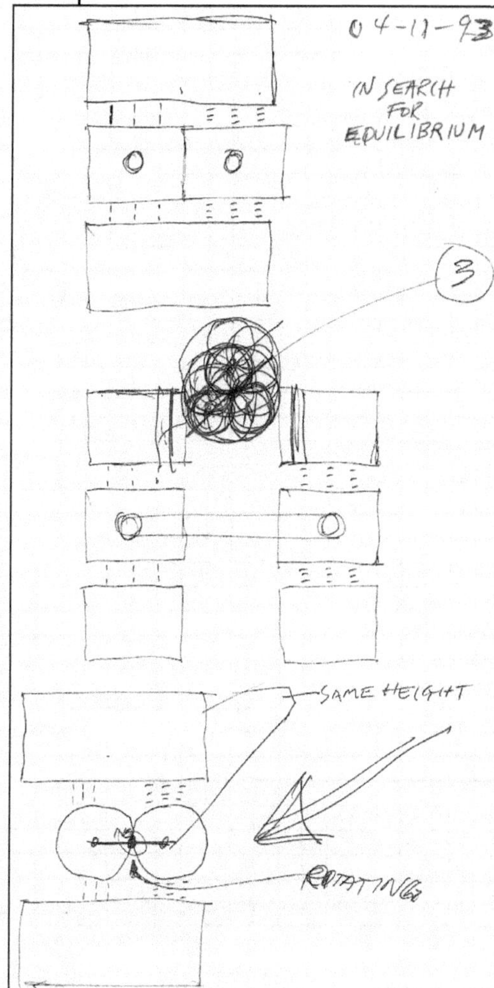
04-11-93
IN SEARCH FOR EQUILIBRIUM
3
SAME HEIGHT
ROTATING

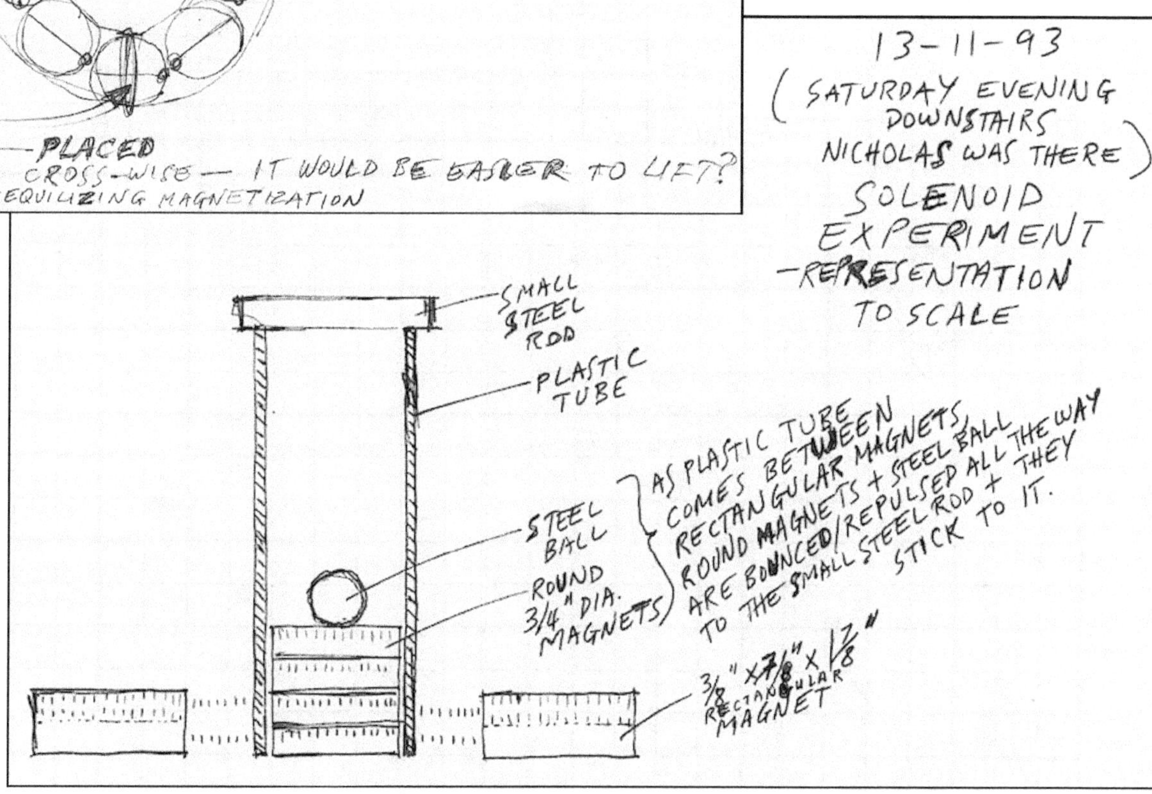
13-11-93
(SATURDAY EVENING DOWNSTAIRS
NICHOLAS WAS THERE)
SOLENOID EXPERIMENT
-REPRESENTATION TO SCALE
SMALL STEEL ROD
PLASTIC TUBE
STEEL BALL
ROUND 3/4" DIA. MAGNETS
AS PLASTIC TUBE COMES BETWEEN RECTANGULAR MAGNETS, ROUND MAGNETS + STEEL BALL ARE BOUNCED/REPULSED ALL THE WAY TO THE SMALL STEEL ROD + THEY STICK TO IT.
3/8" X 7/8" X 1/2" RECTANGULAR MAGNET

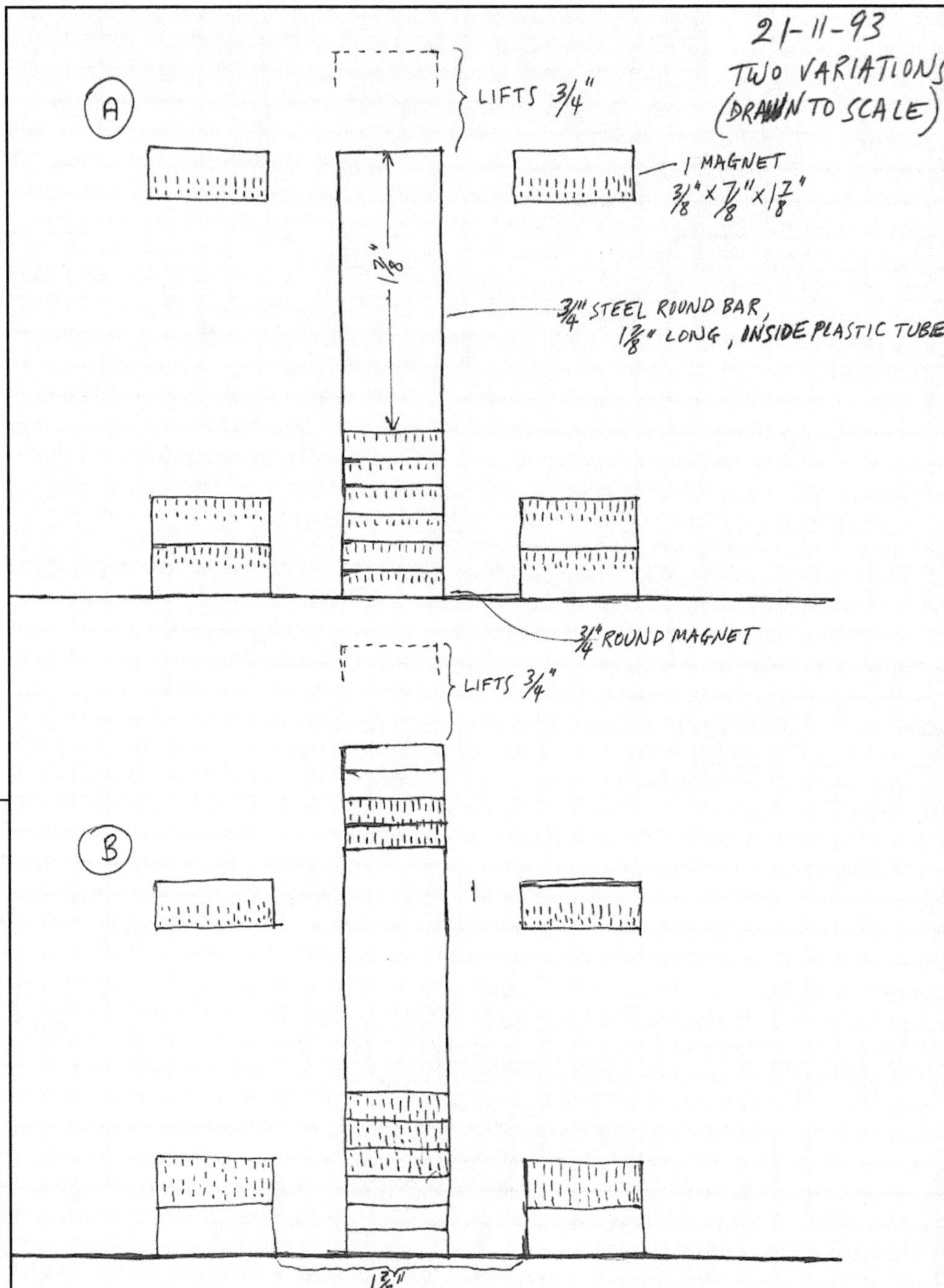
21-11-93
TWO VARIATIONS
(DRAWN TO SCALE)
A
LIFTS 3/4"
1 MAGNET
3/8" x 7/8" x 1 7/8"
3/4" STEEL ROUND BAR,
1 7/8" LONG, INSIDE PLASTIC TUBE
3/4" ROUND MAGNET
B
LIFTS 3/4"
1 3/8"

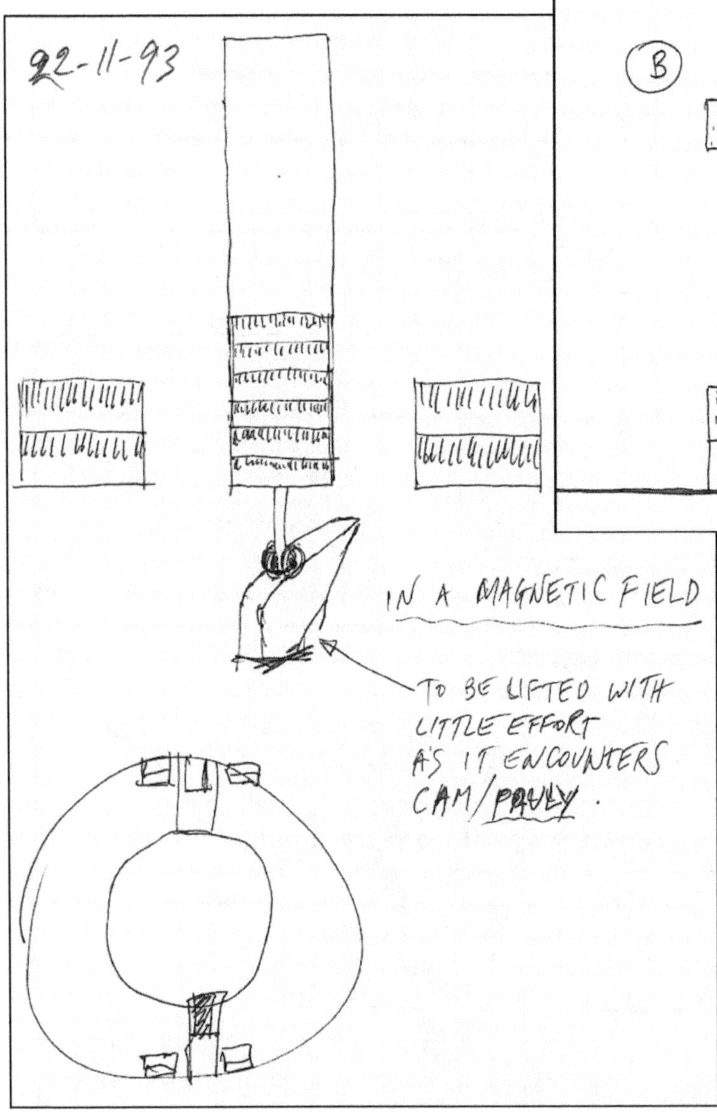
22-11-93
IN A MAGNETIC FIELD
TO BE LIFTED WITH
LITTLE EFFORT
AS IT ENCOUNTERS
CAM/

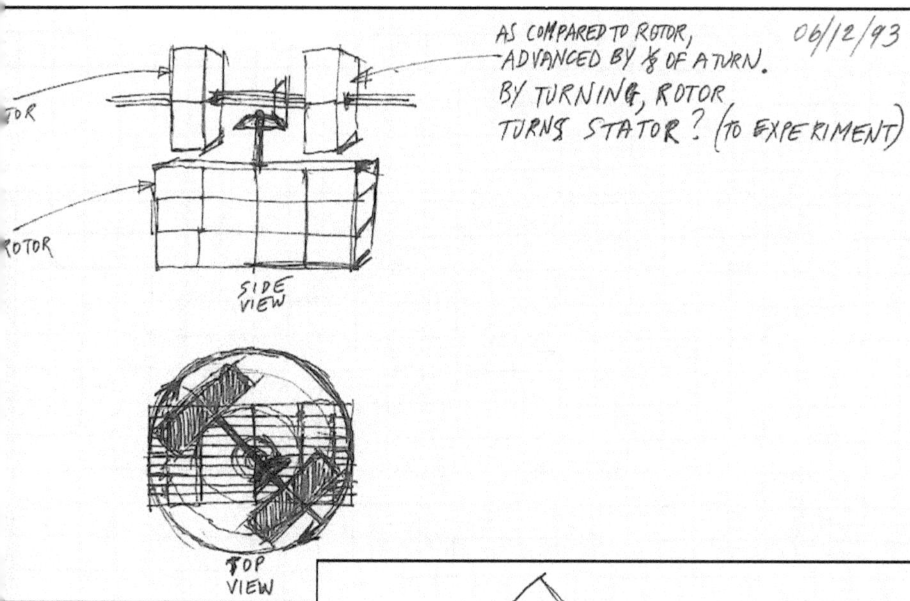
AS COMPARED TO ROTOR,
ADVANCED BY 1/8 OF A TURN.
BY TURNING, ROTOR
TURNS STATOR? (TO EXPERIMENT)
06/12/93
SIDE VIEW
TOP VIEW

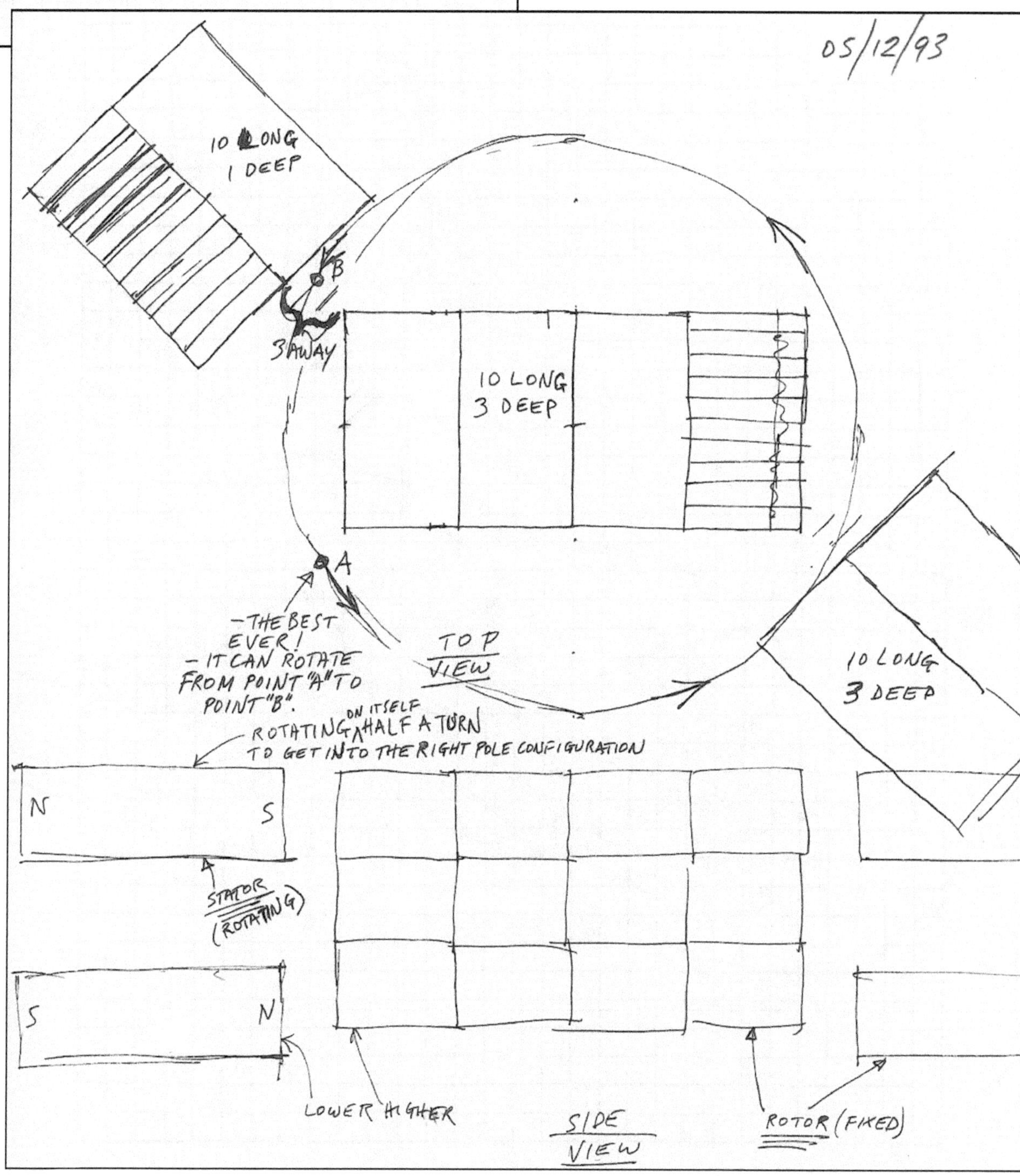
05/12/93
10 LONG
1 DEEP
B
3 AWAY
10 LONG
3 DEEP
A
- THE BEST EVER!
- IT CAN ROTATE FROM POINT "A" TO POINT "B".
TOP VIEW
10 LONG
3 DEEP
ROTATING ON ITSELF HALF A TURN
TO GET INTO THE RIGHT POLE CONFIGURATION
N
S
STATOR (ROTATING)
S
N
LOWER HIGHER
SIDE VIEW
ROTOR (FIXED)

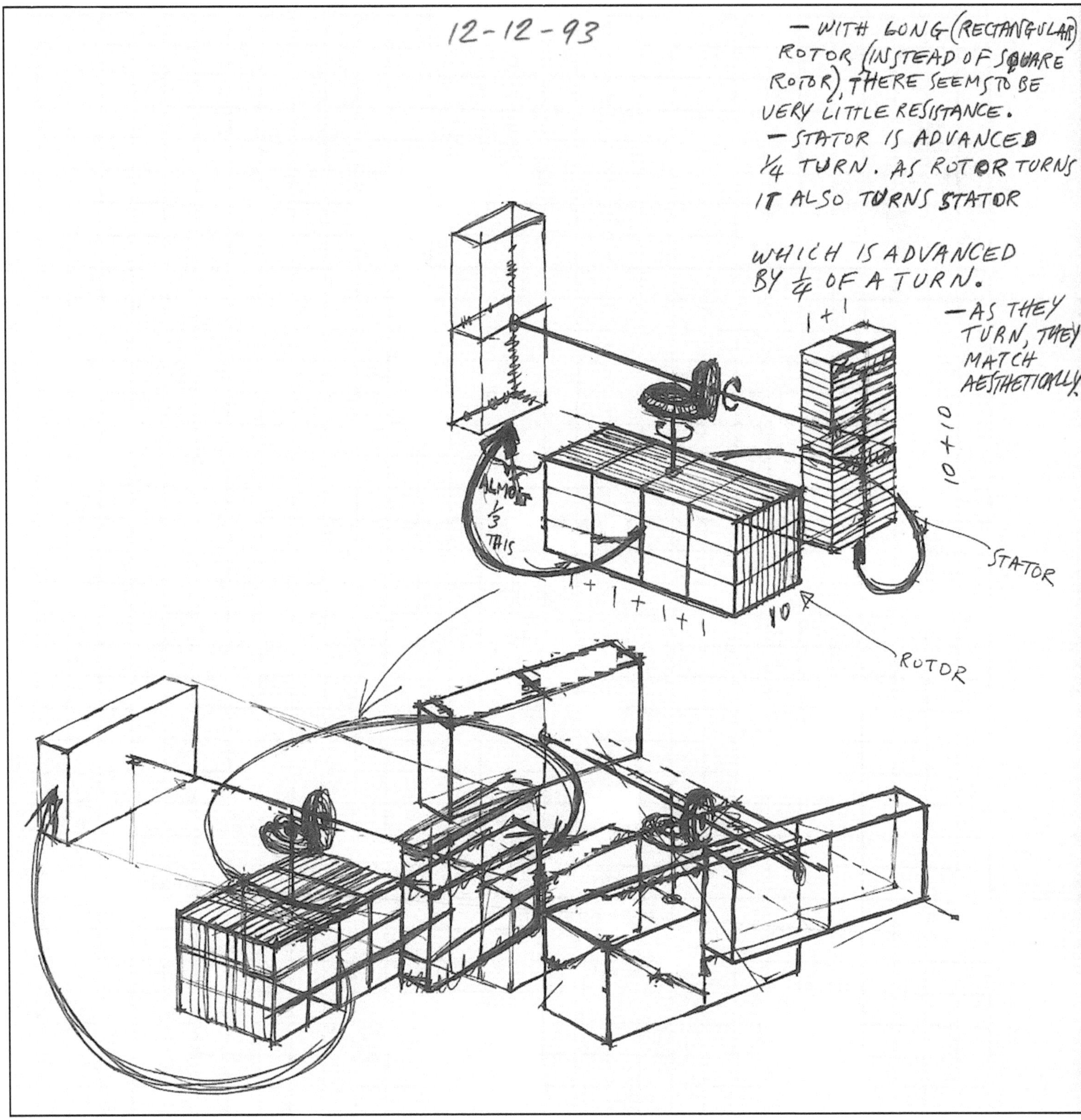
12-12-93
—WITH LONG (RECTANGULAR) ROTOR (INSTEAD OF SQUARE ROTOR), THERE SEEMS TO BE VERY LITTLE RESISTANCE.
—STATOR IS ADVANCED 1/4 TURN. AS ROTOR TURNS IT ALSO TURNS STATOR
WHICH IS ADVANCED BY 1/4 OF A TURN.
—AS THEY TURN, THEY MATCH AESTHETICALLY.
1+1
10+01
ALMOST 1/3 THIS
1+1+1+1
10
STATOR
ROTOR

20-12-93

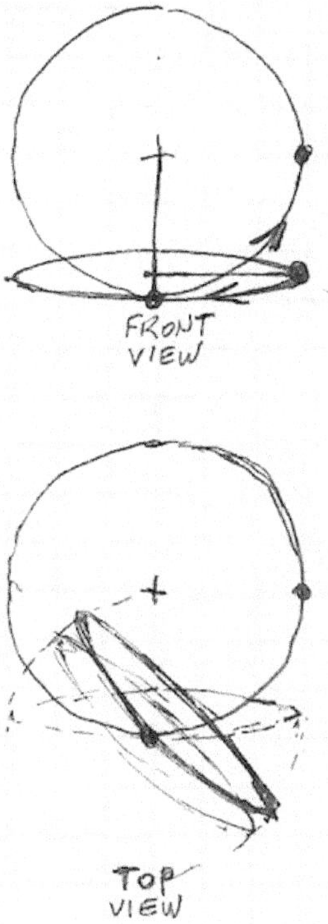

- BY BUILDING PROTOTYPE WITH BEVEL GEARS + UPON. TRIAL OUT, I DISCOVERED/FOUND OUT IT NOT NOT WORK. FLAW IS BASED ON ON THE FACT THAT I USED ONE PART-STATOR AS COMPARED TO TWO-PART-STATOR OF ORIGINAL IDEA. BACK TO THE ORIGINAL IDEA.

19-12-93

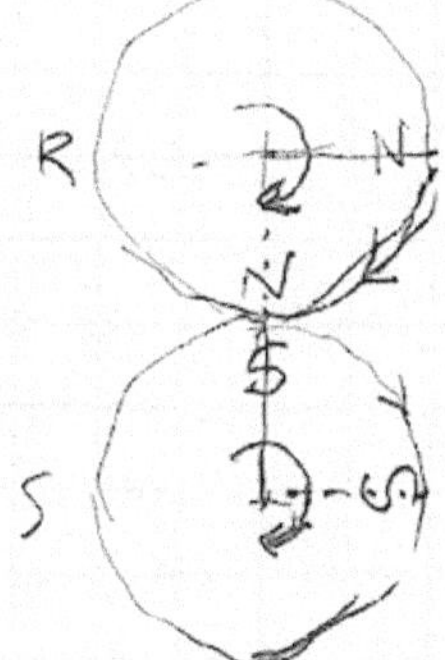

- R (ROTOR) TURNS TO MEET S, WHEREAS S (STATOR) TURNS AWAY FROM IT. THIS IS WHY R IS STRONGER.

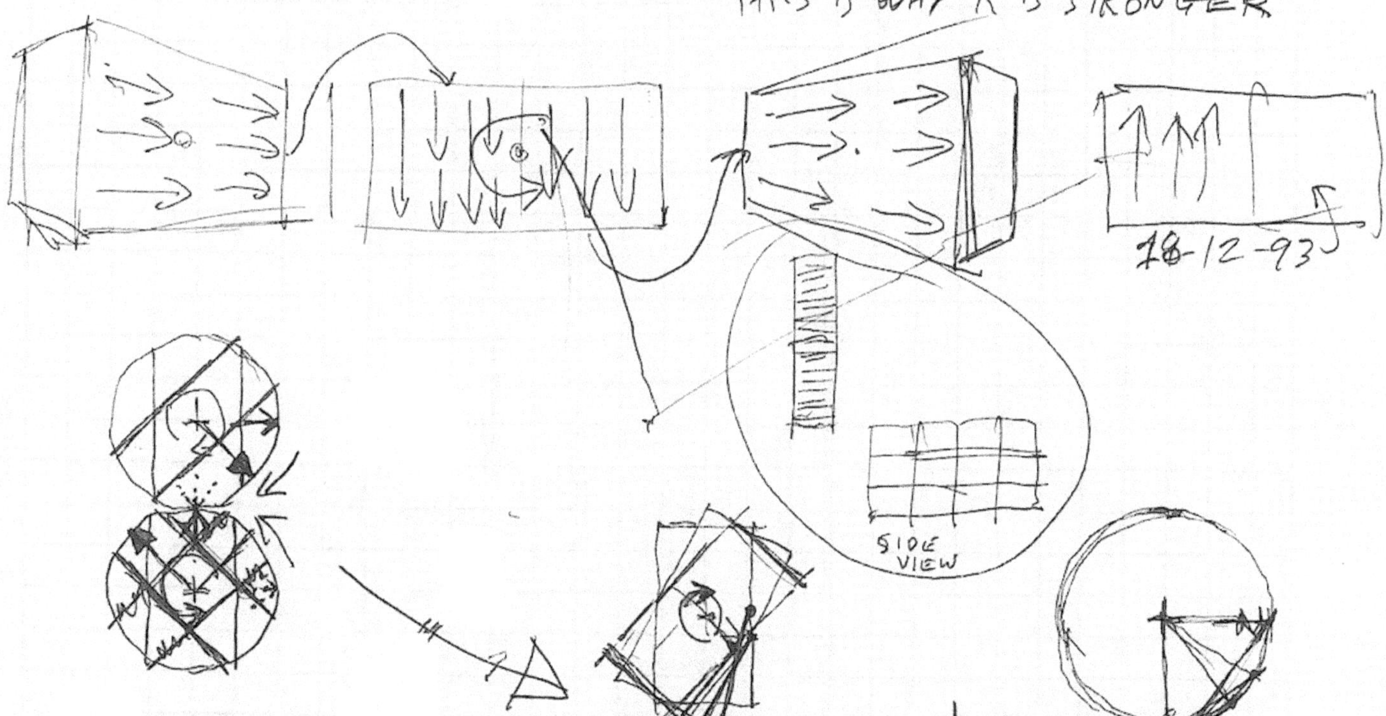

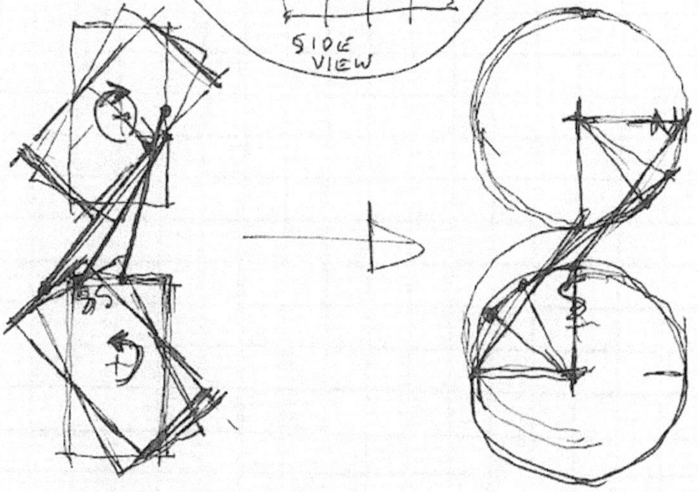

- IT IS BY THE WAY TWO INTERLOCKED GEARS TURN, THAT NORTH + SOUTH POLES STAY ALWAYS AT THE SAME DISTANCE + THUS KEEP THE MAGNETIC PRESSURE ON (CONSTANTLY CONTINUOUSLY).

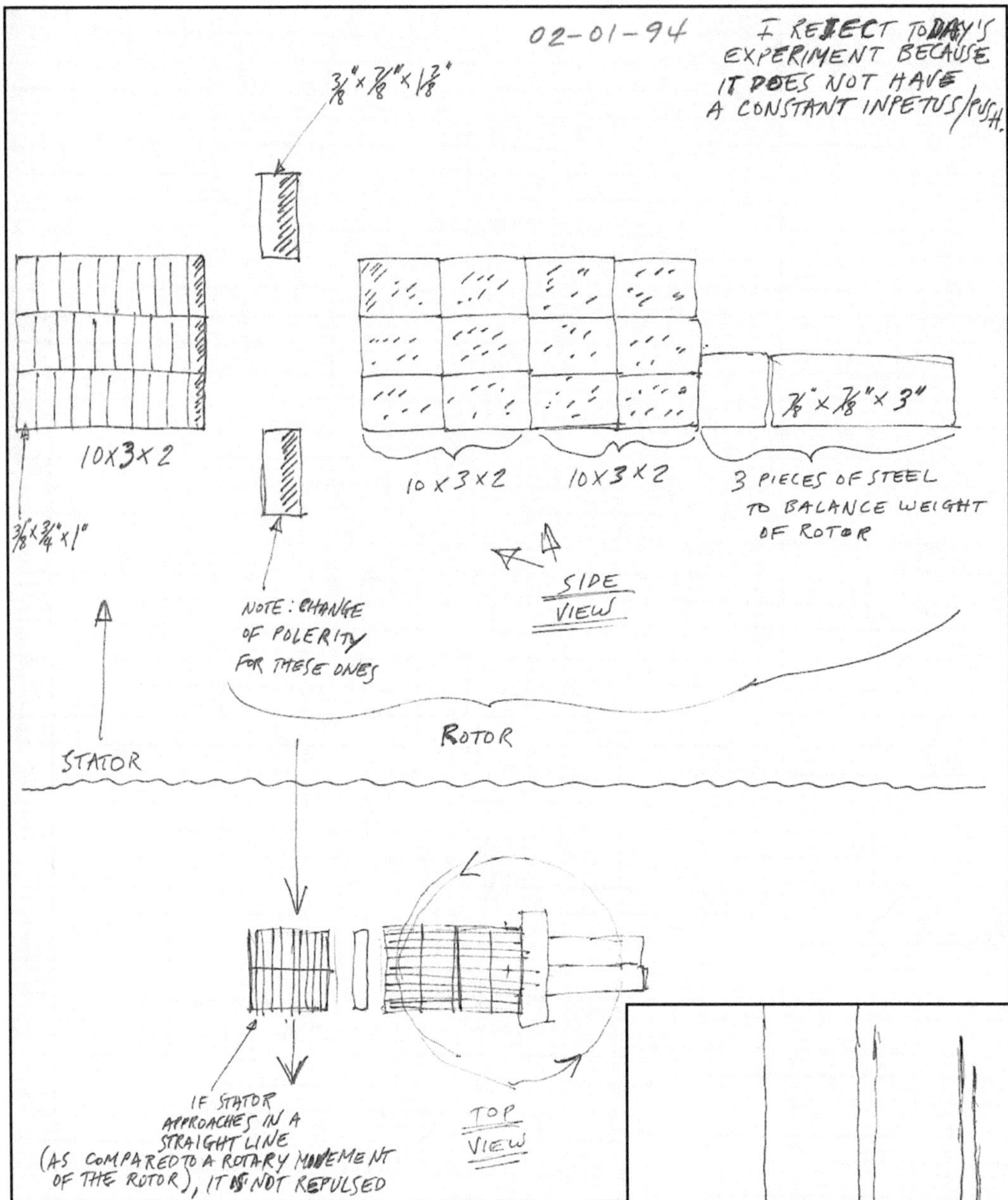
02-01-94
I REJECT TODAY'S EXPERIMENT BECAUSE IT DOES NOT HAVE A CONSTANT INPETUS/PUSH.
3/8"x 7/8"x 1 2/8"
10X3X2
3/8"x3/4"x1"
10x3x2
10x3x2
7/8"x 7/8"x 3"
3 PIECES OF STEEL TO BALANCE WEIGHT OF ROTOR
SIDE VIEW
NOTE: CHANGE OF POLERITY FOR THESE ONES
ROTOR
STATOR
IF STATOR APPROACHES IN A STRAIGHT LINE (AS COMPARED TO A ROTARY MOVEMENT OF THE ROTOR), IT IS NOT REPULSED
TOP VIEW

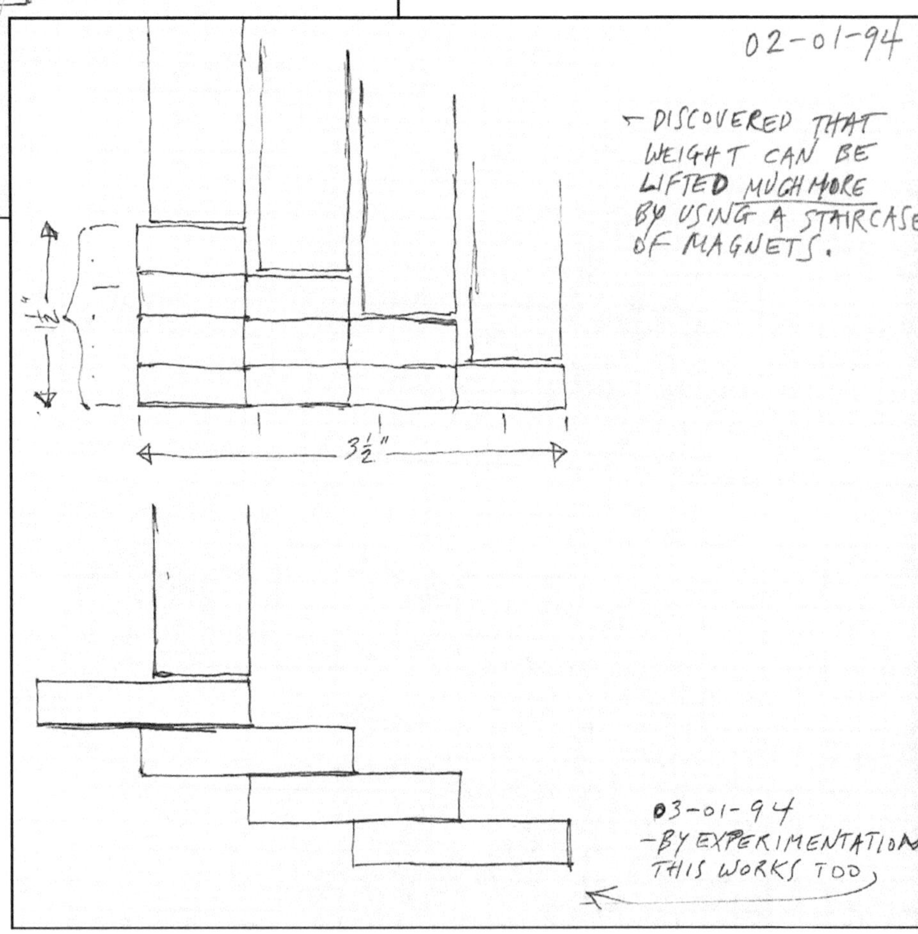
02-01-94
← DISCOVERED THAT WEIGHT CAN BE LIFTED MUCH MORE BY USING A STAIRCASE OF MAGNETS.
1 IN
3 1/2"
03-01-94
-BY EXPERIMENTATION THIS WORKS TOO

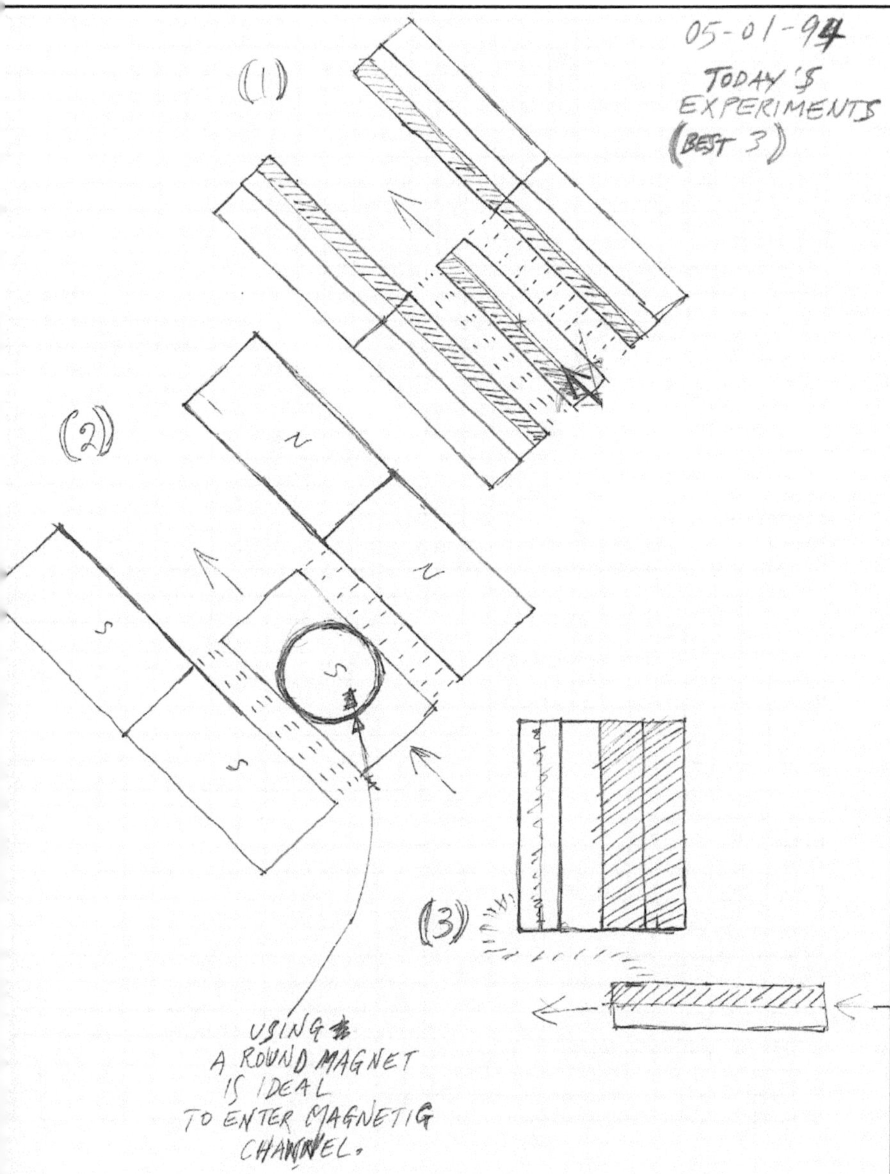
05-01-94
TODAY'S
EXPERIMENTS
(BEST 3)
(1)
(2)
N
N
S
S
S
(3)
USING
A ROUND MAGNET
IS IDEAL
TO ENTER MAGNETIG
CHANNEL.

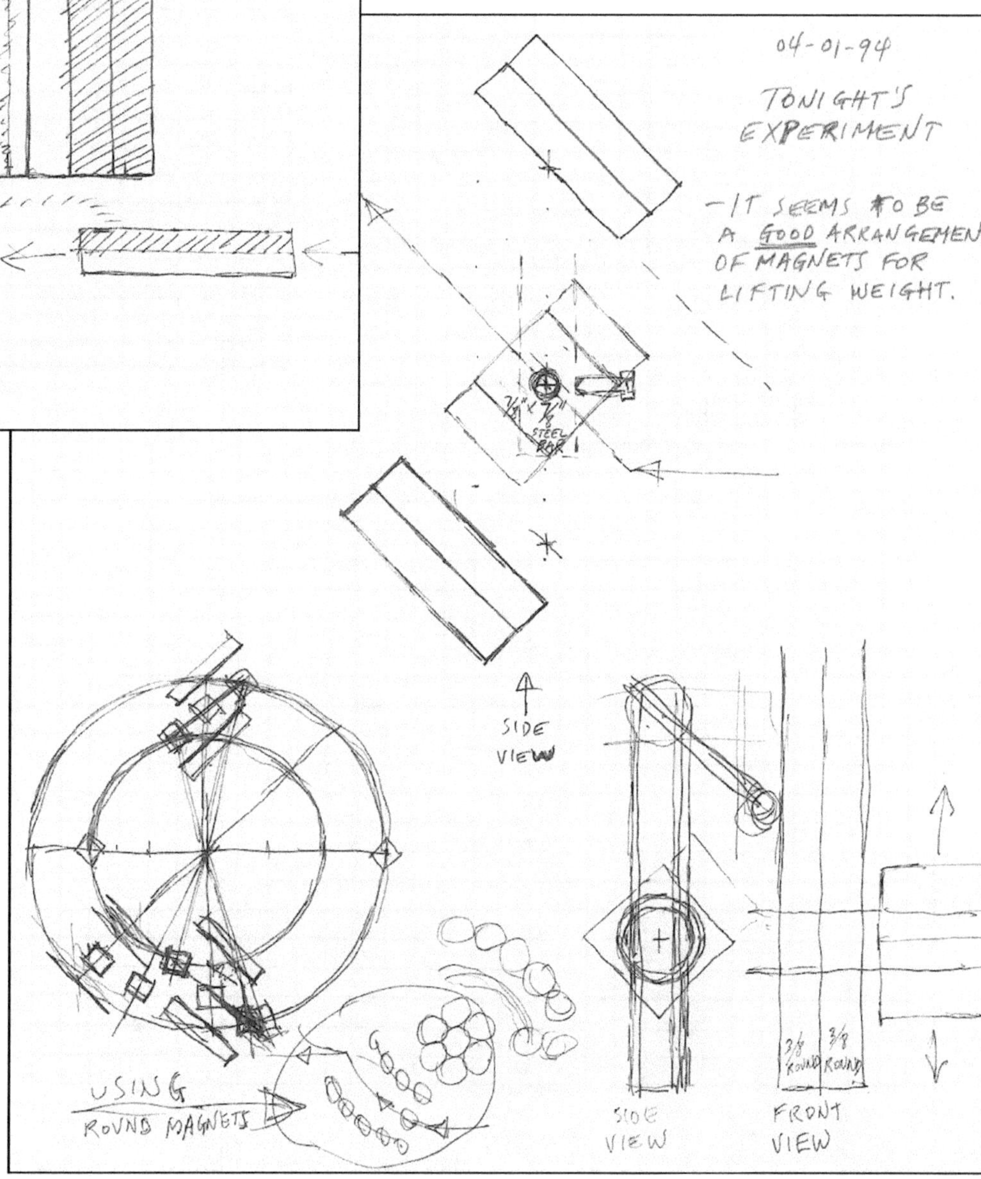
04-01-94
TONIGHT'S
EXPERIMENT
-IT SEEMS TO BE
A GOOD ARRANGEMENT
OF MAGNETS FOR
LIFTING WEIGHT.
7/8" x 7/8"
STEEL
BAR
SIDE
VIEW
USING
ROUND MAGNETS
SIDE
VIEW
3/8
ROUND
3/8
ROUND
FRONT
VIEW

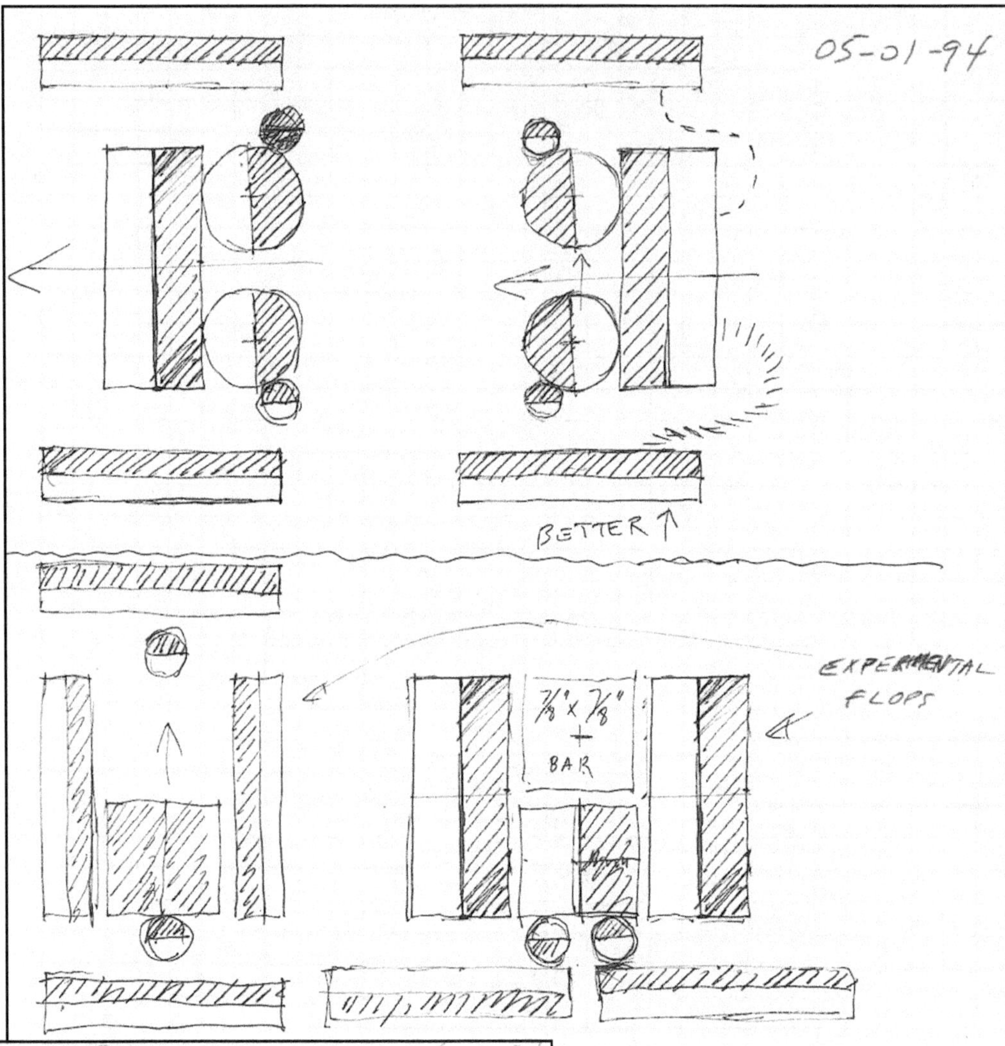
05-01-94
BETTER ↑
EXPERIMENTAL FLOPS
7/8" x 7/8"
BAR

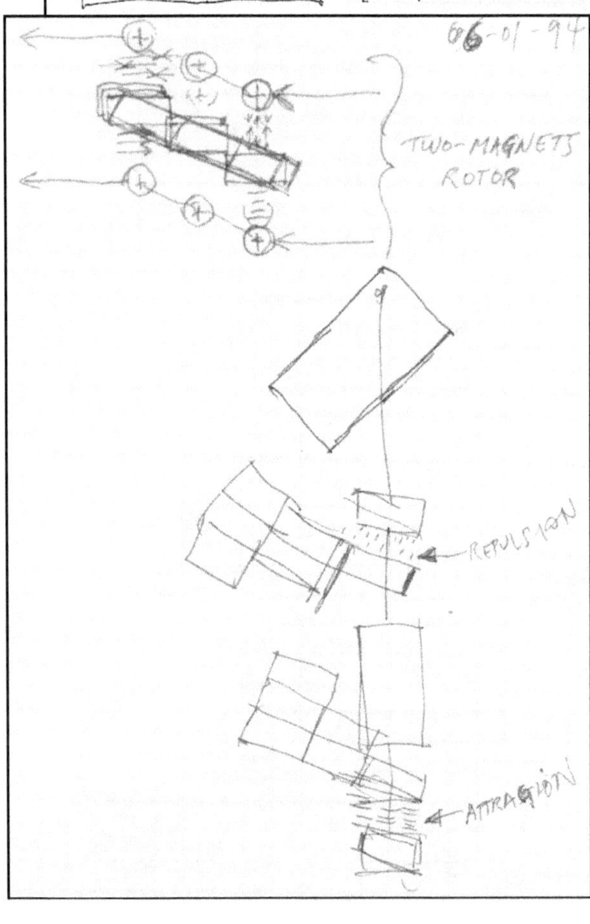
06-01-94
TWO-MAGNETS ROTOR
REPULSION
ATTRACTION

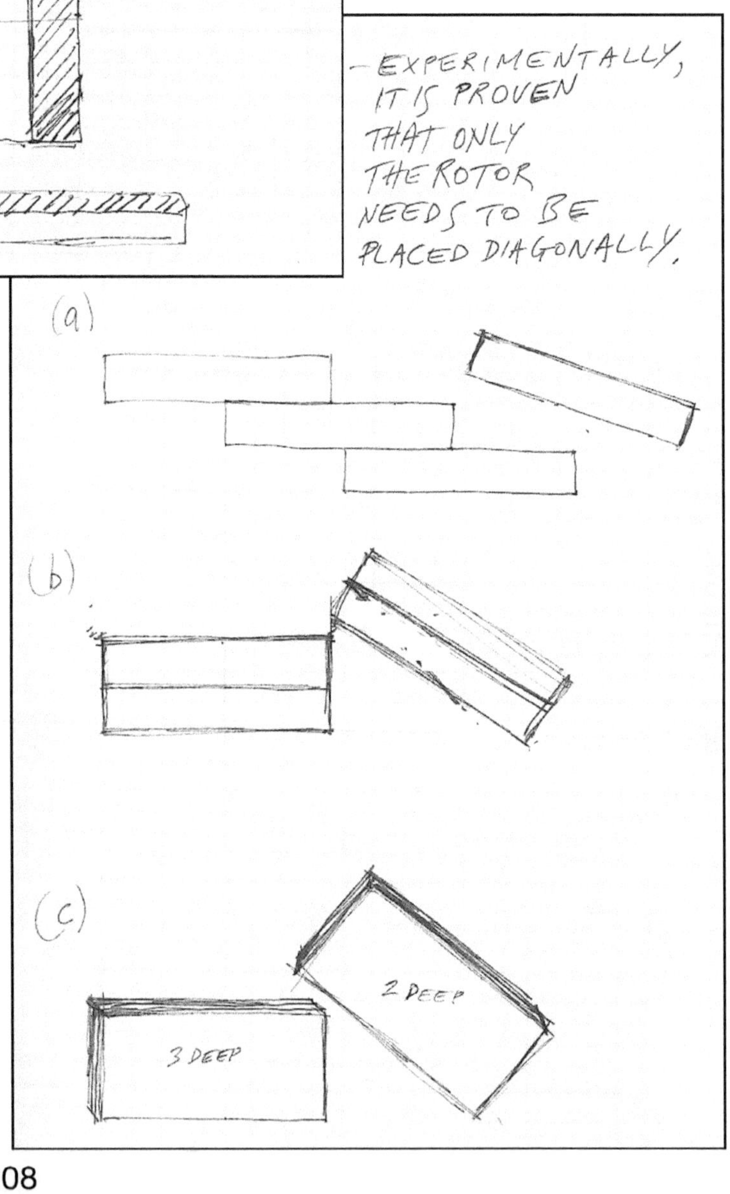
- EXPERIMENTALLY, IT IS PROVEN THAT ONLY THE ROTOR NEEDS TO BE PLACED DIAGONALLY.
(a)
(b)
(c)
3 DEEP
2 DEEP

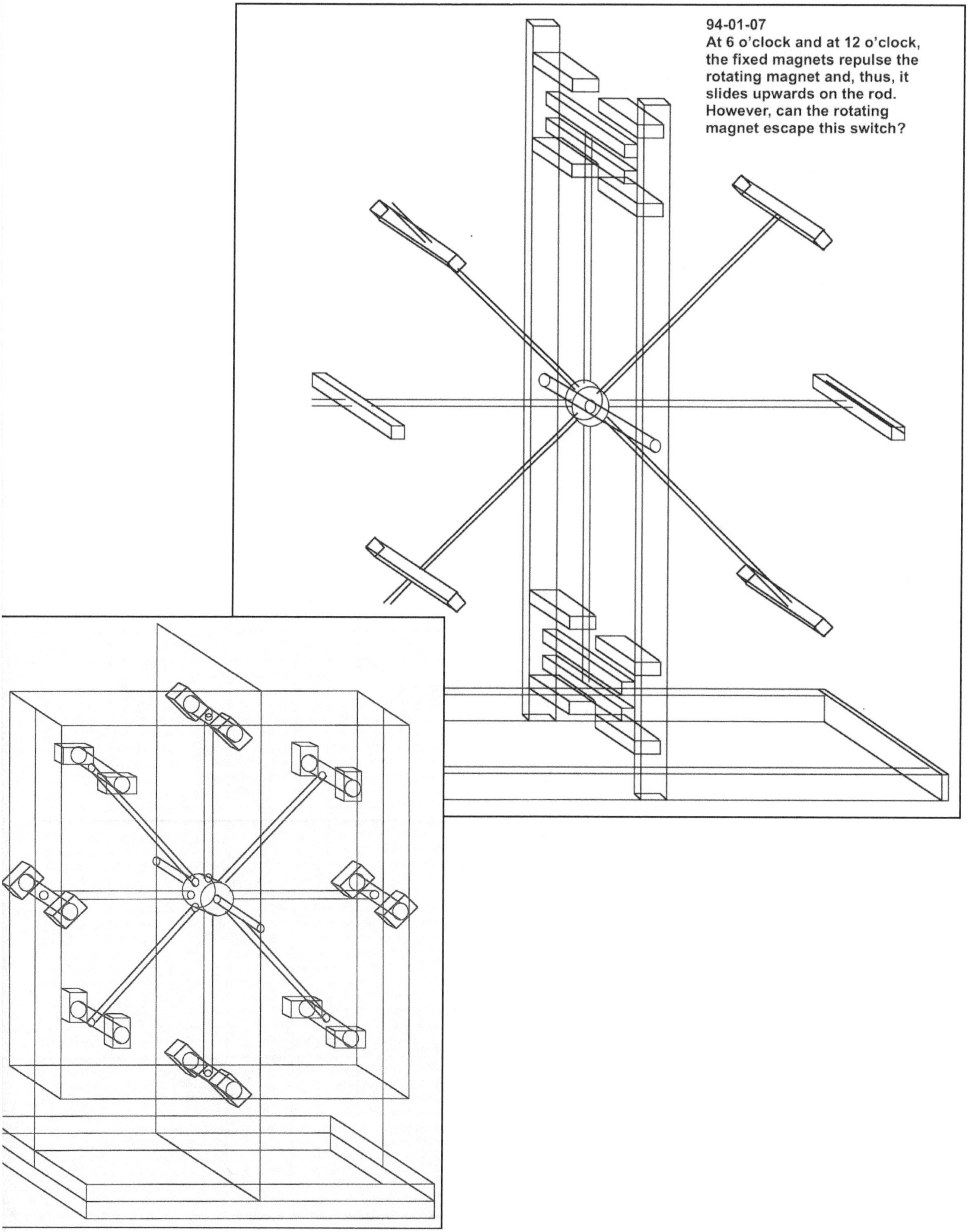

94-01-07
At 6 o'clock and at 12 o'clock, the fixed magnets repulse the rotating magnet and, thus, it slides upwards on the rod. However, can the rotating magnet escape this switch?

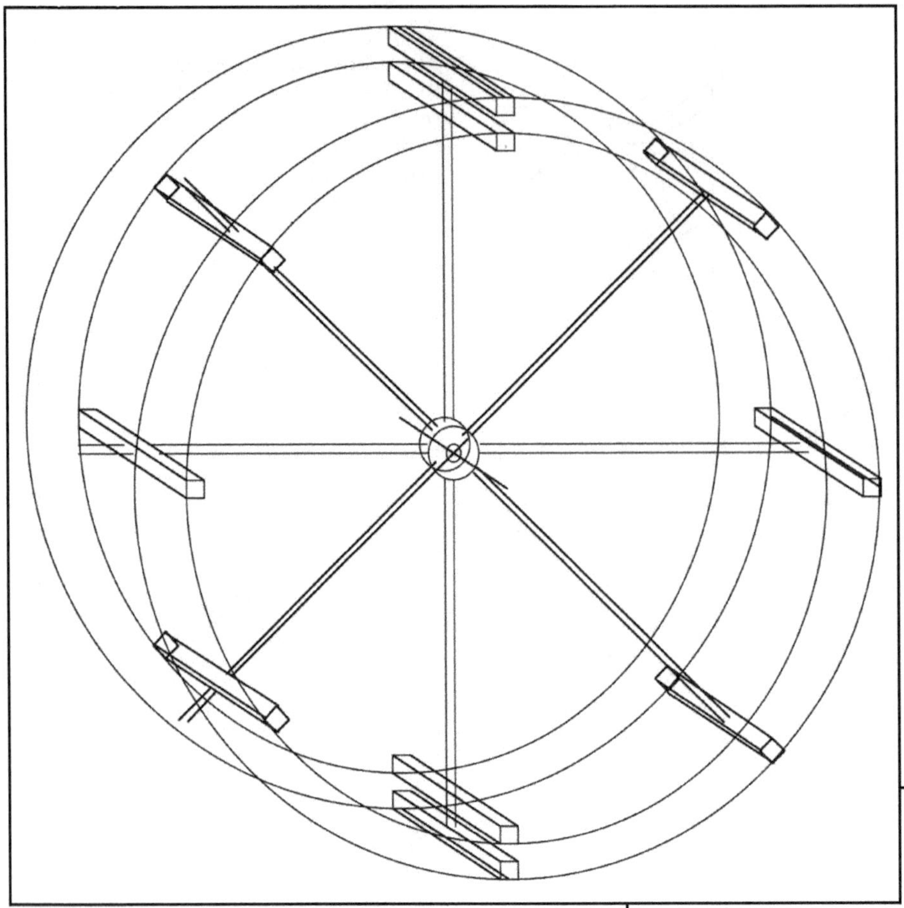

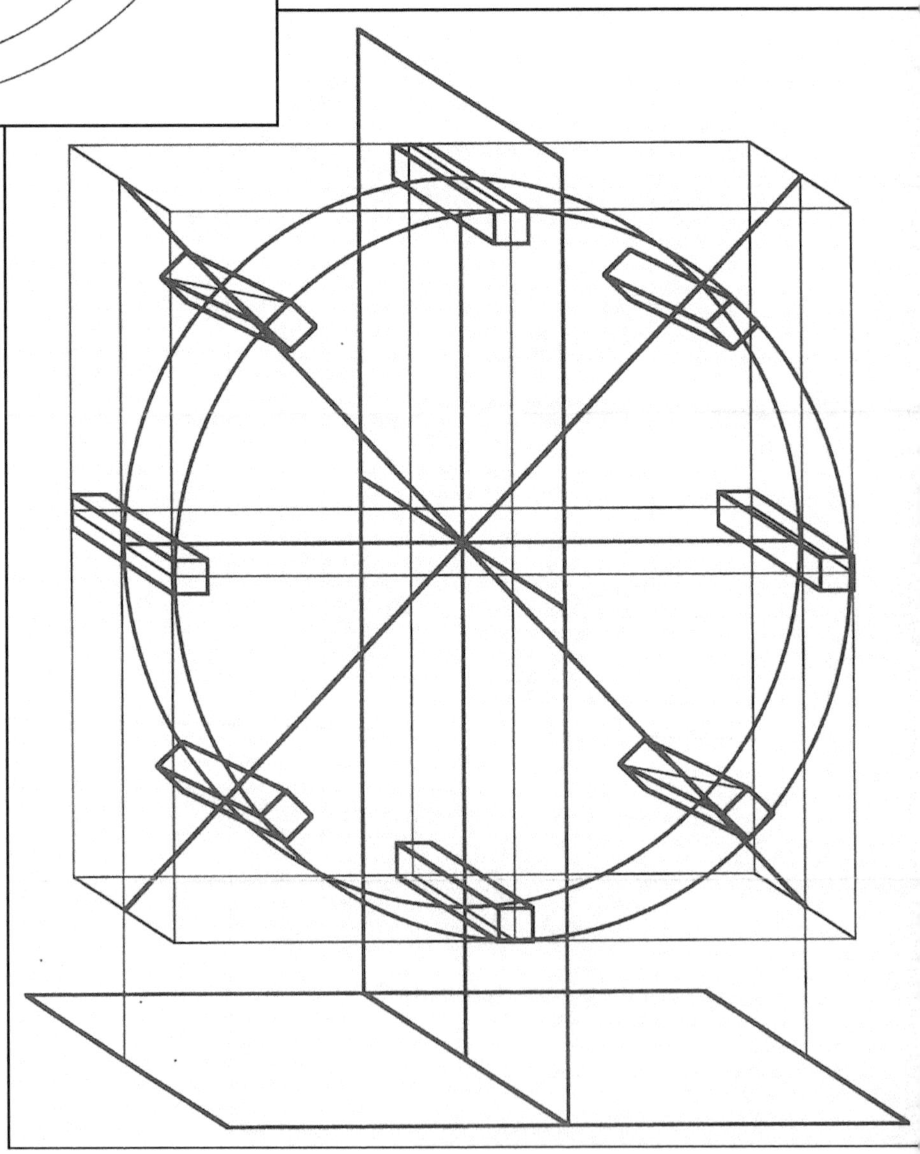

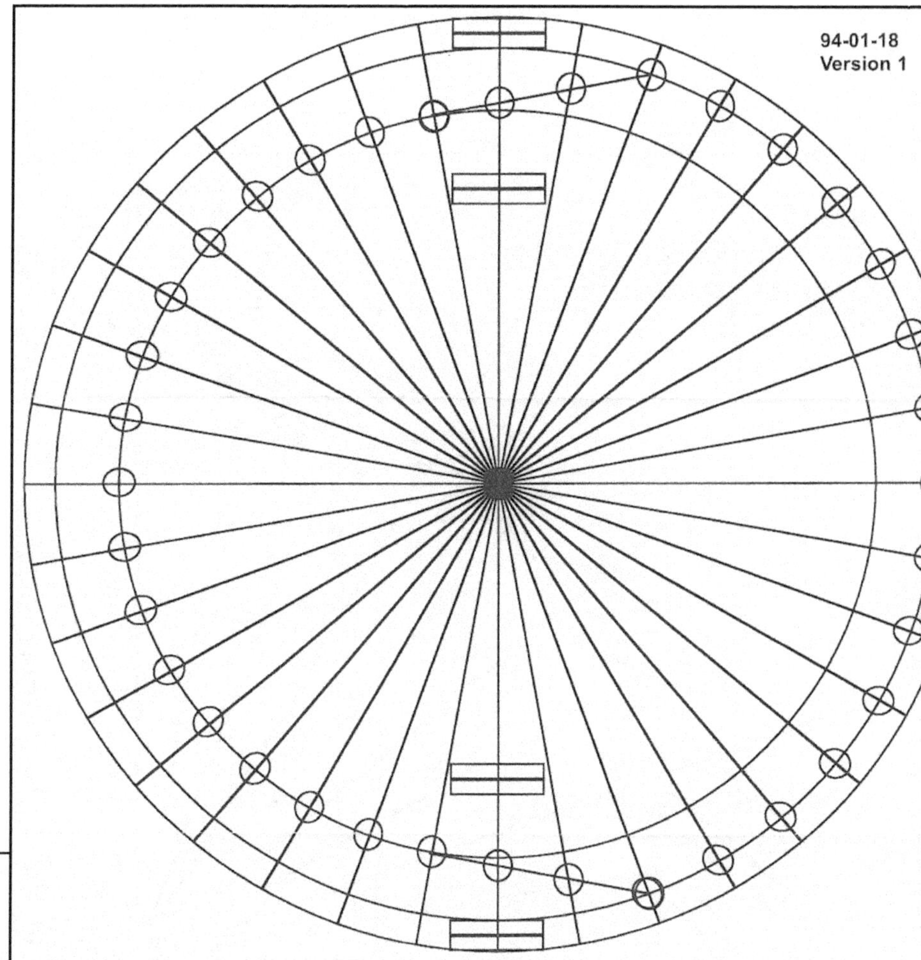
94-01-18
Version 1

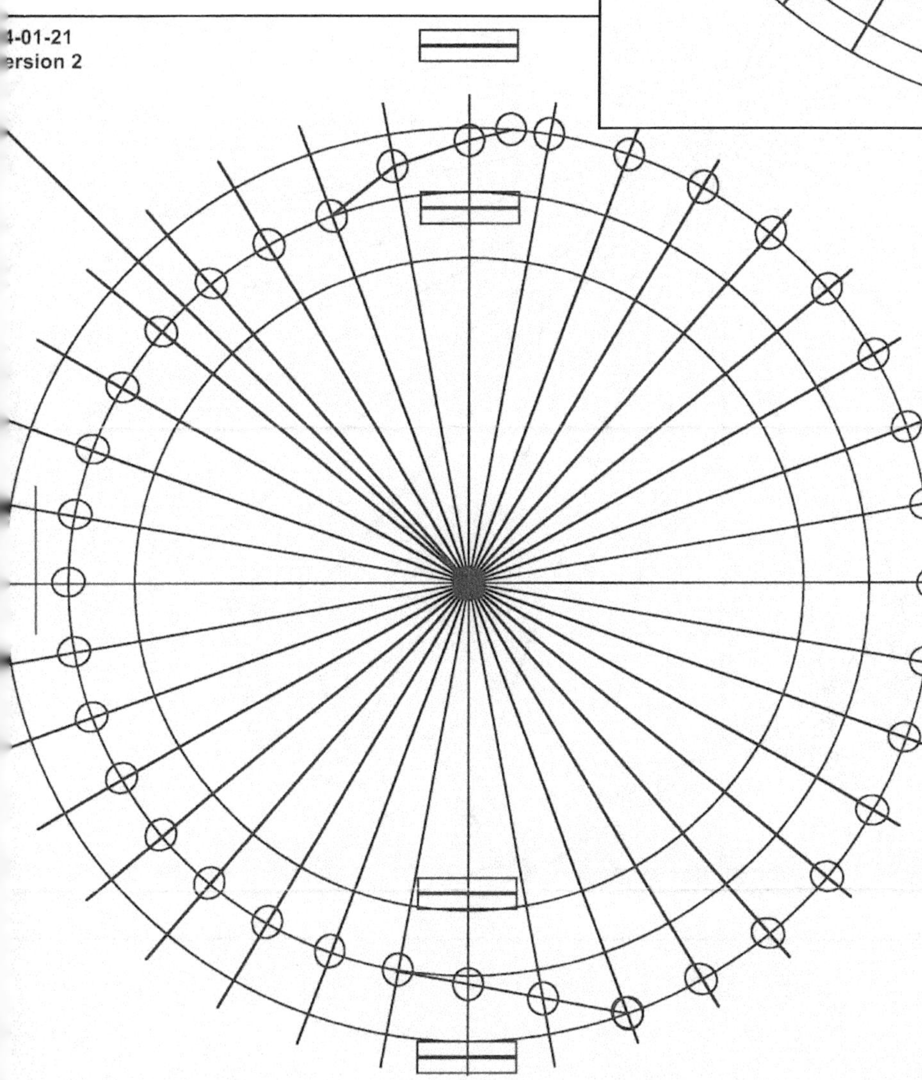
4-01-21
ersion 2

94-01-18
The rotation of the rotor-magnet creates the disequilibrium.

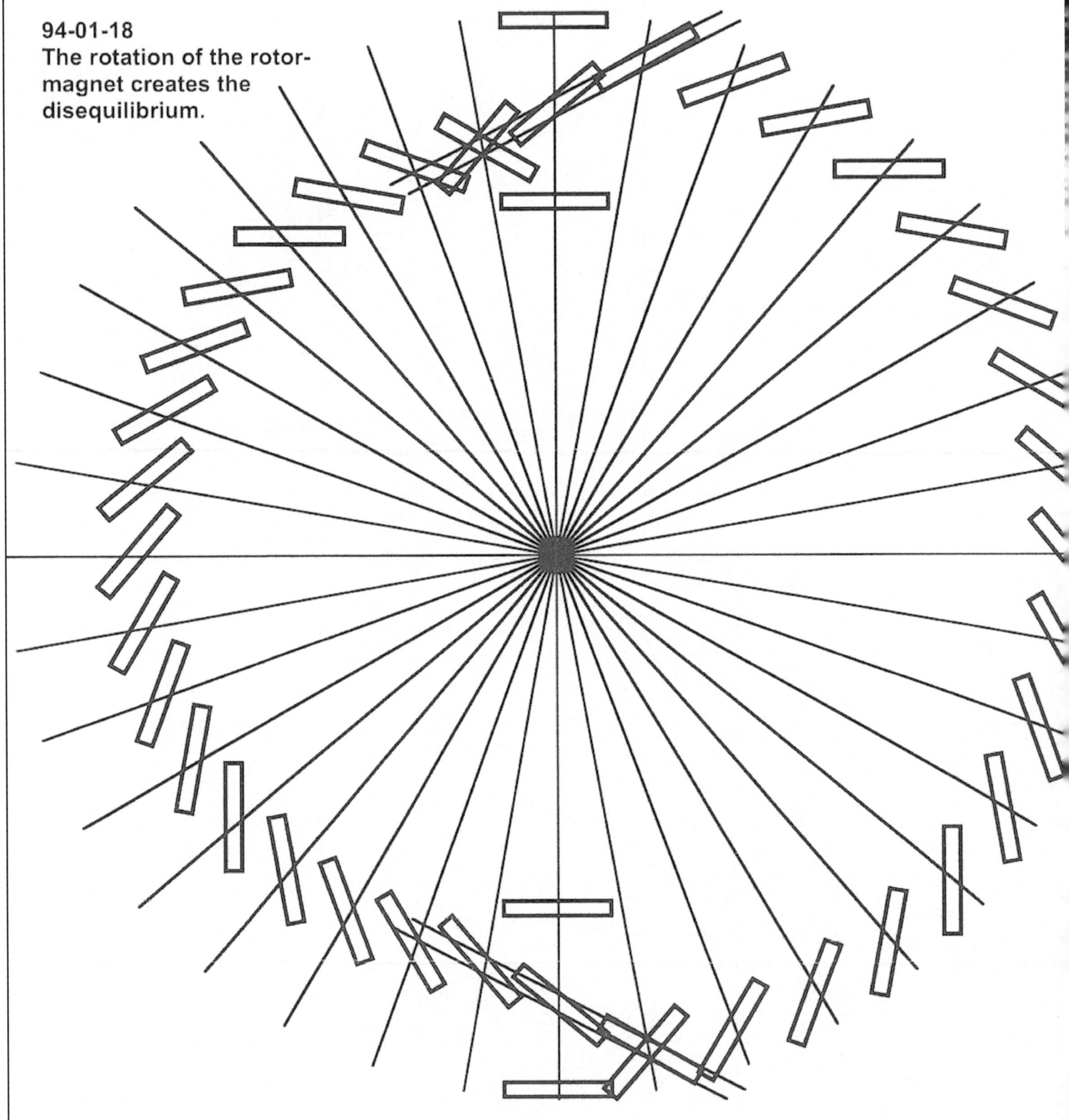

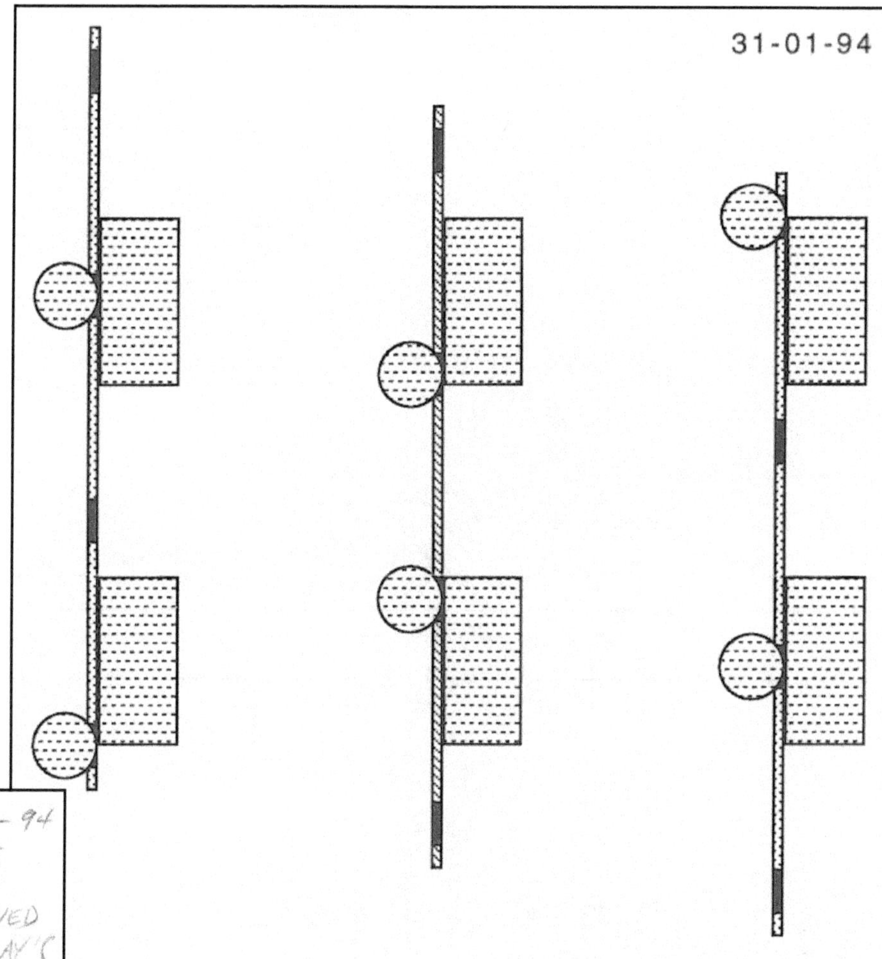
31-01-94

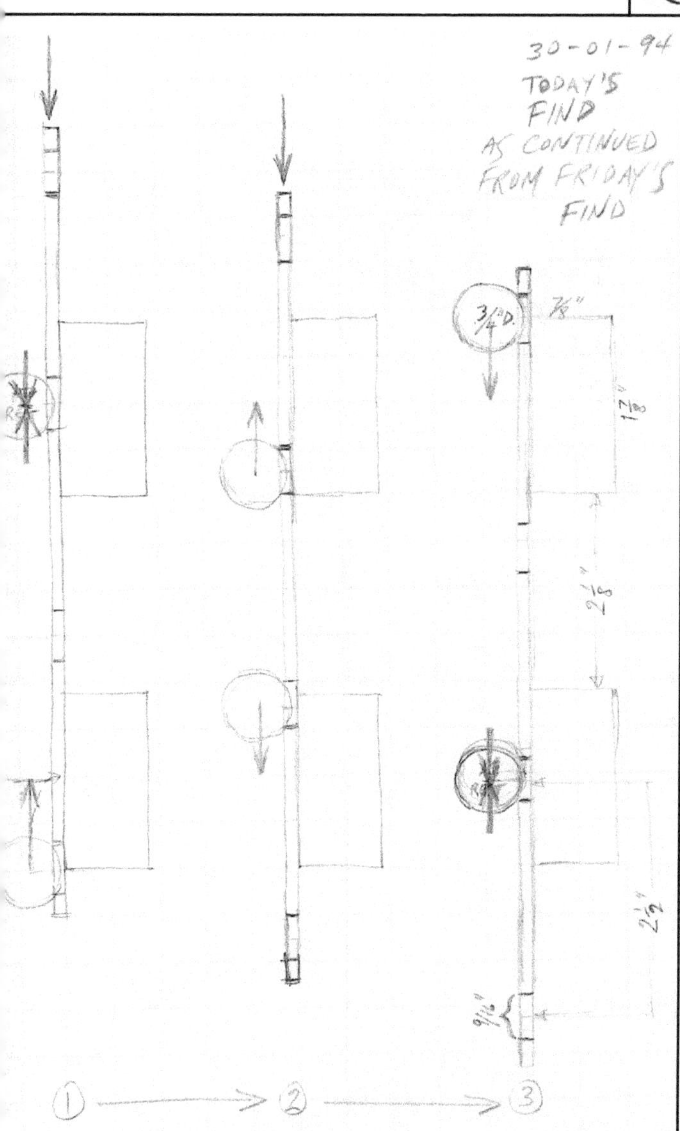
30-01-94
TODAY'S FIND
AS CONTINUED
FROM FRIDAY'S
FIND
3/4"D.
7/8"
2 1/8"
2 1/2"
9/16"
1
2
3

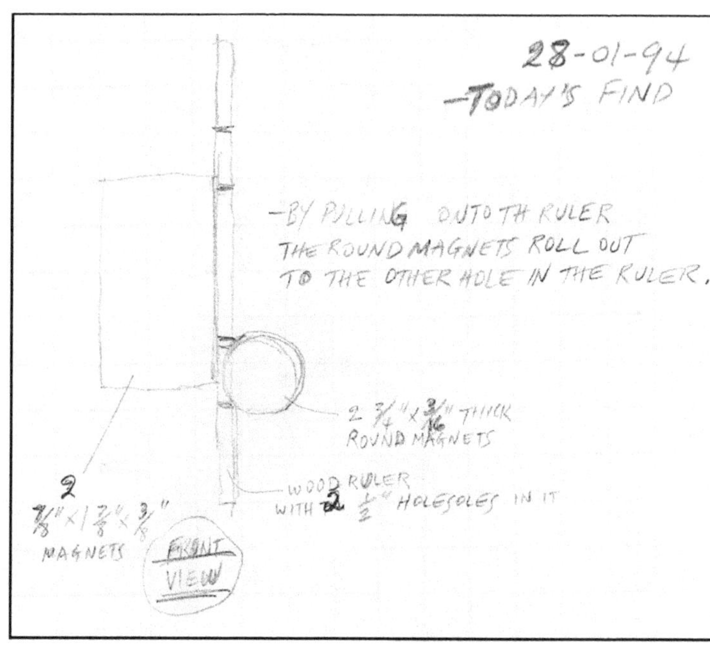
28-01-94
TODAY'S FIND
BY PULLING OUT TH RULER
THE ROUND MAGNETS ROLL OUT
TO THE OTHER HOLE IN THE RULER.
2 3/4" x 3/16" THICK
ROUND MAGNETS
WOOD RULER
WITH 2 1/2" HOLES IN IT
2
7/8" x 1 7/8" x 3/8"
MAGNETS
FRONT
VIEW

01-02-94

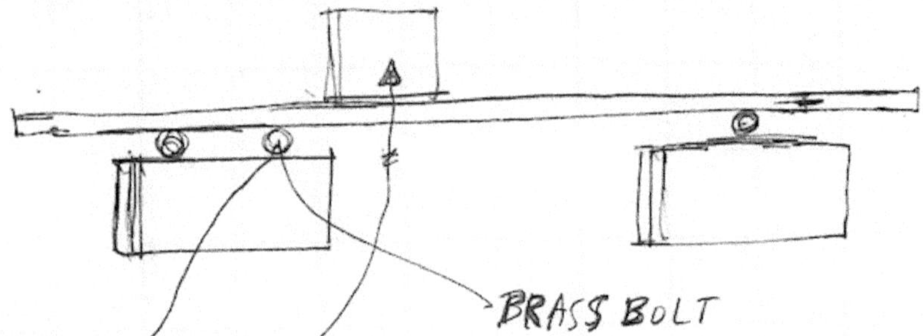

TONIGHT, BY EXPERIMENTATION,
I DEMONSTRATED THAT,
BY USING ROLLERS, A WEIGHT,
EVEN A SQUARE WEIGHT, CAN ROLL ON A MAGNET
AND CAN, PROBABLY, BE LIFTED

01-02-94

THE SOLENOID IDEA

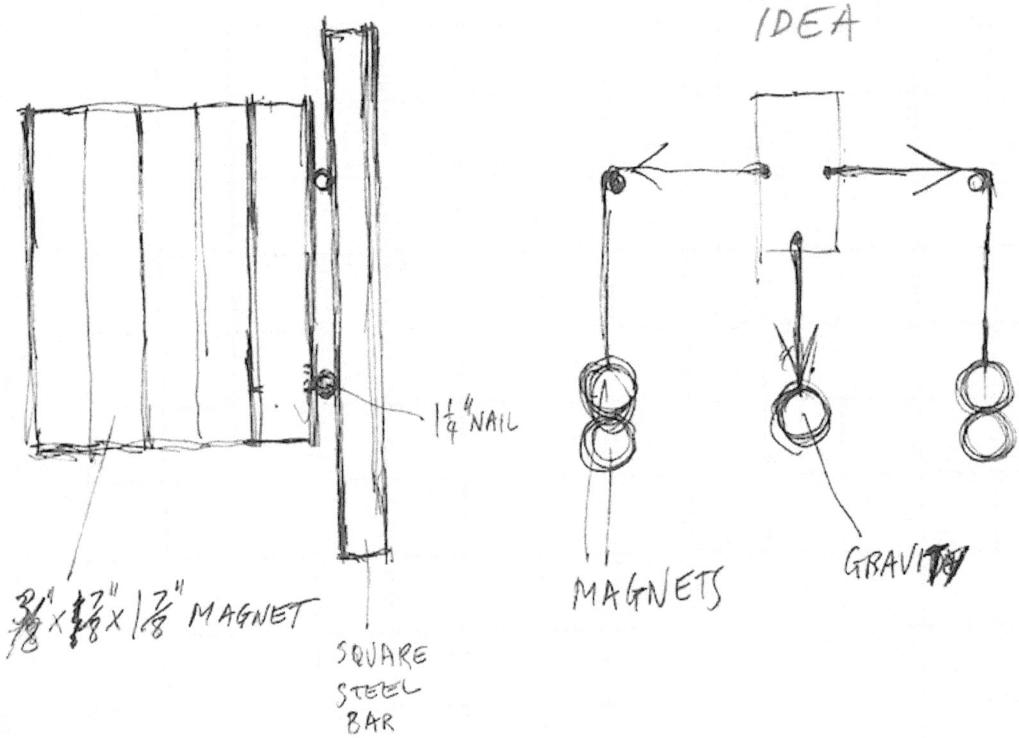

—NOT TO GET MAGNETICALLY
LOCKED, THE SOLENOID-WEIGHT
SHOULD NOT BE PUSHED
ALL THE WAY AT EITHER WAY.

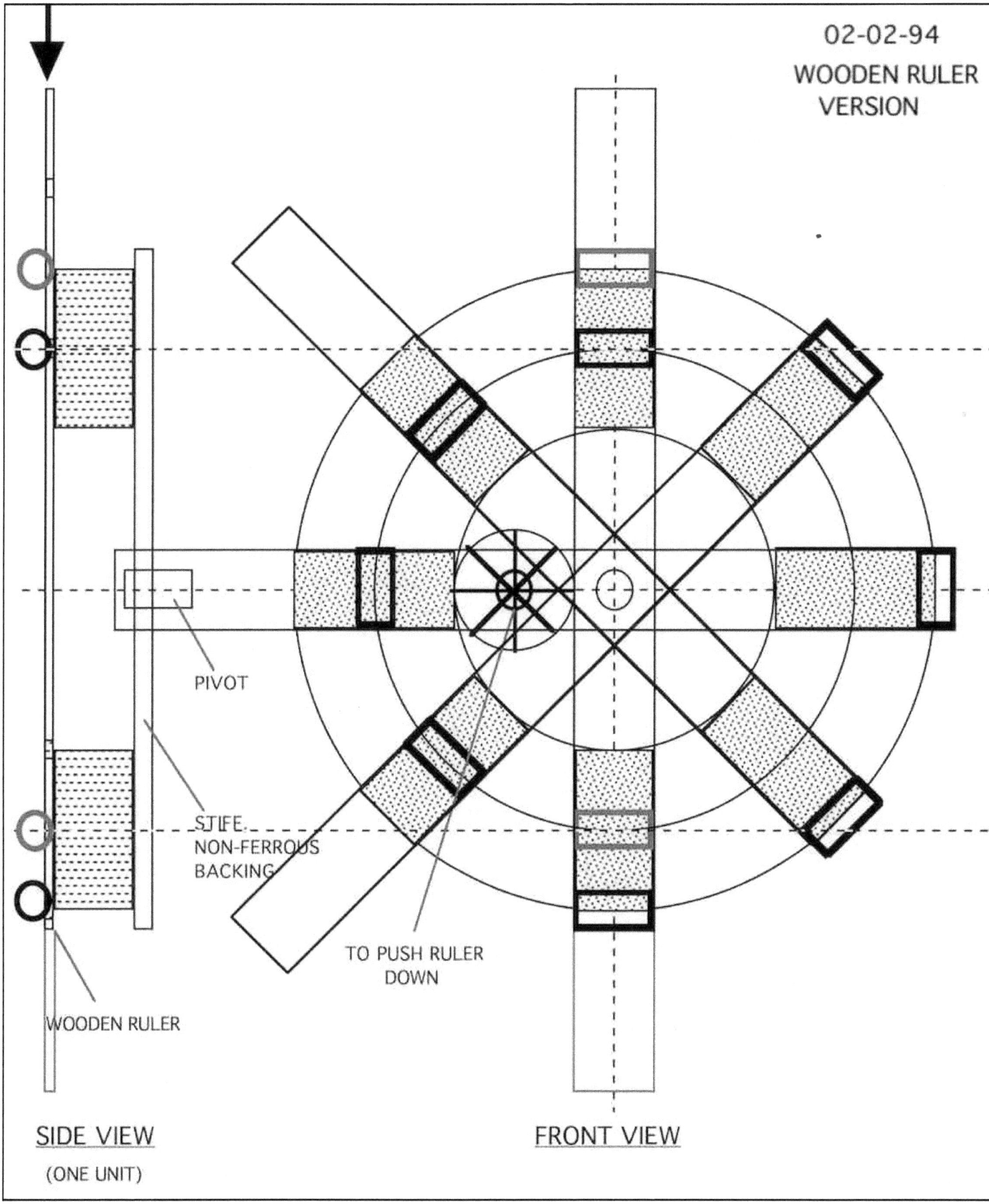
02-02-94
WOODEN RULER
VERSION
PIVOT
STIFF
NON-FERROUS
BACKING
TO PUSH RULER
DOWN
WOODEN RULER
SIDE VIEW
(ONE UNIT)
FRONT VIEW

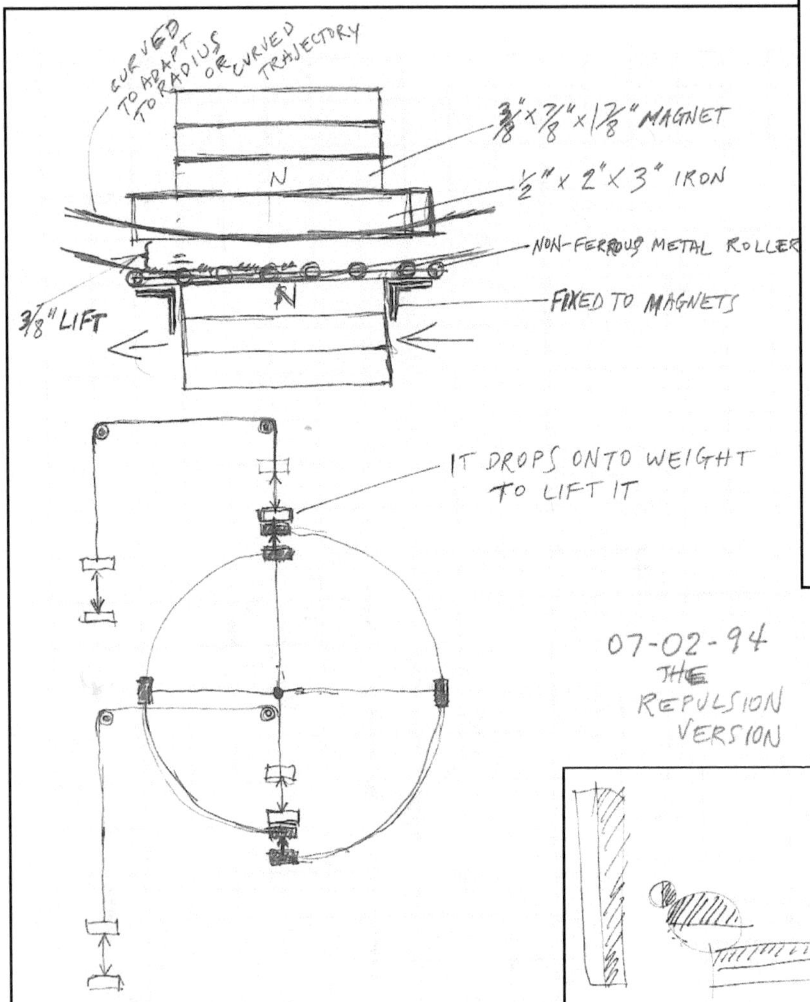

— RE.: PERPETUAL MOTION MACHINE 08-02-94

— IT IS UTTERLY NONSENSICAL TO THINK THAT AN EFFORT (THE EFFORT BY GRAVITY OF A FALLING WEIGHT OF, SAY, ONE KILO) OF METAL OR WATER OR... COULD LIFT ITSELF (AN EFFORT OF THE SAME DEGREE) BY SHEER USAGE OF OR REDESTRIBUTION OF EFFORT BY LEVERAGE, INCLINE, DROP (STRAIGHT OR PENDULUM), OR...

— THE GREATEST DISSERVICE WAS DONE BY LEONARDO DA VINCI WITH HIS DEFINITION/EXPLANATION OF THE LEVER, I.E., THAT UNEQUAL DISTANCE FROM THE FULCRUM MAKES A DIFFERNCE IN THE ORGANIZATION OF AFFECTING EFFORT (LIFTING WEIGHT) AND AFFECTED EFFORT (LIFTED WEIGHT).

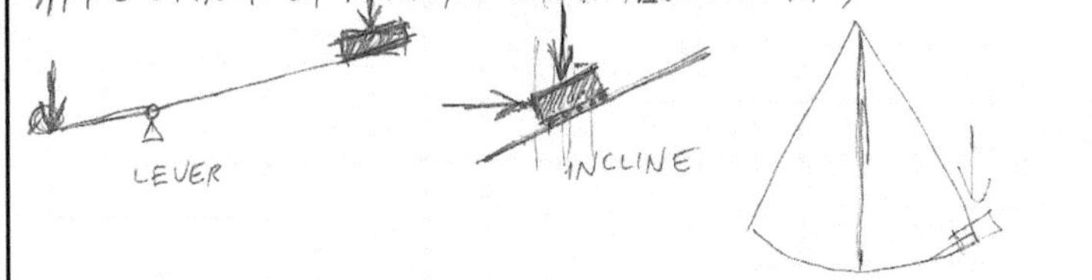

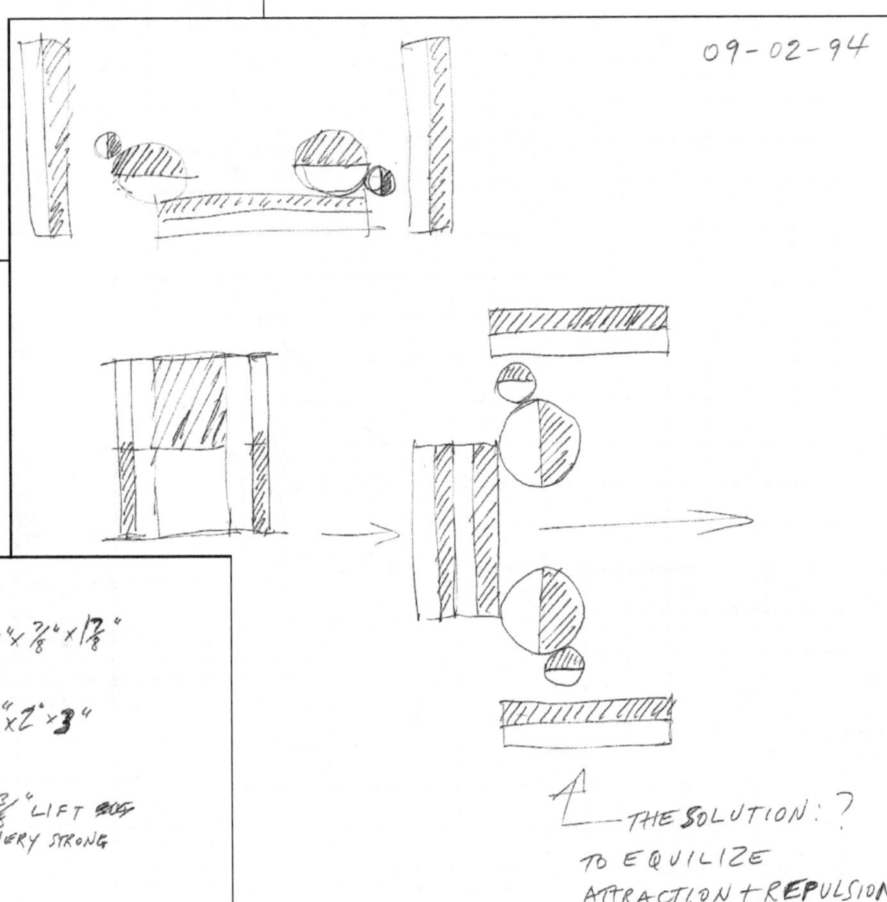

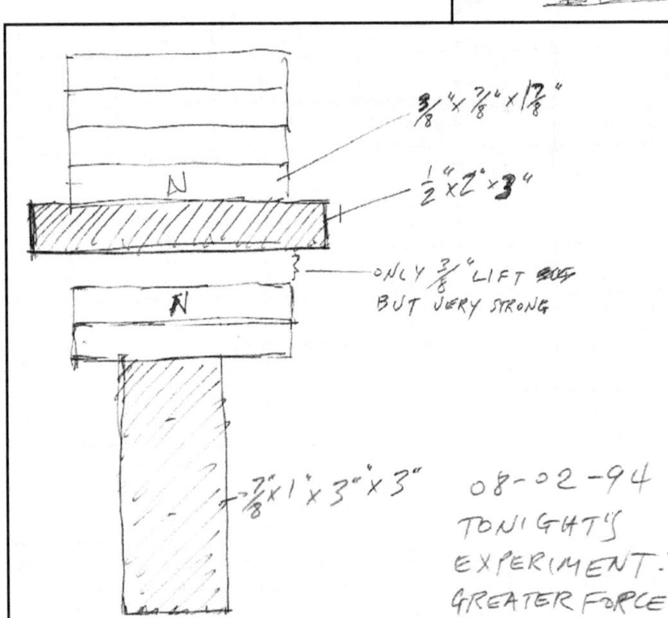

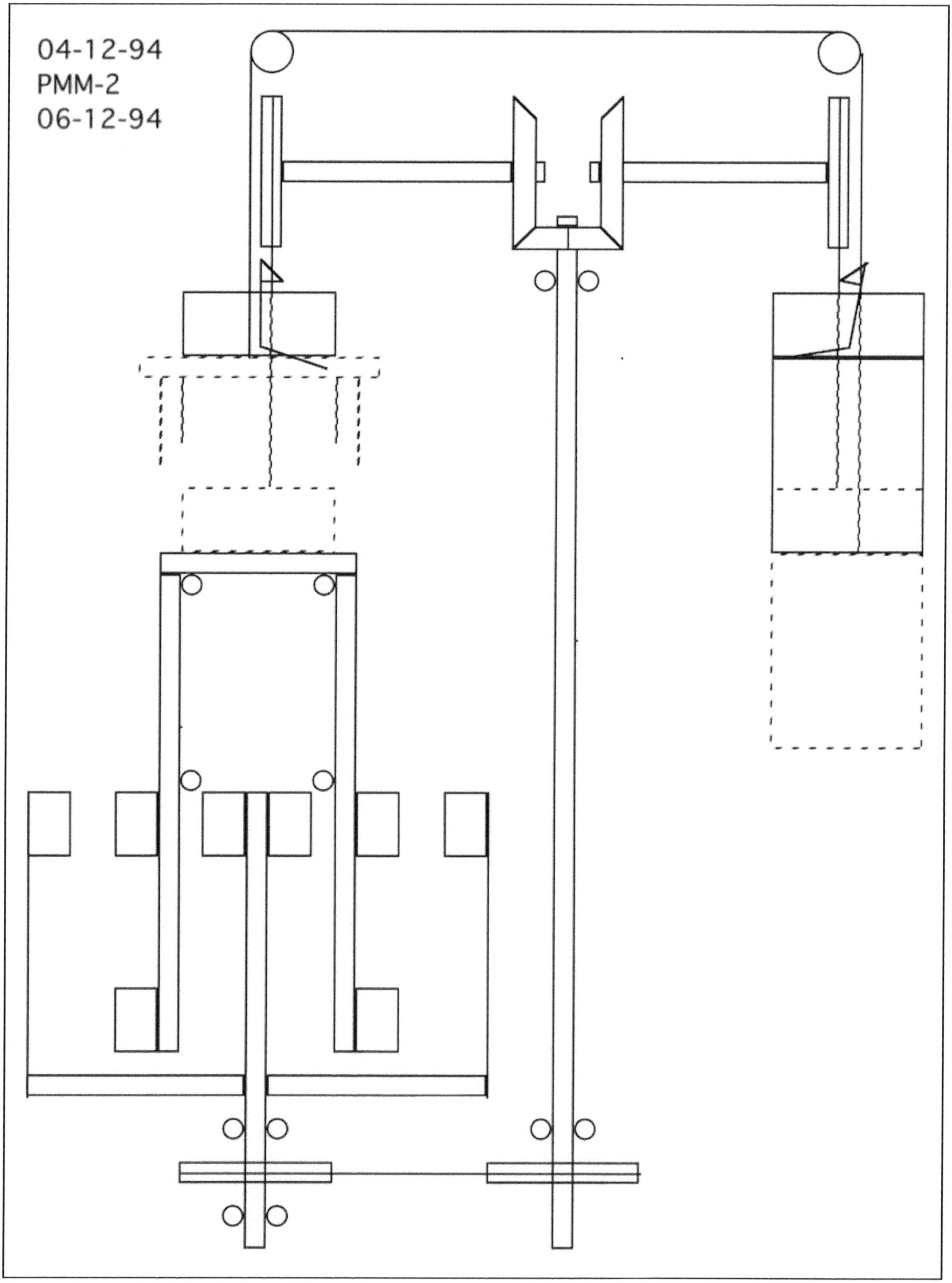
04-12-94
PMM-2
06-12-94

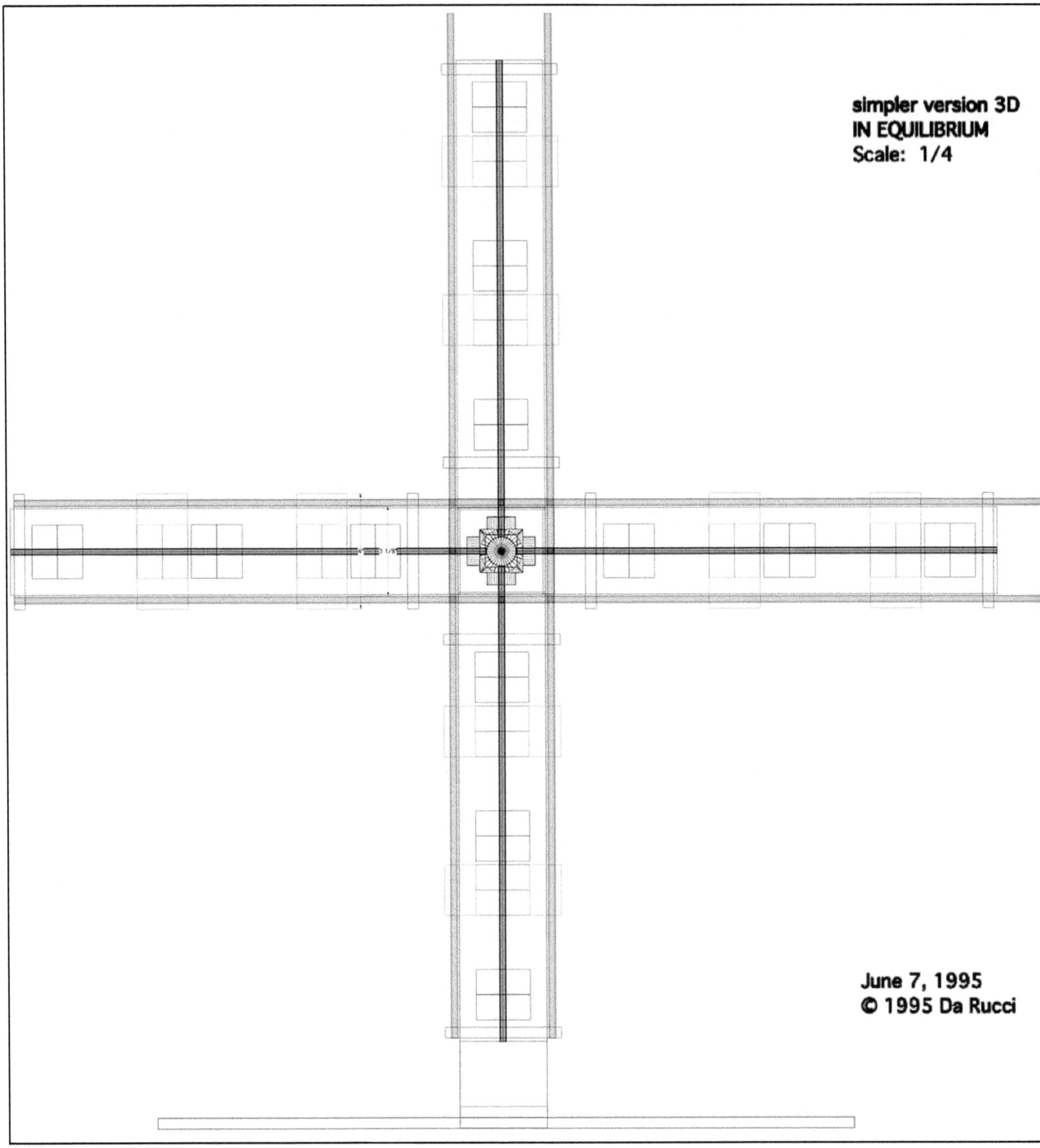
simpler version 3D
IN EQUILIBRIUM
Scale: 1/4
June 7, 1995
© 1995 Da Rucci

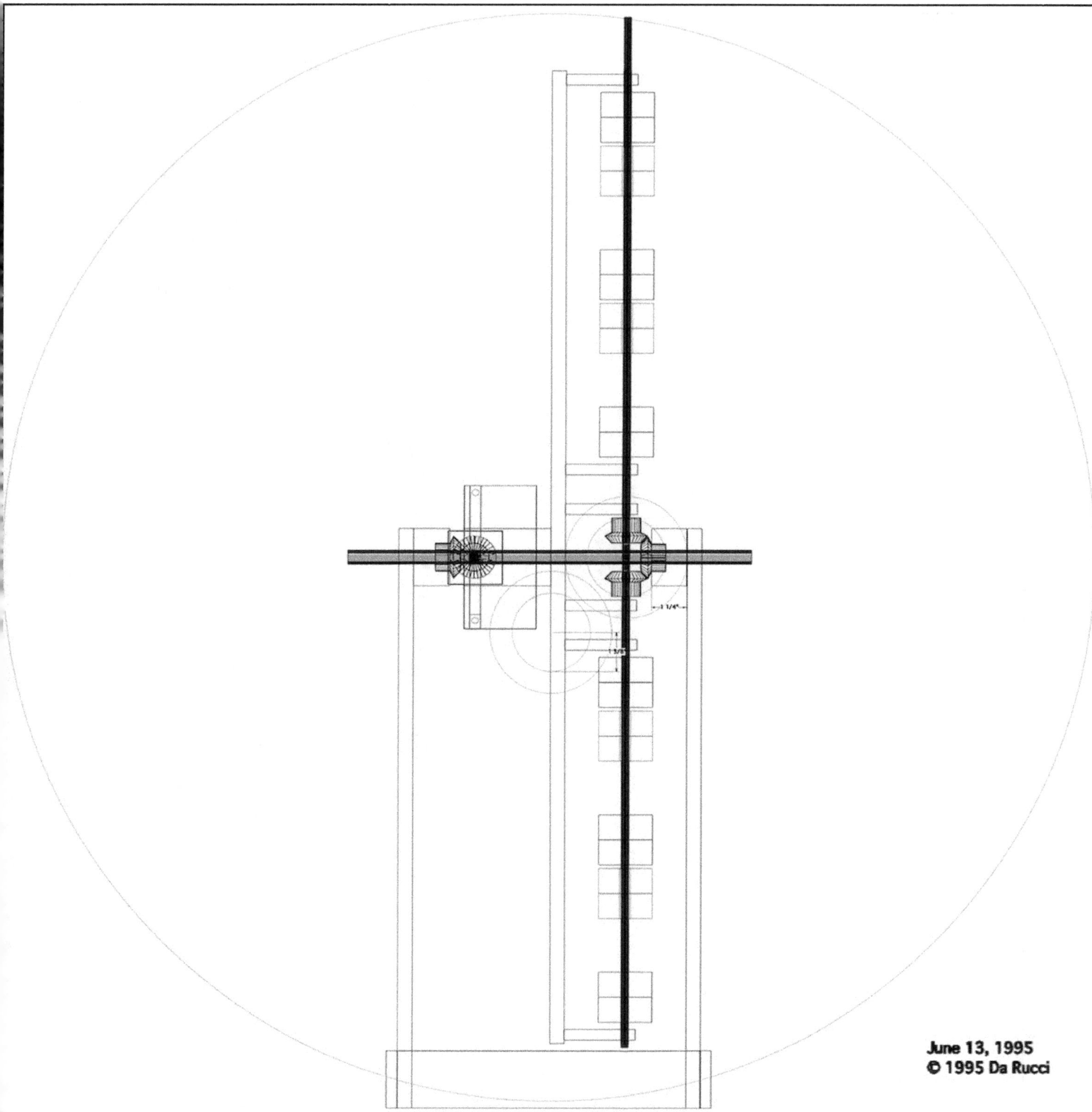
June 13, 1995
© 1995 Da Rucci

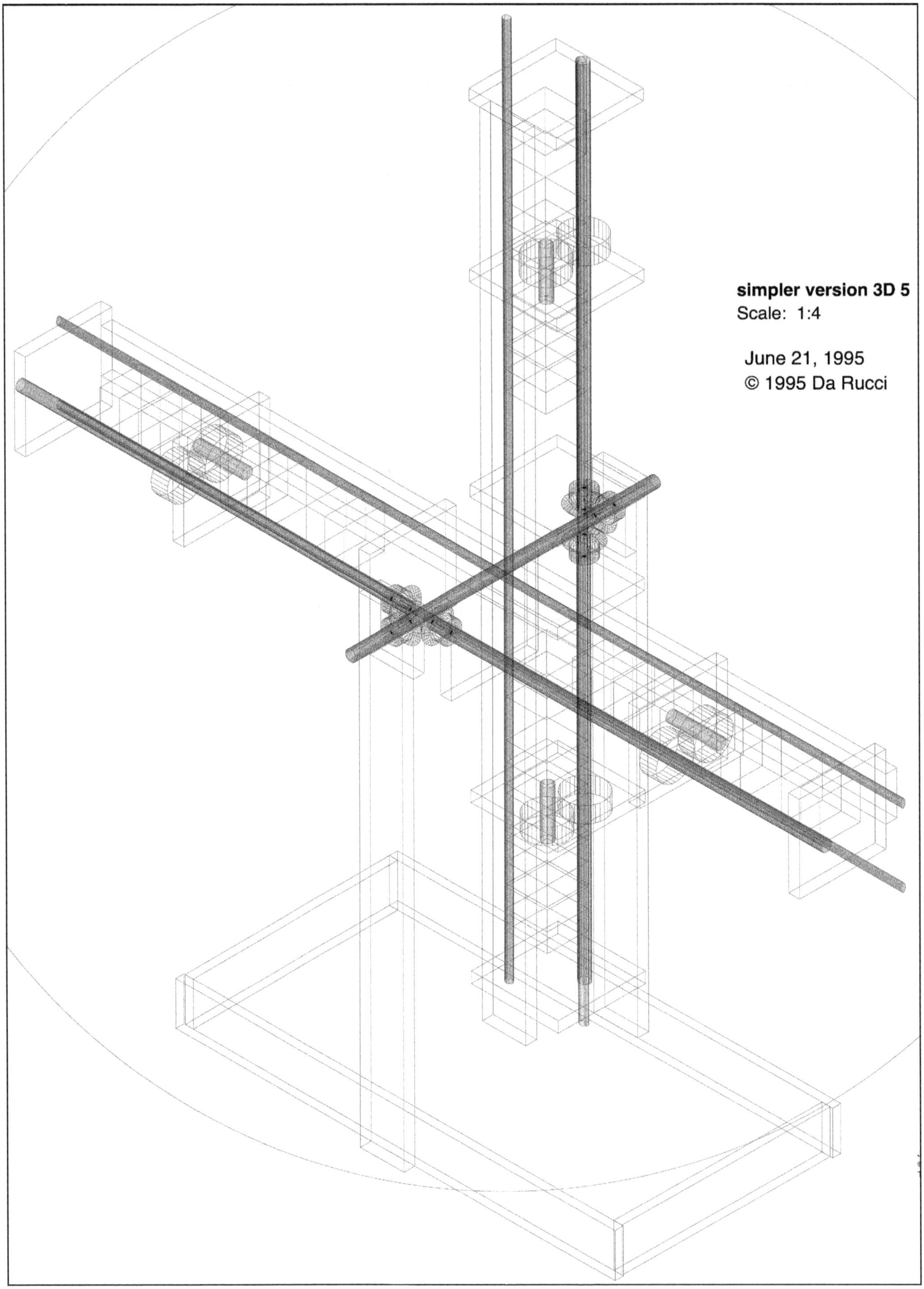

simpler version 3D 5
Scale: 1:4

June 21, 1995

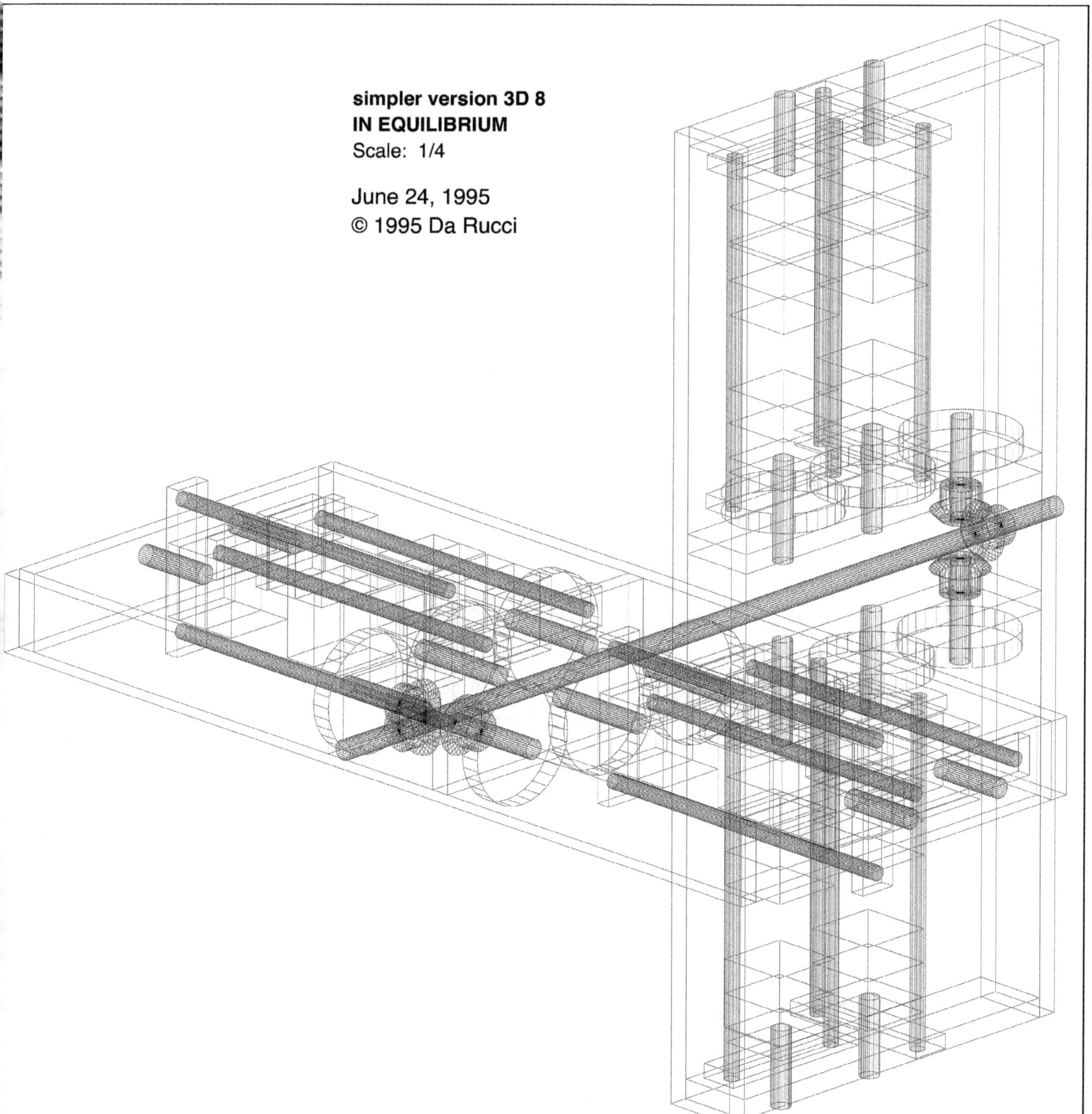
simpler version 3D 8
IN EQUILIBRIUM
Scale: 1/4
June 24, 1995
© 1995 Da Rucci

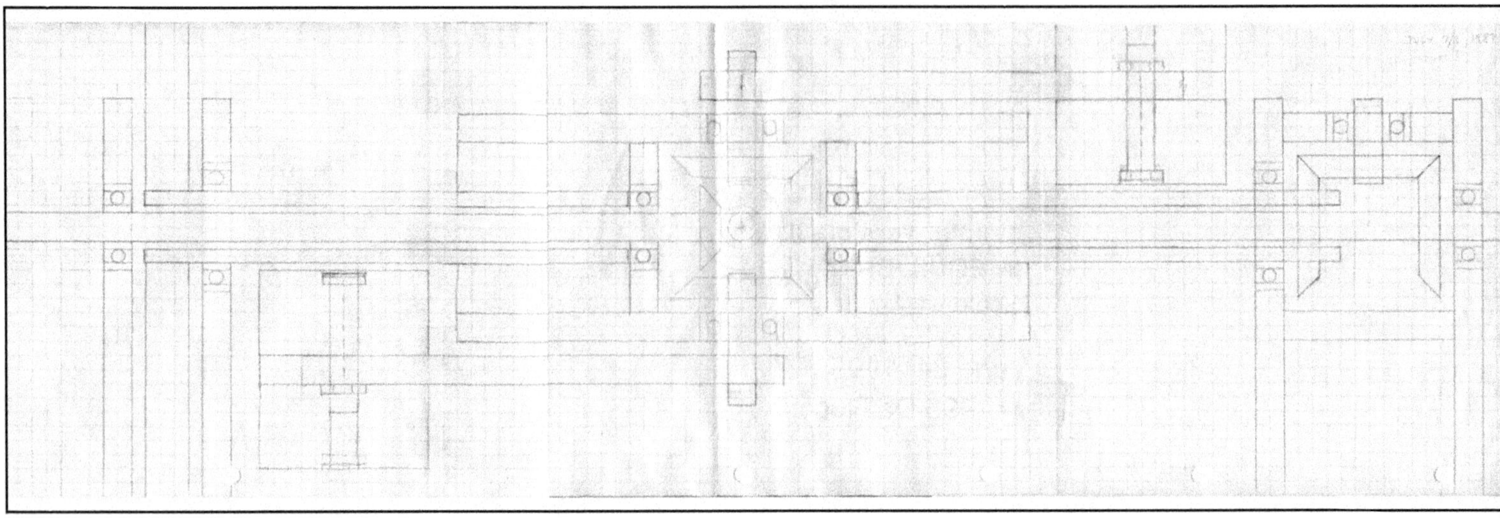

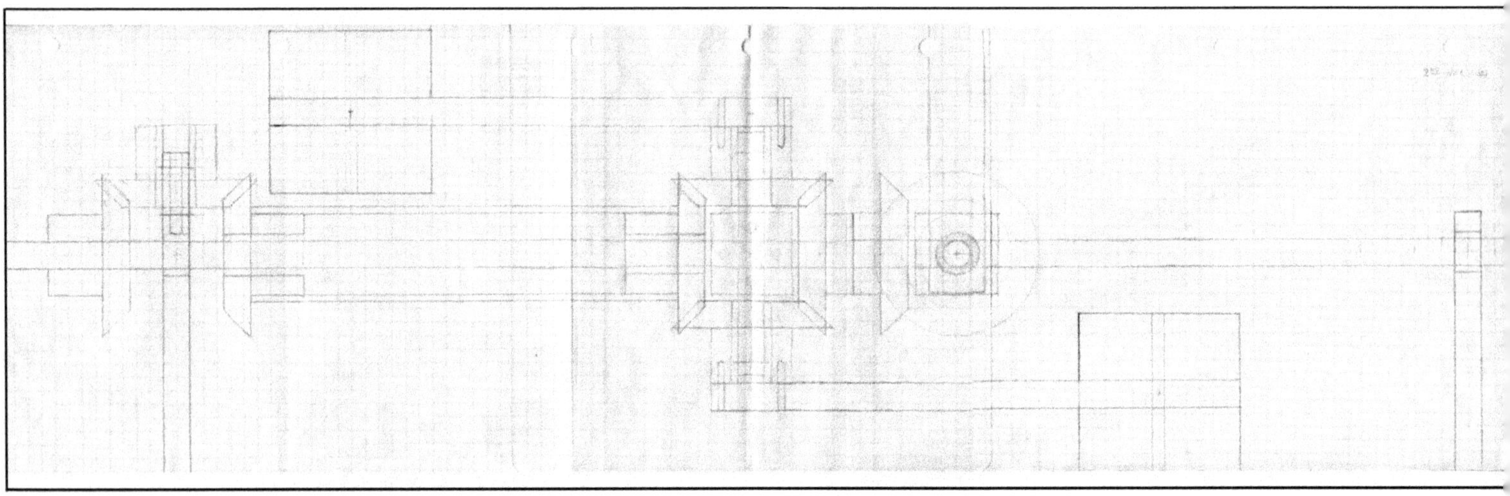

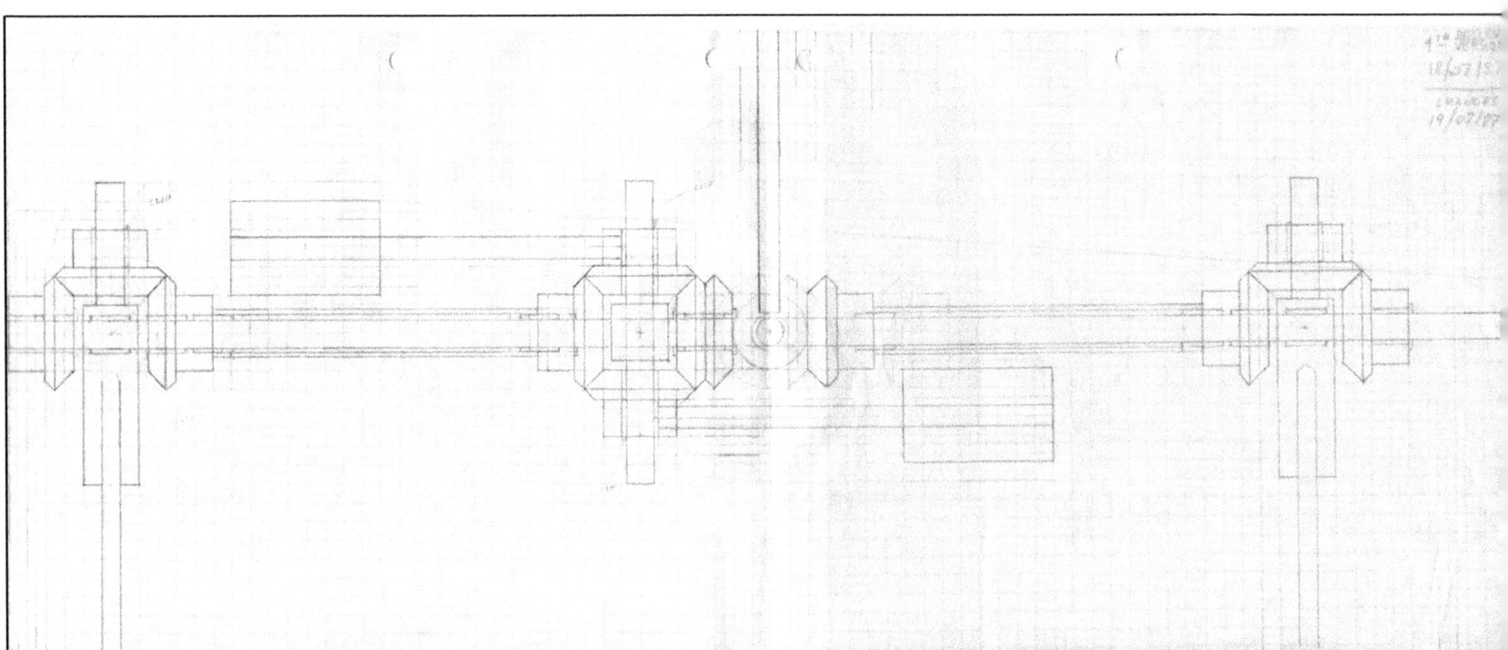

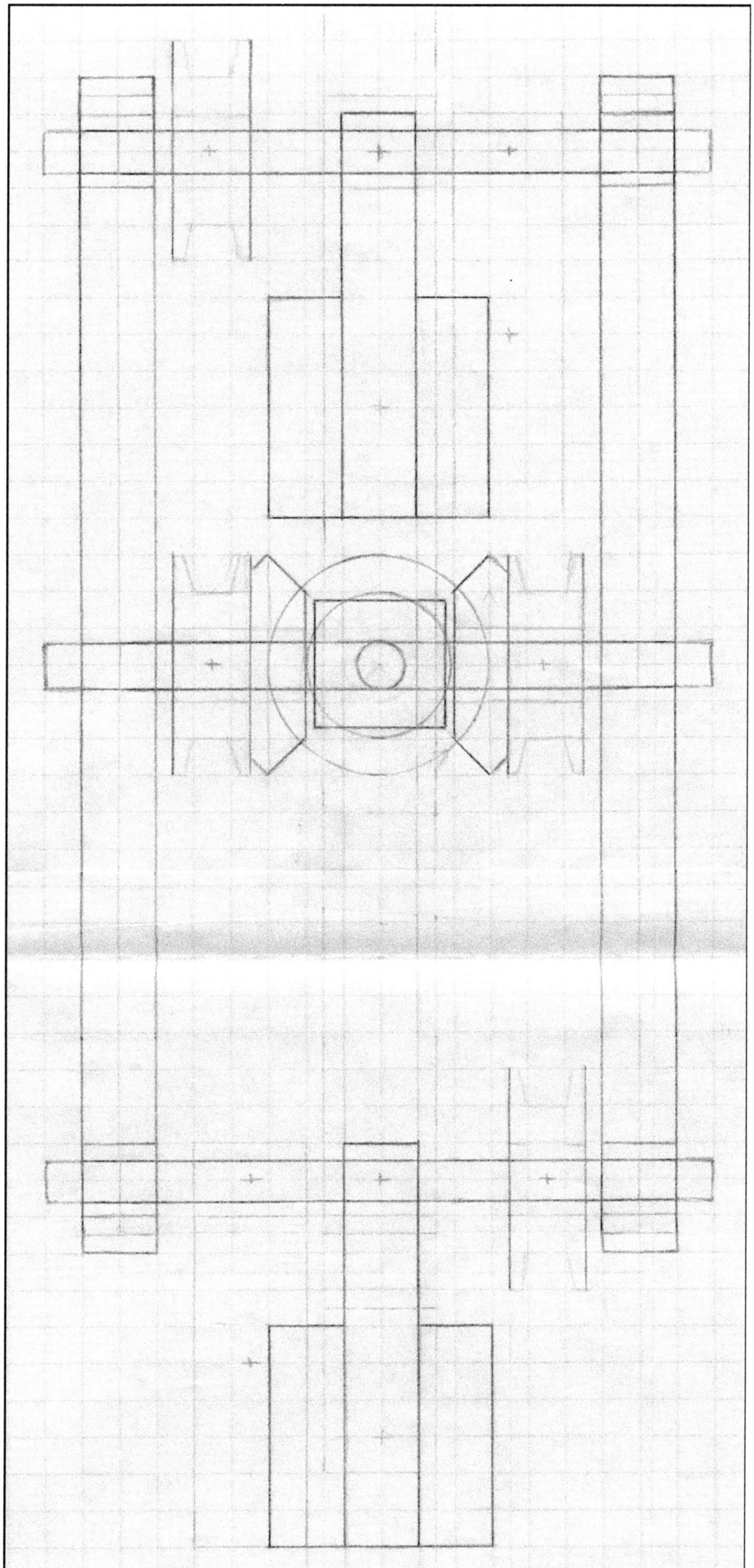

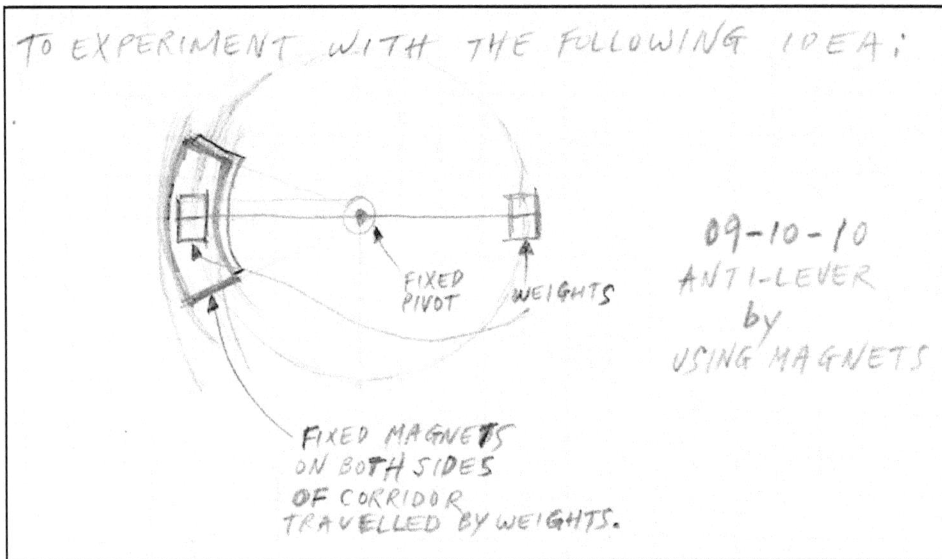
TO EXPERIMENT WITH THE FOLLOWING IDEA;
FIXED PIVOT
WEIGHTS
09-10-10
ANTI-LEVER
by
USING MAGNETS
FIXED MAGNETS
ON BOTH SIDES
OF CORRIDOR
TRAVELLED BY WEIGHTS.

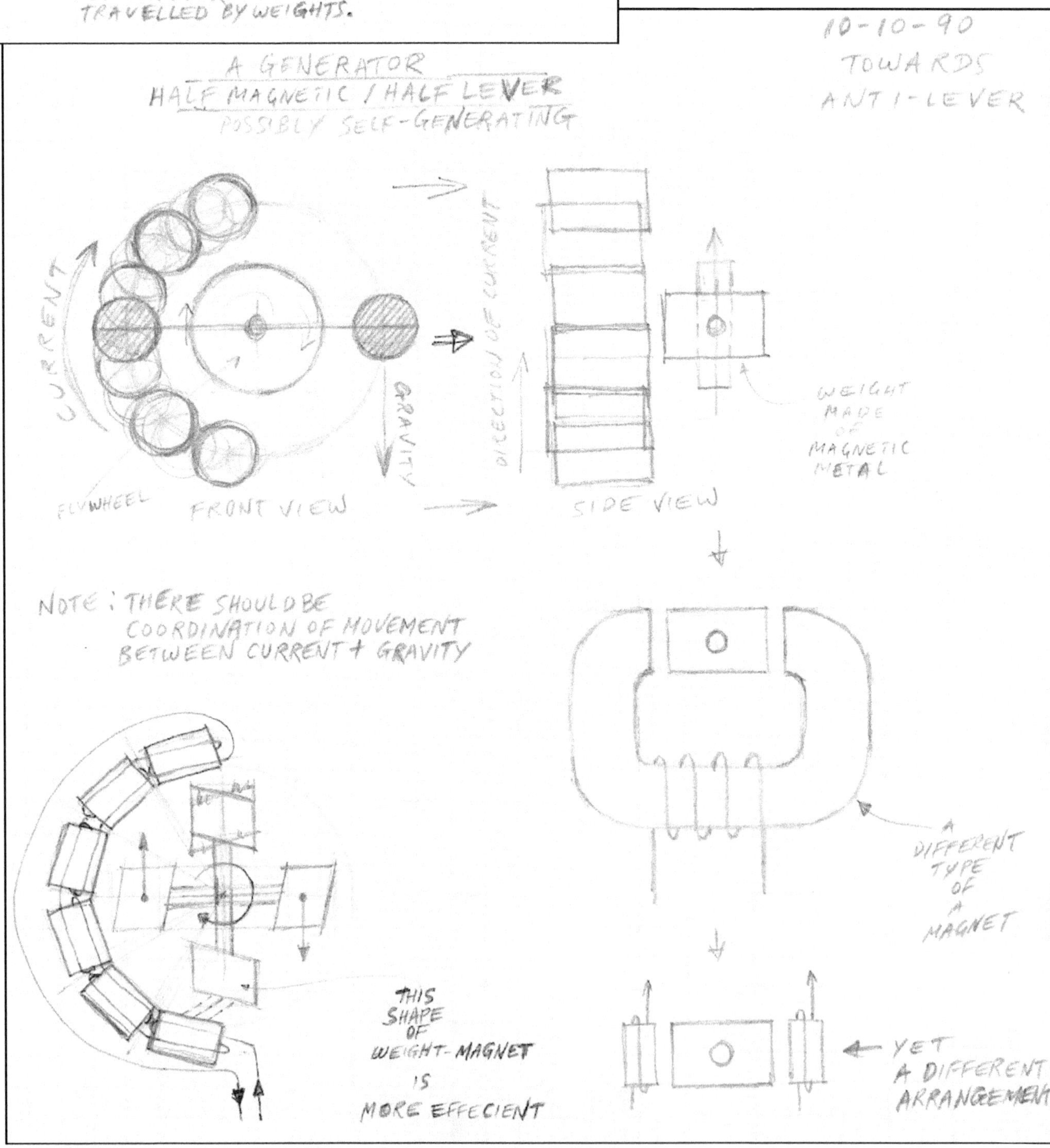
10-10-90
TOWARDS
ANTI-LEVER
A GENERATOR
HALF MAGNETIC / HALF LEVER
POSSIBLY SELF-GENERATING
CURRENT
DIRECTION OF CURRENT
GRAVITY
FLYWHEEL
FRONT VIEW
SIDE VIEW
WEIGHT
MADE
OF
MAGNETIC
METAL
NOTE: THERE SHOULD BE
COORDINATION OF MOVEMENT
BETWEEN CURRENT + GRAVITY
A
DIFFERENT
TYPE
OF
A
MAGNET
THIS
SHAPE
OF
WEIGHT-MAGNET
IS
MORE EFFECIENT
YET
A DIFFERENT
ARRANGEMENT

8E. ELECTRIC MOTOR

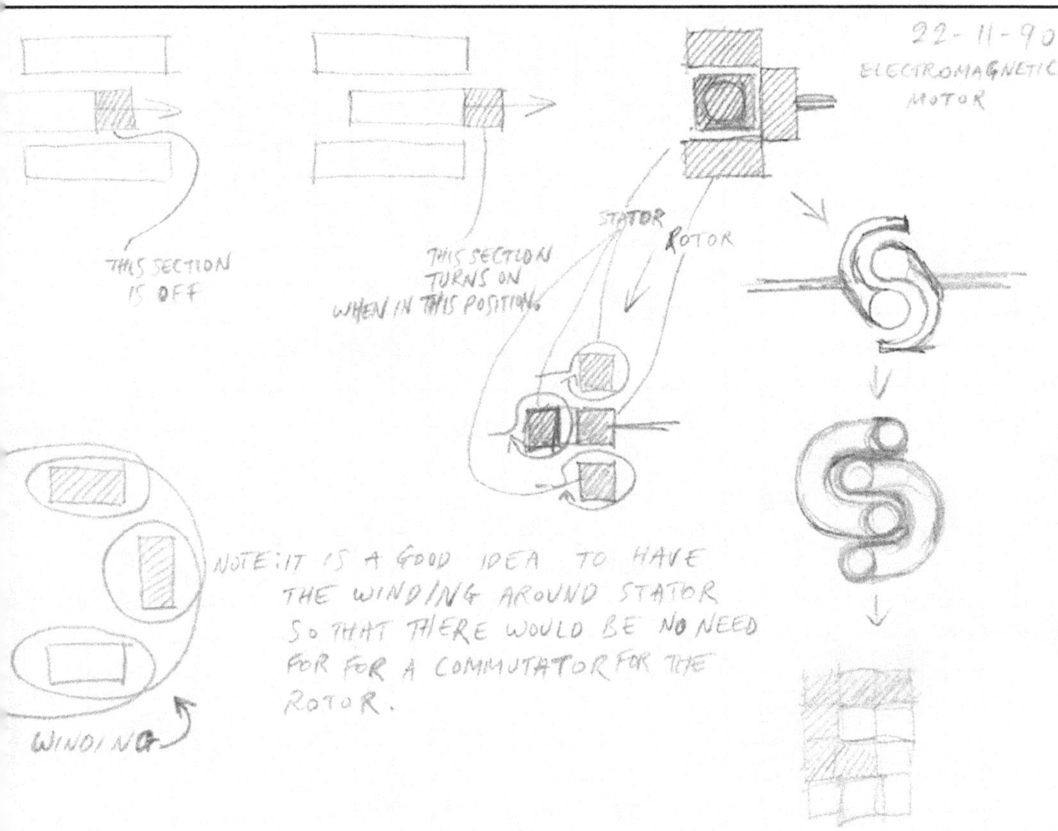

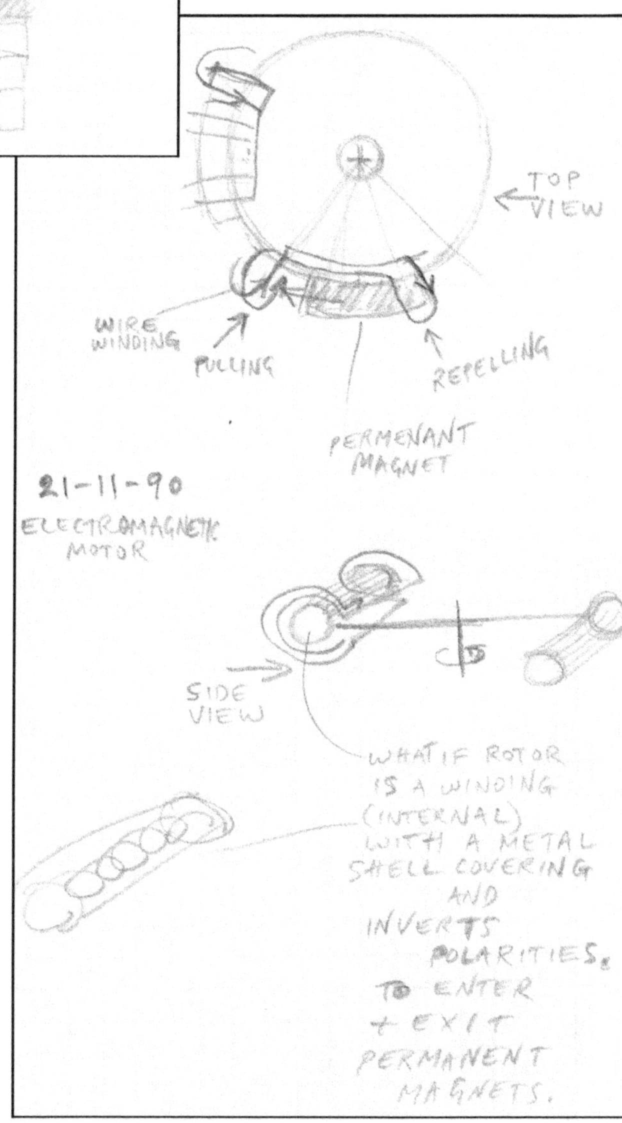

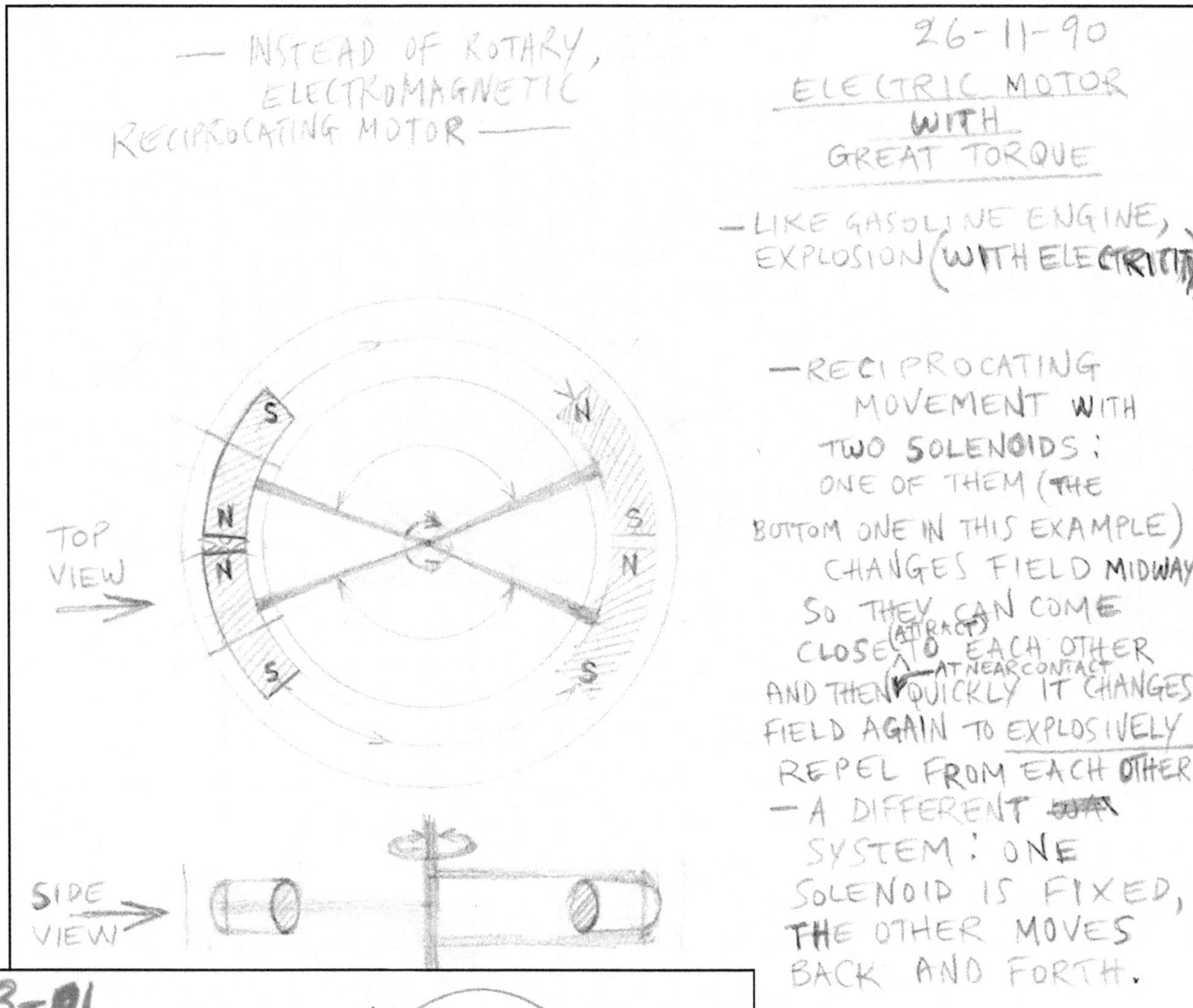

26-11-90
ELECTRIC MOTOR WITH GREAT TORQUE
— INSTEAD OF ROTARY, ELECTROMAGNETIC RECIPROCATING MOTOR —
—LIKE GASOLINE ENGINE, EXPLOSION (WITH ELECTRICITY)
—RECIPROCATING MOVEMENT WITH TWO SOLENOIDS; ONE OF THEM (THE BOTTOM ONE IN THIS EXAMPLE) CHANGES FIELD MIDWAY SO THEY CAN COME CLOSE TO (ATTRACT) EACH OTHER AND THEN (AT NEAR CONTACT) QUICKLY IT CHANGES FIELD AGAIN TO EXPLOSIVELY REPEL FROM EACH OTHER.
—A DIFFERENT SYSTEM: ONE SOLENOID IS FIXED, THE OTHER MOVES BACK AND FORTH.
TOP VIEW
SIDE VIEW
S
N
N
S
N
S
N
S

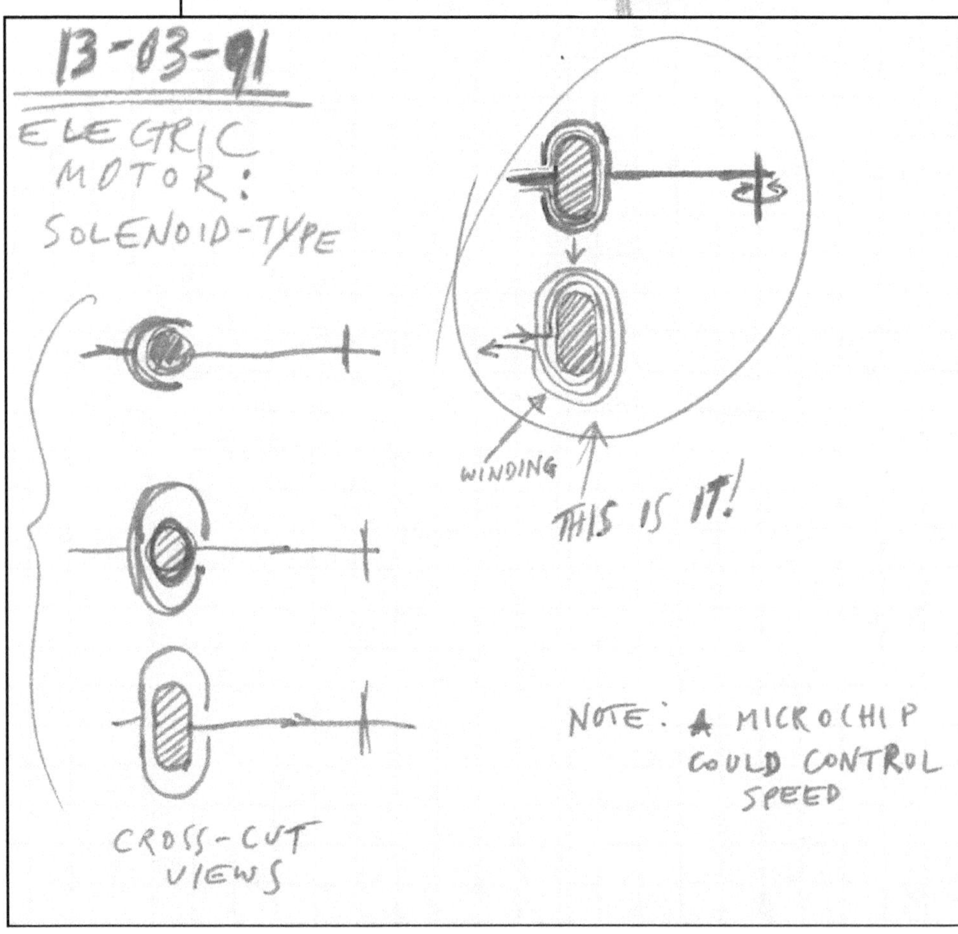

13-03-91
ELECTRIC MOTOR: SOLENOID-TYPE
WINDING
THIS IS IT!
CROSS-CUT VIEWS
NOTE: A MICROCHIP COULD CONTROL SPEED

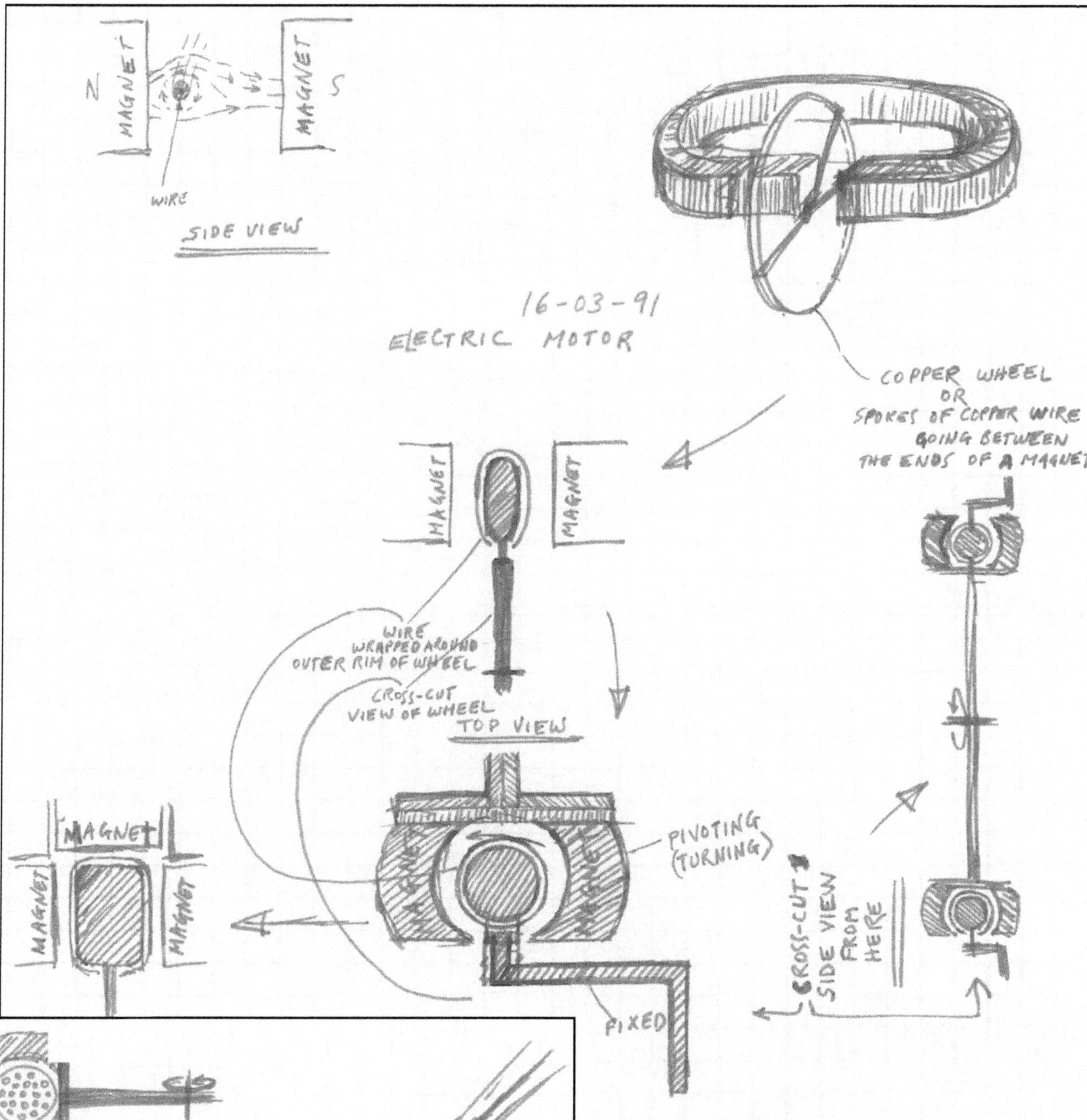
N
MAGNET
MAGNET
S
WIRE
SIDE VIEW
16-03-91
ELECTRIC MOTOR
COPPER WHEEL
OR
SPOKES OF COPPER WIRE
GOING BETWEEN
THE ENDS OF A MAGNET
MAGNET
MAGNET
WIRE
WRAPPED AROUND
OUTER RIM OF WHEEL
CROSS-CUT
VIEW OF WHEEL
TOP VIEW
MAGNET
MAGNET
MAGNET
PIVOTING
(TURNING)
CROSS-CUT
SIDE VIEW
FROM
HERE
FIXED

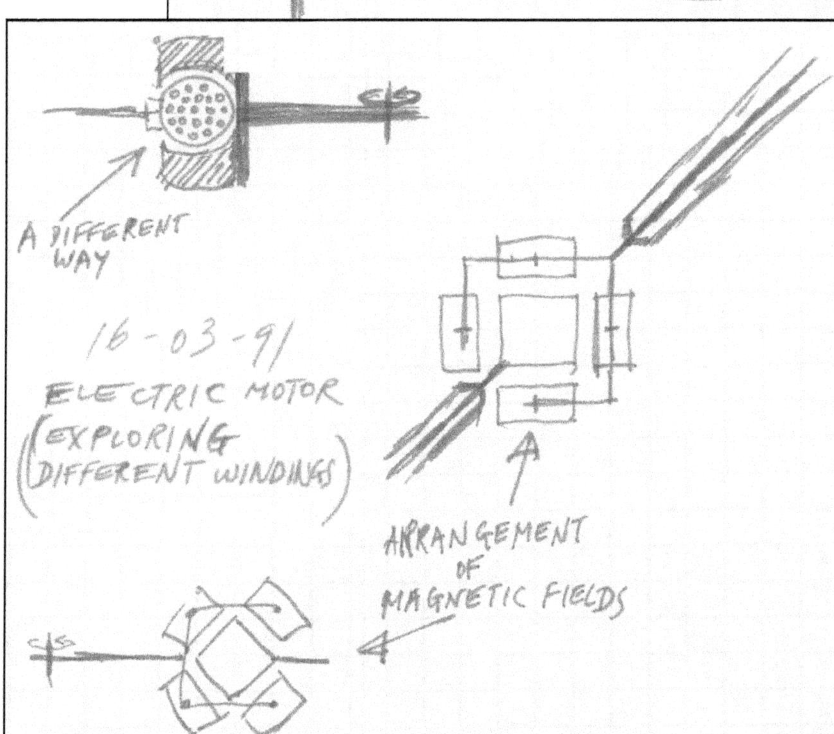
A DIFFERENT
WAY
16-03-91
ELECTRIC MOTOR
(EXPLORING
DIFFERENT WINDINGS)
ARRANGEMENT
OF
MAGNETIC FIELDS

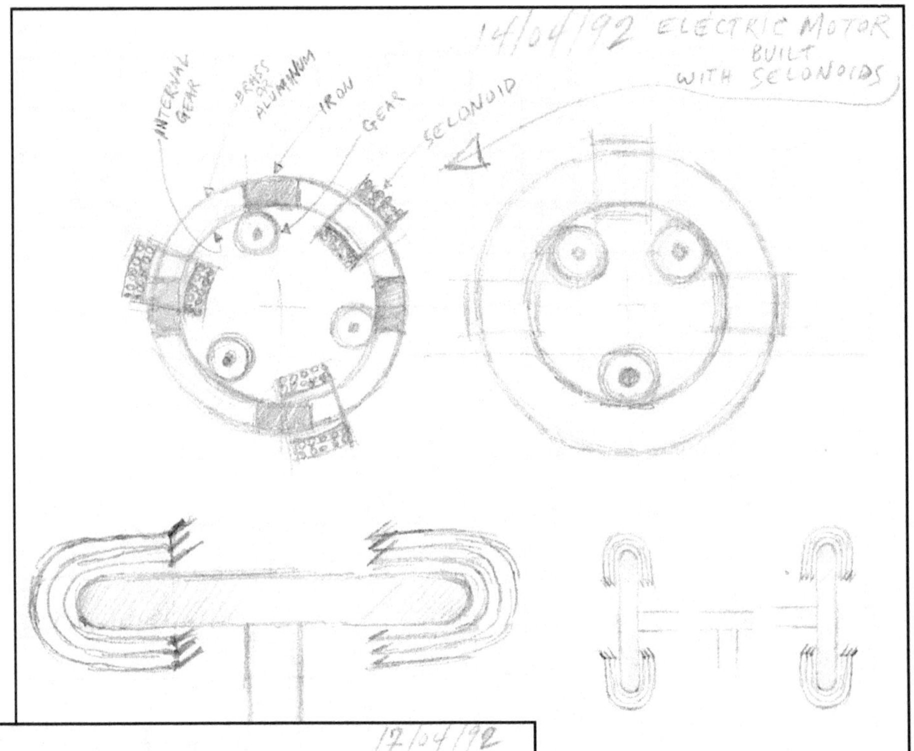
14/04/92 ELECTRIC MOTOR
BUILT
WITH SOLENOIDS
INTERNAL GEAR
BRASS OR ALUMINUM
IRON
GEAR
SOLENOID

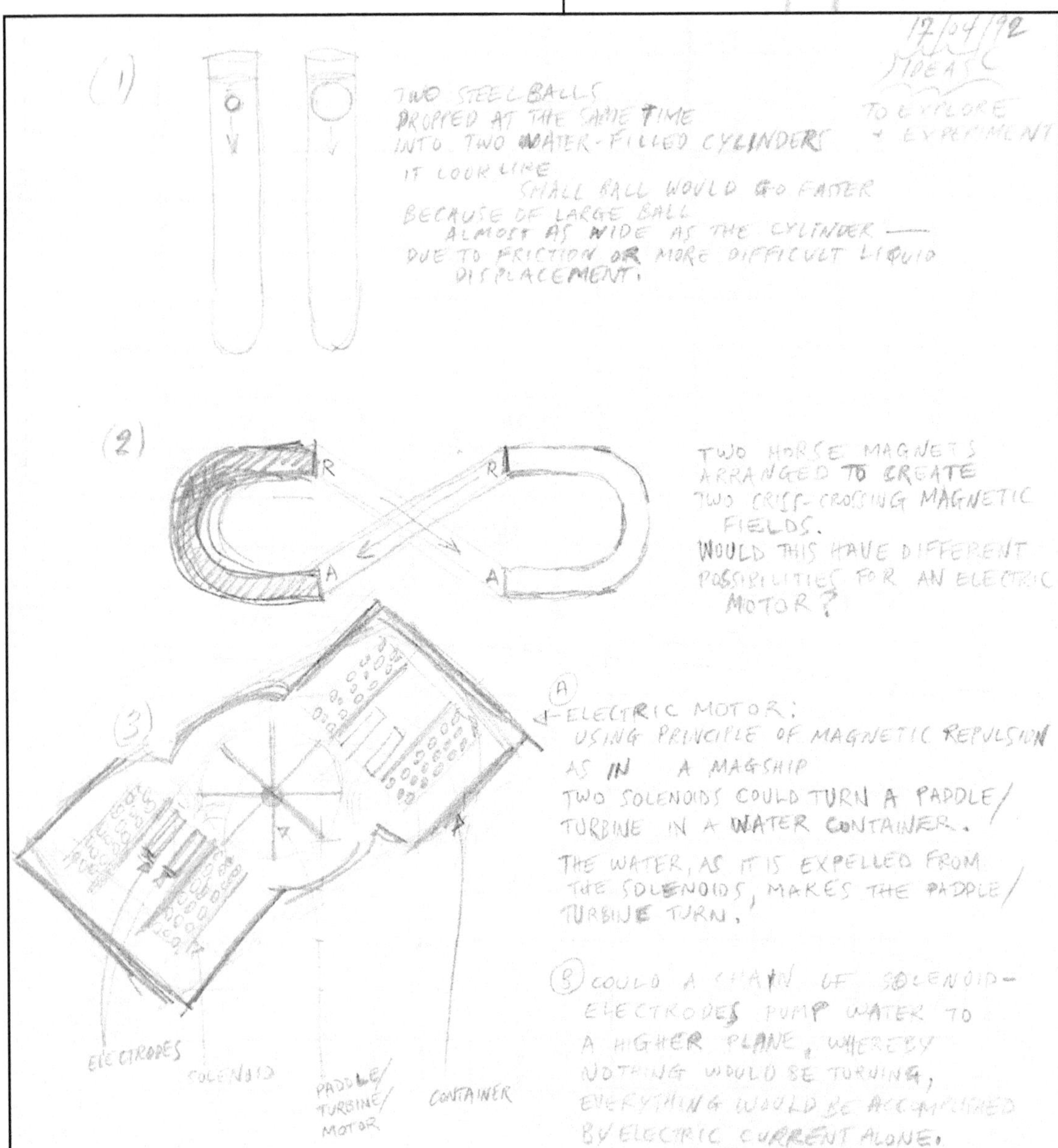
17/04/92
IDEAS
TO EXPLORE
+ EXPERIMENT
(1)
TWO STEEL BALLS
DROPPED AT THE SAME TIME
INTO TWO WATER-FILLED CYLINDERS
IT LOOK LIKE
SMALL BALL WOULD GO FASTER
BECAUSE OF LARGE BALL
ALMOST AS WIDE AS THE CYLINDER —
DUE TO FRICTION OR MORE DIFFICULT LIQUID
DISPLACEMENT.
(2)
R
R
A
A
TWO HORSE MAGNETS
ARRANGED TO CREATE
TWO CRISS-CROSSING MAGNETIC
FIELDS.
WOULD THIS HAVE DIFFERENT
POSSIBILITIES FOR AN ELECTRIC
MOTOR?
(3)
A
A
ELECTRIC MOTOR:
USING PRINCIPLE OF MAGNETIC REPULSION
AS IN A MAGSHIP
TWO SOLENOIDS COULD TURN A PADDLE/
TURBINE IN A WATER CONTAINER.
THE WATER, AS IT IS EXPELLED FROM
THE SOLENOIDS, MAKES THE PADDLE/
TURBINE TURN.
ELECTRODES
SOLENOID
PADDLE/
TURBINE/
MOTOR
CONTAINER
(B) COULD A CHAIN OF SOLENOID-
ELECTRODES PUMP WATER TO
A HIGHER PLANE, WHEREBY
NOTHING WOULD BE TURNING,
EVERYTHING WOULD BE ACCOMPLISHED
BY ELECTRIC CURRENT ALONE.

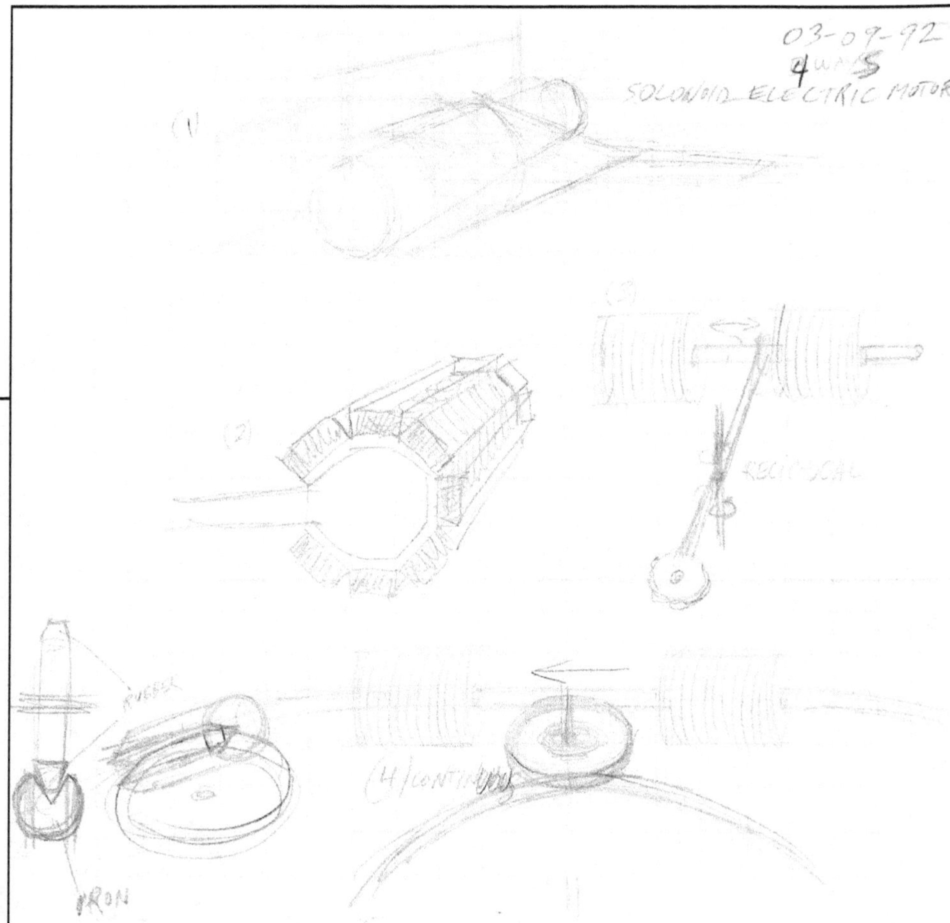
03-09-92
SOLONOID ELECTRIC MOTOR
RUBBER
IRON

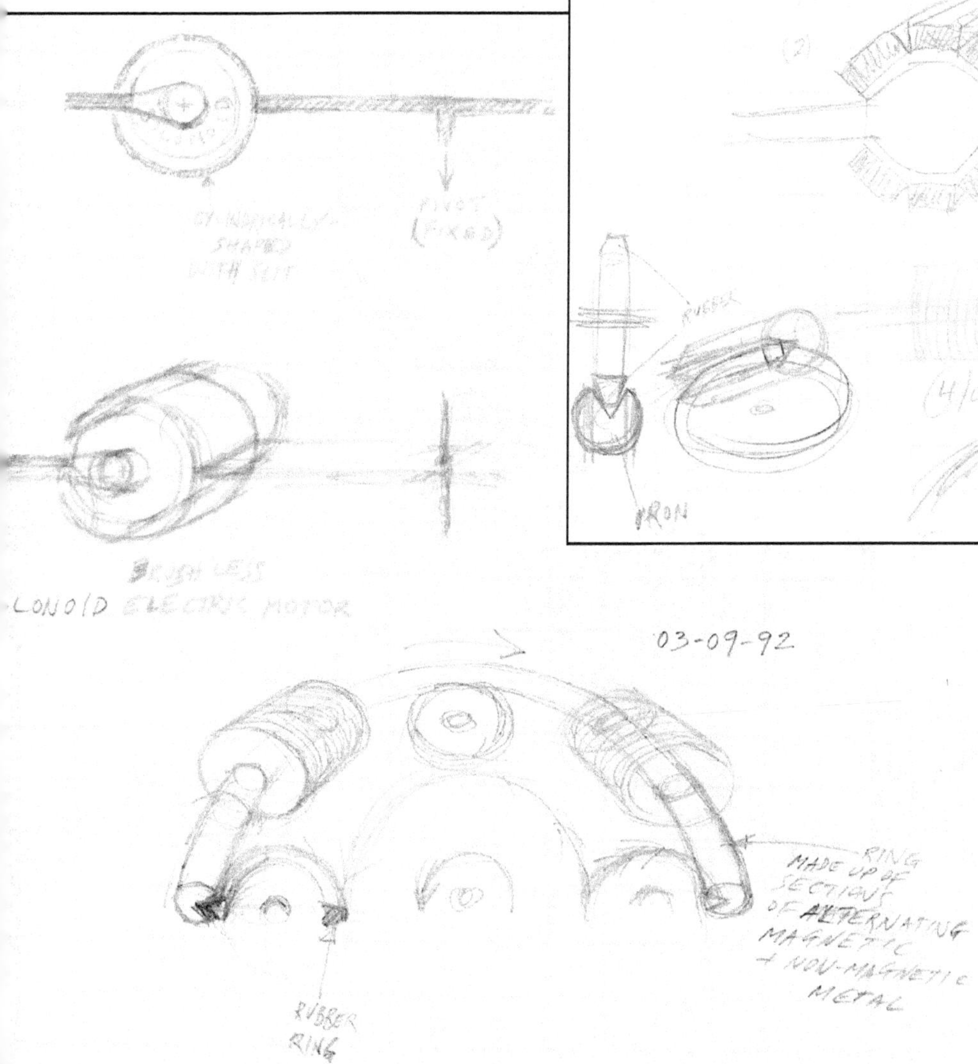
LONOID ELECTRIC MOTOR
03-09-92
RING MADE UP OF SECTIONS OF ALTERNATING MAGNETIC + NON-MAGNETIC METAL
RUBBER RING

23/11/92
A MORE EFFICIENT
ELECTRIC MOTOR?
(RIDING THE WAVE)
COMMUTATOR

9A. MOTION CAMERA

MOTION CAMERA TITLES

— Solid block letter titles animated in the motion camera style. [The title MOTION CAMERA would follow a brief graphic (narrative/minimal) history of the motion picture. The various inventions in the field.] This would be appropriate

— OTHER ALTERNATIVES:

(1) Titles on the faces of a 12-sided solid. So it would be turned under the movie camera insted of having to take still pictures (negatives) first.

(2) Titles on a figure. Animating the figure + zooming in on the titles.

(3) A variation of (2) would be: Superimposition of figure + title. Titles ON A CARD ZOOMING OUT to LONGSHOT as full figure ZOOMING IN to CLOSEUP

(1)

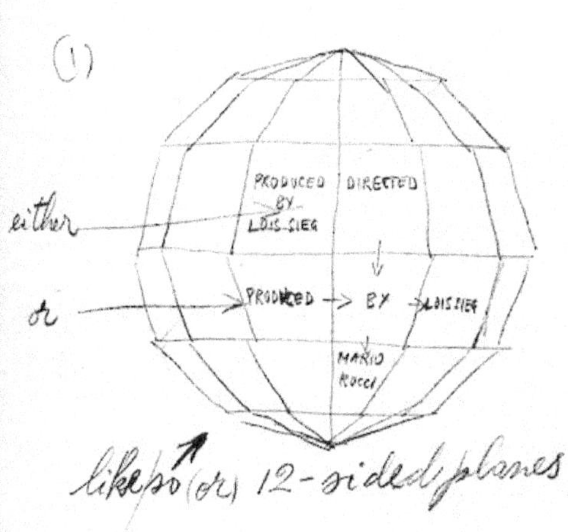

(2)

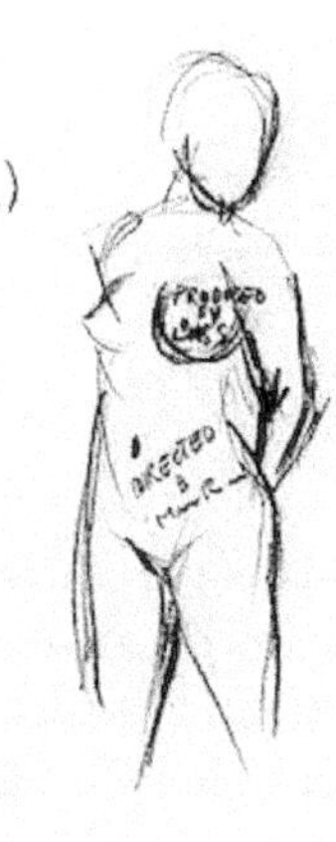

(3)

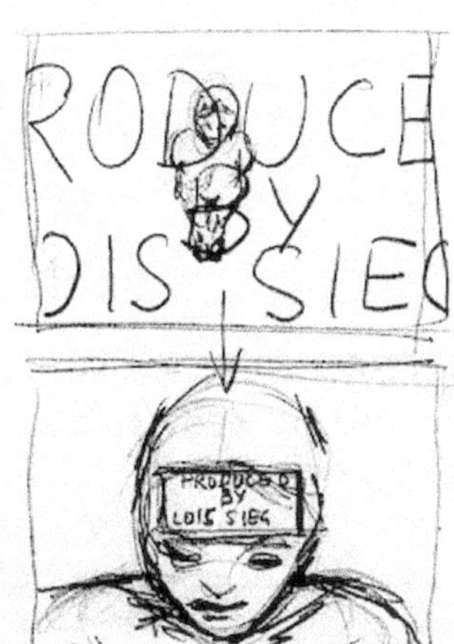

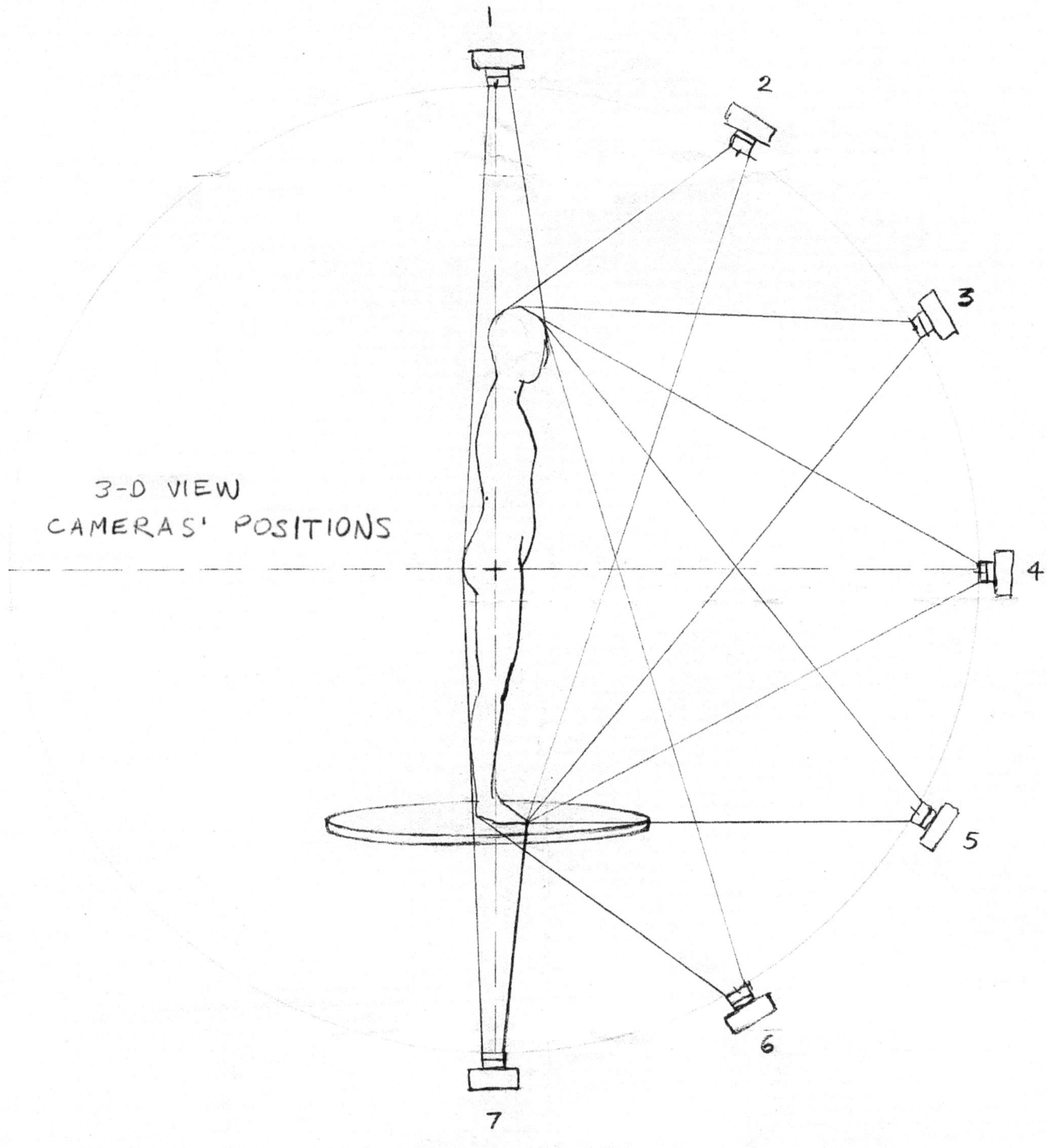

The 3-D CAMERA SYSTEM was my first invention to which I gave serious, long-term consideration. The purpose of this camera system was to take pictures of a model from every point of view. The pictures would be used to produce an animation movie or stone sculptures. I called it MOTION CAMERA when planning the animation movie and READER when planning the stone sculptures.

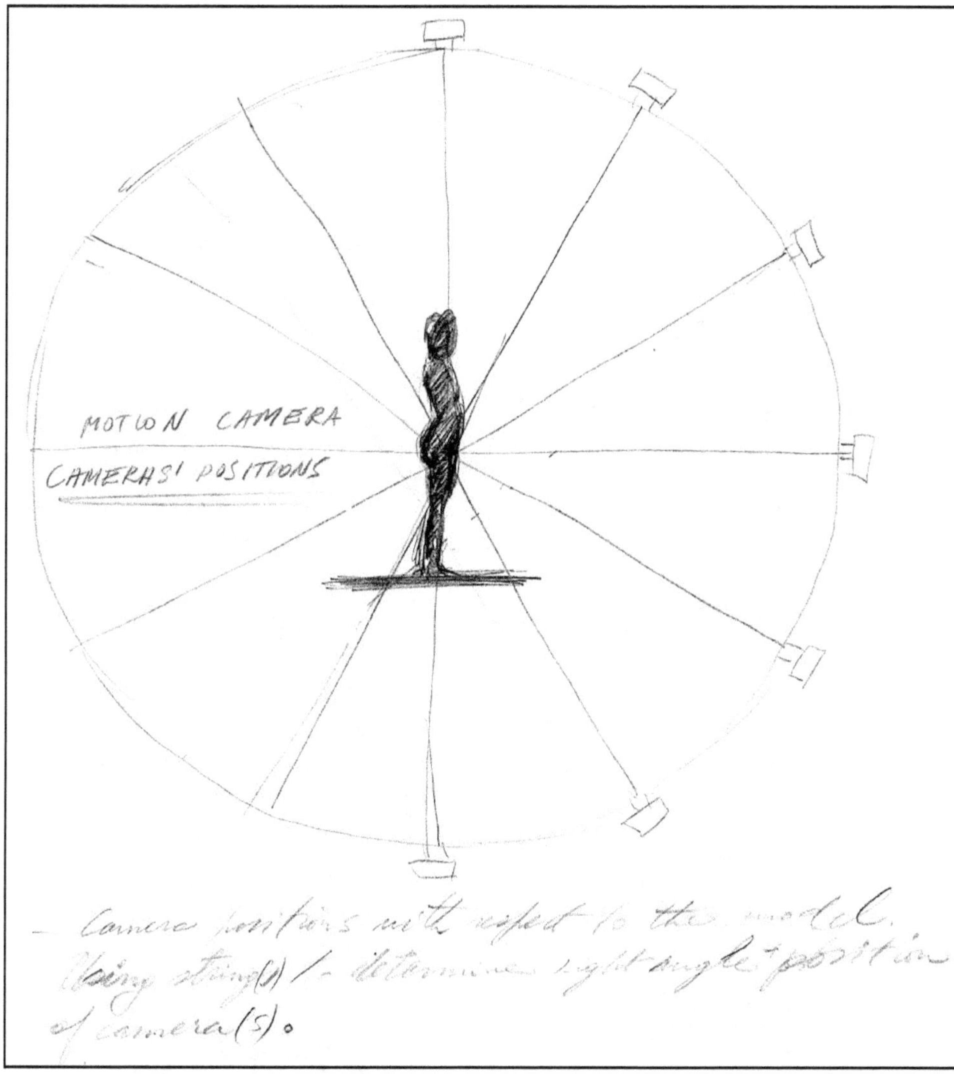
MOTION CAMERA
CAMERAHS' POSITIONS
– Camera positions with respect to the model.
Using string(s) – determine right angle + position
of camera(s).

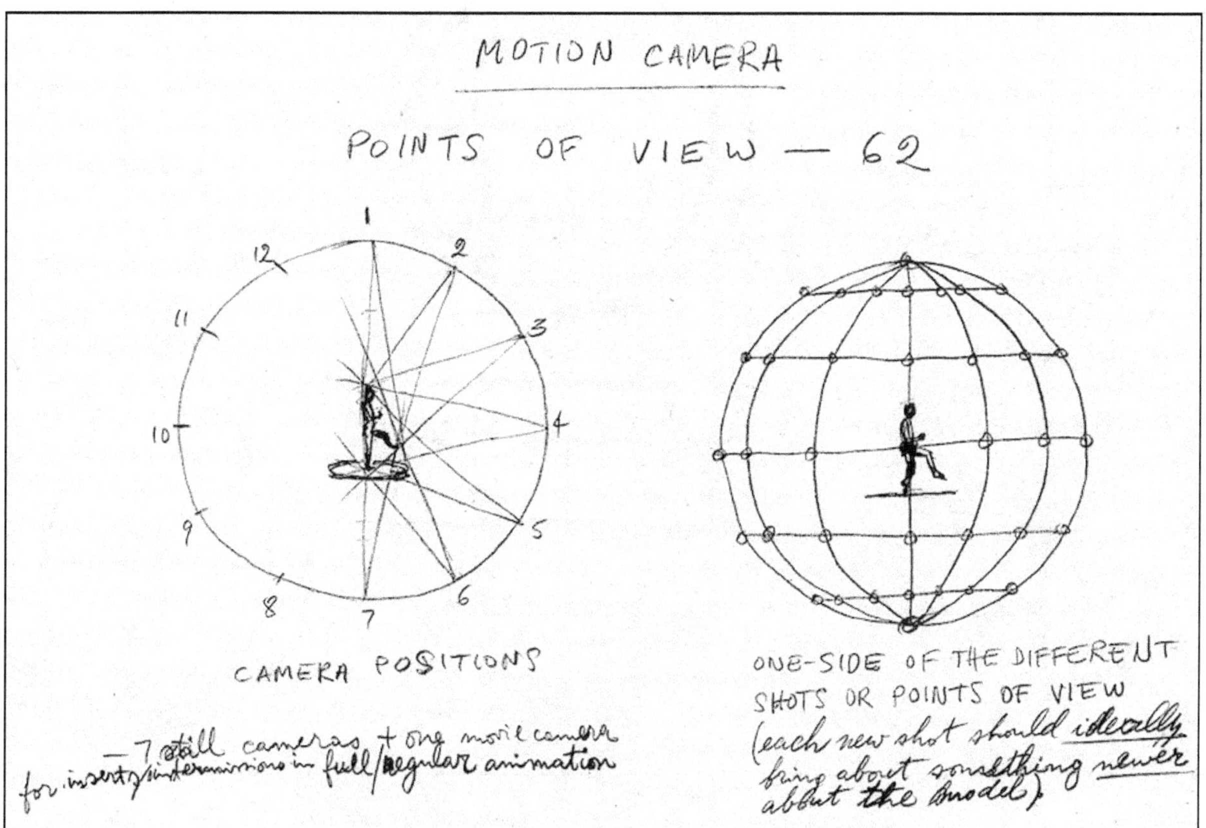
MOTION CAMERA
POINTS OF VIEW — 62
CAMERA POSITIONS
– 7 still cameras + one movie camera
for inserts/intermissions in full/regular animation
ONE-SIDE OF THE DIFFERENT
SHOTS OR POINTS OF VIEW
(each new shot should ideally
bring about something newer
about the model)

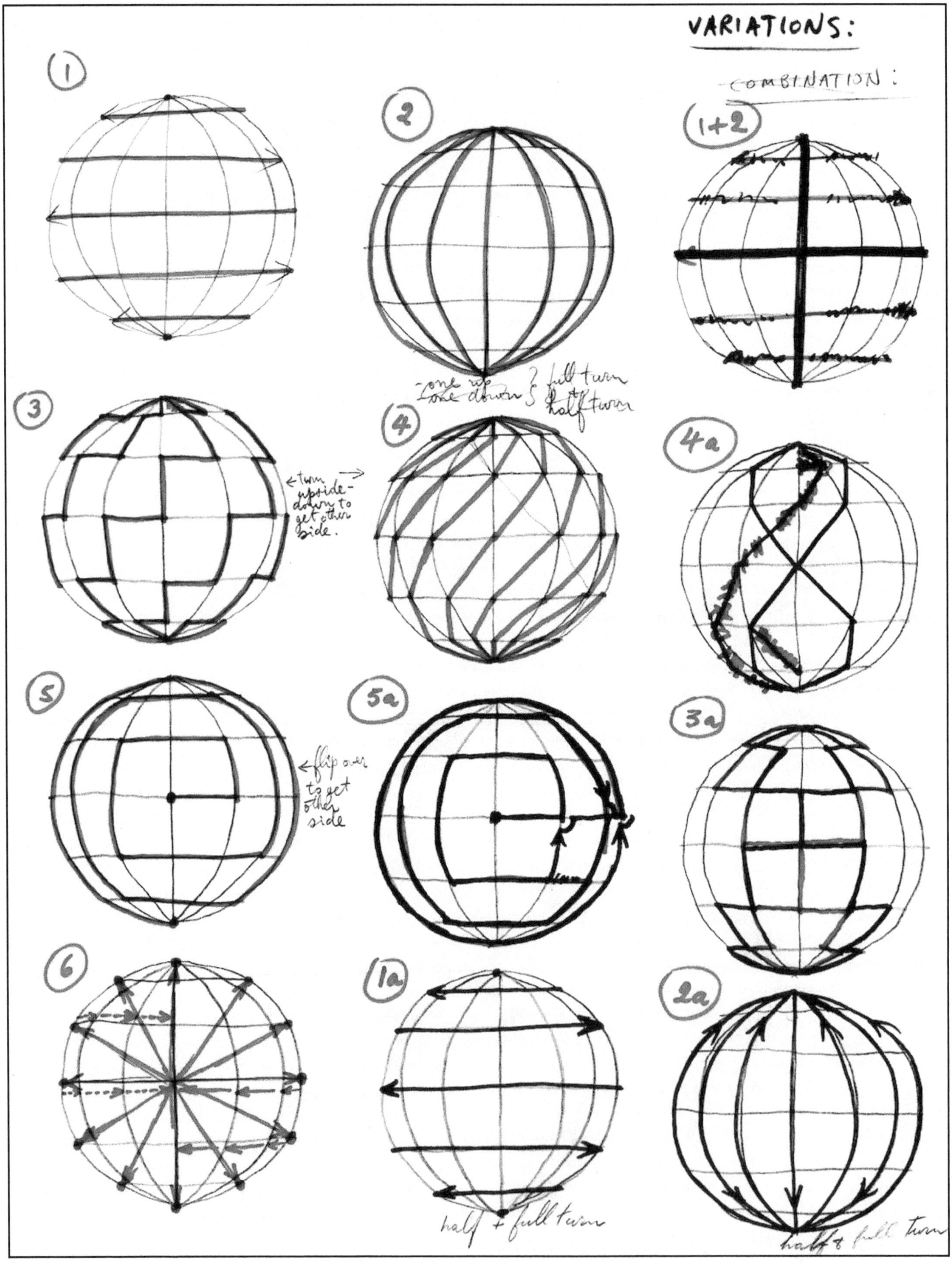
VARIATIONS:
COMBINATION:
1
2
1+2
one up
one down
full turn
half turn
3
turn upside-down to get other side.
4
4a
5
flip over to get other side
5a
3a
6
1a
half + full turn
2a
half & full turn

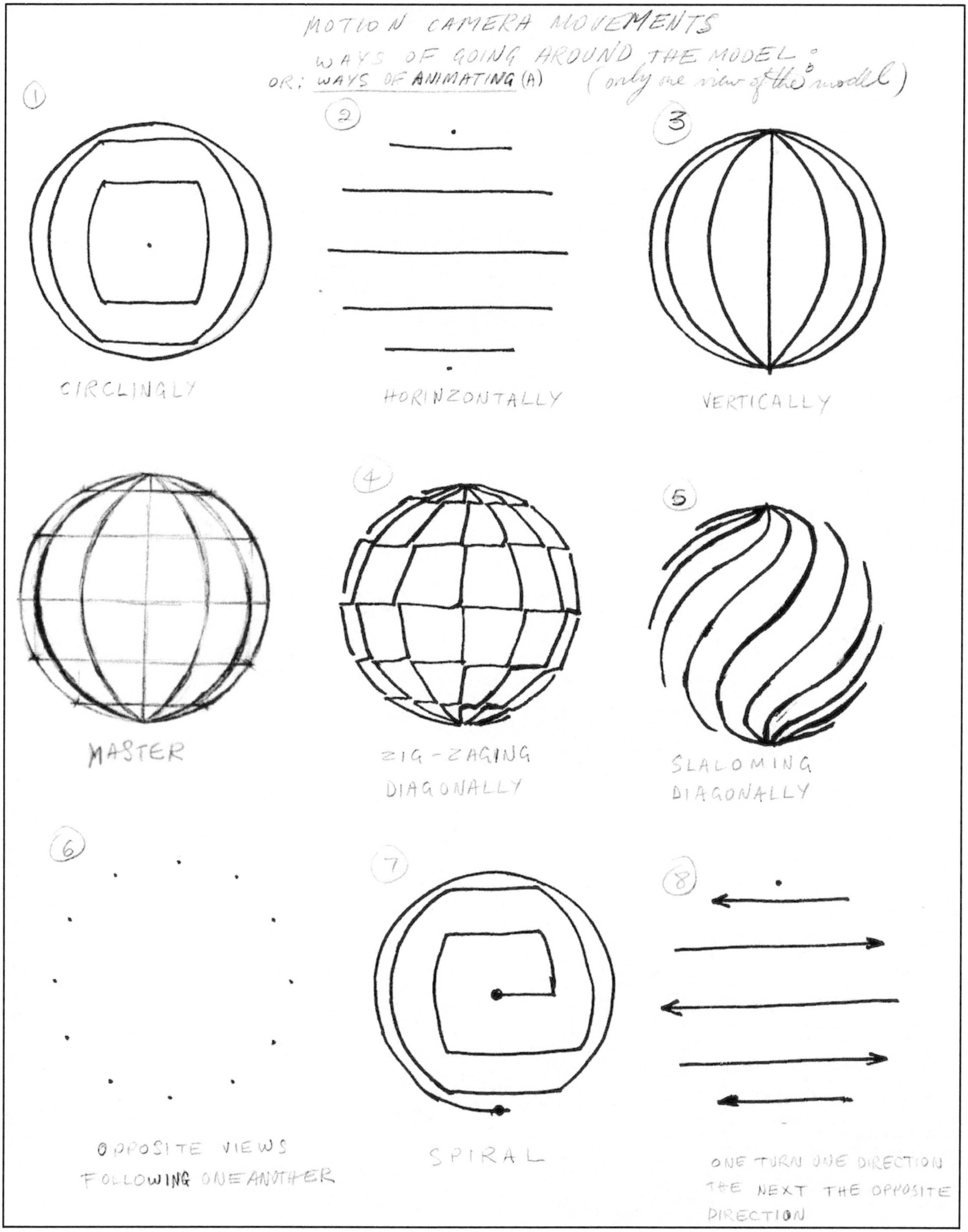
MOTION CAMERA MOVEMENTS
WAYS OF GOING AROUND THE MODEL:
OR: WAYS OF ANIMATING (A)
(only one view of the model)
1
CIRCLINGLY
2
HORINZONTALLY
3
VERTICALLY
MASTER
4
ZIG-ZAGING
DIAGONALLY
5
SLALOMING
DIAGONALLY
6
OPPOSITE VIEWS
FOLLOWING ONE ANOTHER
7
SPIRAL
8
ONE TURN ONE DIRECTION
THE NEXT THE OPPOSITE
DIRECTION

MOTION CAMERA   WAYS OF ANIMATING (8)

SRAIGHT CUT

+/OR

↓

SUPERIMPOSITION

As a transition, super-imposition with every ~~differ~~ picture from a different point of view. Say, two superimposed + two not superimposed. Or better said, 2 clear + 2 overlapping pictures

*

Animating a turn + then cutting + back to animate it again + again. As in the film "Feu Follet" (about the guy who wanted to commit suicide).

*

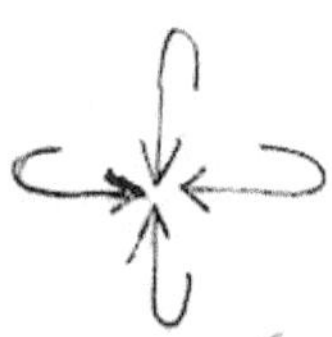

Same principle of animation as explained above, only this time cutting + animating + turning from the opposite direction.

*

Building the object by strips going around it. This is done by cropping on the long shot negative figure

A different filter with every other picture. Then 3 differently-colored filters.

One right, one left, one up, one down

ta! ta! ta! ta! ...

With use of the 3-turret (3 lenses) BOLEX → one w/a lens shot, one normal lens shot, one telephoto lens shot + so on + so forth.

Using a 12-sided parallelogram + with slides in every window. Filming is done as it turns under the movie camera. So to give a feeling 3-D or the feelings of different points of views.

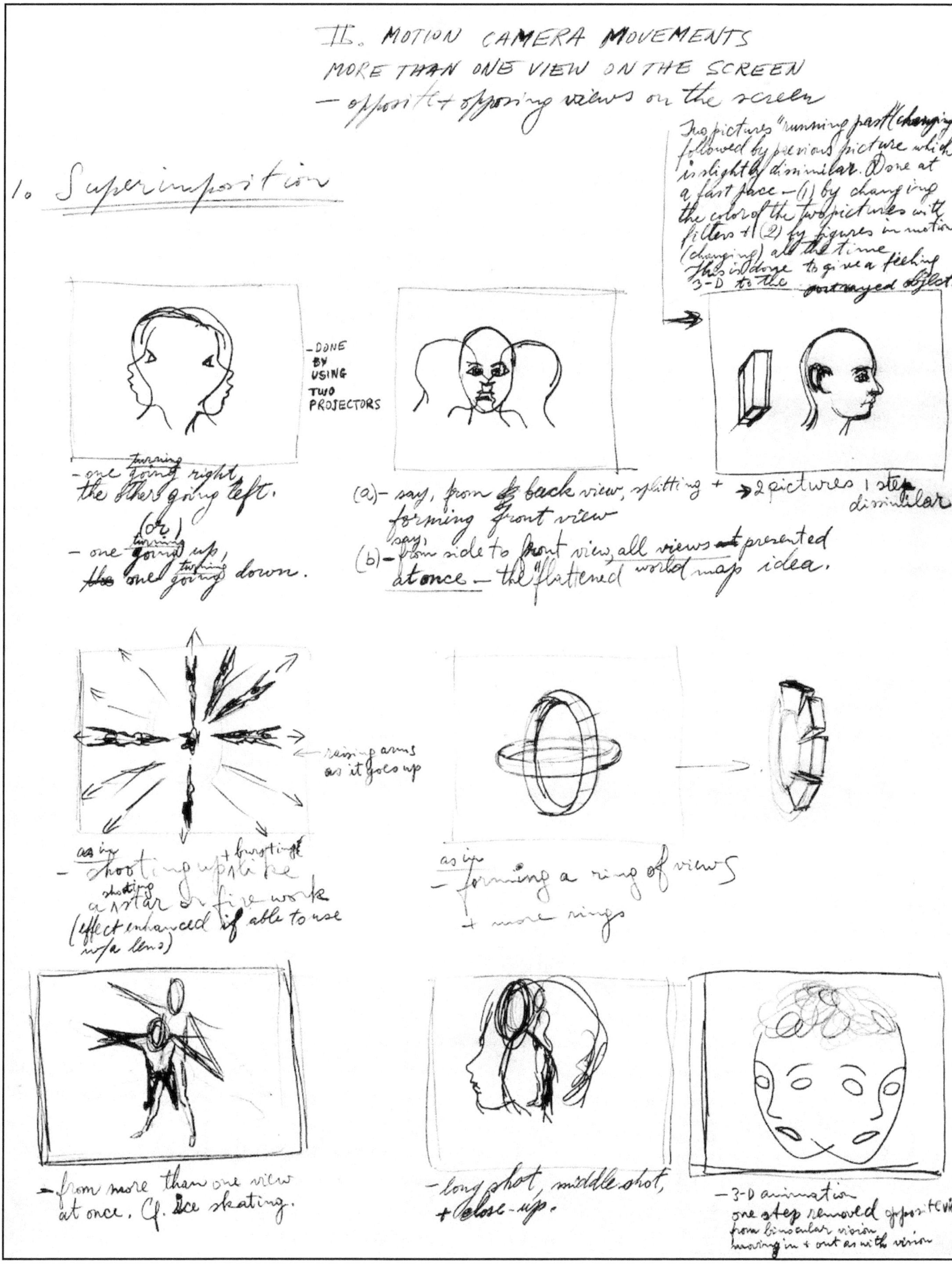
II. MOTION CAMERA MOVEMENTS
MORE THAN ONE VIEW ON THE SCREEN
— opposite + opposing views on the screen
1. Superimposition
Two pictures "running past (changing)" followed by previous picture which is slightly dissimilar. Done at a fast pace — (1) by changing the color of the two pictures with filters + (2) by figures in motion (changing) all the time. This is done to give a feeling 3-D to the portrayed object.
—DONE BY USING TWO PROJECTORS
— one turning right, the other going left.
(or)
— one turning up, the one turning down.
(a) — say, from back view, splitting + forming front view
(b) — say, from side to front view, all views presented at once — the "flattened world map idea.
→ 2 pictures 1 step dissimilar
raising arms as it goes up
— as in shooting up like + bursting of a star or fire work (effect enhanced if able to use w/a lens)
— as in forming a ring of views + more rings
— from more than one view at once. Cf. Ice skating.
— long shot, middle shot, + close-up.
— 3-D animation one step removed opposite view from binocular vision moving in + out as with vision

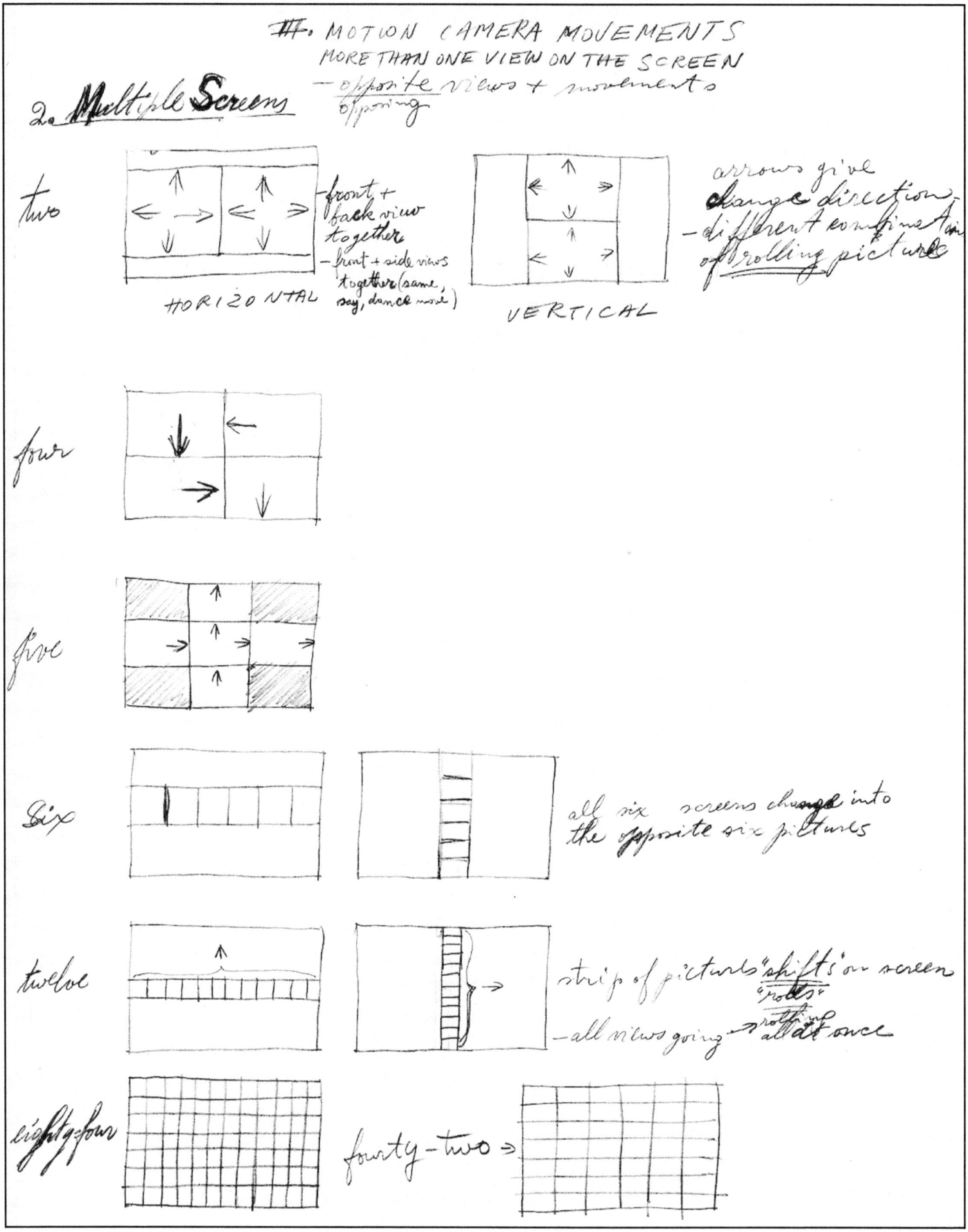
III. MOTION CAMERA MOVEMENTS
MORE THAN ONE VIEW ON THE SCREEN
– opposite views + movements
opposing
2. Multiple Screens
two
front + back view together
– front + side views together (same, say, dance move)
HORIZONTAL
VERTICAL
arrows give change direction –
– different confine of rolling picture
four
five
Six
all six screens change into the opposite six pictures
twelve
strip of pictures "shifts" on screen
"rolls"
– all views going → rotating all at once
eighty-four
fourty-two →

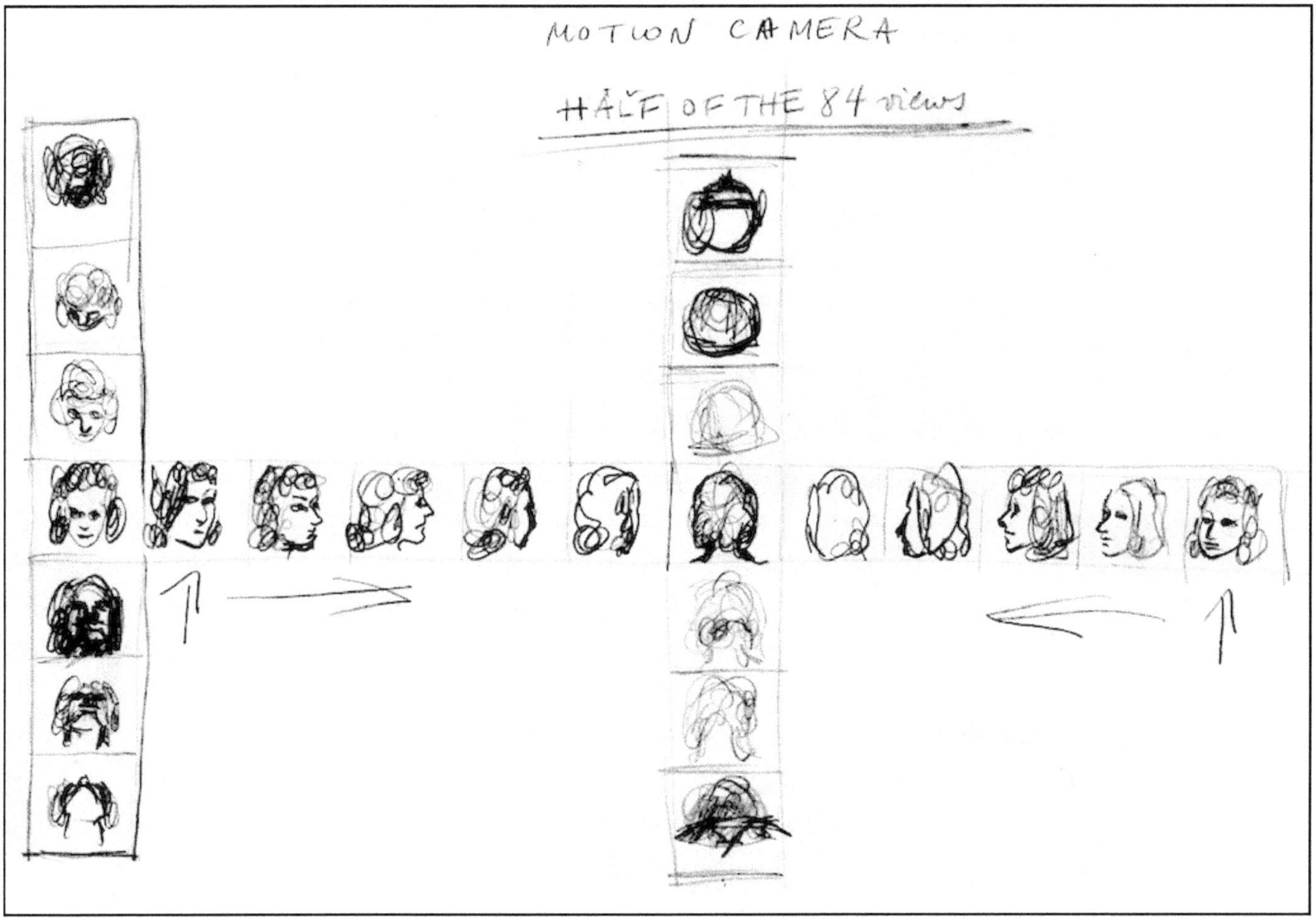
MOTION CAMERA
HALF OF THE 84 views

from side (view) going around to front
+ then circling above to back
+ going below —
zooming in + show buttocks movement in walk
THE WALK
with every 2-3 frames
showing the other-side's view

MOTION CAMERA
a "spread-out" view of one side
(all views from one side)
follows by spread-out views of other side (back).

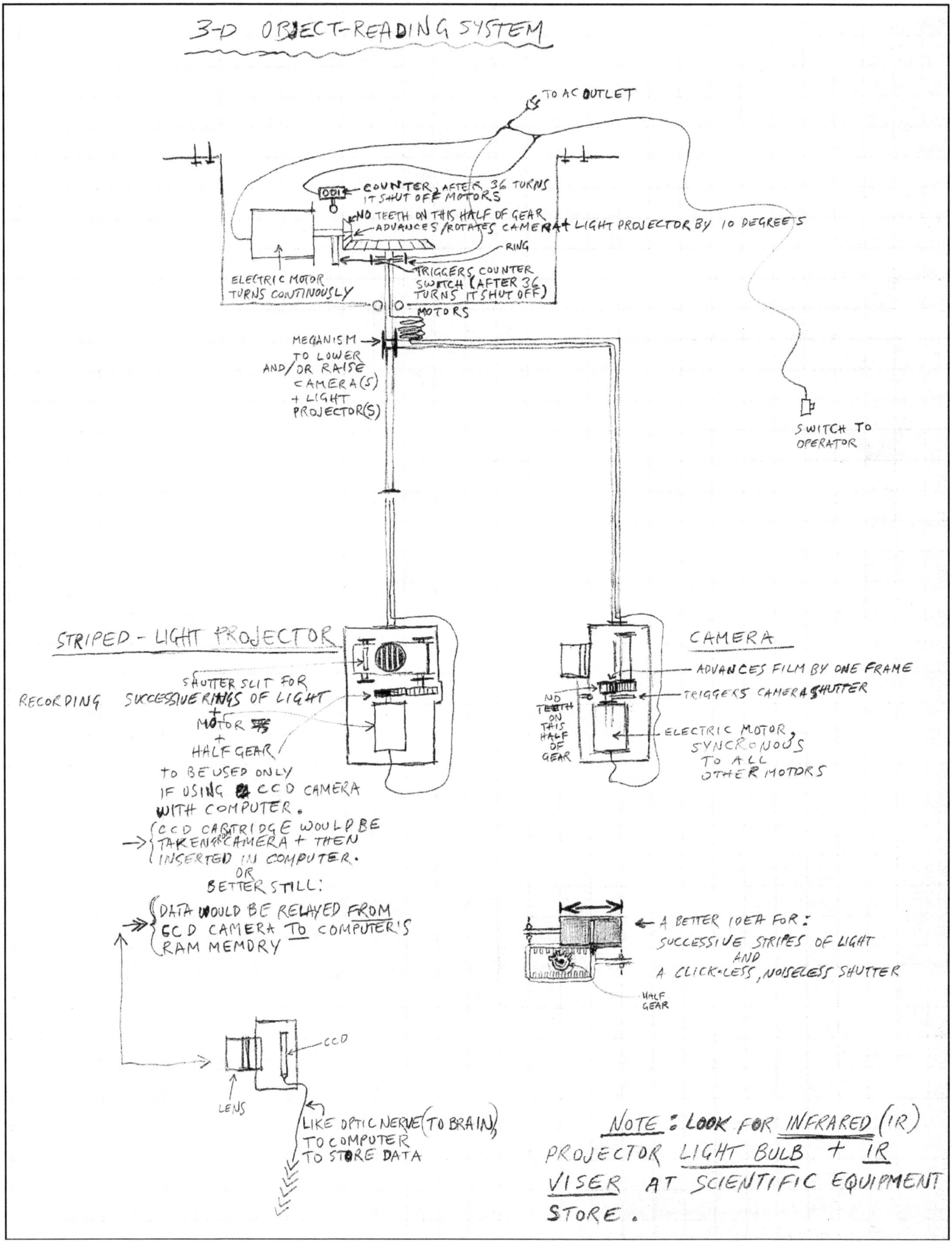

3-D OBJECT-READING SYSTEM
TO AC OUTLET
COUNTER, AFTER 36 TURNS IT SHUT OFF MOTORS
NO TEETH ON THIS HALF OF GEAR
ADVANCES/ROTATES CAMERA + LIGHT PROJECTOR BY 10 DEGREES
RING
TRIGGERS COUNTER SWITCH (AFTER 36 TURNS IT SHUT OFF) MOTORS
ELECTRIC MOTOR TURNS CONTINOUSLY
MECANISM TO LOWER AND/OR RAISE CAMERA(S) + LIGHT PROJECTOR(S)
SWITCH TO OPERATOR
STRIPED-LIGHT PROJECTOR
SHUTTER SLIT FOR RECORDING SUCCESSIVE RINGS OF LIGHT + MOTOR + HALF GEAR
TO BE USED ONLY IF USING CCD CAMERA WITH COMPUTER.
CCD CARTRIDGE WOULD BE TAKEN FROM CAMERA + THEN INSERTED IN COMPUTER.
OR
BETTER STILL:
DATA WOULD BE RELAYED FROM CCD CAMERA TO COMPUTER'S RAM MEMORY
CAMERA
ADVANCES FILM BY ONE FRAME
TRIGGERS CAMERA SHUTTER
NO TEETH ON THIS HALF OF GEAR
ELECTRIC MOTOR, SYNCRONOUS TO ALL OTHER MOTORS
A BETTER IDEA FOR: SUCCESSIVE STRIPES OF LIGHT AND A CLICK-LESS, NOISELESS SHUTTER
HALF GEAR
CCD
LENS
LIKE OPTIC NERVE (TO BRAIN), TO COMPUTER TO STORE DATA
NOTE: LOOK FOR INFRARED (IR) PROJECTOR LIGHT BULB + IR VISER AT SCIENTIFIC EQUIPMENT STORE.

OTHER PROBLEMS TO TACKLE

SET
– pedestal construction (turntable) →

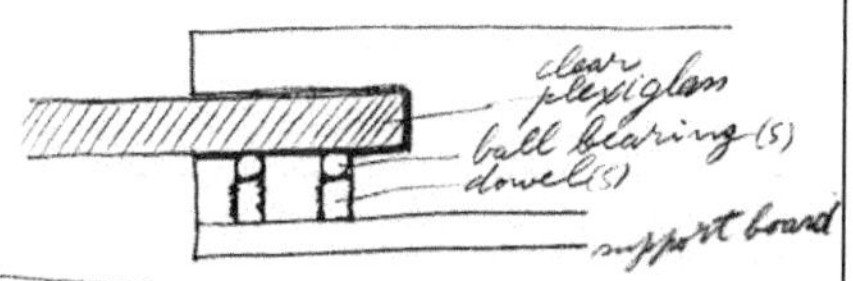

– registration marks on black background seemless paper →

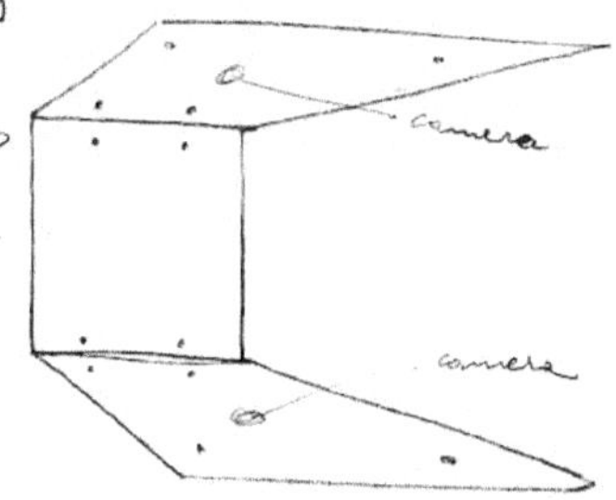

FILMING
on the Oxberry stand →

– budget

– models' "appearance"
– by herself or in group
– dressed (skirt + worn-out jeans) ballet dancer suit, +/or nude

– personnel (crew)
– models
– photographers

WAYS OF SHOOTING

– Double exposure of 2 or "blended" poses in possible views + then animated. As on p. 23 in "Photolighting Techniques"

round screen instead of rectangular screen

which is for panoramic/horizontal "follow-action" takes.

WAYS OF LIGHTING

— A graphic representation: like a drawing: simplified representation

— One light source. This will give the sensation that the model is turning as of the light source coming from one direction only.

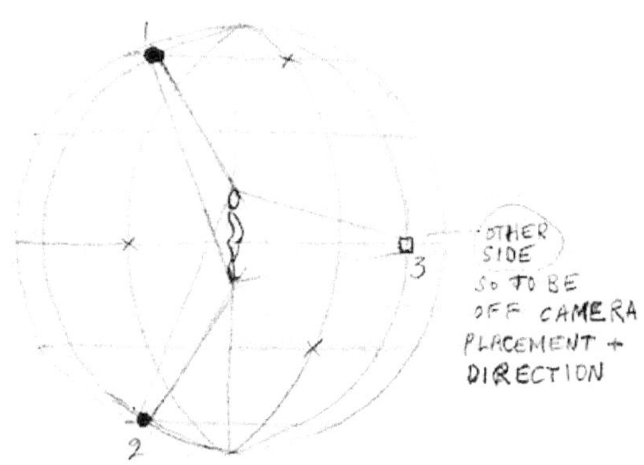

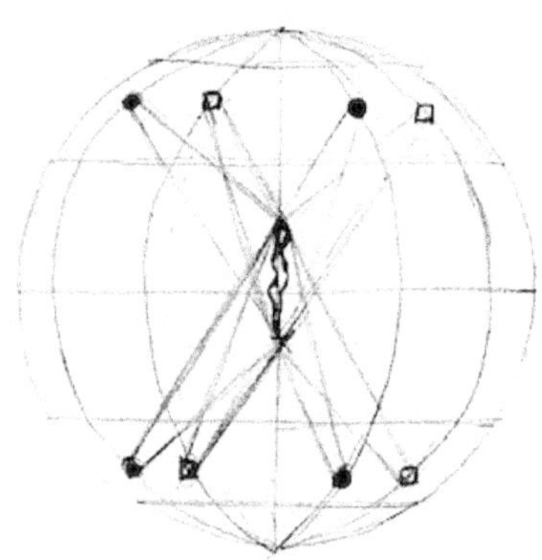

— Three lights. ~~lighting~~
OR: 6 lights. (Positions)
(X)

— Eight lights.
NOTE: Square lights coming from other side of globe.

MOTION CAMERA

POSES (A series of interrelated poses — for consistancy/unity + economy — for all or quasi all visual effects to be created.)

ONE STILL

- standing with weight on one leg
- standing with left knee + elbow bent at right angle
- standing straight (also used for "dressing" technique — going up around model till she's completely "dressed" or revealed)
- leaning on stand as in sketchbook "Take Me Along" {using clear plexiglass supports}
- sitting as in Playboy centerfold on wall.
- close-ups of face, breasts, buttocks, hands, + feet.

- ice skating poses
- ballet dancing poses.

A SERIES OF STILLS

FROM SAME POINTS OF VIEW + THEN TURNING A STOP.

- dance steps (to be animated to the sound of music — probably different pieces of music & yet using the same steps)
- gymnastics movements (for all possible movements of the body + to also have model face points of view as well)
- walking phases

EVERY TIME FROM A DIFFERENT POINT OF VIEW. (moving at every shot)

- walking phases (one turn of the motion camera with every change/phase) (2) full walking cycles from every point of view.
- gymnastics movements (changing one point of view with one change of the figure)

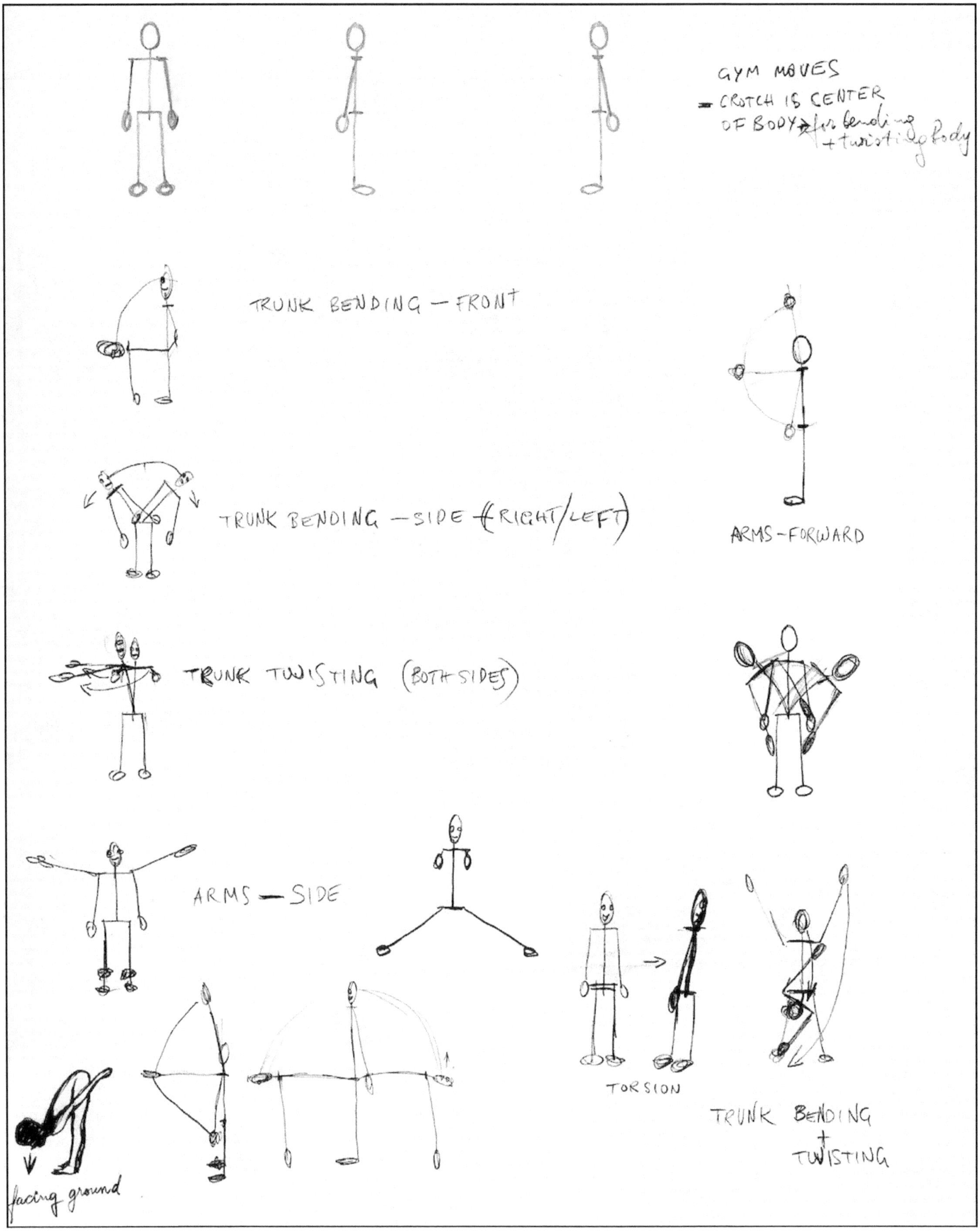
GYM MOVES
- CROTCH IS CENTER OF BODY → for bending + twisting body
TRUNK BENDING — FRONT
ARMS-FORWARD
TRUNK BENDING — SIDE (RIGHT/LEFT)
TRUNK TWISTING (BOTH SIDES)
ARMS — SIDE
TORSION
TRUNK BENDING + TWISTING
facing ground

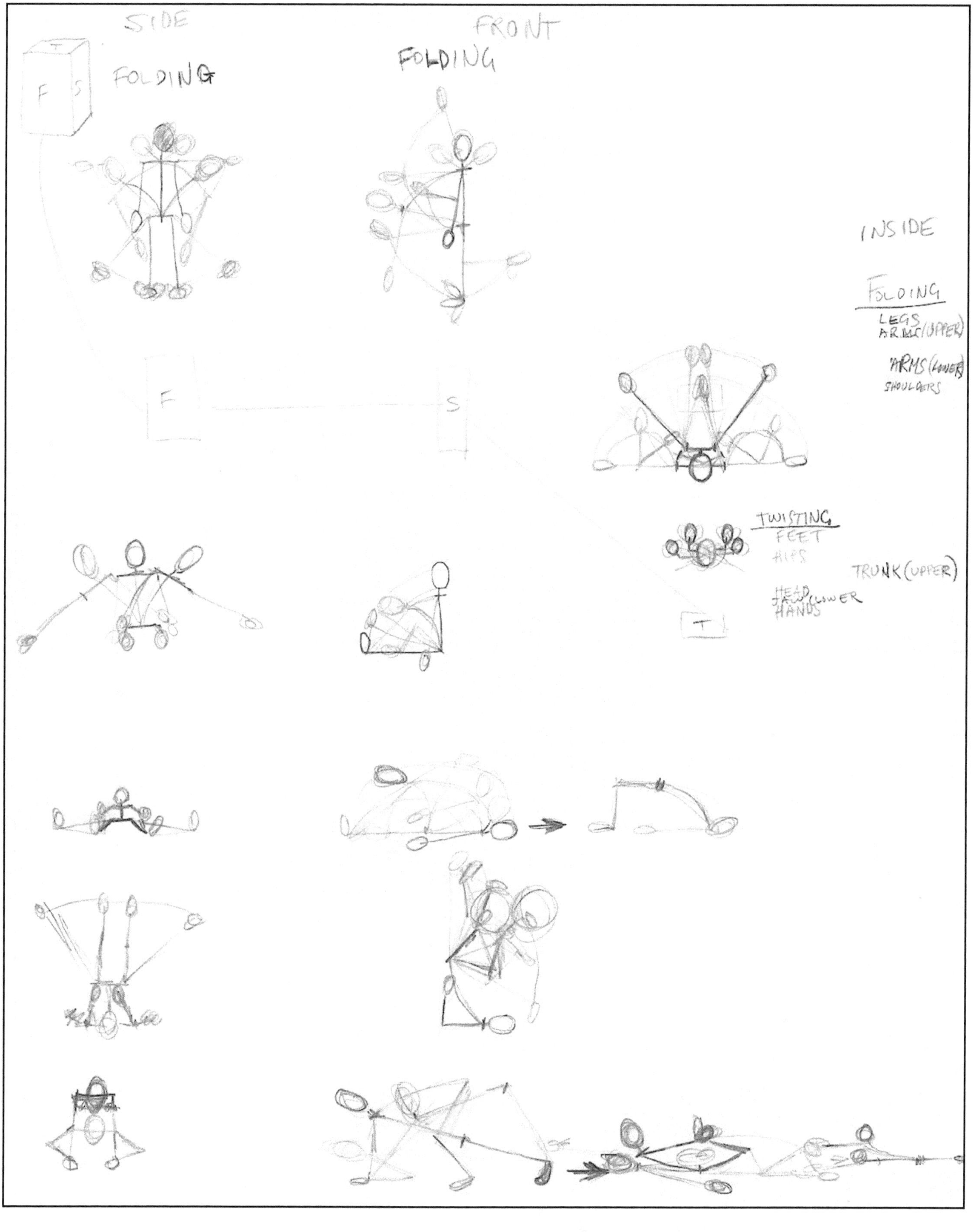

9B. PORTRAIT CAMERA (READER)

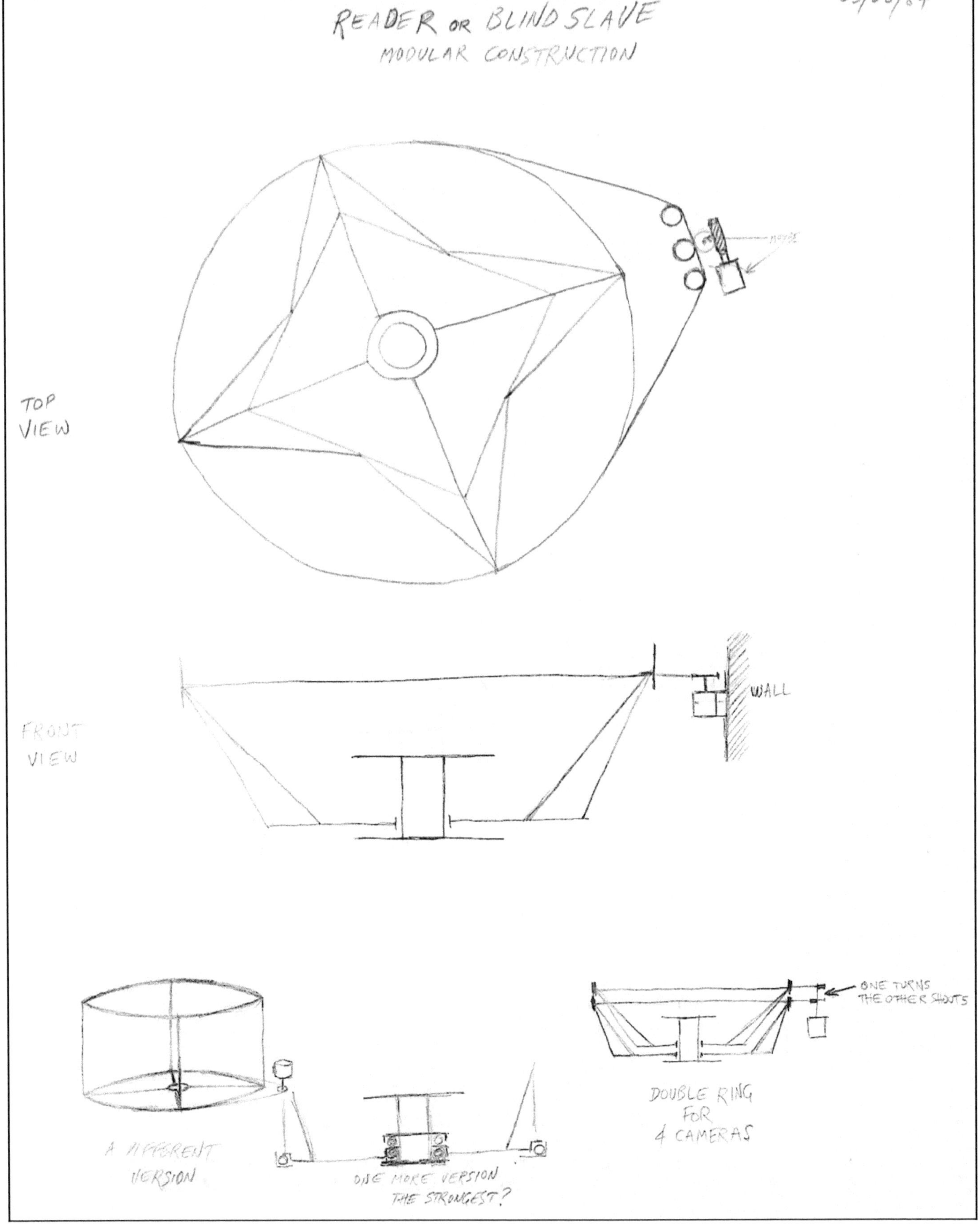
05/08/84
READER OR BLIND SLAVE
MODULAR CONSTRUCTION
TOP VIEW
MAYBE
FRONT VIEW
WALL
ONE TURNS
THE OTHER SHOUTS
DOUBLE RING
FOR
4 CAMERAS
A DIFFERENT
VERSION
ONE MORE VERSION
THE STRONGEST?

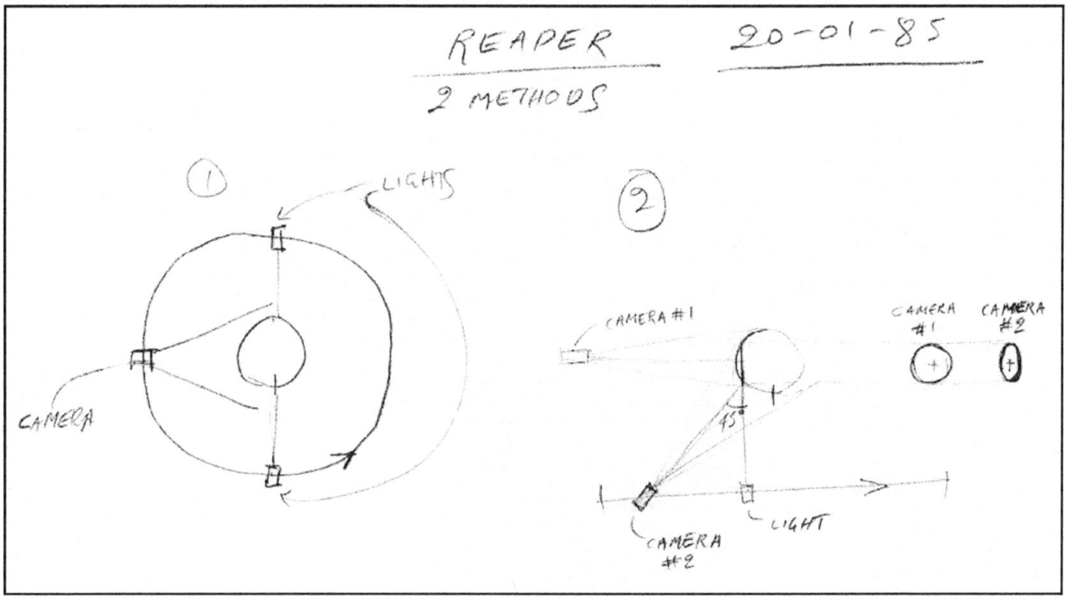
REAPER
20-01-85
2 METHODS
1
LIGHTS
CAMERA
2
CAMERA #1
CAMERA #1
CAMERA #2
45°
CAMERA #2
LIGHT

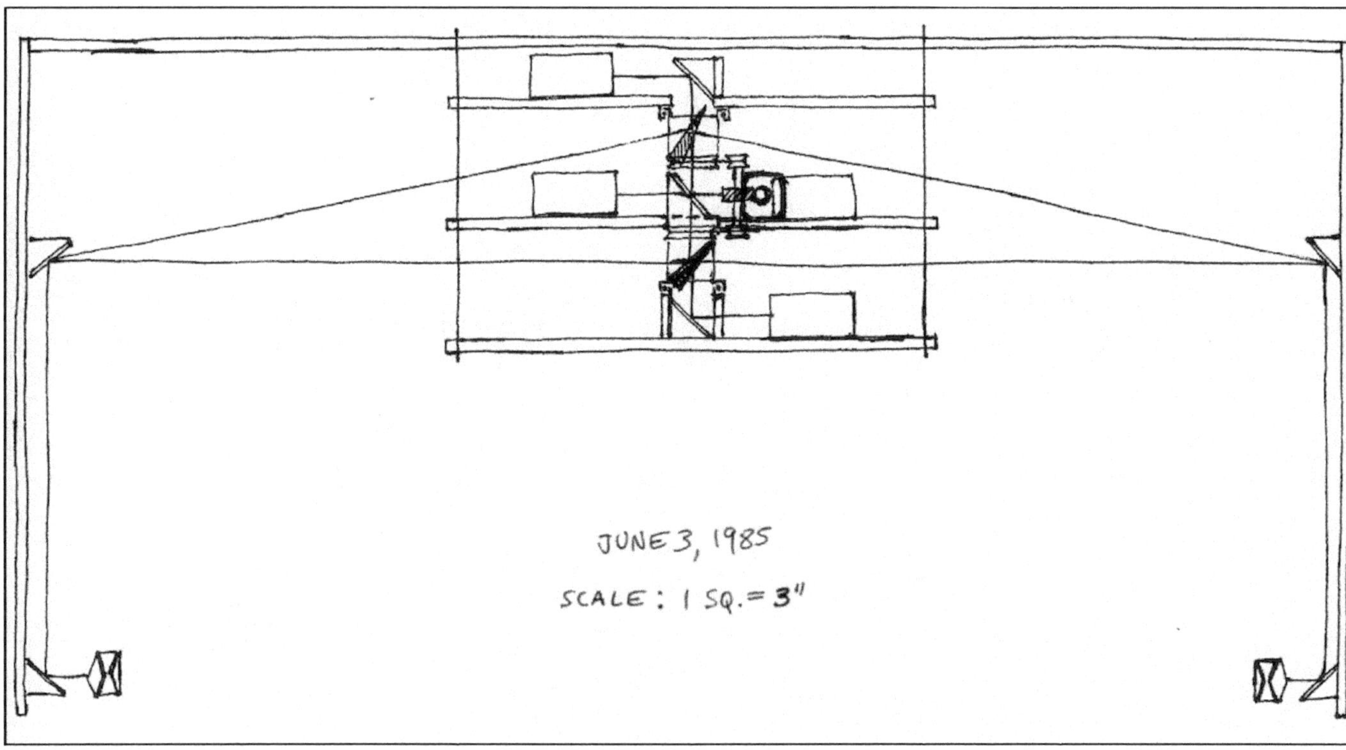
JUNE 3, 1985
SCALE: 1 SQ. = 3"

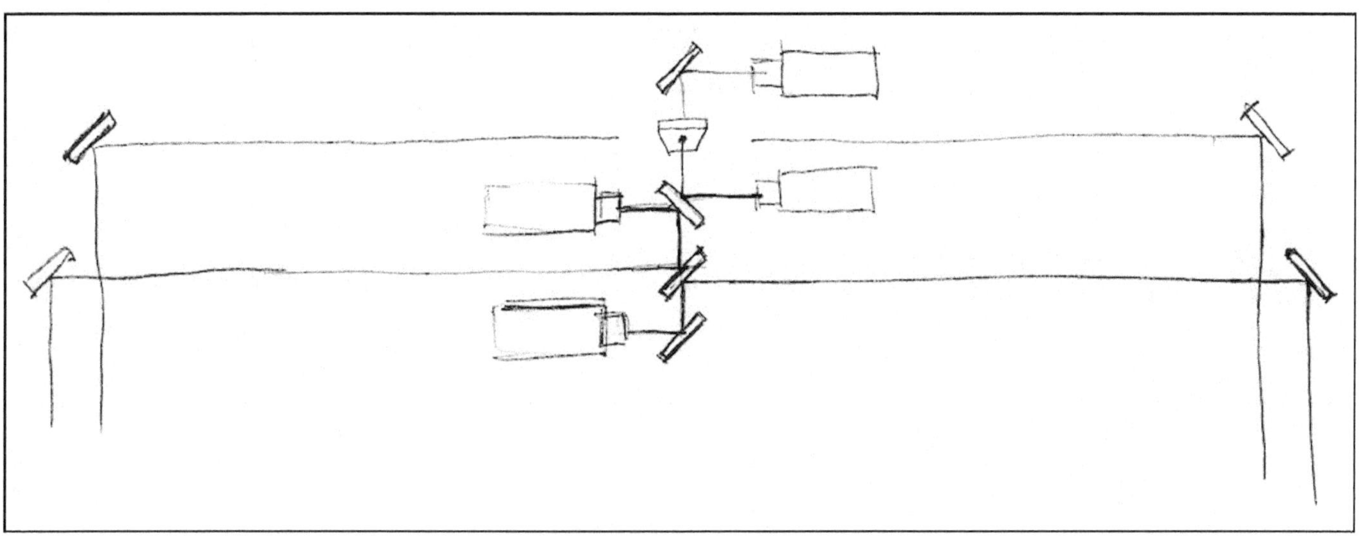

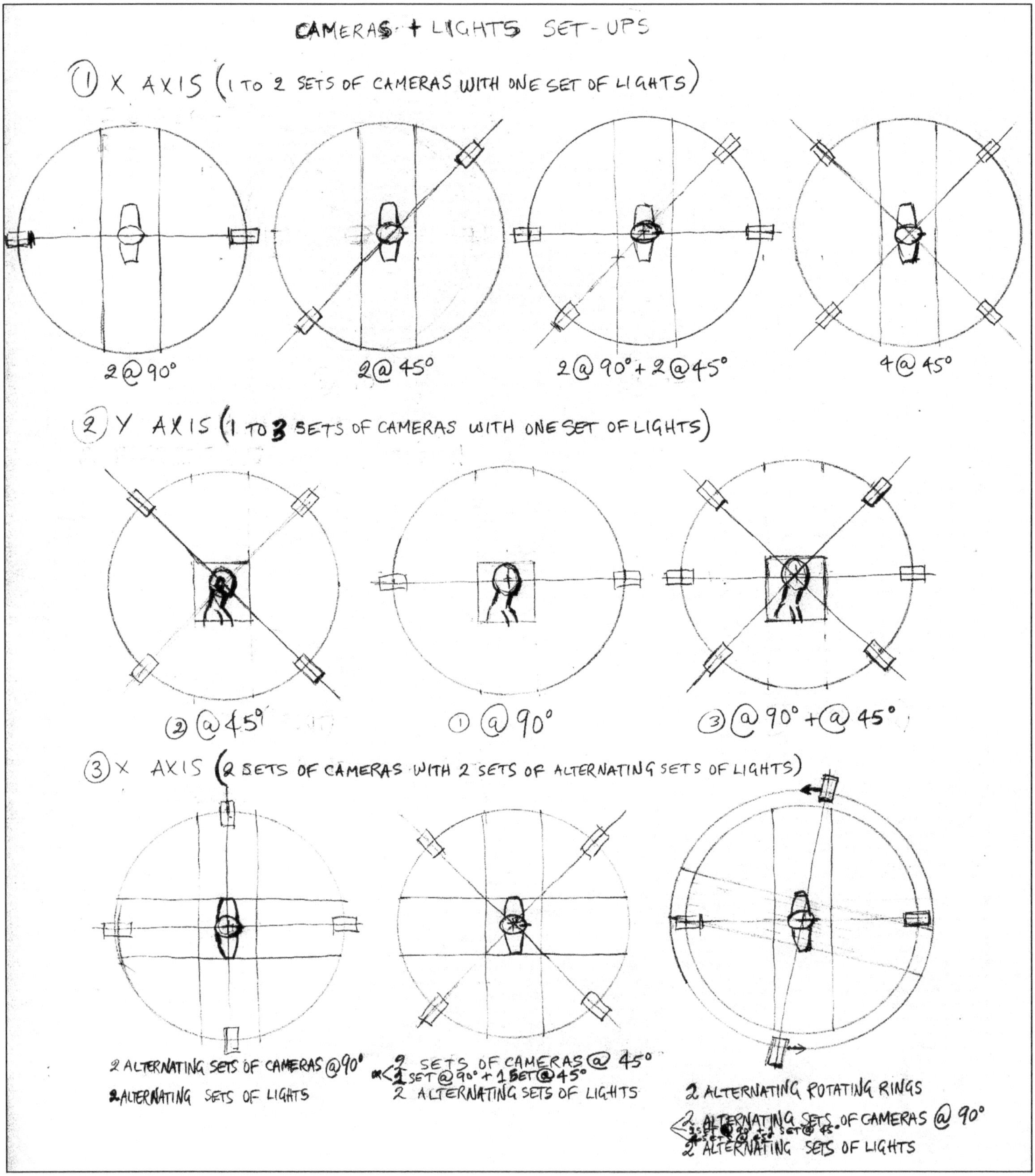
CAMERAS + LIGHTS SET-UPS
1 X AXIS (1 TO 2 SETS OF CAMERAS WITH ONE SET OF LIGHTS)
2@90°
2@45°
2@90°+2@45°
4@45°
2 Y AXIS (1 TO 3 SETS OF CAMERAS WITH ONE SET OF LIGHTS)
2 @45°
1 @90°
3 @90°+@45°
3 X AXIS (2 SETS OF CAMERAS WITH 2 SETS OF ALTERNATING SETS OF LIGHTS)
2 ALTERNATING SETS OF CAMERAS @90°
2 ALTERNATING SETS OF LIGHTS
2 SETS OF CAMERAS @ 45°
1 SET @ 90° + 1 SET @ 45°
2 ALTERNATING SETS OF LIGHTS
2 ALTERNATING ROTATING RINGS
2 ALTERNATING SETS OF CAMERAS @ 90°
2 ALTERNATING SETS OF LIGHTS

CAMERAS SET-UPS

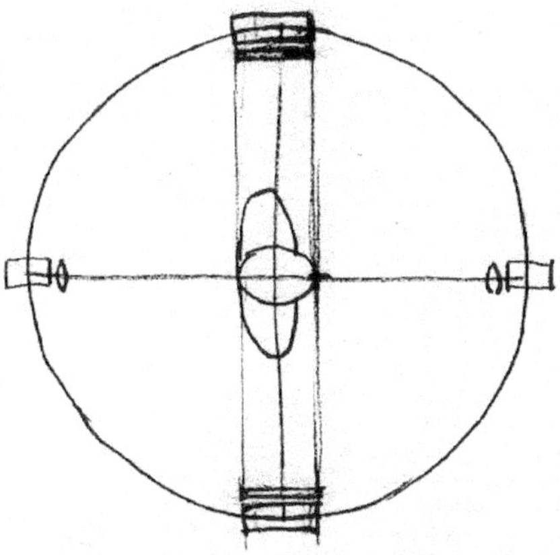

2 CAMERAS @ 90°
1 SET OF LIGHTS
→ 90° SHOTS

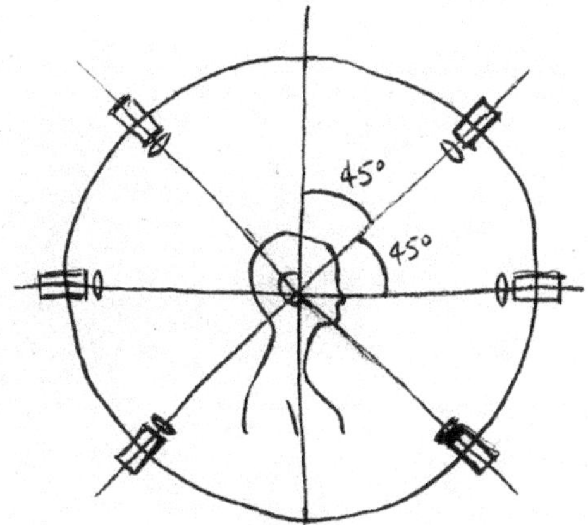

POSSIBLE SET-UPS OF CAMERAS
SIDE VIEW
2, 4, 6, 8, OR 12 CAMERAS
→ 90° OR 45° SHOTS

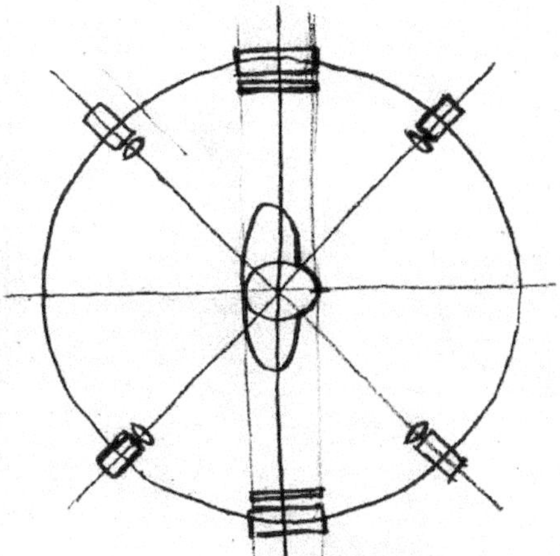

4 CAMERAS
1 OR 2 SETS OF LIGHTS
→ 45° SHOTS

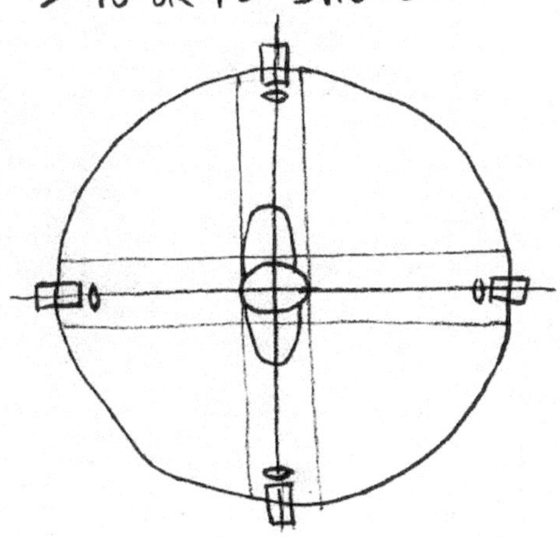

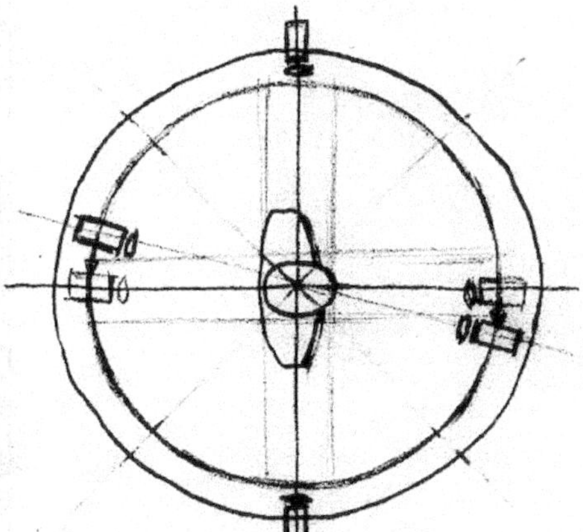

2 RINGS
4 CAMERAS
2 SETS OF LIGHTS

READER (JUNE 10, 1985)

LIGHTING FROM 5 DIFFERENT DIRECTIONS
WITH 5 DIFFERENT COLORING FILTERS;
1 OR 2 CAMERAS

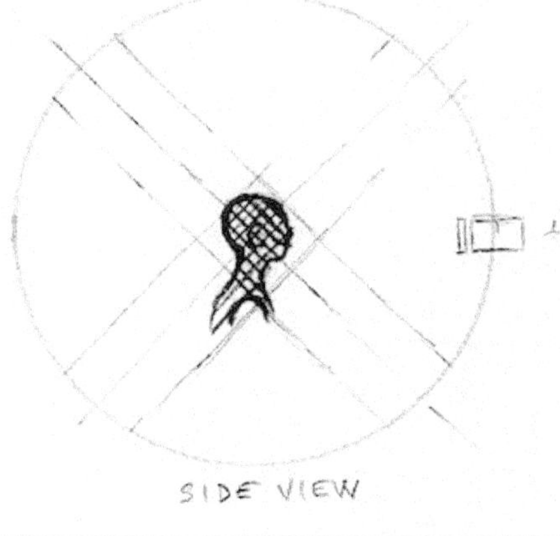

SIDE VIEW

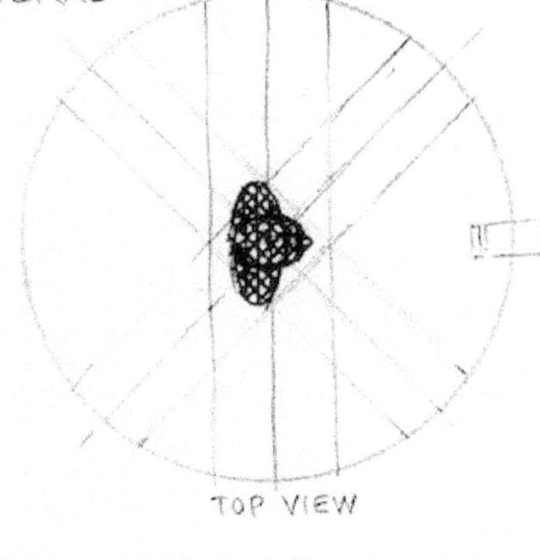

TOP VIEW

NEW IDEAS RE. READER: MAY 9, 1986

(1) TO TURN MIRRORS FOR NEXT FRAME, END OF FIELD/FRAME WOULD BE PICKED UP FROM TV SCREEN WITH LIGHT PEN.

(2) TO READ FROM EITHER A HIGHER OR LOWER ANGLE, MIRRORS WOULD MOVE UP + DOWN AS A WHOLE UNIT (A RING) ON 4, 6, OR 8 PILLAR-BARS. COUNTERWEIGHTS WOULD REDUCE WEIGHT TO ZERO.

(3) TO ANGLE MIRRORS TOGETHER ALL AT ONCE, THEY WOULD BE LINKED WITH RODS LINKED, IN TURN, BY UNIVERSAL JOINTS.

(1)

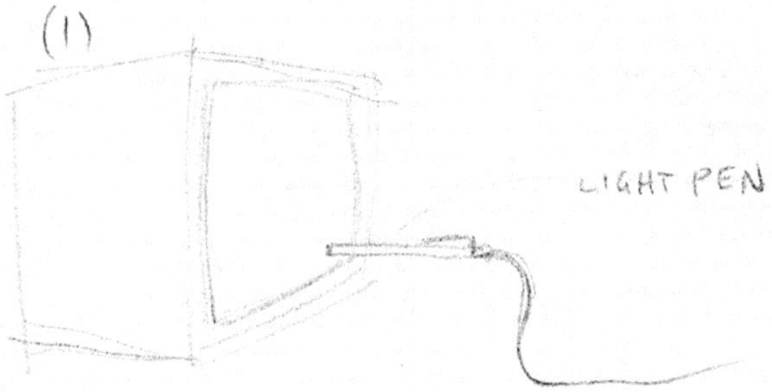

(2)

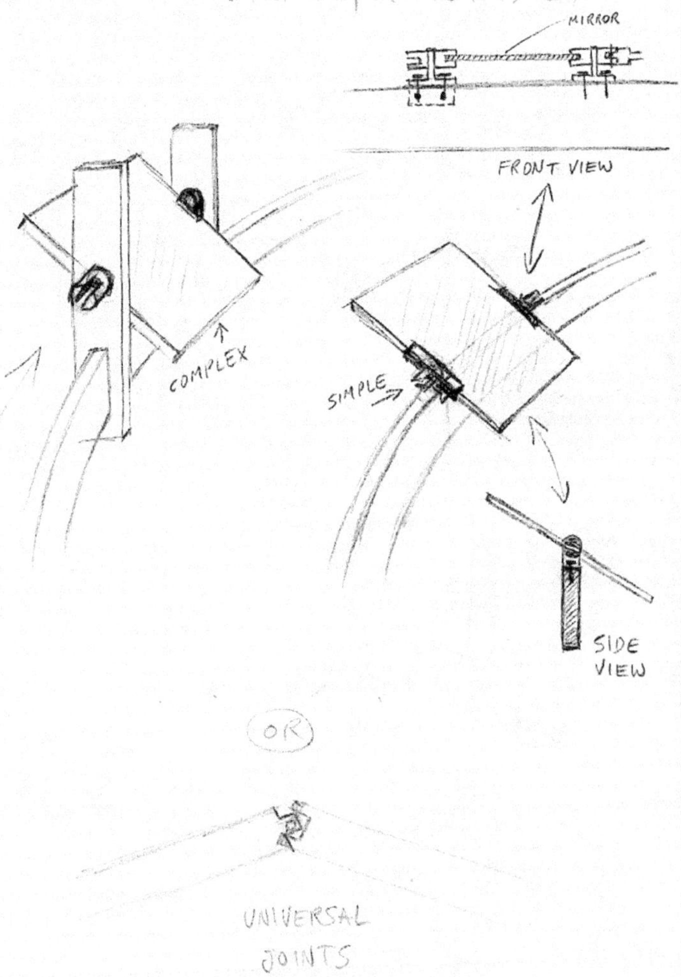

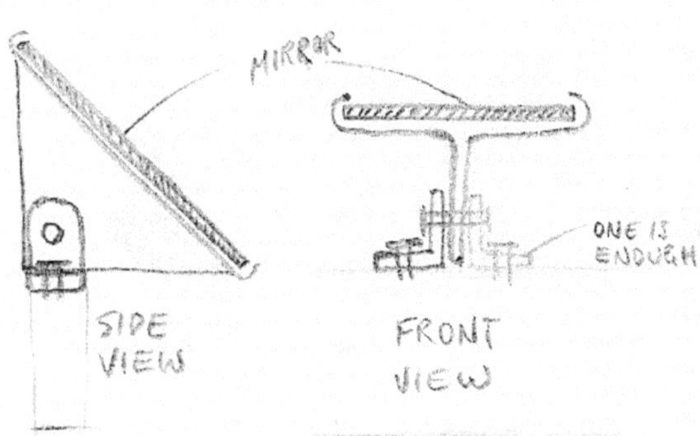

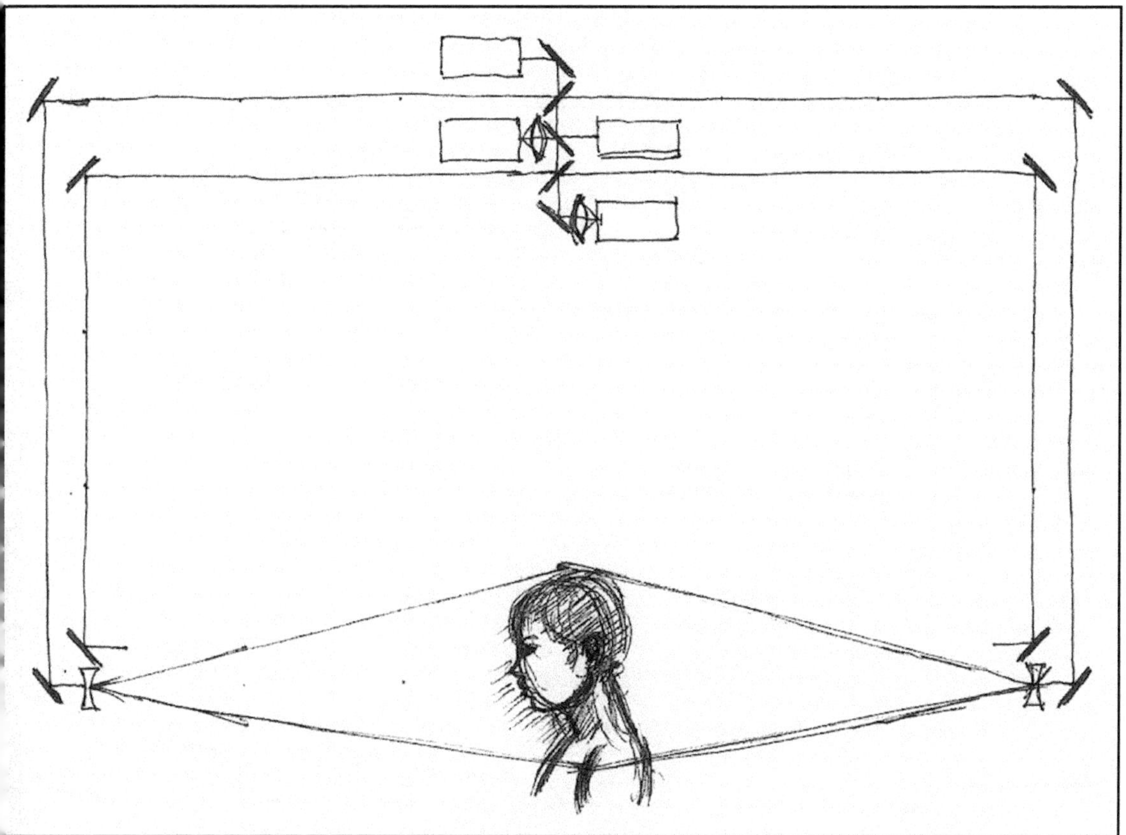

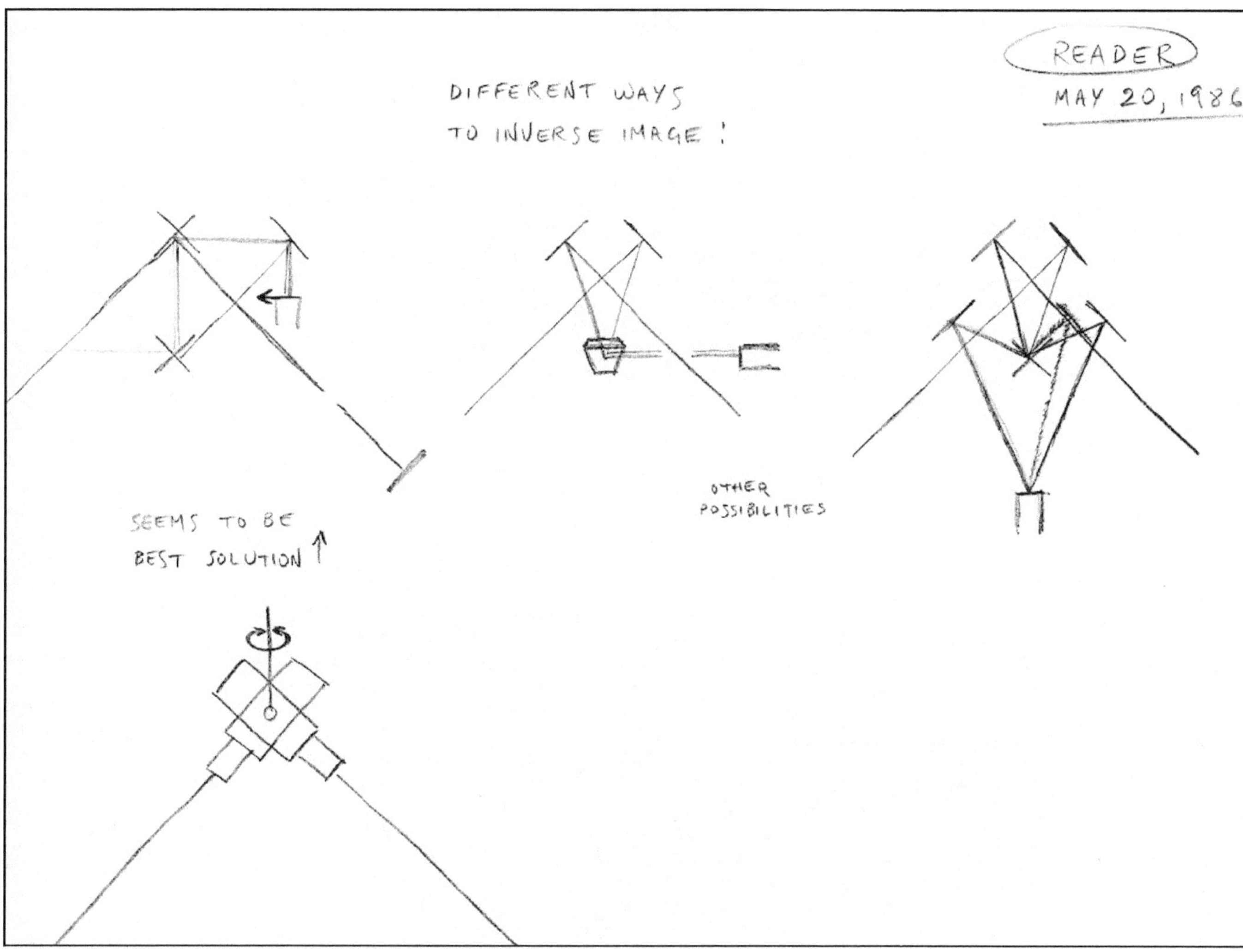
READER
MAY 20, 1986
DIFFERENT WAYS
TO INVERSE IMAGE !
OTHER
POSSIBILITIES
SEEMS TO BE
BEST SOLUTION ↑

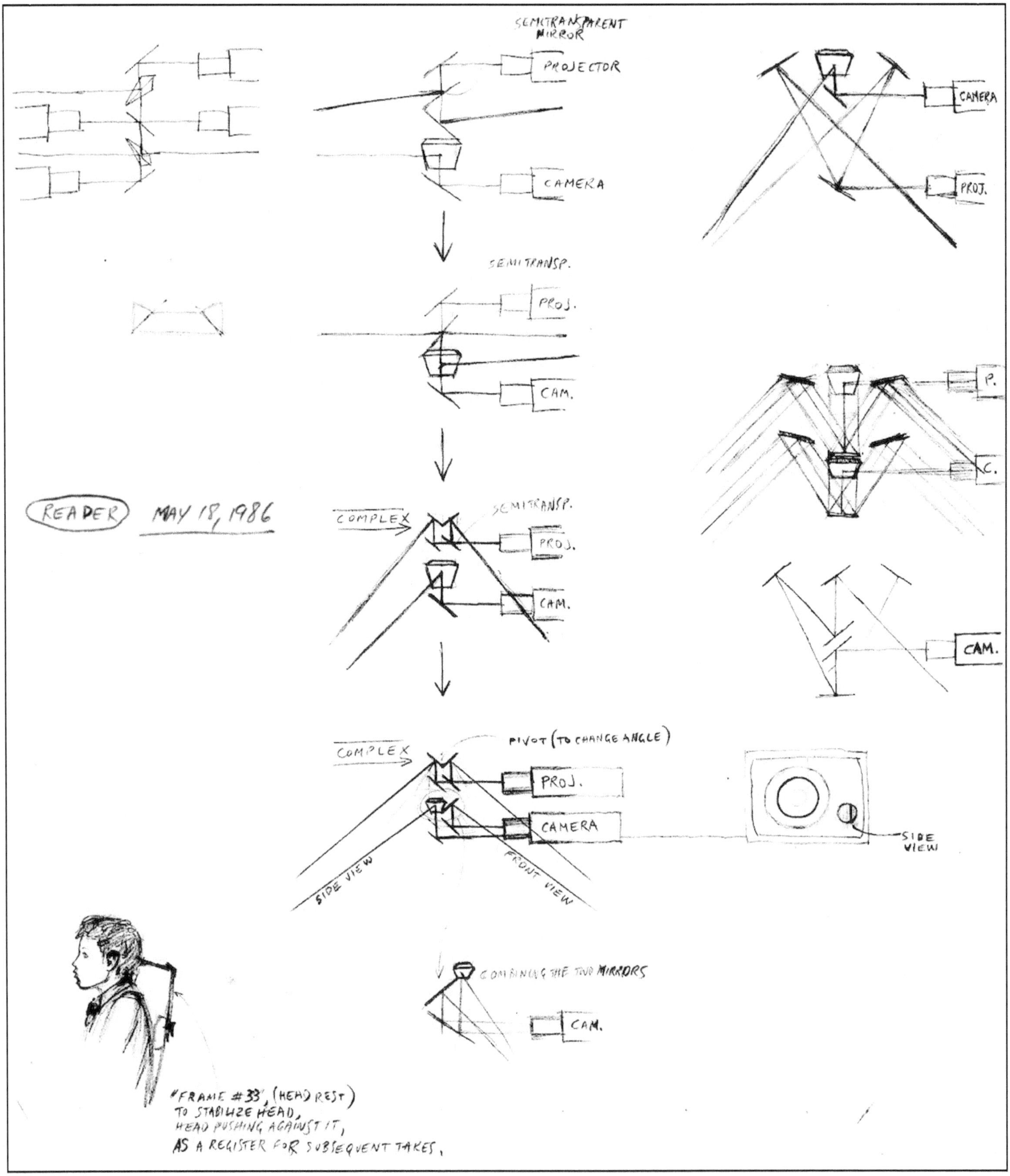
SEMITRANSPARENT MIRROR
PROJECTOR
CAMERA
CAMERA
PROJ.
SEMITRANSP.
PROJ.
CAM.
P.
C.
READER
MAY 18, 1986
COMPLEX
SEMITRANSP.
PROJ.
CAM.
CAM.
COMPLEX
PIVOT (TO CHANGE ANGLE)
PROJ.
CAMERA
SIDE VIEW
SIDE VIEW
FRONT VIEW
COMBINING THE TWO MIRRORS
CAM.
"FRAME #33", (HEAD REST)
TO STABILIZE HEAD,
HEAD PUSHING AGAINST IT,
AS A REGISTER FOR SUBSEQUENT TAKES,

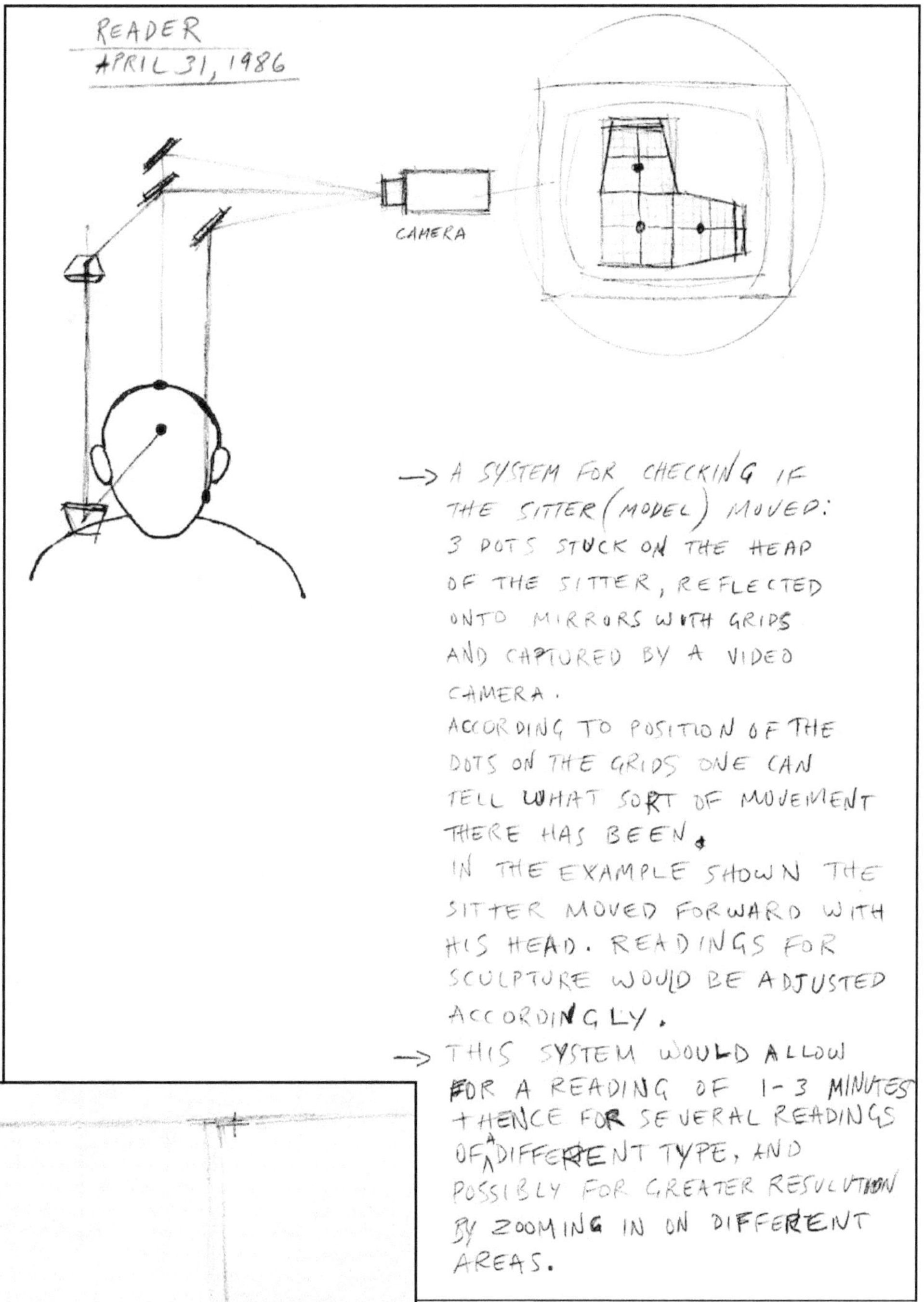
READER
APRIL 31, 1986
CAMERA
→ A SYSTEM FOR CHECKING IF THE SITTER (MODEL) MOVED: 3 DOTS STUCK ON THE HEAD OF THE SITTER, REFLECTED ONTO MIRRORS WITH GRIDS AND CAPTURED BY A VIDEO CAMERA.
ACCORDING TO POSITION OF THE DOTS ON THE GRIDS ONE CAN TELL WHAT SORT OF MOVEMENT THERE HAS BEEN.
IN THE EXAMPLE SHOWN THE SITTER MOVED FORWARD WITH HIS HEAD. READINGS FOR SCULPTURE WOULD BE ADJUSTED ACCORDINGLY.
→ THIS SYSTEM WOULD ALLOW FOR A READING OF 1-3 MINUTES + HENCE FOR SEVERAL READINGS OF A DIFFERENT TYPE, AND POSSIBLY FOR GREATER RESOLUTION BY ZOOMING IN ON DIFFERENT AREAS.

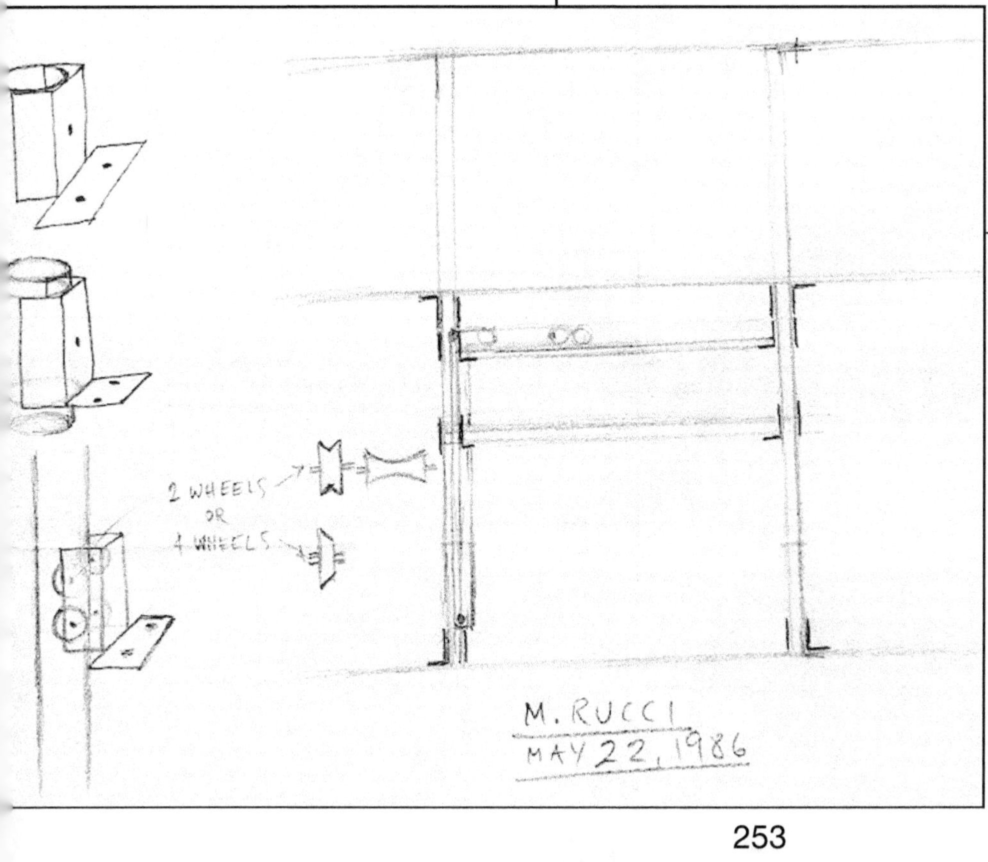
2 WHEELS
OR
4 WHEELS
M. RUCCI
MAY 22, 1986

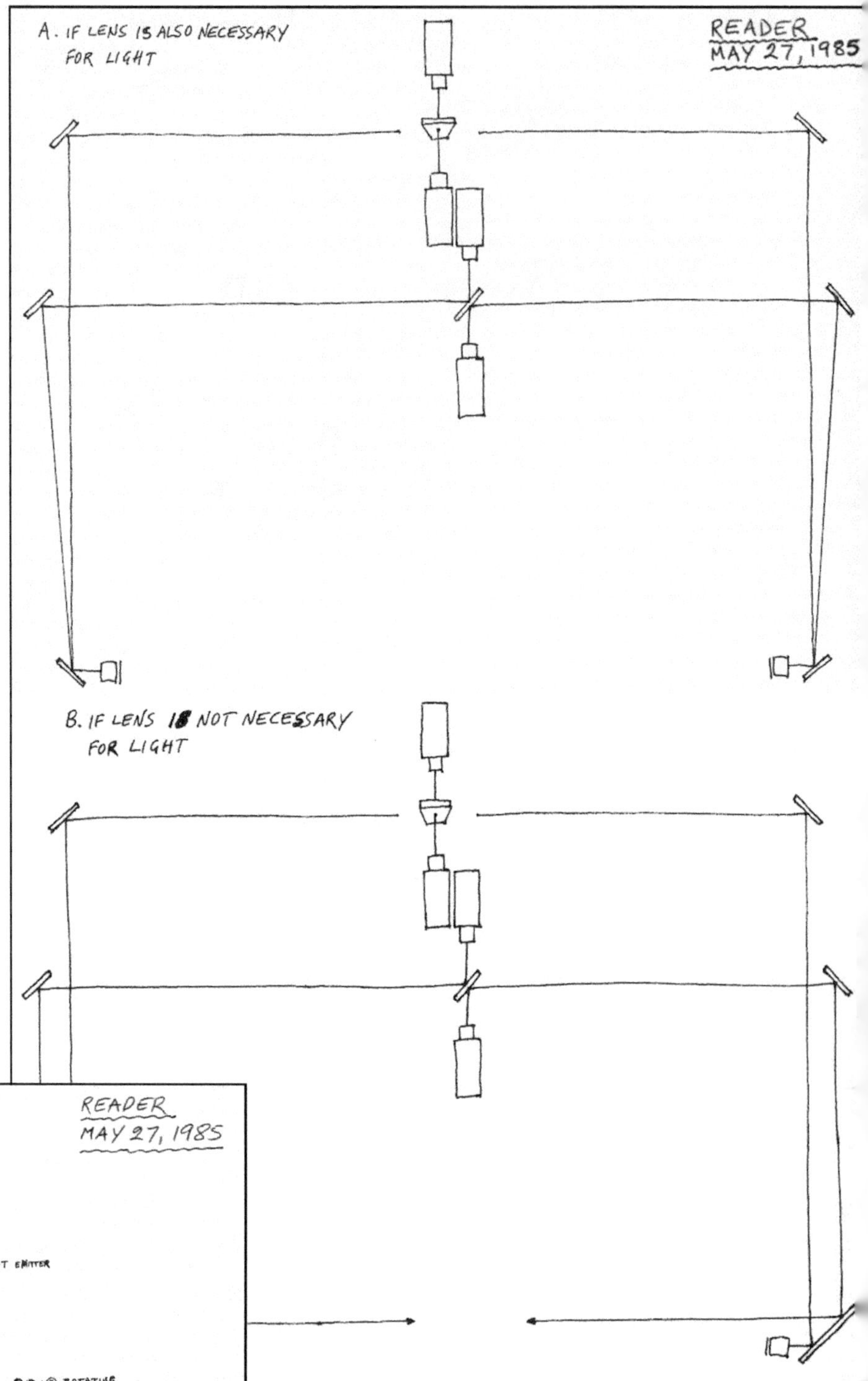
A. IF LENS IS ALSO NECESSARY
FOR LIGHT
READER
MAY 27, 1985
B. IF LENS IS NOT NECESSARY
FOR LIGHT

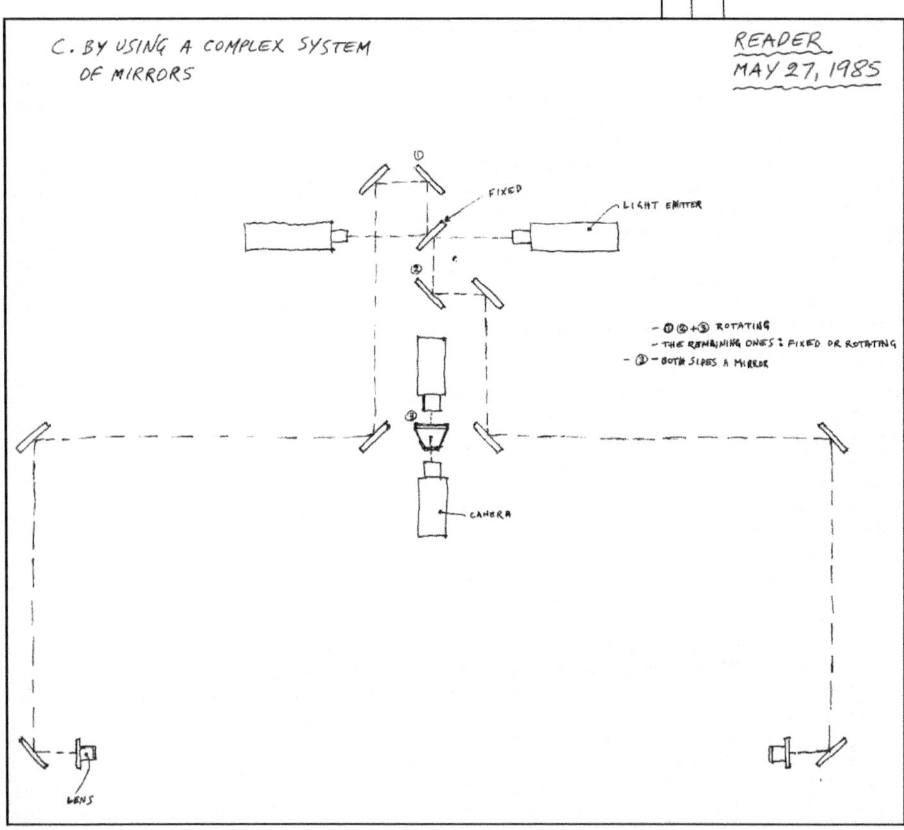
C. BY USING A COMPLEX SYSTEM
OF MIRRORS
READER
MAY 27, 1985
FIXED
LIGHT EMITTER
- ①②+③ ROTATING
- THE REMAINING ONES : FIXED OR ROTATING
- ② - BOTH SIDES A MIRROR
CAMERA
LENS

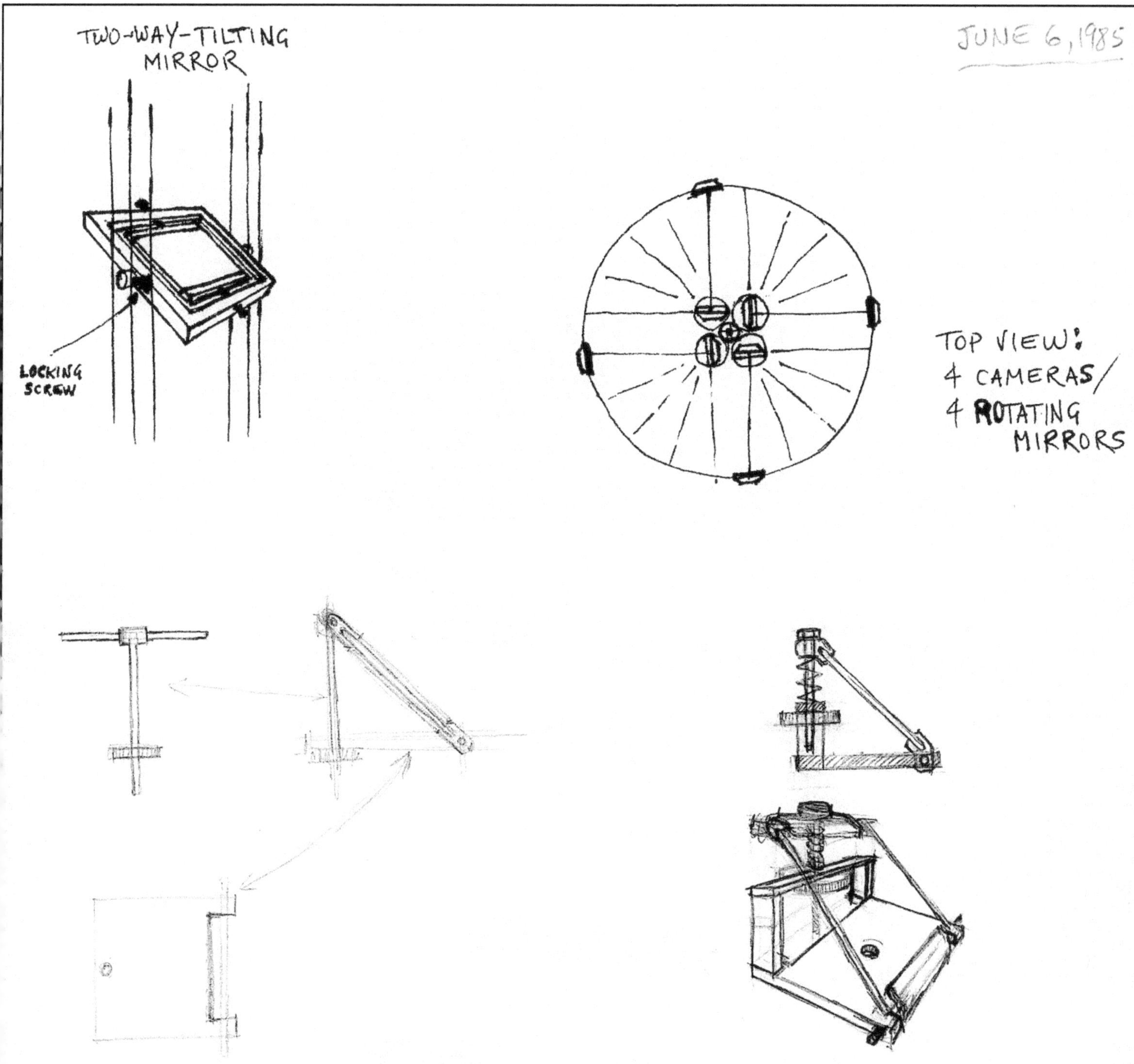
TWO-WAY-TILTING MIRROR
JUNE 6, 1985
LOCKING SCREW
TOP VIEW:
4 CAMERAS/
4 ROTATING MIRRORS

10A. CAD

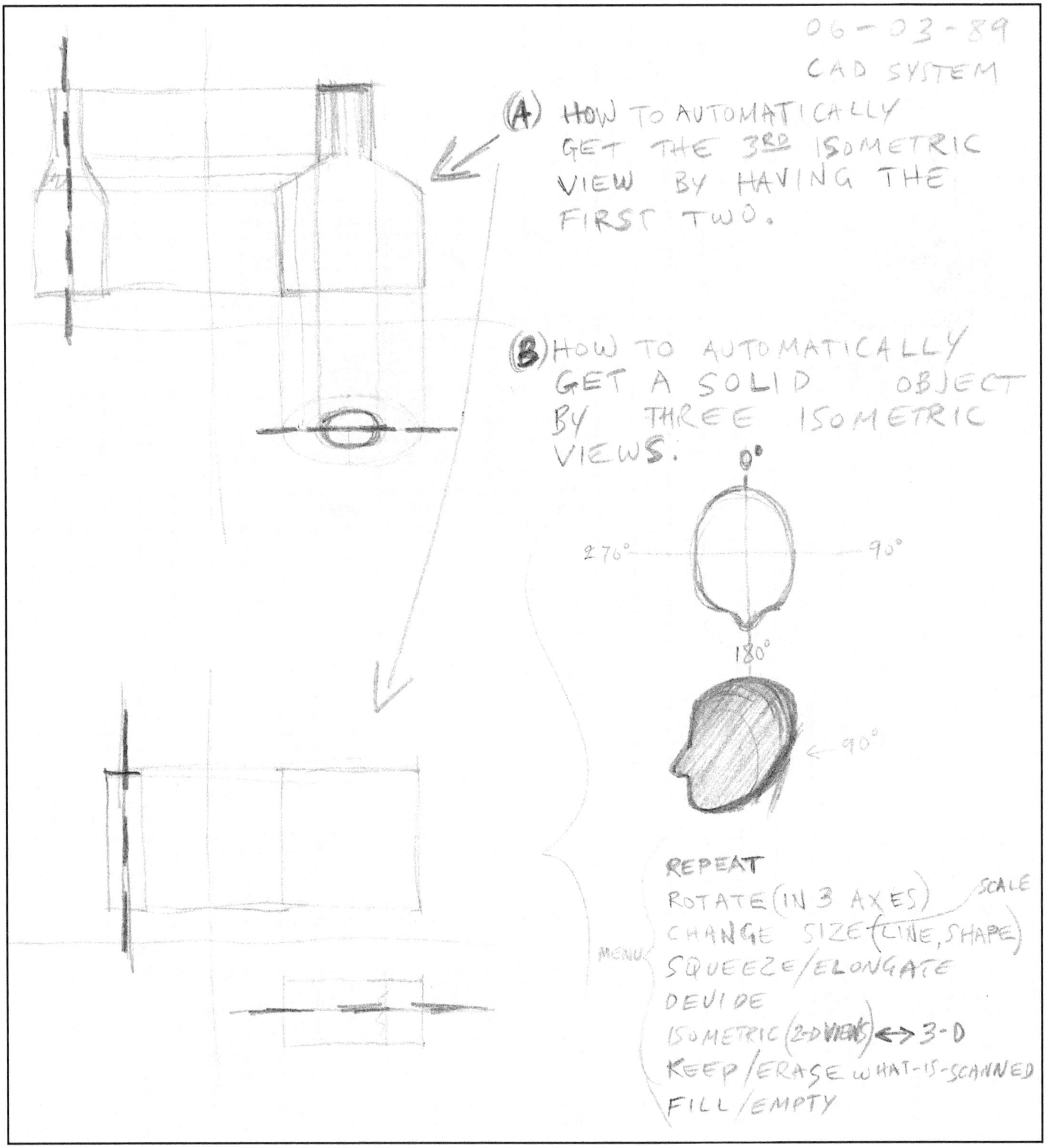
06-03-89
CAD SYSTEM
(A) HOW TO AUTOMATICALLY GET THE 3RD ISOMETRIC VIEW BY HAVING THE FIRST TWO.
(B) HOW TO AUTOMATICALLY GET A SOLID OBJECT BY THREE ISOMETRIC VIEWS.
0°
270°
90°
180°
90°
REPEAT
ROTATE (IN 3 AXES)
SCALE
CHANGE SIZE (LINE, SHAPE)
MENU
SQUEEZE/ELONGATE
DEVIDE
ISOMETRIC (2-D VIEWS) ↔ 3-D
KEEP/ERASE WHAT-IS-SCANNED
FILL/EMPTY

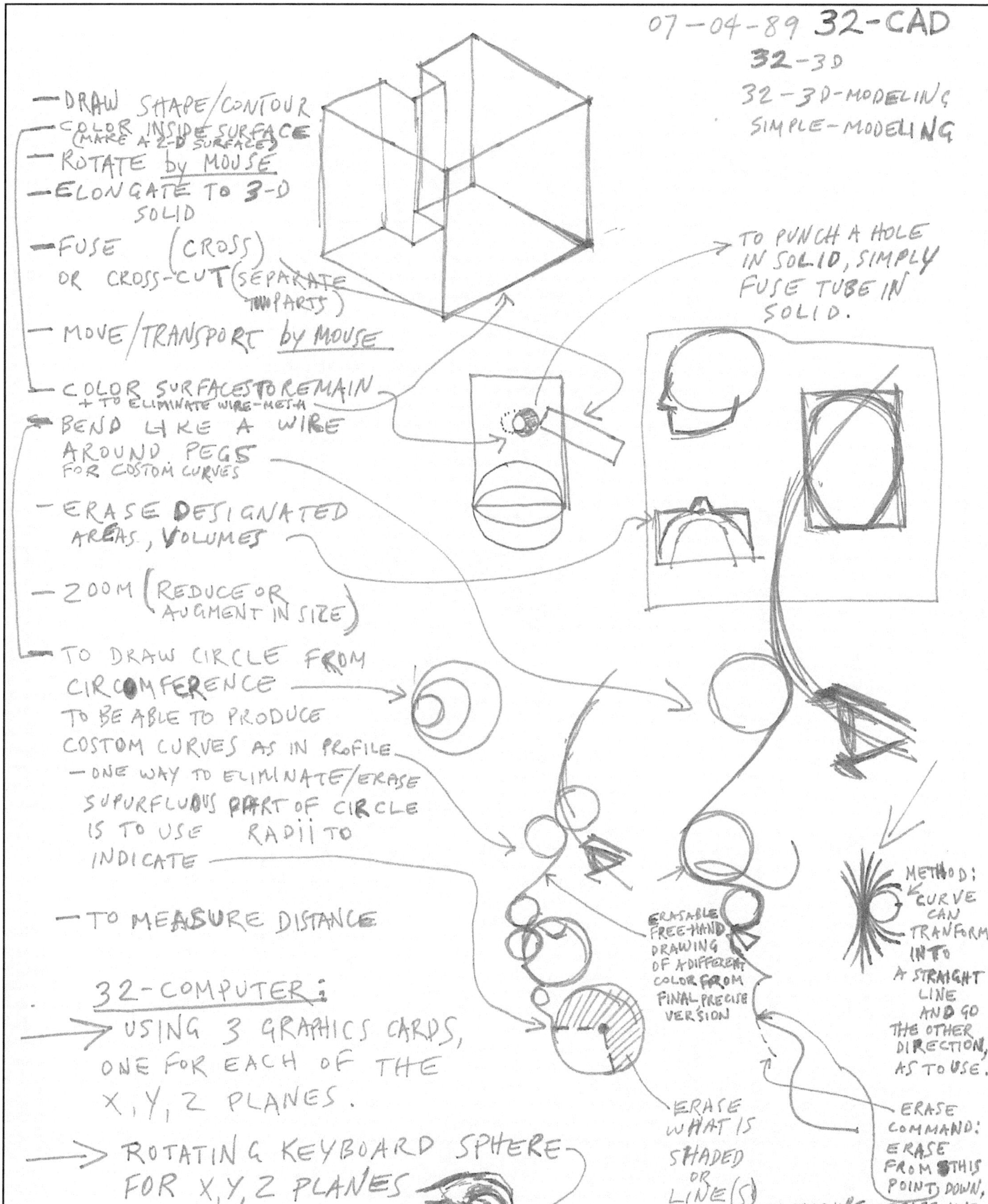

07-04-89 32-CAD
32-3D
32-3D-MODELING
SIMPLE-MODELING
— DRAW SHAPE/CONTOUR
— COLOR INSIDE SURFACE (MAKE A 2-D SURFACE)
— ROTATE by MOUSE
— ELONGATE TO 3-D SOLID
— FUSE (CROSS) OR CROSS-CUT (SEPARATE TWO PARTS)
— MOVE/TRANSPORT by MOUSE
— COLOR SURFACES TO REMAIN + TO ELIMINATE WIRE-MESH
— BEND LIKE A WIRE AROUND PEGS FOR COSTOM CURVES
— ERASE DESIGNATED AREAS, VOLUMES
— ZOOM (REDUCE OR AUGMENT IN SIZE)
— TO DRAW CIRCLE FROM CIRCOMFERENCE
TO BE ABLE TO PRODUCE COSTOM CURVES AS IN PROFILE
— ONE WAY TO ELIMINATE/ERASE SUPURFLUOUS PART OF CIRCLE IS TO USE RADII TO INDICATE
— TO MEASURE DISTANCE
TO PUNCH A HOLE IN SOLID, SIMPLY FUSE TUBE IN SOLID.
ERASABLE FREE-HAND DRAWING OF A DIFFERENT COLOR FROM FINAL PRECISE VERSION
METHOD: CURVE CAN TRANFORM INTO A STRAIGHT LINE AND GO THE OTHER DIRECTION, AS TO USE.
ERASE WHAT IS SHADED OR LINE(S) TOUCHED BY SHADING
ERASE COMMAND: ERASE FROM THIS POINT, DOWN, ERASE LINE FROM THIS POINT ON IN THIS DIR.
32-COMPUTER:
→ USING 3 GRAPHICS CARDS, ONE FOR EACH OF THE X, Y, Z PLANES.
→ ROTATING KEYBOARD SPHERE FOR X, Y, Z PLANES

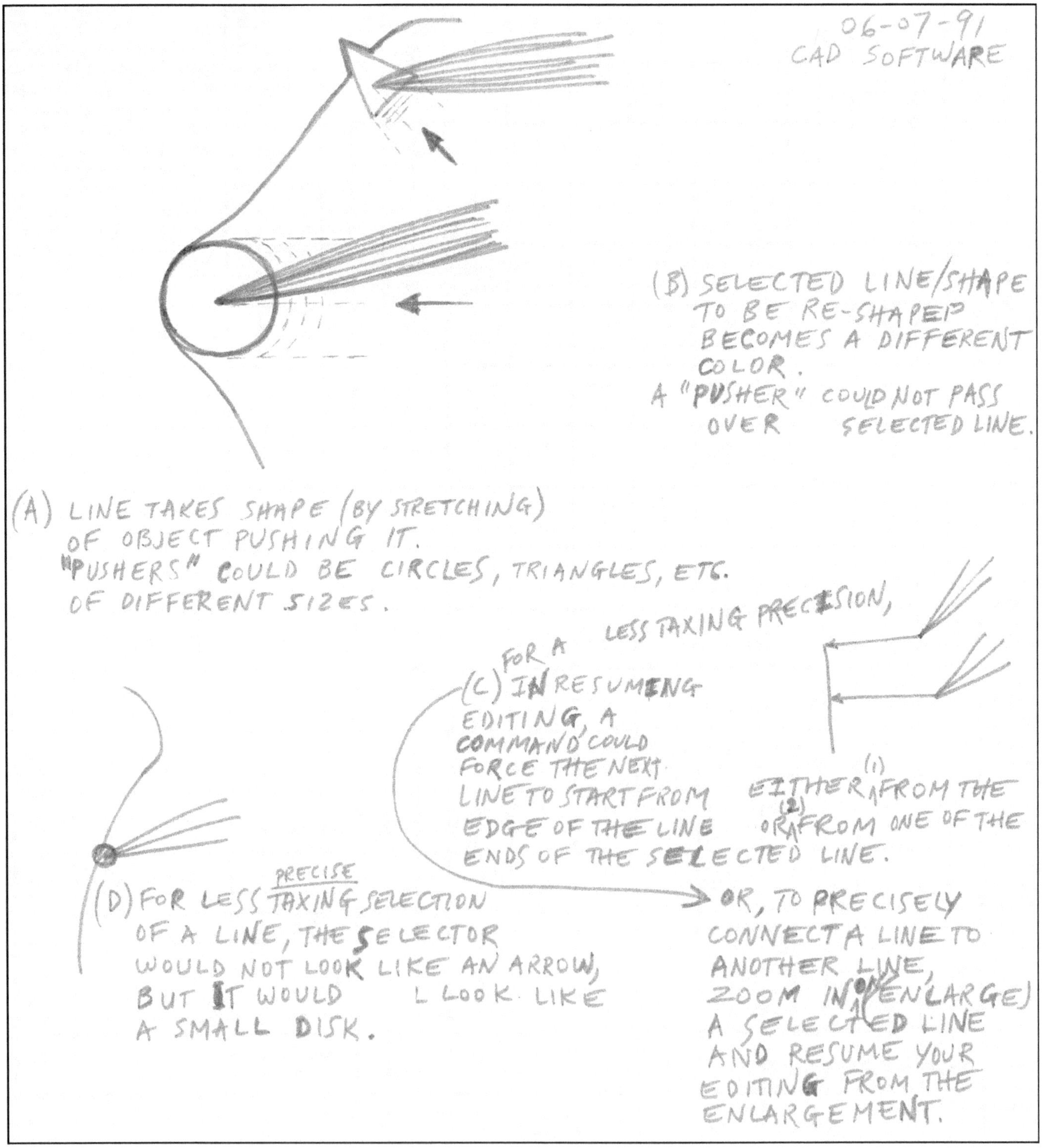
06-07-91
CAD SOFTWARE
(B) SELECTED LINE/SHAPE
TO BE RE-SHAPED
BECOMES A DIFFERENT
COLOR.
A "PUSHER" COULD NOT PASS
OVER SELECTED LINE.
(A) LINE TAKES SHAPE (BY STRETCHING)
OF OBJECT PUSHING IT.
"PUSHERS" COULD BE CIRCLES, TRIANGLES, ETC.
OF DIFFERENT SIZES.
(C) FOR A LESS TAXING PRECISION, IN RESUMING
EDITING, A
COMMAND COULD
FORCE THE NEXT
LINE TO START FROM EITHER (1) FROM THE
EDGE OF THE LINE OR (2) FROM ONE OF THE
ENDS OF THE SELECTED LINE.
OR, TO PRECISELY
CONNECT A LINE TO
ANOTHER LINE,
ZOOM IN ON (ENLARGE)
A SELECTED LINE
AND RESUME YOUR
EDITING FROM THE
ENLARGEMENT.
(D) FOR LESS PRECISE TAXING SELECTION
OF A LINE, THE SELECTOR
WOULD NOT LOOK LIKE AN ARROW,
BUT IT WOULD LOOK LIKE
A SMALL DISK.

10B. MEASURING TELESCOPE

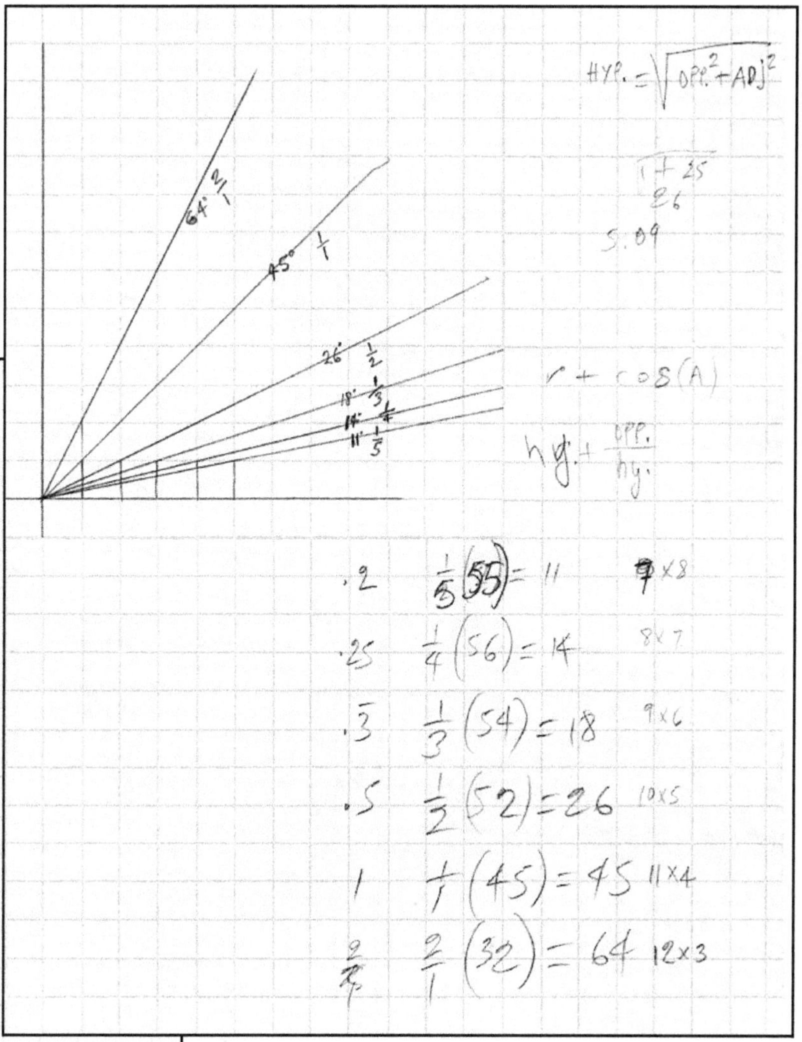

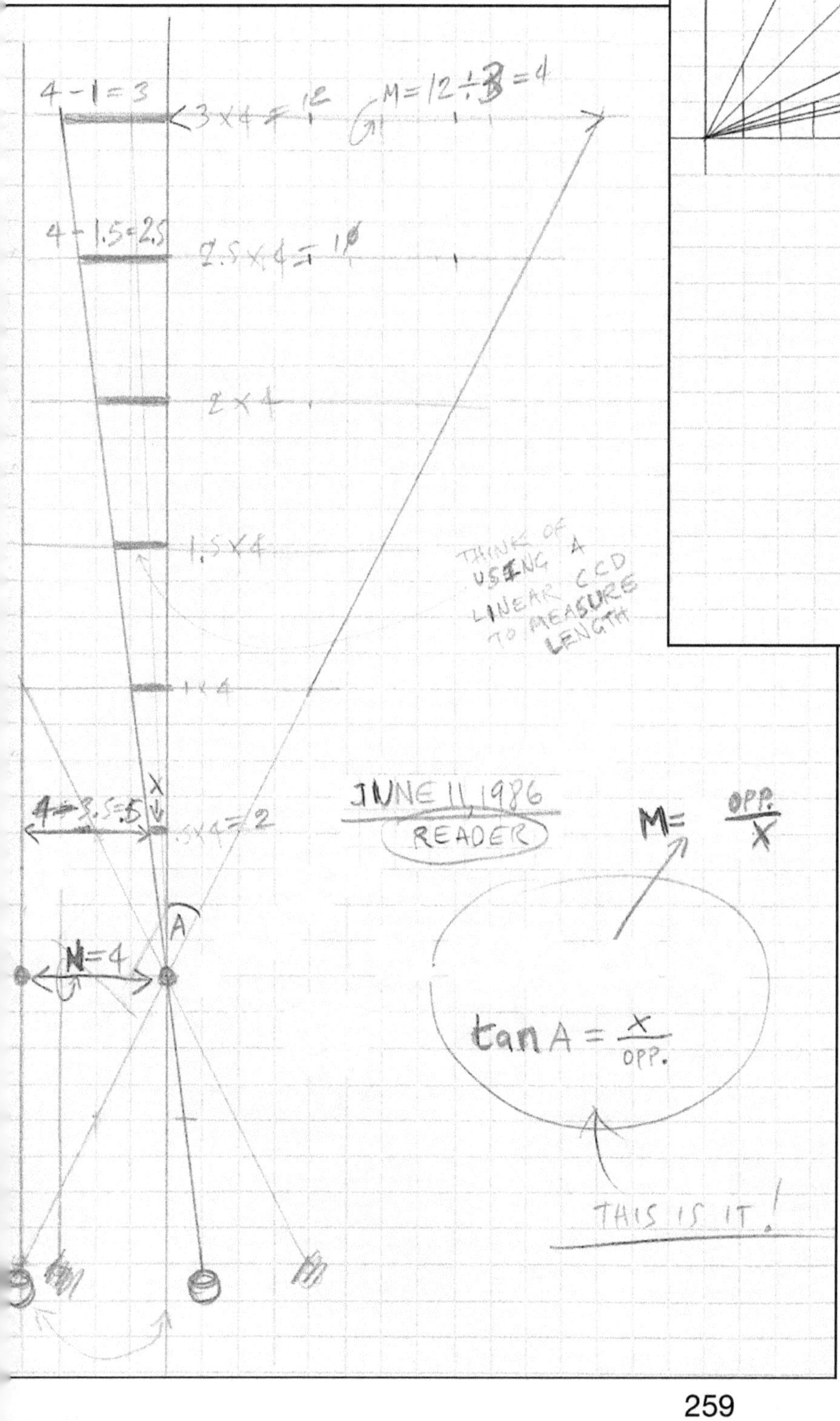

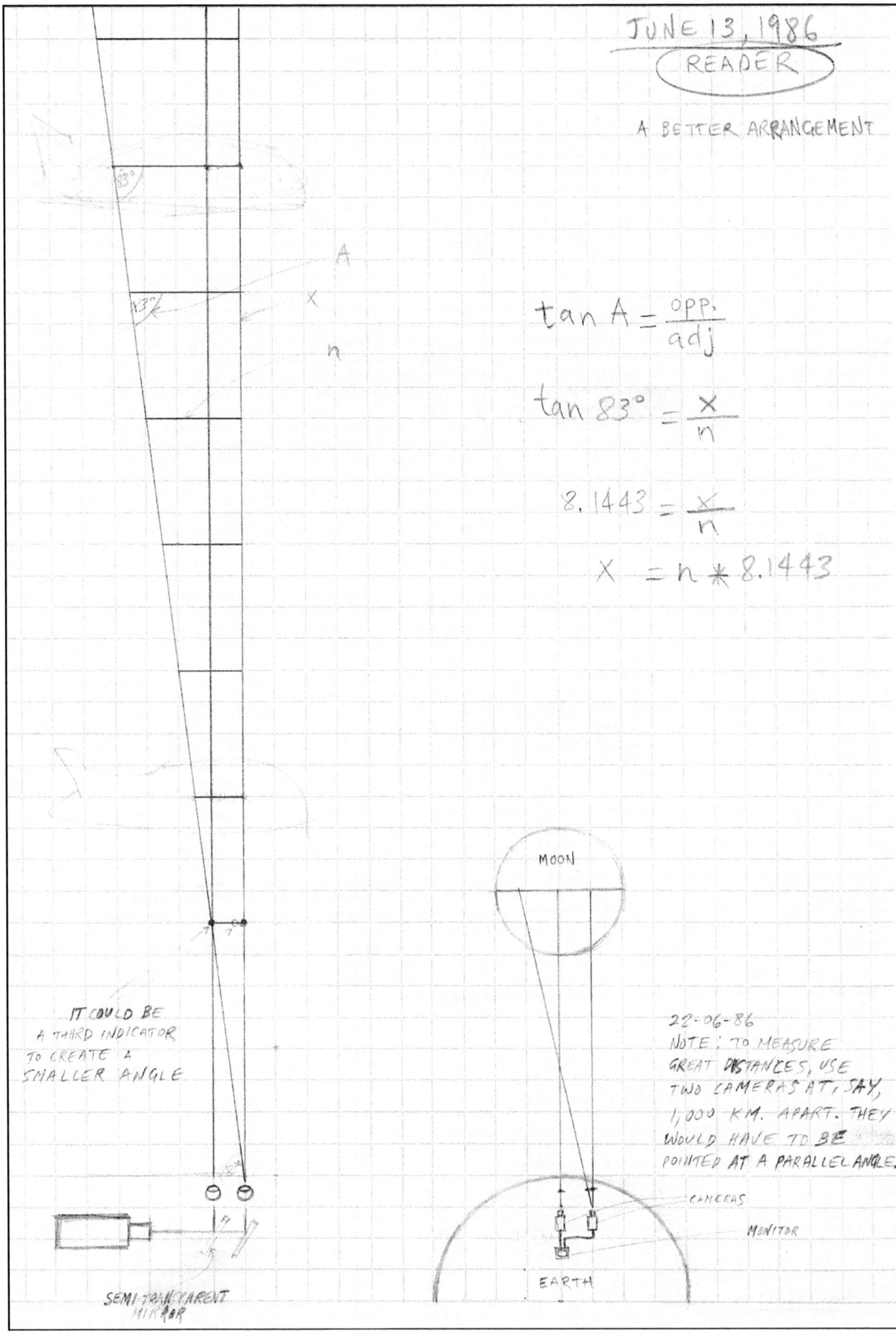

JUNE 13, 1986

READER

A BETTER ARRANGEMENT

$$\tan A = \frac{opp.}{adj}$$

$$\tan 83° = \frac{x}{n}$$

$$8.1443 = \frac{x}{n}$$

$$x = n * 8.1443$$

IT COULD BE A THIRD INDICATOR TO CREATE A SMALLER ANGLE

22-06-86
NOTE: TO MEASURE GREAT DISTANCES, USE TWO CAMERAS AT, SAY, 1,000 KM. APART. THEY WOULD HAVE TO BE POINTED AT A PARALLEL ANGLE.

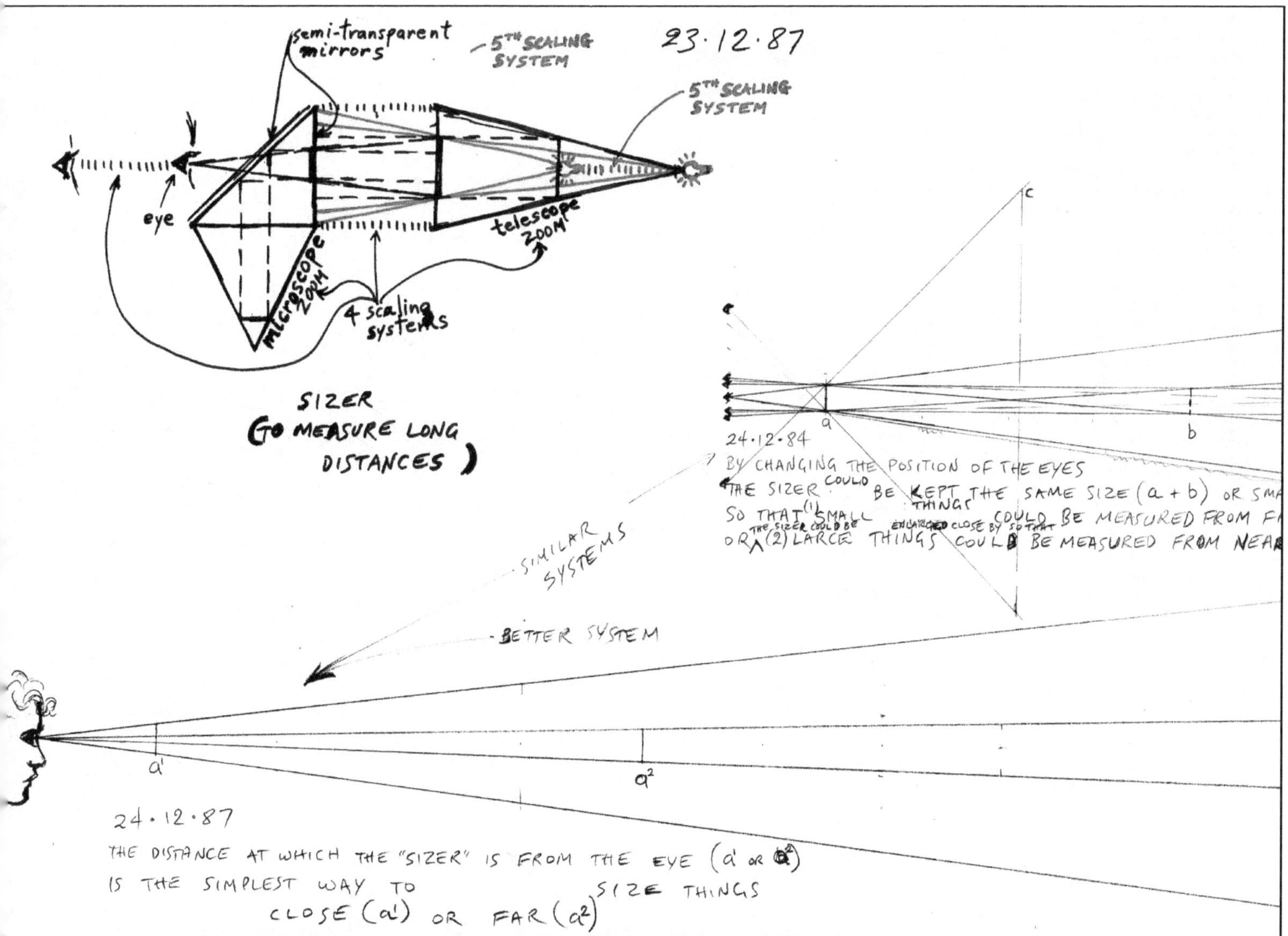

semi-transparent mirrors
5TH SCALING SYSTEM
23·12·87
5TH SCALING SYSTEM
eye
telescope ZOOM
microscope ZOOM
4 scaling systems
SIZER (TO MEASURE LONG DISTANCES)
c
a
b
24·12·84
BY CHANGING THE POSITION OF THE EYES
THE SIZER COULD BE KEPT THE SAME SIZE (a + b) OR SMA
SO THAT (1) SMALL THINGS COULD BE MEASURED FROM FA
OR (2) LARGE THINGS COULD BE MEASURED FROM NEAR
SIMILAR SYSTEMS
BETTER SYSTEM
a¹
a²
24·12·87
THE DISTANCE AT WHICH THE "SIZER" IS FROM THE EYE (a¹ OR a²)
IS THE SIMPLEST WAY TO SIZE THINGS
CLOSE (a¹) OR FAR (a²)

A SYSTEM TO MEASURE DEPTH ①

SIZER

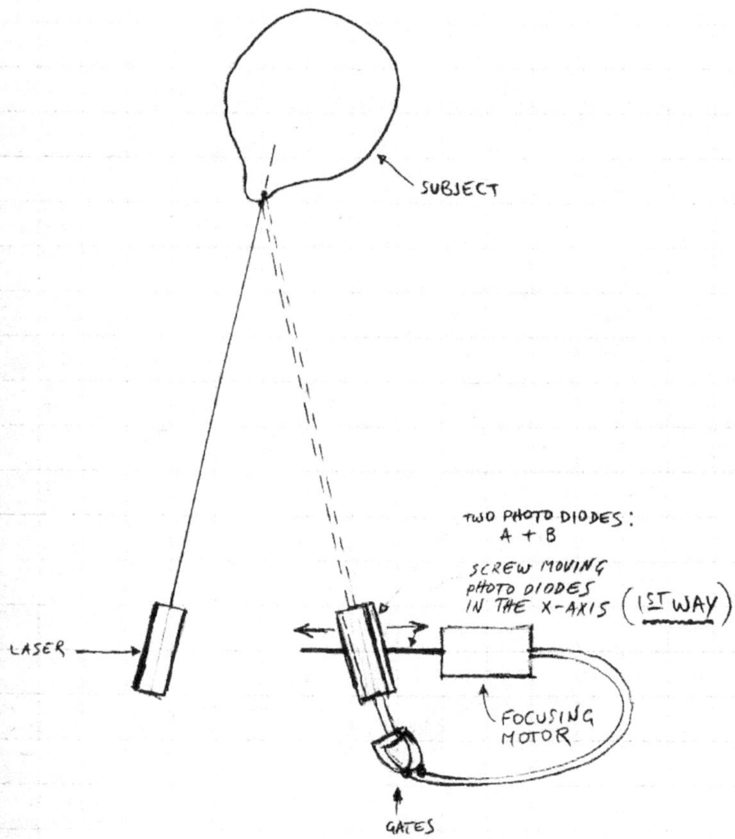

TRUTH TABLE
ONLY POSSIBILITIES

A	B	OUT
1	Ø	Ø
1	1	1 (MOTOR DRIVES FORWARD)
Ø	Ø	1 (MOTOR DRIVES BACKWARDS)

A 2ND WAY FOR FOCUSING IS TO TURN A MIRROR

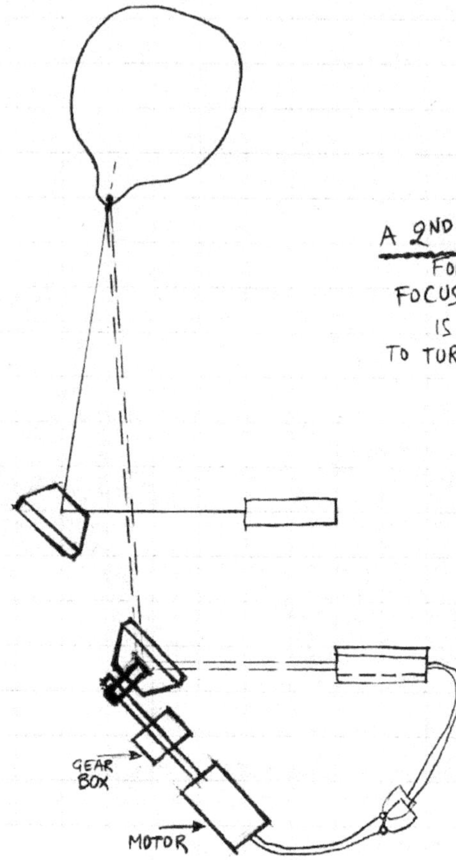

A 3RD WAY WOULD BE TO FOCUS WITH ONE OR TWO LENSES

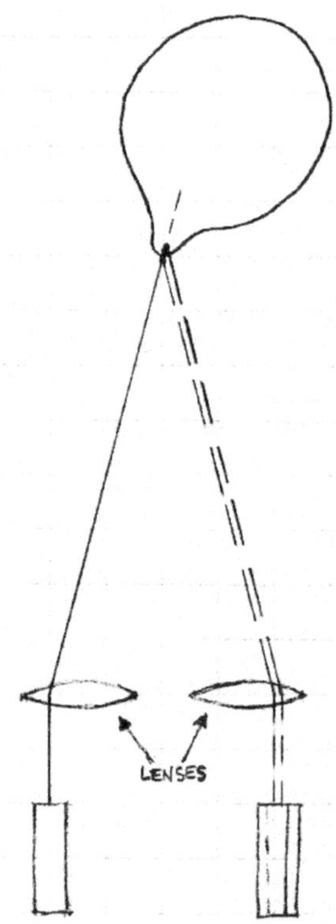

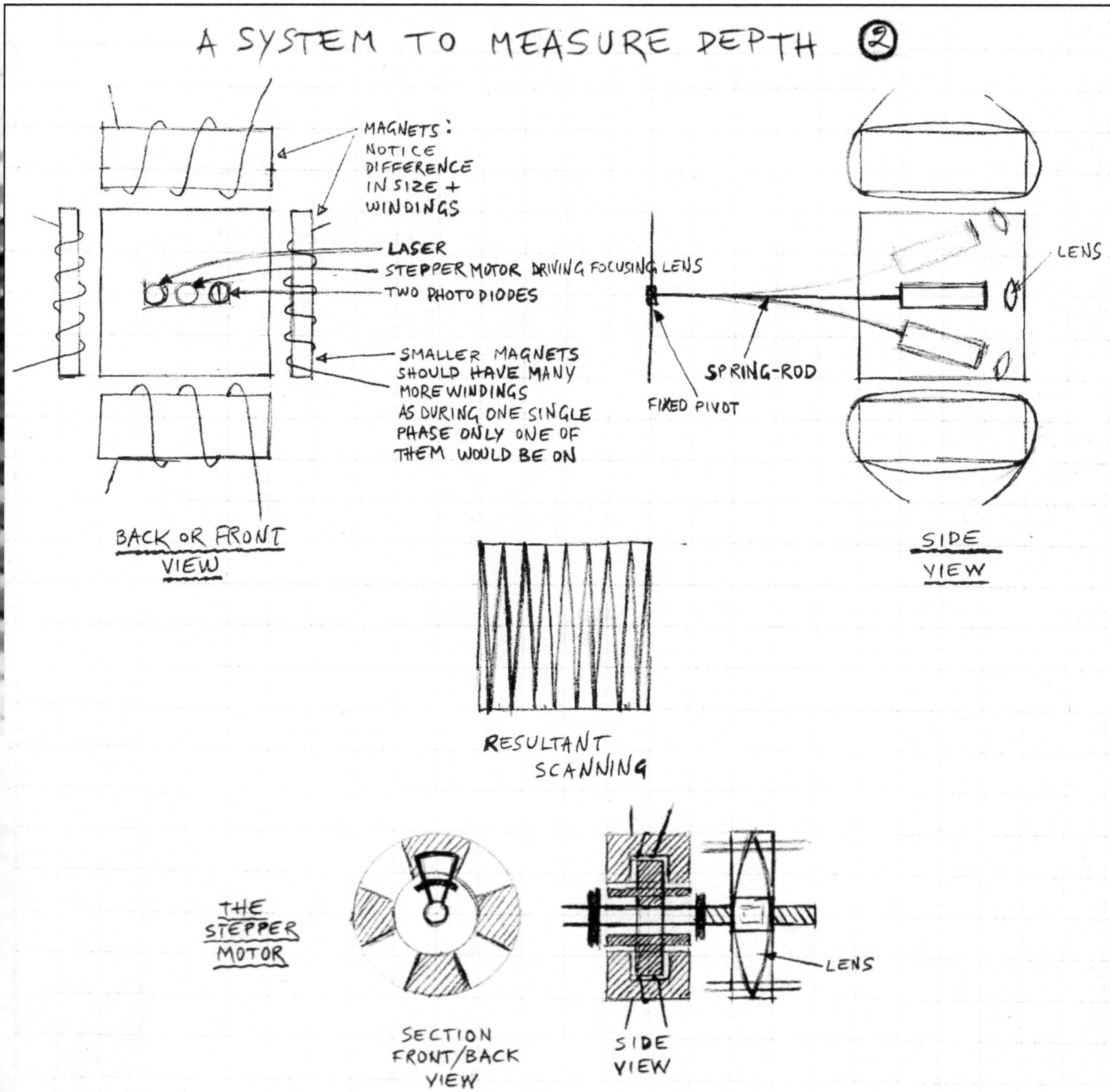
A SYSTEM TO MEASURE DEPTH ②
MAGNETS: NOTICE DIFFERENCE IN SIZE + WINDINGS
LASER
STEPPER MOTOR DRIVING FOCUSING LENS
TWO PHOTODIODES
SMALLER MAGNETS SHOULD HAVE MANY MORE WINDINGS AS DURING ONE SINGLE PHASE ONLY ONE OF THEM WOULD BE ON
BACK OR FRONT VIEW
SPRING-ROD
FIXED PIVOT
LENS
SIDE VIEW
RESULTANT SCANNING
THE STEPPER MOTOR
SECTION FRONT/BACK VIEW
SIDE VIEW
LENS

A SY SYSTEM TO MEASURE DEPTH ③

SPEED OF LIGHT (ELECTRICITY) : 186,000 MILES/SECOND
299,500 KM/SECOND

1 TURN OF LENS MOTOR = .5 ft. of electric wire = 1 mm.

1 TURN OF LARGER MOTORS = .5 ft. X 75 mm. (3 inches) = 37.5 ft.

AREA TO BE READ : 450 mm. X 450 mm. = 202,500 moves

202,500 X 37.5 ft. of electric wire for 1 move of larger motor = 7,593,750

7,593,750 ft. ÷ 1,760 ft. (1 mile) = 4,314.6306 miles

186,000 ÷ 4,314.63 = 43 → 1/43 OF A SECOND

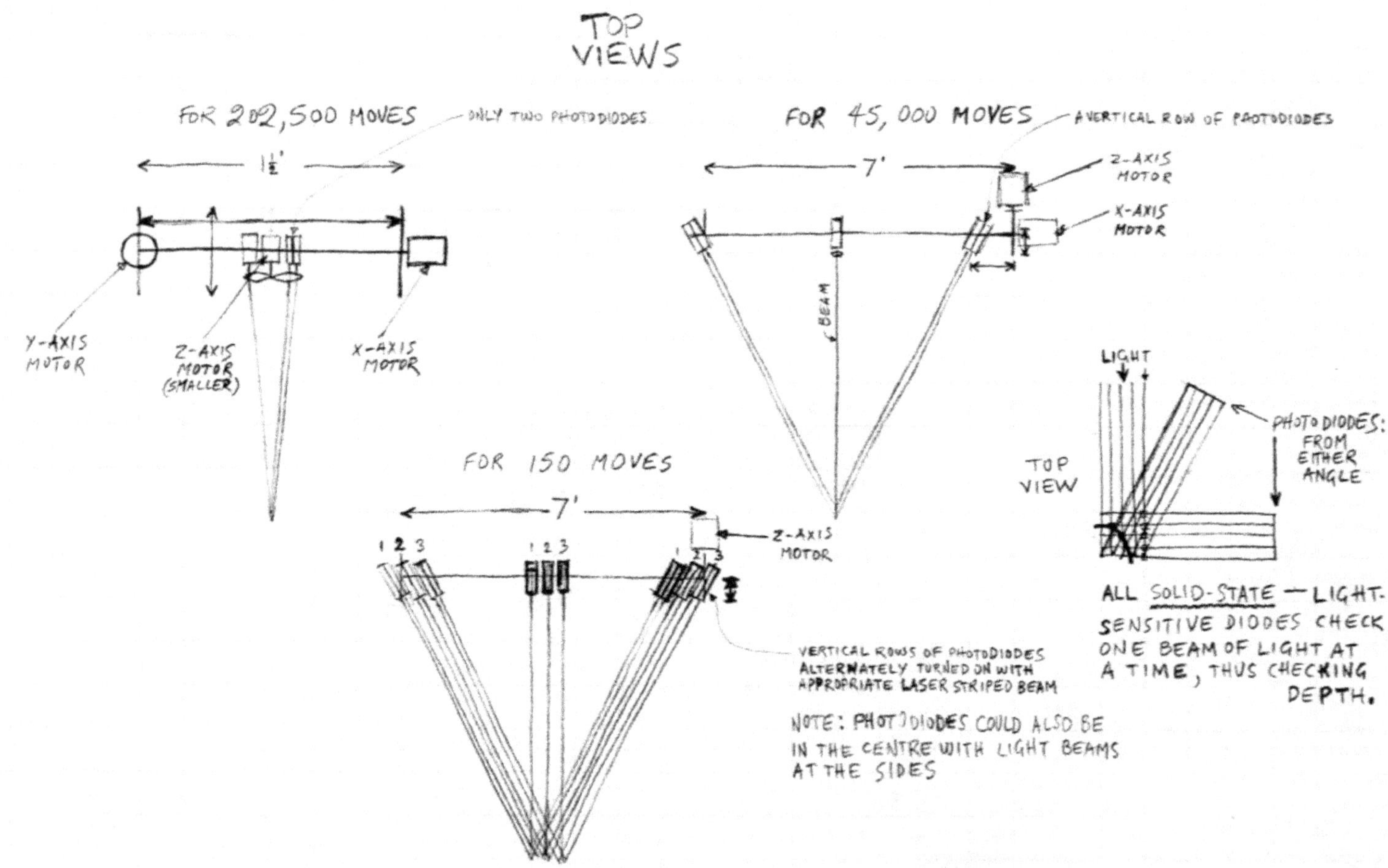

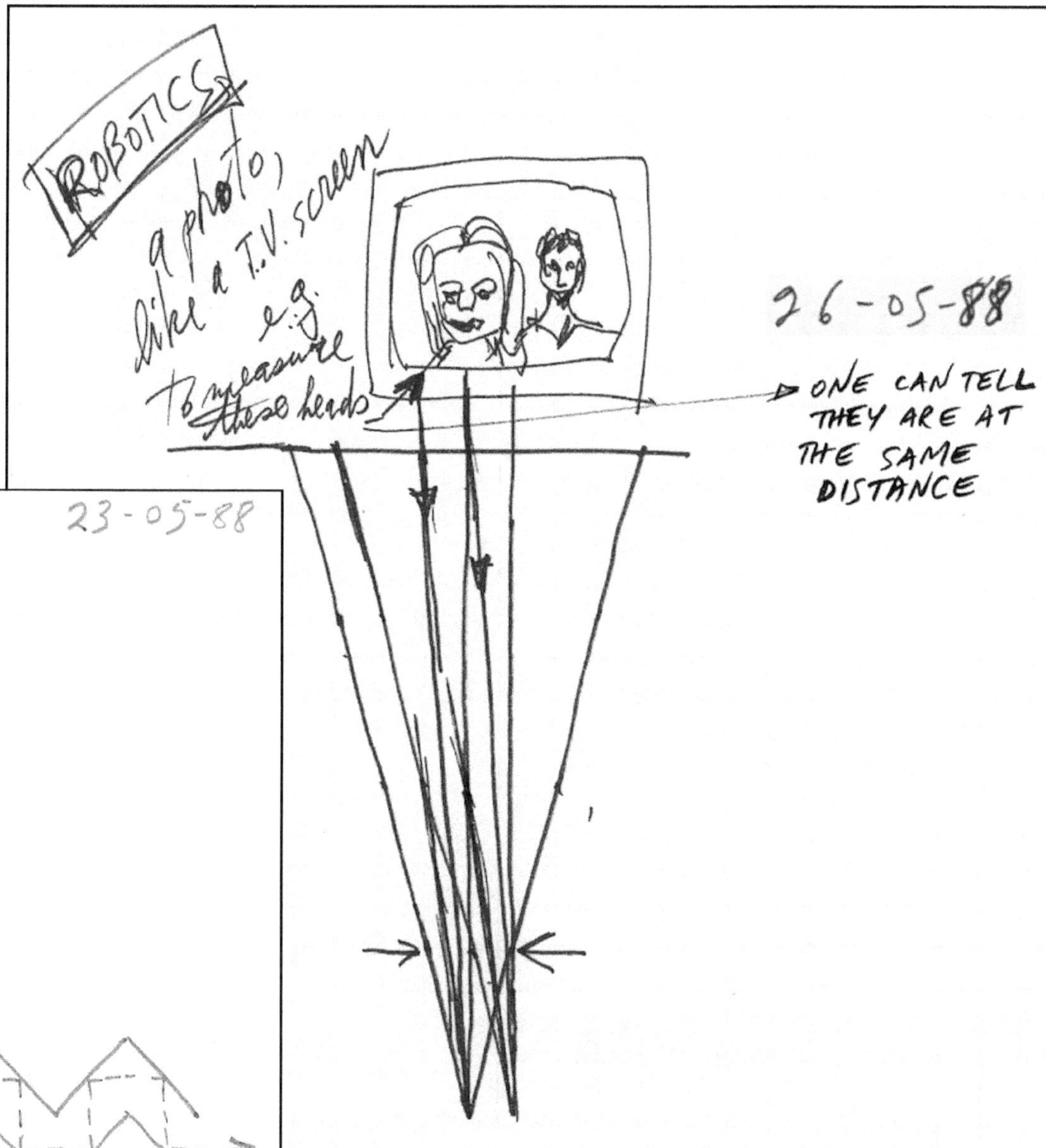
ROBOTICS
a photo, like a T.V. screen e.g.
To measure these heads
26-05-88
ONE CAN TELL THEY ARE AT THE SAME DISTANCE

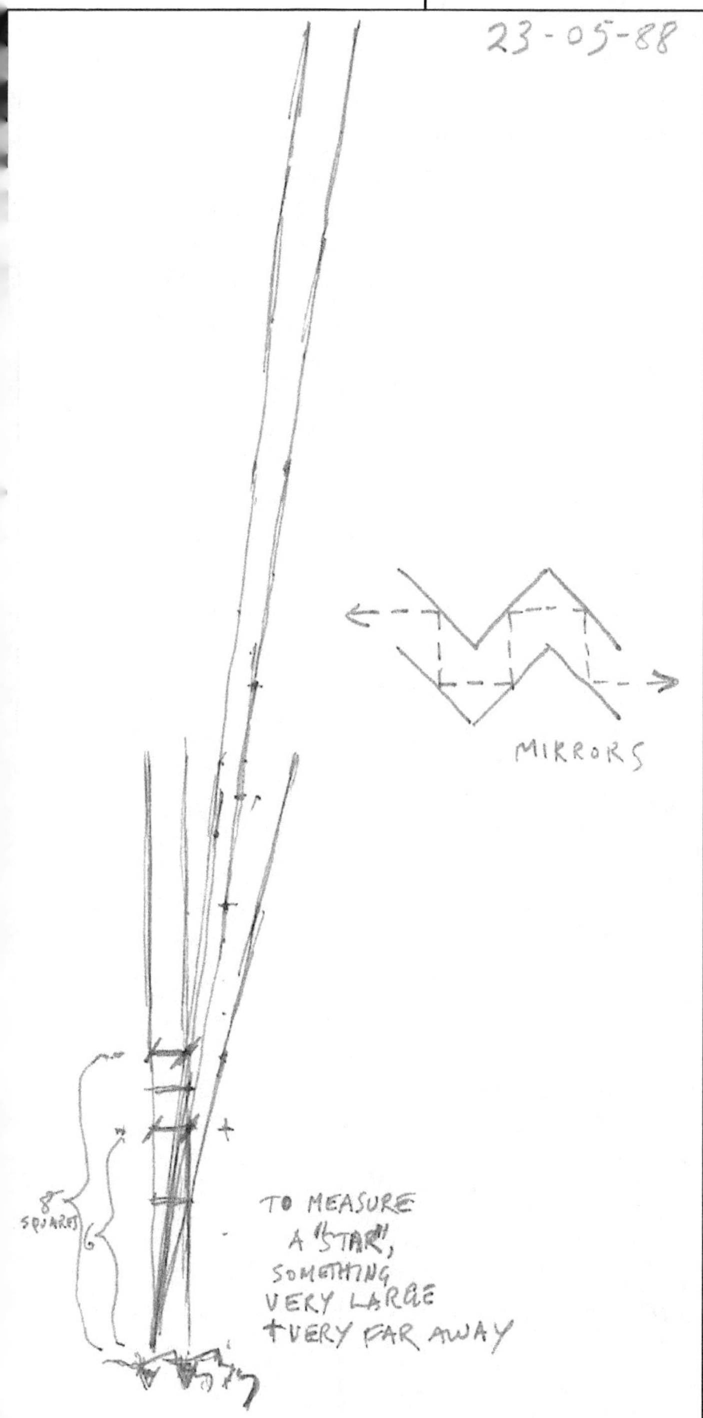
23-05-88
MIRRORS
8 SQUARES
TO MEASURE A "STAR", SOMETHING VERY LARGE + VERY FAR AWAY

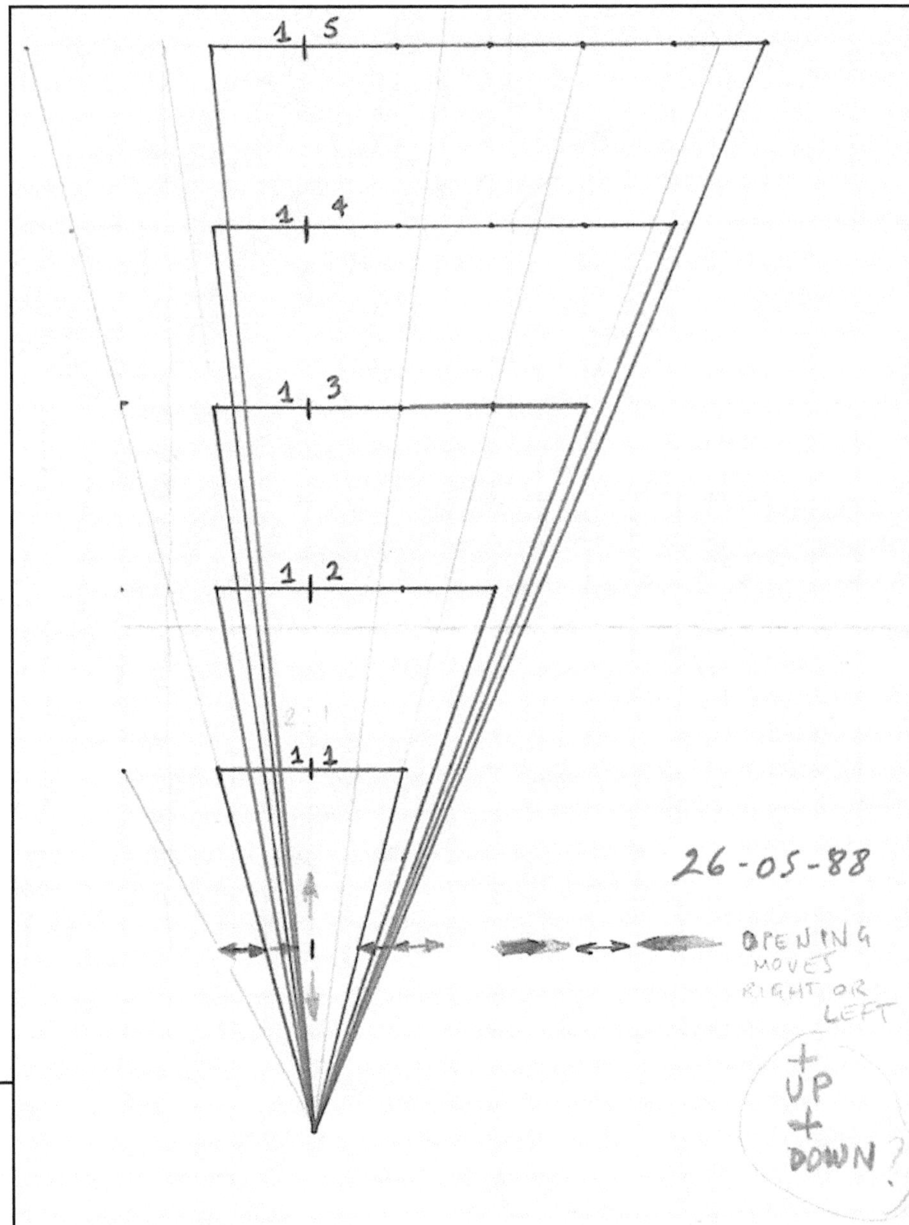

1 5
1 4
1 3
1 2
1 1
26-05-88
OPENING MOVES RIGHT OR LEFT
+ UP + DOWN?

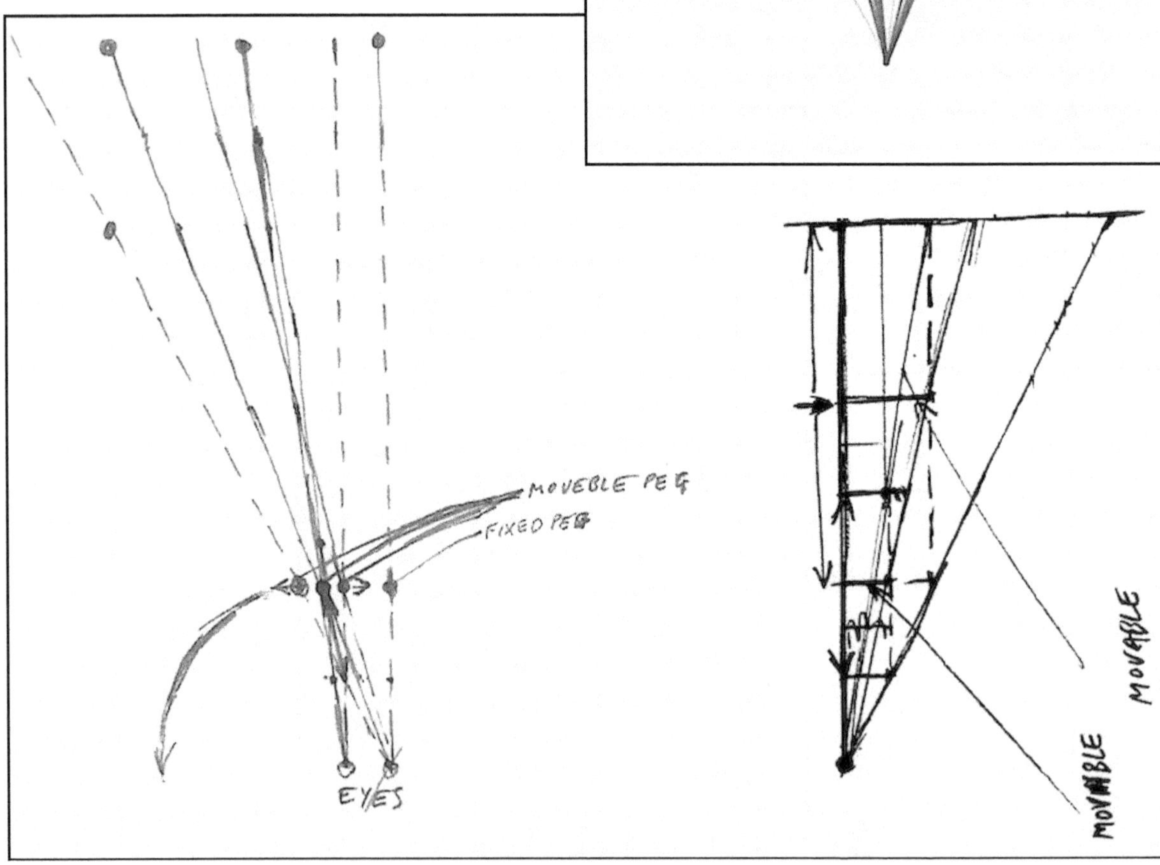

MOVEBLE PEG
FIXED PEG
EYES
MOVABLE
MOVABLE

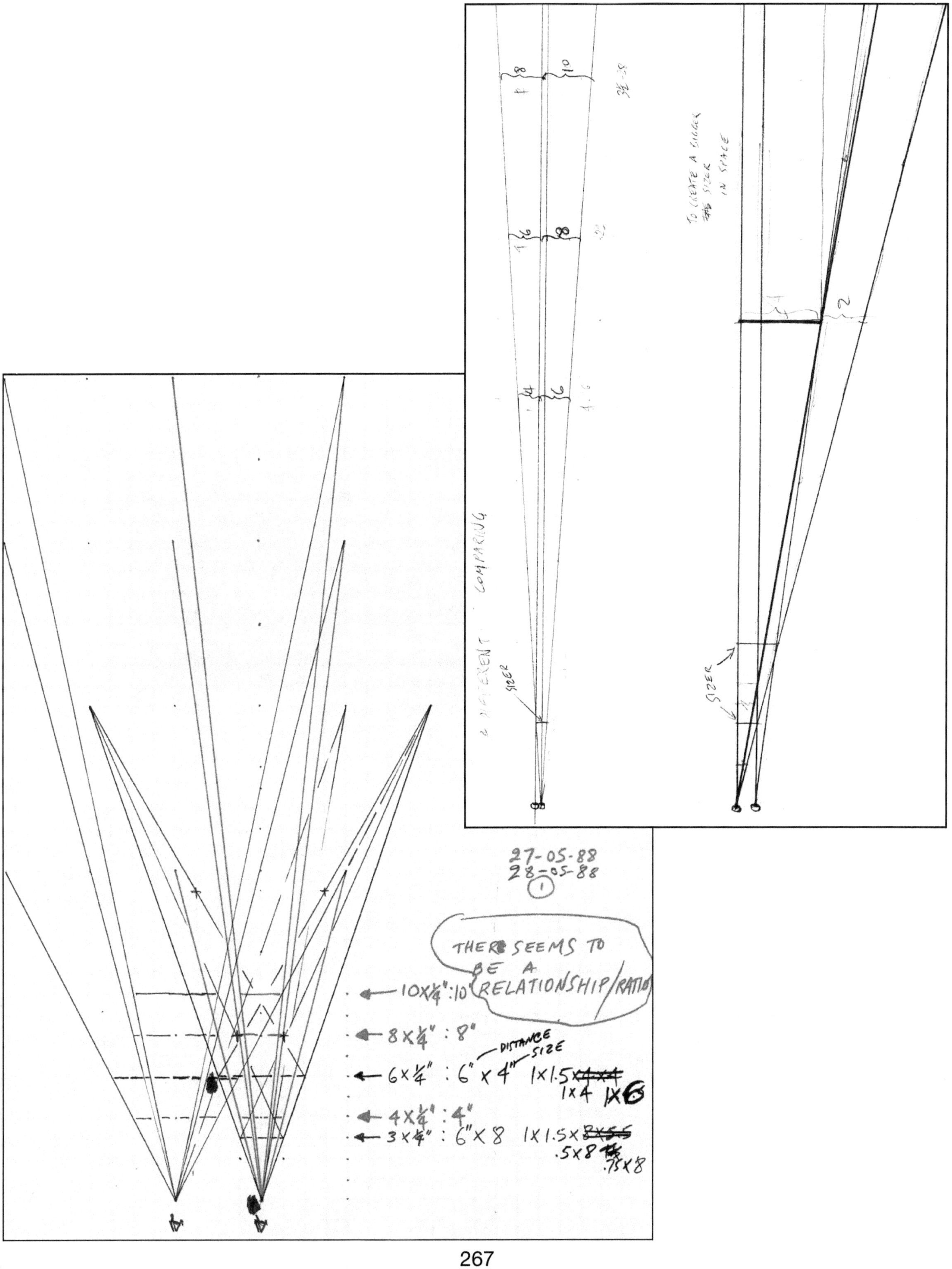
27-05-88
28-05-88
1
THERE SEEMS TO BE A RELATIONSHIP/RATIO
10X¼" : 10"
8X¼" : 8"
DISTANCE
SIZE
6X¼" : 6" X 4"
1X1.5X
1X4
1X6
4X¼" : 4"
3X¼" : 6" X 8
1X1.5X
.5X8
.75X8
COMPARING
SIZER
SIZER
TO CREATE A BIGGER SIZER IN SPACE

05-06-88
"SIZER"
PUTTING TO REST "SIZER"
(FINALIZED)

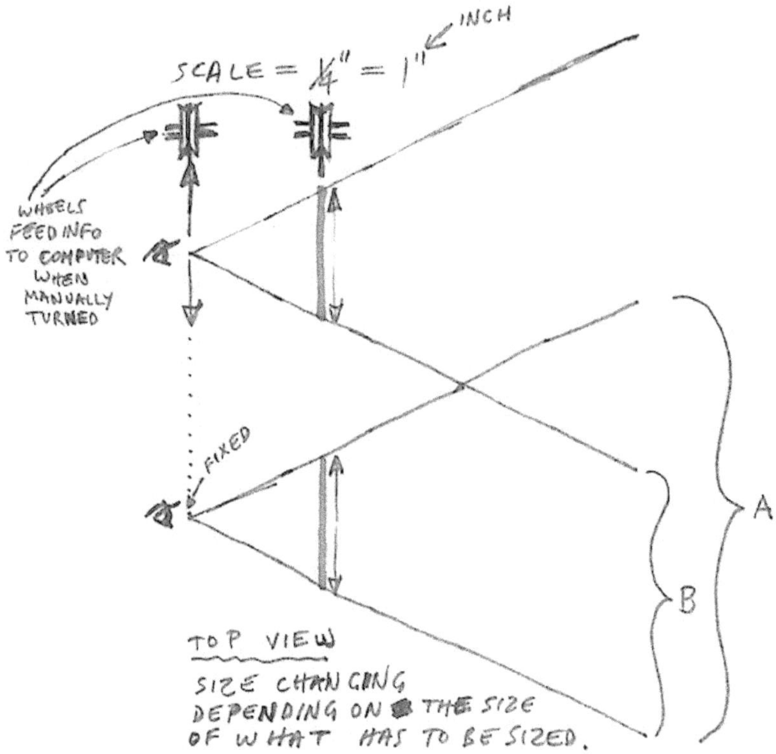

$D = A \div B$
↑
DISTANCE

TOP VIEW

SIZE CHANGING DEPENDING ON THE SIZE OF WHAT HAS TO BE SIZED.

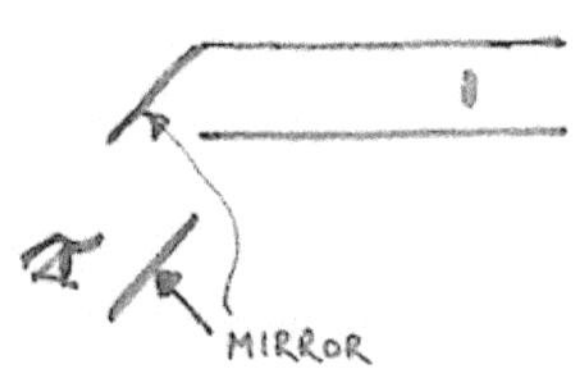

SIDE VIEW

(2) THIS ONE-TIME POINTING SYSTEM IS IDEAL IN PHOTOGRAPHY + ROBOTICS WHERE THE AIM IS TO ANALIZE THE IMAGE OR SCENE INSTANTANEOUSLY.

(3) THE AIM OF STEREOVISION IS NOT SOLELY TO DISCOVER THE 3-DIMENSIONALITY OF OBJECTS, BUT, MORE IMPORTANTLY ITS IS TO DISCOVER OR TELL UNCONSCIOUSLY OF SPACE OR DISTANCE OF OBJECTS FROM THE "STEREO-ONLOOKER"

SENSUALLY, NOT MATHEMATICALLY

(1)
NOTE: THIS IS A BETTER ONE-TIME SYSTEM, WHEREAS, IN TRIGONOMATRY ONE HAS TO POINT TWICE AT THE OBJECT TO FIND ITS SIZE, I.E.:

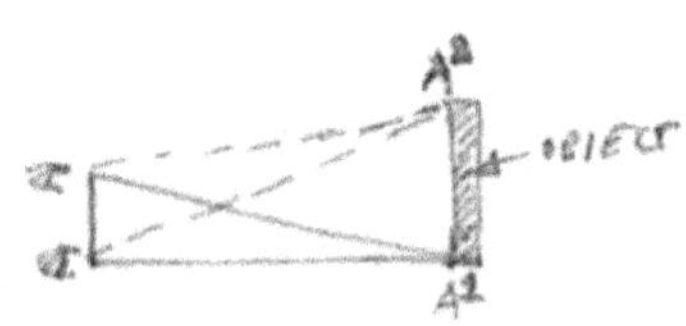

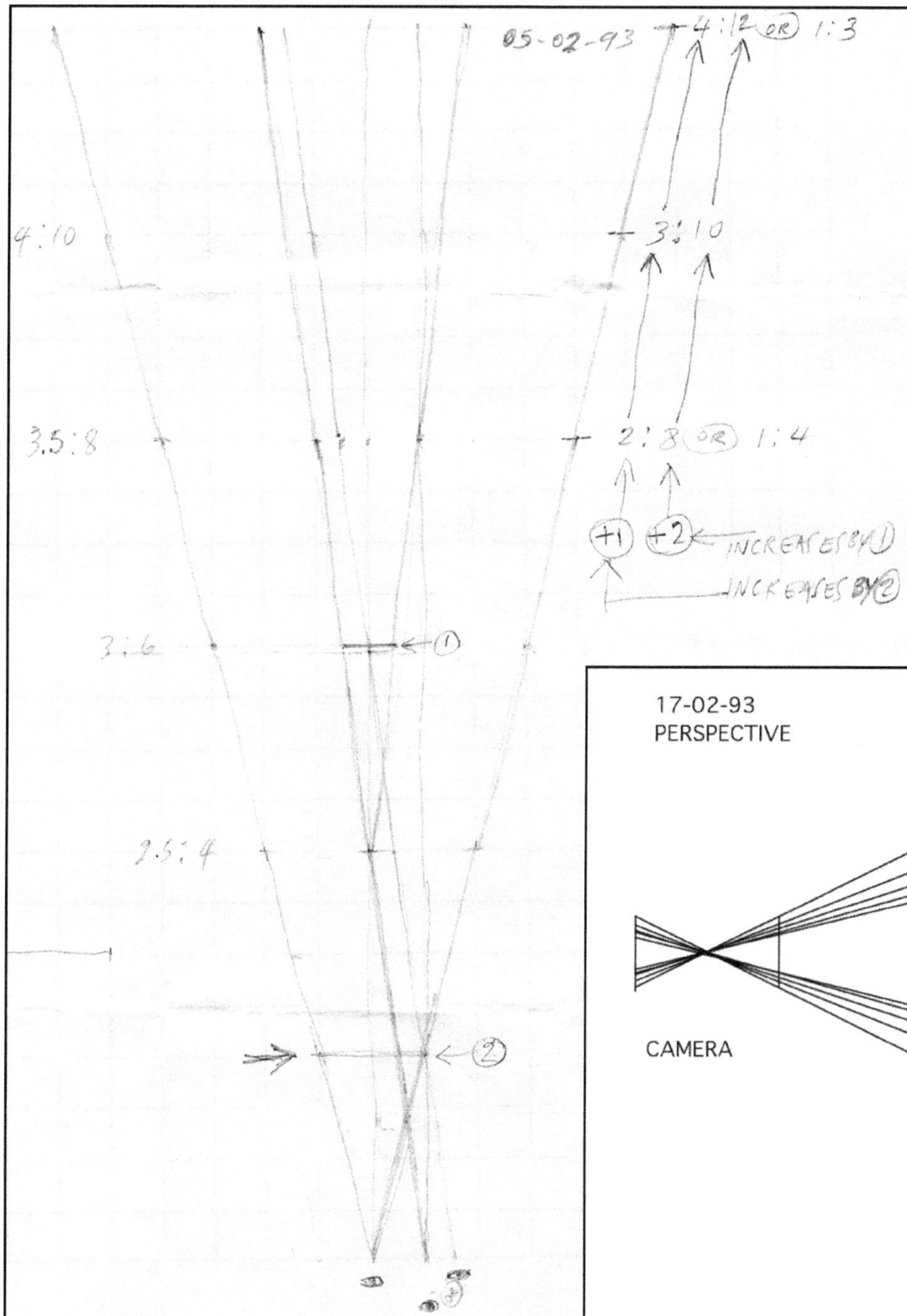
05-02-93
4:12 OR 1:3
4:10
3:10
3.5:8
2:8 OR 1:4
+1
+2
INCREASES BY 1
INCREASES BY 2
3:6
2.5:4

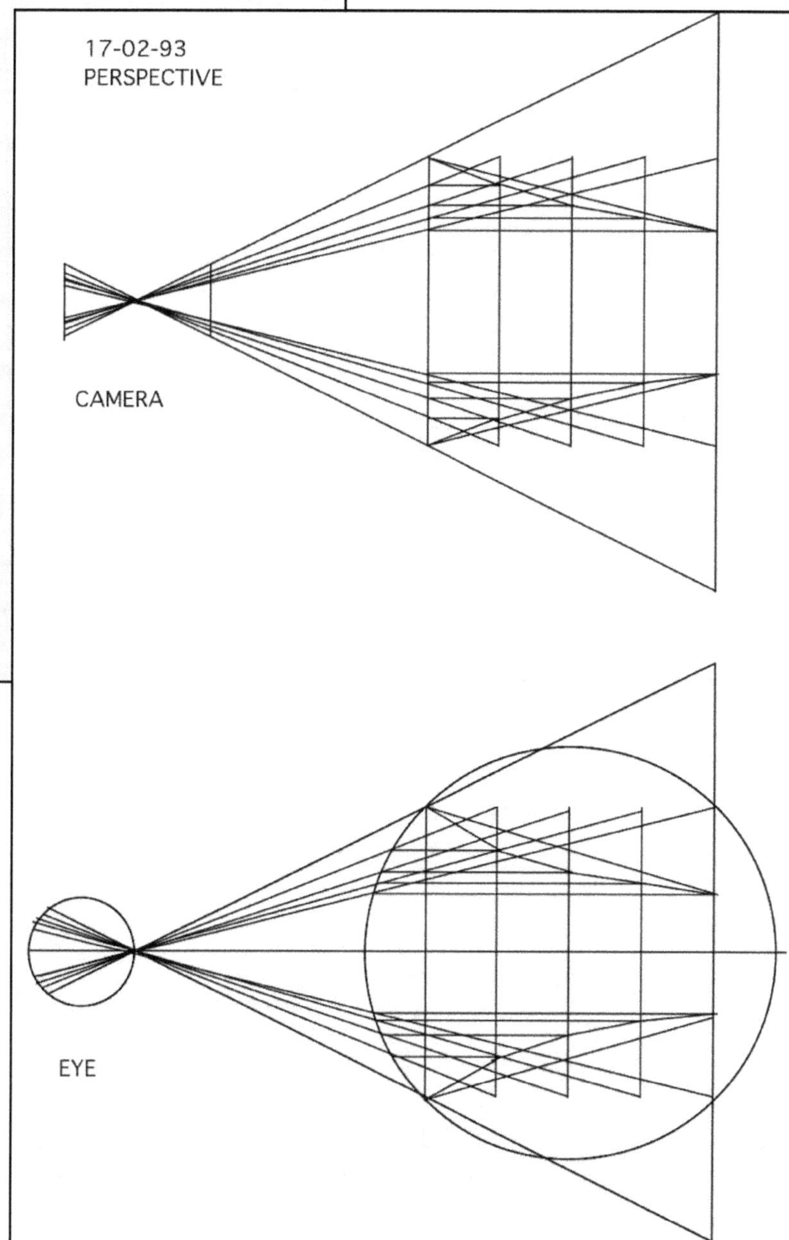
17-02-93
PERSPECTIVE
CAMERA
EYE

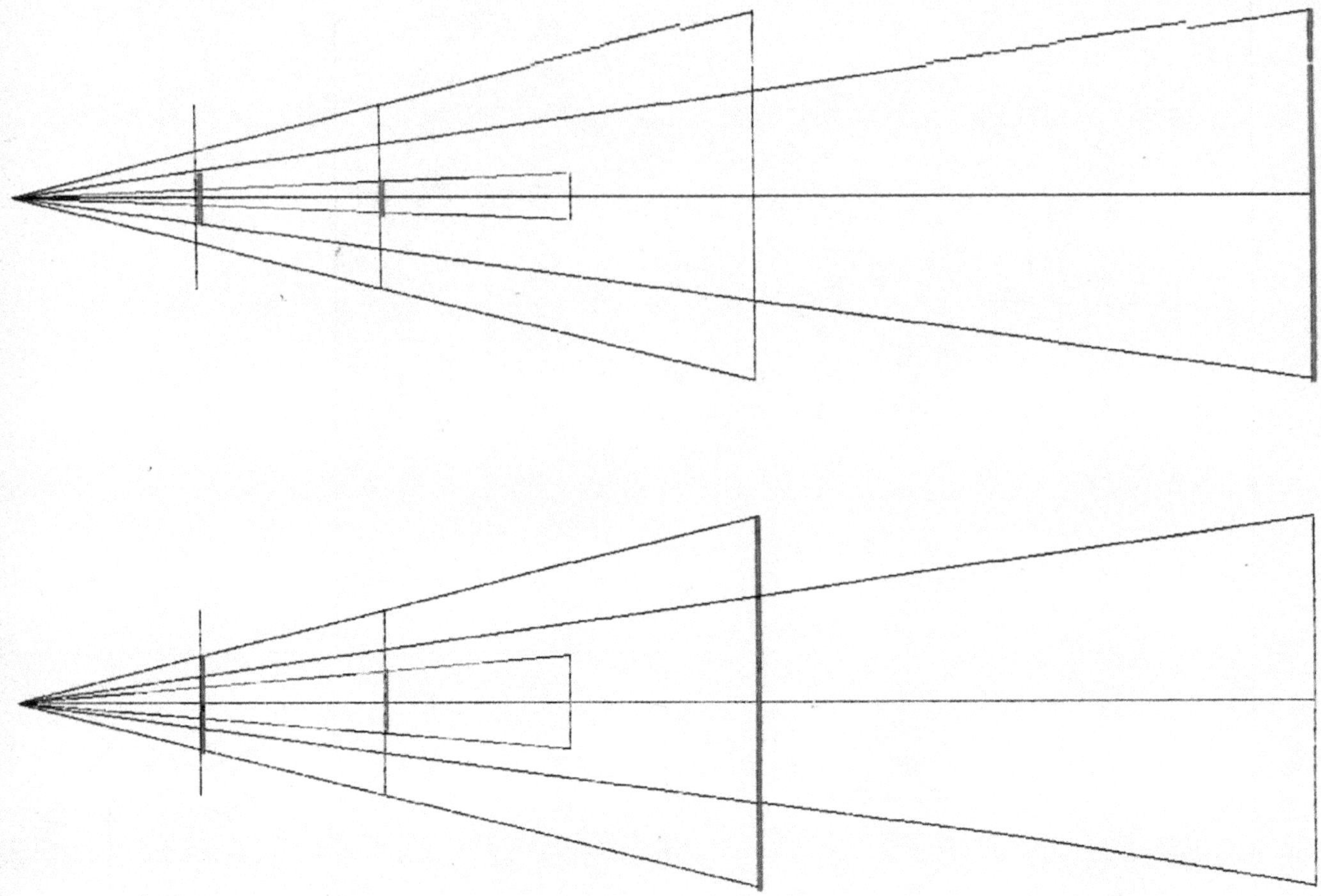

2:4
3:5

PERSPECTIVE
25-02-93

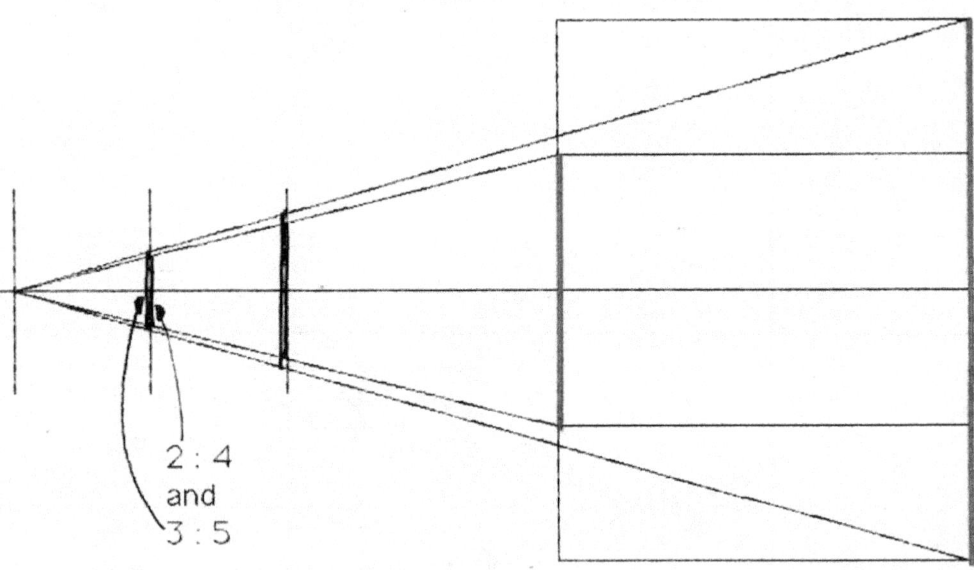

PERSPECTIVE
25-02-93

Epiphany: Many mirrors could be used to measure large distances the same as many levers are used to measure large loads like a heavy truck.

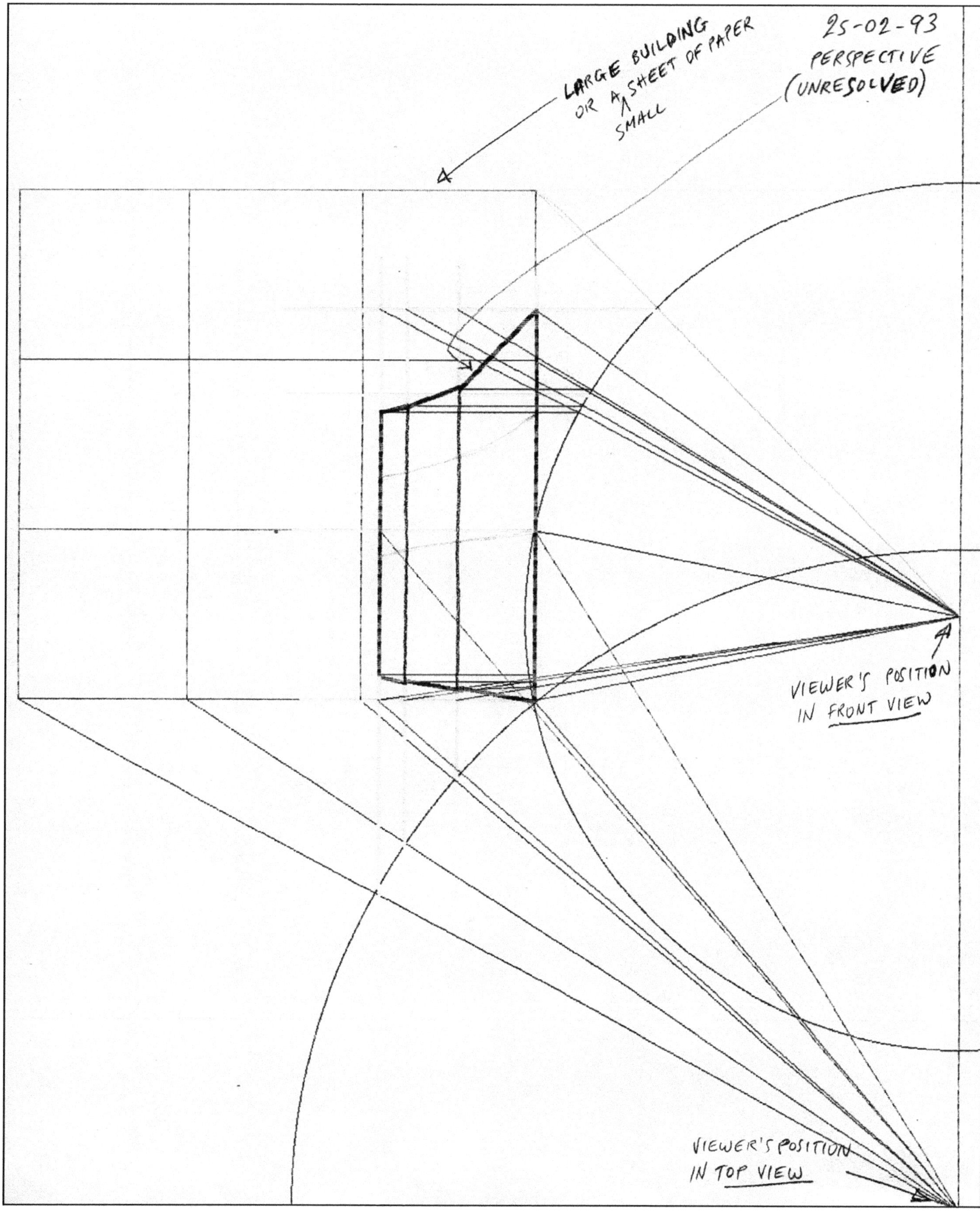
25-02-93
PERSPECTIVE
(UNRESOLVED)
LARGE BUILDING
OR A SMALL SHEET OF PAPER
VIEWER'S POSITION
IN FRONT VIEW
VIEWER'S POSITION
IN TOP VIEW

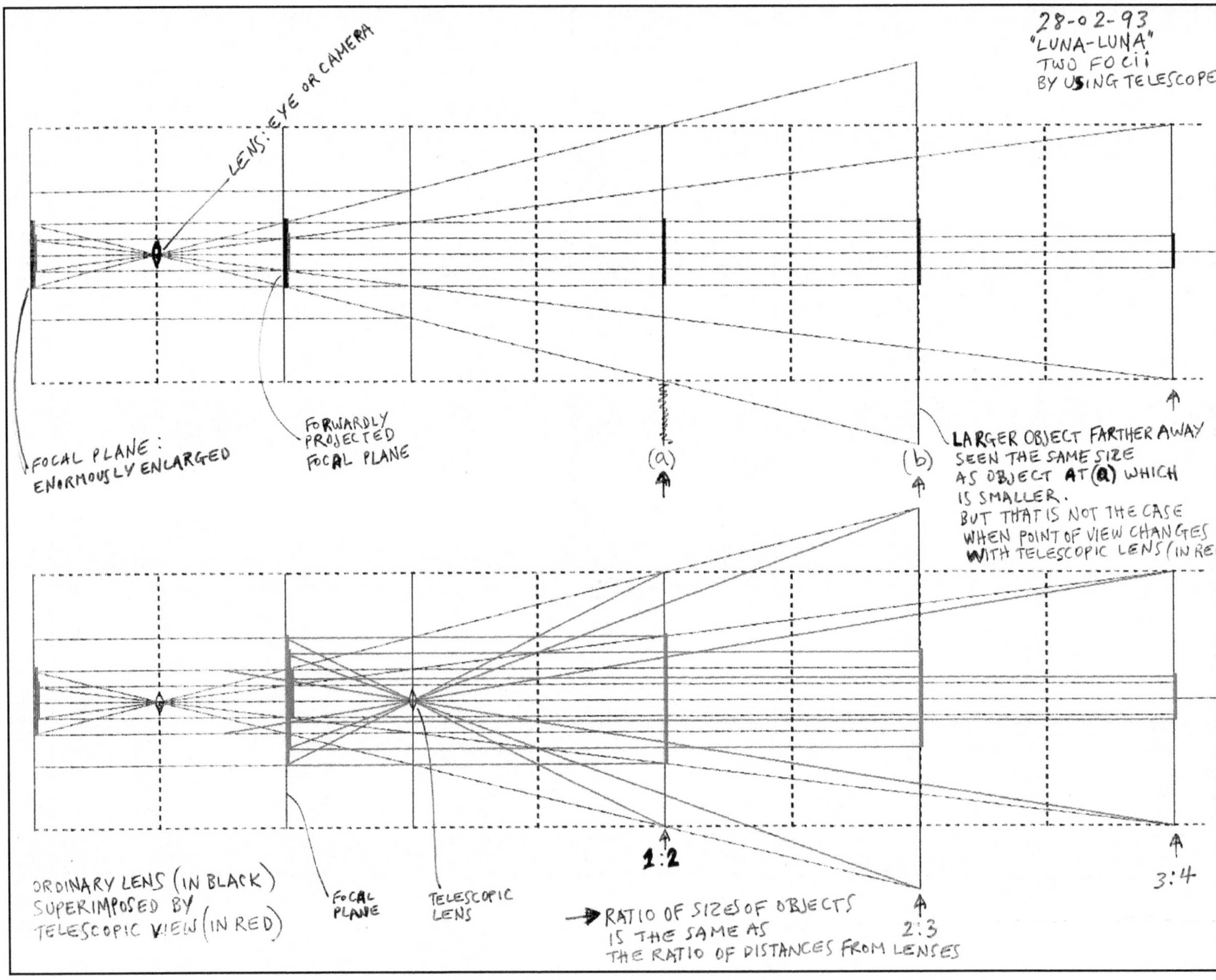

28-02-93
"LUNA-LUNA"
TWO FOCII
BY USING TELESCOPE
LENS: EYE OR CAMERA
FOCAL PLANE :
ENORMOUSLY ENLARGED
FORWARDLY
PROJECTED
FOCAL PLANE
(a)
(b)
LARGER OBJECT FARTHER AWAY
SEEN THE SAME SIZE
AS OBJECT AT (a) WHICH
IS SMALLER.
BUT THAT IS NOT THE CASE
WHEN POINT OF VIEW CHANGES
WITH TELESCOPIC LENS (IN RED
ORDINARY LENS (IN BLACK)
SUPERIMPOSED BY
TELESCOPIC VIEW (IN RED)
FOCAL
PLANE
TELESCOPIC
LENS
1:2
2:3
3:4
RATIO OF SIZES OF OBJECTS
IS THE SAME AS
THE RATIO OF DISTANCES FROM LENSES

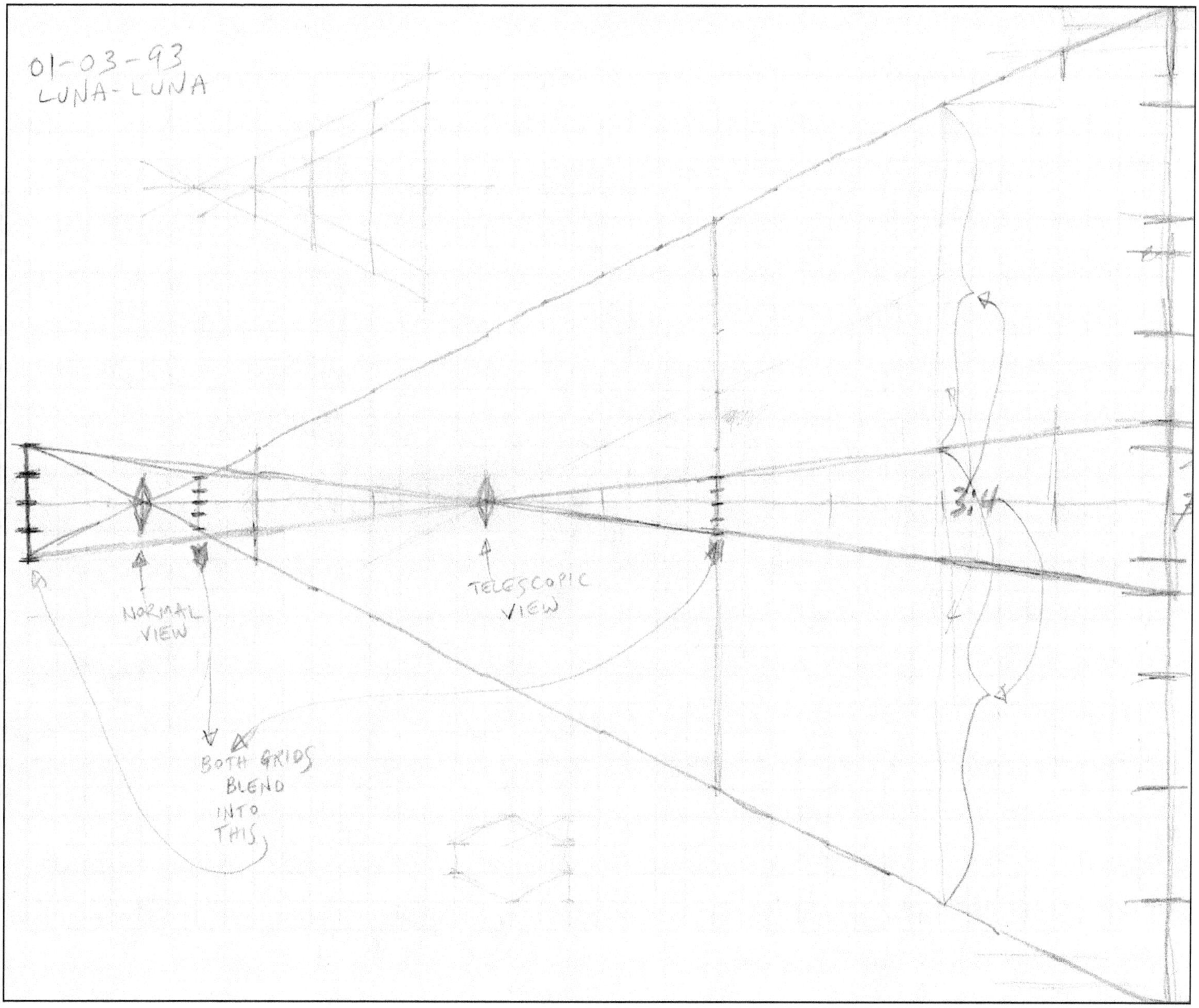
01-03-93
LUNA-LUNA
NORMAL VIEW
TELESCOPIC VIEW
BOTH GRIDS BLEND INTO THIS
3:4

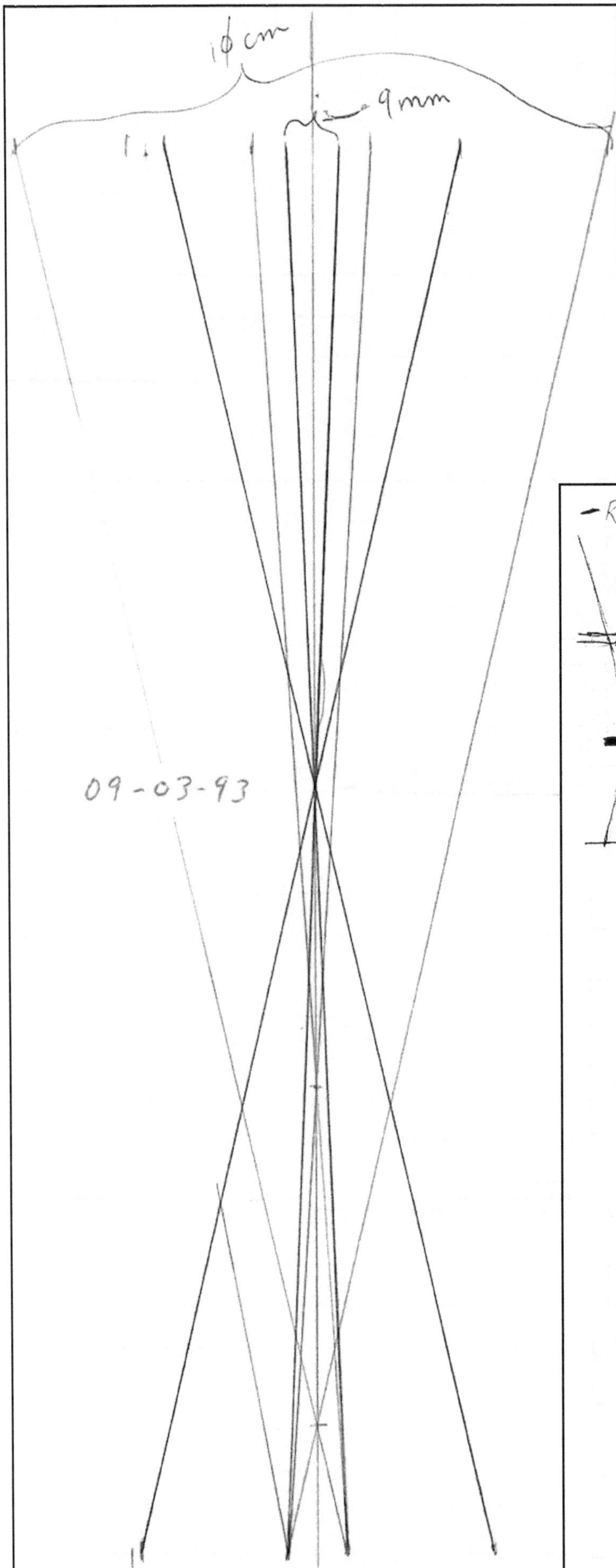
10 cm
9 mm
09-03-93

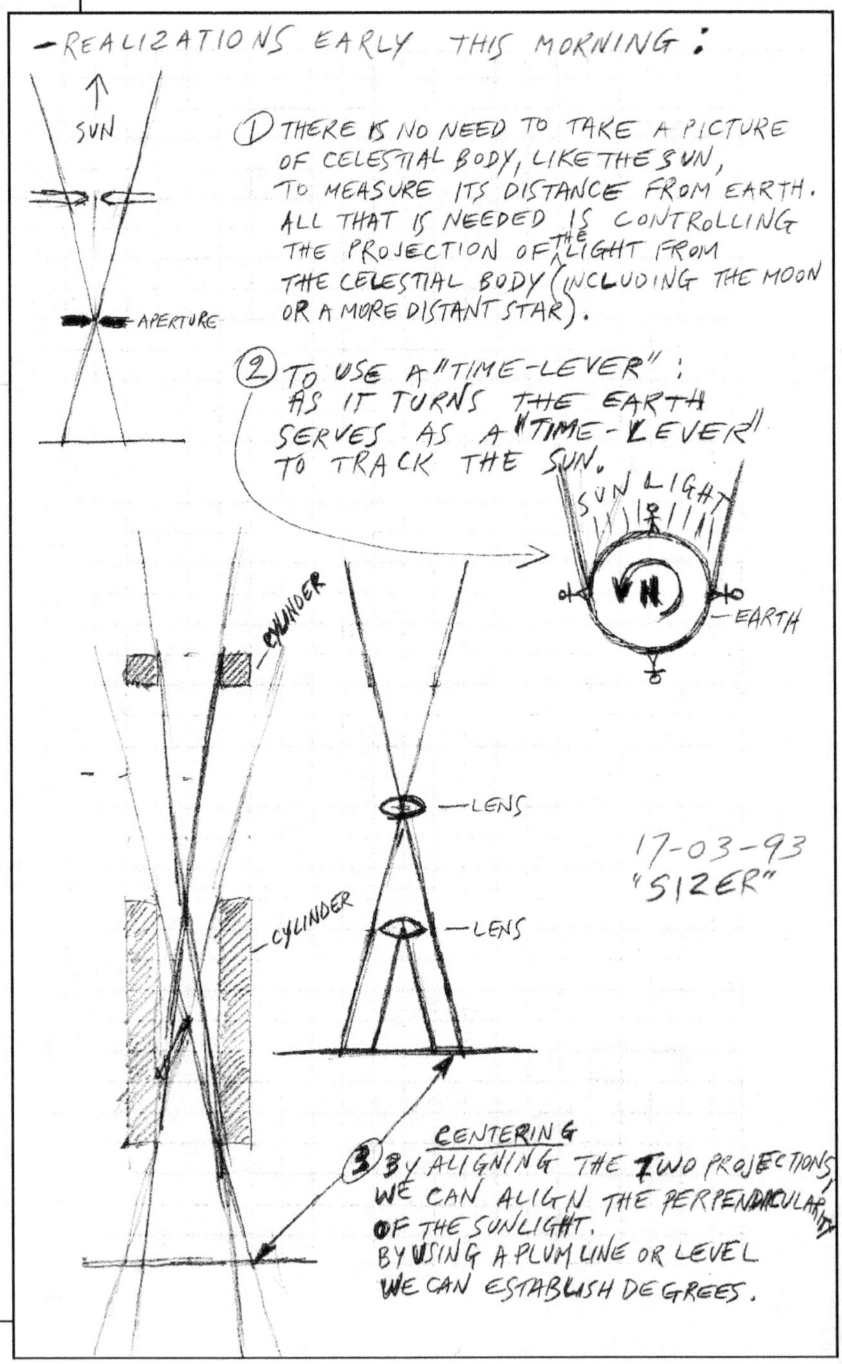
-REALIZATIONS EARLY THIS MORNING:
SUN
APERTURE
① THERE IS NO NEED TO TAKE A PICTURE OF CELESTIAL BODY, LIKE THE SUN, TO MEASURE ITS DISTANCE FROM EARTH. ALL THAT IS NEEDED IS CONTROLLING THE PROJECTION OF THE LIGHT FROM THE CELESTIAL BODY (INCLUDING THE MOON OR A MORE DISTANT STAR).
② TO USE A "TIME-LEVER": AS IT TURNS THE EARTH SERVES AS A "TIME-LEVER" TO TRACK THE SUN.
SUN LIGHT
EARTH
CYLINDER
CYLINDER
LENS
LENS
17-03-93
"SIZER"
CENTERING
③ BY ALIGNING THE TWO PROJECTIONS, WE CAN ALIGN THE PERPENDICULARITY OF THE SUNLIGHT.
BY USING A PLUM LINE OR LEVEL WE CAN ESTABLISH DEGREES.

LUNA-LUNA 25-03-93
(YESTERDAY I WAS
45 YEARS OLD)

PRESENT SITUATION
IN MY EFFORT
TO UREVEL THE SOLAR SYSTEM.
WHAT FOLLOWS ARE THE 3 STUMBLING BLOCKS:

(1) THE EARTH MUST TURN MORE THAN ONE COMPLETE CIRCLE A DAY (I.E., 1 MORE DEGREE) TO AGAIN FACE THE SUN THE WAY IT WAS THE DAY BEFORE.

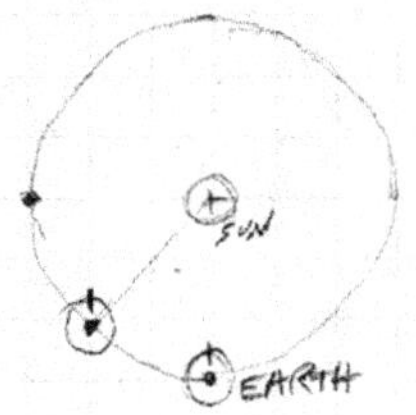

(2) THE LENGTH OF DAYLIGHT FROM WINTER TO SUMMER IS DUE TO THE TILT OF THE EARTH, BUT THE HEIGHT OF THE SUN ON THE HORIZON DURING WINTER DOES NOT SEEM TO MAKE SENSE.

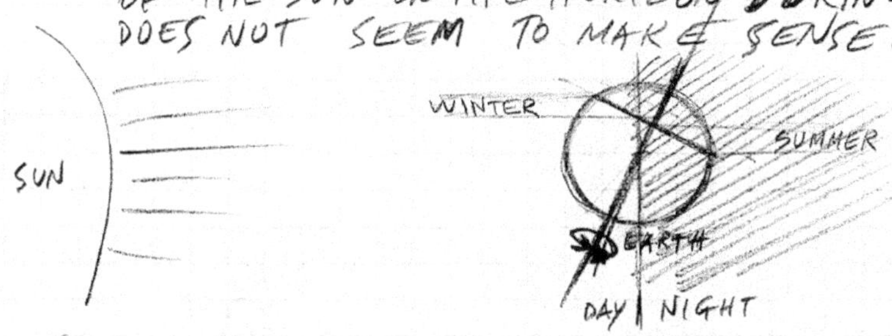

(3) TO MEASURE THE DISTANCE THE EARTH TRAVELS AS IT ROTATES AROUND THE SUN, BY OBSERVING THE SUN.
UNRESOLVED.

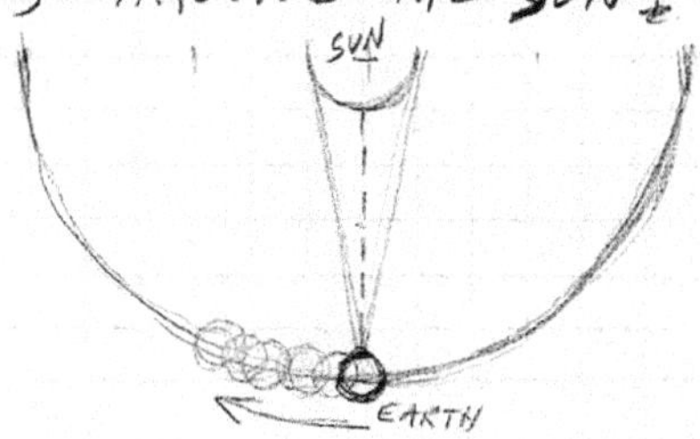

25-03-93
THE SOLAR
SYSTEM

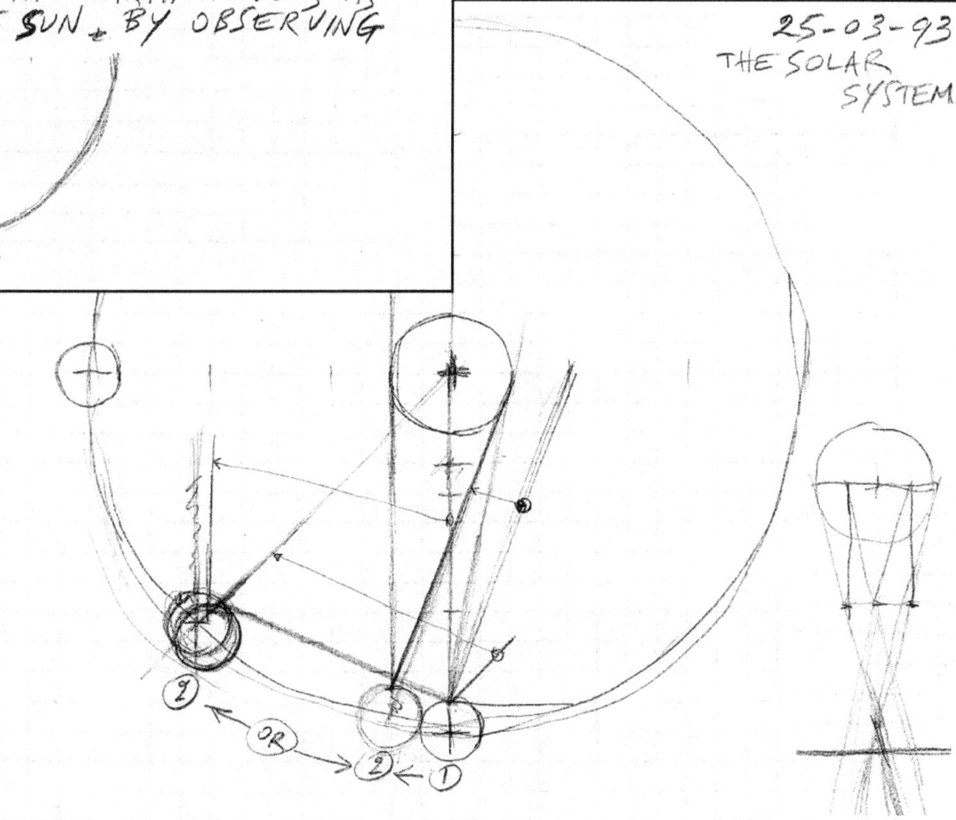

- CALCULATING DISTANCE TRAVELLED BY THE EARTH AS IT MOVES AROUND THE SUN.

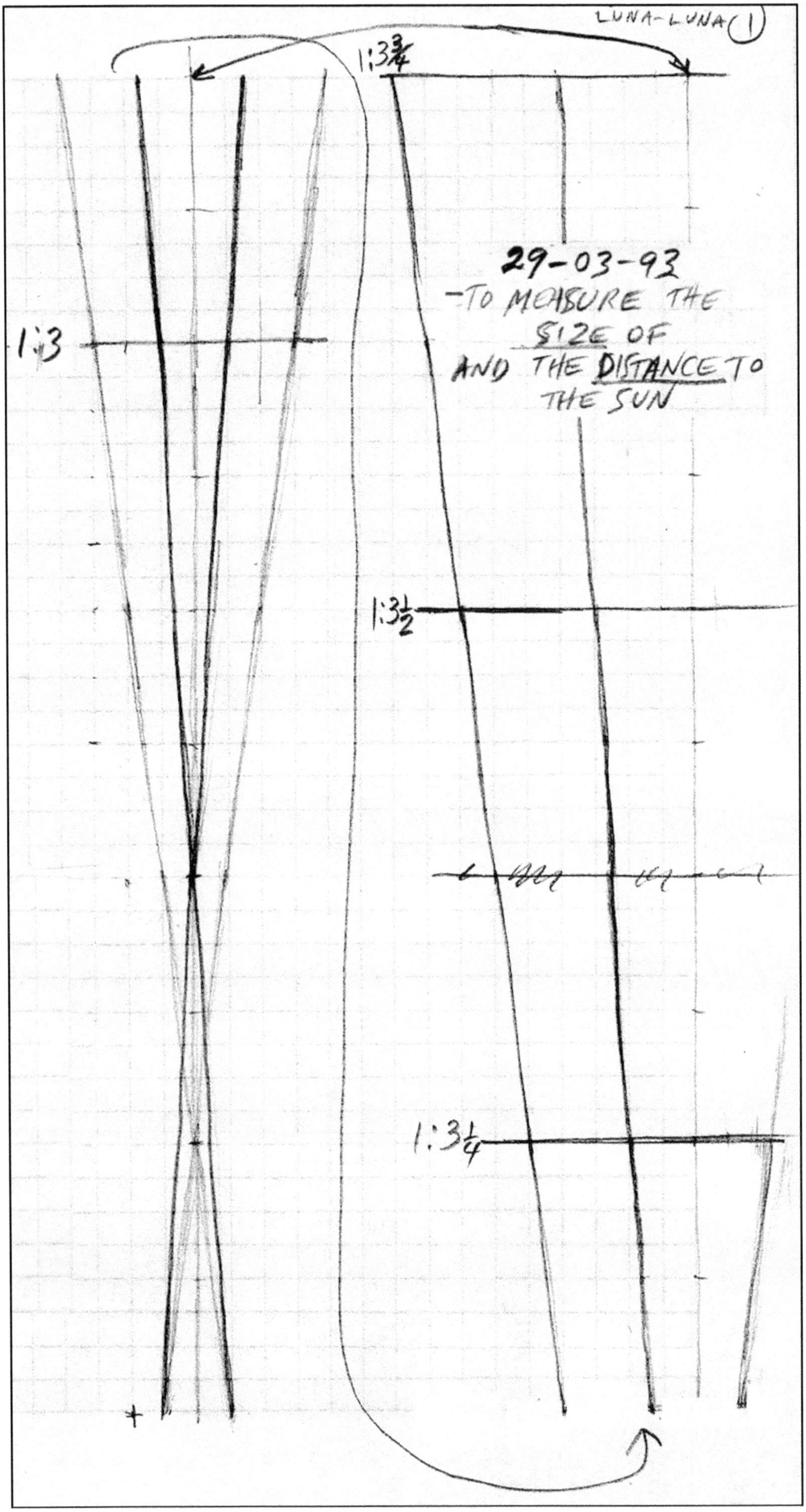
LUNA-LUNA (1)
1:3¾
29-03-93
-TO MEASURE THE SIZE OF AND THE DISTANCE TO THE SUN
1:3
1:3½
1:3¼

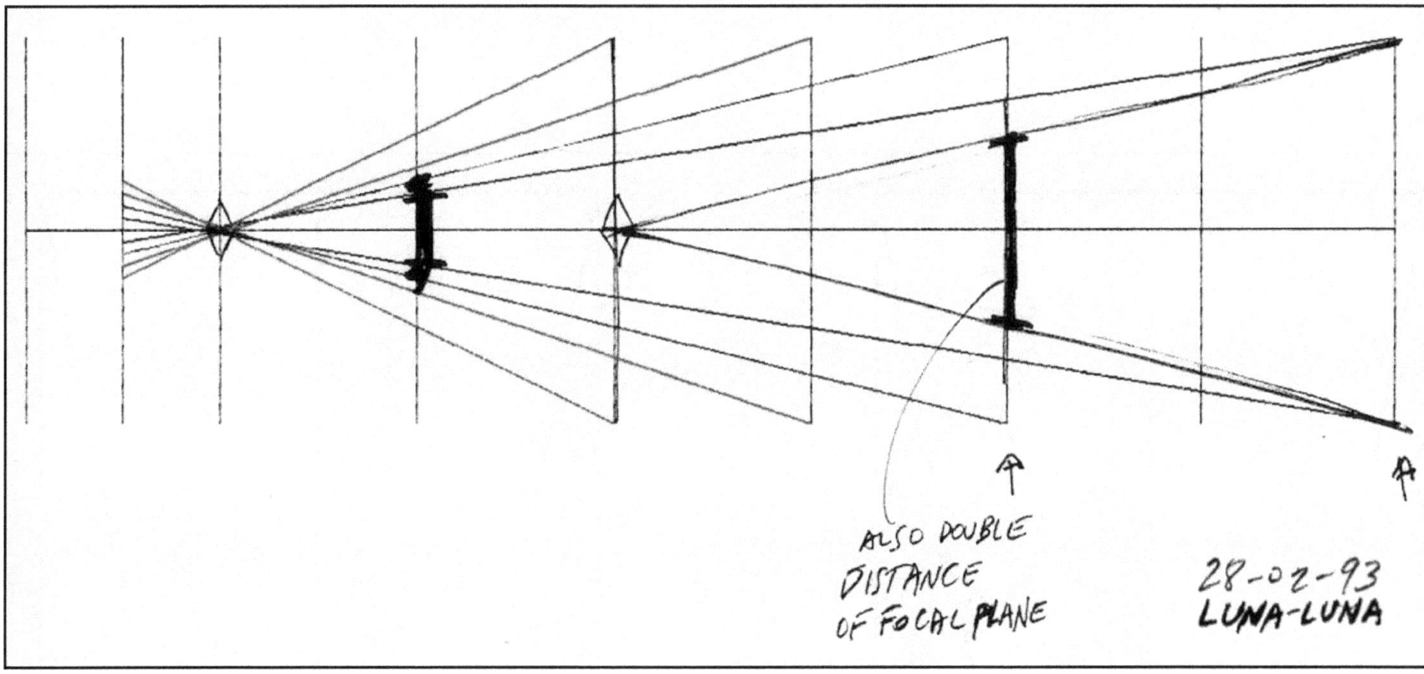

Measuring the stars.
01-01-95

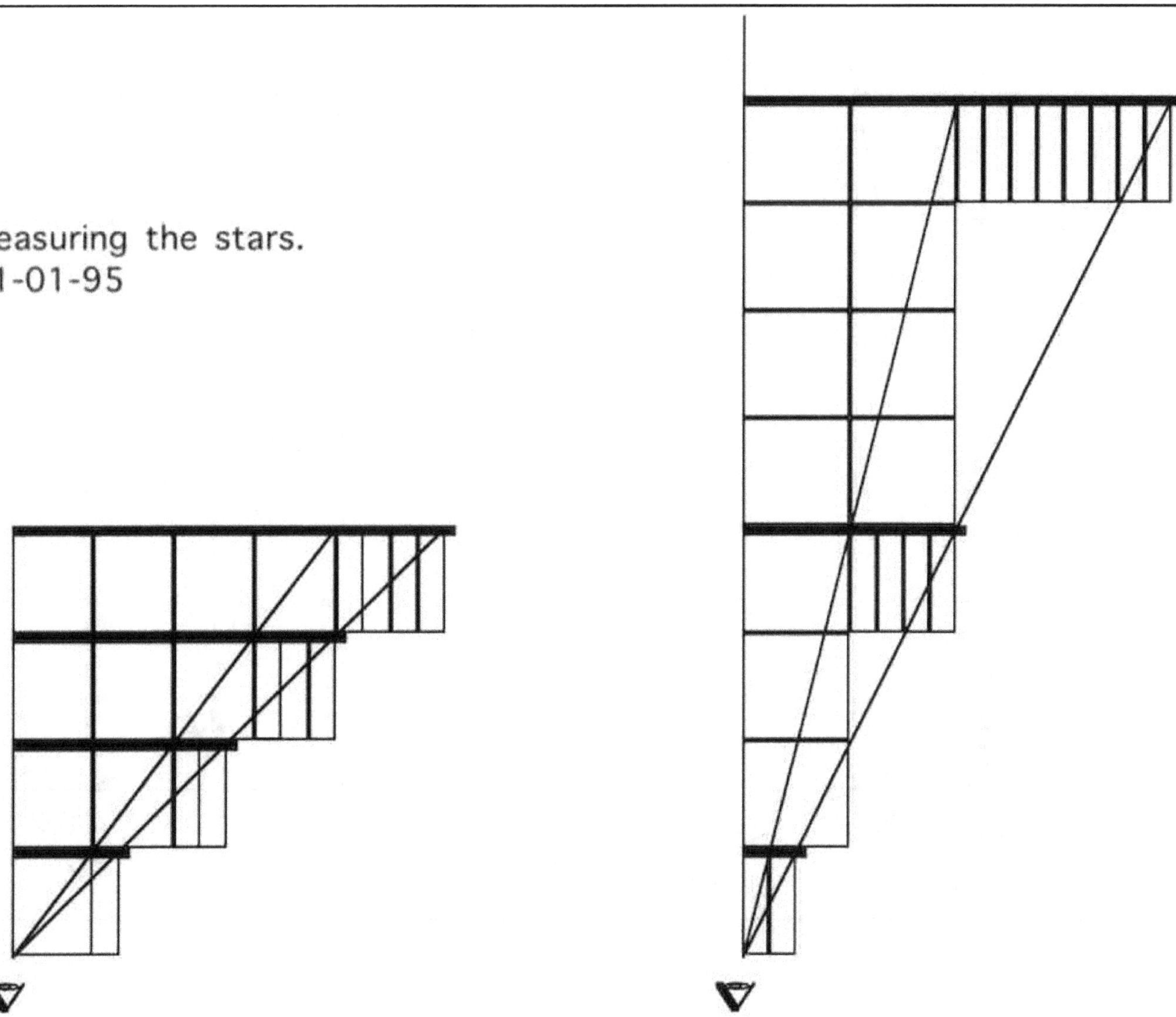

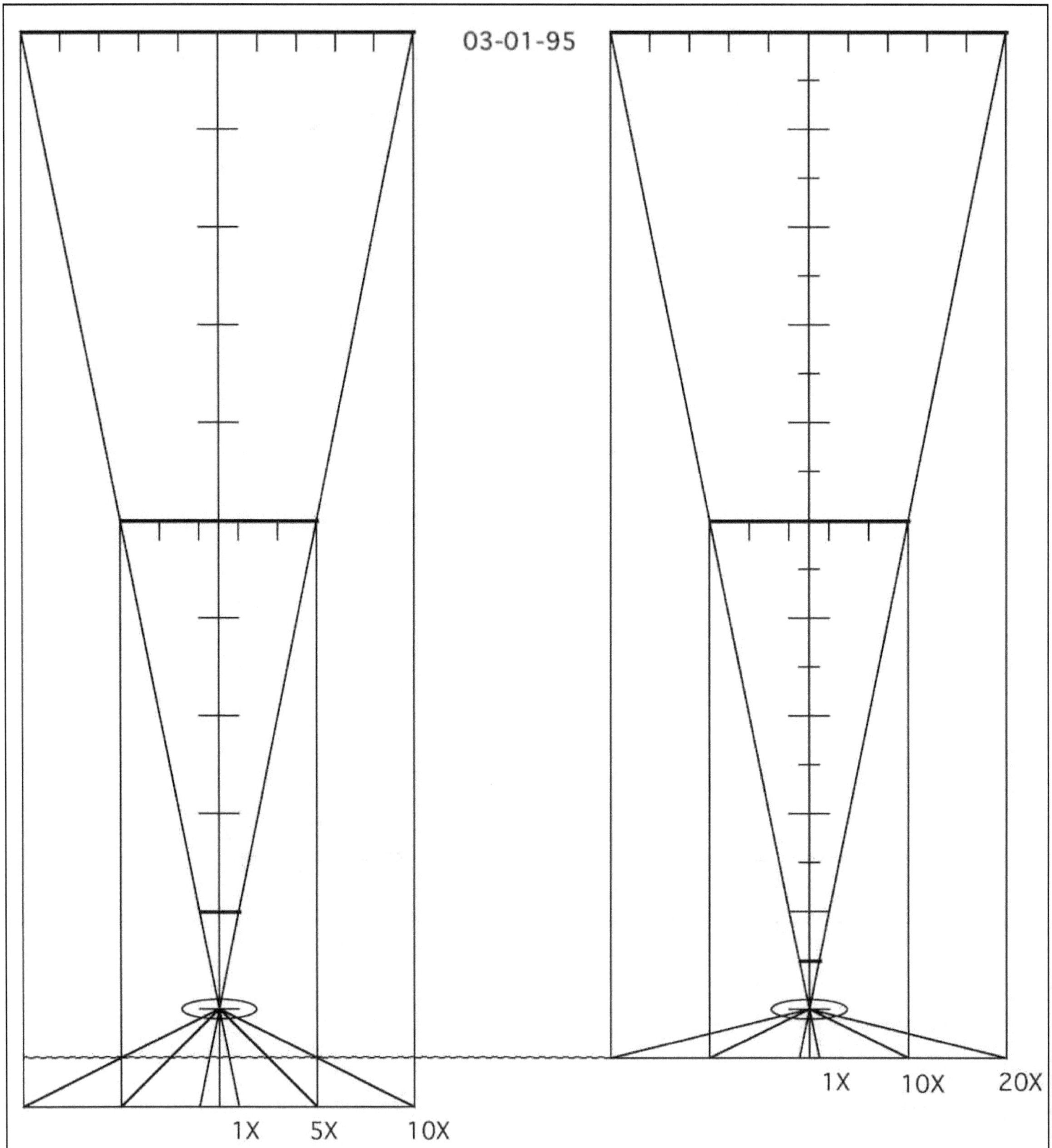
03-01-95
1X
5X
10X
1X
10X
20X

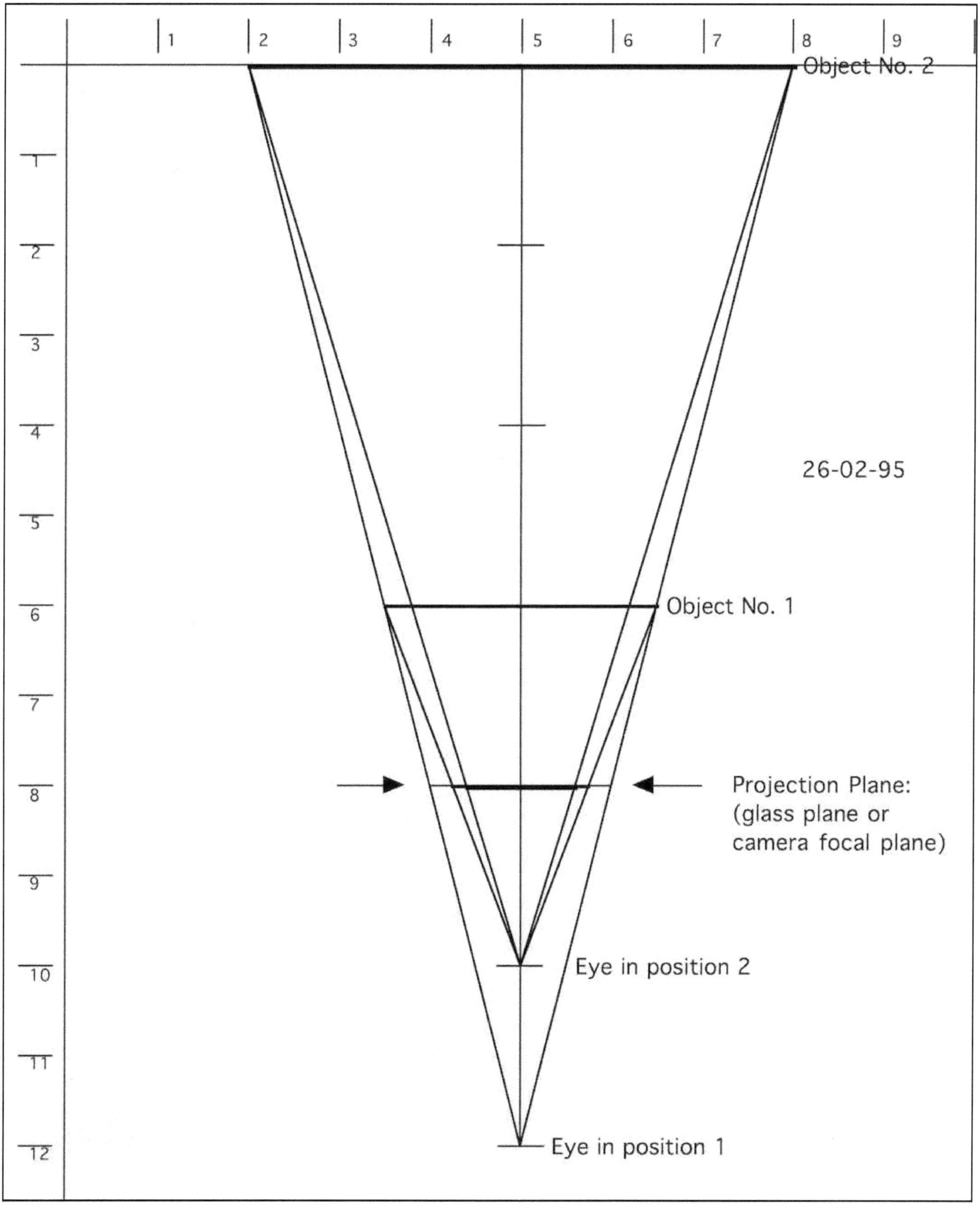
1
2
3
4
5
6
7
8
9
Object No. 2
1
2
3
4
5
6
7
8
9
10
11
12
26-02-95
Object No. 1
Projection Plane:
(glass plane or
camera focal plane)
Eye in position 2
Eye in position 1

10C. MEASURING PEN

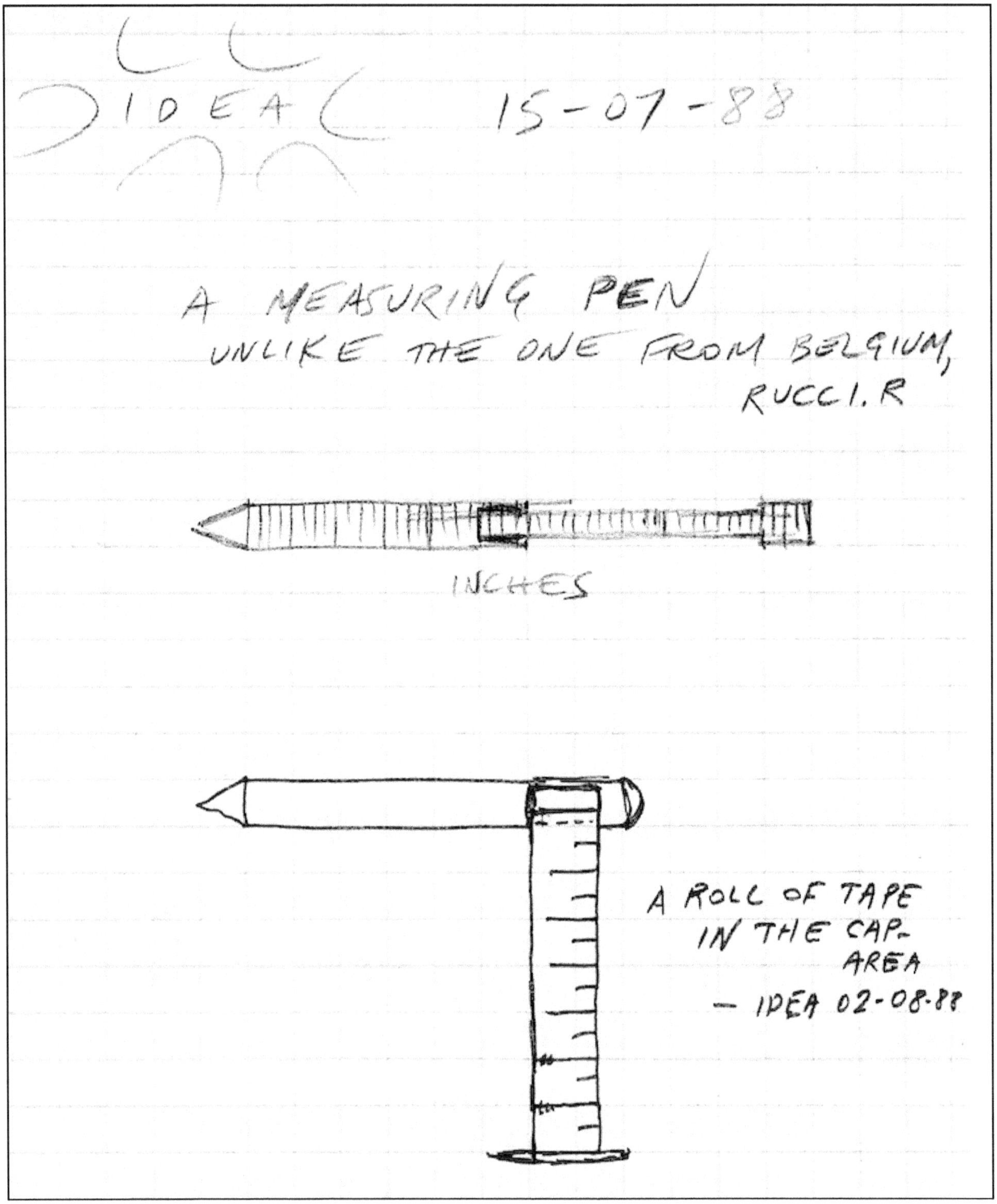

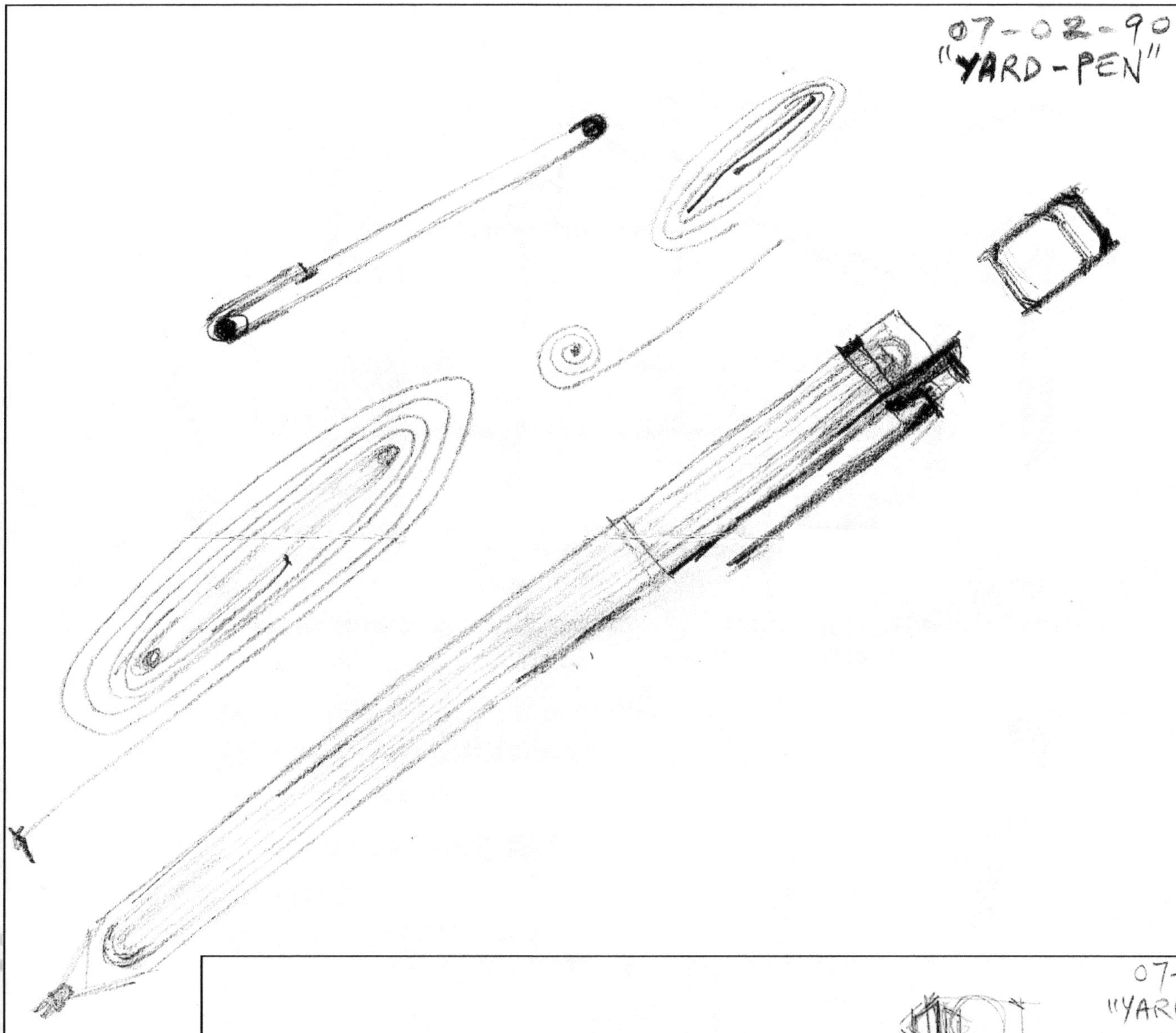
07-02-90
"YARD-PEN"

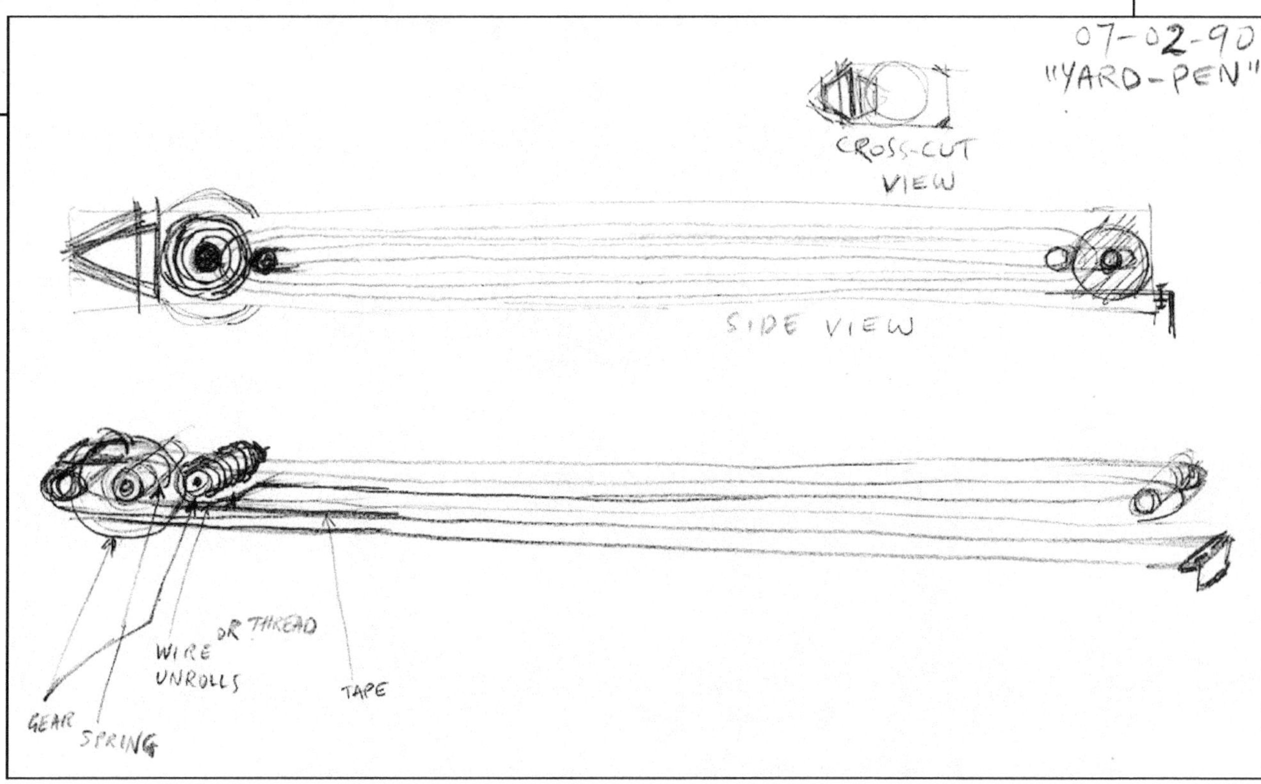
07-02-90
"YARD-PEN"
CROSS-CUT VIEW
SIDE VIEW
WIRE OR THREAD UNROLLS
TAPE
GEAR
SPRING

10D. FIGURE ABSTRACTION

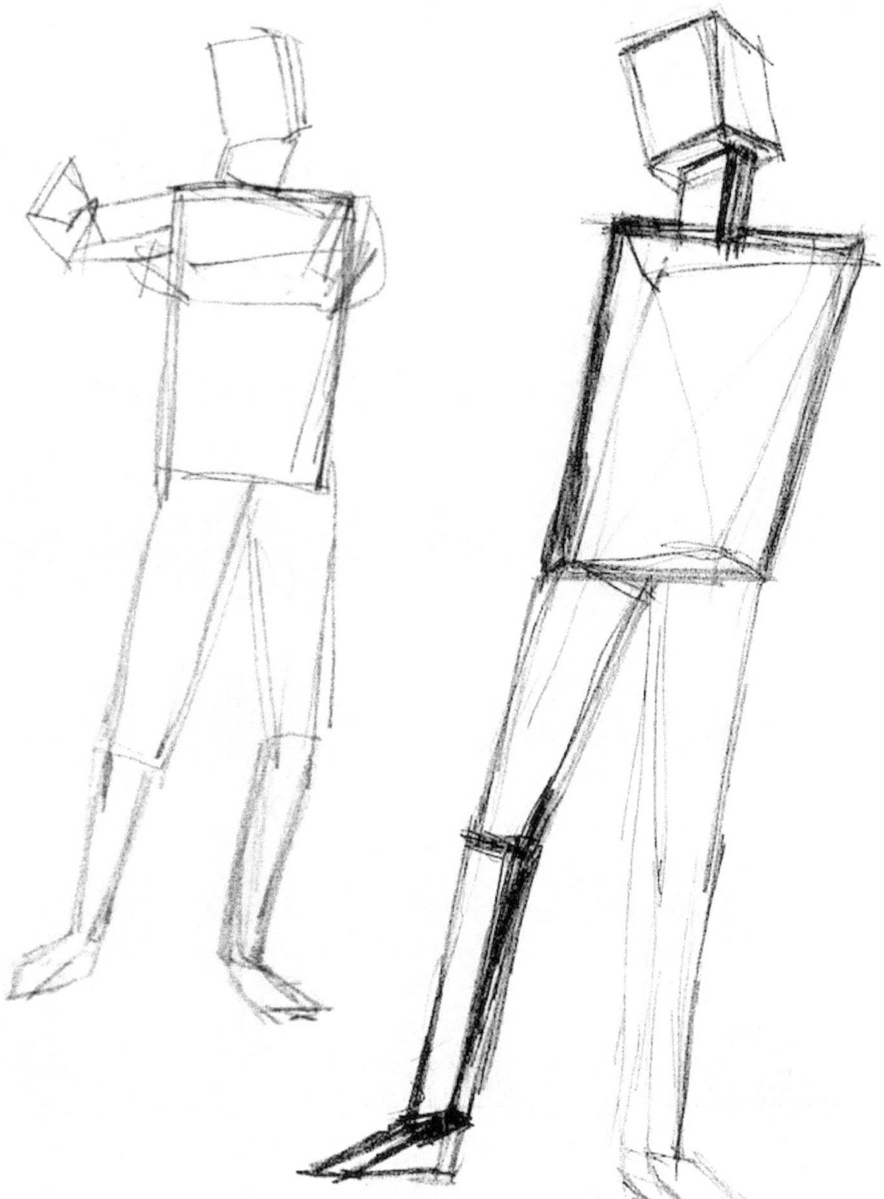

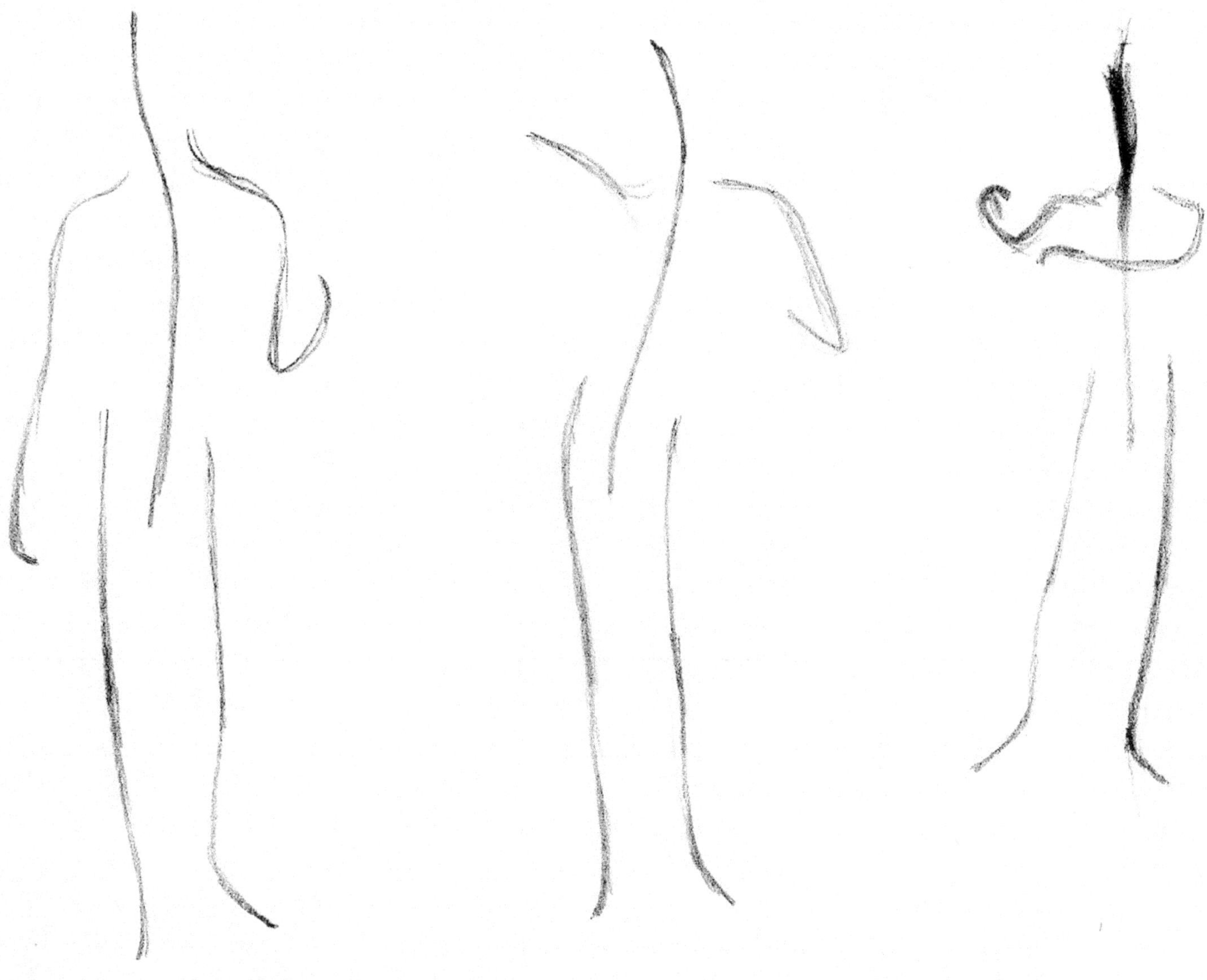

11. COMPUTER

3 IDEAS TO MAKE INPUT TO A COMPUTER DECIMAL.

(1) FIBER GLASS WIRES OF DIFFERENT COLORS. EACH COLOR A DIFFERENT NUMBER OR LETTER.

(2) TO INPUT PULSES NOT MERELY HORIZONTALLY BUT ALSO VERTICALLY SO TO HAVE TEN LEVELS OF CIRCUIT BOARDS.

(3) TO READ AS A DIFFERENT NUMBER OR LETTER A PULSE OF HIGHER OR LOWER VOLTAGE.

04-08-83

PAGE 5
26-12-85

AI READER

FUNCTION READER

ONCE IT HAS ABSTRACTED THE BASIC SHAPE OF THE OBJECT BY ZOOMING-OUT AND/OR IN, THE MACHINE COMPARES WITH BASIC SHAPES STORED ON A LASER DISK TO FIND OUT FUNCTION (e.g. SEAT: ONE SITS ON IT (OR) DOOR: TO OPEN HOUSE/ROOM), GIVE THE DIFFERING VARIATIONS OF AN OBJECT, + HENCE IDENTIFY IT

→ WHAT I REASONED: IT IS FUNCTION THAT IDENTIFIES OBJECT, e.g. THINK OF THE MANY FORMS OF A CHAIR

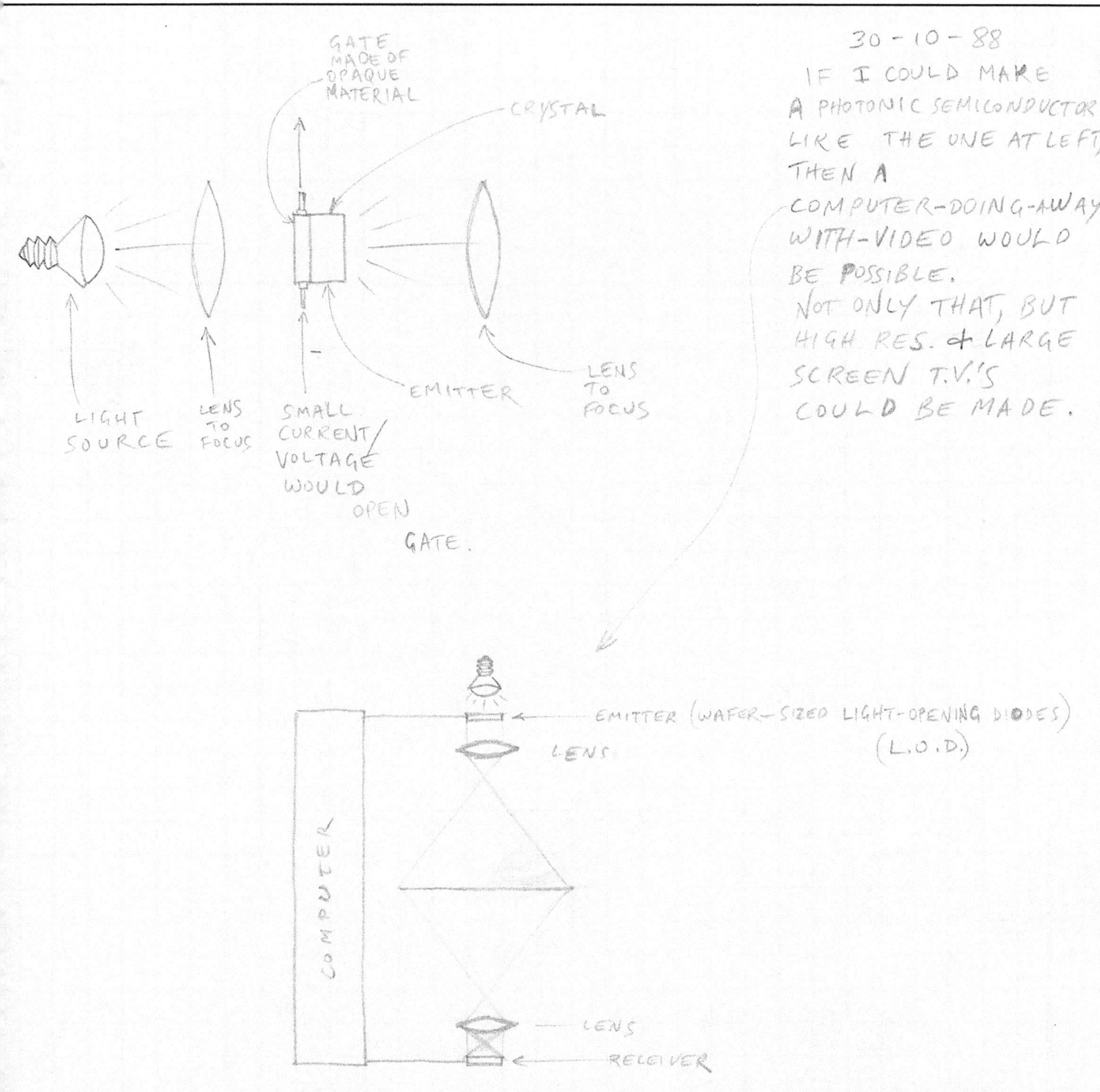
GATE MADE OF OPAQUE MATERIAL
CRYSTAL
LIGHT SOURCE
LENS TO FOCUS
SMALL CURRENT/VOLTAGE WOULD OPEN GATE.
EMITTER
LENS TO FOCUS
30-10-88
IF I COULD MAKE A PHOTONIC SEMICONDUCTOR LIKE THE ONE AT LEFT, THEN A COMPUTER-DOING-AWAY WITH-VIDEO WOULD BE POSSIBLE.
NOT ONLY THAT, BUT HIGH RES. + LARGE SCREEN T.V.'S COULD BE MADE.
EMITTER (WAFER-SIZED LIGHT-OPENING DIODES) (L.O.D.)
LENS
COMPUTER
LENS
RECEIVER

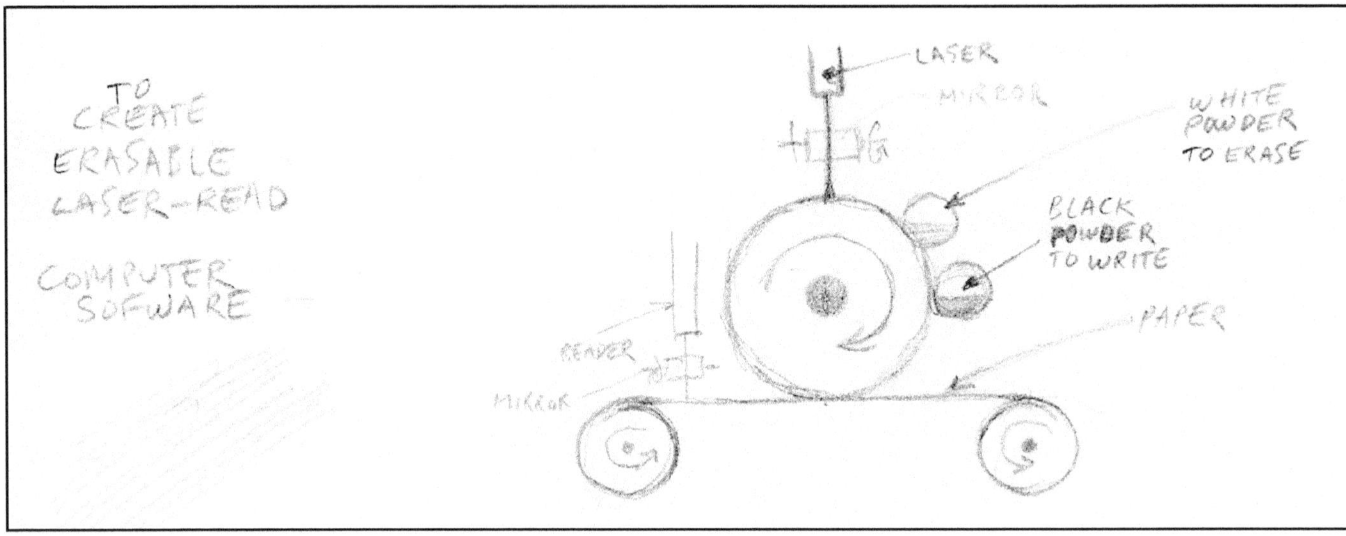
TO CREATE ERASABLE LASER-READ COMPUTER SOFWARE
LASER
MIRROR
WHITE POWDER TO ERASE
BLACK POWDER TO WRITE
PAPER
READER
MIRROR

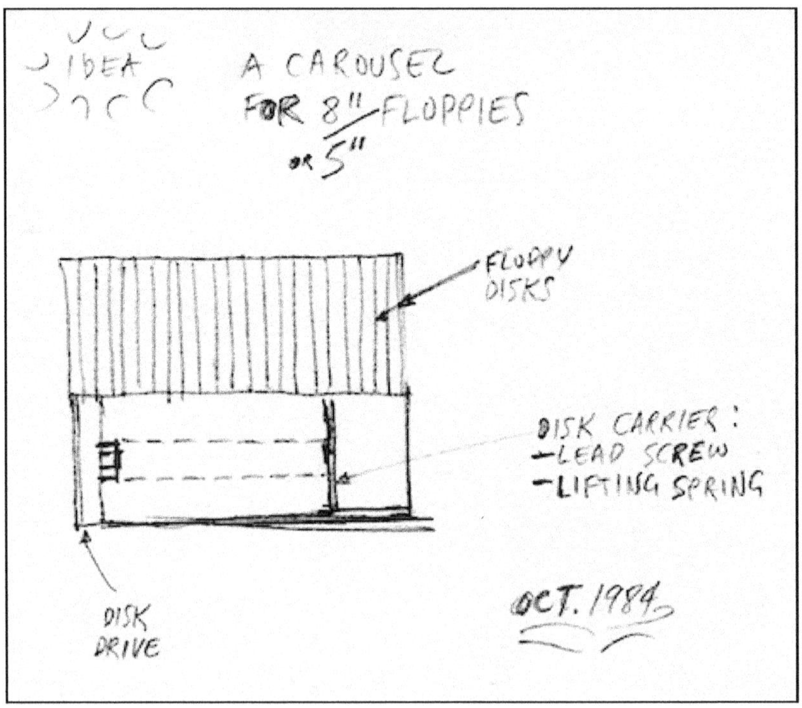
IDEA
A CAROUSEL FOR 8" FLOPPIES or 5"
FLOPPY DISKS
DISK CARRIER:
-LEAD SCREW
-LIFTING SPRING
DISK DRIVE
OCT. 1984

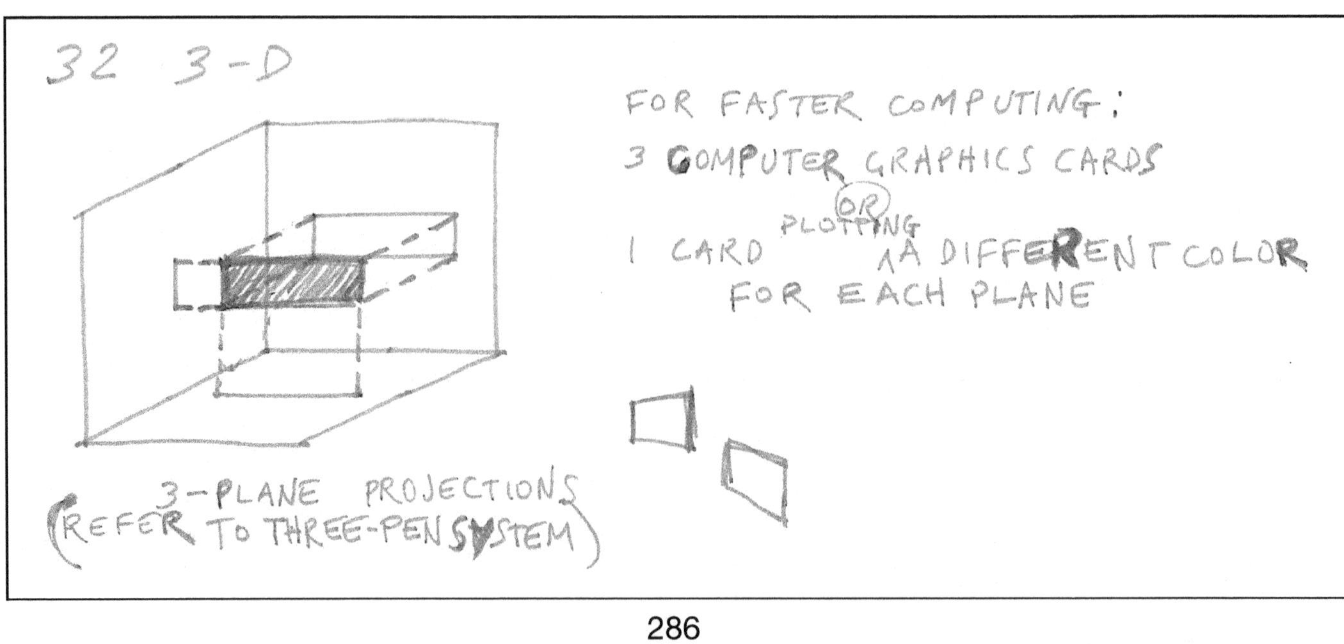
32 3-D
FOR FASTER COMPUTING:
3 COMPUTER GRAPHICS CARDS
OR
1 CARD PLOTTING A DIFFERENT COLOR FOR EACH PLANE
3-PLANE PROJECTIONS
(REFER TO THREE-PEN SYSTEM)

12. KEYBOARD

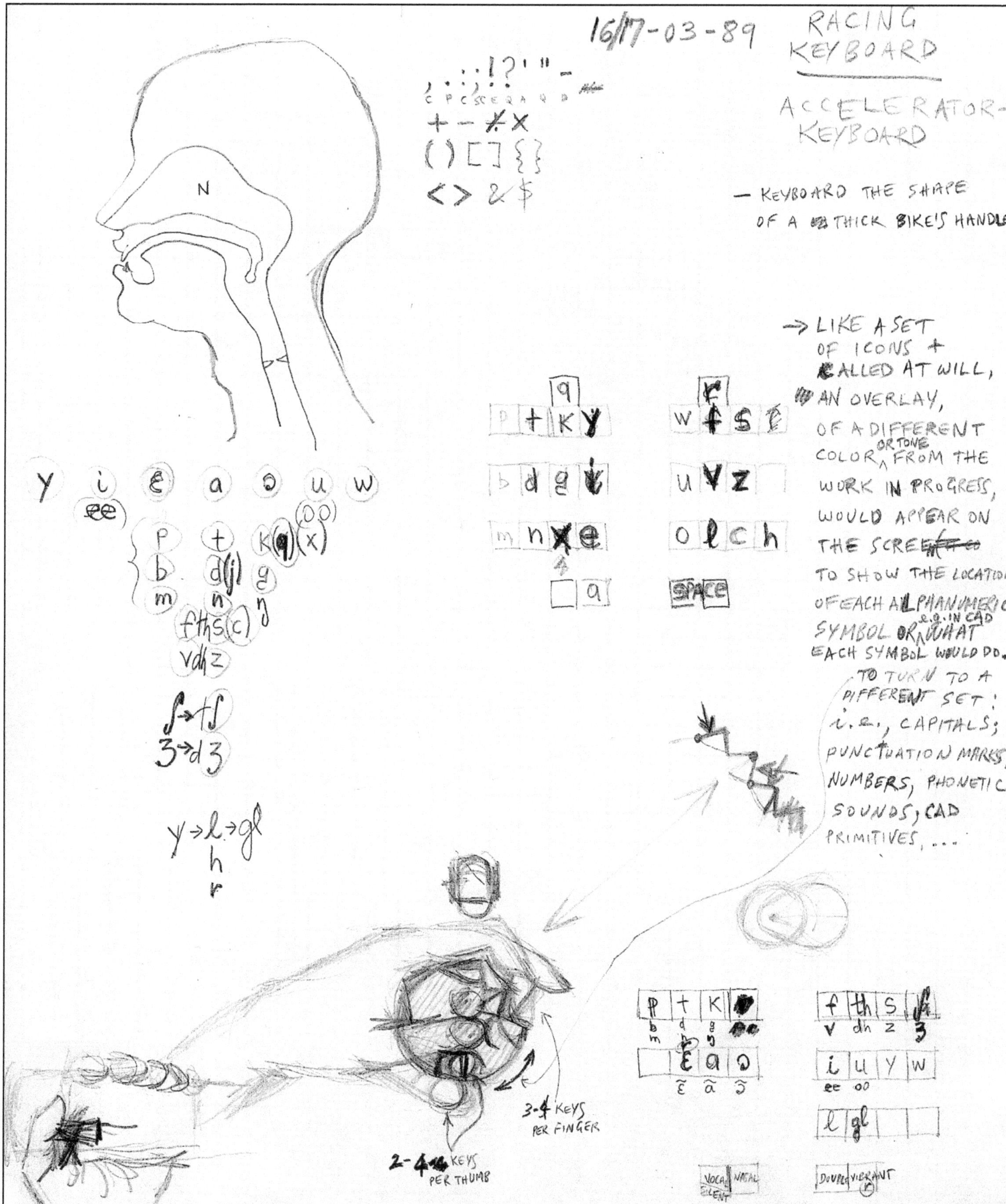
16/17-03-89
RACING KEYBOARD
ACCELERATOR-KEYBOARD
— KEYBOARD THE SHAPE OF A THICK BIKE'S HANDLE
→ LIKE A SET OF ICONS + CALLED AT WILL, AN OVERLAY, OF A DIFFERENT COLOR OR TONE FROM THE WORK IN PROGRESS, WOULD APPEAR ON THE SCREEN TO SHOW THE LOCATION OF EACH ALPHANUMERIC SYMBOL OR e.g. IN CAD WHAT EACH SYMBOL WOULD DO.
TO TURN TO A DIFFERENT SET: i.e., CAPITALS; PUNCTUATION MARKS, NUMBERS, PHONETIC SOUNDS, CAD PRIMITIVES, ...
SPACE
3-4 KEYS PER FINGER
2-4 KEYS PER THUMB
VOCAL ELEMENT
NASAL
DOUBLE VIBRANT

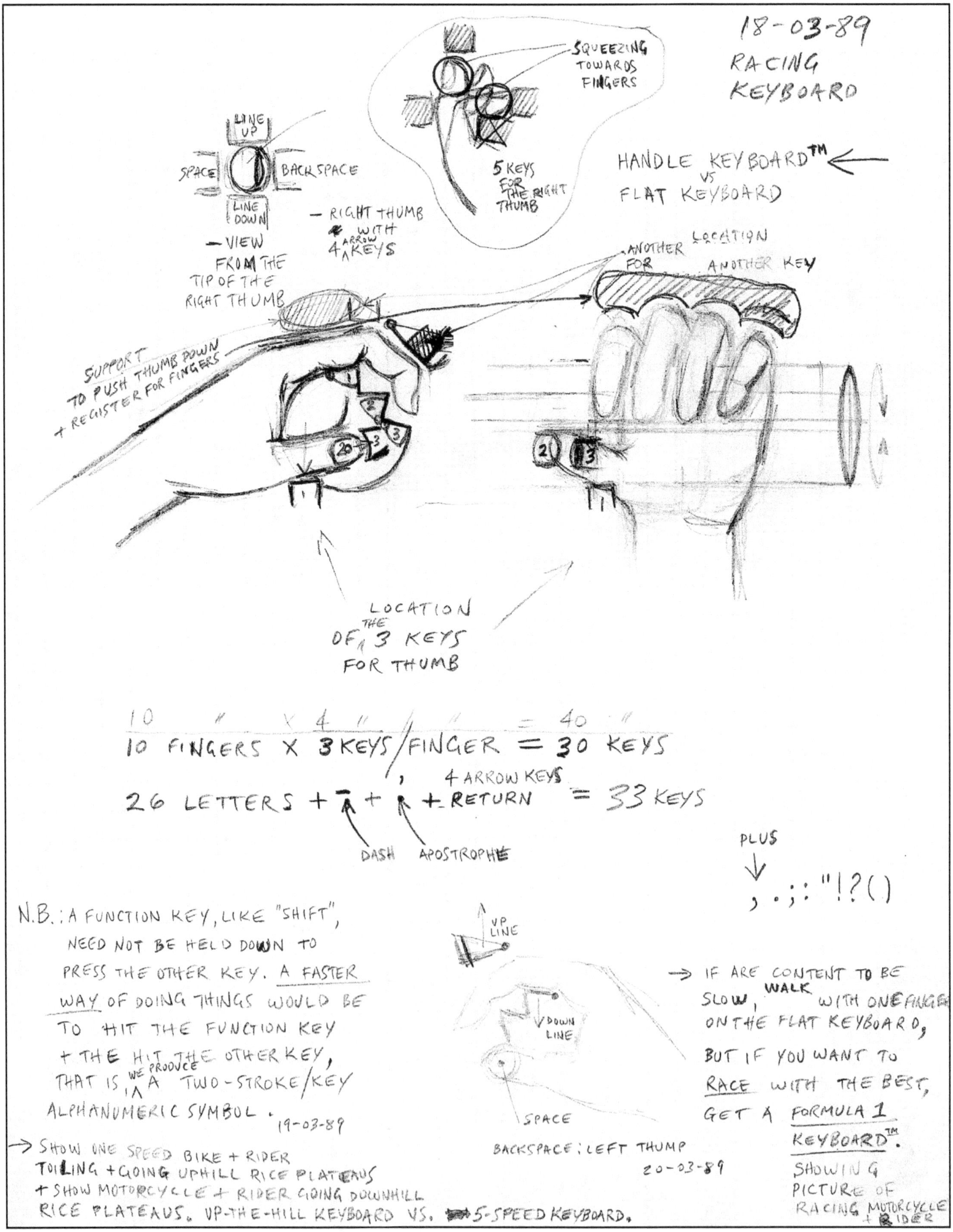
18-03-89
RACING KEYBOARD
SQUEEZING TOWARDS FINGERS
5 KEYS FOR THE RIGHT THUMB
LINE UP
SPACE
BACKSPACE
LINE DOWN
– VIEW FROM THE TIP OF THE RIGHT THUMB
– RIGHT THUMB WITH 4 ARROW KEYS
HANDLE KEYBOARD™ VS FLAT KEYBOARD
ANOTHER LOCATION FOR ANOTHER KEY
SUPPORT TO PUSH THUMB DOWN + REGISTER FOR FINGERS
LOCATION OF THE 3 KEYS FOR THUMB
10 " X 4 " / " = 40 "
10 FINGERS X 3 KEYS/FINGER = 30 KEYS
26 LETTERS + - + ' + 4 ARROW KEYS + RETURN = 33 KEYS
DASH
APOSTROPHE
PLUS
, . ; : " ! ? ( )
N.B.: A FUNCTION KEY, LIKE "SHIFT", NEED NOT BE HELD DOWN TO PRESS THE OTHER KEY. A FASTER WAY OF DOING THINGS WOULD BE TO HIT THE FUNCTION KEY + THE HIT THE OTHER KEY, THAT IS, WE PRODUCE A TWO-STROKE/KEY ALPHANUMERIC SYMBOL. 19-03-89
UP LINE
DOWN LINE
SPACE
BACKSPACE: LEFT THUMP
20-03-89
→ IF ARE CONTENT TO BE SLOW, WALK WITH ONE FINGER ON THE FLAT KEYBOARD, BUT IF YOU WANT TO RACE WITH THE BEST, GET A FORMULA 1 KEYBOARD™.
SHOWING PICTURE OF RACING MOTORCYCLE + RIDER
→ SHOW ONE SPEED BIKE + RIDER TOILING + GOING UPHILL RICE PLATEAUS + SHOW MOTORCYCLE + RIDER GOING DOWNHILL RICE PLATEAUS. UP-THE-HILL KEYBOARD VS. 5-SPEED KEYBOARD.

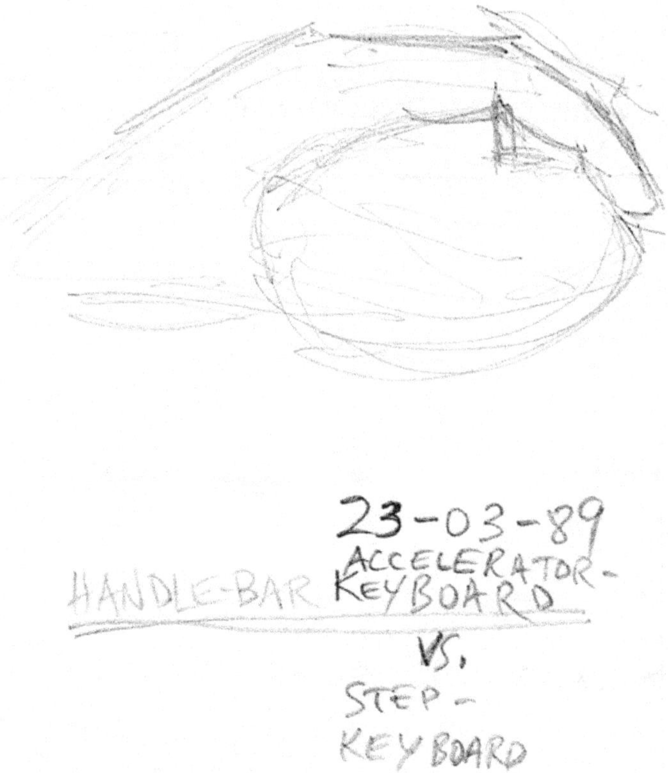

Typing would be by squeezing key, not by hitting it. And, to help the process, to make certain that the letter was typed, a clicking sound would accompany the squeezing of the key (when the squeezing is enough). This way the typist would know the letter has been typed, the same as with the production of speech sounds.

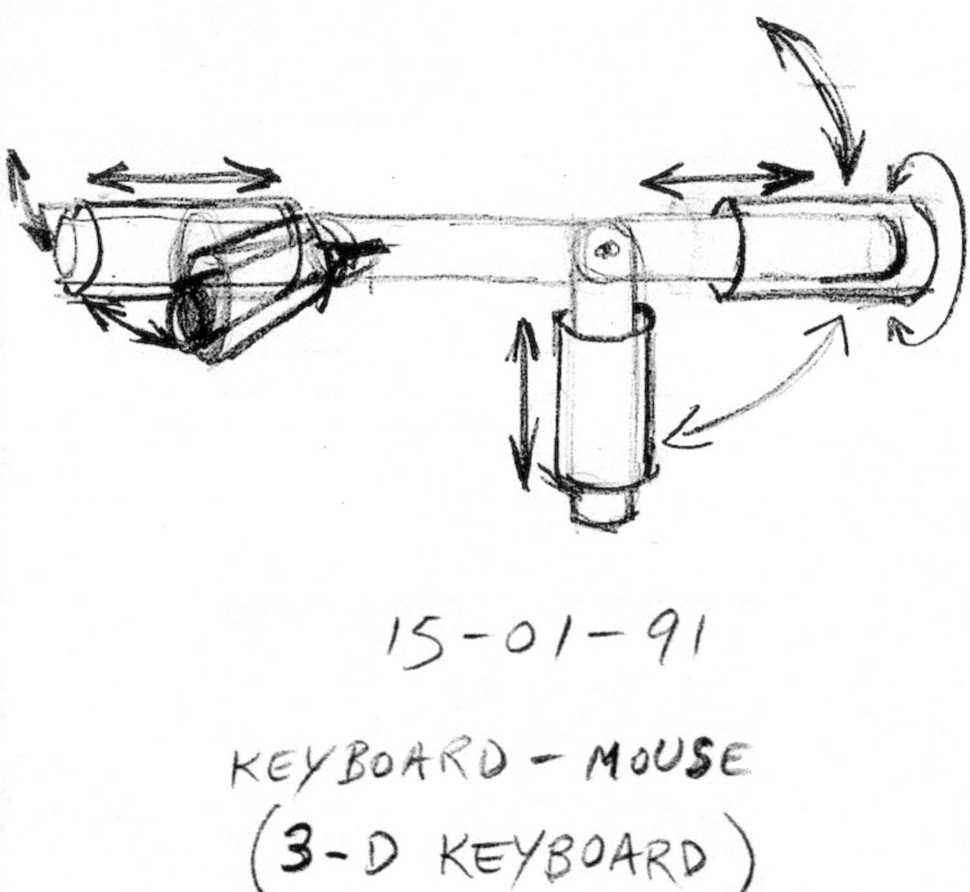

13. OTHER IDEAS

Connected Car (Knowing What The Other Cars Are Doing)

IDEAS MARCH 31, 1984

A CAR ENDOWED WITH ELECTRONIC SIGHT ROUTE WOULD BE COMPUTERIZED. THREE LANES: ONE FAST, ONE SLOW TO GET ONTO THE FAST ONE, AND ONE TO PARK.

HIGHWAY COULD ALSO WORK LIKE AN ELEVATOR. AN EMPTY CAR WOULD COME OUT OF THE FAST LANE TO THE PARKING LANE WHEN CALLED, WITHOUT HAVING THE OTHER CARS-TO-FURTHER-DESTINATIONS ALSO STOPPING.

Little Robot Ball As Security Guard

IDEAS MARCH 1987

BALL

- SECURITY GUARD
  - A LITTLE ROBOT BALL (IT COULD BE AS TINY AS IT COULD ROLL UNDER THE COUCH)
    - SELF-PROPELLED
    - SELF-BOUNCING (TO ATTACK OR TO GET HIGH PLACES OR UP + DOWN THE STAIRS)
    - SELF-VACUUM

07-04-89

IDEAS

DOOR INSTALLER

ALUMINUM FRAME TO PUT UP DOOR FRAME + DOOR

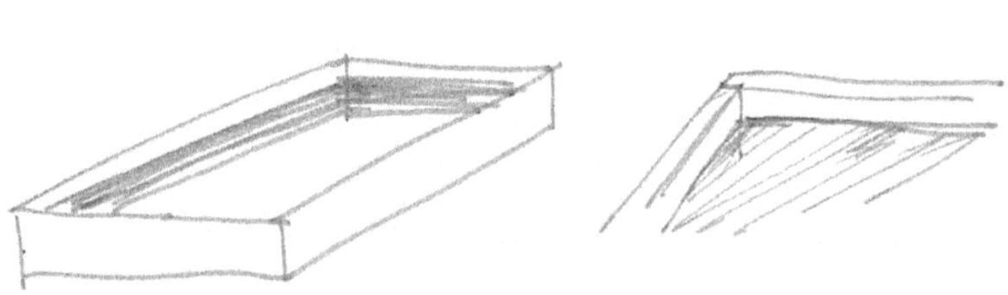

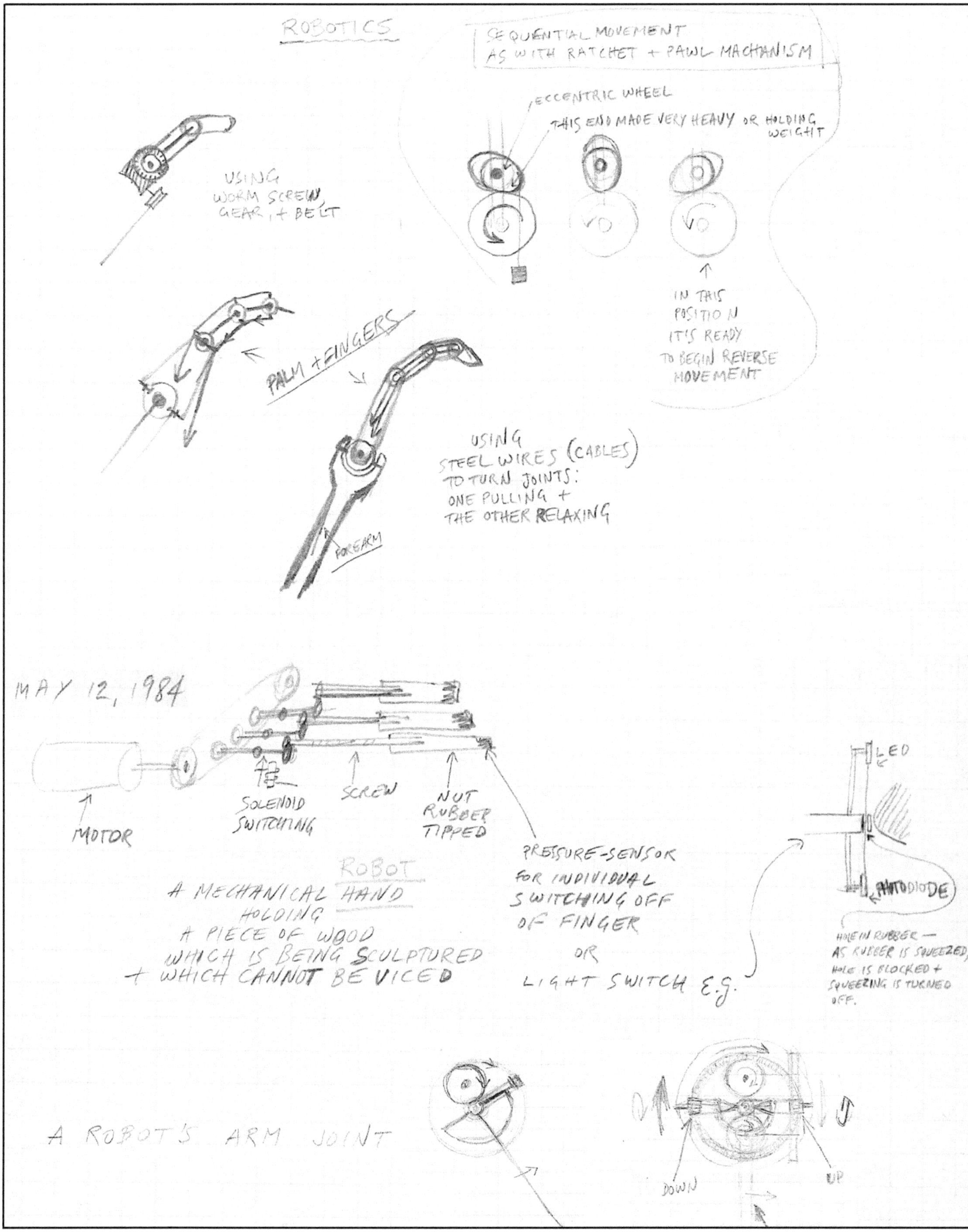
ROBOTICS
SEQUENTIAL MOVEMENT
AS WITH RATCHET + PAWL MACHANISM
ECCENTRIC WHEEL
THIS END MADE VERY HEAVY OR HOLDING WEIGHT
USING
WORM SCREW,
GEAR, + BELT
IN THIS
POSITION
IT'S READY
TO BEGIN REVERSE
MOVEMENT
PALM + FINGERS
USING
STEEL WIRES (CABLES)
TO TURN JOINTS:
ONE PULLING +
THE OTHER RELAXING
FOREARM
MAY 12, 1984
MOTOR
SOLENOID
SWITCHING
SCREW
NUT
RUBBER
TIPPED
LED
PHOTODIODE
ROBOT
A MECHANICAL HAND
HOLDING
A PIECE OF WOOD
WHICH IS BEING SCULPTURED
+ WHICH CANNOT BE VICED
PRESSURE-SENSOR
FOR INDIVIDUAL
SWITCHING OFF
OF FINGER
OR
LIGHT SWITCH E.G.
HOLE IN RUBBER —
AS RUBBER IS SQUEEZED,
HOLE IS BLOCKED +
SQUEEZING IS TURNED
OFF.
A ROBOT'S ARM JOINT
DOWN
UP

20/08/84

— INVENTIONS —

→ COMPUTER HAND

TWO SENSORS:
(1.) VISUAL (AWAY): THE MEETING OF THE LINE ON MONITOR SCREEN
(2.) TACTILE (THERE): THE SENSING OF PRESSURE/OBSTACLE

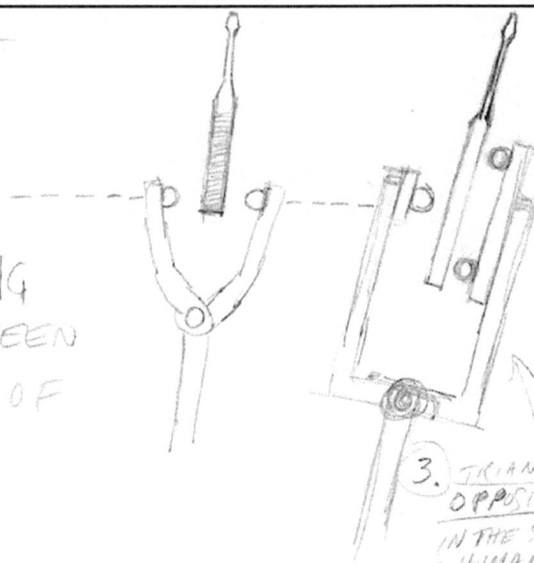

(3.) TRIANGULAR-OPPOSITION: IN THE SAME WAY A HUMAN HAND HOLDS A PENCIL

→ THE SPIDER ROBOT

· LIKE A SPIDER, A MULTILEGGED MACHINE THAT WALKS/MANOUVERS ON THE WALL.
COULD ALSO LEGS FUNCTION AS A HELICOPTER'S BLADES IN CASE OF A FALL?

THE SOLUTION SHOULD BE FOUND IN SOMETHING SIMILAR TO TRIANGULAR OPPOSITION (ABOVE) OR A WALL SHELF.

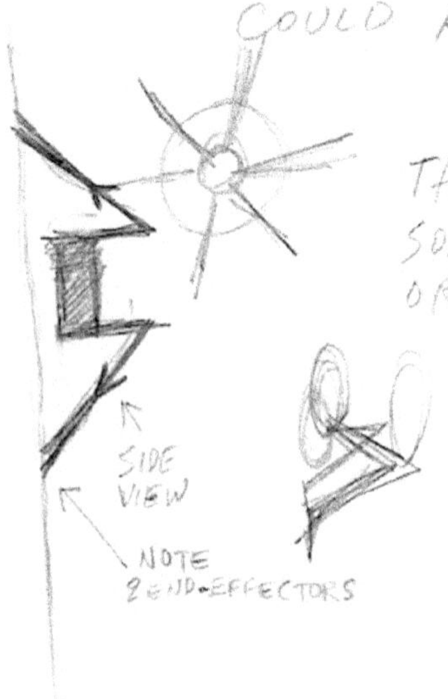

→ GRAPHICS TABLET

THAT WOULD GIVE LENGTH OF LINE LINEARLY (WHAT IS NATURAL) → JUST BY LOOKING, NOT BY NUMBERS (WHAT IS ARTIFICIAL)

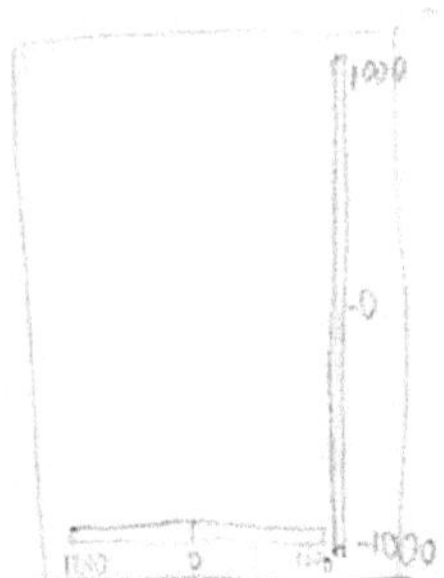

Zooming With Mirrors

PAGE A
25-12-85

THIS IS THE ANSWER

ZOOMING OUT OR IN TO QUICKLY SCAN FOR LOCATION, SIZE (ABSTRACTION), AND DIRECTION OF THE WHOLE OR PARTS OF THE SHAPE cf. p. 3 AND IDENTIFY IT

→ THE IDEA STEMS FROM "A LINE IS MADE UP OF DOTS"

EX.

a) ....

b) ●●●●●●●●

c)

TO FOLLOW THE FLOW, THE COMPUTER CAN DEAL WITH (b) BUT IT CANNOT DEAL WITH (c).

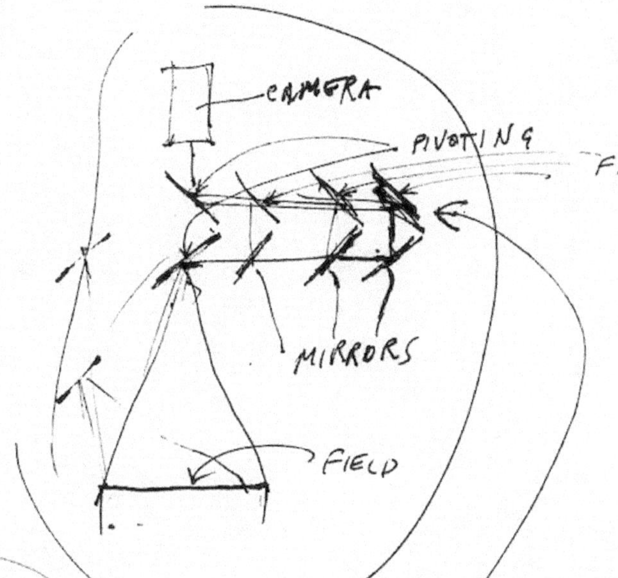

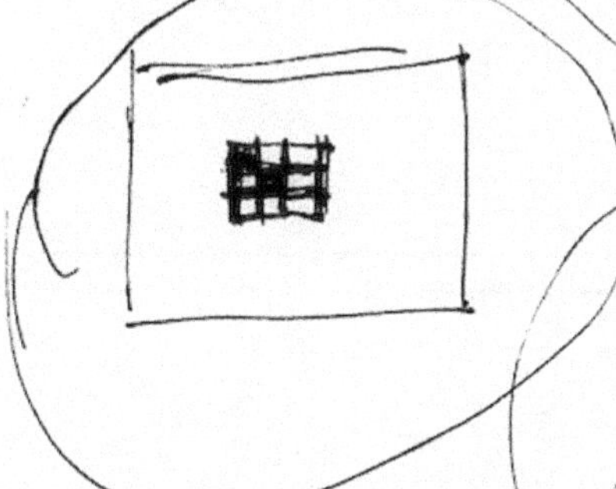

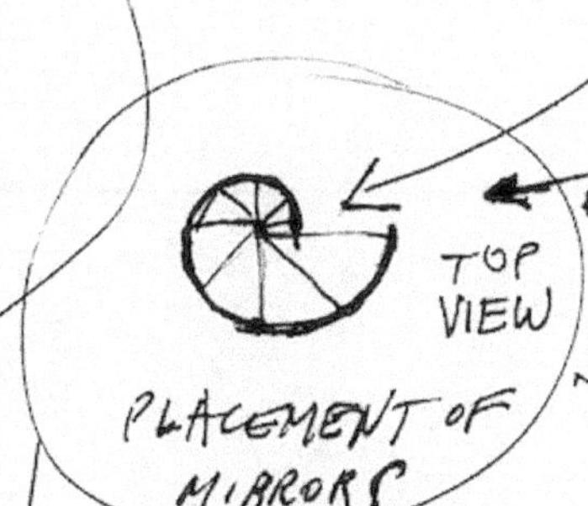

MUCH FASTER + POWERFUL THAN A ZOOM LENS

TO QUICKLY ZOOM OUT OR IN (ONE PIXEL TO 480x640)

26-12-85

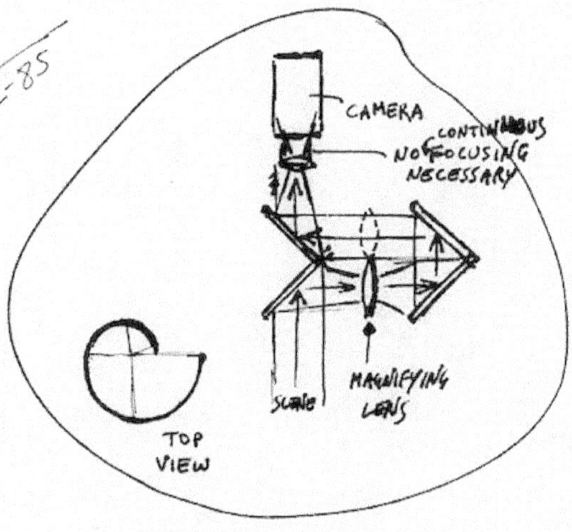

MAKING A ZOOM LENS WITH A SNAIL SHELL ARRANGEMENT OF MIRRORS:

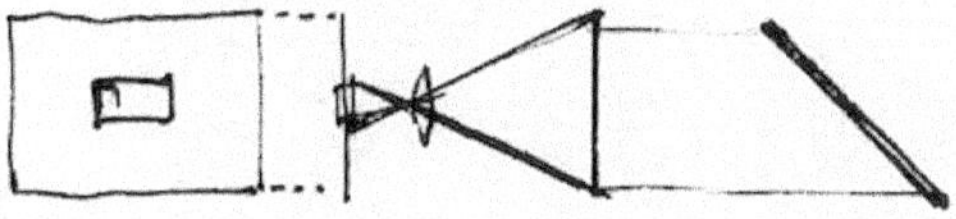

Computer Workstation

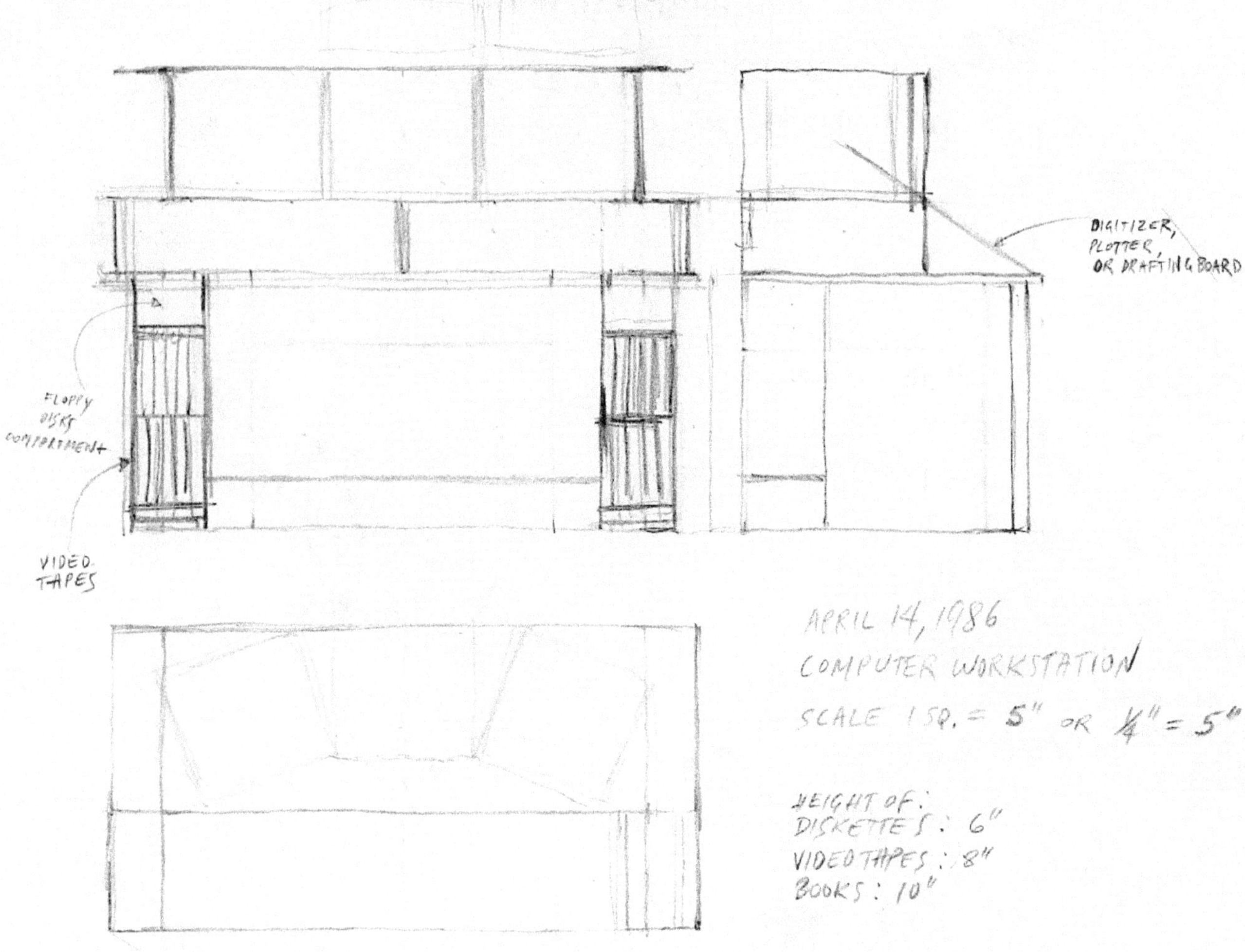

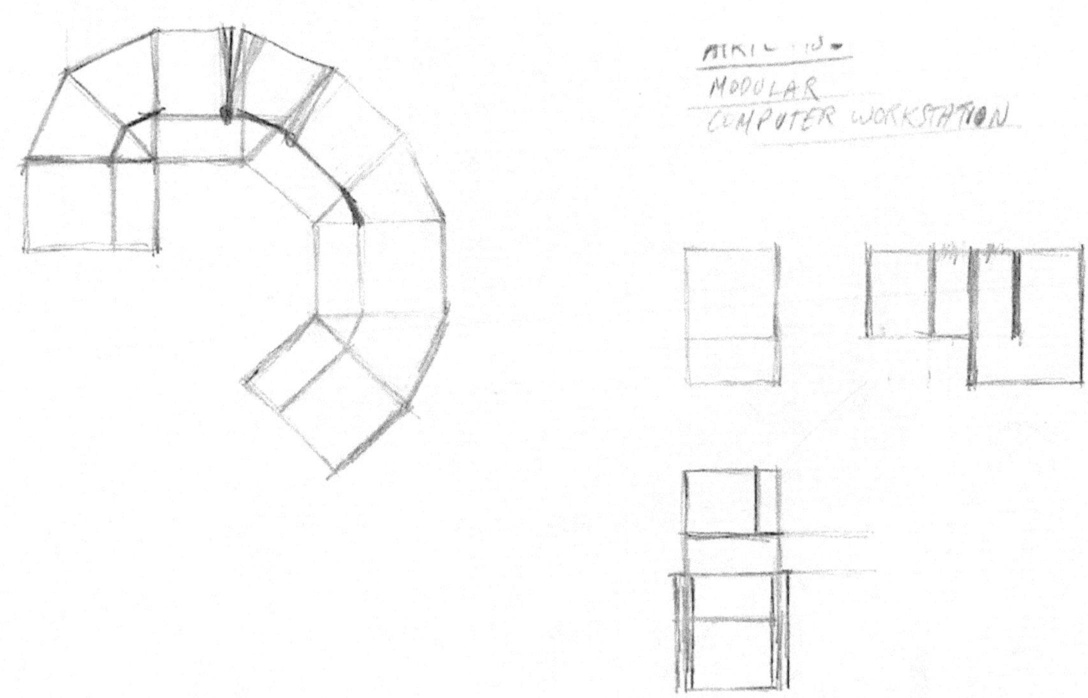

IDEAS 23·02·88

Lever Bed / Bed Hammock

① A BED THAT DISTRIBUTES MUCH MORE BODY WEIGHT (FIGHTING GRAVITY):

ⓐ

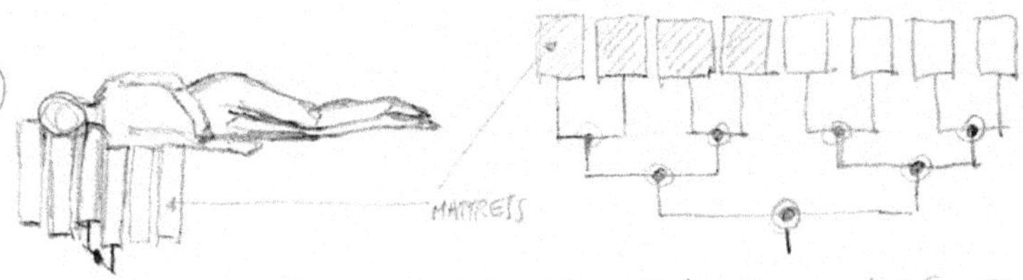

CROSS-SECTIONS ON A SYSTEM OF LEVERS SO THAT THE MOVE AS TO THE SHAPE OF THE BODY.

ⓑ

← A SINGLE RESILIENT STRING THAT, INSTEAD OF GOING LENGTHWISE, LIKE FOR THE "AMACA", GOES CROSS-WISE.

② A DISHWASHER → INTO WHICH DISHES ARE LOADED AS A STACK,

→ THAT USES BRUSHES TO SCRUB DISHES, LIKE A CARWASH,

→ THUS DOES NOT NEED TO USE JUST HOT WATER

→ THAT COULD WASH JUST ONE SINGLE DISH WITHOUT WASTE OF ENERGY (WATER), THAT IT HAS TO GO THROUGH A FULL LOAD

→ THAT IT IS OPEN AT THE TOP OF THE COUNTER, LIKE A COUNTER TOP DISH RACK.

Centering Device for Drill Press
A battery-powered penlight, above drill bit, would project crosshairs onto the workpiece under drill bit.

04·03·88

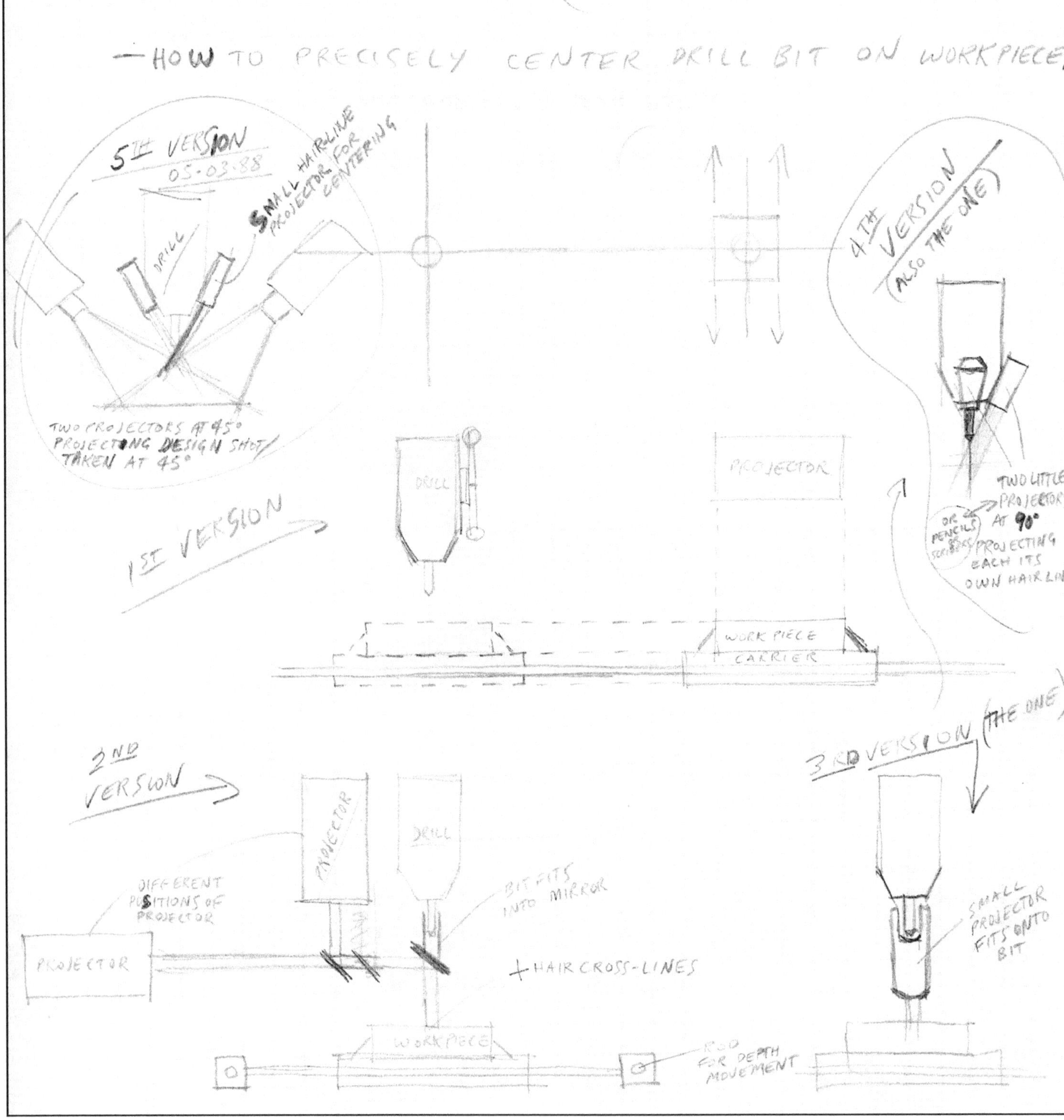

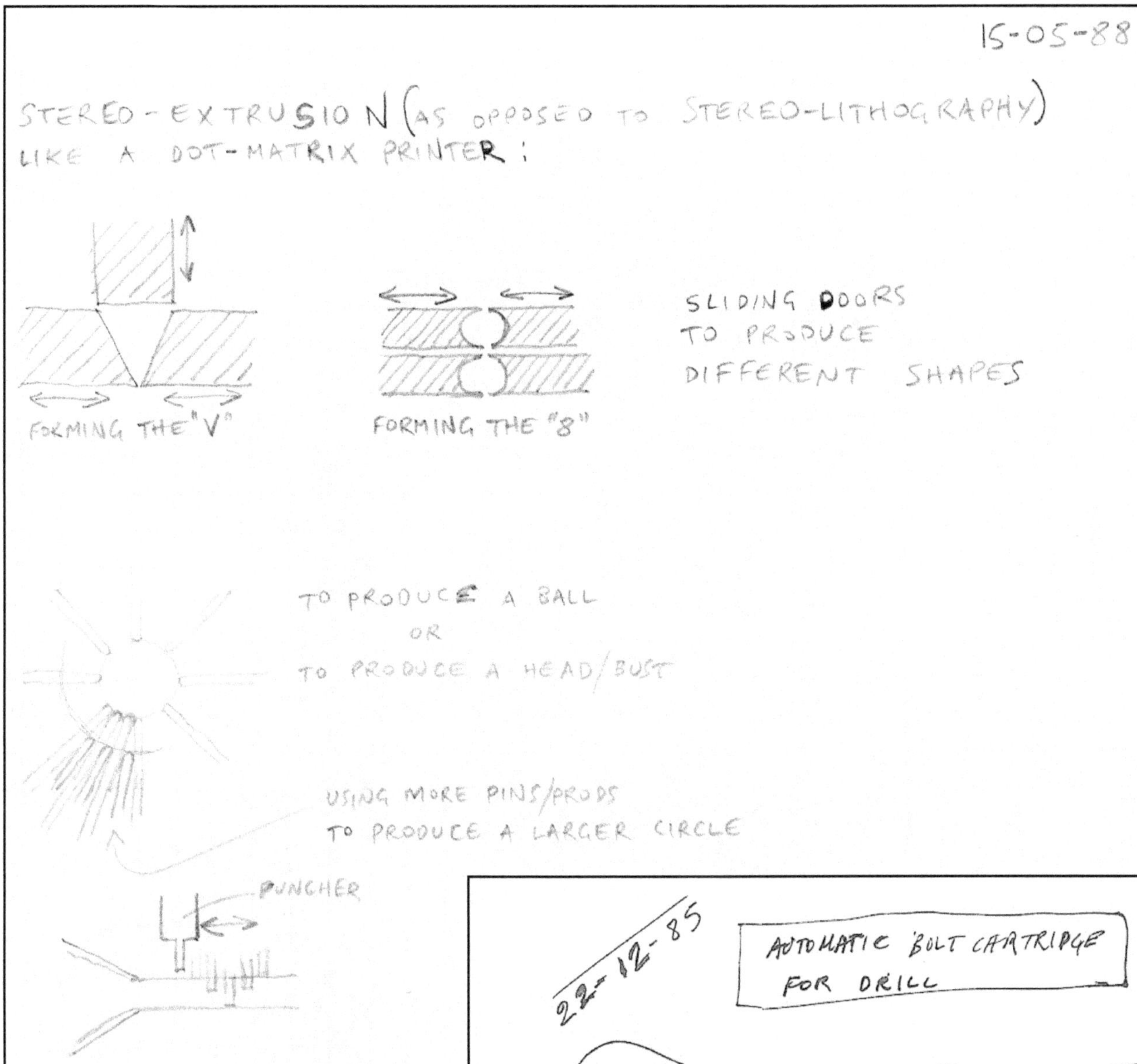
15-05-88
STEREO-EXTRUSION (AS OPPOSED TO STEREO-LITHOGRAPHY)
LIKE A DOT-MATRIX PRINTER:
FORMING THE "V"
FORMING THE "8"
SLIDING DOORS TO PRODUCE DIFFERENT SHAPES
TO PRODUCE A BALL
OR
TO PRODUCE A HEAD/BUST
USING MORE PINS/PRODS TO PRODUCE A LARGER CIRCLE
PUNCHER

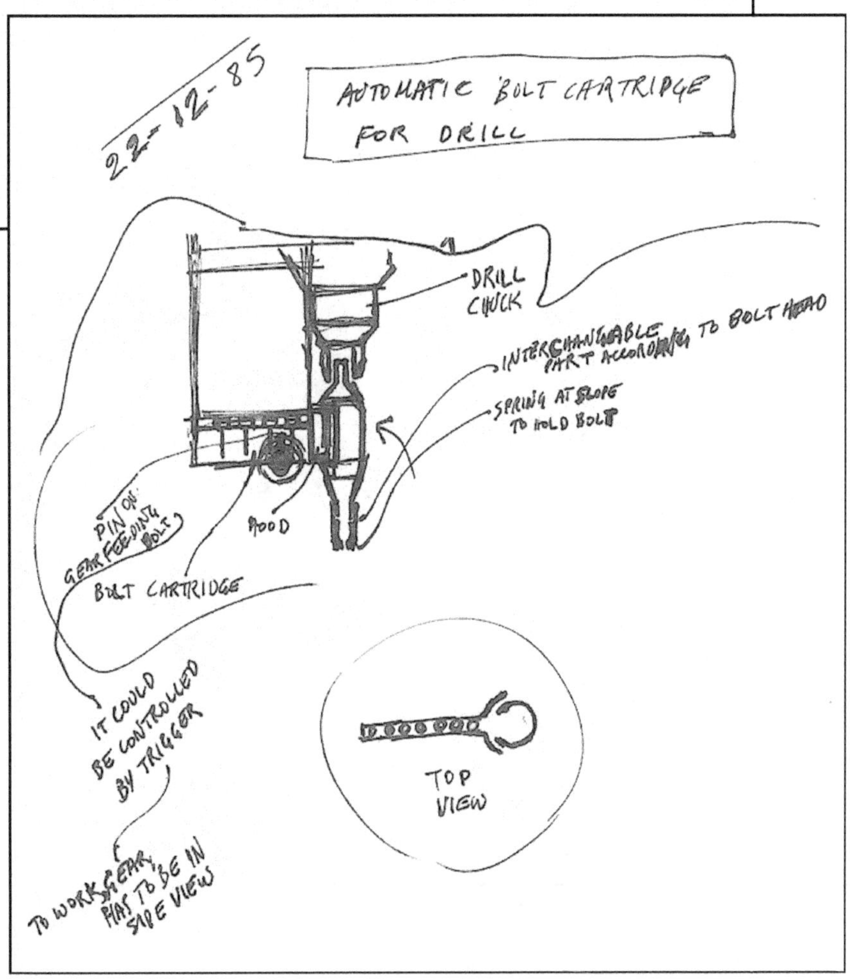
22-12-85
AUTOMATIC BOLT CARTRIDGE FOR DRILL
DRILL CHUCK
INTERCHANGEABLE PART ACCORDING TO BOLT HEAD
SPRING AT SLOPE TO HOLD BOLT
PIN ON GEAR FEEDING BOLT
BOLT CARTRIDGE
IT COULD BE CONTROLLED BY TRIGGER
TO WORK GEAR, HAS TO BE IN SIDE VIEW
TOP VIEW

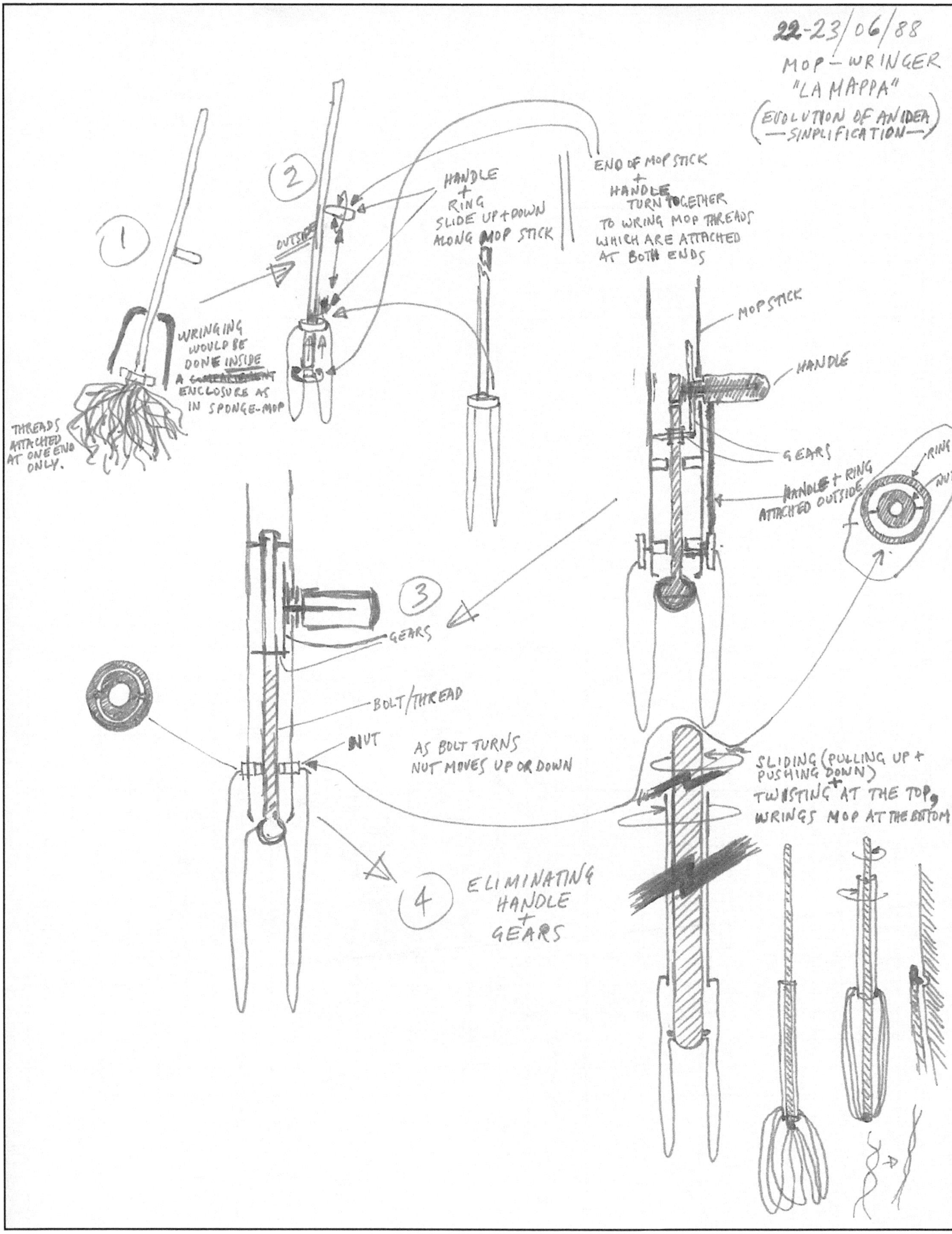

22-23/06/88
MOP-WRINGER
"LA MAPPA"
(EVOLUTION OF AN IDEA
—SIMPLIFICATION—)
1
2
OUTSIDE
WRINGING WOULD BE DONE INSIDE A ENCLOSURE AS IN SPONGE-MOP
THREADS ATTACHED AT ONE END ONLY.
HANDLE + RING SLIDE UP + DOWN ALONG MOP STICK
END OF MOP STICK + HANDLE TURN TOGETHER TO WRING MOP THREADS WHICH ARE ATTACHED AT BOTH ENDS
MOP STICK
HANDLE
GEARS
HANDLE + RING ATTACHED OUTSIDE
RING
NUT
3
GEARS
BOLT/THREAD
NUT
AS BOLT TURNS NUT MOVES UP OR DOWN
SLIDING (PULLING UP + PUSHING DOWN) + TWISTING AT THE TOP, WRINGS MOP AT THE BOTTOM.
4
ELIMINATING HANDLE + GEARS

August 13, 1983

THE SOLAR HOUSE (PASSIVE)

RUN BY WATER-PRESSURE TO RISE + BY GRAVITY TO DROP

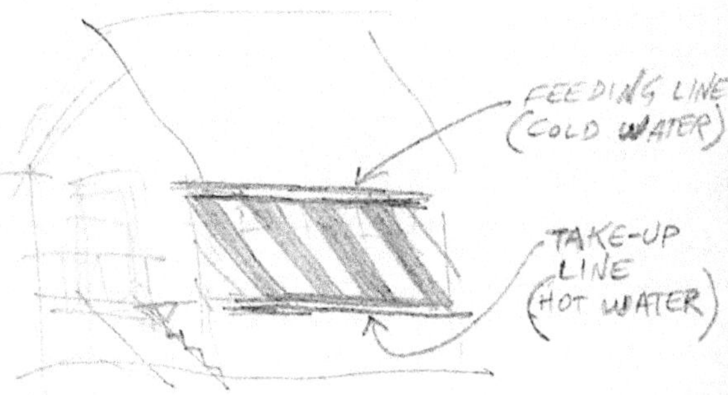

HOUSE COVERED WITH THIN DIAGONAL

SOLAR PANELS— 1 MM-THIN PLASTIC SANDWICHES FILLED WITH WATER.

AT "BOILING" TEMPERATURE TAKE-UP LINE OPENS (BY THERMOMETER) TO DROP HOT WATER ONTO LARGE CHAMBER OF STONES AND/ OR GRAVEL.

HEAT RADIATION FROM ROCKS IS DUCTED THROUGHOUT THE HOUSE.

STEAM FROM DROP (ON ROCKS) COULD BE USED TO WORK A STEAM ENGINE AND HENCE A GENERATOR.

EXCESS HOT WATER COULD BE DROPPED IN HOT WATER TANK

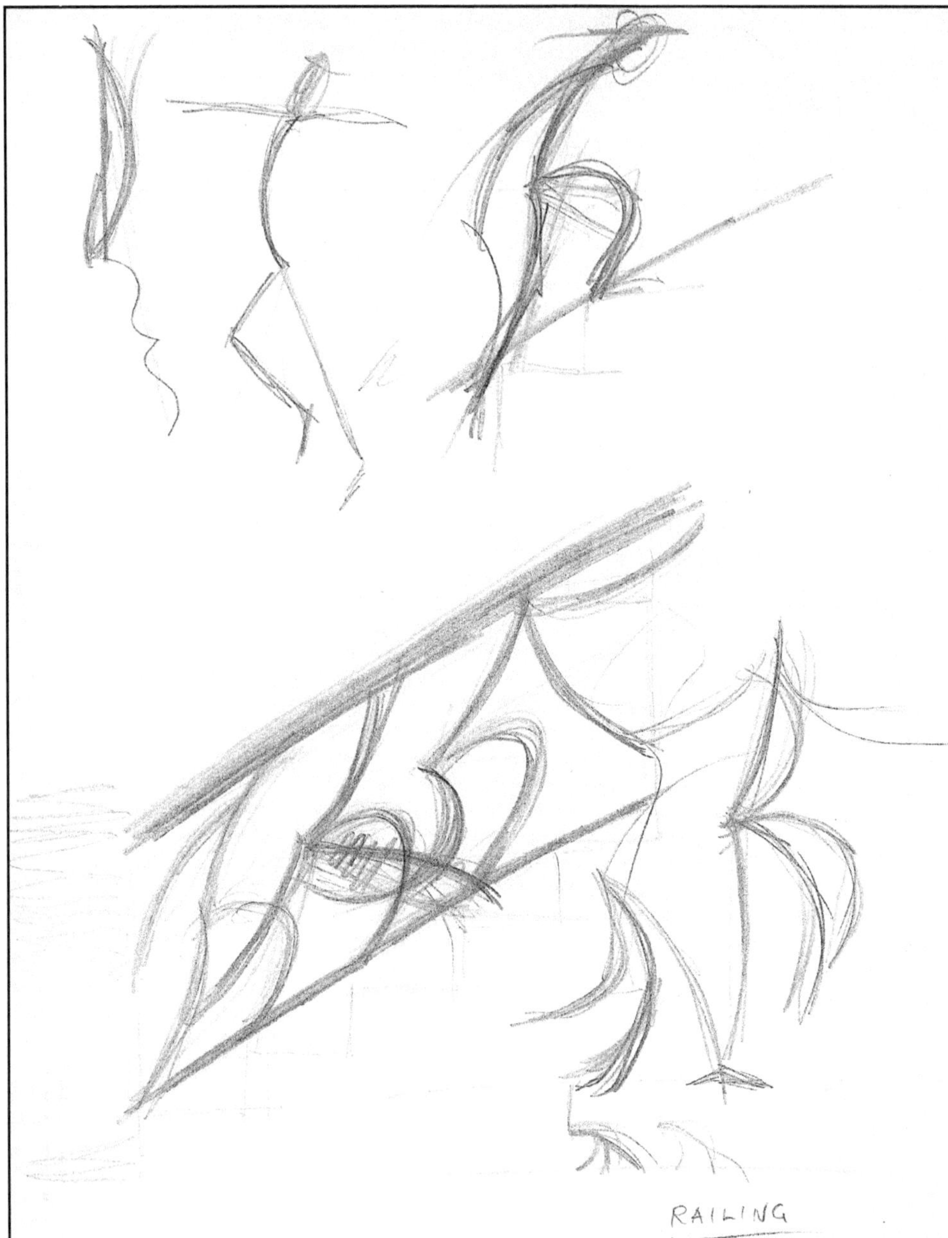
RAILING

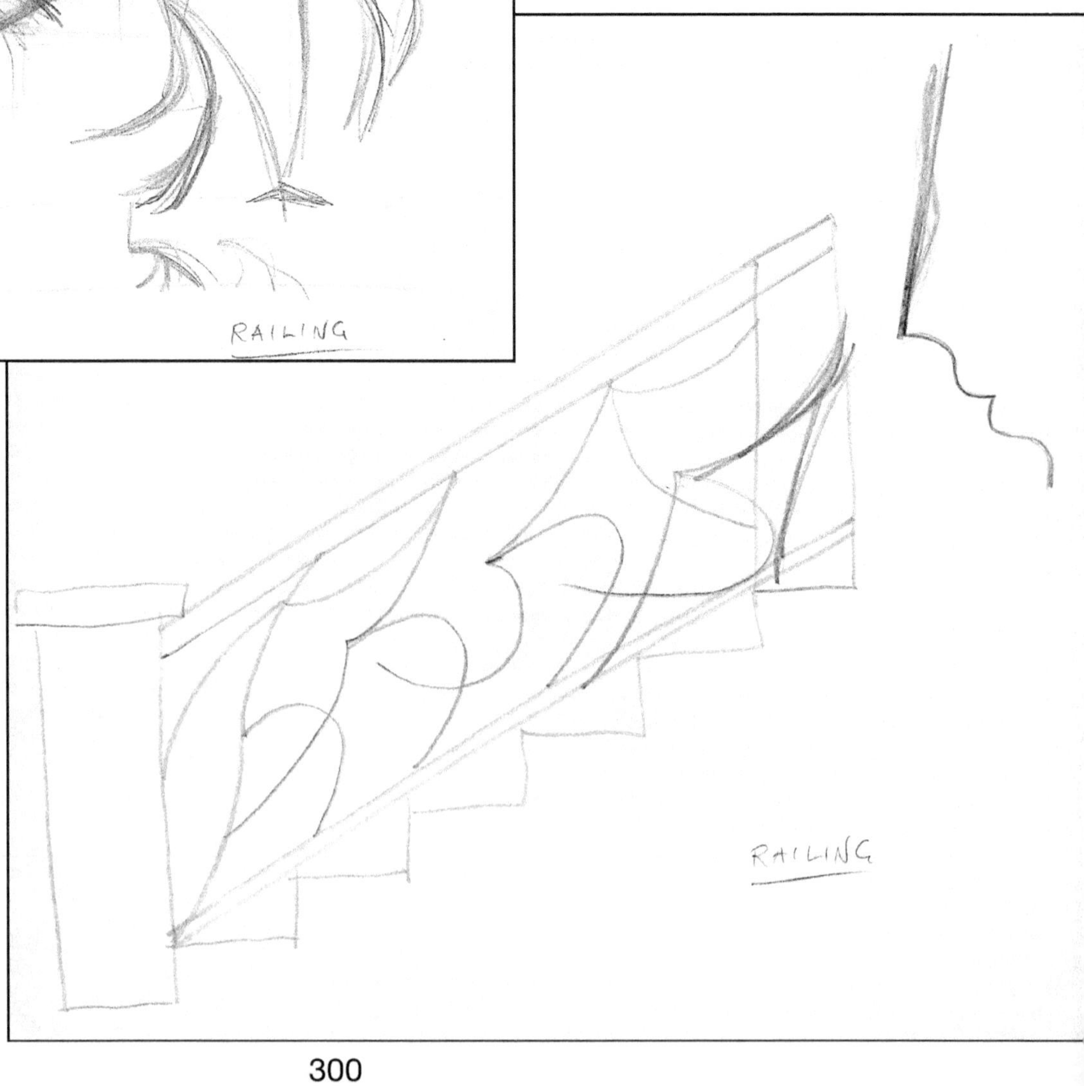
RAILING

IDEA

24-10-88

ENERGY-SAVER POT

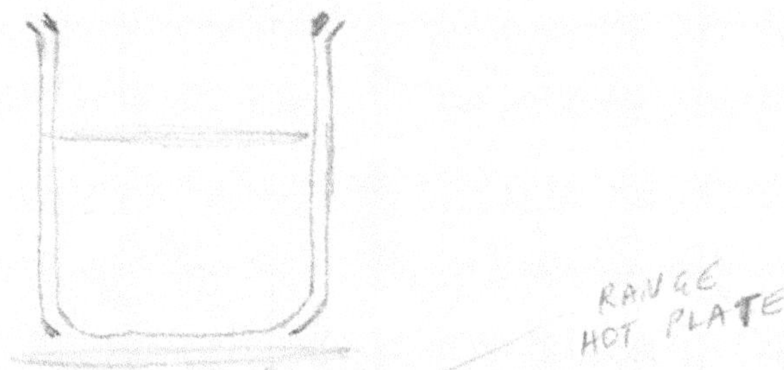

A POT WITH A DOUBLE WALL SO HEAT RADIATING FROM THE POT DOES NOT DISSIPATE. THUS TRAPPED, HEAT RISES UP + CONTRIBUTES TO A FASTER WARMING UP OF THE CONTENTS OF THE POT.

OTHER WAYS:

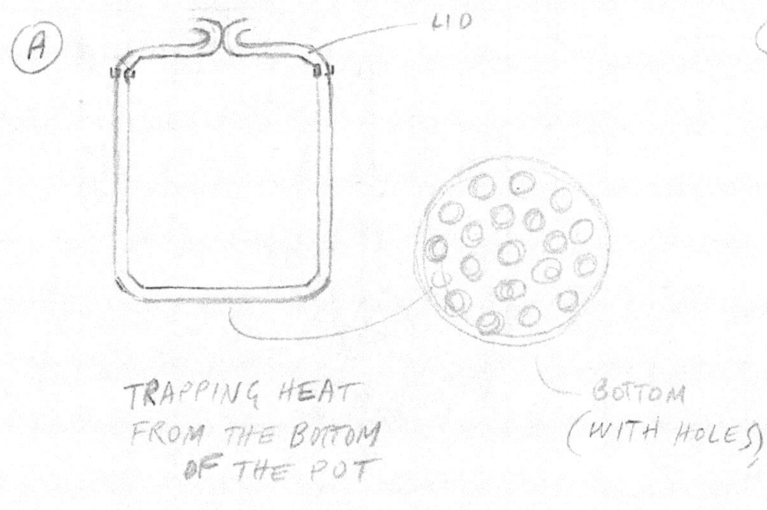

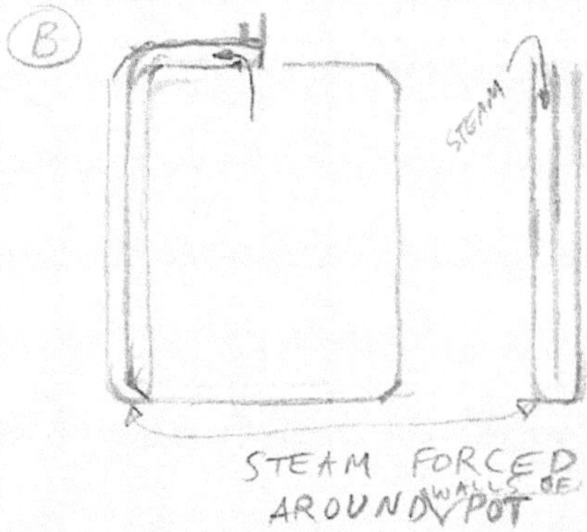

(C) A COMBINATION OF (A) + (B)

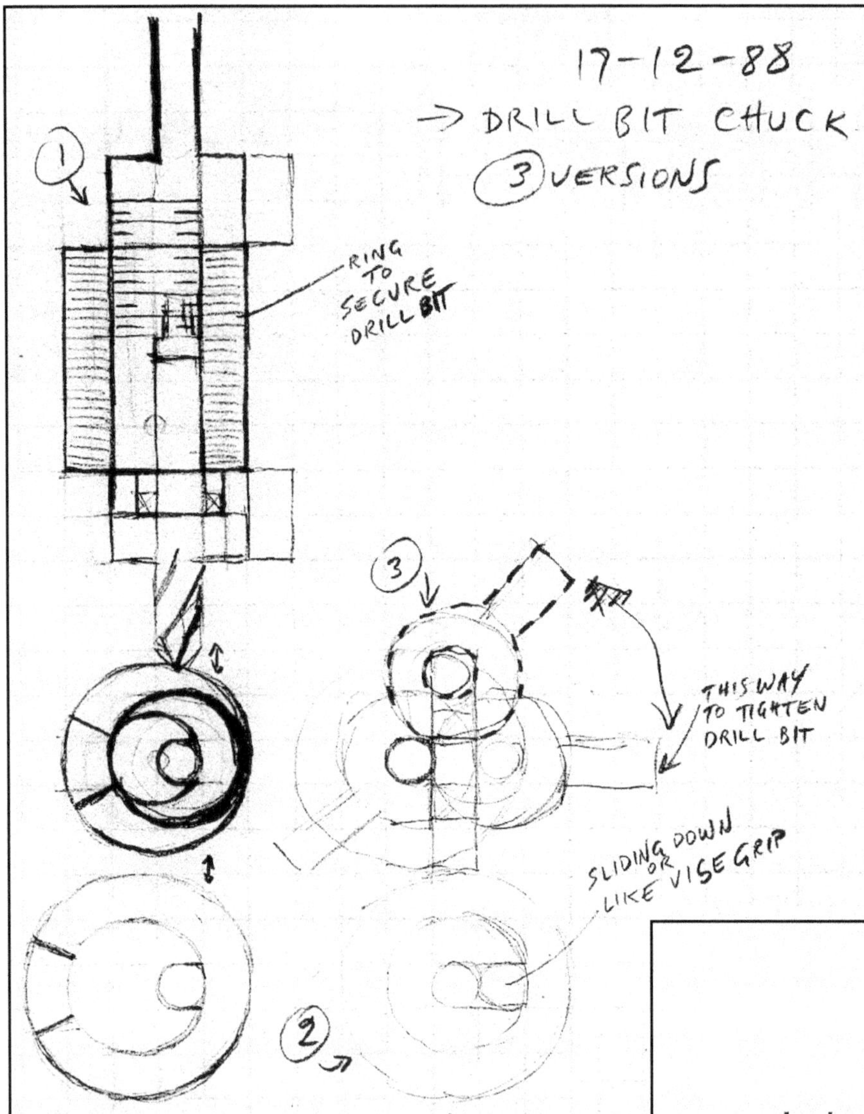
17-12-88
→ DRILL BIT CHUCK
3 VERSIONS
1
RING TO SECURE DRILL BIT
3
THIS WAY TO TIGHTEN DRILL BIT
SLIDING DOWN OR LIKE VISE GRIP
2

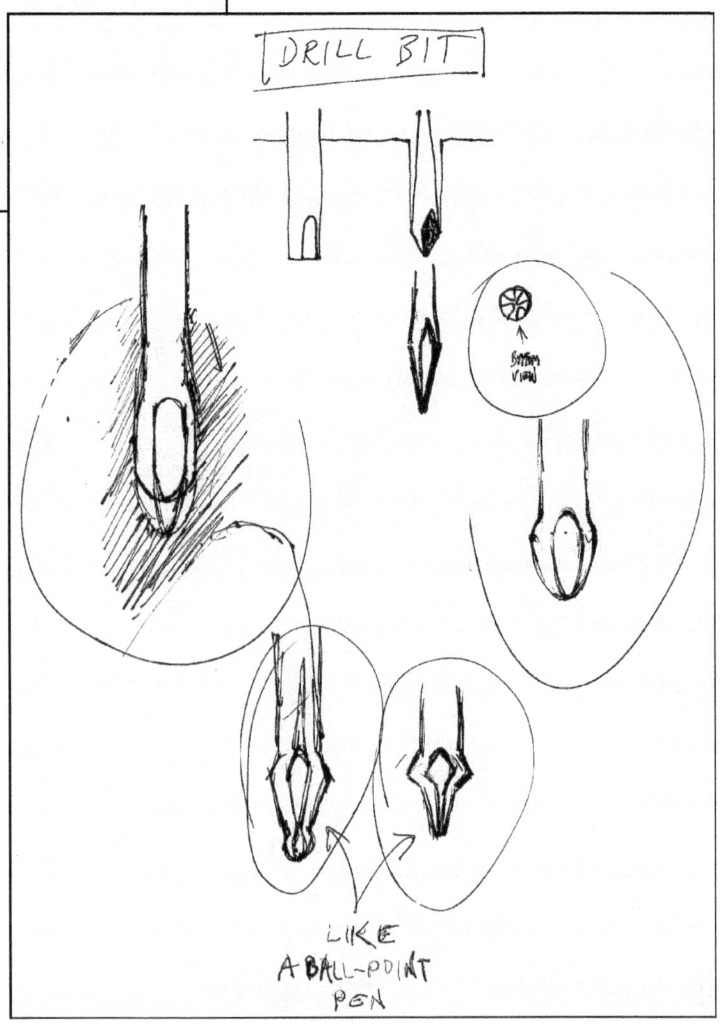
DRILL BIT
LIKE A BALL-POINT PEN

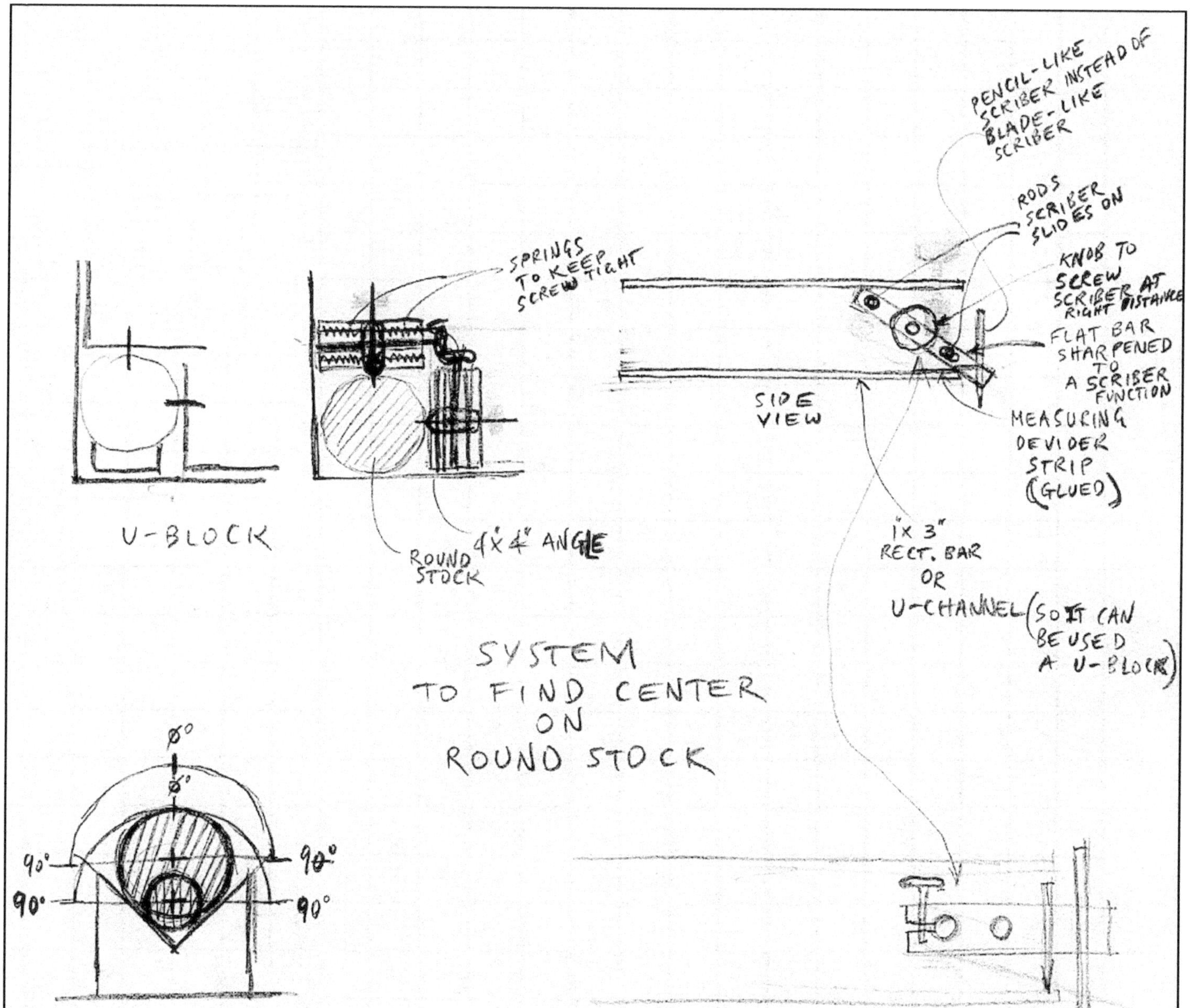
PENCIL-LIKE SCRIBER INSTEAD OF BLADE-LIKE SCRIBER
RODS SCRIBER SLIDES ON
KNOB TO SCREW SCRIBER AT RIGHT DISTANCE
FLAT BAR SHARPENED TO A SCRIBER FUNCTION
MEASURING DEVIDER STRIP (GLUED)
SPRINGS TO KEEP SCREW TIGHT
SIDE VIEW
U-BLOCK
ROUND STOCK
4"x4" ANGLE
1"x 3" RECT. BAR OR U-CHANNEL (SO IT CAN BE USED A U-BLOCK)
SYSTEM TO FIND CENTER ON ROUND STOCK
Ø°
Ø°
90°
90°
90°
90°
USING A SQUARE BLOCK AND AN ANGLE

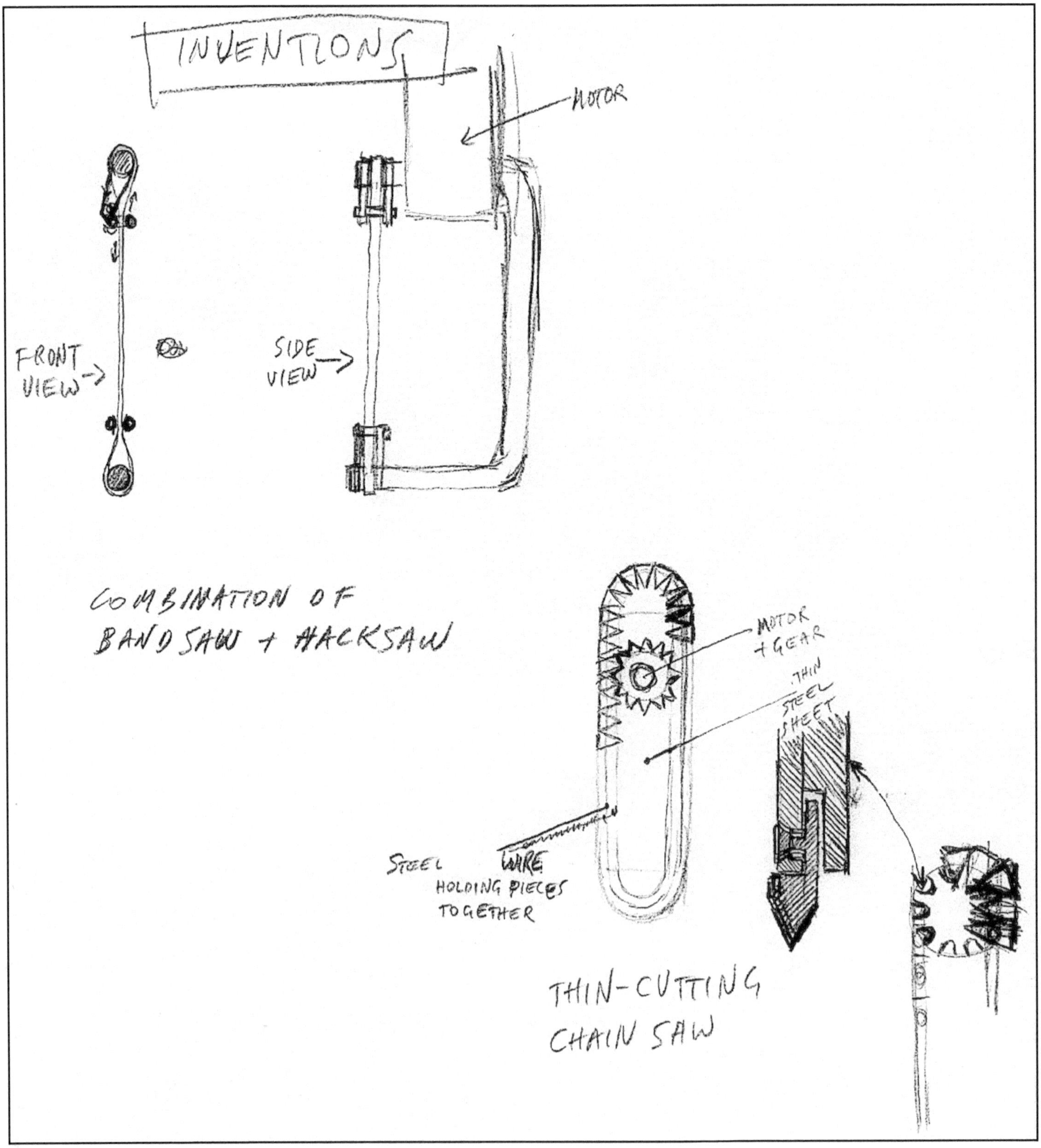
INVENTIONS
MOTOR
FRONT VIEW
SIDE VIEW
COMBINATION OF
BANDSAW + HACKSAW
MOTOR
+GEAR
THIN
STEEL
SHEET
STEEL WIRE
HOLDING PIECES
TOGETHER
THIN-CUTTING
CHAIN SAW

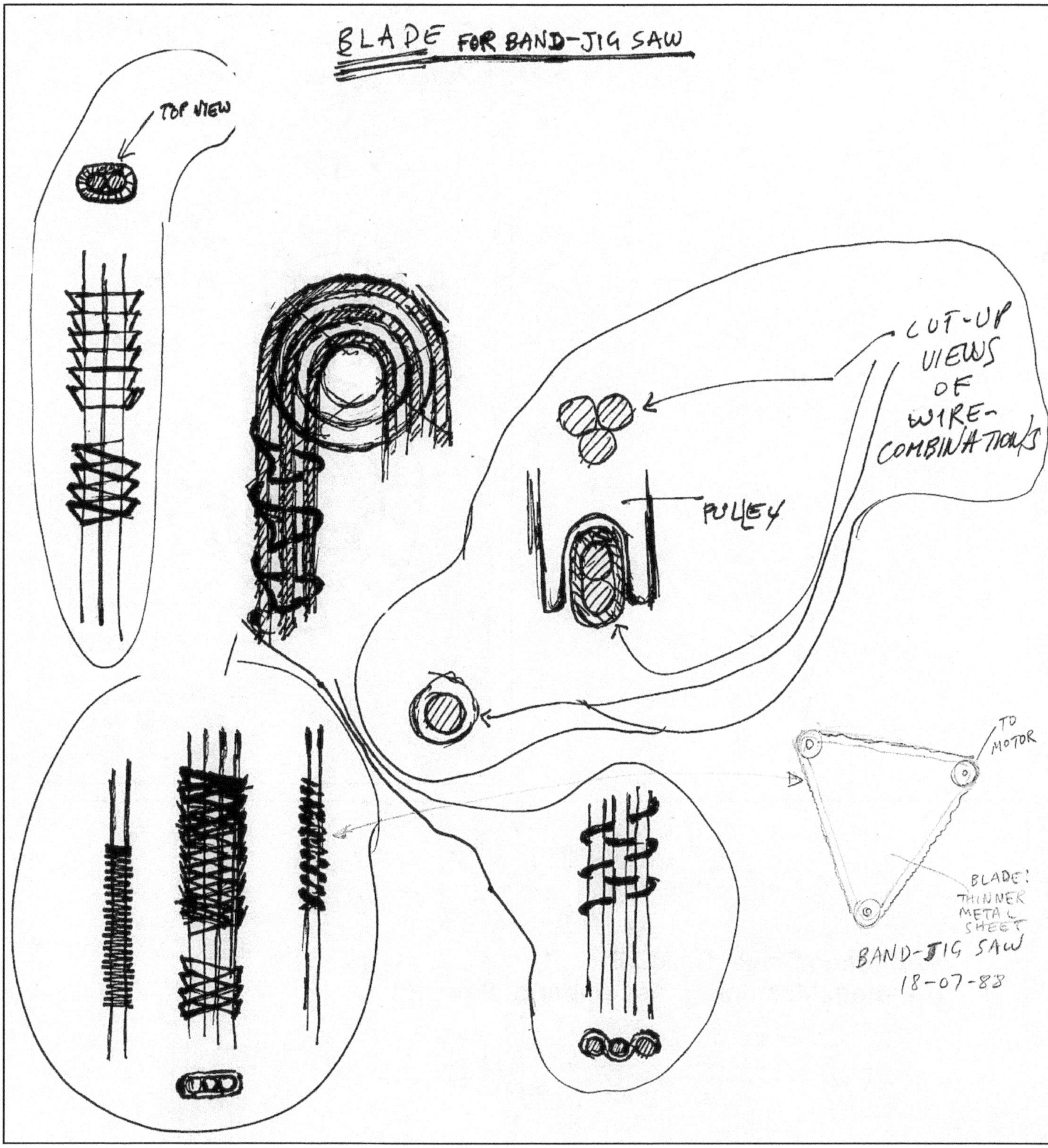
BLADE FOR BAND-JIG SAW
TOP VIEW
CUT-UP VIEWS OF WIRE-COMBINATIONS
PULLEY
TO MOTOR
BLADE: THINNER METAL SHEET
BAND-JIG SAW
18-07-88

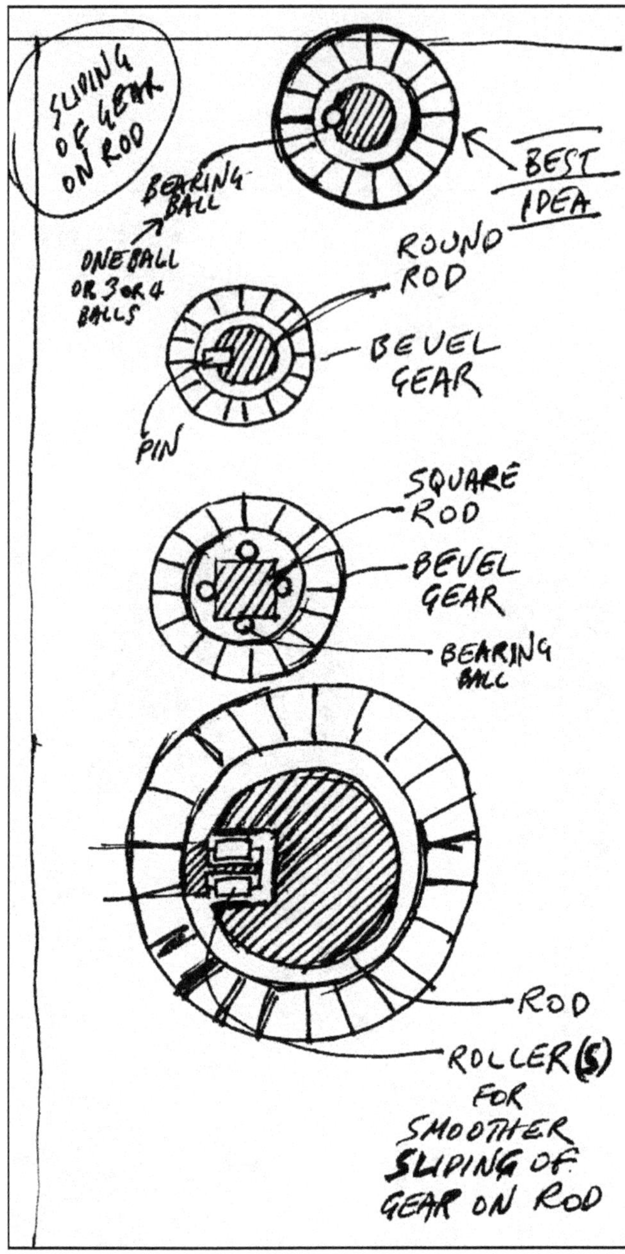

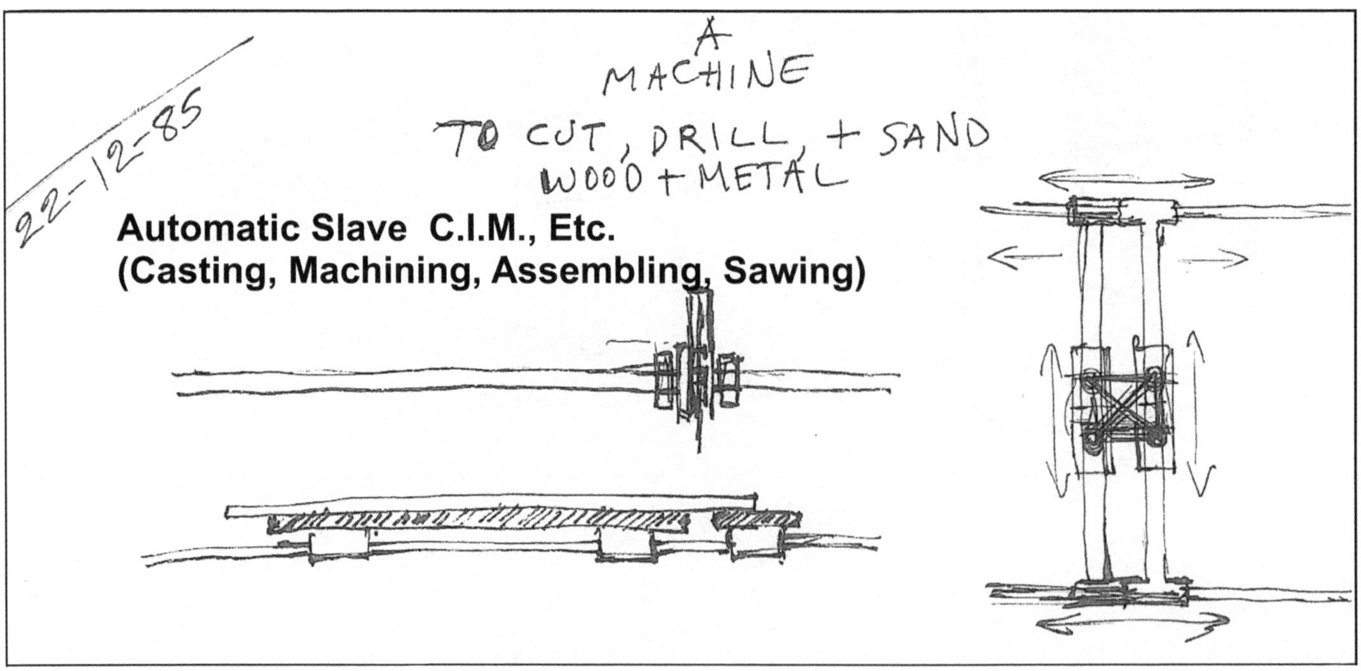

Automatic Slave C.I.M., Etc.
(Casting, Machining, Assembling, Sawing)

Soda Can Spout/Straw

22-02-89 1/2
"SPOUT-BEC"

A BETTER WAY TO DRINK
FOR CANNED BEVERAGES — COKE, PEPSI, BEER, ETC.

A SPOUT THAT POPS/SPRING UP WHEN PART OF THE TOP OF THE CAN IS PULLED UP.

ADVANTAGES:

(1) CLEAN SPOUT TO DRINK FROM. NOT HAVING TO CLEAN CAN AREA (WHERE) YOU ARE GOING TO TOUCH WITH YOUR LIPS/MOUTH.

(2) EASIER WAY TO "SUCK" LIQUID FROM CAN. COMPARE IDEAL/LOGICAL SHAPE OF SPOUT TO ODD/UNCOMFORTABLE PRESENT-DAY SHAPE OF TOP OF THE CAN AND ITS DRINKING OPENING (SHARP EDGE + FAR AWAY FROM THE SIDE OF CAN WITH DEPRESSION/FURROW BETWEEN THE SIDE OF CAN AND THE DRINKING OPENING).

(3) DUE TO PRESENT-DAY DRINKING HOLE, DRINKER SUCKS A LOT OF AIR-BUBBLES WITH THE BEVERAGE — THUS CHANGING "TEXTURE" OF BEVERAGE.

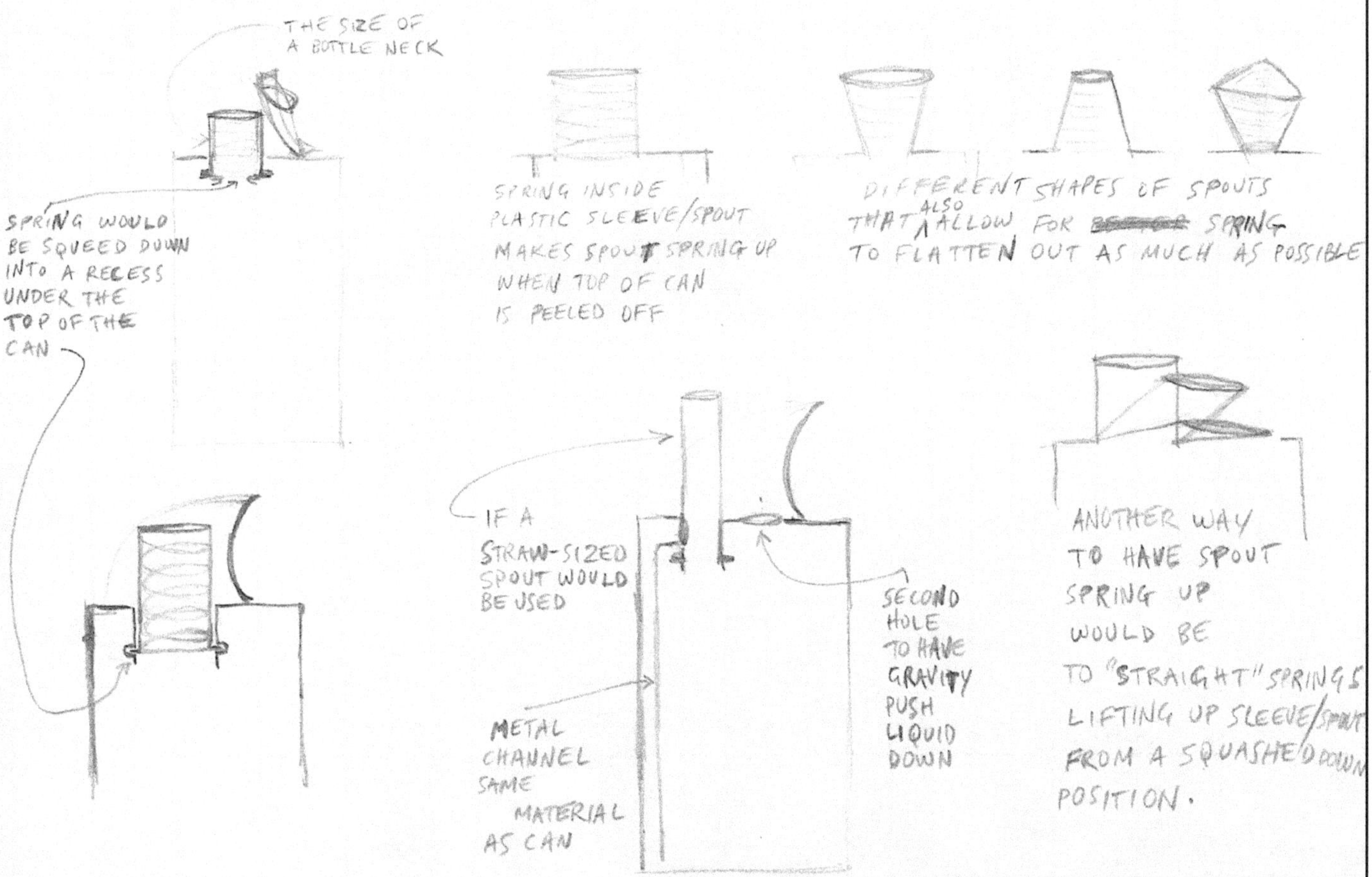

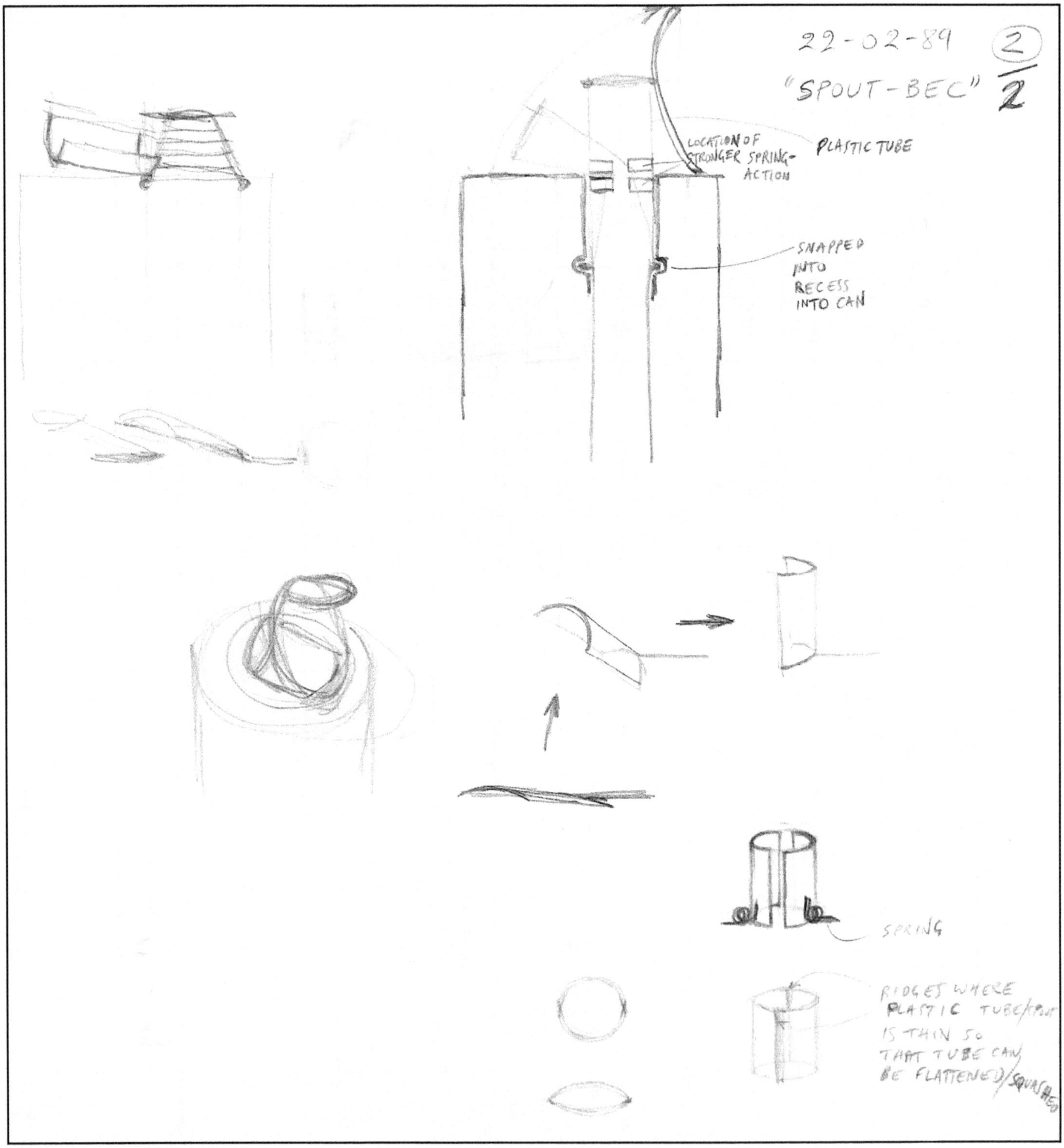

22-02-89
2
"SPOUT-BEC"
2
LOCATION OF STRONGER SPRING-ACTION
PLASTIC TUBE
SNAPPED INTO RECESS INTO CAN
SPRING
RIDGES WHERE PLASTIC TUBE/ IS THIN SO THAT TUBE CAN BE FLATTENED/SQUASHED

24-02-89
"SPOUT-BEC"
ELASTIC MEMBRANE
ELASTIC MEMBRANE
SPRING

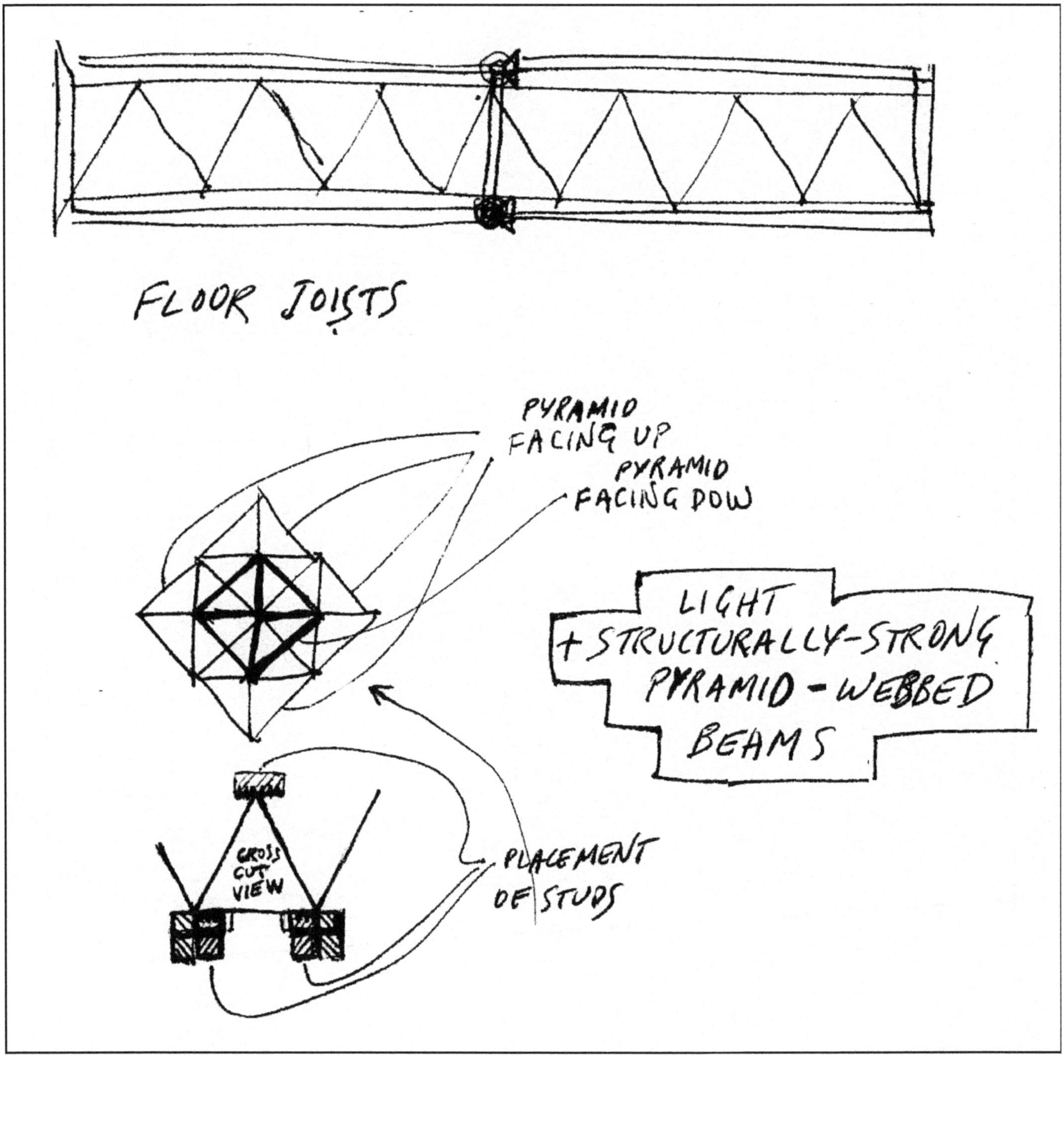
FLOOR JOISTS
PYRAMID FACING UP
PYRAMID FACING DOW
LIGHT + STRUCTURALLY-STRONG PYRAMID - WEBBED BEAMS
CROSS CUT VIEW
PLACEMENT OF STUDS

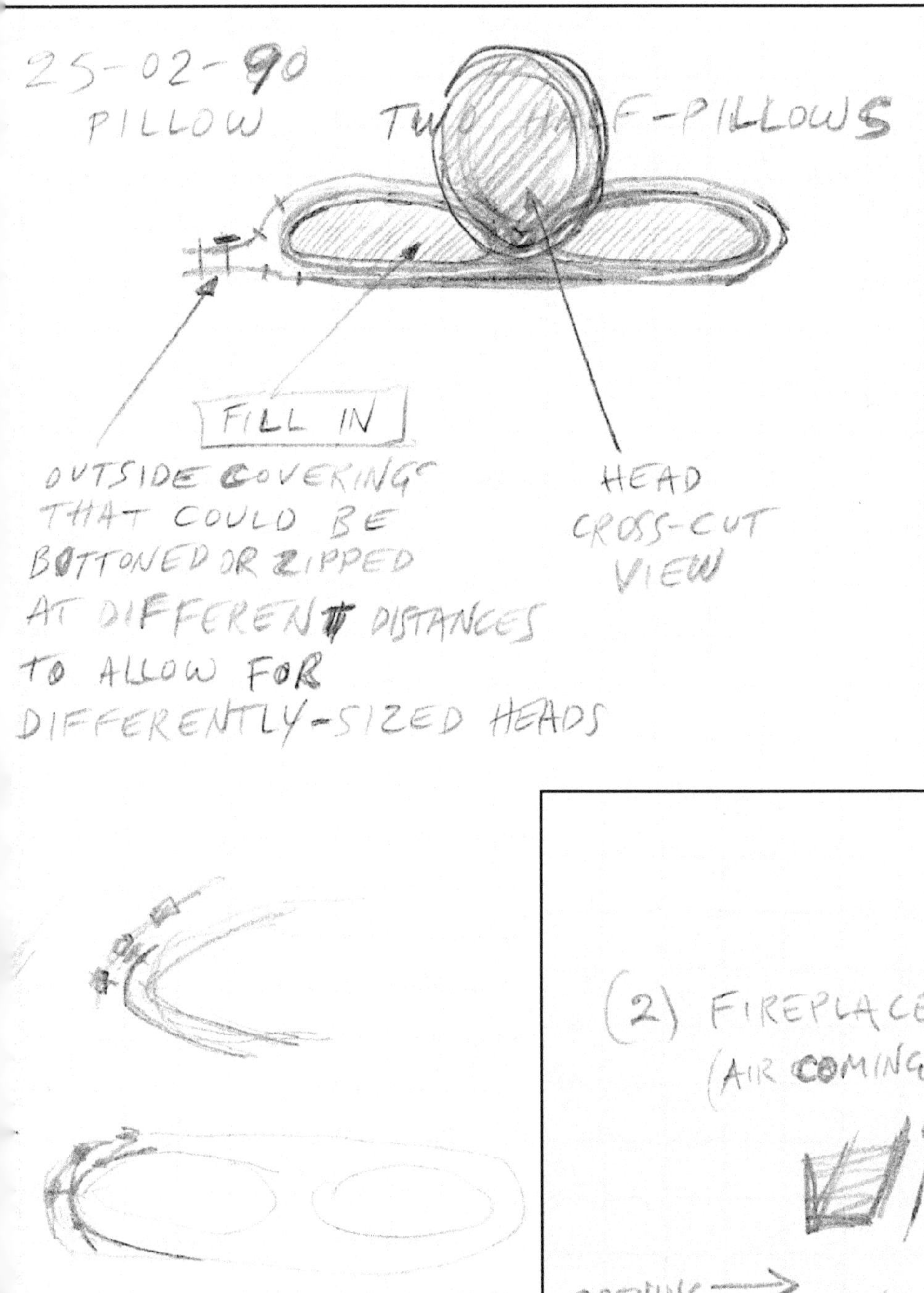
25-02-90
PILLOW
TWO HALF-PILLOWS
FILL IN
OUTSIDE COVERINGS THAT COULD BE BOTTONED OR ZIPPED AT DIFFERENT DISTANCES TO ALLOW FOR DIFFERENTLY-SIZED HEADS
HEAD CROSS-CUT VIEW

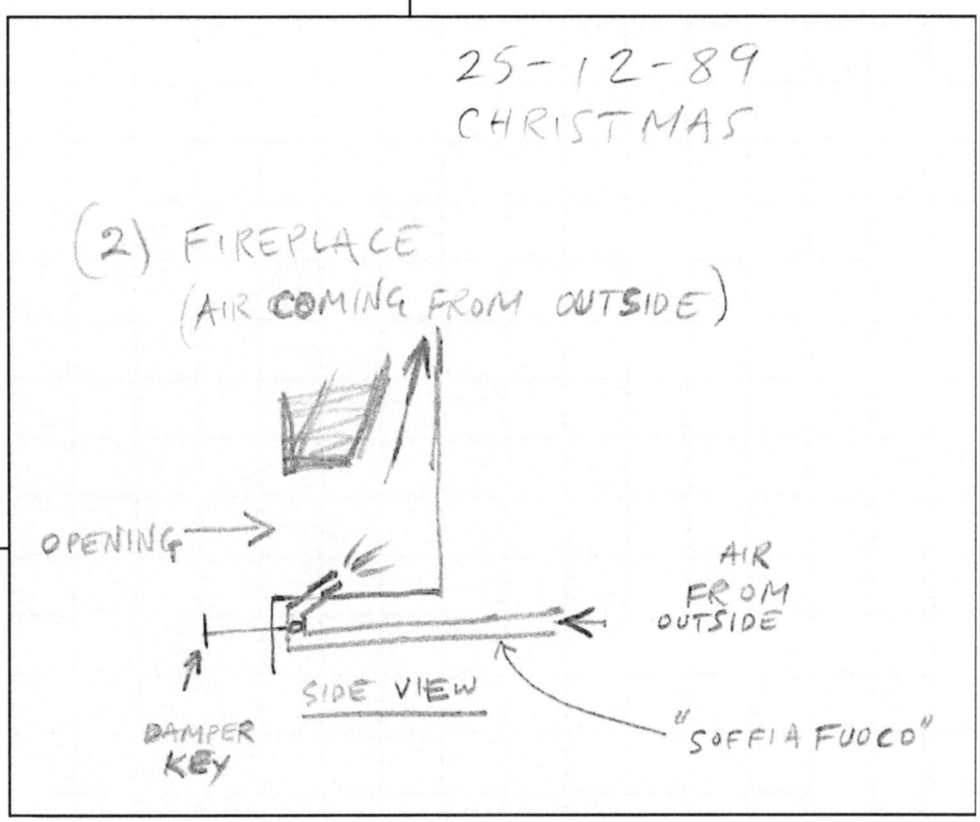
25-12-89
CHRISTMAS
(2) FIREPLACE
(AIR COMING FROM OUTSIDE)
OPENING
AIR FROM OUTSIDE
SIDE VIEW
DAMPER KEY
"SOFFIA FUOCO"

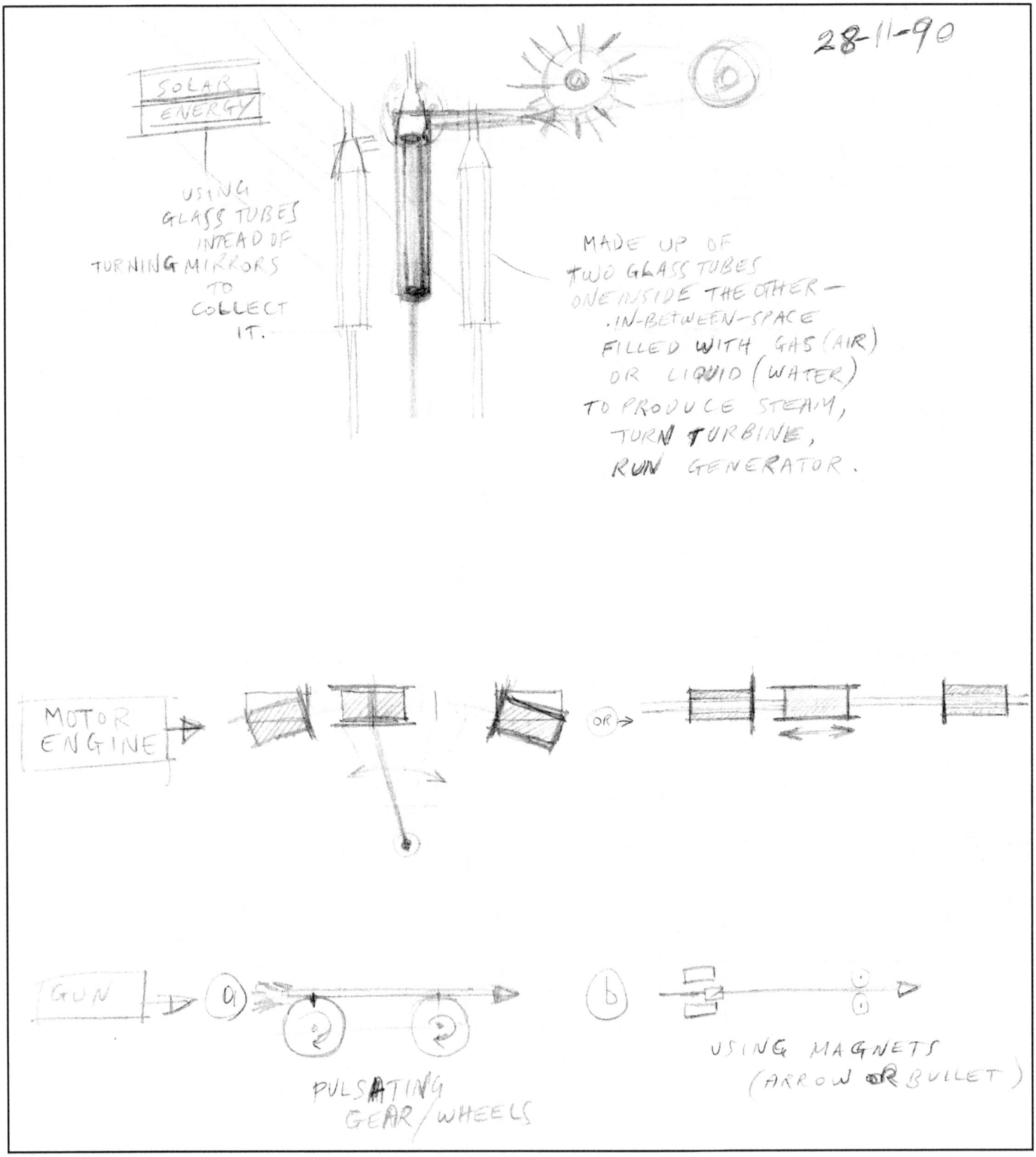
28-11-90
SOLAR ENERGY
USING GLASS TUBES INTEAD OF TURNING MIRRORS TO COLLECT IT.
MADE UP OF TWO GLASS TUBES ONE INSIDE THE OTHER — .IN-BETWEEN-SPACE FILLED WITH GAS (AIR) OR LIQUID (WATER) TO PRODUCE STEAM, TURN TURBINE, RUN GENERATOR.
MOTOR ENGINE
OR
GUN
a
b
PULSATING GEAR/WHEELS
USING MAGNETS (ARROW OR BULLET)

26-01-91
WAR IN THE GULF

— FIRE – EXTINGUISHER

— AN ASBESTOS PARACHUTE (HUGE) ~~THAT~~ SHOT BY AN AIRCRAFT, OPENS, ~~AN~~ HOMED IN BY LASER BEAM, DROPS ON FIRE + SMOTHERS IT.

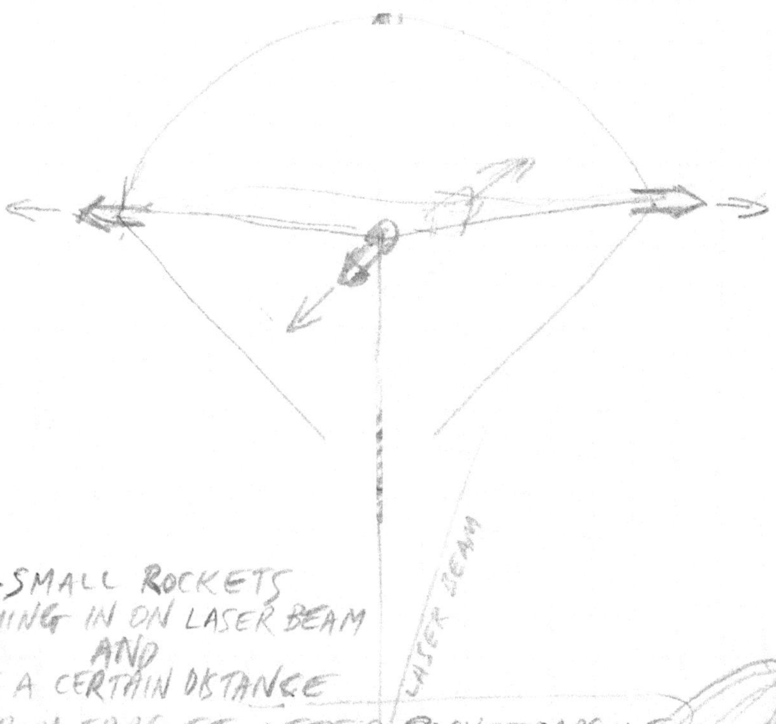

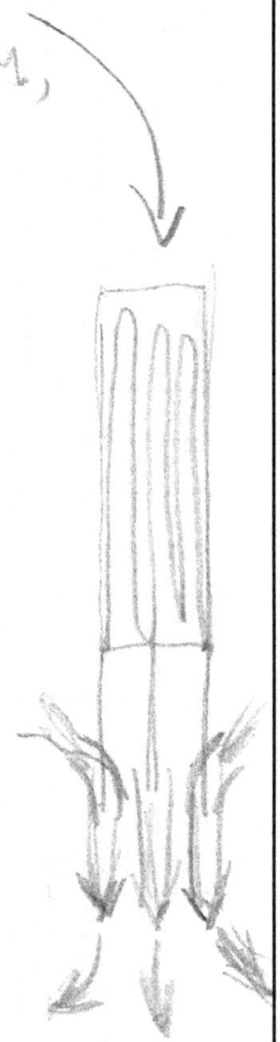

4 SMALL ROCKETS HOMING IN ON LASER BEAM AND AT A CERTAIN DISTANCE FROM TARGET, AFTER ROCKET CAPSULE OPENS, THEY SEPARATE + GO SIDEWAYS TO OPEN PARACHUTE + TEAR AWAY WHEN WITH THE PARACHUTE NOW OPEN INERTIA/RESISTANCE BUILDS.

FOLDING THE PARACHUTE

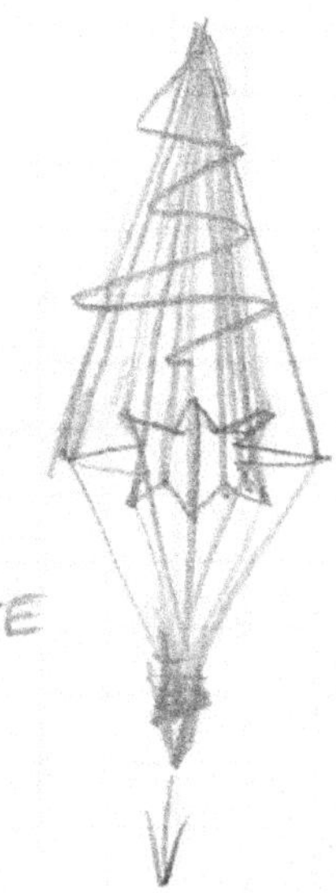

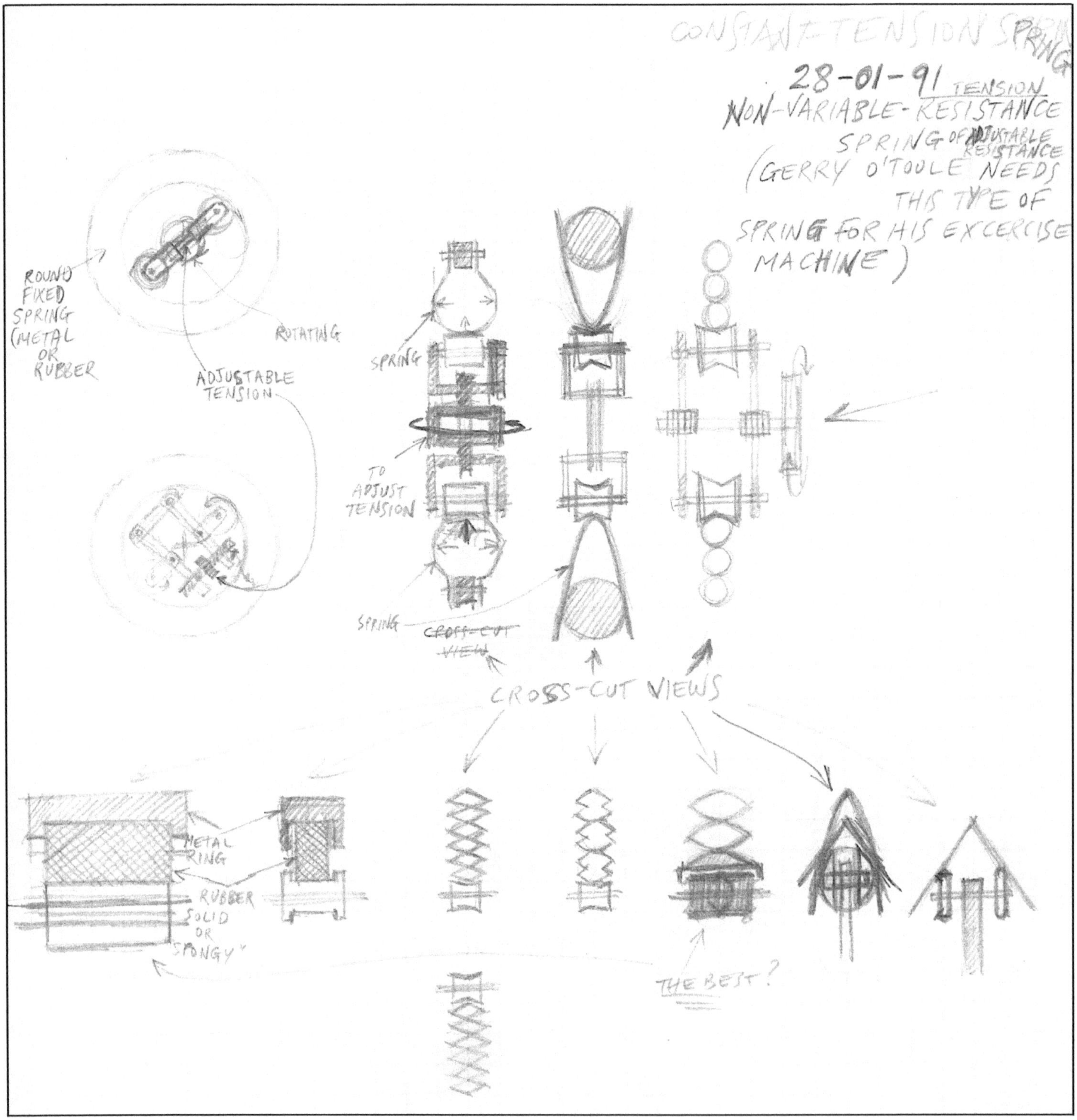
CONSTANT-TENSION SPRING
28-01-91
NON-VARIABLE-RESISTANCE SPRING OF ADJUSTABLE TENSION RESISTANCE
(GERRY O'TOOLE NEEDS THIS TYPE OF SPRING FOR HIS EXCERCISE MACHINE)
ROUND FIXED SPRING (METAL OR RUBBER
ROTATING
ADJUSTABLE TENSION
SPRING
TO ADJUST TENSION
SPRING
CROSS-CUT VIEW
CROSS-CUT VIEWS
METAL RING
RUBBER SOLID OR "SPONGY"
THE BEST?

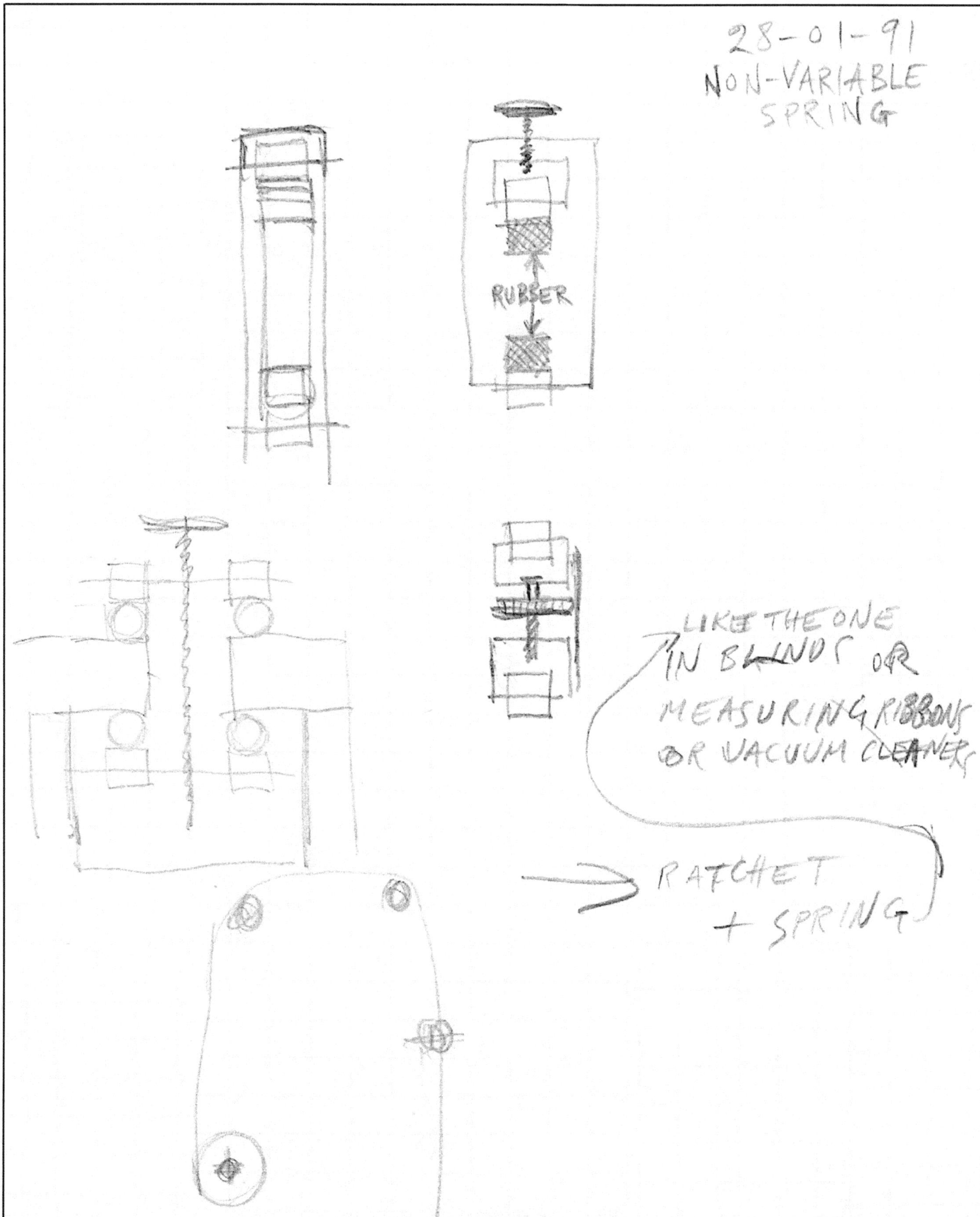
28-01-91
NON-VARIABLE SPRING
RUBBER
LIKE THE ONE IN BLINDS OR
MEASURING RIBBONS
OR VACUUM CLEANERS
RATCHET + SPRING

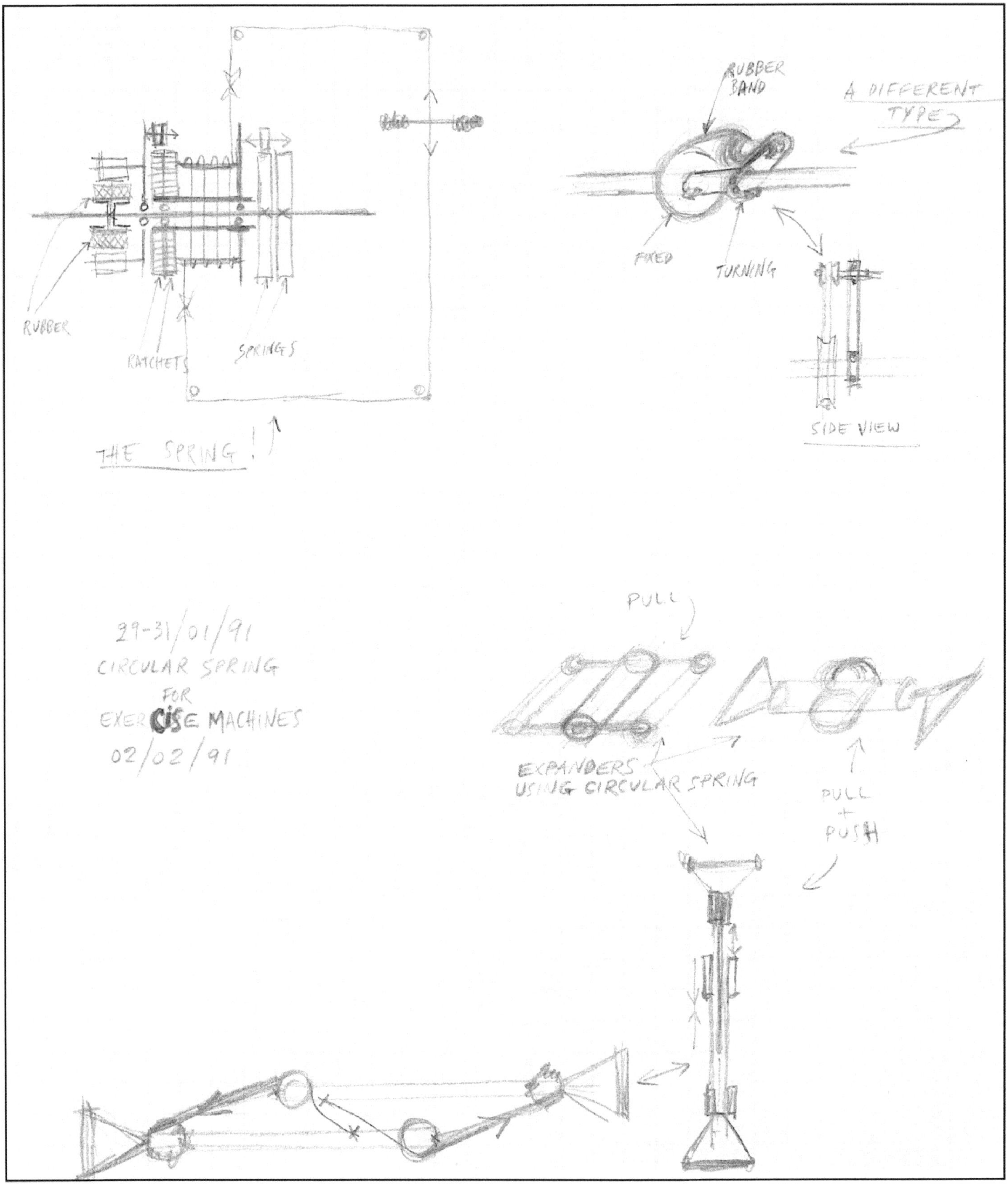
RUBBER
RATCHETS
SPRINGS
THE SPRING!
RUBBER BAND
A DIFFERENT TYPE
FIXED
TURNING
SIDE VIEW
29-31/01/91
CIRCULAR SPRING
FOR
EXERCISE MACHINES
02/02/91
PULL
EXPANDERS
USING CIRCULAR SPRING
PULL
+
PUSH

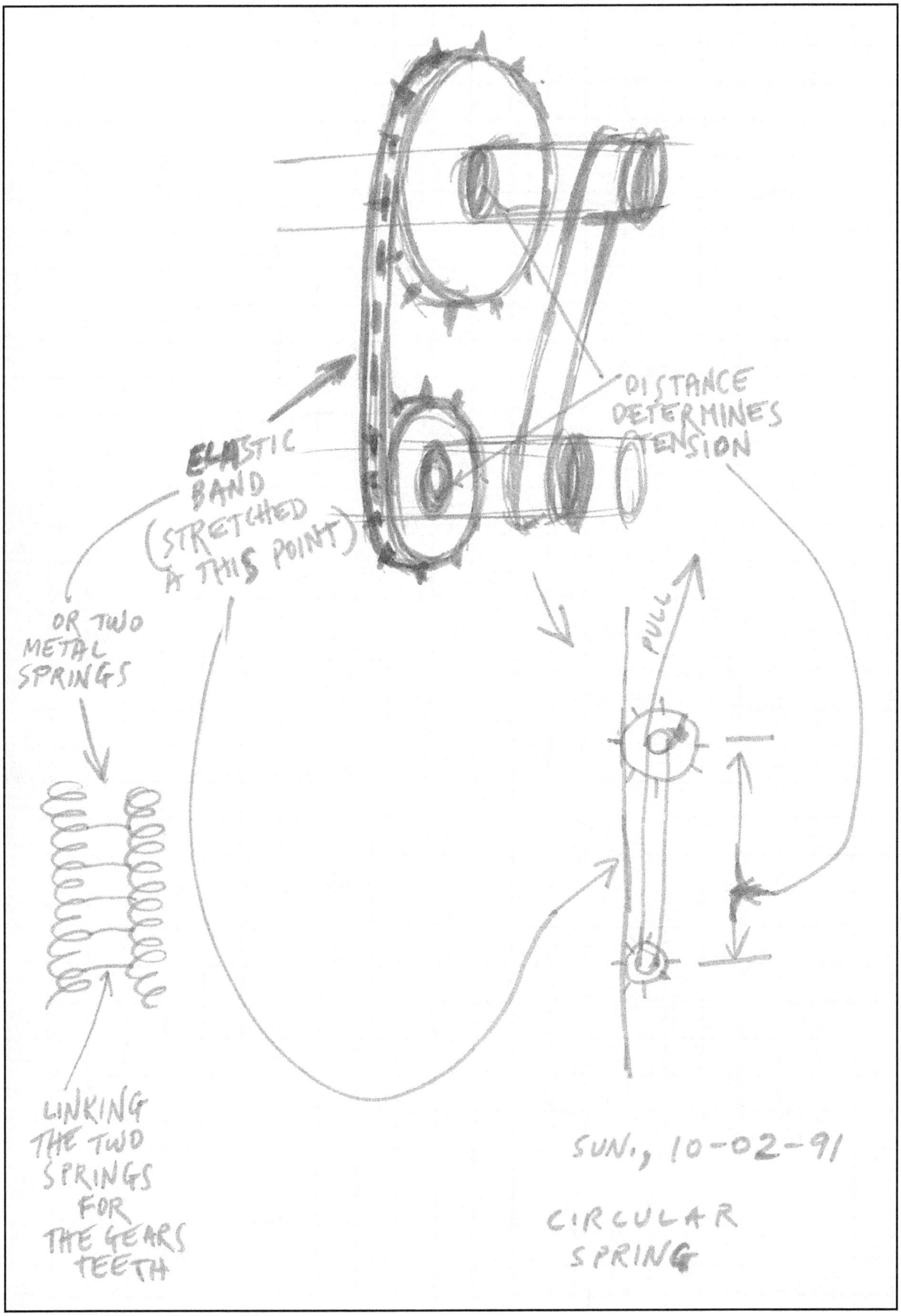
ELASTIC BAND (STRETCHED AT THIS POINT)
DISTANCE DETERMINES TENSION
OR TWO METAL SPRINGS
PULL
LINKING THE TWO SPRINGS FOR THE GEARS TEETH
SUN., 10-02-91
CIRCULAR SPRING

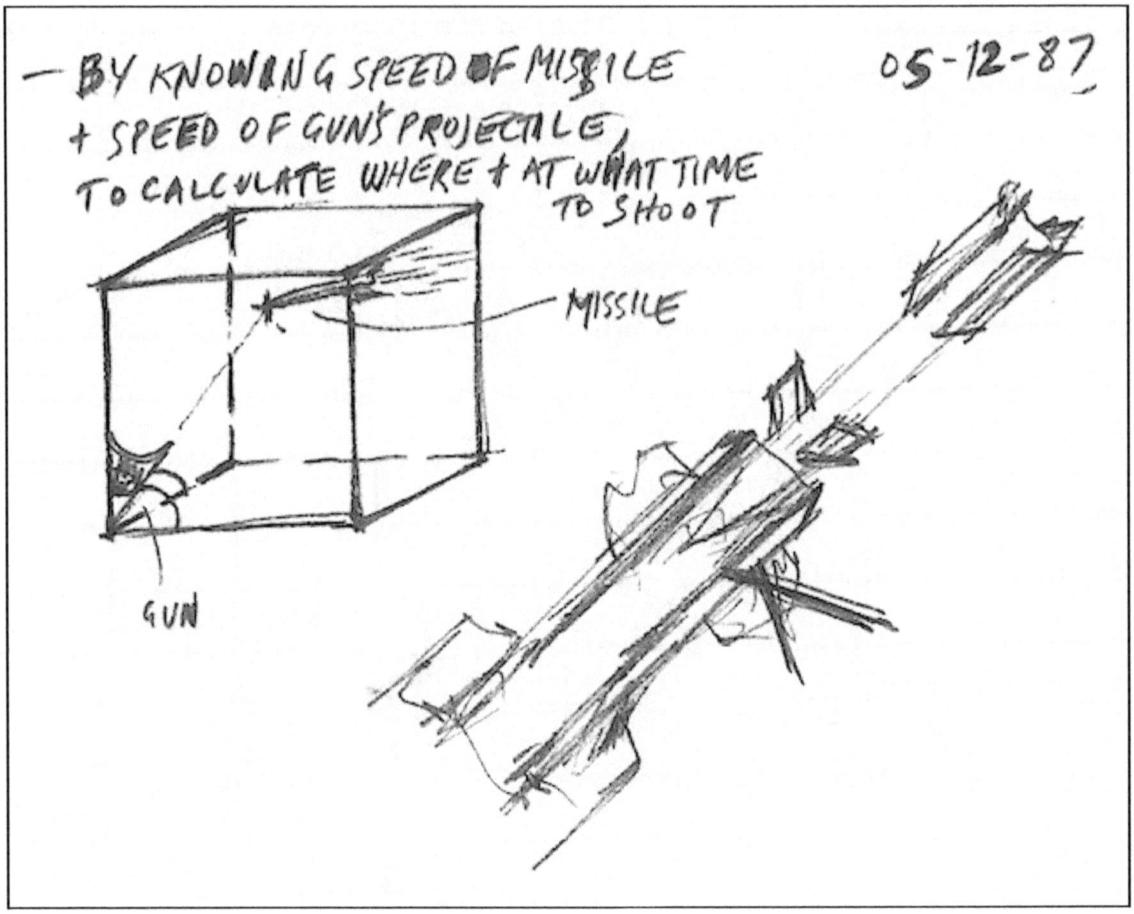
- BY KNOWING SPEED OF MISSILE
+ SPEED OF GUN'S PROJECTILE,
TO CALCULATE WHERE + AT WHAT TIME
TO SHOOT
05-12-87
MISSILE
GUN

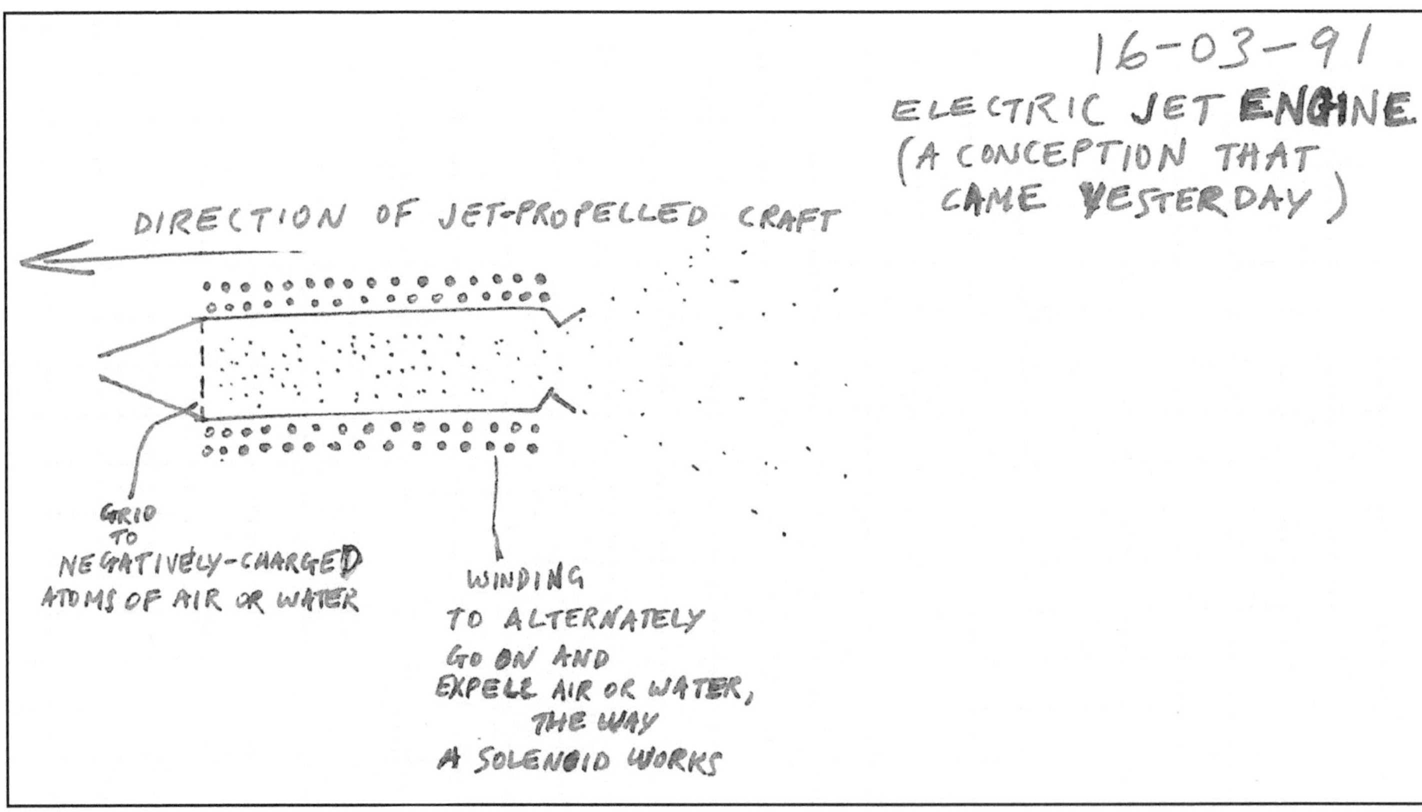
16-03-91
ELECTRIC JET ENGINE
(A CONCEPTION THAT
CAME YESTERDAY)
DIRECTION OF JET-PROPELLED CRAFT
GRID
TO
NEGATIVELY-CHARGED
ATOMS OF AIR OR WATER
WINDING
TO ALTERNATELY
GO ON AND
EXPELL AIR OR WATER,
THE WAY
A SOLENOID WORKS

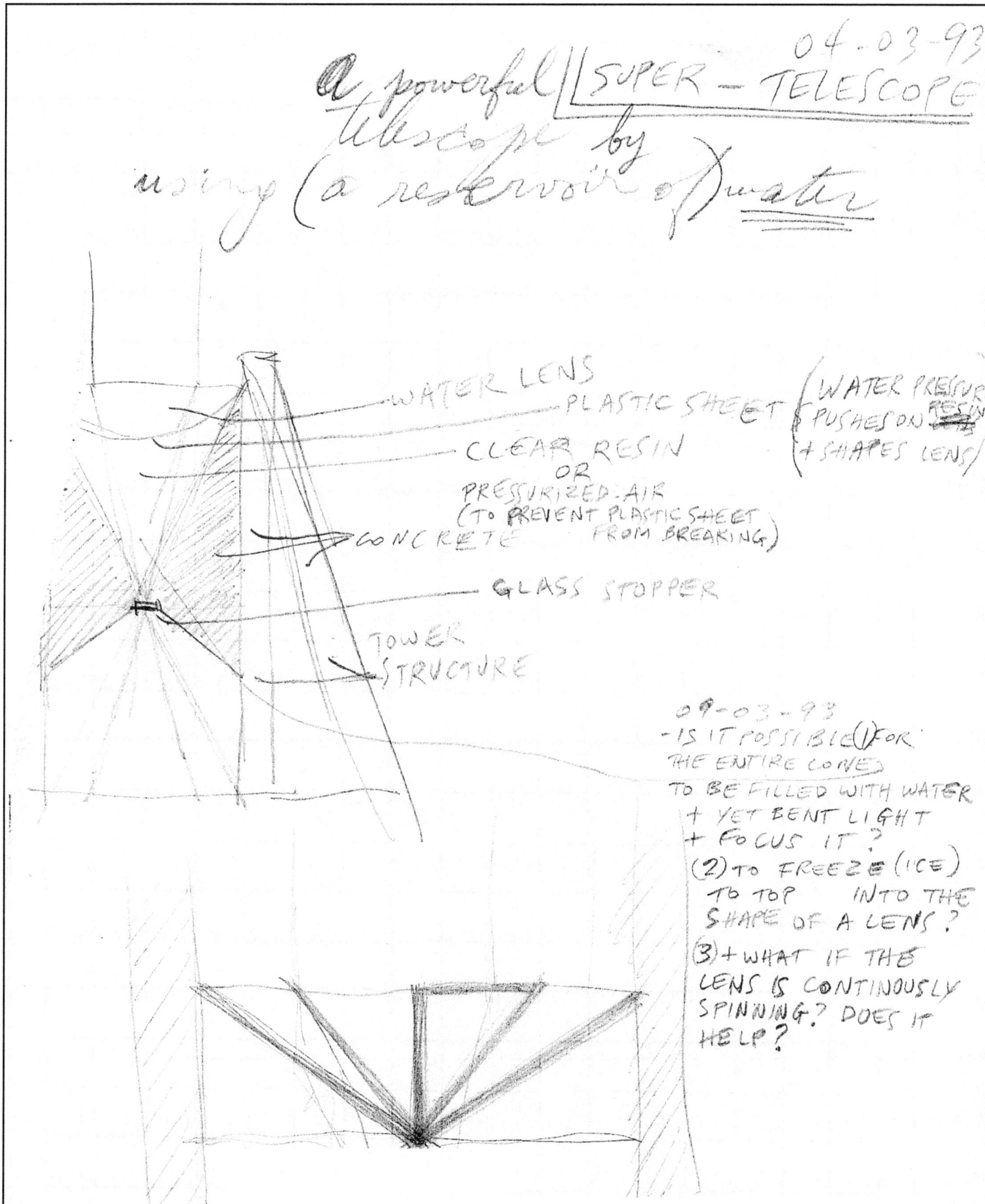
04-03-93
a powerful SUPER - TELESCOPE
telescope by
using (a reservoir of) water
WATER LENS
PLASTIC SHEET
(WATER PRESSURE PUSHES ON RESIN + SHAPES LENS)
CLEAR RESIN
OR
PRESSURIZED AIR
(TO PREVENT PLASTIC SHEET FROM BREAKING)
CONCRETE
GLASS STOPPER
TOWER STRUCTURE
09-03-93
-IS IT POSSIBLE (1) FOR THE ENTIRE CONE TO BE FILLED WITH WATER + YET BENT LIGHT + FOCUS IT?
(2) TO FREEZE (ICE) TO TOP INTO THE SHAPE OF A LENS?
(3) + WHAT IF THE LENS IS CONTINOUSLY SPINNING? DOES IT HELP?

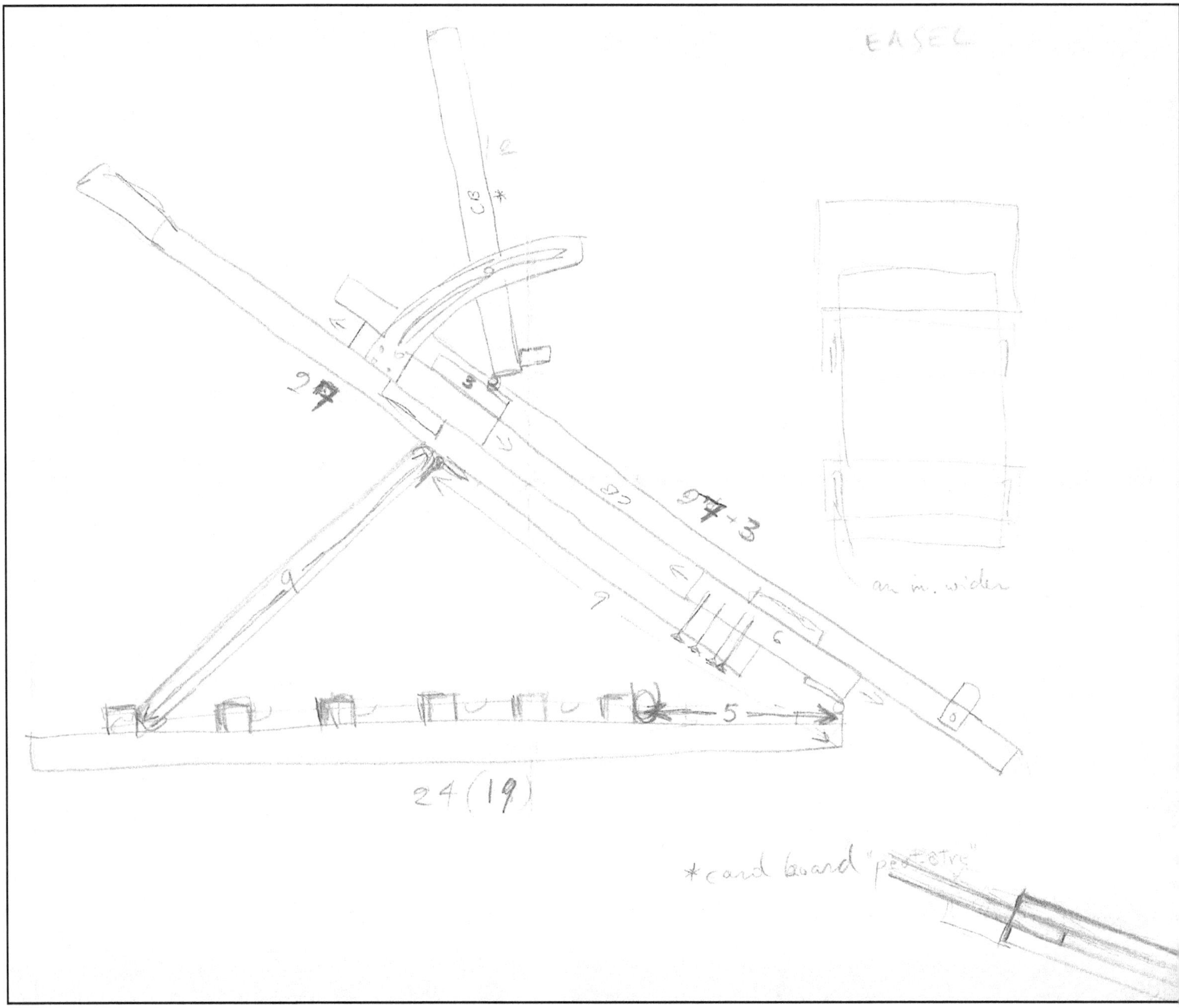
EASEL
an in. wider
5
24(19)

IDEAS TO BE ILLUSTRATED

Winter Pants with Zippered Legs
One-Dish Dish Washer
Floppy Disks Carousel
Stacked Laser Disks With Windows
Stacked Cd's Searchable By Paper Flaps
Pre-Fabricated House System
Building Block For Children
Water Faucet Gage To Permanently Regulate Temperature
Online Food Distribution Center Where You Can See And Pick The Products

www.ingramcontent.com/pod-product-compliance
Lightning Source LLC
LaVergne TN
LVHW080641011026
849160LV00014B/1282